Fertilizer Technology and Use

Third Edition

Fertilizer Technology and Use

Third Edition

O. P. ENGELSTAD, *editor*

Editorial Committee

F. C. Boswell | **T. C. Tucker**
L. F. Welch | **O. P. Engelstad, chair**

Managing Editor: RICHARD C. DINAUER
Associate Editor: SUSAN ERNST
Editor-in-Chief SSSA Publications: JOHN J. MORTVEDT

Published by
Soil Science Society of America, Inc.
Madison, Wisconsin USA

1985

Cover design by Ron Koontz, National Fertilizer Development Center, TVA, Muscle Shoals, AL

Soil Science Society of America, Inc.
677 South Segoe Road, Madison, WI 53711 USA

Library of Congress Cataloging-in-Publication Data

Fertilizer technology and use.

Includes bibliographies and index.
1. Fertilizers and manures. 2. Fertilizer industry—United States. I. Engelstad, Orvis P. II. Boswell, F. C. III. Soil Science Society of America.
S633.F43 1985 631.8′1 85-32051
ISBN 0-89118-779-0

Printed in the United States of America

CONTENTS

FOREWORD

We depend on soils to produce most of our food, feed, and fiber, but their tremendous variation presents problems. They differ by their locations on the continental land masses, elevations, latitudes, parent materials, and ages. As a result, they have different chemical composition, nutrient availabilities, biological colonization, physical properties, exposure to moisture supplies, organic matter, and other properties. All of these characteristics affect a soil's productivity but most cannot be changed significantly. Variability in the essential mineral element content for efficient plant growth is one of the factors that can be changed relatively easily to accommodate specific crop needs by adding soil amendments.

The origin of amending soils to make them better media for plant growth is shrouded in history. Applying soil amendments to correct for specific essential element deficiencies dates from the 19th century. But it was not until the 20th century when almost all of the knowledge about plant nutrient requirements developed into a well documented and generally understood basis for fertilizer use.

The recognized need for and usefulness of additional nutrient elements has led to the development of a large heavy chemical industry sector supplying plant-based agricultural activities. This has led to voluminous research on the science and technology of fertilizer production, distribution, needs, and economic use. From time to time, these new ideas should be made available for use by practitioners in the extended field of fertilizers, soil fertilization, and plant nutrition. This volume addresses that purpose.

Other issues surrounding fertilizer production and use have surfaced during the last 10 or 20 years. One of these is the exhaustible nature of the prime resources for fertilizer manufacture. This reality makes it important that fertilizers be used efficiently and effectively. Another issue is that fertilizers are implicated more and more in environmental problems, especially those concerning both surface and subsurface water supplies. Demonstrated prudent use of fertilizers is one step in avoiding unnecessary contamination. An informed public is more likely to deal rationally with positive and negative factors associated with production technology, so this volume has a significant potential role beyond production agriculture alone.

December 1985

JOHN PESEK, *president*
Soil Science Society of America

PREFACE

The Soil Science Society of America sponsored shortcourses on fertilizer technology and use at the University of Maryland in 1950 and at Purdue University in 1962. As a result of the second shortcourse, the Society published a very successful book entitled *Fertilizer Technology and Usage*. Since technology continued to advance, the Society published a second edition entitled *Fertilizer Technology and Use* in 1971. This edition also was a highly successful and popular book.

This book, the third edition of the series, embodies the latest developments in fertilizer technology and use. The authors present current information on fertilizer markets; soil and tissue testing; fertilizer-plant interactions in both acid and alkaline soils; behavior of plant nutrients in flooded soils; the production, marketing, and use of macronutrient, secondary and micronutrient fertilizers in solid, solution, and suspension forms; slow-release and inhibitor-amended nitrogen fertilizers; fertilizer use in relation to the environment, including concerns for nonpoint pollution effects; and the nutritional quality of crop plants in relation to fertilizer use. A chapter dealing with organic sources of nutrients replaces an earlier chapter on human and animal wastes as fertilizers. New chapters in this edition are entitled "Agronomic and Statistical Evaluation of Fertilizer Response" and "Modern Techniques in Fertilizer Application." The latter is especially relevant to the increasingly popular conservation tillage practices.

All of these presentations by recognized experts should be of great interest and utility to university staff and students, to fertilizer industry representatives, and to the public at large with interest in agriculture.

The Editorial Committee gratefully acknowledges the excellent and patient cooperation of all contributing authors. The editors also wish to extend their special thanks to Ms. Susan Ernst and Mr. Richard Dinauer of the Society headquarters for their editorial diligence and moral support.

December 1985

The Editorial Committee

O. P. ENGELSTAD, chair
National Fertilizer Development Center (TVA)

F. C. BOSWELL
University of Georgia

T. C. TUCKER
University of Arizona

L. F. WELCH
University of Illinois

CONTRIBUTORS

Frank P. Achorn	Chief Chemical Engineer, National Fertilizer Development Center, Tennessee Valley Authority, Muscle Shoals, Alabama
W. H. Allaway	Visiting Fellow, Department of Agronomy, Cornell University, Ithaca, New York
Hubert L. Balay	Chemical Engineer, National Fertilizer Development Center, Tennessee Valley Authority, Muscle Shoals, Alabama
Stanley A. Barber	Professor of Agronomy, Department of Agronomy, Purdue University, West Lafayette, Indiana
James D. Beaton	Northwest Director, Potash & Phosphate Institute, Cochrane, Alberta, Canada
Fred C. Boswell	Professor of Soil Science, Agronomy Department, University of Georgia, Experiment, Georgia
F. E. Broadbent	Professor of Soil Microbiology, Department of Land, Air, and Water Resources, University of California, Davis, California
Ned L. Case	Director of Agronomy (retired), Phillips Petroleum Company. Current address: Pavillion, Wyoming
Gary W. Colliver	Chief Agronomist, Farmland Industries, Kansas City, Missouri
J. T. Cope	Professor (Emeritus), Department of Agronomy and Soils, Auburn University, Auburn, Alabama
F. R. Cox	Professor, Department of Soil Science, North Carolina State University, Raleigh, North Carolina
W. B. Dancy	Director of Development, International Minerals & Chemical Corporation, Carlsbad, New Mexico. Deceased 10 August 1985
R. Ellis, Jr.	Professor of Agronomy, Kansas State University, Manhattan, Kansas. Deceased 9 September 1982
Robert L. Fox	Professor of Soil Science, Department of Agronomy and Soil Science, University of Hawaii, Honolulu, Hawaii
Charles D. Foy	Research Soil Scientist, Agricultural Research Service, U.S. Department of Agriculture, Beltsville Agricultural Research Center, Beltsville, Maryland
J. W. Gilliam	Professor of Soil Science, Soil Science Department, North Carolina State University, Raleigh, North Carolina
David L. Grunes	Soil Scientist, Agricultural Research Service, U.S. Department of Agriculture, U.S. Plant, Soil, and Nutrition Laboratory, Ithaca, New York
John J. Hanway	Professor, Agronomy Department, Iowa State University, Ames, Iowa
Edwin A. Harre	Supervisor, Marketing and Distribution Economics Section, National Fertilizer Development Center, Tennessee Valley Authority, Muscle Shoals, Alabama
Roland D. Hauck	Soil Scientist, National Fertilizer Development Center, Tennessee Valley Authority, Muscle Shoals, Alabama
Milton B. Jones	Agronomist, Department of Agronomy and Range Science, University of California, Hopland, California

Eugene J. Kamprath Professor of Soil Science, Soil Science Department, North Carolina State University, Raleigh, North Carolina

David E. Kissel Professor of Agronomy, Department of Agronomy, Kansas State University, Manhattan, Kansas

Terry J. Logan Professor of Agronomy, Agronomy Department, The Ohio State University, Columbus, Ohio

J. J. Meisinger Soil Scientist, Agricultural Research Service, U.S. Department of Agriculture, Beltsville Agricultural Research Center, Beltsville, Maryland

Duane S. Mikkelsen Professor and Director of International Programs, Department of Agronomy and Range Science, University of California, Davis, California

John J. Mortvedt Soil Chemist, National Fertilizer Development Center, Tennessee Valley Authority, Muscle Shoals, Alabama

Robert D. Munson Northcentral Director, Potash & Phosphate Institute, St. Paul, Minnesota

Larry A. Nelson Professor of Statistics, Department of Statistics, North Carolina State University, Raleigh, North Carolina

R. I. Papendick Soil Scientist, Agricultural Research Service, U.S. Department of Agriculture, Washington State University, Pullman, Washington

W. H. Patrick, Jr. Boyd Professor, Laboratory for Wetland Soils and Sediments, Louisiana State University, Baton Rouge, Louisiana

John Pesek Professor and Head, Agronomy Department, Iowa State University, Ames, Iowa

J. F. Power Research Leader, Agricultural Research Service, U.S. Department of Agriculture, Department of Agronomy, University of Nebraska, Lincoln, Nebraska

Gyles W. Randall Soil Scientist and Professor, University of Minnesota, Southern Experiment Station, Waseca, Minnesota

D. H. Sander Professor of Agronomy, Department of Agronomy, University of Nebraska, Lincoln, Nebraska

Regis D. Voss Professor of Agronomy, Agronomy Department, Iowa State University, Ames, Iowa

L. Fred Welch Professor of Soil Fertility, Department of Agronomy, University of Illinois, Urbana, Illinois

B. R. Wells Professor, Department of Agronomy, University of Arkansas, Fayetteville, Arkansas

K. L. Wells Extension Professor (Soils), Department of Agronomy, University of Kentucky, Lexington, Kentucky

D. G. Westfall Professor of Agronomy, Department of Agronomy, Colorado State University, Fort Collins, Colorado

William C. White Senior Vice President, Member Services, The Fertilizer Institute, Washington, D.C.

David A. Whitney Professor, Agronomy Department, Kansas State University, Manhattan, Kansas

Ronald D. Young Staff Chemical Engineer (retired), National Fertilizer Development Center, Tennessee Valley Authority, Muscle Shoals, Alabama

Conversion Factors for SI and non-SI Units

To convert Column 1 into Column 2, multiply by	Column 1 SI Unit	Column 2 non-SI Unit	To convert Column 2 into Column 1 multiply by
		Length	
0.621	kilometer, km (10^3 m)	mile, mi	1.609
1.094	meter, m	yard, yd	0.914
3.28	meter, m	foot, ft	0.304
1.0	micrometer, μm (10^{-6} m)	micron, μ	1.0
3.94×10^{-2}	millimeter, mm (10^{-3} m)	inch, in	25.4
10	nanometer, nm (10^{-9} m)	Angstrom, Å	0.1
		Area	
2.47	hectare, ha	acre	0.405
247	square kilometer, km^2 (10^3 m)^2	acre	4.05×10^{-3}
0.386	square kilometer, km^2 (10^3 m)^2	square mile, mi^2	2.59
2.47×10^{-4}	square meter, m^2 (10^3 m)^2	acre	4.05×10^3
10.76	square meter, m^2 (10^3 m)^2	square foot, ft^2	9.29×10^{-2}
1.55×10^{-3}	square millimeter, m^2 (10^{-6} m)^2	square inch, in^2	645
		Volume	
6.10×10^4	cubic meter, m^3	cubic inch, in^3	1.64×10^{-5}
2.84×10^{-2}	liter, L (10^{-3} m^3)	bushel, bu	35.24
1.057	liter, L (10^{-3} m^3)	quart (liquid), qt	0.946
3.53×10^{-2}	liter, L (10^{-3} m^3)	cubic foot, ft^3	28.3
0.265	liter, L (10^{-3} m^3)	gallon	3.78
33.78	liter, L (10^{-3} m^3)	ounce (fluid), oz	2.96×10^{-2}
2.11	liter, L (10^{-3} m^3)	pint (fluid), pt	0.473
1.06	liter, L (10^{-3} m^3)	quart (liquid), qt	0.946
9.73×10^{-3}	meter^3, m^3	acre-inch	102.8
35.7	meter^3, m^3	cubic foot, ft^3	2.80×10^{-2}

continued on next page

Conversion Factors for SI and non-SI Units

To convert Column 1 into Column 2, multiply by	Column 1 SI Unit	Column 2 non-SI Unit	To convert Column 2 into Column 1 multiply by
		Mass	
2.20×10^{-3}	gram, g (10^{-3} kg)	pound, lb	454
3.52×10^{-2}	gram, g	ounce (avdp), oz	28.4
2.205	kilogram, kg	pound, lb	0.454
10^{-2}	kilogram, kg	quintal (metric), q	10^{2}
1.1×10^{-3}	kilogram, kg	ton (2000 lb), T	907
1.102	megagram, Mg (tonne)	ton (U.S.), T	0.907
		Yield and Rate	
0.893	kilogram per hectare, kg ha^{-1}	pound per acre, lb $acre^{-1}$	1.12
7.77×10^{-2}	kilogram per cubic meter, kg m^{-3}	pound per bushel, lb bu^{-1}	12.87
1.49×10^{-2}	kilogram per hectare, kg ha^{-1}	bushel per acre, 60 lb	67.19
1.59×10^{-2}	kilogram per hectare, kg ha^{-1}	bushel per acre, 56 lb	62.71
1.86×10^{-2}	kilogram per hectare, kg ha^{-1}	bushel per acre, 48 lb	53.75
0.107	liter per hectare, L ha^{-1}	gallon per acre	9.35
893	megagram per hectare, Mg ha^{-1}	pound per acre, lb $acre^{-1}$	1.12×10^{-3}
0.446	megagram per hectare, Mg ha^{-1}	ton (2000 lb) per acre, T $acre^{-1}$	2.24
2.10	meter per second, m s^{-1}	mile per hour	0.477
		Specific Surface	
10	square meter per kilogram, m^2 kg^{-1}	square centimeter per gram, cm^2 g^{-1}	0.1
10^{3}	square meter per kilogram, m^2 kg^{-1}	square millimeter per gram, mm^2 g^{-1}	10^{-3}

		Pressure	
9.90	megapascal, MPa (10^6 Pa)	atmosphere	0.101
10	megapascal, MPa (10^6 Pa)	bar	0.1
1.00	megagram per cubic meter, Mg m^{-3}	gram per cubic centimeter, g cm^{-3}	1.00
2.09×10^{-2}	pascal, Pa	pound per square foot, lb ft^{-2}	47.9
1.45×10^{-4}	pascal, Pa	pound per square inch, lb in^{-2}	6.90×10^3
		Temperature	
1.00	Kelvin, K	Celsius, °C	1.00 (°C + 273)
(9/5 °C) + 32	Celsius, °C	Fahrenheit, °F	5/9 (°F −32)
		Energy, Work, Quantity of Heat	
9.52×10^{-4}	joule, J	British thermal unit, Btu	1.05×10^3
0.239	joule, J	calorie, cal	4.19
10^7	joule, J	erg	10^{-7}
0.735	joule, J	foot-pound	1.36
2.387×10^{-5}	joule per square meter, J m^{-2}	calorie per square centimeter (langley)	4.19×10^4
10^5	newton, N	dyne	10^{-5}
1.43×10^{-3}	watt per square meter, W m^{-2}	calorie per square centimeter minute (irradiance), cal cm^{-2} min^{-1}	698
		Transpiration and Photosynthesis	
3.60×10^{-2}	milligram per square meter second, mg m^{-2} s^{-1}	gram per square decimeter hour, g dm^{-2} h^{-1}	27.8
5.56×10^{-3}	milligram (H_2O) per square meter second, mg m^{-2} s^{-1}	micromole (H_2O) per square centimeter second, μmol cm^{-2} s^{-1}	180
10^{-4}	milligram per square meter second, mg m^{-2} s^{-1}	milligram per square centimeter second, mg cm^{-2} s^{-1}	10^4
35.97	milligram per square meter second, mg m^{-2} s^{-1}	milligram per square decimeter hour, mg dm^{-2} h^{-1}	2.78×10^{-2}
		Angle	
57.3	radian, rad	degrees (angle) °	1.75×10^{-2}

continued on next page

Conversion Factors for SI and non-SI Units

To convert Column 1 into Column 2, multiply by	Column 1 SI Unit	Column 2 non-SI Unit	To convert Column 2 into Column 1 multiply by
	Electrical Conductivity		
10	siemen per meter, S m^{-1}	millimho per centimeter, mmho cm^{-1}	0.1
	Water Measurement		
9.73×10^{-3}	cubic meter, m^3	acre-inches, acre-in	102.8
9.81×10^{-3}	cubic meter per hour, m^3 h^{-1}	cubic feet per second, ft^3 s^{-1}	101.9
4.40	cubic meter per hour, m^3 h^{-1}	U.S. gallons per minute, gal min^{-1}	0.227
8.11	hectare-meters, ha-m	acre-feet, acre-ft	0.123
97.28	hectare-meters, ha-m	acre-inches, acre-in	1.03×10^{-2}
8.1×10^{-2}	hectare-centimeters, ha-cm	acre-feet, acre-ft	12.33
	Concentrations		
1	centimol per kilogram, cmol kg^{-1} (ion exchange capacity)	milliequivalents per 100 grams, meq 100 g^{-1}	1
0.1	gram per kilogram, g kg^{-1}	percent, %	10
1	megagram per cubic meter, Mg m^{-3}	gram per cubic centimeter, g cm^{-3}	1
1	milligram per kilogram, mg kg^{-1}	parts per million, ppm	1
	Plant Nutrient Conversion		
	Elemental	***Oxide***	
2.29	P	P_2O_5	0.437
1.20	K	K_2O	0.830
1.39	Ca	CaO	0.715
1.66	Mg	MgO	0.602

5 March 1986

1

Fertilizer Market Profile

Edwin A. Harre

National Fertilizer Development Center
Tennessee Valley Authority
Muscle Shoals, Alabama

William C. White

The Fertilizer Institute
Washington, DC

Since World War II, increased world agricultural production to feed a growing population has been one of the greatest success stories ever documented. World agriculture, responding to this pressure, has consistently outperformed the population growth rate. From 1970 to 1981 the index of food production registered a dramatic increase of > 28%, raising the per capita index by more than five points (FAO, 1983b). World population increases are expected to average 1.8%/yr growth by 1994, creating an ever increasing demand for food (FAO, 1981a).

The precariousness of the world food chain was dramatically illustrated in 1983. Pressured at the beginning of the year by high inventory levels of agricultural commodities, low prices, and low farm incomes, the USA initiated a cropland reduction program. Farmer response was greater than anticipated, resulting in a record-setting reduction in production and in reduced inventory level expectations for cereal crops and soybeans [*Glycine max* (L.) Merr.]. However, a severe drought throughout most of the country brought further yield reductions and reduced stock levels, both in the USA and worldwide, to a point where reserves would have been inadequate if similar growing conditions occurred again in 1984.

The specter of world famine emerges from time to time, only to be deterred by dramatic gains in food supplies. The question remains, however, as to whether world agriculture can continue to meet this challenge until long-term population control programs begin to take effect. To meet this challenge, worldwide agriculture must depend more on increased per-hectare yields as land for expanded crop production becomes increasingly scarce. And, with greater dependence on higher yields, improved soil fertility will be the primary option for increasing food supplies.

This chapter will emphasize the primary plant nutrients and review the history of fertilizer supply and use, products used and marketing forms, and relationships between fertilizers and other farm inputs. A review of past trends should help indicate the future role of fertilizers in the world food chain.

I. THE IMPORTANCE OF FERTILIZERS IN THE WORLD FOOD CHAIN

Fertilizer has been in the forefront of the struggle to increase world food production and perhaps, more than any other input, has been largely responsible for the success that has been attained. Only fertile soils are productive soils. Where plant nutrients are deficient, soil productivity and crop yields are low. Thus, by supplying plant nutrients essential for high crop yields, fertilizers have become vital for crop production.

Many attempts have been made to quantify the contribution of fertilizer to agricultural production. Estimates of the contribution fertilizer makes to increased food production range as high as 50 to 75% or more in some developing countries. Most studies indicate that fertilizers, while not the only factor, are regarded as the leading contributor (FAO, 1981b). The gain in food production is obviously the result of a combination of factors, any one of which is ineffective without the others. In the USA, 30 to 40% of total crop production is attributable to fertilizer use (TVA, 1983). Worldwide estimates show a 20 to 25% contribution from fertilizers.

Not only have great strides been made in fertilizer development, but advances also have occurred in farm management, water supplies, crop protection chemicals, and new crop varieties. Studies by Allan in Kenya (Wortman & Cummings, 1978) illustrate that a complete package of improved management practices, including hybrid seed and fertilizer, increased yields by >300%, hybrid seed and fertilizer without the other inputs were responsible for a 65% gain. There is no substitute for good farming practices; combining them with adequate inputs produces substantial yield gains.

As new advances flow from both public and private research in crop production, there is reason to believe that production records will continue to be broken and that food supplies will be available to meet world population needs. Since there are only limited prospects for expanding cropland significantly, most food production increases will be through increased yields made possible with improvements in plant nutrition.

Until 1940, agricultural production in the USA was expanded primarily by increasing the land area used for cultivation and grazing. Land, however, is a limited resource and cannot sustain high crop production levels without replacing the nutrients removed by crops. Farmers can no longer farm until the soil is depleted and yields decline. They must maintain soil fertility from year to year. As population growth increases, so does the demand for land resources and commercial fertilizers. Fertilizers are, in effect, a substitute for land—they are an increasingly important factor as available land for food production becomes more scarce. While the role of fertilizers in world agricultural production has been substantial, fertilizers will have an even greater role to play if our limited land area is to supply the needed food production.

Like land, fertilizer resources are limited. Mineral reserves, while ample under current production levels, are finite. Also, trends in future N fertilizer production will depend largely on energy supplies and costs. Thus, the challenge to the fertilizer industry is not only to continue to produce an adequate supply of nutrients, but also to increase the efficiency of these nutrients in crop production.

II. FERTILIZER MARKET CHARACTERISTICS

Modern farming places high demands on N, especially in warm, humid regions. Requirements for N exceed any other plant nutrient, and only very rarely do soils have enough available N to produce high yields of nonlegumes. Organic matter is a source of N for crops, but today's efficient crop production depends not only on restoring and maintaining the amount of organic matter, but also supplementing it with N fertilizers. World N fertilizer use has expanded more rapidly than any of the other primary plant nutrients because results of increased N applications are easily seen and measured in units of increased profits.

Unlike N, levels of P can be increased in soil. With the recent high rates of fertilizer applications in many areas of the world, P levels of many soils have been dramatically increased. Recently, however, P rates in some countries have remained about the same. This is causing concern that P reserves in some soils may be depleted to a point where the soil will no longer support high yields without the addition of P fertilizers. Potassium use continues to gain in importance, and in many areas of the world exceeds the use of P. Future consumption patterns for K and N may be similar as opposed to the past pattern of similarity of K with P.

A. The Major Fertilizer Materials

Any material applied to the soil to supply nutrients for plant growth can be considered a fertilizer. The term *fertilizer* can be used to include green cover crops, manure, or similar materials. Recently, however, it has come to denote *commercial fertilizers* as materials containing one or more plant nutrients, used primarily for their plant nutrient content, and designed for use or claimed to have value in promoting plant growth. In the USA, fertilizer production is one of the nation's major industries. Such products as ammonia (NH_3), urea, phosphoric acid (H_3PO_4), and sulfuric acid (H_2SO_4), ammonium phosphates, and K salts are high on the list of tonnage rankings for chemical production and mineral processing.

Ammonia, phosphoric acid, and potassium chloride (KCl) are the basic raw materials for fertilizers in the USA. In other areas of the world, superphosphate manufacture and the use of ground phosphate rock still

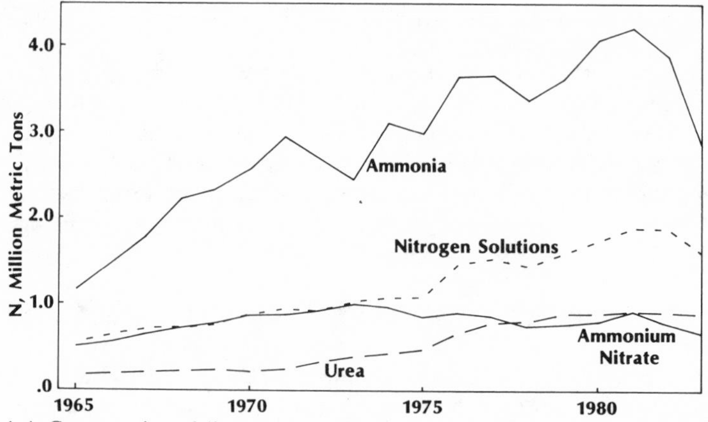

Fig. 1–1. Consumption of direct-application N materials (Bridges, 1983; USDA, 1983).

contribute significantly to the P supply. The following is a brief description of the fertilizer materials used in the USA and their consumption patterns since 1965. Similar product market trends are occurring in the international fertilizer market. Several countries, however, do not make market information available on a product basis.

1. Ammonia

Ammonia is not only used in the manufacture of all N fertilizers used in the USA, but is also applied directly to the soil. Almost 40% of the total N fertilizer used in the USA has been in the form of direct-application ammonia (Fig. 1–1) (USDA, 1983). In world markets, however, the use of ammonia as a direct-application material has been limited to certain areas of western Europe and Latin America. The primary reason for the limited acceptance of this practice is the relatively small farm size in other countries compared with the large areas involved in the USA and the specialized equipment required. Soil type and climate are also important factors in determining the direct application of ammonia.

2. Nitrogen Solutions

Second in importance as a N source in the USA, are N solutions, usually a combination of ammonium nitrate (NH_4NO_3), urea, and water. Use of N solutions has been expanding faster than that of ammonia or some of the solid N materials. In 1982, N solutions accounted for about 18% of the total U.S. market of 10 million metric tons of N. Nitrogen solutions have been used more extensively in other countries than has ammonia; however, they do not have the large market share that prevails in the USA.

3. Ammonium Nitrate

First introduced as a high-analysis N fertilizer after World War II, ammonium nitrate followed ammonia as the leading source of N until the mid-1970s when it was replaced by N solutions. It remained the leading solid N fertilizer until 1978, when it was replaced by urea. Since reaching its peak tonnage in 1973, its market share has slowly declined.

4. Urea

Urea has the highest N content of the solid N fertilizers. Since its introduction as a fertilizer material, its use, especially in international markets, has grown rapidly. Urea became the world's leading N source in the early 1970s and it followed ammonia and N solutions in importance in the U.S. market by 1980. With one-third of the N content of N solutions coming from urea, the relative value of urea as a fertilizer material is more important than market data indicate. As world fertilizer production patterns change and world N trade continues to expand, urea's importance will grow even faster.

5. Ammonium Sulfate

At one time, ammonium sulfate [$(NH_4)_2SO_4$] was the leading N fertilizer. Significant tonnages are still applied in some areas of the world. It is also a source of S. The true importance of ammonium sulfate is masked by the fact that much of this material is used to manufacture mixed fertilizers, and no accurate accounting of actual use in this form is available. While ammonium sulfate continues to decline in relative importance as a N fertilizer, it will continue to be used because it is a by-product of the steel industry and the production of caprolactum and other industrial chemicals.

6. Normal Superphosphate

Once the mainstay of the P fertilizer market, normal superphosphate has almost disappeared from the market in many countries. In the USA, ammoniated normal superphosphate accounted for much of the mixed fertilizer used until the 1960s, but now represents only 2% of the total P supply (Fig. 1–2).

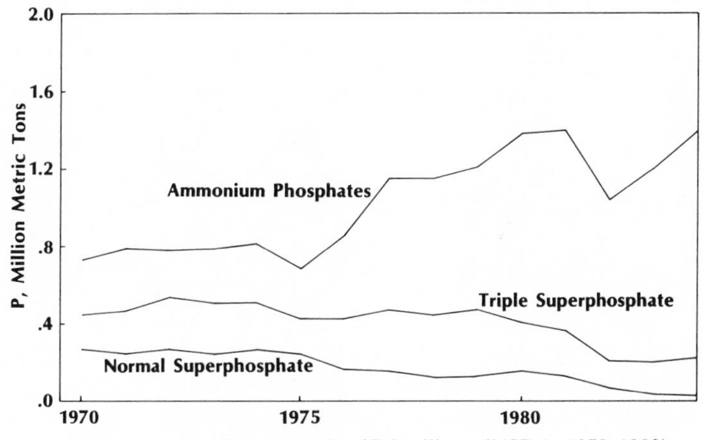

Fig. 1–2. United States supply of P fertilizers (USDA, 1970–1983).

7. Triple Superphosphate

Triple superphosphate's (TSP) market share also has eroded slowly to about 10% of the total P supply in the USA. Triple superphosphate is still used extensively to manufacture fertilizer grades containing little or no N and grades with high P/N ratios. Large amounts of TSP enter world trade, especially to those countries with N production but no P reserves.

8. Ammonium Phosphates

Diammonium phosphate (DAP) and monoammonium phosphate (MAP) grades have become the cornerstones of the P fertilizer industry in the USA and in international trade. These two products now represent >50% of total U.S. P supply and should increase in importance. With the advent of ammoniation-granulation technology and the subsequent production of DAP, no other fertilizer material matches the impact of the ammonium phosphates on the marketing and distribution system for fertilizers in the USA and in many other areas of the world. Monoammonium phosphate, while still only a small percentage of the total ammonium phosphate production, is increasing in importance because of its use as either a solid material or in the manufacture of suspension grade fluid fertilizers. Similar to TSP, MAP can be made from sludge phosphoric acid and used to manufacture grades with high P/N ratios.

9. Nitric Phosphates

Since nitric phosphate grades are produced by reacting nitric acid (HNO_3) rather than sulfuric or phosphoric acid with phosphate rock, they are more popular in areas lacking large reserves of S or facilities for S recovery. At one time, several nitric phosphate plants operated in the USA; however, they were abandoned or converted to ammonium phosphate or ammoniation-granulation plants as the new technology became available. Nitric phosphates still predominate in western Europe, but some manufacturers have shifted to imported P materials for the blending of similar grades.

10. Potash

The demand for K is supplied primarily by potassium chloride. Smaller amounts of potassium sulfate (K_2SO_4) and compounds containing secondary nutrients combined with K make up the remainder. Potassium chloride is mined in only a few areas of the world. The market is dominated by production from Canada and USSR. Other major producers are in western and eastern Europe, USA, Israel, and Jordan. Mining projects are underway where significant K reserves have been found in Thailand and Brazil. Similar mining operations, primarily brine recovery, have also been proposed in other developing countries.

Many other materials contain plant nutrients. Small amounts are used in the areas where they exist as a unique supply or by-product and are com-

Table 1-1. Nutrient content of principal fertilizer materials (Young & Johnson, 1982).

Material	N	P	K	Ca	S
	%				
Nitrogen					
Anhydrous ammonia	82				
Aqua ammonia	16-25				
Ammonium nitrate	33.5				
Ammonium nitrate-lime	20.5			7	
Ammonium sulfate	21				24
Ammonium sulfate-nitrate	26				15
Calcium cyanamide	21			39	
Calcium nitrate	15			19	
Nitrogen solutions	21-49				
Sodium nitrate	16				
Urea	46				
Phosphate					
Basic slag		4-5		29	
Normal superphosphate		8-9		20	12
Concentrated superphosphate		18-22		14	1
Phosphoric acid		23-26			
Superphosphoric acid		30-33			
Potash					
Potassium chloride			50-51		
Potassium sulfate			42		18
Potassium-magnesium sulfate			18		23
Multinutrient materials					
Diammonium phosphate	16-21	20-23			
Monoammonium phosphate	10-11	21-24			
Nitric phosphates	14-22	4-10		8-10	0-4
Potassium nitrate	13		37		

petitive with the more common sources of plant nutrients. The primary materials used as fertilizers and their nutrient content are listed in Table 1-1. More details on the use and characteristics of these materials are found in later chapters.

B. Fertilizer Distribution by Class

In addition to the large numbers of materials available, there are several forms or classes in which fertilizer materials are marketed. These include ammonia, solids, liquids, suspensions, homogenous mixtures, or blending of two or more materials to a specific grade or ratio. Since 1960, distribution of fertilizers by class has changed because of new product developments that changed the traditional producer-wholesaler-retailer system to a more direct producer-retailer production flow (Hargett & Pay, 1980).

Total consumption of fertilizer materials in the USA has increased from 29 million metric tons in 1965[1] to 47 million metric tons in 1981—the record high level of consumption to date (USDA, 1983). During this time

[1]All U.S. fertilizer consumption statistics are for the fiscal year July–June. The year mentioned represents the year-ending date.

Table 1-2. United States consumption of fertilizer by class (Bridges, 1983; Hargett & Berry, 1985).

Year	Total fertilizer					Mixtures		
	Bagged	Bulk	Fluid	NH_3	Total	Bagged	Bulk	Fluid
	metric tons × 10^6							
1965			3.4	1.4	28.9			0.9
1966				1.8	31.3			
1967	14.1	11.0	4.8	2.2	32.0	11.0	6.5	1.7
1968	12.6	13.0	4.9	2.8	33.2	9.8	7.5	1.8
1969	11.9	13.8	5.3	2.8	33.8	9.0	8.0	2.0
1970	11.0	14.4	5.9	3.1	34.4	8.3	8.1	2.3
1971	10.4	15.4	6.5	3.6	35.9	7.9	8.8	2.7
1972	10.2	15.8	6.5	3.3	35.9	7.7	8.9	2.8
1973	9.6	18.3	7.0	3.1	37.9	7.2	10.4	2.9
1974	9.5	19.8	7.7	3.8	40.8	7.3	11.4	3.1
1975	8.2	17.6	7.4	3.6	36.8	6.3	9.6	2.9
1976	7.8	21.2	9.1	4.5	42.5	6.0	11.5	3.3
1977	7.5	22.8	9.8	4.5	44.5	5.8	12.5	3.6
1978	6.8	21.3	9.2	4.1	41.4	5.2	11.3	3.6
1979	6.5	23.4	10.4	4.4	44.8	4.9	12.6	4.1
1980	6.1	23.5	11.2	5.0	45.8	4.7	12.3	4.2
1981	5.7	24.2	12.0	5.2	47.0	4.3	12.6	4.5
1982	5.0	21.3	11.4	4.6	42.3	3.8	10.9	4.1
1983	4.5	19.1	9.8	3.5	36.9	3.4	10.1	3.5
1984	4.7	22.7	12.3	4.3	44.0	3.5	11.6	4.1

there was a rapid change in distribution systems. Dry-bagged materials and mixtures have steadily declined from > 14 million metric tons to 4.5 million, or only 12% of the total market. Fluid mixtures, including suspensions, which made up < 1 million metric tons of material in 1965 now account for almost 4 million metric tons; fluid materials, primarily ammonia and N solutions, represent almost 16 million metric tons, or 38% of the market. At the same time, bulk distribution and handling of fertilizers more than doubled from 11 million metric tons in 1967 to 24 million metric tons in 1982 (Table 1–2).

Since 1965, consumption of mixed fertilizers has changed very little. In 1965, almost 17 million metric tons of mixed fertilizers were distributed to farmers; in 1982, total distribution was nearly 19 million metric tons. During this time, however, there was a dramatic shift in the market share of the three major systems used in the USA to distribute mixed fertilizers. In the early 1960s, ammoniation-granulation plants accounted for almost 80% of the mixed fertilizer market. Fluid mixtures represented < 5%, while bulk blending—the physical mixing of two or more materials to get the desired nutrient ratio—made up the remaining 15% (Fig. 1–3). In 1976, and as a result of the growth of bulk blending, it became the leading form of mixed fertilizer distribution. Bulk blending now represents about 53% of all fertilizer mixtures, ammoniation-granulation makes up 27%, and fluid mixtures make up the remaining 20%. The change in market share for solid forms of fertilizer mixtures has been declining, while that for fluids increased.

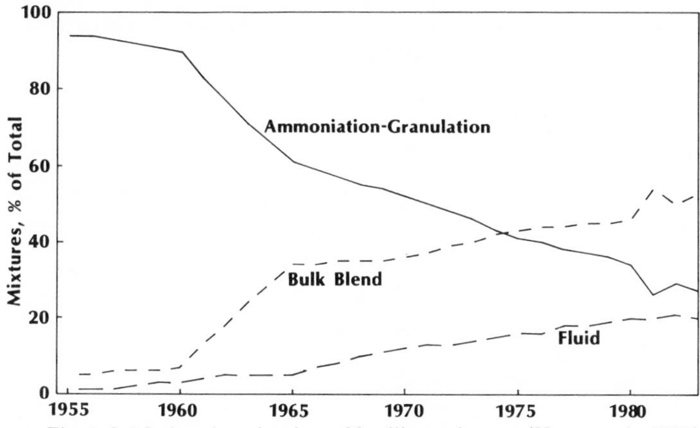

Fig. 1–3. Market share by class of fertilizer mixtures (Harre et al., 1983).

Bulk blending and liquid mixing have changed the traditional marketing system for fertilizers in the USA by transferring part of the functions of wholesale distributors to basic producers and to retail dealers. Basic producers now sell directly to the retailer and are required to store large quantities of materials in the off-season. Blenders also store large amounts of materials and blend these materials to specifications and soil test requirements. Retail distributors in the traditional system did little if any mixing or blending and sold only a minimum number of grades. There is some indication that the market situation in the USA is bringing wholesalers back into the U.S. fertilizer distribution system.

Each of the major manufacturing-marketing systems has a place in the U.S. market. Each must be evaluated in its particular market location, raw material supply, size of plant, total investment, and the type of agriculture within the market area. A manufacturing-marketing system designed for one area may not have the same advantages in another area or under different circumstances (Harre et al., 1983).

Ammoniation-granulation plants continue to be a major marketing channel for mixed fertilizers. However, these plants have changed from production of relatively low-analysis grades based on normal superphosphate to high-analysis materials based on the ammoniation of phosphoric and sulfuric acids to produce granular high-analysis grades. Many producers are installing pipe-cross reactors to further increase production levels and to reduce energy requirements. Granulation plants also provide an excellent way to incorporate micronutrients into mixed fertilizers and to use by-product raw materials when economically advantageous.

In 1983, there were 81 ammoniation-granulation plants in the USA, of which almost half were in the Southeast. Total production from these plants was > 3.8 million metric tons/yr. Many produce materials for the nonfarm fertilizer market as well as for agriculture.

Compared with other manufacturing systems, the ammoniation-granulation plant is usually in or near large cities rather than in rural areas, and its larger capacity serves a much larger market area. Average throughput from ammoniation-granulation units is > 120 000 metric tons/yr. Only about half of the operating plants are exclusively ammoniation-granulation units—the remainder also produce DAP and MAP for sales to blenders and retailers in their area.

Bulk blends are produced from granular high-analysis materials and are often custom-manufactured to meet specific soil test requirements. The physical mixture is generally sold directly to farmers in bulk form rather than through wholesalers or conventional dealer outlets. The product is seldom stored, which distinguishes the bulk blender from the larger, conventional ammoniation-granulation plants. Most bulk blenders serve a market within a 5- to 30-km radius of the plant. In the present U.S. marketing system, the blender is primarily a direct marketer of fertilizers to farmers and provides adequate product storage as close to that market area as possible.

There are about 5000 bulk blend plants in the USA. The typical bulk blend plant has an annual throughput of 4900 metric tons/yr. About 2800 metric tons of this total distribution is dry bulk mixtures. Blenders also distribute significant tonnages of solid direct application materials as well as N solutions and ammonia. The average analysis of blended mixtures is generally higher than for other manufacturing systems.

While the average throughput of a U.S. blend plant is > 4500 metric tons/yr, about half of the plants distribute only 1000 to 3500 metric tons/yr. This reflects the fact that most of these plants distribute fertilizer within a very small market area.

Storing raw materials is an important function of the blender-retailer because it allows basic producers to operate plants at relatively constant production levels throughout the year. In a typical fertilizer season, blenders fill more than two-thirds of their storage during the winter months. Since there is very little farmer storage of fertilizers, blenders generally provide storage for almost half of their total annual distribution.

The number of fluid-fertilizer-mix plants has also grown at a remarkable rate during the past 15 yr. However, production of liquid and suspension fertilizers has been hampered at times by limited supplies of suitable P materials. In the 1950s, more elemental P and furnace phosphoric acid was available than was used in the industrial market. This surplus was neutralized with low-cost ammonia to produce liquid fertilizer materials. Production of high-N, nonpressure urea–ammonium nitrate solutions (UAN) further promoted liquid fertilizers, as these solutions can be directly applied or used to manufacture fluid fertilizers. Advantages of fluid fertilizers include ease of mixing, ease of incorporating additives and securing homogeneity of the mixture, convenience of mechanical handling, and high reliability of fluid application systems.

Consumption of all fluid mixtures (liquids and suspensions) for 1979 was 4 million metric tons. In 1979, five states—Georgia, Illinois, Indiana, Iowa, and Texas—accounted for 50% of all fluid mixtures used. Suspensions comprise 40% of all fluid mixtures, up from 25% in 1974 and 33% in 1976.

A typical liquid mix fertilizer plant has a 3500 metric ton average annual throughput. A typical plant distributes 1900 metric tons of liquid mixtures, 720 metric tons of ammonia, 1170 metric tons of N solutions, and 650 metric tons of liquid direct application materials. Analysis for the grades distributed is usually below that of dry mixtures. These plants may also distribute bulk and bagged dry mixtures and materials.

Suspensions, relatively new on the marketing scene, are rapidly expanding in the U.S. market. *Suspensions* are defined as liquids in which salts are suspended by a suspending agent. This allows a wider range of phosphate materials to be used than with nonsuspension products. Suspension plants have an average throughput of 3250 metric tons/yr, including 1800 metric tons of mixtures. Almost 79% of these suspension mixtures is custom-applied, compared with 44% of the bulk blending mixtures. Average analysis is higher than for ammoniation-granulation products, but lower than bulk blends. Most of the 1100 suspension units in the USA are in rural areas and follow the same storage patterns as the bulk blend units.

No one marketing system dominates all others. With a changing product mix possible because of the depletion of high-quality P resources and the growing importance of imported materials in the U.S. fertilizer market, it is possible that there will be a shift from one system to another or that still a new system will emerge. One requirement remains, regardless of the system used—to get the fertilizer to the farmer at the right time and at the lowest cost.

C. Criteria for Nutrient Guarantees

In the early 1800s, limited knowledge of plant nutrition regarding the role of the soil solution in plant growth resulted in setting "water solubility" as a criterion for "available" nutrients. Initial attention centered on phosphates because the first patents for commercial fertilizers were for superphosphates, beginning with the work of H. W. Kohler for processing bones in Pilsen, Czechoslovakia in 1831. Early principles established in P nutrition of plants showed that most P uptake by the plant was via the soil solution.

Following these developments, commercial phosphates from acidified bones or from phosphatic slag became the first points of focus for government standards in the fertilizer industry. The first meeting of the Associations of Agricultural Chemists (now the Association of Official Analytical Chemists, AOAC) in 1884 was to establish official methods of determining available phosphoric acid, expressed as P_2O_5. At that meeting the official water solubility test was adopted, as well as the citrate insoluble test. A combination of different tests continued until 1949 when the official water and citrate solubility tests were adopted for phosphate.

In the latter part of the past century, German interest in promoting basic slag led to the citrate test. Basic slag has very little water-soluble phosphate, but German agronomists measured crop responses to levels of P in the slag soluble in ammonium citrate ($NH_4C_6H_5O_7$). Refinements in this fraction led to designating neutral 1 *M* ammonium citrate as part of the official method in use today (formally adopted by AOAC in 1950).

In this official method, monocalcium phosphate [$Ca(H_2PO_4)_2$] present in normal and triple superphosphates, and the P in ammonium phosphates are almost completely extracted with water. The citrate fraction (neutral ammonium citrate heated to 60°C) includes some dicalcium phosphate ($CaHPO_4$), such as in superphosphates, and a fraction of the P in basic slag and bone meal. That remaining as citrate-insoluble is not included in the guarantee for commercial fertilizers.

Methods of analyzing for N have concentrated on total N, including NO_3^-, NH_4^+, and organic forms. Guarantees for N percentage in fertilizers usually are as the single number, percent total N. For some specialty crops, such as tobacco (*Nicotiana tabacum* L.) and vegetables, NH_4^+, and NO_3^- percentages may be included. Suppliers are required to indicate the percentage of ammonium N (NH_4^+–N), nitrate N (NO_3^-–N), water-insoluble N, and other recognized and determinable forms of N for specialty fertilizers.

The criterion for K guarantees in fertilizer is soluble potash, or that part soluble in boiling ammonium oxalate [$(NH_4)_2C_2O_4$]. Potassium in fertilizers is as inorganic salts, i.e., potassium chloride or potassium-sulfate and is water-soluble.

Such developments have resulted in today's rules and language for describing commercial compound or mixed fertilizers. These are encapsulated in the *grade,* or guaranteed analysis stated in whole numbers as the minimum percentage by weight of total N, available phosphate (P_2O_5), and soluble potash (K_2O). Guarantees of secondary and micronutrients are expressed as the minimum percentage by weight of the element, and with few exceptions these nutrients are expressed as the total form with no solubility criteria.

Most state fertilizer laws and regulations require commercial fertilizers to have a total nutrient content of at least 16 to 20% (N, P_2O_5, and K_2O). Exceptions may exist for specialty, or nonfarm, fertilizers. The Uniform State Fertilizer Bill, developed by the Association of American Plant Food Control Officials (AAPFCO), establishes minimum levels of secondary and micronutrients if claimed or guaranteed (Table 1–3).

Table 1–3. Minimum percentage for secondary and micronutrients if claimed or guaranteed (Young & Johnson, 1982).

Element	Percent
Ca	1.00
Mg	0.50
S	1.00
B	0.02
Cl	0.10
Co	0.0005
Cu	0.05
Fe	0.10
Mn	0.05
Mo	0.0005
Na	0.10
Zn	0.05

Most state fertilizer laws have developed in conjunction with methods of analyses and standards for quality control. With these provisions, administrative procedures also have evolved covering requirements for registering grades, licensing plants or firms, payment of fees, and tonnage reporting. Fertilizer laws exist in all states except Alaska and Hawaii. No federal law exists and the fertilizer industry has supported state controls. Industry has also actively supported uniformity among states in these statutes based on the Uniform State Fertilizer Bill of AAPFCO.

Important elements of state fertilizer laws include: licensing of firms or product registration with corresponding fees, inspection fees, tonnage reports, quality control (investigational allowances and penalties), regulations for the handling and storage of ammonia, and mobile equipment regulations. State departments of agriculture and, in a few cases, state land grant universities administer these laws.

III. FERTILIZER CONSUMPTION AND PRODUCTION

A. World Fertilizer Market Trends

World plant nutrient consumption in 1982 exceeded 93 million metric tons. This was a 36 million ton increase over 1972 and four times the plant nutrient use in 1962. The world fertilizer market has increased steadily, with an annual growth rate of 7.2% from 1965 to 1985. Only twice in this period—1975 and 1982—has the use of plant nutrients worldwide failed to register a gain over the previous year. Five-year forecasts indicate that increases in world plant nutrient use will continue, but at somewhat lower annual growth rates. As world plant nutrient consumption approaches 120 million metric tons by the end of the 1980s, the growth rate will average 3.5 to 4.0%/yr (Fig. 1–4).

If these forecasts are correct, they signal changes in the trend of the ratios of plant nutrient use (Fig. 1–5). After the growing N/P ratio, forecasts indicate that this ratio will stabilize at about a 4.5:1 relationship. Lit-

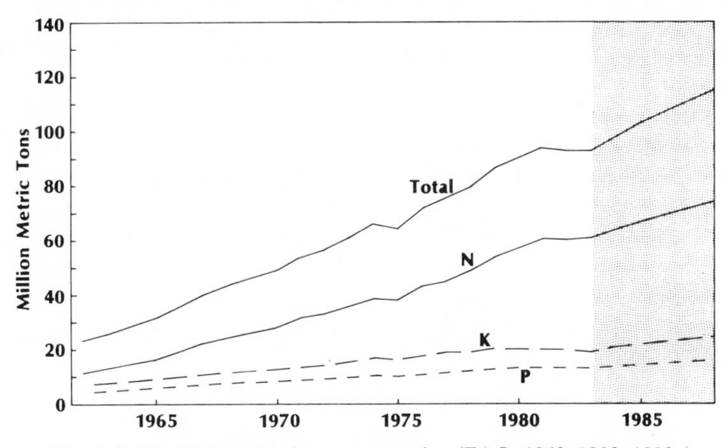

Fig. 1–4. World plant nutrient consumption (FAO, 1962–1983, 1983a).

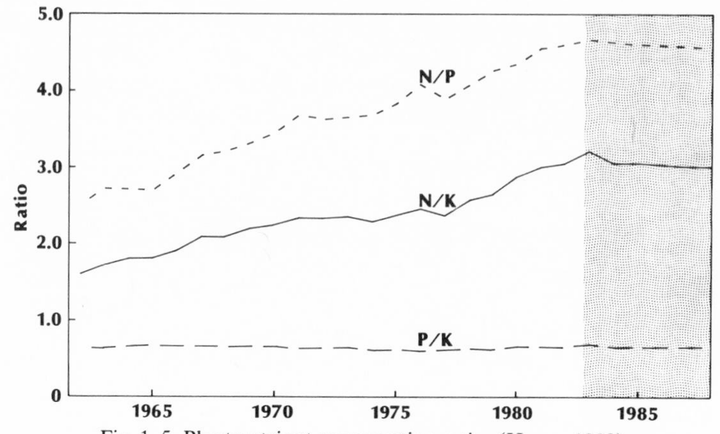

Fig. 1–5. Plant nutrient consumption ratios (Harre, 1983).

tle change is expected between P and K; N/K ratios will also begin to stabilize.

No N supply shortages are expected until 1987 as the world expansion of N capacity begun in the early 1980s nears completion. However, with few production units scheduled beyond 1987, world N demand will begin to pressure supply capabilities, and shortages could occur.

Phosphoric acid capacity and P demand are expected to increase at about the same rate. The large potential supply of P appears likely to outrun demand for several years after the plant building cycle ends. A similar situation exists for K; however, the magnitude of the surplus is smaller than that in the P sector.

1. Nitrogen

Buried in the changing supply-demand outlook are several market developments with international trade implications as changing market share trends begin to stabilize. In 1980 the Centrally Planned Economies (CPE) emerged as the leading users of N fertilizers in the world, and their share has continued to expand. However, forecasted market growth indicates that both the CPE nations and the developed nations will begin to lose some of their market share to the developing countries (Fig. 1–6).

A regional analysis of market shares for supply and demand presents an indication of future N supply sources. The developed regions, with continuing declines in market share in both N consumption and ammonia capacity, will emerge with similar shares for both supply and demand. They will meet their own needs but will not remain the large exporters that they were before 1980.

Nitrogen market shares in developing regions continue to indicate a supply deficit through 1988. The gap continues to narrow, however, as more and more capacity is being built in these countries.

In 1977 the CPE nations moved from a deficit market share in N supply-demand to a surplus. Through 1988 they will continue to have a sur-

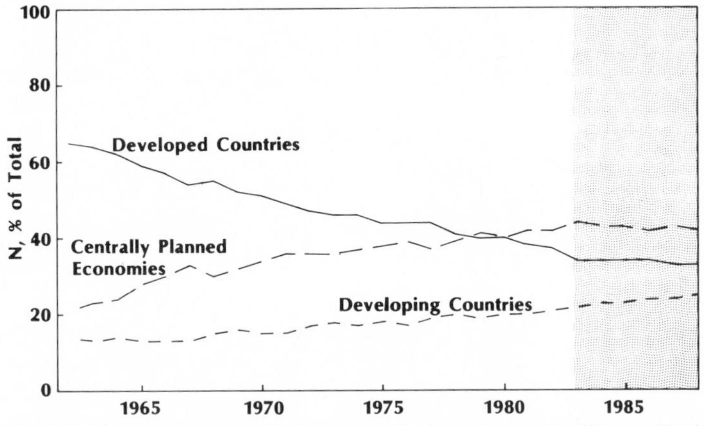

Fig. 1–6. Nitrogen consumption by economic class of countries (Harre, 1983).

plus, which will allow them to maintain a prominent role in international trade and replace some of the traditional, developed world suppliers.

Concentrating production facilities at raw material sites rather than in the marketplace will continue. This trend is already true in K and, to a great extent, in the world P market as North Africa and the USA dominate world trade. This change is rapidly taking place in the N industry as more and more ammonia/urea complexes are located in the natural gas–rich areas of the world.

Nitrogen production has been located primarily in the developed regions—North America, western Europe, and Japan. However, the rapidly changing world energy supply situation and rising energy costs are shifting the location of N supplies. As indicated in Table 1–4, natural gas is

Table 1-4. Estimated proved reserves of natural gas (Harre & Young, 1983).

Region	Natural gas	Share of world natural gas reserves	Share of ammonia capacity		Ammonia capacity based on natural gas (1980)
			1980	1985	
	$m^3 \times 10^{12}$	%			
North America	7.8	11	19	18	96
Latin America	4.5	6	4	6	87
Western Europe	4.4	6	17	14	67
Eastern Europe	26.1	35	32	35	85
Africa	5.8	8	2	3	45
Asia	24.3	33	25	27	35
Oceania	1.0	1			42
Developed countries	13.2	18	41	36	75
Developing countries	33.9	46	17	21	58
Central Planned Economies	26.8	36	42	43	72
World	73.9				71

the most common feedstock for ammonia production. Developed regions, however, account for < 20% of world reserves; the developing regions are estimated to control almost 50% of all natural gas reserves. The developing nations have < 20% of the total world ammonia capacity, but their share will increase as the developed nations continue to decline in relative importance. This shift in N capacity to areas with large natural gas reserves will lead to an increased level of N trade as greater amounts of both ammonia and urea are shipped to developed regions.

2. Phosphate

In 1983 the developed regions and the CPE countries reached an equal share of world P consumption. Neither group is expected to drastically change market share by 1988. At the same time, the developing regions will slowly become more important in the world phosphate market.

In 1967 the developed countries had almost 90% of the world's phosphoric acid capacity; by 1988 this capacity level will have declined to about 50%. Since 1967 the developed regions' share of total P demand has declined at about the same rate; thus, the developed regions continue to have excess capacity and remain a strong factor in world P trade.

While only accounting for about 25% of the total P market, the developing nations have steadily increased their share of world phosphoric acid capacity. They will become a larger factor in international P trade, as capacity expansions grow at a much faster rate than expected demand.

The CPE countries will continue to have a large deficit in phosphoric acid capacity. Since they still rely heavily on production of ground phosphate rock, normal superphosphate and other products not based entirely on phosphoric acid, the total deficit will not be as large as indicated. The region as a whole, however, will continue to be a net importer of P if building plans proceed as scheduled.

3. Potash

Potash market share for all economic classes of countries will change very little by 1988. The leading market share will remain in the developed regions, followed by the CPE nations. The developing world, while increasing K use more rapidly than other areas of the world, will only increase its share of the market by slightly more than 2%.

Since only a small addition to K capacity is expected in the developing regions, they will remain large net importers through the 1980s. The developed regions will remain net exporters. In 1976 the market share of K capacity exceeded the market share of K demand in the CPE nations for the first time. With capacity increasing faster than demand, the CPE nations will be in a position to expand their future export markets.

B. Fertilizers in the USA

While world fertilizer use was increasing at a 7% annual rate from 1962 to 1982, fertilizer consumption in the USA was increasing at a rate of just below 5%/yr. Between 1962 and 1972, U.S. consumption increased by an average annual rate of 7.4%, but the average for 1973–1983 was at 2.3%/yr (Hargett & Berry, 1985). This significantly lower rate is mainly because of the 9% decline in use from 1981 to 1982. A further 14% decline occurred in 1983, reflecting a large cut in planted area and generally depressed world economic market conditions. In 1984 the U.S. fertilizer market recovered most of its losses of the past 2 yr. Whether it returns to rates of increase similar to the trend of the previous 10 yr will depend on the agricultural and economic recovery in general and on agricultural exports in particular.

With slower rates of growth in fertilizer consumption, the U.S. share of world fertilizer use has steadily declined. In 1962 the USA accounted for 25% of world plant nutrient use; in 1982 this share was 17%. The USA still dominates the world market as it ranks first among all countries in terms of total plant nutrient consumption.

Through the 1980s, U.S. fertilizer market growth rates should remain below the world average. This trend has significant implications to the U.S. market structure. As its share declines, the U.S. market is influenced more by world events and trends and less by domestic market developments. The U.S. farmer thus not only competes with his neighbor but also with farmers worldwide for his fertilizer supply. His position is further jeopardized by his operating in a free market system, while > 70% of the world market is either under direct government control or subject to a wide range of subsidies that effectively reduce the competitive nature of the world fertilizer market. In the world trade sector, more and more fertilizer sales will be on a government-to-government basis with prices being dictated rather than allowed to fluctuate with changes in supply and demand.

1. U.S. Agriculture

The U.S. farmer has been among the most productive in the world. The USDA index of total agricultural production showed a gain of > 35 points from 1970 through 1981 when farmers set a new record in total crop production. This achievement was based on an area about equal to that under cultivation in the 1930s, but with the use of record levels of plant nutrients and other farm inputs (Fig. 1–7). To fully understand the role of fertilizers in agriculture, it is necessary to review recent agricultural market trends since the fertilizer market expansion began in 1970.

The recent growth in farm output was spurred more by a rapidly expanding export market for grains than by growing domestic use. Volume of agricultural exports increased from 63 million metric tons in 1970 with a value of $7 billion to > 162 million metric tons in 1981 worth almost $44 billion (USDA, 1982). About 33% of all cropland planted in 1981 was to produce crops for the export market.

Record crop production levels have been a mixed blessing for the farmer. Crop prices declined steadily as domestic and export demand failed to keep pace with supply, boosting crop inventory levels, and depressing farm prices. Repeatedly, farmers have reacted to favorable mar-

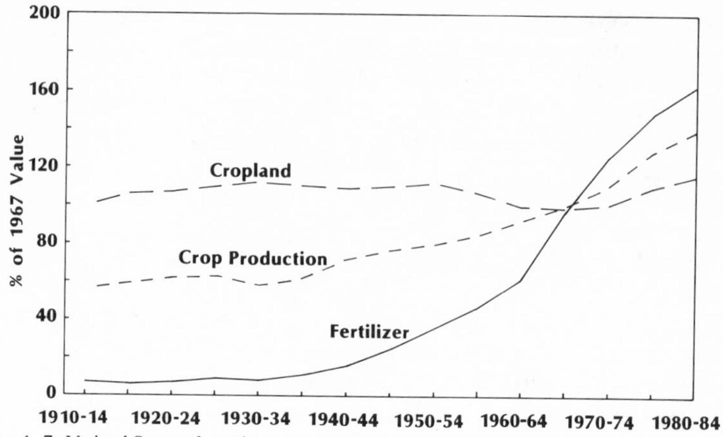

Fig. 1–7. United States farm input-output index values (USDA, 1976, 1983, 1984).

ket conditions by rapidly increasing the use of farm inputs to increase production, only to see commodity prices falter. In periods of low prices, farmers also intensify use of farm inputs in an attempt to earn maximum income from every hectare to cover their high level of fixed costs and investment. This condition has generally led to some form of production control program by USDA, which usually limits planted area and thus the use of other farm inputs.

The farmer constantly faces a cost-price squeeze, and the farm input/cost input ratio remains a key factor in the fertilizer demand equation. Fertilizer remains a vital input for agriculture because it continues to have a substantial benefit/cost ratio. As long as this ratio remains favorable, fertilizer use will be maintained at high levels and will expand as the nation's agricultural markets grow.

Each of the major plant nutrients has experienced a unique market pattern from 1965 to 1985 and supply/demand relationships will continue to change.

2. Nitrogen

In the USA, N use has exceeded that of P or K by a wide margin. Over the 1960–1982 period, N use increased an average of 6.5%/yr compared with $<3\%$ for P and 4.5% for K. In 1982 total N use was almost 10 million metric tons compared with 2.5 million tons in 1960 and 6.8 million tons in 1970 (Table 1–5). On a regional basis, N use is centered in the East and West North Central states where average annual growth rates have consistently exceeded the national average. In 1982 $>50\%$ of all N applied in the USA was used in these two regions. Corn (*Zea mays* L.), wheat (*Triticum aestivum* L.), and soybeans accounted for 69% of the cropland harvested.

Nitrogen production, however, is centered in the Gulf Coast region and Great Plains. Louisiana accounts for 33% of U.S. ammonia capacity, followed by Oklahoma with 11% and Texas with 9%. Other major producing states are Mississippi, Iowa, Alaska, and Arkansas. These seven states

Table 1-5. United States consumption of fertilizers and plant nutrients (Bridges, 1983; Hargett & Berry, 1985).

Year	Total fertilizer	N	P	K	Total nutrients
	metric tons × 10^6				
1960	22.6	2.5	1.0	1.6	5.1
1961	23.2	2.7	1.0	1.6	5.4
1962	24.1	3.1	1.1	1.7	5.9
1963	26.2	3.6	1.2	1.9	6.7
1964	27.8	3.9	1.3	2.1	7.3
1965	28.9	4.2	1.4	2.1	7.7
1966	31.3	4.8	1.5	2.4	8.8
1967	33.6	5.5	1.7	2.7	9.9
1968	35.1	6.2	1.8	2.9	10.8
1969	35.3	6.3	1.8	2.9	11.1
1970	35.9	6.8	1.8	3.0	11.6
1971	37.3	7.4	1.9	3.2	12.5
1972	37.4	7.3	1.9	3.3	12.5
1973	39.3	7.5	2.0	3.5	13.0
1974	42.7	8.3	2.0	3.8	14.2
1975	38.5	7.8	1.8	3.4	12.9
1976	44.6	9.4	2.1	3.9	15.4
1977	46.8	9.7	2.2	4.4	16.3
1978	43.1	9.0	2.0	4.2	15.2
1979	46.7	9.7	2.2	4.7	16.7
1980	47.9	10.3	2.2	4.7	17.2
1981	49.0	10.8	2.2	4.8	17.7
1982	44.2	10.0	1.9	4.2	16.1
1983	38.4	8.3	1.6	3.7	13.6
1984	45.5	10.1	2.0	4.4	16.5

comprised about 70% of the total U.S. ammonia capacity in 1980 (Bridges, 1983).

The ability of the USA to supply all of its N requirements will continue to undergo change as costs of natural gas increase. Since capacity reached 20 million metric tons of ammonia in 1978, there have been many plant closings. United States ammonia capacity in 1983 was slightly more than 15 million metric tons, because the nation looked to imports rather than pay high prices for natural gas feedstocks. The result has been a steady increase in N imports into the USA and a significant increase in the N trade deficit (Fig. 1–8).

With high reserves of coal in the central and western areas, the USA has the potential to produce enough N to meet its needs. The large investment in coal gasification facilities for N production makes the economic feasibility of a shift in feedstock for N unlikely until natural gas prices increase significantly. A shift to coal gasification for N production in the USA would also require a dramatic change in the distribution system for N materials.

Regardless of feedstock source, there will be a general upward movement in N production costs and in retail prices paid by farmers. It will be

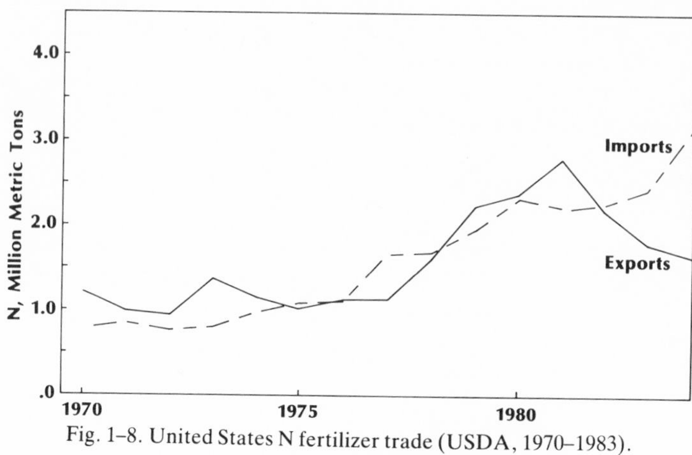

Fig. 1–8. United States N fertilizer trade (USDA, 1970–1983).

more difficult to maintain the favorable crop/N price ratio previously enjoyed, and efforts will be made to increase the efficiency of N fertilizer use through development of new products or those containing nitrification or urease inhibitors.

3. Phosphate

Gains in P consumption averaged < 1% from 1972 to 1982. In 1982 total P use was about 1.9 million metric tons compared with 1.0 million metric tons in 1960. Total usage in 1972 and 1982 was almost identical as P use reached its peak of 2.2 million metric tons in 1979. Phosphorus use should continue to increase slowly. However, with many soils testing adequate in available P levels and requiring little more than maintenance applications, additional growth must come as a result of higher per-hectare crop yields that remove more P. By the end of the 1980s, P use may not equal its 1979 peak. Because more than half of the nation's P use is centered in the central Corn Belt, future consumption patterns will depend on the market for corn and soybeans—the major crops in this area.

For many years, P production was based on normal superphosphate produced in large plants located primarily in the southeastern USA. With development of ammoniation-granulation technology and DAP, production shifted to high-analysis granular materials, but the primary production units remained in the market area. After hybrid corn was introduced, P demand in the Midwest began to expand and production facilities were inadequate to meet demand. Most new plants were located primarily at phosphate rock mines to take advantage of the freight savings by shipping a finished fertilizer material with a higher analysis instead of raw materials. With the introduction of DAP, the blending and distribution of bulk materials changed the structure of the market. Today, production is located at the raw material site. Of the > 200 normal superphosphate plants operating in the 1950s, only 25 exist today (USDC, 1959, 1984). Mixed fertilizers produced through the ammoniation-granulation process has declined from > 7.7 million metric tons in 1979 to about 4.1 million in 1982.

The U.S. phosphate fertilizer industry is based almost entirely on the production of phosphoric acid made from reacting phosphate rock with sulfuric acid. Wet-process phosphoric acid capacity is almost 4.0 million tons of P—twice the level needed to meet domestic P demands. Since 1975, the U.S. industry has become the major world supplier of P fertilizers, with almost 60% of all P entering world trade originating at U.S. production points. With foreign trade taking > 40% of the total U.S. phosphate production, the 4 million tons of capacity available in the USA has been able to run at a high operating rate and has allowed the construction of large-scale plants competitive with producers in other areas of the world.

With world demand growing faster than domestic use and the continuing participation of U.S. suppliers in world trade, P exports will account for a greater share of total U.S. production than domestic use. Farmers in the USA will be competing with farmers and governments throughout the world for their P supplies, because world market events rather than domestic changes will dictate fertilizer prices and supply levels.

4. Potash

Potash markets in the USA and Canada have been combined because 75% of the total K used in the USA is produced in Canada. Total K consumption in 1982 was 4.2 million metric tons in the USA and 285 000 tons in Canada. Potash has continued to show an increase in use and has ranked second in total nutrient use in North America since 1975. Potash should continue to expand in use at a rate between N and P consumption rates.

Development of the mines in Canada in the early 1960s ensured North America with an adequate supply of K to replace the reserves mined for many years in Carlsbad, NM. Canada also has become a major supplier of potash to the world market with exports accounting for 33% of total North American K use. With the continued expansion of potash capacity and the relatively slow growth in domestic demand, exports will become a much greater factor in the supply-demand picture in North America in the 1980s.

C. Importance of World Trade

In the past 20 yr, world fertilizer trade patterns have changed significantly. While the overall percentage of total world consumption represented by trade has only changed from 26 to 28% in this period, the pattern for the individual nutrients has shifted drastically.

Nitrogen trade as a percent of total N consumption has decreased from 26% in 1962 to 19% in 1982, which indicates the trend toward the establishment of production units in developing nations with natural gas reserves. This trend also reflects the withdrawal from trade of some of the traditional suppliers, because their feedstock source no longer allows them to be competitive with these new production areas. As most of the new N plant construction is taking place in regions with substantial natural gas reserves, but with only a small domestic demand base, N trade will once again contribute to a greater share of world consumption. When ammonia trade is included in the overall trade statistics, this trend is much more pronounced because of the recent dramatic growth of this market.

Phosphate trade has moved from only 12% of the total consumption in 1962 to 21% in 1982. When phosphoric acid is included, the shift to a greater reliance on trade is even more pronounced. Phosphate trade is centered in the USA and North Africa, which have extensive reserves of phosphate rock. These areas have established production units to upgrade phosphate rock to intermediate or finished fertilizers for export. With Morocco expanding its phosphate rock and phosphoric acid capacity, further increases in the share of total P consumption met by trade are expected.

The most dramatic shift in trade patterns has been with potash. In 1982 about 62% of the K used worldwide was supplied through trade. Since there are relatively few large-scale mines, most countries must meet their K needs through international trade. Canada, USSR, and eastern and western Europe remain the major suppliers with most of the recent production expansion coming in Canada and the USSR. While several projects are being considered in the developing countries, they are not expected to have any significant impact on trade patterns.

Fertilizer trade and the interdependence of producing and consuming nations will be important in ensuring an adequate supply of fertilizer. Inconsistent buying patterns, leading to wide variations in prices from season to season, tend to curtail an adequate inflow of investment capital for new facilities. In the long-run, an unrealistic price level can lead to short supplies. Suppliers must also realize that pricing policy and an overreaction to brief shortages will not be profitable if the result is a decrease in demand because the farmer cannot afford the fertilizer.

The desire to ensure an adequate fertilizer supply will expand world trade as the industry continues to move production facilities toward raw material reserves. The future should see traditional suppliers investing in these new production areas if economic and investment conditions allow an adequate return. Without these guarantees, producers would be forced to use high-cost feedstocks. This would increase production costs and lead to higher farm production costs. As world population growth continues to pressure the world's food supply, it is imperative that the inputs to expand food production remain in adequate supply and available to all nations.

IV. SUMMARY

Commercial fertilizers, materials sold primarily for their plant nutrient content, are essential to maintain and increase world crop production. At present rates of use, they account for 20 to 30% of the world's crop production, with the percentages much higher in developed countries. As pressures continue to increase on arable land, the role of fertilizers as a land substitute will increase in importance.

An underlying purpose in all fertilizer production processes is to produce materials containing plant nutrients in an available form. Plants absorb nutrients primarily from the soil solution, hence, the necessity of producing fertilizers with soluble forms of nutrients. This condition is particularly important for P production where the major role of production processes is that of solubilizing forms of insoluble calcium phosphate materials.

The principal aim in the N production processes is to chemically fix N in forms (NO_3^- or NH_4^+) that plant roots can absorb. This process is very energy intensive, requiring 55.5 to 64.8 MJ/kg of N. The major role in K production is that of mining and refining potassium chloride and potassium sulfate minerals and removing impurities.

A natural obstacle to producing fertilizers worldwide is the scattered nature of many fertilizer resources. This particularly applies to the mineral deposits of P and K. Many countries have no known resources of these nutrients. As for N production, countries with low-cost natural gas sources have distinct comparative advantages as basic producers.

International trade becomes the distributor, with the aim of abolishing uneven distribution of these fertilizer resources. Increasingly, exports and imports of fertilizers among countries is the means of supplying individual farmers with N, P, and K. Brine from several major lakes, such as the Dead Sea, and underground deposits of K salts in Canada and the USSR are distributed worldwide by trade. Sedimentary deposits of phosphates in the USA and Morocco are the principal sources of P materials.

The role of the fertilizer industry is changing rapidly from a producing industry to that of a distribution industry. Distribution in turn depends on market development and trends in fertilizer consumption. Some of the developing countries have expanded consumption rapidly, with annual average increases in tonnage ranging from 15 to 20%. In contrast, in many of the developed countries the market has been increasing at a much slower rate, in the range of 2 to 5% annually.

The fertilizer market has international dimensions—fertilizer is truly an international commodity. Developments in the U.S. market will depend increasingly on market trends in agriculture as well as the fertilizer industry.

REFERENCES

Bridges, J. D. 1983. Fertilizer trends 1982. TVA Rep. TVA/OACD-83/6, Bull. Y-176. Tennessee Valley Authority, Muscle Shoals, AL.

Food and Agriculture Organization of the United Nations. 1962–1983. FAO fertilizer yearbooks. Food and Agriculture Organization of the United Nations, Rome.

Food and Agriculture Organization of the United Nations. 1981a. Agriculture: Toward 2000. Food and Agriculture Organization of the United Nations, Rome.

Food and Agriculture Organization of the United Nations. 1981b. Crop production levels and fertilizer use. FAO Fert. and Plant Nutr. Bull. 2. Food and Agriculture Organization of the United Nations, Rome.

Food and Agriculture Organization of the United Nations. 1983a. Current world fertilizer situation and outlook 1981/82–1978/88. Food and Agriculture Organization of the United Nations, Rome.

Food and Agriculture Organization of the United Nations. 1983b. FAO indices of food and agricultural production. Monthly bulletin of agricultural economics and statistics. 6 (11):9. Food and Agriculture Organization of the United Nations, Rome.

Hargett, N. L., and J. T. Berry. 1985. 1984 fertilizer summary data. TVA Rep. TVA/OACD-85/10, Bull. Y-189. Tennessee Valley Authority, Muscle Shoals, AL.

Hargett, N. L., and R. Pay. 1980. Retail marketing of fertilizers in the U.S. p. 82–95. *In* A. Spillman (ed.) Proceedings of the 30th annual meeting fertilizer round table 1980. Fertilizer Industry Round Table, Glen Arm, MD.

Harre, E. A. 1983. World fertilizer market trends. *In* IFDC fertilizer marketing management training program—1983. International Fertilizer Development Center, Muscle Shoals, AL.

Harre, E. A., N. L. Hargett, and R. J. Williams. 1983. The economics of bulk blending in the United States. *In* The economic aspects of bulk blends. Int. Fert. Ind. Assoc. Rep. A/F/83/57. International Fertilizer Industry Association, Paris.

Harre, E. A., and R. D. Young. 1983. Nitrogen compounds. p. 961–988. *In* S. J. Lefond (ed.) Industrial minerals and rocks, 5th ed., Vol. 2. American Institute of Mining, Metallurgical, and Petroleum Engineers, Society of Mining Engineers, New York.

Tennessee Valley Authority. 1983. The impact of TVA's national fertilizer program. NFDC Rep. TVA/OACD-83/5, Cir. Z-145. National Fertilizer Development Center, Muscle Shoals, AL.

U.S. Department of Agriculture. Agricultural Stabilization and Conservation Service. Emergency Preparedness Branch. 1970–1983. The fertilizer supply (annual report). U.S. Department of Agriculture, Washington, DC.

U.S. Department of Agriculture. Economic Research Service. National Economic Analysis Division. 1976. Changes in farm production and efficiency. USDA Stat. Bull. 561. U.S. Department of Agriculture, Washington, DC.

U.S. Department of Agriculture. Economic Research Service. 1982. U.S. foreign agricultural trade statistical report, fiscal year 1981, a supplement to foreign trade of the United States. U.S. Department of Agriculture, Washington, DC.

U.S. Department of Agriculture. Statistical Reporting Service. Crop Reporting Board. 1983. Commercial fertilizers, consumption for year ended June 30, 1983. USDA SpCr 7 (11-83). U.S. Department of Agriculture, Washington, DC.

U.S. Department of Agriculture. Statistical Reporting Service. Crop Reporting Board. 1984. Crop production, 1983 summary. USDA CrPr 2-1 (84). U.S. Department of Agriculture, Washington, DC.

U.S. Department of Commerce. Bureau of the Census. Industry Division. 1959. Facts for industry, superphosphate and other phosphatic fertilizers, summary for 1958. USDC M28D-08. U.S. Department of Commerce, Washington, DC.

U.S. Department of Commerce. Bureau of the Census. Industry Division. 1984. Current industrial reports, inorganic chemicals. USDC MA-28A (82)-1. U.S. Department of Commerce, Washington, DC.

Wortman, S., and R. W. Cummings, Jr. 1978. To feed this world. The Johns Hopkins University Press, Baltimore, MD.

Young, R. D., and F. J. Johnson. 1982. Fertilizer products. p. 45–68. *In* W. C. White and D. N. Collins (ed.) The fertilizer handbook. The Fertilizer Institute, Washington, DC.

2

Prescribing Soil and Crop Nutrient Needs

David A. Whitney
Kansas State University
Manhattan, Kansas

J. T. Cope
Auburn University
Auburn, Alabama

L. Fred Welch
University of Illinois
Urbana, Illinois

The primary objective behind most lime and fertilizer decisions is to apply the rates of lime and fertilizer that will realize the greatest profit from the crops produced. Alternative objectives exist such as crop quality, maximum yields, soil conservation, and environmental quality. The lime and fertilizer rates necessary to meet these latter objectives may be different from those for maximum profit. To meet any of these objectives, a similar approach for prescribing nutrients can be used.

Fitts (1959) suggested that yield could be described as a function of four factors: (i) the crop grown, (ii) the soil characteristics, (iii) the climatic conditions, and (iv) the management of the producer. Other breakdowns of plant growth have been suggested, but all illustrate that many variables are involved in determining the yield potential of the crop and ultimately in making the optimum lime and fertilizer rate decisions. Some of these variables can be controlled by management, while others, such as seasonal rainfall and distribution, cannot be controlled. Because of the strong interaction between plant response to fertilizer and some of these uncontrollable factors, prescribing the exact optimum rate of lime and fertilizer prior to growing the crop is difficult. This gives added incentives for using all information available in striving to meet fertilization objectives.

The discussion in this chapter will center on soil testing for prescribing current nutrient needs and the use of soil and plant analysis to monitor the pH and fertility status of soils for adjustment of future lime and fertilizer practices.

I. SOIL ANALYSIS

A soil test program has been described by various workers as having four distinctive parts, namely, (i) sample collection, (ii) chemical analysis of the sample, (iii) interpretation of the chemical test, and (iv) lime and fertilizer recommendations for the crop(s) to be grown. Although it is beyond the scope of this chapter to review and report all research information available, we will focus on some factors important to a sound soil test program.

A. Soil Sampling Techniques

Soil sample collection has often been described as the "weak link" in a soil test program. With the tremendous inherent variability in many soils and the additional variability caused by fertilizer and tillage practices, collection of representative soil samples is difficult. However, samples taken with proper care can be quite meaningful in making lime and fertilizer decisions.

1. Field Sampling

The objective for taking the soil sample should be considered before sampling a field. If the objective is to have an unbiased estimate of the mean fertility level of the field, then a single composite sample is justified (Cline, 1944). However, a single composite sample does not yield any information about variability within the field, which may be quite important in assessing production differences that exist between areas within the field.

A number of different sampling plans or designs have been suggested. Each has merit for particular situations. Petersen and Calvin (1965) discuss the merits of four sampling plans—judgement sampling, simple random sampling, stratified random sampling, and systematic sampling. Many sampling instructions use a combination of these sampling plans. Irrespective of what sampling technique is followed, there is unanimous agreement that a single core or shovel slice of soil is not a representative sample because of the heterogeneity that exists in most fields. However, the sampling plan must be simple and cost-effective for acceptance by those doing the sampling.

Numerous publications and soil test laboratory instructional guides exist on soil sampling. Most of these instructional guides advise that the field be divided into areas of visually uniform soil types with similar past management. They also recommend that a composite sample of 10 to 30 subsamples (cores) be taken at random over the area.

A survey was made in 1951 by the Soil Test Work Group of the National Soil and Fertilizer Research Committee on soil sampling procedures recommended by various states. A standard sampling procedure was proposed suggesting that a farm be divided into areas of uniform soil type and past management not larger than 2 to 4 ha each. If an area or a field was uniform in appearance, production, and past treatment, the sample may represent as much as 8 ha (Reed et al., 1953). These instructions are essentially those being used today by those advising farmers on proper sampling techniques. With an increase in the size of farm equipment and the concurrent increase in field size, limiting sample areas to 2 to 4 ha may not be practical. Cline (1945) suggested the minimum sample area be an area that a farmer can be expected to fertilize as a unit. In many cases it becomes the judgement of the sampler as to the number of subdivisions to make in the field.

James and Dow (1972) stated that a systematic sampling plan is superior because it provides estimates of the average soil fertility, its variability, and its distribution. Dow et al. (1973) showed that in most cases soil variation was not random and would lend itself to mapping and differential fertilization. In contrast to the above results, a systematic sampling of two 16-ha tracts of relatively uniform and nonuniform mapped soil types on 25 m grid by Illinois researchers indicated no pattern to P test values as related to a soil type or a specific area of the field (Peck & Melsted, 1973). These data illustrate that variability within a field may or may not show a consistent trend across an area. Illinois soil sampling instructions call for systematic subdividing of fields so that one composite sample is taken from every 1 to 2 ha. Five cores are suggested for the composite sample (Ill. Agron. Handb., 1982).

If the results from systematic sampling identify areas of low or high fertility, fertilizer rates can then be adjusted to the specific areas to lessen such variations and perhaps allow future sampling by a composite sample. Although detailed systematic sampling is an excellent technique, the time and expense for collection of samples will likely not be practiced except in research plots or fields being developed for production of high cash-value crops.

The ideal number of subsamples (cores) required for the composite sample to represent the sample area will depend on the nutrient being tested, soil variability within the area, sampling depth, past management, and the sampling plan being followed for the field. Statistical methods for determining the optimum number of subsamples for various sampling schemes have been published (Cline, 1944; Peterson & Calvin, 1965) and will not be presented here.

Sampling studies often reveal considerable variability among individual cores collected within a visually uniform soil area (Peck & Melsted, 1973; Hooker et al., 1976). The need for taking a sufficient number of cores for making the composite sample cannot be overemphasized. Hooker et al. (1976) found that to obtain a level of ± 3 mg/kg of soil test P, 14 to 20 cores per composite sample were required on two soil types studied. Other studies have shown that even much larger numbers of cores may be needed, depending on the soil type. The variability has also been shown to be different for various tests.

Thomas and Hanway (1968) suggested that two or more composite samples from the same area be taken as a field method for testing the adequacy of sampling. If the soil test results agree, then the area has been characterized rather well by the sampling technique used. Wide disagreement among composite samples indicates the need to reevaluate sampling, either by subdividing the sample areas or by increasing the number of cores.

More than one field should be evaluated in this manner before using the results to establish a general sampling guideline.

2. Sampling Depth

In addition to concern about variations in nutrient availability among areas of a field, the sampler should also be aware of vertical differences in nutrient availability. Most soil test development research has been done using a specific sampling depth. This makes it imperative that the sampler follow laboratory instructions on sampling depth or be able to adjust the interpretations for the altered sampling depth.

Subsoil fertility can contribute significantly to the available nutrient supply. Differences in the supply of available nutrients in subsoils may be a reflection of parent material, degree of weathering, crop rooting pattern, or restrictions in root development. Several states have classified soil types or broad soil association areas within the state as to their nutrient supply and have adjusted soil test interpretations accordingly (Schulte et al., 1980; Ill. Agron. Handb., 1982). Periodic testing of subsoils may prove to be helpful to individual producers. Additional research is needed to further define nutrient contributions by subsoils in most states.

Traditionally, suggested sampling depth on cultivated land has been the tillage depth. This has normally been a depth of 15 to 30 cm. On noncultivated land, such as pastures, a sampling depth of 5 to 15 cm has been suggested. With the rapid increase in the use of reduced or no-tillage seedbed preparation methods, a reevaluation of traditional sampling depths need to be made. However, any adjustment in traditional sampling depths must be calibrated into the interpretation of soil test results.

Greater soil test P gradients within the top 15 cm of soil have been shown on fields cropped with no-till seedbed preparation compared with more conventional methods (Welch & Fitts, 1956; Walker et al., 1970). With broadcast application of P remaining in the top 2 to 4 cm of soil with no-till planting, efficient plant utilization is a question that has not been extensively researched. Some recent research in the eastern USA suggests that P availability and utilization is quite good when broadcast on the soil surface. Kunishi et al. (1982) showed that corn made good utilization of P broadcast on no-till planted plots in Maryland, even though P remained in the top 3.75-cm layer. Broadcast P fertilizer showed little movement below 7.5 cm except at high application rates with disk and no-till seedbed preparation in Georgia (Touchton et al., 1982). Their conclusion was that P recommendations for disk and no-till systems should be based on soil samples taken from the upper 7.5 cm of soil rather than the more common 0- to 20-cm depth. In addition to concern with P accumulation within the top few centimeters of soil, a sharp decline in surface pH has been observed with no-till as compared with conventionally tilled fields (Blevins et al., 1977). This has implications for effectiveness of certain herbicides (Kells et al., 1979).

The above observations would suggest that splitting of the traditional tillage layer sample into short increments would be advantageous on fields that have been no-till farmed for several years. The results from vertical-increment sampling would reveal fields with nutrient accumulation or low pH in the top few centimeters. However, research data is needed to develop the best recommendations for succeeding crops.

Sampling depths of 60 to 120 cm are recommended by several of the Great Plains states (North Dakota, South Dakota, Nebraska, Kansas, and Oklahoma) for their available N test (nitrates). Annual rainfall in this area is normally insufficient to leach soluble nitrates below the rooting depth of most crops. In such situations, the available N test has good predictive value. Recommended sampling depth varies among states and within a state for specific crop and production practices. It would seem logical that deeper sampling would also be advantageous in testing for other soluble and mobile nutrients, such as sulfate-S. Further research is needed in this area.

3. Sampling Equipment

Sampling equipment used should provide a sample that is (i) uncontaminated, (ii) uniform in cross-section to the desired depth, and (iii) reproducible (Cline, 1944). A number of sampling tools, such as shovels, soil tubes, or augers, can meet these criteria if properly used. Welch and Fitts (1956) showed no difference in test values when sampling surface soil with a soil tube, trowel, or spade, but lower pH and organic matter results were probably due to loss of dry soil from the surface when sampling with an auger. No tool will be superior for all sampling situations.

There have been mechanical adaptations of the basic soil tubes and augers to aid in better and easier soil sampling. Hydraulic probes mounted in four-wheel drive vehicles have become quite common in the Great Plains states where sampling to depths greater than the traditional tillage layer is recommended for nitrate testing. Other types of vehicles with floatation tires can be used to transport the sampler over the field, with perhaps more representative samples being collected. However, any mechanization of sampling should be carefully examined to ascertain that the sampling objectives are realized.

4. Handling of Samples

Soil sample collection in the field should be done so that contamination does not occur from the collection container or other sources. Of special concern is contamination of the sample for micronutrient testing. If micronutrient analyses are to be performed, it is essential that all surfaces coming in contact with the soil be stainless steel, plastic, or wood, preferably in the order listed (Eik et al., 1980).

Thorough mixing of the composited cores and subsequent reduction of volume to that suggested by the testing laboratory should be done to ensure that a representative subsample is delivered to the laboratory and used in the testing. Laboratory processing of samples is geared to handle a specific volume of soil, and improper subsampling of large volume samples may lead to errors.

The sample should be handled so that minimal change occurs in the chemical or physical characteristics in the field. Ideally, testing of the soil as it exists in the field would be best. However, for ease of handling, soil samples are normally dried and pulverized by the testing laboratory before analysis. Drying of samples causes a marked change in the exchangeable K levels of many midwestern soils. The soil testing laboratory at Iowa State University had developed a system to handle moist samples (Eik et al., 1980). Immediate drying is necessary on samples analyzed for nutrients whose availability is primarily microbially related, such as nitrates. Most drying instructions suggest drying at no more than 35 to 40° C. Sample handling instructions are given by most laboratories; their instructions should be followed because their test correlation and interpretation are based on their prescribed handling of the samples.

5. Seasonal Change and Sampling Frequency

Seasonal trends in soil test values have been reported in the literature (Keough & Maples, 1972; Fitts & Nelson, 1956). Soluble salts and CO_2 may affect pH, while crop removal often affects the soil equilibria for P and K. The seasonal trend has been most consistent for K. Generally, samples taken during the growing season are lower in exchangeable K than those taken in the winter or early spring. Illinois adjusts their K soil test results on samples taken between 30 September and 1 May (Ill. Agron. Handb., 1972). In most cases, the seasonal trends have not been strong and sampling is encouraged whenever it can be done. Sampling at the same time each year would be best for research studies to assure that seasonal variation does not influence the results.

Crop removal and fertilizer additions are usually such that most routine soil test measurements do not change rapidly. Therefore, most recommendations suggest sampling every 2 to 4 yr for routine tests. The frequency of testing is adjusted in some areas by the intensity of agricultural practices and the buffering capacity of the soil. On irrigated cropland, some laboratories recommend sampling on a yearly basis.

Yearly samples taken for 2 to 3 consecutive years or multiple samples taken within a year are needed to establish the true fertility status of the area represented by the sample. This type of data base allows comparison of present and future tests to past test results in evaluating changes due to fertilization and liming. This also requires good records with field maps of areas represented by current and past samples. In the future, soil tests may be used more to monitor fertility levels in the field, making consistent sampling techniques even more important.

B. Methodology in Laboratory Operation

Variability associated with laboratory analytical tests are normally quite small when compared with field sampling variability. Most laboratories have a quality control system built into their operation. This often includes control samples that are run repeatedly in the daily operation. Several states have quality control programs to help laboratories operating in the state maintain quality control. This consists of periodically sending "check" samples to all laboratories. The question on analytical methods is more selection of the most appropriate method for the soil and climatic conditions than the analytical accuracy for the test within the laboratory.

1. Analytical Methods

The method used in the laboratory must be one that reliably measures nutrient availability in the sample. It must also lend itself to a routine soil test operation that handles a large volume of samples, and it must be inexpensive.

Discussion of individual analytical methods is beyond the scope of this chapter and the reader is referred to other references on analytical methods (Dahnke, 1980; Sabbe & Breland, 1974; Jones, 1980; Black et al., 1965; Page et al., 1982). As pointed out in these publications, modification of the method from the published procedure can alter the results, thus invalidating published interpretations. Any modification, even if slight, should be identified and thoroughly researched before incorporation into a soil test operation. Such modifications should be emphasized in publications reporting use of reference procedures.

2. Reliability of Test Methods

Technological advances in current instrumentation allows for quick and accurate detection of most essential elements. These instrumentation advancements have far exceeded the correlation and interpretation research necessary for development of a reliable soil test for several nutrients. Extreme care must be taken not to extrapolate test methods developed in one region into other regions where the need for application of the nutrient has not been documented through field research. There is also the temptation to analyze and report results for nutrients where no correlation research exists. Understanding the chemical reactions involved with a particular extraction procedure and knowledge of soil chemistry of the test area can be helpful in ascertaining the potential feasibility of the procedure. But this soil chemistry knowledge should not be the sole basis for adoption of the test method.

In addition to the capability to quickly and accurately detect most essential elements, instruments are used in testing laboratories to detect two or more elements simultaneously or in rapid succession in the same extract. This has led to research into universal extractants. Care should be taken that test accuracy is not sacrificed for laboratory convenience.

C. Correlation and Test Calibration Requirements

The native fertility of virgin soils is determined by the amount and form of nutrient elements contained in the parent material and subsequent changes by soil-forming factors, such as weathering and growth of native vegetation. Fertility status of these virgin soils can be determined by relatively simple field experiments designed to determine which elements are inadequate. When soils have been classified into types based on parent material, texture, site characteristics, etc., fertilizer recommendations can be based on soil type and crops to be grown, using data from simple rate experiments.

After soils have been put into production, fertility status is likely to be changed by fertilization, crop removal, erosion, tillage, and other practices. Under good management and fertilizer programs, soils generally improve in fertility, while those with poor management and inadequate fertilization will likely be depleted. Since management may cause similar soils to differ greatly in fertility status, general fertilizer recommendations based on soil type may no longer be satisfactory for maximum economic returns from the use of fertilizers. Depletion and/or buildup of specific nutrients have increased the need for soil tests to determine the need for fertilization of individual fields. This is demonstrated by data from a long-term experiment in Alabama (Fig. 2–1), which shows buildup of both P and

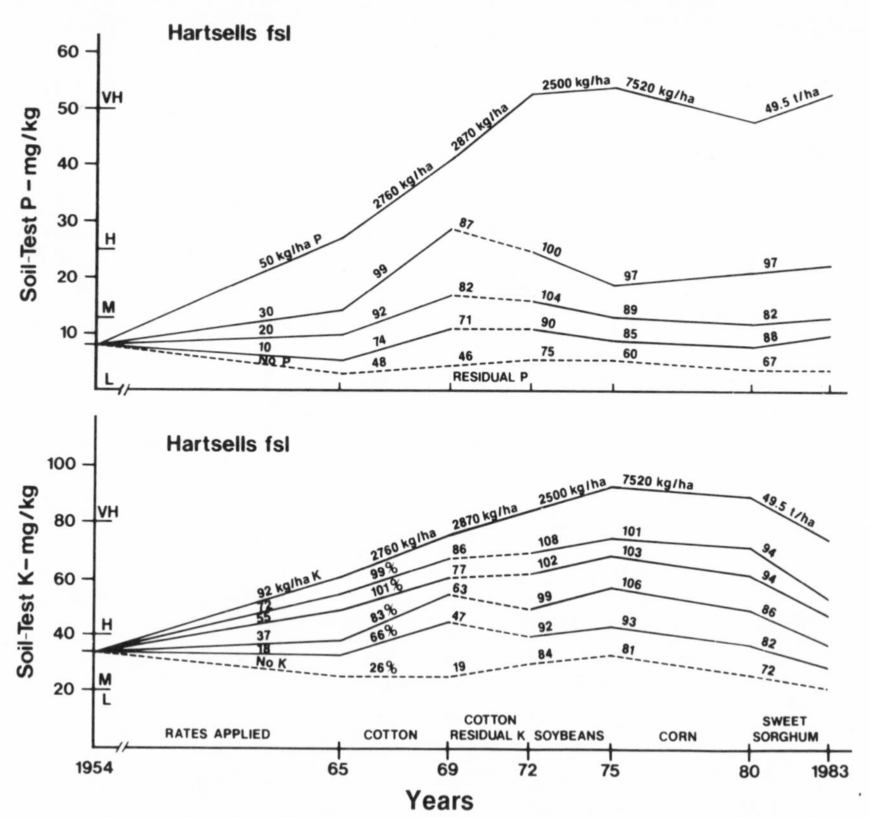

Fig. 2–1. Soil test P and K (Mehlich I) and relative yields of cotton, soybeans, corn, and sweet sorghum from 1954 through 1982 in fertilizer rates experiment on Hartsells fine sandy loam (dashed lines indicate residual period from 1970 through 1972 when none was applied). (Unpublished data.)

K on a Hartsells fine sandy loam (Typic Hapludult), with cation exchange capacity (CEC) of 6 cmol (+)/kg, from relatively low rates of fertilization applied over a 29-yr period. Data on soils over a 50-yr period showed that buildup or depletion of P and K were dependent on rates applied, but were quite consistent among these low CEC soils.

1. Collection of Data

To be useful in making fertilizer recommendations, soil tests must be calibrated against yield responses in field experiments. This calibration requires large numbers of experiments to develop reliable data on the numerous soil types and crops. Simple "some and none" experiments can be used for the purpose of determining soil test levels above which response is not expected. Such experiments can be used to determine if fertilizer is needed, but do not give much information on amounts that should be recommended. More detailed rate experiments on areas known to be deficient are necessary for determining optimum rate. Accuracy of data from rate experiments is improved by keeping plots on the same site for several crop years to permit buildup or depletion effects to be accurately expressed (Fig. 2–1).

Experiments for soil test calibration should be designed to determine the "adequate level" of a nutrient in a soil. This is the soil test level of a nutrient that is adequate for $\geq 95\%$ of maximum yield without addition of the element. This means that crops grown on soils having this soil test level of a nutrient would not generally be expected to respond to nutrient applications. At this soil test level, some laboratories still recommend small amounts as maintenance applications; many laboratories have ceased to make maintenance recommendations, as their data show that economic returns from applications at high soil test levels are unusual.

Percentage yield or relative yield is normally used in soil test calibration research for plotting data from numerous experiments conducted on different soils to normalize for individual location effects. Soils vary widely in soil test levels required for adequacy and should be grouped into categories showing similar response. Such groupings may be based on CEC or on soil associations such as Coastal Plain or Piedmont soil regions, which would serve the same purpose. Crops also vary in fertility requirements, so experiments should be conducted with as many crops as possible so that crops can be grouped according to response characteristics. This is demonstrated in Fig. 2–1, where relative yields of cotton can be compared with those of soybeans, corn, and sorghum, where P or K is deficient.

2. Calibration Experiments

Initial calibration can best be accomplished using large numbers of simple experiments on soils covering a range of fertility levels. Only two rates of a nutrient are required to determine if a soil will respond, but four or more rates are generally needed to determine the optimum amount needed on responsive soils. Experiments should include rates of all nutrients suspected to be deficient. This can be determined by soil tests prior to site selection. Prior to the establishment of experiments, soil samples should be taken from at least each replication to determine within-site variation.

Selection of treatments should begin with a standard treatment including what is considered to be adequate amounts of all applied nutrients. Yield from this treatment is assigned a value of 100% and yields from other treatments are expressed as percentages of this treatment. Response to other nutrients is determined by omitting one at a time, while applying adequate amounts of other nutrients that may be deficient. Several rates of a nutrient may be used, but other practices, including the use of lime or micronutrients, should be constant and adequate. Such experiments do not require an unfertilized check, because the law of the minimum makes it impossible to accurately determine the need for one nutrient when another is limiting the yield. Field experiments of this type are seldom precise enough to accurately determine interactions among nutrients where more than one is limiting. When yields are seriously limited by drought, poor stands, or other uncontrolled factors, data should not be used in calibration.

3. Statistical Methods

Randomized block designs with three to five replications have been used most frequently in calibration work with large numbers of simple experiments on similar soils at different sites. Many procedures have been reported for analyzing and partitioning data for use in calibration. Many response equations have been developed, but few have been used extensively in making fertilizer recommendations from soil tests. Detailed calculations for making recommendations for individual sites are generally not justified because the entire process of making fertilizer recommendations from soil tests involves a series of approximations. These include sampling variation in the field; analytical variation in the laboratory, which is generally much less; experimental variation incurred in field calibration experiments; variation in yield level and response among experiments on different soils grouped together; variation among crops in fertility requirements; and safety factors associated with suggested use of higher fertilizer rates than that shown to be needed by research data. Farmers frequently increase applications above recommended rates because of inaccuracy in application rates and/or to compensate again for the factors above. Even with all of these limitations, a good soil test recommendation program is the best and most reliable basis for determining fertilizer needs.

Perhaps the most frequently used calibration procedure is demonstrated in Fig. 2–2, where each point plotted represents one experiment. Regression curves from such data may be calculated as described by Bray (1948), Melsted and Peck (1977), and others. Cate and Nelson (1971), Fitts and Hanway (1971), and Nelson and Anderson (1977) described a simplified system for partitioning data based on probability of response. This system lacks precise definition for the "critical level" which may be needed in making recommendations. Hauser (1973) points out that in many cases

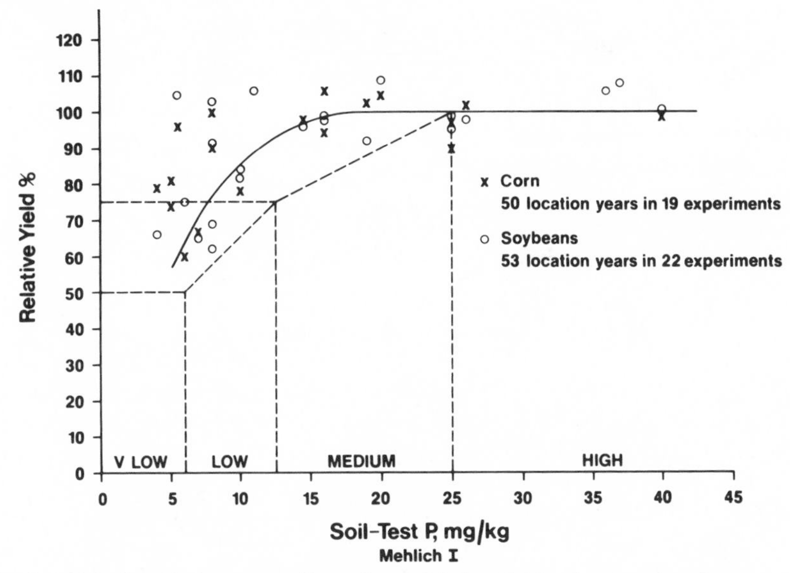

Fig. 2–2. Relationship between soil test P and relative yields of corn and soybeans on low CEC soils in Alabama, 1976–1981. (Unpublished data.)

hand-drawn curves may be adequate. These can be fitted into a system such as described by Cope and Rouse (1973), where the curve is divided into sections based on definitions of soil test ratings for percentage crop yields predicted from the curve (Fig. 2–2). Percent yield refers to that yield from plots receiving none of the nutrient, relative to that from an adequate supply, while other nutrients and inputs are optimum. The points plotted in Fig. 2–2 are in most cases 2- to 5-yr averages from long-term experiments where soil test levels of check plots were well established and quite stable. About two-thirds of these plots had not received P applications for 25 yr. The solid line curve was hand-drawn through the data points. The dashed line is the interpretation now in use in the Auburn University Soil Testing Laboratory on which P recommendations are based for corn and soybeans. These recent yield data show that very little response was obtained on "medium" fertility level soils in these long-term experiments. The discrepancy between the two curves shows that present P recommendations are more than adequate. These data indicate that a large safety factor is built into this system and that consideration should be given to lowering soil test levels required for "medium" and "high" ratings. The data suggest a need for constant and continued updating of calibration programs that are based on old data. New experiments should be conducted to keep calibration current with increasing yield levels and changes in prices of fertilizer and crops produced.

4. Greenhouse Studies

Greenhouse studies are useful in comparing soils and evaluating their relative nutrient-supplying capacity by growing plants in the same environment, except for soil differences. Total nutrient uptake can be measured by analyzing plant materials. Different extractants can be compared by determining correlation between amounts of nutrients extracted and nutrient uptake. This can often be done more quickly and accurately in the greenhouse than in field experiments. Data from greenhouse experiments are of limited value in calibrating nutrient test levels for fertilizer recommendations. These calibrations must be conducted in field experiments, but greenhouse and laboratory studies can be helpful in locating field experiments on soils that are likely to be responsive.

D. Recommendation Philosophy

The primary objective of a soil testing program should be to make lime and fertilizer recommendations that will produce maximum economic returns. This involves consideration of many factors that contribute to production of profitable yields. Various laboratories use widely different procedures and philosophies in arriving at recommendations. Their effectiveness and reliability should be judged on the accuracy of the recommendation in producing economic returns rather than on how it is reached. Profits are reduced by use of fertilizer in excess of optimum rates, but this loss is generally less than that from insufficient fertilizer applications. Growers can use either insufficient or excessive amounts of most nutrients for several years without being conscious of economic loss. Growers and fertilizer suppliers, using laboratories that recommend excessive rates, will be satisfied until research is done to determine soil tests levels and fertilizer rates associated with maximum economic returns (Olson et al., 1982).

1. Nutrient Accumulation and Removal

Increases in soil test P levels in the plow layer of some fertilized soils have been generally recognized for many years. Because P is not leached by passage of water through the soil, application of rates higher than amounts removed in harvested portions of crops results in P accumulation. The P content of most crops is usually in the range of 0.2 to 0.4%, so that 1000 kg of dry plant material contains only 2 to 4 kg of P. On the other hand, removal of K may be much greater when all of the plant material is removed, often being as high as 2 to 4% in nongrain portions of many crops. Potassium is subject to some leaching, particularly on extremely sandy soils. Cope (1981) reported that on six soils in Alabama over the 50-yr period 1928–1978, average annual application of 14 to 18 kg/ha P increased soil test P from an average of 19 mg/kg in 1928 to 33 mg/kg by 1957. Application of 28 kg/ha K over the 50 yr in a cotton-corn rotation produced top yields and increased the level of soil test K in five of the six soils.

Research data and soil test summaries from many states have shown increasing levels of both P and K under current fertilizer practices. Regular soil testing by reliable laboratories will promote efficient use of fertilizer and increase profits. Rates of P and K that produce maximum returns have been shown to also increase soil test levels. Therefore, recommendations may be limited to rates from which response can be predicted on the next crop or within the rotation. As long as the release rate of a nutrient from the soil is adequate to supply the needs of a crop without reduction in yield or quality, application of the nutrient is not justified.

2. Cation Ratio vs. Sufficiency Levels in Making Recommendations

Soils vary tremendously in their capacity to supply P and K to growing crops. Some of the factors that affect the nutrient-supplying capacity of soils are amount and type of clay, organic matter content, total nutrient content, and amounts of extractable nutrient ions. Soils with similar amounts of chemically extractable P or K may vary widely in nutrient-supplying capacity to crops. Therefore, these differences in supplying power and response characteristics must be considered in making soil fertility ratings and recommendations based on soil tests.

McLean (1977) described two concepts of soil test interpretation that have been widely used. These are the sufficiency levels of available nutrients (SLAN) and basic cation saturation ratio (BCSR) methods. The SLAN concept involves assigning soil test sufficiency levels based on response data from field experiments. This concept is illustrated by Fig. 2–2 and is sometimes referred to as "fertilizing the crop." Most laboratories using this interpretation concept adjust sufficiency levels for crop and soil association effects. This concept assumes little or no interaction among nutrients in their effect on needed fertilization and applies to all nutrients.

The BCSR concept is used by many commercial laboratories to adjust for interactions among cations so that a specific cation ratio or base saturation is achieved. This involves determining the CEC and making recommendations to achieve specific ratios among K, Mg, and Ca ions in the soil. Recent research reported by McLean (1981) and others has shown that the values for BCSR per se are completely independent of crop yields. Use of this concept results in recommendations that are higher than can be justified by most recent studies.

Some state laboratories have also recognized a need for classifying soils into groups for making P recommendations. Soils with low CEC release P more readily to extracting solutions and therefore require higher levels for adequacy than do soils with more clay or organic matter. Laboratories in Alabama, Georgia, Florida, and South Carolina separate the sandy soils of the Coastal Plain from those of the Piedmont or others that have more clay or organic matter. This relationship for P is reversed from that with the cations as indicated by Cope and Rouse (1973).

3. Reporting Results and Making Recommendations

Many systems of reporting soil test results to growers are used. Most laboratories use interpretive ratings ranging from very low to very high to identify the level of sufficiency in the soil. These ratings are generally based on anticipated yield without application of the element. High and very high are usually above the adequate level. Many laboratories recommend maintenance applications at these levels, but some are discontinuing this practice as research data show that they are generally not needed. Medium is normally the area of moderate response and low or very low is where response should be expected. The recommended rate of application to obtain optimum yields at these soil test levels will result in a gradual soil test increase.

Most laboratories recognize differences among crops in their fertility requirements. Some laboratories rate soils differently based on crops to be grown, such as vegetable and other horticultural crops; forage crops, which remove large amounts of nutrients; cotton; and crops grown for silage, which have higher requirements for P and K than do corn for grain, rice, most other grasses, and peanuts (Cope et al., 1981). These crops with low P and K requirements can frequently be grown without direct applications when grown in rotation with other well-fertilized crops.

In addition to ratings, most laboratories use some method of reporting results more precisely, especially for use in record keeping and fertility monitoring by growers. Some report kilograms per hectare or parts per million extracted, but these data are often confusing to growers, because each element has a different level for a specific degree of adequacy. For example, the adequate or critical level for one soil may be 25 mg/kg P, 120 mg/kg K, 200 mg/kg Ca, and 30 mg/kg Mg. Adequate levels in other soils and from other extracting procedures would likely be different for each element.

To simplify reporting and make it more easily understood by growers, some laboratories have developed fertility indexes for use on soil test reports. Use of computers in making soil test recommendations has simplified reporting of index values. No indexing system has been universally accepted. Some form of a percent sufficiency index, as reported by Cope and Rouse (1973), offers promise of simplified reporting and record keeping. This system assigns the adequate level of each element for each soil a value of 100. Below this level the index follows the yield curve and above is a straight line relationship. It has been particularly helpful in emphasizing to growers the degree of buildup of P under continued applications.

II. PLANT ANALYSIS

Plant analysis, like soil tests, can be a valuable aid to producers in making efficient use of lime and fertilizer. Recommendations based on soil tests are made prior to seeding; however, plant samples for analysis cannot be taken until the corp is growing, resulting in a delay that may prevent corrective action for the current crop. The value of plant analysis may therefore be more for the future than for current crops, especially for annual crops.

According to Ulrich and Hills (1967), *plant analysis* is defined as "the concept that the concentration of a nutrient within the plant at any particular time is an integrated value of all the factors that have influenced the nutrient concentration up to the time the plant sample is taken." Plant tissue can be analyzed for the total quantity of the nutrient or for only a soluble fraction of the nutrient in the plant sap. The latter is commonly referred to as tissue testing and normally measurements are taken in the field for only N, P, and K in freshly extracted plant sap. Our discussion will focus on plant analysis for total quantity of nutrient present in the plant; however, much of the information on sampling and interpretation also applies to tissue testing.

A plant analysis program can be thought of as having four parts: (i) collection of the sample in the field, (ii) chemical analysis of the plant tissue, (iii) interpretation of the analytical results, and (iv) recommendations based on the interpretation and supplemental information provided with the sample. Our discussion will center on these four factors.

A. Plant Sampling Techniques

Collection of proper plant tissue samples is essential to a sound plant analysis program. Not only do we have to concern ourselves with field division and sample handling in plant analysis, but we must also concern ourselves with which part of the plant and at what growth stage to sample.

1. Field Division

Research information is limited on methods of subdividing fields for collection of plant samples. Likewise, little research is reported on the sampling pattern to follow in the field. The purpose for sampling will influence the choice of sampling procedure. Plant samples are normally taken either to monitor the nutrient status of the crop or to help diagnose poor growth. Our discussion on field division will relate to these two objectives for sampling.

Field division and sampling patterns to monitor nutrient status of a crop should be similar to those used for soil sampling in the field. This allows the information from the soil and plant analysis to be used in combination for a better interpretation. The person taking the plant samples should be alert for abnormal growth areas that were not apparent when soil sampling. If soil samples were not recently collected from the field, then they should be taken at the same time as the plant samples. For the diagnostic approach to plant analysis, normal and abnormal growth areas should be sampled separately for comparison of results.

The sampler for both approaches should avoid plants of different maturity due to planting date differences that might exist within the field. Likewise, sample each cultivar separately if several have been planted in the field to avoid differences in test results due to genetics. A separate composite sample should be taken from different maturity or cultivar areas to minimize plant size and maturity effects (Jones & Steyn, 1973). As with soil sample collection, avoid unusual soil areas when plant sampling to monitor overall nutrient status of the field. The number of subsamples required to characterize the field will vary with crop and soil conditions. In general, the larger the number of subsamples composited, the more likely the test sample will reflect the condition in the sampled area.

The sampler should observe the general condition of the crop for such things as drought stress, insect damage, disease damage, and nutrient deficiency, because each can affect nutrient content of the plant. Information about crop stress is valuable to the person who interprets the results. Before going to the field, it is advisable to contact the laboratory for detailed sampling instructions, information sheets, and sample bags.

An inspection of the plant root system and the general condition of the soil root zone should be made, especially in those fields showing abnormal growth. Most analytical laboratories have forms to record such information. Fertilizer and pesticide application history is also requested on most information sheets. This total package of observations and field history can be extremely helpful in pinpointing problems when coupled with plant and soil analyses.

2. Plant Part Selection and Sampling Time

Numerous studies have compared different plant maturities and various plant parts for nutrient concentration. Results have generally shown that both plant part and maturity have a marked effect on the concentration of one or more nutrients. Bates (1971), in a review of important factors affecting critical nutrient concentrations, not only identifies age and choice of tissue as important factors, but also discusses cultivars, nutrient interactions, environmental conditions, and the specific form of a nutrient. Therefore, strict attention should be given to taking plant samples as prescribed by the laboratory if their interpretations for critical levels are to be used.

General guidelines have been developed and prescribed by several investigators for various crops (Chapman, 1964; Jones et al, 1971; Jones & Steyn, 1973). Jones et al. (1971), working as a committee from The Council on Soil Testing and Plant Analysis and The Fertilizer Institute, developed sampling guides for a number of plants. Table 2–1 gives plant sampling guidelines for some common field crops.

Several general observations can be made about plant sampling of field crops based on the experience of the authors and published information. Six of these general observations follow:

1. Sampling should be done at or prior to the reproductive stage. Samples taken later than this on annual crops are of little value for monitoring nutrient adequacy.
2. On immature plants, collect whole plant samples or take the last mature leaf near the top of the canopy.
3. Samples should be taken when a nutrient deficiency first appears rather than waiting for a specific growth stage. A plant sample should be collected from the area of the field showing the deficiency symptom(s) and another from the normal growth area of the field. Plant tissue of similar maturity should be taken from both areas.
4. Avoid sampling tissue that is diseased, insect damaged, mechanically damaged, or in severe moisture stress. Likewise, do not sample dead tissue.

Table 2-1. Suggested stage of growth, plant part, and number of plants to sample for several crops (adapted from Jones et al., 1971 and other sources).

Crop	Stage of growth	Plant part to sample	Number of plants
Corn (*Zea mays* L.)			
	Seedling stage (<30 cm)	Whole plants	20–30
	Prior to tasseling	Last fully developed leaf below whorl	15–25
	Tassel to silking	Ear leaf or immediately below	15–25
	(Sampling after silking not recommended)		
Sorghum [*Sorghum bicolor* (L.) Moench]			
	Seedling stage (<20 cm)	Whole plants	20–30
	Prior to boot	Last fully developed leaf below whorl	15–25
	Boot to heading	Flag leaf or leaf immediately below	15–25
	(Sampling after heading not recommended)		
Wheat (*Triticum aestivum* L.)			
	Seedling stage (<30 cm)	Whole plants	50–100
	Prior to heading	Uppermost leaves	50–100
		Whole plants	20–30
	(Sampling after heading not recommended)		
Rice (*Oryza sativa* L.)			
	(Similar to wheat)		
Soybeans [*Glycine max* L. (Merr.)]			
	Seedling stage (<30 cm)	Whole plants	20–30
	Prior to or during initial flowering	Fully developed leaves at top of the plant	20–30
	(Sampling after pods begin to set not recommended)		
Cotton (*Gossypium* spp.)			
	Prior to or at first bloom or when first square appears	Youngest fully mature leaves on main stem	30–40
Peanuts (*Arachis hypogaea* L.)			
	Prior to or at first bloom	Youngest fully mature leaves	40–50
Alfalfa (*Medicago sativa* L.)			
	Prior to tenth bloom	Mature leaves or	50–100
		top 10–15 cm	30–40

5. Although there is no general agreement on the magnitude of total nutrient concentration flux with time of day, samples collected at a similar time of day are preferred. This is especially true if a soluble fraction of the nutrient is tested.
6. Any sampling guide developed for grower use should be easy to follow.

3. Handling of Samples

Care should be exercised when handling plant samples to prevent deterioration of the samples before they reach the laboratory. Most laboratories specify in their sampling instructions how they prefer to have samples handled, packaged, and shipped.

Exposed plant tissues often have some contamination with dust. Spray residues from pesticides and fertilizer application to the crop can also cause surface contamination. This surface contamination could lead to high results, especially for the micronutrients if not removed. As an example, Jacques et al. (1974) compared nutrient content of unwashed and washed sorghum plant tissue parts taken throughout the growing season. Iron was the only nutrient consistently higher for unwashed than for washed samples. The differences in Fe content were much greater between unwashed and washed samples for exposed tissue such as leaf blades than for the culm. Other studies have shown differences for nutrients other than Fe for unwashed compared with washed samples. Much of the differential between washed and unwashed samples will be governed by the degree and type of contamination.

Different opinions exist on how best to handle samples to minimize the effect of surface contamination. Bowen (1978) suggests that loose soil and dust be removed by shaking and gently brushing. To remove additional contamination, the samples should be rinsed in flowing tap water with gentle scrubbing followed by a gentle rinse through a dilute phosphate-free soap solution and, finally, rinsed with distilled water. Schulte and Simson (1981) recommend that the sample only have the dust and loose soil removed by gently brushing and that no washing of the samples be done. Another recommendation suggests wiping samples with a damp cloth to remove surface contamination (Jones et al., 1971).

Washing plant samples is difficult for many producers and field scouts due to the time involved and logistics of transporting samples. There is general agreement that if plant samples are to be washed, the washing must be done while the plant tissue is still fresh. Washing must be done without prolonged immersion of the samples in the wash solution to minimize leaching of soluble nutrients from the tissue. In many cases, if plants heavily covered with dust and spray residue are avoided, a gentle brushing of the tissue is adequate to minimize contamination. However, the sampler should be aware of the potential for contamination. If considerable contamination is present on the tissue surface, then washing the tissue seems advisable.

Fresh plant tissue tightly packed into containers (plastic or paper) will begin to deteriorate rather quickly. Most laboratory instructions recommend that samples be spread out in a clean area and air-dried before shipment to the laboratory. They should then be packaged in proper mailing containers to further reduce the potential for contamination and rotting in transit to the laboratory. If possible, samples should be shipped the first part of the week to assure processing by the laboratory prior to sample deterioration due to extended holdover periods, especially if not dried. If facilities are available, the samples should be oven-dried before shipment to the laboratory. Jones and Steyn (1973) state that drying of plant material must be at a sufficiently high temperature to destroy enzymes responsible for decomposition, but not so high as to cause appreciable thermal decomposition. Most research results on drying report an ideal temperature for drying in a forced-air oven to be in the 60 to 70° C range.

Jones and Steyn (1973) made an extensive review of research on sample grinding and laboratory preparation. The two major concerns are that the sample be ground to a homogeneous state, and that contamination from the grinder not occur. The entire sample collected in the field should be processed. For large samples this may mean first grinding through a mill or shredder with a coarse screen and then taking from the thoroughly mixed coarse material a subsample for final grinding. Finer grinding ensures a more homogenous particle size; however, the vascular tissue tends to be more difficult to grind and tends to segregate. Samples should be mixed again after grinding and no material should be left in the grinding chamber.

Sample contamination is of real concern for micronutrients. To minimize contamination, all grinding surfaces should be of tooled or stainless steel or of agate. All new grinders should be checked for potential contamination by grinding several different plant materials and comparing analytical results to those of the same plant material ground in a manner that does not cause contamination.

4. Methodology in Laboratory Operation

Plant analysis, as routinely conducted by most laboratories, is for the total amount of the nutrient present in the plant tissue. There are exceptions where some fraction of the total is measured, such as nitrates in petiole samples of cotton or sugarbeets (*Beta vulgaris* L.). The tissue is most often digested by one of several wet- or dry-ashing procedures to get the nutrients into solution. Once in solution, detection of most essential elements is relatively simple with modern instrumentation. However, each laboratory needs a reliable quality control system to ensure analytical accuracy.

Most laboratories have a quality control system in which check samples are included routinely with the unknown samples. In addition, plant sample exchanges are carried out among laboratories. Several different plant tissue standards are available from the U.S. National Bureau of Standards. Quality control is one of the biggest concerns in most laboratories.

Jones and Steyn (1973) point out that it is difficult to select one analytical system that is superior to others. In many cases, equipment available and nutrients for which tests are desired will dictate the procedure. Dry-ashing of samples will not allow measurement of N and S. It is beyond the scope of this chapter to discuss analytical methods in any detail. However, note that measurement of total quantity of essential nutrients is relatively easy with modern instrumentation and results among laboratories for total concentration should be comparable. As with soil testing, the analytical test is probably the most accurate part of plant analysis.

C. Correlation and Calibration of Results

Considerable progress has been made in identifying physiological stages of plant growth. Good descriptions are available for defining specific vegetative and reproductive stages of corn (Ritchie & Hanway, 1982), soybeans (Ritchie et al., 1982), wheat (Large, 1954), sorghum (Vanderlip, 1972), and other crops. Researchers and growers need to be more diligent in acquainting themselves with these physiological stages. This knowledge should be used to identify the growth stage at which plant samples are collected for nutrient analysis. Many studies have provided valuable information on plant analysis because the samples were taken at a time that was easily identifiable, i.e., at silk emergence of corn. However, samples taken prior to tassel emergence have not always been defined for physiological stage so other researches could duplicate the sampling in future years.

1. Collection of Data

The expense of collecting useful plant analysis data may be greatly reduced by close cooperation among researchers. Field research designed to provide data for correlating soil tests with plant growth is also a good potential source of data for establishing the relationship between plant nutrient concentrations and yields. If the soil test correlation research is to be conducted, the only additional effort required would be that for plant sample collection and laboratory analyses.

There are also numerous experiments, other than those concerned with soil testing, that might provide valuable plant analysis data. For example, many soil testing laboratories do not analyze for N because of their failure to correlate a N soil test. Nevertheless, yield evaluations and plant N data from N rate experiments provide a source of data on the effect of N concentration in the plant on yields.

Scientists interested in collecting information on plant nutrient concentrations and yields should be aware of fertility experiments being conducted by other scientists. The scientist should be alert to the possibility of gaining additional information over the primary objective of other research efforts by helping identify secondary objectives, especially those concerned with plant analysis and crop yields. A willingness to share the work and expense and a cooperative spirit on the part of the plant analysis researcher will expedite the collection of valuable data that otherwise might go uncollected.

2. Utilizing Collected Data

Plant analysis becomes a useful diagnostic and predictive tool only when a given nutrient concentration can be related to the nutrient concentration necessary for near-maximum yields. (Maximum yield implies the yield attained when the nutrient in question is not limiting, even though some other growth factors might be limiting.) After establishing the relationship between a given concentration of nutrient and plant yield, the information becomes even more valuable when one suggests a treatment that will result in producing near-maximum yields. Both the nutrient concentration–yield relationship and the interpretation are necessary to make plant analysis a viable tool for growers.

Through appropriate statistical techniques, one can correlate nutrient concentrations with plant yields. This may involve both linear and curvilinear relationships. Once this relationship has been established, the researcher can then quantify the yield loss that occurs because of a particularly low concentration of a nutrient. After the absolute magnitude of yield loss is known, prices can be assigned per unit of crop yield and the economic loss calculated. Once this is done, growers are able to determine the dollars that might profitably be spent to correct the low nutrient concentration.

The second phase necessary for giving value to plant analysis data is to determine the treatment necessary to go from an initial yield to the desired yield. Assume that P concentration in the corn leaf at tassel is such that only 80% of maximum yield will be obtained. How much P fertilizer will be required to remove P as a limiting growth factor? Researchers determine this by using rates of P fertilizer and measuring corn yields. Once the data are collected and analyzed, one would then suggest to growers the rate of P fertilizer that was necessary for maximum yields. Growers can easily compute the cost of the P fertilizer treatment necessary to increase yields from 80% to near 100%. In addition, the amount of P fertilizer required for between 80 and 100% yields also would be obtained as the rate experiments were being interpreted. That is, the P fertilizer application rate necessary for 85%, 95%, or any other percentage between 80 and 100% would become known after appropriate experimentation had been conducted. Adequate data will be available only if sufficient rates of fertilizer are used to establish a curve that expresses the relationship between fertilizer and crop yields.

This brief description of data necessary to make plant analysis meaningful may be misleading to those who are unfamiliar with field research. Considerable time, effort, and expense are involved in collecting and interpreting the data necessary for various crop species and soil conditions. But without such information, plant analysis and its relationship to crop production can never become the quantitative science that is desired.

3. Quantifying Other Growth Factors

The quantity of nutrient in a known quantity of plant tissue is determined in the laboratory. This is expressed as a nutrient concentration. The concentration has often been reported in terms such as percent, parts per million, or grams per kilogram. In either case, nutrient concentration is a ratio that is calculated by dividing quantity of nutrient by quantity of plant tissue. Nutrient concentration is therefore dependent on two factors: (i) nutrients accumulated up to the time of sample collection, and (ii) growth that has occurred up to the point of sample collection. It follows then that any factor that affects either nutrient uptake or plant growth affects nutrient concentration. This makes it important that researchers attempt to better understand the many factors that affect the two components of nutrient concentration.

Even if the same quantity of nutrient was absorbed by the crop, the concentrations of the nutrient in plant tissue could vary widely between fields, and would be directly proportional to the amount of growth differential. That is, if growth was restricted, nutrient concentration would be higher than if growth flourished. There are many examples where factors other than nutrient availability from the soil and added fertilizer have influenced nutrient concentration. Drought may restrict plant growth such that a nutrient concentration may be much higher in the plant than during seasons of adequate rainfall. The difficulty in attempting to interpret plant analysis during droughty conditions is that nutrient concentration may not be the factor that is most limiting for growth. It is impossible to properly utilize plant analysis unless there is a known cause-and-effect relationship between plant analysis and crop yield. Knowing that a nutrient concentration is not limiting growth is important information for analytical interpretation.

Other sections of this chapter have discussed the importance of time of sampling. Time per se is not a factor that affects nutrient concentration. Time is important only in that growth occurs over time. Additional growth dilutes nutrient concentration unless nutrient uptake remains directly proportional to growth over time. In most cases, nutrient concentration in older plants is less than in younger plants because of the dilution effect due to growth (Martin & Matocha, 1973).

To be most meaningful, plant analyses data should be evaluated simultaneously with quantitative evaluations of other factors that affect growth. Plant diagnosticians will render the greatest service when they are familiar with both the factors that affect nutrient availability and uptake from the soil system and factors that affect growth. Researchers need to provide users of plant analysis data with the proper relationship between nutrient concentration and other factors that might be quantified during the growing season. For plant growth factors that are easily controlled by the researcher, maintenance at levels that provide optimum crop yield is preferred. If rates of P fertilizers are being studied, then the researcher should be sure that other nutrients like N, K, etc. are present in optimum concentrations in the soil. For those interactive factors that are not controlled by the researcher under field conditions, they should at least be measured and reported. Precipitation is not easily controlled, but is easy to measure and report. Water supply is very significant because it has a profound effect on growth and consequently nutrient concentration.

D. Interpretation and Recommendation Philosophy

Researchers and/or those making recommendations often do not agree on plant analysis interpretations. One reason is because they often are using different sets of data collected under different growing conditions. Even if a number of investigators were interpreting the same data set, there is still ample opportunity for philosophical differences among individuals in interpretations and recommendations. Researchers should agree on the relationship between nutrient concentration and plant yield as expressed by the data. There could be disagreement, though, in determining what nutrient concentration should be defined as sufficient. One investigator might prefer that sufficiency be defined as the concentration that results in at least 98% of maximum yield, while another may prefer to define sufficiency as the concentration that produces 90% of maximum yield. If this occurred, it provides the basis for later disagreement among individuals on suggested corrective treatments.

One disadvantage often cited for plant analysis is that for annual crops there may be insufficient time for corrective action after the diagnosis is made. For this reason, it would be desirable to use plant analysis from the previous crop to prevent nutrient shortages from occurring in the present crop. Plant analysis is used to monitor nutrient sufficiency.

1. Cost of Being Wrong

Plant analyses data are used to make decisions that have economic consequences. Management decisions by growers may tend to put more emphasis on the cost of taking action rather than proper evaluation of the cost of inaction. The cost of corrective treatment is something that is quite evident to the grower. However, the cost of inaction may not be so evident. Growers may be more impressed by the supposed saving if corrective action is not taken than by the less evident reduced crop loss and profits that will occur due to not taking corrective action.

Decisions are much easier to make under severe nutrient deficiencies than under only moderate nutrient deficiencies. The probability that a large profit will be made by correcting severe deficiency is so high that it cannot be easily ignored by growers. If the nutrient concentration is well within the sufficiency range, but not excessive, then the decision to not take corrective action is also easily made. However, as the nutrient deficiency becomes less severe, the ratio between increased value of crop yield due to treatment and cost of treatment will become narrower. As the potential for increased profit becomes less, the decision becomes more difficult and is dependent on several other factors. Borderline cases give rise to disagreement among those who interpret and make recommendations from plant analysis.

The direct cost of corrective treatment can easily be determined. However, the number of crops to which the treatment costs might be appropriately assigned can differ depending on the particular corrective treatment. If N is deficient for corn in Illinois, then corrective action and its associated cost likely will benefit only the current crop. This is because in many years carryover N is not high. If N fertilizer is not utilized by the immediate crop, there may not be opportunity for it to benefit future crops. On the other hand, immobile nutrients like P and K can be counted on to benefit more than the immediate crop where application rates are greater than that removed in the current crop. Growers should not require as favorable a return from the immediate crop for corrective action involving immobile nutrients as compared with mobile nutrients. Any carryover from treatment that will benefit future crops should have some of the initial treatment cost appropriated to those future crops.

Some corrective treatments cost little, while the potential for increased crop output may be high. Such situations are attractive even to growers who normally are not in a financial position to assume great risk. Grower outlook toward risk is greatly modified by their cash flow position and whether they are tenants or land owners. Because of such diverse outlooks toward risk, it is not surprising that some growers opt for corrective treatment, while others decide to let the situation go uncorrected, even though it was not optimum for high crop yields.

2. Nutrient Sufficiency

Interpretation of plant analysis is accomplished by comparing the nutrient concentration of the plant sample with predetermined nutrient concentrations known to be sufficient. The recommendation based on plant analysis depends on whether the sample nutrient concentration falls within the predetermined sufficiency range. As discussed earlier, nutrient sufficiency and recommendations can be appropriately determined only from field research. Nutrient deficiency must occur before one can establish the sufficiency level. Unless nutrient shortages are such that there are yield increases when the particular nutrient is applied, it is impossible to calibrate plant nutrient concentrations. Until one demonstrates through field research that a yield increase occurs from application of a particular nutrient, one can only assume that the initial nutrient concentration in the soil is sufficient. There may be an increase in plant nutrient concentration from the applied nutrient, but without the concomitant increase in yield such information is of little value for calibration purposes.

Rankings for different concentrations of N in the corn ear leaf sampled in Ohio at initial silk stage are shown in Table 2–2 (Jones, 1967). These data serve as an example of the concept used in making plant analysis worthwhile. This table provides nutrient levels that help the diagnostician determine the adequacy of a given nutrient concentration. The data, as presented, do not indicate the amount of N fertilizer that would have to be applied to increase the N concentration from 2.45 to 3.50%. All interpreters would not use the same percent of maximum yield to represent a sufficient nutrient level. For example, the Ohio scale defined sufficient as any N concentration in the leaf that would result in at least 90% of maximum yield; other researchers might prefer defining sufficient as that concentration that results in 98% of maximum yield—the percent of maximum

Table 2–2. Interpretation of N analyses of corn ear leaf samples at initial silking for Ohio (Jones, 1967).

Percent N in leaf	Nutrient level category	Percent of maximum yield obtained
<2.45	Deficient	≤80
2.46–2.75	Low	80–90
2.76–3.50	Sufficient	90–100
3.51–3.75	High	100
>3.75	Excess	<100

yield defined as sufficient will vary with the value of the crop and the cost of corrective treatment. If the cost of corrective treatment is low and the value of increased crop yield high, then one would want to approach 100% of maximum yield. While there might be justification for some disagreement as to what corrective action should be taken when the nutrient concentration in the plant is near sufficiency, there would surely be almost unanimous agreement that corrective action should be taken when plant analysis reveals that the nutrient concentration would be categorized as deficient. Growers who habitually attain only 80% of the yield possible for their soil and climatic situation will likely have economic problems because of unrealized profit.

Much research has been conducted to relate nutrient concentration in plants to yield. Data similar to that given above for N on corn are available for other nutrients and other crops (Jones, 1967; Melsted et al., 1969). Plant analysis research continues to be a viable area of investigation (Dow & Roberts, 1982; Jones & Bowen, 1981; Ware et al., 1982).

3. Monitoring of Nutrient Levels

The concept of plant analyses to let the plant "speak" is sound. The nutrient concentration in the tissue represents the integration of nutrient uptake and growth up to the point of sample collection. An adequate concentration of a nutrient in early stages of growth does not guarantee that an adequate level will be maintained through physiological maturity. But, a low nutrient concentration at early growth stages does indicate that the full potential for yield, that is afforded by other growth factors, cannot be realized because of nutrient shortage. One should view plant analysis as a "checkup" for the crop. Avoiding problems is often more economical than attempting to correct problems after they occur. Annual leaf sample analyses are a good way to determine if nutrients are being adequately supplied. Even though corrective action may not be possible during the current season, depending on the particular nutrient deficiency and crop, borderline nutrient concentrations provide the grower an opportunity to adjust fertilizer practices prior to seeding future crops.

Plant analysis has been extremely useful for "troubleshooting." Plant analysis results can greatly improve the confidence of the diagnostician when preliminary diagnosis of the problems and their solution are confirmed by chemical analyses. The diagnostician should be aware of crop species, soils, and environmental conditions where specific nutrient deficiencies are likely to appear, based on prior experience for the area. The better the soil-plant system is characterized and understood, the more likely problems can be avoided or at least judiciously corrected.

4. Multinutrient Deficiencies

The simplest situation is when only one nutrient is deficient. If there are interactions between plant nutrient concentrations and crop yields, then interpretation becomes more complex when more than one nutrient is deficient. Multiple regression techniques have been useful in evaluating the importance of various nutrients, provided there are adequate ranges of the various nutrient concentrations to allow mathematical coefficients to be assigned to nutrient concentrations (Hanway, 1973). The diagnosis and recommendation integrated system (DRIS) attempts to determine the most limiting nutrient with respect to plant yield. This system evaluates the concentration of a number of nutrients simultaneously (Beaufils & Sumner, 1976).

REFERENCES

Bates, T. E. 1971. Factors affecting critical nutrient concentrations in plants and their evaluation: A review. Soil Sci. 112:116–130.

Beaufils, E. R., and M. E. Sumner. 1976. Application of the DRIS approach for calibrating soil and plant factors in their effects on yield of sugarcane. Proc. Annu. Congr. S. Afr. Sugar Technol. Assoc. 50:118–124.

Black, C. A., D. D. Evans, J. L. White, L. E. Ensminger, and F. E. Clark (ed.). 1965. Methods of soil analysis, part 1. Agronomy 9.

Blevins, R. L., G. W. Thomas, and P. L. Cornelius. 1977. Influence of no-tillage and nitrogen fertilization on certain soil properties after 5 years of continuous corn. Agron. J. 69:383–386.

Bowen, J. E. 1978. Plant tissue analysis: Costly errors to avoid. Crops Soils 31(3):6–10.

Bray, R. H. 1948. Correlation of soil tests with crop response to added fertilizers and with fertilizer requirement. p. 53–86. *In* H. B. Kitchen (ed.) Diagnostic techniques for soils and crops. American Potash Institute, Washington, DC.

Cate, R. B., Jr., and L. A. Nelson. 1971. A simple statistical procedure for partitioning soil test correlation data into two classes. Soil Sci. Soc. Am. Proc. 35:658–659.

Chapman, H. D. 1964. Foliar sampling for determining the nutrient status of crops. World Crops 16:35–46.

Cline, M. G. 1944. Principles of soil sampling. Soil Sci. 58:275–288.

Cline, M. G. 1945. Methods of collecting and preparing soil samples. Soil Sci. 59:3–5.

Cope, J. T. 1981. Effects of 50 years of fertilization with phosphorus and potassium on soil test levels and yields at six locations. Soil Sci. Soc. Am. J. 45:342–347.

Cope, J. T., C. E. Evans, and H. C. Williams. 1981. Soil test fertilizer recommendations for Alabama crops. Alabama Agric. Exp. Stn. Circular 251.

Cope, J. T., and R. D. Rouse. 1973. Interpretation of soil test results. p. 35–54. *In* L. M. Walsh and J. D. Beaton (ed.) Soil testing and plant analysis. Soil Science Society of America, Madison, WI.

Dahnke, W. C. (ed). 1980. Recommended chemical soil test procedures for the North Central Region. North Dakota Agric. Exp. Stn. Bull. 499 (Revised).

Dow, A. I., D. W. James, and T. S. Russell. 1973. Soil variability in central Washington and sampling for soil fertility tests. Washington Agric. Exp. Stn. Bull. 788.

Dow, A. I., and S. Roberts. 1982. Proposal: Critical nutrient ranges for crop diagnosis. Agron. J. 74:401–403.

Eik, K., E. L. Hood, and J. J. Hanway. 1980. Soil sample preparation. *In* W. C. Dahnke (ed.) Recommended chemical soil test procedures for the North Central Region. North Dakota Agric. Exp. Stn. Bull. 499 (Revised).

Fitts, J. W. 1959. Research + extension = higher farming profits. Plant Food Rev. 5(2):10–12.

Fitts, J. W., and J. J. Hanway. 1971. Prescribing soil and crop nutrients needs. p. 57–79. *In* R. A. Olson et al. (ed.) Fertilizer technology and use, 2nd ed. Soil Science Society of America, Madison, WI.

Fitts, J. W., and W. L. Nelson. 1956. The determination of lime and fertilizer requirements of soil through chemical tests. Adv. Agron. 8:241–282.

Hanway, J. J. 1973. Experimental methods for correlating and calibrating soil tests. p. 55–56. *In* L. M. Walsh and J. D. Beaton (ed.) Soil testing and plant analysis. Soil Science Society of America, Madison, WI.

Hauser, G. F. 1973. The calibration of soil tests for fertilizer recommendations. FAO Soils Bull. 18, Food and Agriculture Organization of the United Nations, Rome.

Hooker, M. L., G. A. Peterson, and D. H. Sander. 1976. Soil Samples: How many do you need? Univ. of Nebraska Agric. Exp. Stn., Farm, Ranch and Home Quarterly 23(2):8–11.

Illinois Agronomy Handbook 1983–84. 1982. Ill. Coop. Ext. Service Circular 1208. University of Illinois, Urbana.

Jacques, G. L., R. L. Vanderlip, D. A. Whitney, and R. Ellis, Jr. 1974. Nutrient contents of washed and unwashed grain sorghum plant tissues compared. Commun. Soil Sci. Plant Anal. 5:173–182.

James, D. W., and A. I. Dow. 1972. Source and degree of soil variation in the field: The problem of sampling for soil tests and estimating soil fertility status. Washington Agric. Exp. Stn. Bull. 749.

Jones, C. A., and J. E. Bowen. 1981. Comparative DRIS and crop log diagnosis of sugarcane tissue analyses. Agron. J. 73:941–944.

Jones, J. B., Jr. 1967. Interpretation of plant analysis for several agronomic crops. p. 49–58. *In* Soil testing and plant analysis, part 2. Spec. Pub. 2. Soil Science Society of America, Madison, WI

Jones, J. B., Jr. (ed.). 1980. Handbook on reference methods for soil testing. Council on Soil Testing and Plant Analysis, University of Georgia, Athens.

Jones, J. B., Jr., and W. J. A. Steyn. 1973. Sampling, handling, and analyzing plant tissue samples. p. 249–270. *In* L. M. Walsh and J. D. Beaton (ed.) Soil testing and plant analysis. Soil Science Society America, Madison, WI.

Jones, J. B. Jr., R. L. Large, D. B. Pfleider, and H. S. Klosky. 1971. The proper way to take a plant sample for tissue analysis. Crops Soils 23(8):15–18.

Kells, J. J., C. E. Rieck, R. L. Blevins, and C. H. Slack. 1979. Relationship of weed control and soil pH to no-tillage corn yields. Agron. Notes 12:2 Department of Agronomy, University of Kentucky, Lexington.

Keough, J. L., and R. Maples. 1972. Variations in soil test results as affected by seasonal sampling. Arkansas Agri. Exp. Stn. Bull. 777.

Kunishi, H. M., V. A. Bandel, and F. R. Mulford. 1982. Measurement of available soil phosphorus under conventional and no-till management. Commun. Soil Sci. Plant Anal. 13:607–618.

Large, E. C. 1954. Growth stages in cereals. Illustrations of the Feekes scale. Plant Pathol. 3:128–129.

Martin, W. E., and J. E. Matocha. 1973. Plant analysis as an aid in the fertilization of forage crops. p. 393–426. *In* L. M. Walsh and J. D. Beaton (ed.) Soil testing and plant analysis. Soil Science Society of America, Madison, WI.

McLean, E. O. 1977. Contrasting concepts in soil test interpretation: Sufficiency levels of available nutrients versus basic cation saturation ratios. p. 39–54. *In* T. R. Peck et al. (ed.) Soil testing: Correlating and interpretating the analytical results. Spec. Pub. 29. American Society of Agronomy, Madison, WI.

McLean, E. O. 1981. Chemical equilibrations with soil buffer systems as bases for future soil testing programs. *In* Eighth Soil-Plant Analyst's Workshop Proceedings, Cleveland, OH. 4 Nov. 1981. The Council on Soil Testing and Plant Analysis, Athens, GA.

Melsted, S. W., H. L. Motto, and T. R. Peck. 1969. Critical plant nutrient composition values useful in interpreting plant analysis data. Agron. J. 61:17–20.

Melsted, S. W., and T. R. Peck. 1977. The Mitscherlich-Bray growth function. p. 1–18. *In* T. R. Peck et al. (ed.) Soil testing: Correlating and interpretating the analytical results. Spec. Pub. 29. American Society of Agronomy, Madison, WI.

Nelson, L. A., and R. L. Anderson. 1977. Partitioning of soil test-crop response probability. p. 19–38. *In* T. R. Peck et al. (ed.) Soil testing: Correlating and interpretating the analytical results. Spec. Pub. 29. American Society of Agronomy, Madison, WI.

Olson, R. A., K. D. Frank, P. H. Grabowski, and G. W. Rehm. 1982. Economic and agronomic impacts of varied philosophies of soil testing. Agron. J. 74:492–499.

Page, A. L., R. H. Miller, and D. R. Keeney (ed.). 1982. Methods of soil analysis, part 2, 2nd ed. Agronomy 9.

Peck, T. R., and S. W. Melsted. 1973. Field sampling for soil testing. p. 67–75. *In* L. M. Walsh and J. D. Beaton (ed.) Soil testing and plant analysis. Soil Science Society of America, Madison, WI.

Petersen, R. G., and L. D. Calvin. 1965. Sampling. *In* C. A. Black et al. (ed.) Methods of soil analysis, part 1. Agronomy 9:54–72.

Reed, J. F., J. W. Fitts, J. J. Hanway, L. T. Kardos, W. T. McGeorge, and L. A. Dean. 1953. Sampling soils for chemical tests. Better Crops Plant Food 37(8):13–18.

Ritchie, S. W., and J. J. Hanway. 1982. How a corn plant develops. Iowa Coop. Ext. Service Spec. Rep. 48.

Ritchie, S. W., J. J. Hanway, and H. E. Thompson. 1982. How a soybean plant develops. Iowa Coop. Ext. Service Spec. Rep. 53.

Sabbe, W. E., and H. L. Breland (ed). 1974. Procedures used by state soil testing laboratories in the Southern Region of the United States. Alabama Agric. Exp. Stn. Southern Coop. Series. Bull. 190.

Schulte, E. E., E. A. Liegel, C. R. Simson, and K. A. Kelling. 1980. Optimum soil test levels for Wisconsin. Wisconsin Coop. Ext. Service, Bull. A2030.

Thomas, G. W., and J. J. Hanway. 1968. Determining fertilizer needs. *In* L. B. Nelson et al. (ed.) Changing patterns in fertilizer use. Soil Science Society of America, Madison, WI.

Touchton, J. T., W. L. Hargrove, R. R. Sharpe, and F. C. Boswell. 1982. Time, rate, and method of phosphorus application for continuously double-cropped wheat and soybeans. Soil Sci. Soc. Am. J. 46:861–864.

Ulrich, A., and F. J. Hills. 1967. Principles and practices of plant analysis. p. 11–24. *In* Soil testing and plant analysis, part 2. Spec. Pub. 2. Soil Science Society of America, Madison, WI.

Vanderlip, R. L. 1972. How a sorghum plant develops. Kansas Coop. Ext. Service Bull. C-447.

Walker, W. M., J. C. Siemens, and T. R. Peck. 1970. Effect of tillage treatments upon soil test for soil acidity, soil phosphorus and soil potassium at three soil depths. Commun. Soil Sci. Plant Anal. 1:367–375.

Ware, G. O., K. Ohki, and L. C. Moon. 1982. The Mitscherlich plant growth model for determining critical nutrient deficient levels. Agron. J. 74:88–91.

Welch, C. D., and J. W. Fitts. 1956. Some factors affecting soil sampling. Soil Sci. Soc. Am. Proc. 20:54–56.

3

Agronomic and Statistical Evaluation of Fertilizer Response

Larry A. Nelson
North Carolina State University
Raleigh, North Carolina

Regis D. Voss
Iowa State University
Ames, Iowa

John Pesek
Iowa State University
Ames, Iowa

The existing trend for farming operations to involve larger land and capital commitments requires unbiased estimates of the input/output relationships in fertilizing various crop-soil combinations. These relationships can best be estimated by conducting fertilizer experiments in the field over a series of sites and years. These experiments should be planned by an interdisciplinary team of agronomists, statisticians, and economists because the validity of statistical and economic interpretations depends on optimum design of experiments and reliable data that result from carefully applied experimental technique. The economic interpretations also depend on proper application of statistical techniques such as those for model selection and for parameter estimation. For comprehensive discussions of how the agronomic, statistical, and economic aspects of fertilizer response research relate to one another, see Baum (1956, 1957) and *Status and Methods of Research in Economic and Agronomic Aspects of Fertilizer Response and Use* published by the National Academy of Sciences–National Research Council (1961). Much of the material in these three references is relevant to the topic of this chapter, so our major goal will be to amplify and to update the information that was presented in them.

Sites for fertilizer response experiments must be chosen in a manner that makes them representative of the soils and environmental conditions under study. Yield responses need to be related to both edaphic and climatic variables in order to permit extrapolation of experimental results to individual farm situations. This implies that sites be selected to provide the necessary ranges in these properties, which will ensure that a regression relationship between yields and the variables can be estimated. It is also necessary to measure these variables during the period in which the experiments are being conducted.

Fertilizer recommendations are developed using response surface functions to estimate the input-output relationships. These functions, which usually incorporate variables representing controlled variables and those representing random uncontrolled (but measured) variables, are often called covariance models. These functions take on a variety of forms depending on assumed relationships between yield, fertilizer nutrients and edaphic climatic variables. The statistical assumptions also vary with models, but in general, the usual multiple regression assumptions apply, i.e., the X_i's (fertilizer variables) are fixed and measured without error, the deviations from regression are normally and independently distributed with mean 0 and constant variance, σ^2. The response surface that is estimated from the data is comprised of average values of the Y variables (yield) estimated from given combinations of the X_i.

Knowledge of the response functions is very useful for determining fertilizer rates for unlimited and limited capital situations and for linear programming to determine how resources should be allocated among fertilization and other enterprises in the overall farm program.

I. DESIGN OF FERTILIZER EXPERIMENTS

A. Importance of Planning

Planning is one of the most important and yet neglected phases of fertilizer response experimentation. It assures that the population of sites within a given soil about which inferences are drawn is the population of interest and also that statistical analyses will not be more complicated than necessary. In addition, it assures that probability statements made about parameters estimated from the experimental data will be correct or nearly so. Planning assures that the derived fertilizer response surface relating yield to fertilizer rate is an unbiased estimate of the true surface and that an estimate of experimental error is available from the experimental data. Planning also helps to assure that the size of an individual experiment and a series of experiments will be appropriate for a particular situation. Undersized experiments often result in wide confidence intervals for estimates of model parameters or for the response surface itself. Oversized experiments are costly.

B. Steps in Planning

1. Selection of a Series of Sites at Which Experiments will be Conducted

Baird and Fitts (1957) discussed various agronomic aspects of designing and conducting a series of fertilizer trials involving corn (*Zea mays* L.) in North Carolina. Some important points concerning these aspects will be discussed here. It is necessary to establish the universe to be explored before selecting a series of sites on which to conduct experiments. This would usually be a soil or a soil association in a region of a state or country. Soil samples would then be taken from potential sites within that soil area and would be characterized for chemical properties. This characterization information would be helpful in making the final selection of sites on which to conduct fertilizer experiments. In order that a series of experiments be most useful in describing the response to an added nutrient, it is desirable to select sites that represent a range of soil levels of that nutrient from the most responsive to just where no response occurs. Selection of sites representing only low or a narrow range of soil nutrient levels will not provide an accurate estimate of the coefficient for the interaction of soil by fertilizer source of nutrient; thus, the predicted yield response for other soil nutrient levels will be biased high or low depending on the narrow range selected. The site should show sizable response to more than one nutrient to be useful in describing the production function. One would use existing information on the correlation of yields with soil test results in the selection of potential sites.

Once a set of sites (which appear likely to respond to fertilizer) has been identified, one should consider the distribution of edaphic properties other than those used to predict fertilizer response. In this connection, it is useful to construct a multiway table showing ranges of predetermined edaphic properties exhibited by sites that might be selected for the sample and visually determine what combinations of property values would best cover the universe. Sites possessing these combinations of properties would then be purposely selected for the series.

Certain principles are helpful in selecting the number of sites and the number of replications per site. First, it is necessary to use enough sites to provide stable estimates of the parameters for the relationship between yield and the most important site variables and to provide suitable degrees of freedom (df) for estimating the standard errors of the parameter estimates. If possible, the number of site variables should be $< 25\%$ of the number of sites. If there were four important site variables, perhaps 12 sites would be adequate. It is also necessary to provide stable estimates of the interaction between the added nutrient and site variables, although presumably this would automatically be satisfied if one used the rule of thumb given above for selecting the number of sites based on the number of important site variables. Another consideration is that the sample size is adequate to provide stable estimates of error terms that will be used for testing hypotheses about parameters or placing confidence limits on them or on predicted yields. A minimum of 10 to 15 df for each of three estimates appears necessary. The width of the confidence intervals for the contours representing sections through the response surface is also affected by the number of sites. If a reasonably good estimate of experimental error is available and if the number of replications per site has been established (see below), one is able to calculate the widths of the confidence intervals for the yield contours for different numbers of sites to see how they are affected by this variable.

It is usually advisable to have a self-contained experiment at each site so replication at each site is necessary to estimate experimental error. For a fixed resource level, a greater number of replications per site comes at the expense of fewer sites. Therefore it becomes important to weigh the relative importance of having some self-containment within sites against the possibility of having more precision on the estimates obtained over all sites. Three replications per site would seem to be a reasonable compromise, but in some situations this may be prohibitive in cost and time. In such cases, two would be used.

2. Selection of Experimental Design

An essential step in planning is the selection of a suitable experimental design. It is important to use a design that is common to the entire series and also of the simplest form that will achieve the objectives desired. The design should be compatible with the development of a general model that contains both controlled and random uncontrolled variables. The design should provide for error control, which should provide a good estimate of experimental error. For most fertilizer experiments, the randomized complete block design seems to be a reasonable choice. It is simple and flexible and the blocking provides adequate precision. Missing plots do not cause undue difficulty in the analysis of the data.

There are several design-model relationships that should be taken into consideration when choosing a design. First, one needs to consider what questions are to be answered from the study. If interactions among factors are to be studied, a factorial arrangement is suggested. The complete factorial is also appropriate if the data are to be used to confirm the original choice of model.

If prediction is most important, the variances of the estimates of the parameters of the model are not as important as if the objective of the study is to determine what variables belong in the general fertilizer response model. For the predictive model case, treatment design would not necessarily be chosen on the basis of minimizing variances of the individual regression coefficients.

Many fertilizer experiments have been conducted using factorial arrangement of treatments. With only two levels per factor, an exploratory 2^n fertilizer response experiment may be conducted to obtain the absence or presence of effects of a number of nutrients. This would help eliminate factors that are not relevant and to provide information about the range in rates for more intensive future experiments. With a minimum of three levels per factor, it is possible to estimate a response surface. Parameter estimates for the response function, however, may not have an optimal degree of precision with such a limited number of levels. For example, the pure quadratic terms (β_{11}, β_{22}, etc.), which are often of major interest, are estimated with less precision than the mixed quadratic (second-order) terms (β_{12}, β_{13}, etc.).

Factorial experiments have the advantage of hidden replication (i.e., the other factor(s) serve(s) as a source of replication for the first). Interpretation of results from complete factorials that contain four or more levels of each factor also have the advantage that it is possible to evaluate the appropriate model throughout the entire design space because all rows and columns have the same number of levels of the other factor(s). It is also possible with factorials of this size to give a better estimate of the optimal fertilizer rates than with the 3^n system, for which it is difficult to choose and space the levels to provide good estimates of the regression coefficients in the response function. With four or more levels of each factor, the number of treatments becomes large if there are several factors. The block size problem may ultimately decrease the precision of the experiment. This is the greatest disadvantage of factorial experiments. This problem may be obviated to a certain extent by using incomplete block designs that do provide more error control for large experiments (say number of treatments ≥ 27) at the expense of partial or complete confounding of certain higher-ordered interactions with blocks. Single or fractionally replicated factorial experiments are other possibilities. These reduce the ratio of the number of plots in the experiment to the number of effects being estimated, and yet the precision of these estimates is not greatly reduced. A defined estimate of error is not available, but by making certain assumptions and using certain generally accepted result-guided procedures, a reasonably good estimate of experimental error can be obtained. See Cochran and Cox (1957) for a comprehensive discussion of factorial experiments including the single and fractional replication versions and the incomplete block designs.

Split-plot designs are often used for fertilizer response investigations because it is desirable to incorporate into the experiment a second factor such as crop variety or tillage with a different optimum plot size than that for a fertilizer treatment. From the standpoint of precision, fertilizer rate treatments should be placed into the subplots, although mechanical considerations may dictate otherwise. Generally, the whole-plot treatments are arranged in randomized complete blocks.

There is a class of incomplete factorial designs that were developed primarily for exploring polynomial response surfaces for two or more controlled variables. The designs for exploring second-order surfaces are appropriate for fertilizer response evaluation. The response surface designs have the advantage over complete factorials in that they require considerably less experimental material. The coefficients of the response model are estimated with the same or even greater degree of precision (expressed on a per-observation basis) than those estimated from a complete factorial. But one limitation of these designs is that the precision is redistributed due to the fact that there are fewer points at the extremes of the permissible ranges of the factors. They were originally developed for use in industrial experiments, but were later adapted for use in fertilizer response studies in the field. Basically, the response surface designs often consist of two geometric figures such as a cube and an octahedron having the same center, hence the term *composite design.* Many of the industrial experiments were replicated only on the center treatment, but for field use, a randomized complete block design usually has been used, each block of which accommodates a complete set of response surface treatments, there being only a single center treatment per block. Experimental error is then estimated from the complete set of treatments rather than just from the center treatment. This provides a more stable estimate of experimental error, which represents the entire design space and which is based on adequate degrees of freedom for a field experiment.

Box and Hunter (1957) introduced the concept of a response surface design having a "spherical variance function," meaning that the variance of estimated response at a given point has a value that is dependent only on the distance of that point from the center of the design and not on the direction. Such designs are called *rotatable designs* because they are insensitive to rotation with respect to the original coordinate axes. There seems to be a case for using a design that has at least some degree of rotatability, especially if there is uncertainty where the optimum combination of fertilizer levels will be within the design space. Rotatability may be achieved by proper choice of spacing of octahedral points from the origin. It is also possible to assure that the variance of the estimated response value is approximately uniform over the design region by proper choice of the number of replications of the center treatment. See Cochran and Cox (1957) for axial spacings that give rotatability and numbers of center points that give uniform variance for various numbers of controlled variables used in connection with certain standard designs. Hader et al. (1957) gave a description of the experimental design and statistical methods for characterizing a response surface for a set of data from a biological experiment. A typical response surface design for fertilizer response investigations is shown in Fig. 3–1.

A different and basically intuitive approach to the selection of incomplete factorials in fertilizer work has been employed in recent years. Cady and Laird (1973) described a number of different incomplete factorials that would not be classed as composite response surface designs, but that do give a good exploration of the response surface space. One, which has desirable spatial characteristics, low bias error, and low variance error, and which has been used by workers in Latin America, is the 13-treatment de-

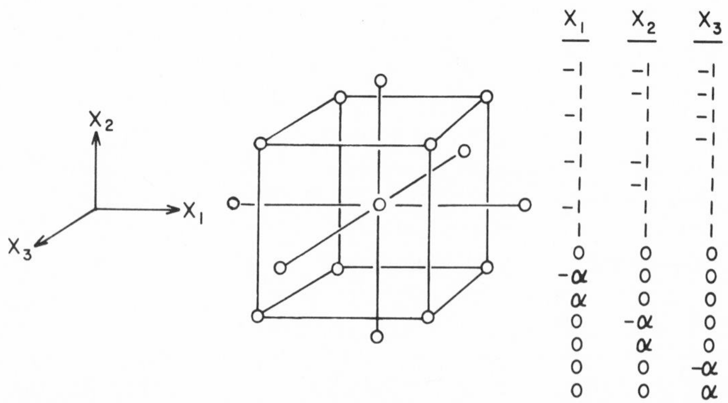

X_1	X_2	X_3
−1	−1	−1
1	−1	−1
−1	1	−1
1	1	−1
−1	−1	1
1	−1	1
−1	1	1
1	1	1
0	0	0
−α	0	0
α	0	0
0	−α	0
0	α	0
0	0	−α
0	0	α

Fig. 3–1. The central composite design that is formed from a cube plus an octahedron and is frequently used in fertilizer response investigation. The α's are coordinate values for axial points; α for rotatability $= 2^{k/4} = 1.68$ for $k =$ three factors.

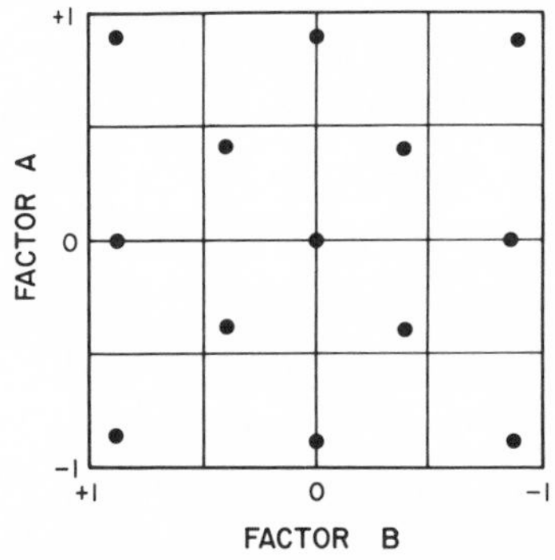

Fig. 3–2. Treatment design for 5^2 partial factorial modification by Escobar (1967).

sign developed by Escobar (1967), which consists of 13 selected treatments from a 5 × 5 factorial. The spatial location of the treatments of this design is shown in Fig. 3–2. Because two or three levels of each factor are studied at each of five levels of the other factor, a rapid graphic evaluation of the effects of the two factors can be obtained. This is especially useful in interpreting data from individual sites before doing a combined analysis.

3. *Selection of Treatments*

Another step in planning a fertilizer experiment is the selection of treatments. One purpose of treatments is to provide evidence of the absence or presence and, if present, the magnitude of the fertilizer effect. Often, however, they are used to provide an estimate of the response surface. In view of these two purposes, treatments should be chosen so that they best estimate the fertilizer effect or the response surface.

Several aspects of treatment design need to be taken into consideration when planning a fertilizer response investigation. *Treatment design* refers to how a limited number of points (fertilizer rate combinations) should be distributed within a given factor space. The object is to select that particular treatment design that will best give the specific information on plant response to fertilizer. Cady and Laird (1973) gave the following criteria for choosing a treatment design:

1. Interpretable data without extensive analysis
2. Relatively small number of treatment combinations
3. Low variance of estimated coefficients and, if hypothesis testing is important, independent estimates
4. Variance of the predicted values small over the central part of the factor space
5. Variance of the estimated response function slopes small over the central part of the factor space
6. Bias of the predicted values small over at least the central part of the factor space
7. Measure of lack of fit available
8. Check plot included in the design

9. The combinations of the zero level of one factor and high level of another factor are excluded

They thoroughly discussed minimization of variance (4) and bias (6). From a purely statistical point of view, the treatment designs may be altered to give either minimum variance or minimum bias error by changing the numbers of levels and their spacings. *Bias error* refers to the failure of the model being used to predict the response curve over the entire range of the controlled variable. The treatment designs that produce the minimum of one type of error are not necessarily the ones that produce the minimum of the other type of error. Designs were chosen on the basis of variance considerations in the past because it is possible to obtain a good idea of the variance before the experiment is run by examining the properties of the inverse matrix; also, the true model has to be known to study bias, a situation not found in practice.

It is very important to be careful in choosing the fertilizer rates to be used in an experiment. The question of the optimal number of incremental applications and their placement when considering rates to be used must also be addressed. The lowest and highest rates of each fertilizer nutrient are completely at the discretion of the researcher, but they should be chosen so that the responsive portion of the range is bracketed and there is at least one rate beyond the responsive range. The center of the range should be in the vicinity of the anticipated optimum rate for that particular fertilizer nutrient. This is because, statistically, the variance of predicted yield is smallest at the center.

Centering of the range in the vicinity of the anticipated optimum may force one to use a rate other than zero for the lowest rate. In this case, a check treatment will need to be included in addition to the response surface design points. The spacing of the fertilizer rates will be determined by the particular treatment design being used.

II. FIELD PLOT TECHNIQUE

A. Decisions Related to Use of Randomized Complete Block Design

The randomized complete block design is a good choice for fertilizer response investigations because the blocks control site variation, thereby increasing precision. To minimize the extent of the area involved and thus keep the experimental area relatively uniform, block size should be held to 15 to 20 plots if possible. The blocks should be square or nearly square in shape and should be oriented in such a way that, although the blocks may differ considerably from one another, the units within blocks are relatively uniform. Usually the blocks are placed side by side so that they control variation along a fertility or moisture gradient.

B. Randomization of Treatments

Randomization is employed in connection with use of the randomized complete block design. Randomization refers to the process of assigning treatments to the plots in such a way that all treatments have an equal chance of being assigned to a particular plot. Randomization provides assurance that a treatment is not continually favored or handicapped in various replications due to some extraneous source of variation, known or unknown.

Randomization for each experiment in a series of experiments should be performed independently. This will assure that no systematic biases will occur throughout the series and that estimates of site, year, site × treatment, year × treatment, and site × year × treatment effects will be valid.

C. Size and Shape of Plots

The plot size and shape that will give minimum error variance, and hence can be called *optimal,* will vary depending on the crop being studied. In general, plots should be long and narrow rather than square. The long dimension of the plot should be parallel to the fertility gradient if one exists. Four-row plots 6 to 10 m long with only the center two rows harvested for experimental purposes are usually adequate for row crops such as corn and soybean [*Glycine max* (L.) Merr.]. One should refer to the literature to find appropriate field plot technique recommendations for a particular crop. An example is the handbook on rice (*Oryza sativa* L.) experimentation by Gomez (1972). There are quantitative methods for determining optimal plot size empirically based on the original work of Smith (1938) and later that of Hatheway and Williams (1958). Optimal plot sizes for various crops, which were determined using their techniques, have been reported in the literature.

Mechanical considerations will often dictate what size and shape will be used, in spite of what is considered to be an ideal plot size and shape. Uniform spacing of plants within the row should reflect the plant population desired as well as results of past investigations concerning spacing × fertilizer interaction. Often the plants will need to be sampled before maturity to determine quality and plant tissue composition. This can best be accomplished by taking a systematic sample with a random start. To implement this, one would choose a plant at one end of the plot randomly and then take every *k*th plant (e.g., every 10th plant) along the row to include in the sample. See Federer (1955) for more detailed information on size and shape of plots as well as the statistical aspects of sampling within plots.

D. Use of Border Rows

Border rows are necessary in situations where the researcher considers that fertilizer treatments imposed on one plot might affect neighboring plots. A correct judgment of the degree of this influence is important. Providing insufficient border area can produce interplot interference that may cause representational or cryptic error. Having too much area in border rows would result in wasted space and an increase in area in the experiment, which could result in a large error variance.

E. Use of Uniform Experimental Technique Throughout the Series

Overall precision can be increased by using a uniform experimental technique throughout the series of experiments. Some ways of standardizing technique are to:

1. Write out procedures for conducting various phases of the experiments and a time schedule for their execution
2. Make all personnel dealing with the treatments, plots, and data aware of the various sources of error and the need for good technique
3. Apply the treatments uniformly
4. Exercise sufficient control over external influences so that every treatment produces its effect under controlled, comparable conditions. For example, having each of three persons harvest an individual replication is better than having all three harvest all replications as a team. Still better is to have one person harvest all three replications. If it is impossible to control environmental conditions, observations on major environmental variables should be taken and used as covariables
5. Devise suitable unbiased measures of the effects of treatments
6. Prevent gross errors

F. Recording of Site Data for Predetermined and Random Uncontrolled Variables

It is necessary to characterize each site used for experimentation for principal soil and meteorological characteristics that might affect yield, quality, and plant composition. Selection of site variables to be measured should be based on agronomic criteria. These site data are very important in a fertilizer response study; care should be taken in both the planning and measurement phases so that the relevant and reliable site data will be available at the time of data analysis. Usually one will need to take readings on many more variables than will appear as site variables in the general model. The most critical stages in the growth of a particular plant species are the times at which it will be important to make observations on specific variables (such as temperature). The dates of specific physiological development stages such as anthesis and maturity can be very important in explaining treatment effects. An incorrect judgment concerning timing could render the variable useless in the data analysis phase. Observations on random uncontrolled events such as pest attacks, flooding, etc. should also be made and carefully recorded.

G. Measurement of Soil Properties of Individual Plots

If there is a potential for sizable variability in chemical and physical properties within a site, the site should be characterized by individual plot. It is then possible to estimate the within-site interactions involving the soil properties and the added nutrients in addition to among-site interactions.

III. PHYSICAL RESPONSE MEASUREMENTS

Although yield is the most common physical response measured, measurement of certain other characteristics may provide useful information on the mechanism of plant response. Variables such as chemical composition of plant tissue, both grain and vegetative, are useful in this connection. In some cases, complete biomass yields are also measured because they provide more complete information on utilization of nutrients. Physical and biological quality measurements of the product are also important because they affect its nutritive and economic value.

Plant survival is another response that is measured and used for conducting a covariance analysis to adjust yield to a constant stand. Causes of stand losses also must be recorded.

IV. STATISTICAL ANALYSIS OF FERTILIZER RESPONSE DATA

A. Adjustment of Yield Data to a Specified Moisture Content

Grain yields are often adjusted to a standard moisture content to make them comparable from one treatment or study to another and relate them to the market. This is accomplished by use of the following formula:

$$\text{Yield at standard moisture content} = W[(100-A)/(100-M)]$$

where W = weight of harvested grain, A = actual moisture content of grain (%), and M = standard moisture content of grain (%).

As an example, suppose one plot yielded 25 kg corn that had a moisture content of 20.0%. This yield in kilograms corn per plot at 15.5% moisture is given by:

$$\text{Yield at 15.5\% moisture} = 25\,[100\%-20.0\%)/(100\%-15.5\%)]$$

$$= 23.7 \text{ kg/plot.}$$

The moisture adjustment must be done separately for each plot in an experiment.

B. Analysis of Variance for Individual Site-Year Data

One type of data analysis that should be routine is an analysis of variance of data for each individual site-year. The analysis normally would include single degree of freedom components representing terms of a second-order response surface either in the original X scale or on a transformed scale as subdivisions of the treatment sum of squares. A typical analysis of variance key-out for a N, P, and K fertilizer experiment using a randomized complete block design and fitting a second-order response sur-

Table 3–1. Analysis of variance key-out for a 4 × 4 × 4 N, P, and K fertilizer experiment according to a randomized complete block design with five blocks, assuming the fitting of a second-order response surface.

Source	df		
Blocks	4		
Treatments	63		
N		3	
N_{lin}			1
N_{quad}			1
Other			1
P		3	
P_{lin}			1
P_{quad}			1
Other			1
NP		9	
$N_{lin} \times P_{lin}$			1
Other			8
K		3	
K_{lin}			1
K_{quad}			1
Other			1
NK		9	
$N_{lin} \times K_{lin}$			1
Other			8
PK		9	
$P_{lin} \times K_{lin}$			1
Other			8
NPK		27	
Error	252		

face is shown in Table 3–1. F tests of significance are conducted for treatments, and all single degree of freedom components. The sources labeled Other and NPK are usually pooled to give a "Lack of Fit" source of variation and this mean square is tested with the error mean square to check for inappropriateness of the model. For the example given in Table 3–1, the Lack of fit has 54 df. The results give an indication as to which nutrients the crop responded and how well the model fits.

C. Combined Analysis of Variance Over Locations and/or Years

A combined analysis of variance for data from all sites and/or years is conducted after performing the individual site-year analyses of variances. This analysis is necessary to test for treatment × environment interactions. It also allows a test of the treatment main effects averaged over all sites and/or years. In addition, it provides estimates of the error terms that will be used in deriving errors for testing various terms in the response function. For example, the error term for testing site variable terms is usually different from that used to test fertilizer main effect terms.

The format for an analysis of variance of data combined over sites within a year is shown in Table 3–2. An analysis of variance of data combined over sites and years has the form shown in Table 3–3. Treatments for both of the analyses are subdivided in the same manner as shown in the analysis of variance key-out for the individual site. The interaction terms

Table 3-2. Format for an analysis of variance key-out for data combined over sites within a year.†

Source	df
Sites (S)	$s - 1$
Replications in site	$s(r - 1)$
Treatment (T)	$t - 1$
$S \times T$	$(s - 1)(t - 1)$
Error	$s(r - 1)(t - 1)$

† Capital letter abbreviations refer to the name of the effect. Lower case letter abbreviations refer to the number of levels.

Table 3-3. Format for an analysis of variance key-out for data combined over sites and years.†

Source	df
Sites (S)	$s - 1$
Years (Y)	$y - 1$
$S \times Y$	$(s - 1)(y - 1)$
Replications in ($S \times Y$)	$sy(r - 1)$
Treatment (T)	$t - 1$
$T \times S$	$(t - 1)(s - 1)$
$T \times Y$	$(t - 1)(y - 1)$
$T \times Y \times S$	$(t - 1)(y - 1)(s - 1)$
Error	$sy(t - 1)(r - 1)$

† Capital letter abbreviations refer to the name of the effect. Lower case letter abbreviations refer to the number of levels.

may also be subdivided to show how individual components of treatments interact with sites and years.

Error terms from the individual site-years are pooled to obtain the sum of squares for Error in both of the analyses. Prior to pooling, there should be a reasonable indication that the variances from the individual site-year analyses are homogeneous. This can be checked with a test for homogeneity of variance such as Bartlett's (1937) or Hartley's (1950) tests. There will often be a few site-years within a series that should be considered as outliers and deleted from the combined data set. The results of one of these tests of homogeneity of variance as well as examination of the data scatter will help to identify these sites.

The appropriate mean square to use as the denominator in the F ratio having a given mean square as numerator may be inferred by writing out expectations of the mean squares for the various sources of variation. Rules for writing these are given by Schultz (1955) and Steel and Torrie (1980). One should allow room for some flexibility in methodology in testing procedures based on the unique set of circumstances connected with a particular series of experiments.

D. Analysis of Covariance in Analyzing Fertilizer Response Data

There are two situations in which analysis of covariance may be used to advantage in the analysis of fertilizer response data. This technique is sometimes used to provide estimates of missing plot values that result in unbiased estimates of treatment sums of squares. It is also used to adjust yield or other response variables for differences in the level of some observed uncontrolled variables such as plant population per plot. In connection with the second type of application, the precision of the analysis is often improved considerably. Heady and Pesek (1957) found the relationship between plant population and corn yield to be highly significant in the analysis of data from an experiment on a Haynie soil (coarse-silty, mixed [calcareous], mesic Mollic Udifluvent) in Iowa; therefore, they included this as a term in the response function derived from that particular set of data.

E. Response Surface Fitting to Data From Individual Site-Years

The treatment sum of squares should be subdivided in the analysis of variance into components for terms of the response surface model (see section B above). Our discussion here deals with the choice of an appropriate model that will be used for fitting data from the individual site-years and ultimately the general response function. Mead and Pike (1975) stated that the choice of a particular response function in the analysis of data is often determined by several conflicting objectives. These usually include a purely descriptive reduction in the data, a short-term prediction for particular combinations of factor levels, a long-term prediction for the general pattern of response, and a simple interest in the "true" pattern of response. Models have been classified into two broad categories, *empirical* and *biological.* The parameters of an empirical model do not necessarily have biological meaning, whereas those of a biological model are considered to be approximations of some aspect of a biological process. At this point the distinction between empirical and biological models seems more tenuous than it was previously considered to be. Perhaps the problem rests more in the difficulty with defining the complex underlying biological processes and their interactions than in dealing with their mathematical quantification.

The most frequently used response function in fertilizer response studies is the polynomial of degree, *p*. Many workers consider this an empirical model, although Niklas and Miller (1927) contended that there was a biological basis for use of a second-order polynomial model. Polynomial models are easily fitted by least squares methods and are easily generalized to a multinutrient relationship represented by $\hat{Y}$ regressed on several X variables. Mead and Pike (1975) state that apart from a straight line, the quadratic ($p = 2$) is the most commonly used polynomial. It has the following form for a single nutrient:

$$\hat{Y} = b_0 + b_1X + b_{11}X^2. \qquad [1]$$

In particular, for fitting a response surface for N, P, and K the form is as follows:

$$\hat{Y} = b_0 + b_1\mathrm{N} + b_2\mathrm{P} + b_3\mathrm{K} + b_{11}\mathrm{N}^2 + b_{22}\mathrm{P}^2 + b_{33}\mathrm{K}^2 + b_{12}\mathrm{NP} + b_{13}\mathrm{NK} + b_{23}\mathrm{PK}. \qquad [2]$$

These authors listed the following reasons for the general popularity of the quadratic model: (i) it involves merely the addition of an extra term to the straight line relationship which, for most people, makes it the simplest curvilinear relationship; (ii) it has a simply defined maximum at $X = -b_1/2b_{11}$, and (iii) the method of least squares produces estimates of parameters without complex calculations. Another advantage is that the scope of yield response patterns that may be fitted within the polynomial family using least squares procedures is broad due to the possibility of making various transformations primarily of the X variables.

One disadvantage of the quadratic model is that it does not allow for asymmetry around the optimum in the yield response pattern. This asymmetry often occurs in actual practice. Also extrapolation outside of the experimental range of X values is impossible.

Some have used models containing polynomials in transformed X variables. The most commonly used transformation model is the square root model (X is replaced by $X^{1/2}$ in the quadratic model). The equation for this model in the single nutrient case is

$$\hat{Y} = b_0 + b_1X^{1/2} + b_{11}X. \qquad [3]$$

The curve reaches a maximum at $X = (-b_1/2b_{11})^2$ and then declines. It is asymmetrical around the maximum. When fit to the same data, predicted values from the square root equation rise more rapidly than those for the quadratic because the slope of the square root curve tends to infinity as X tends to zero. The curve for the predicted values of the square root model often tapers off to a broad area in the region of the maximum that has less slope than the quadratic curve predicted from the same data. The physical maximum may be at either a higher or lower value than that for the quadratic, depending on the yield pattern for the high rate points.

Fitting a response surface for N, P, and K by the mathematical form of the square root response surface gives:

$$\hat{Y} = b_0 + b_1\mathrm{N}^{1/2} + b_2\mathrm{P}^{1/2} + b_3\mathrm{K}^{1/2} + b_{11}\mathrm{N} + b_{22}\mathrm{P}$$
$$+ b_{33}\mathrm{K} + b_{12}\mathrm{N}^{1/2}\mathrm{P}^{1/2} + b_{13}\mathrm{N}^{1/2}\mathrm{K}^{1/2} + b_{23}\mathrm{P}^{1/2}\mathrm{K}^{1/2}. \qquad [4]$$

The square root model, mentioned by Heady et al. (1961) for fertilizer recommendation purposes, was also favored by Abraham and Rao (1965). More recently it has been used by the Australian National Soil Fertility Project for the analysis of field data (Colwell, 1979), and it was the principal model discussed in Colwell's (1978) book on design and analysis of field trial data.

Heady et al. (1961) used a polynomial model having a linear term, but also a 1.5 or 3/2 power on the second term. Anderson and Nelson (1971) used models of the type $\hat{Y} = b_1(X + d) + b_2(X + d)^h$, where X is added nutrient, d is soil nutrient (estimated from the data), and h is an exponent that was given each of the following values: 0.50 (square root model), 0.75,

0.95, 1.25, 1.50, 1.75, and 2.00 (quadratic model). In another approach to the estimation problem, $\hat{h}$, an optimal value of h, was determined simultaneously with $\hat{d}$ for each site-replication using iteration techniques. A restriction that $0.25 \leq \hat{h} \leq 2.00$ was imposed. The criterion for optimality was minimum residual sum of squares. The value of $\hat{h}$ varied considerably within a site, implying that the same polynomial model would not be appropriate for all replications, even at the same site. There was almost as much variation in $\hat{h}$ within sites as among sites.

Nelder (1966) developed a group of empirical models called inverse polynomials, which he claimed are more flexible and realistic than ordinary polynomials. These have the general mathematical forms:

$$\widehat{Y^{-1}} = aX^{-1} + b \qquad [5]$$

and

$$\widehat{Y^{-1}} = aX^{-1} + b + cX \qquad [6]$$

for a saturation effect curve and a toxicity effect curve, respectively. They have been used primarily by British scientists. The second curve is analogous to an ordinary quadratic polynomial but is not constrained to be symmetrical. These are somewhat more complicated mathematically than the ordinary polynomials, both from the standpoint of assumptions about the distribution of errors and the fitting procedures.

The power function (Cobb-Douglas) is an empirical model that has been employed in a number of fertilizer response investigations. It has the general form for a single nutrient, such as nitrogen (N)

$$\hat{Y} = a\mathrm{N}^{b}, \qquad [7]$$

where $\hat{Y}$ is the predicted yield, a and b are constants to be estimated from the data, and N is the rate of added nutrient. The equation may be transformed to linear form

$$\widehat{(\log Y)} = \log a + b \log \mathrm{N}. \qquad [8]$$

The Cobb-Douglas function when generalized to the multinutrient case takes on the following form:

$$\hat{Y} = a\mathrm{N}^{b}\mathrm{P}^{c}\mathrm{K}^{d}, \qquad [9]$$

where a, b, and N are as defined above, P is phosphorus, and K is potassium. The constants for P and K are c and d, respectively. The most serious limitation of this function is that it does not provide for a maximum yield and hence cannot adequately describe the complete range of responses observed. It is not even asymptotic to the maximum.

Boyd (1972), Anderson and Nelson (1975), and Boyd et al. (1976) reported the use of intersecting straight-line models, which was prompted by their observation that the quadratic and square root models gave biased estimators of the optimal fertilizer rates when the yield had a plateau response pattern. The advantage is flexibility of the model, i.e., a model or submodel that fits the data well is chosen from a family of models and this is more apt to provide unbiased estimates of optimal fertilizer rates in plateau response situations than would traditional models. Also, the models are readily fit using unsophisticated computing equipment. The reported disadvantages are that these models are mathematically unsatisfying and that the discontinuities that they have at the intersection points are unrealistic. In addition, these models do not lend themselves to situations where interactions between nutrients exist or where a general model that includes site year × added nutrient terms is developed for a combined data set from a series of site years.

Perhaps the model that most workers have traditionally identified with the "biological model' concept is the Mitscherlich equation (Mitscherlich, 1909), which is also called the *negative exponential model.* This equation was based on Mitscherlich's observation that the response to an increment of fertilizer was proportional to the decrement from the maximum yield. He formulated the following differential to express this relationship

$$\delta Y/\delta X = c(A-Y), \quad [10]$$

where $\delta Y/\delta X$ is the differential of yield (Y) with respect to added nutrient (X), c is a response coefficient, and A is the asymptotic maximum yield. When originally developed to apply to plant response in sand culture, the mathematical equation resulting from integration of Eq. [10] was

$$\hat{Y} = A[1-\exp(-cX)], \quad [11]$$

where $\hat{Y}$ = predicted yield and the other terms are as previously defined. This equation was later modified by Baule (1918) for use with a series of soils to include the initial amount of nutrient in the soil, d, which is expressed in fertilizer equivalent units

$$\hat{Y} = A\{1-\exp[-c(X+d)]\}. \quad [12]$$

The two variable generalization of this model is

$$\hat{Y} = A\,\{1-\exp[-c_1(X_1+d_1)]\}\,\{1-\exp[-c_2(X_2+d_2)]\}. \quad [13]$$

Certain attributes of the Mitscherlich equation originally seemed attractive from a biological point of view. The primary one was the concept of a maximum yield, A, which is asymptotically approached as fertilizer is added. There were statistical limitations, however. This equation is more difficult to fit (even with modern digital computers) than models that may be fit by least squares techniques. There may be problems achieving convergence if the data are imprecise or if there are only a few levels of treatments. The Mitscherlich equation is not as flexible as some of the other models such as the quadratic in terms of developing a general model that includes variables to explain variation among site-years.

Some workers contend that the Mitscherlich equation is not really a biological model but rather an empirical model. The fact that there have been two revisions of the mathematical form of the original model based on modified views of the biological processes and how they should be quantified lends support to this view. The first revision resulted from the empirical observation that in many cases the yield decreased beyond a certain X. This so called second approximation, which was formulated by Mitscherlich (1930), took the form

$$\hat{Y} = A[1-\exp(-cX)]\,[\exp(-kX^2)], \qquad [14]$$

where k is the damage factor and other terms are as before.

von Boguslawski and Schneider (1962) proposed a third approximation to the Law of Yield, which was a follow-up to the first two approximations. It has the form

$$\hat{Y} = M10^{-Z\{\log[(X + i)/(m + i)]\}^n},$$

where M is the maximum yield obtainable; m is the corresponding input of X; i can be interpreted with some approximation as the fertilizer equivalent of the soil; Z and n are parameters controlling the shape of the curve; X is the amount of applied nutrient; and $\hat{Y}$ is the predicted yield. This model is extremely flexible but involves a large number of parameters that must be estimated.

Balmukand (1928) developed another so-called biological model that is based on a completely different theory of biological response. Like the Mitscherlich, it has an asymptotic maximum to which the predicted Y converges as the nutrient rate approaches infinity. It has a hyperbolic form, which is related to Nelder's (1966) inverse polynomial models. It assumes a biological analogy to Maskell's electrical resistance formula. The underlying supposition is that each activity of the plant (yield, etc.) is determined by a potential set of resistances, each of which represents one of the external factors.

Balmukand's equation for the relationship of yield to N, P, and K is

$$\widehat{[1/Y]} = c + [a_n/(n + N)] + [a_p/(p + P)]/ + [a_k/(k + K)],$$

where $\widehat{[1/Y]}$ is the predicted reciprocal of the yield for a given combination of added N, P, and K; c is a general constant; n, p, and k are soil levels of the nutrients that are parameters to be estimated from the data; and N, P, and K are added levels of the nutrients. The a_n, a_p, and a_k are constants that express the importance of the three nutrients to the crop. They are estimated from the data.

Although the literature on fertilizer response modeling is extensive, there are few studies dedicated to comparing different models. The reason is that statistically valid comparisons among mathematical models of differing functional form are difficult to make. It is also difficult, if not impossible, to compare models on the basis of correctness of fertilizer recommendations when in fact the true recommendation is never known. It does appear, however, that the choice of model has a considerable effect on the estimates of optimal fertilizer rates.

Cady and Laird (1973) studied the effect of treatment design and postulated model on the bias error. By bias error they meant the integrated area between the true response curve and the curve for the model being used for fitting the data. Of the factors studied, they found that choice of the model had the most important effect on bias error.

Abraham and Rao (1965) compared the following functions as to suitability for fertilizer recommendation purposes: Mitscherlich, Cobb-Douglas, hyperbola, quadratic, and square root transformation of the quadratic. All models fit reasonably well to rice yields. The estimated optimal rates from the quadratic and Mitscherlich were close to each other and their estimated standard errors smaller than those for the other models. On the basis of the ease of fit and results of tests of hypotheses about model parameters, the quadratic model was favored.

Jónsson (1974) studied the effect of varying N ranges on the choice of an optimum model among the quadratic, square root, and an extended hyperbola model. It was concluded that at a narrow range in N, the quadratic and square root equations are similar and superior in describing the data, whereas the hyperbola is inferior. At a wider range of N rates, the superiority of the square root and extended hyperbola over the quadratic was noted. The parameter estimates of the quadratic equation were found to be very dependent on the number of N rates used in the fit. This would suggest that the fertilizer response curves used in practice were not symmetric and this would militate against the quadratic model, which assumes symmetry.

Anderson and Nelson (1975) compared several models with respect to the optimum amount of nutrient to be applied. They generalized by stating that if one is interested in obtaining the optimal fertilizer recommendation (X_0), an estimating procedure that gives a good X_0 estimate, rather than a minimal residual sum of squares, should be developed.

Mombiela and Nelson (1980) compared the quadratic, square root, and Mitscherlich functions on the basis of the functions' recommendation bias, variability of recommendation, and expected economic loss for data generated from an underlying response function for P whose parameters were prespecified. They concluded that response surfaces should be mechanistic rather than empirical if good performance over a wide variety of situations is desired. The quadratic surface was found to be too rigid to accommodate nonsymmetrical shapes and its performance was reported to be greatly affected by price ratio fluctuations. The performance of the square root surface was satisfactory at medium and high price ratios if the soils involved had medium to high soil test values and, thus, small yield responses.

Heady et al. (1961) used corn response data from an Ida silt loam soil (fine-silty mixed [calcareous], mesic Typic Udorthent) to compare the following models with respect to shapes of isoquants and isoclines as well as estimates of economic optima: square root, quadratic, 1.5 power model, and Cobb-Douglas model. They concluded that while each might be appropriate for certain purposes, the square root function appeared to have a slight superiority in statistical efficiency in predicting the production surface for the particular environmental conditions under which the experiment was conducted.

In summary, concerning model selection, one cannot recommend a single model for all situations. The nature of the actual crop response and the relevant area of interest should be considered in choosing a model. It would seem that the square root transformation of the quadratic model would be a reasonable choice in many situations unless there is a definite "plateau" response pattern. In this case, some of the straight-line models would seem applicable. If the quadratic model is being considered, great care should be taken to see that the model will not introduce upward biases into the estimate of the optimal fertilizer rate. Use of the Mitscherlich equation would seem to be impractical from the standpoint of a general model that includes several fertilizer nutrients, several site-years, and interactions between the nutrients and the site-year variables.

Using least squares, one is able to estimate the regression coefficients of the polynomial models and the variances of these estimates. The predicted yield for a given combination of the X_i and its variance are also available. This latter variance is a function of the particular combination of the X_i for which the response is estimated, but it is also dependent on the design used for the experiment.

If we assume the general model is $\hat{Y} = b_0 + b_1X + b_{11}X^2$, then the estimate of optimal fertilizer rate is $(R-b_1)/2b_{11}$, with R being the cost of a unit of fertilizer/price of a unit of product. An exact estimate of the variance of this optimal rate is calculable, but because the denominator, which is very small and unstable, has an important effect on this variance, this variance is often quite large. The regression coefficients and their standard errors are more involved to calculate for the inverse polynomial, the Mitscherlich, and Balmukand Resistance model. The latter two models require iterative fitting techniques.

F. Two-Dimensional Contour Plots of Yield for Individual Site Year Data

Heady and Dillon (1961) have described the properties of the response surfaces generated by the most frequently used response models with two variates. These are the Cobb-Douglas equation, the Mitscherlich-Spillman equation, the resistance equation proposed by Balmukand, and the quadratic and square root transformation of the quadratic equations. It is possible to plot contours in the space of $X_1, X_2, \ldots X_k$ *along which the yields are constant* (called *isoquants*). If the design is rotatable and the contours are circles, spheres, or hyperspheres around the center of the design region the variance of predicted Y values should be equal along a contour meaning that the predictions are equally reliable. In Fig. 3–3 is shown a two-dimensional plot of yield contours for a response surface estimated from a N × P experiment. The visual representation gives an idea as to how

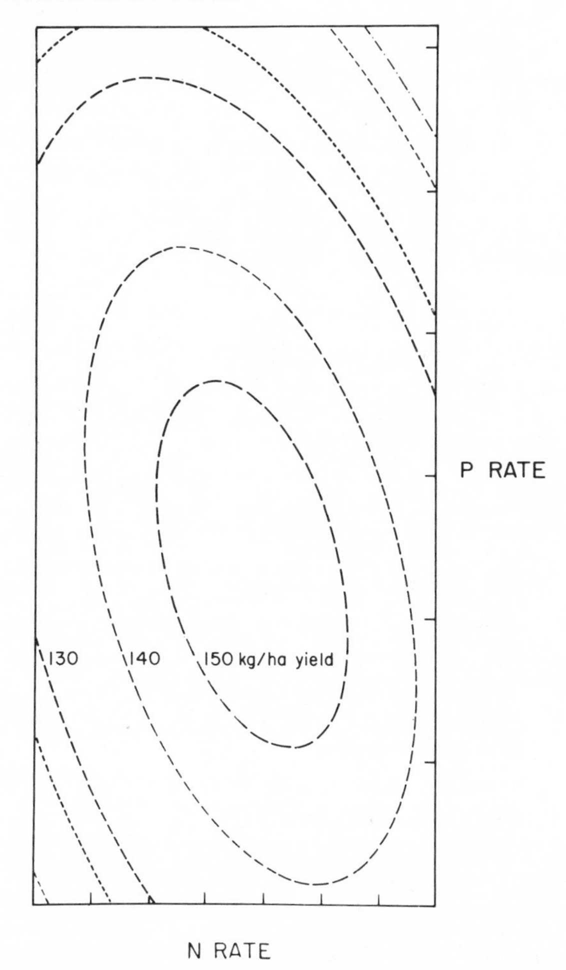

Fig. 3–3. Two-dimensional plot of ellipsoidal yield contours for a response surface estimated from a N × P fertilizer experiment.

responsive the yield measure is to these two fertilizer nutrients and how these nutrients interact with each other. For cases involving three or more nutrients, a two-dimensional plot may be done at each level of the third factor.

There are four fundamental and limiting surfaces generated by a second-degree equation in two dimensions and they have distinctly different appearances (Fig. 3–4). Figure 3–4a is a system of closed ellipsoids, which has a unique maximum in the region investigated. This is a desirable surface to obtain from the standpoint of estimating the optimal fertilizer rate. Figure 3–4b depicts a saddle point (sometimes called a *col* or *minimax*). Such a surface would be difficult to interpret agronomically. Data that give rise to such a surface could result from an experiment that had a poor choice of treatment levels and/or that lacked precision. Substantiation of this type of pattern would be called for before extending the appli-

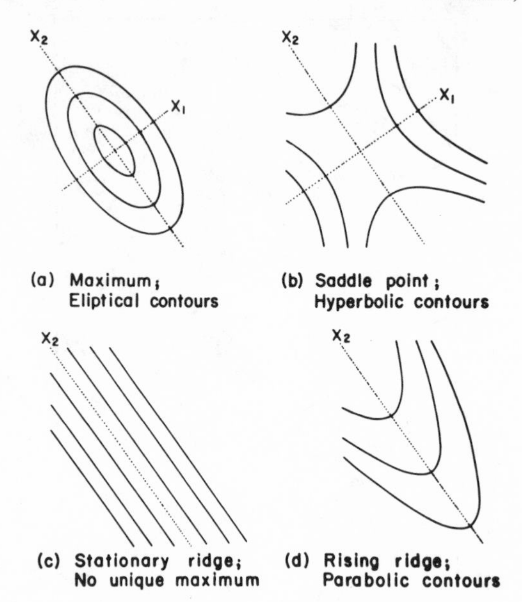

Fig. 3–4. Four fundamental limiting surfaces generated by a second-degree equation in two dimensions.

cation very far. When conducting the substantiation experiment, more treatment combinations in the vicinity of the assumed optimum would seem in order.

Figure 3–4c and 4d represent essentially limiting cases. Figure 3–4c shows a stationary ridge pattern, which might be regarded as a surface such as 4a and 4b, which is infinitely attenuated along the X_2 axis. Figure 3–4d shows a rising ridge in which the contours of the ridge are parabolas. The center is at infinity. For a more complete discussion of the forms of response surfaces see Box (1954).

With modern computing equipment and software packages it is possible to plot the contours in two-dimensions or even to plot a three-dimensional surface. These plots provide information that should supplement the results of the statistical applications such as tests of significance of the regression coefficients of the model.

G. Response Surface Fitting to Combined Data From all Sites and Years

A response surface may be fitted to data from all sites and years using a model that includes the same applied nutrient terms as in the individual site model along with terms to explain among site variation, among year variation, site × nutrient, year × nutrient, and site × year × nutrient variation. This usually is the model that is used for making general recommendations for a crop on a specific soil. The inclusion of site and year terms often causes sizeable adjustments in the applied nutrient regression coefficients so that estimates of optimal fertilizer rates will be very sensitive to the particular site and year variables and their interactions with added nutrient, which are included in the model.

In spite of this model specification difficulty, it would seem that making recommendations from such a general response surface model would be more efficient than obtaining estimates of optima from individual sites and then averaging them. It would also permit tailor-making predictions to individual conditions by substituting values for individual site-year conditions into the model. Obtaining optima estimates from main effect treatment means averaged over sites and years would seem to gloss over the very real possibility of interaction between added nutrients and site and year variables.

Given that a model has been chosen to explain the variation attributable to added nutrients for an individual site-year, there is some uncertainty as to which site-year variables and site-year variable × added nutrient terms should be included and in what form they should be in the model being developed for the combined data.

One question that usually arises is whether to combine soil nutrient with added nutrient or to carry the former in the model as a covariable. For a single nutrient the two forms would be

$$\hat{Y} = b_1(X + d) + b_{11}(X + d)^2 \quad [15]$$

or

$$\hat{Y} = b_0^* + b_1^*X + b_{11}^*X^2 + b_2^*d + b_{22}^*d^2 + b_{12}^*Xd, \quad [16]$$

where X = added nutrient, d = soil nutrient, and the b_i and b_i^* are regression coefficients estimated from the data. Equation [15] would be much more efficient from a statistical and agronomic point of view in terms of low variances of regression coefficients and predicted yields, compactness of the model, and agronomic realism. The exact soil nutrient level is never known, however, so it must either be estimated from the data iteratively in fertilizer equivalent units or else the soil nutrient levels must be estimated from soil tests and then combined with the added nutrient levels. Soil test values are only indexes of nutrient availability and are not equivalent units of the fertilizer source of the nutrient. Hildreth (1957) dealt with the problem of the estimates of soil nutrient obtained by tests having a different fertilization efficiency than the added nutrients. Mombiela et al. (1981) more recently suggested an approach to determine from the data the mathematical form of the relationship between the d estimated from the data and soil test results. Once a reasonably good estimate of the form of this function is known, one can replace d by the function in the model that combines soil and added nutrient.

The form given in Eq. [16] has been used by Voss and Pesek (1967), Voss et al. (1970), Ryan and Perrin (1973), and Colwell (1967, 1968, 1970, 1979). It has the advantage of being convenient and easy to fit by least squares methods. At this point, there seems to be no overwhelming advantage of one form of model over the other.

One may use an agronomic approach to the selection of a general model that would initially involve the controlled variable terms plus other variables that agronomic theory would suggest are important and subsequently those whose relationships with yield are not supported by the data would be removed. The agronomic terms would include the soil nutrient variables as described above. Certain more permanent soil properties such as cation exchange capacity and mineral index (14Å/7Å peak height ratios on x-ray diffraction patterns) were also found by Nelson and McCracken (1962) to be useful for explaining among site variation.

Another class of variables that is necessary in most general models is those that explain weather effects. Sopher et al. (1973) discussed three approaches to reducing daily weather data to a reasonable number of meaningful variables. These included (i) use of seasonal, monthly, bimonthly, or weekly climatic data in the model; and (ii) use of a polynomial regression method devised by Fisher (1924) in which the growing season is divided into n short intervals, a kth degree polynomial is fit to the n intervals, and the k orthogonal polynomial regression coefficients are then entered into the general model as independent variables. Such an approach was used by Hendricks and Scholl (1943), Runge and Odell (1958), and Runge (1968). The third method is employment of the plant-available soil moisture holding capacity, potential evapotranspiration, and rainfall data to calculate a water budget for a specific soil. From the water budget, the number of moisture deficit days in each of several crop growth periods is noted. These moisture deficit days in each growth period are then used as independent variables in a general model. Mason and Cooper (1958), as quoted in Sopher et al. (1973), incorporated the numbers of drought days (van Bavel, 1953) for four selected physiological growth periods in the growing season of corn into a single-variable drought index. This was achieved by regressing yield of corn on the number of drought days in each of the four physiological periods, and then obtaining a weighted average of the number of drought days using the regression coefficients for the four periods as relative weights. The weighted average is the drought index. The index obtained by the above procedure was then used as an independent variable in the general model that was fit to data combined across site-years. Sopher and McCracken (1973) used this drought index in developing yield response models for corn production in the Coastal Plain of North Carolina. Others have found that the stress day concept describes the reaction of plants to climate quite well (Denmead & Shaw, 1962). In general terms, a *stress day* is a day when conditions of atmospheric demand for moisture, water availability in the soil, and stage of crop development are such that adequate moisture for meeting evapotranspiration demands cannot be extracted from the soil by the crop and the plants begin to wilt.

Sopher et al. (1973) were careful to point out that the degrees of freedom associated with the estimation of the drought index should be equal to the number of individual drought components composited in the index plus one for the index itself. They also made the observation that excess moisture and cool temperatures should be included along with drought measurements early in the growing season.

Once it has been established which variables agronomic theory suggests should be included in the general agronomic model, some objective method(s) should be employed to determine which variables are contributing to the prediction of yield and therefore should be retained in the regression equation. In general, those variables whose regression coefficients (b_i) are greater than their standard errors (s_{bi}) are contributing more than "noise" to the prediction of yield. This "sorting out" of variables for which the ratio $= b_i/s_{bi} = t_b \geqslant 1$ is a less conservative approach than making actual tests of significance to determine which variables should be retained in the model. It is for this reason that we recommend this procedure, especially in the early stages of the variable selection process.

The agronomic approach to model selection, although reported by Laird and Cady (1969) to be superior to an empirical approach, has the general limitation that the functional relationship between yield and the combined set of controlled and uncontrolled variables is not known due to the complexity of biological systems. Another choice would be to use a more empirical approach based on fitting alternative models not necessarily having an agronomic basis to the data and evaluating the fit based on criteria such as significance of regression coefficients, minimum mean square error, Mallows C_p statistic, etc. Should several models perform similarly under the previous criteria, the final choice among them is made according to simplicity or convenience. For a discussion of the relative merits of the agronomic and empirical models, see Laird and Cady (1969).

These two authors noted that error structure should be considered in testing the regression coefficients in either type of model. A combined analysis of variance similar to the ones shown in section C above should be conducted first to obtain estimates of the errors (replications in site or replications in site-years and pooled error). The replications in site or replications in site-year should be used to test sites or site-years and their interactions. Added nutrient and added nutrient × site and/or years should be tested by pooled error.

In fitting the empirical models, one would place controlled fertilizer variables into the model and then add additional site-year variables and interaction terms based on their performance using empirical selection procedures. One difficulty in choosing a model and in interpreting the importance of terms contained therein is posed by the correlation (termed *multicollinearity*) of the site-year variables which is often pronounced. There are several fairly standard procedures for building a regression model empirically and these are described in detail in Draper and Smith (1981). Some of the more common of these are the Forward Selection Procedure, which builds a model from a few to a larger number of variables; the Stepwise Regression Procedure, which also builds a model starting with few variables but which reevaluates all previously entered variables as each new variable is entered; the Backward Elimination Procedure in which variables showing little contribution to the prediction of yield are eliminated one-by-one from a "full model," and the Maximum and Minimum R^2 Improvement Procedures in which a new variable is chosen on the basis of producing either maximum or minimum gains in R^2 as it is entered into the model. The danger in using any of these empirical techniques is that because they are more mechanically than agronomically based, a good fit will be obtained to the particular set of data being analyzed, but the equation may not predict well for sites outside of the particular set. There appears to be no easy answer to the question of how to choose a model because the problem of choosing the best function is not soluble using a simple set of rules. The lack of a common definition of the term *best* is one of the problems itself.

Laird and Cady (1969) and Cady and Allen (1972) used a prediction sum of squares approach to the choice of a model rather than a residual sum of squares approach. The principle is that when making an individual prediction, the regression equation should be derived from data that exclude the observed point corresponding to that prediction. The prediction sum of squares is then obtained by squaring the deviations of the observed points from the predicted points, the latter of which have been obtained from all of the observed data except the point being predicted. This should lead to a model that is not confined in applicability to only the available set of data. The prediction sum of squares technique was used by Wood and Cady (1981) in the Benchmark Soils Project of the Universities of Hawaii and Puerto Rico to develop a model for transferring fertilizer response information from one site to another belonging to the same soil family.

Regardless of the selection procedure used, one should not be too optimistic about arriving at a model that is far superior in fit to all other models using empirical techniques. In addition, one can only hope that the variables contained in the "best model" have some agronomic rationale and produce estimates of optimal fertilizer rates that are reasonably free of bias.

As with the individual location models, least squares procedures may be used to estimate the regression coefficients of the general model developed for combined data from all sites and years and to place confidence limits on them. One must remember to use the appropriate error term for calculating the variance of a regression coefficient depending on whether it is a site-year, added nutrient, or site-year × added nutrient variable. One would prefer to keep the number of site-year and site-year × nutrient terms to a minimum to assure that the model is "compact." Extraneous terms in the regression whose regression coefficients are not larger than their standard errors may add "unwanted noise" to the predictions.

As in the case of the individual site-year analyses, the estimate of the optimal fertilizer rate involves a ratio of two regression coefficients and, so again, the variance of this estimate may be quite large.

V. ECONOMIC ANALYSIS OF FERTILIZER USE DATA

A. General Techniques

The economic techniques used to determine optimal rates of fertilizer depend heavily on the model used to fit the data. For example, Anderson and Nelson (1975) described rather simple procedures for economic analysis in connection with the family of linear-plateau models. If the slope of one of the sloping straight lines of the model exceeds the cost of fertilizer/

price of product ratio, one adds fertilizer to the point where that sloping line intersects with another line such as the plateau. In two or three sloping line models there will be two or three slopes to compare with the cost/price ratio. Economics of nondivisible treatments (e.g., plowing, disking) follows a similar "yes-no" decision procedure.

Most of our discussion here, however, will be based on use of a polynomial or polynomial transformation model that provides the basis for using the economic principle of diminishing returns and the added cost-added returns concept. For a much more comprehensive treatment of these concepts and the techniques used in economic analysis see Heady et al. (1955) and Heady and Pesek (1957).

The diminishing returns concept implies that as higher levels of a fertilizer nutrient are applied, the increments of output decrease. A point is reached on the response curve where the cost of an increment of fertilizer applied is equal to the value of its yield return. It is at this point where the maximum profit per unit land area will be achieved. The optimal rate of fertilizer is the quantity of fertilizer that achieves this objective. This then is the rational upper limit of fertilizer use with unlimited capital and without alternative investments. The purpose of the economic analyses then is to find the fertilizer rate where the added returns from an increment of fertilizer just equals the added cost. This is achieved algebraically in the single nutrient case by equating the change in yield per unit change in added nutrient to the cost of fertilizer/price of product ratio, i.e.,

$$\delta Y/\delta X = P_X/P_Y \qquad [17]$$

where Y = yield of crop at fertilizer rate X, P_X = cost of unit of fertilizer, and P_Y = value of an increment of yield of the crop; $\delta Y/\delta X$ is defined as the marginal physical product and it is also the slope of the response function for any particular input.

It can be readily seen that

$$\delta Y \cdot P_Y = P_X \cdot \delta X \qquad [18]$$

obtained from the above relationship is equivalent to the added cost equals added returns identity.

When two or more fertilizer variables are being considered, the principle of substitution must be taken into consideration when determining optimal rates of fertilizer. This occurs with a response surface having iso-yield contours (isoquants). Implied is the fact that there are a number of combinations of input levels of the various fertilizer nutrients that will produce a given yield of the crop being considered. Application of this principle involves comparing costs of different combinations of inputs that could be used in obtaining a specified quantity of production and finding the combination that produces that yield at minimum total per-unit area cost. Applying the procedure of Eq. [17] and solving simultaneously (if needed) gives a unique solution for the optimum of each variable.

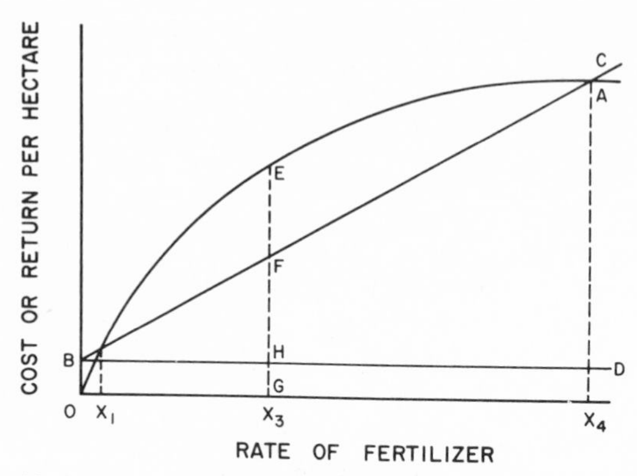

Fig. 3–5. Relationship between the response curve, OEA, the total fertilizer cost line, BFC, and the optimal rate of fertilizer, X_3. Line BHD is the fixed cost of applying fertilizer. (Pesek & Heady, 1958).

1. Graphical Methods of Economic Interpretation of Fertilizer Response Data

The relationships described in the previous paragraphs may be graphed as in Fig. 3–5. The conditions in Eq. [17] are defined graphically by drawing a line parallel to the fertilizer cost line, BFC, (expressed in the same units as Y) and tangent to the yield response curve, OEA. The amount of fertilizer, X_3, needed to produce this magnitude of yield response at this point of tangency is the optimal rate. The vertical distance, EF, between the fertilizer cost line and the yield response curve, is the profit and this distance is greatest at this fertilizer rate.

Economic interpretations for an experiment representing a single site-year and involving two or more nutrients may be obtained readily by plotting the yield contours and then on the same map drawing isoclines equal to particular price ratios. In Fig. 3–6 is shown such a map for a NP fertilizer on corn experiment. The isoquants are calculated by setting the values of yield and fixing the levels of all but one of the nutrients and then solving for the level of the other nutrient. There are several computer packages that have contour plot procedures in which the computer does these calculations and then plots the contours. The slope of a tangent to each isoquant at a given point represents the rate at which one input factor substitutes for or replaces the other in maintaining output at a fixed level. If the isoquants are curved, the rate at which one input factor substitutes for the other diminishes as the output is produced with more of the former and less of the latter. The lines indicated by P_i in Fig. 3–6 are isoclines or expansion paths. They denote the path of optimum (least cost) nutrient combinations as higher yield levels are obtained. They also connect points of equal tangential slopes on successively higher isoquants. Hence, they also denote points on the isoquants that have equal marginal rates of substitution (i.e., the ratios of marginal physical products for the two nutrients are equal). Literally, this is a measure of the relative productivity of the two inputs at the intersection of an isocline with the isoquant.

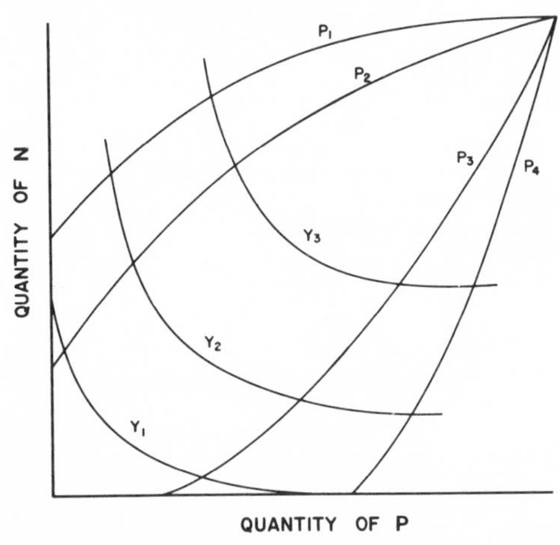

Fig. 3–6. Map showing isoquants and isoclines for graphical economic interpretation.

Use of the graphical procedure for situations where data from a series of experiments are being combined makes it necessary to first substitute values for the site-year variables and their interactions with the nutrients into the general prediction equation. The resulting reduced general equation may then be used to produce isoquant plots as mentioned above.

2. *Algebraic Procedures for Economic Interpretation of Fertilizer Response Data*

For a single nutrient model being fitted to data from a single site-year, the optimal rate will be obtained by taking the derivative with respect to the nutrient and setting it equal to the cost of fertilizer price of product ratio (Eq. [17]). If one has a single experiment with two nutrients, two partial derivatives will need to be taken, one with respect to each of the nutrients. These will be set equal to the respective cost/price ratios. The solution of the two equations will need to be done simultaneously if there is an interaction term in the model. Otherwise, the two equations will be independent. As an example, supposing the following two nutrient model pertains to a set of data from a single site-year for crop yield

$$\hat{Y} = b_0 + b_1\mathrm{N} + b_2\mathrm{P} + b_{11}\mathrm{N}^2 + b_{22}\mathrm{P}^2 + b_{12}\mathrm{NP}. \qquad [19]$$

Because the model contains an NP term, the solution of the optimal rates will need to be done simultaneously, i.e.,

$$\begin{aligned} \delta Y/\delta N &= b_1 + 2b_{11}\mathrm{N} + b_{12}\mathrm{P} = P_n/P_c \\ \delta Y/\delta P &= b_2 + 2b_{22}\mathrm{P} + b_{12}\mathrm{N} = P_p/P_c \end{aligned} \qquad [20]$$

where P_n and P_p are prices of N and P, respectively, and P_c is the price of the crop. These equations reduce to the following:

$$2b_{11}N + b_{12}P = (P_n/P_c) - b_1$$
$$b_{12}N + 2b_{22}P = (P_p/P_c) - b_2 \qquad [21]$$

The set of two simultaneous equations may be readily solved for optimal rates of N and P using a computer assuming the following matrix relationship

$$A\begin{bmatrix} N \\ \underline{P} \end{bmatrix} = \underline{C}, \qquad [22]$$

where A is the matrix of regression coefficients given above,

$$\begin{bmatrix} N \\ \underline{P} \end{bmatrix}$$

is the vector of unknown optimal rates of N and P fertilizer and $\underline{C}$ is the vector of price ratios minus the respective linear regression coefficients.

$$\begin{bmatrix} N \\ \underline{P} \end{bmatrix}$$

may then be solved by the following relationship derived from Eq. [22]:

$$\begin{bmatrix} N \\ \underline{P} \end{bmatrix} = A^{-1}\underline{C}. \qquad [23]$$

Solution of

$$\begin{bmatrix} N \\ \underline{P} \end{bmatrix}$$

could be readily obtained by using Procedure Matrix described in the manual for the Statistical Analysis System (SAS Inst., 1982) or procedures from comparable statistical software packages.

B. Inductive Inferences From the Series of Experiments to Decisions Made by an Individual Farmer

If the response function is a general one developed for a series of site-years, it would contain terms for site-years and site-years × controlled fertilizer variables. For one of the site-years studied, or for an individual farm that was not in the set of experiments studied, one would substitute the levels of the various site-year variables for that particular site into the general equation, which will leave an equation containing only constants and terms for individual controlled variables. This would then be solved for the optimum combination of added fertilizer nutrients, assuming a given set of cost/price ratios and using the procedures given above.

C. Discounting (Cost of Money)

A decision-maker might not use the optimum rate of fertilizer, or even some high rational rate, because he might discount future returns that are subject to uncertainty to a greater degree than the return on a "sure" investment. This discounting characteristic varies among individuals but enters into decision making, although it may not be recognized.

Usually discounting is expressed as a percentage just as in the case of interest. Suppose that the total discounting rate (including return on alternative investments) for an individual is 20%. Under these conditions, the rate of fertilizer that this individual should use is given by

$$\delta Y/\delta X = 1.2\, P_X/P_Y. \quad [24]$$

Pesek et al. (1961) presented examples of how discounting affects fertilizer use in the presence and absence of residual fertilizer effects and given one or more cropping seasons. If compounding returns is considered, and the period of investment in crop growth exceeds the length of one compounding period, then the standard compounding formula must be used to determine the coefficient of the inverse price ratio in Eq. [24]. The formula is

$$C = c(1 + r)^t = P_X \quad [25]$$

where C is the compounded value over time; c is the original cost; r is the interest rate; and t is time in years for each crop return period.

D. Equating Marginal Returns and Limited Capital

Frequently, the resources for buying fertilizer are limited to the extent that not all fields can be fertilized to the level designated in Eq. [17] or even to the level provided for in Eq. [24]. The problem then becomes one of how to allocate the investment in fertilizer among various crops and fields to maximize net return. This goal is achieved when the marginal returns from fertilizer use to all crops or all fields are the same. If the fertilizer response equations for three fields or three crops are given as $f(Y_1)$, $f(Y_2)$, and $f(Y_3)$ and the costs of fertilizer and crops are appropriately designated, the maximum revenue will occur when the following holds:

$$\delta Y_1/\delta X_1 = \delta Y_2/\delta X_2 = \delta Y_3/\delta X_3,$$

and the total of X_1, X_2, and X_3 multiplied by the respective prices does not exceed the limited resources. These equal marginal returns must be greater than the returns in alternative investments.

If resources for fertilizer use are extremely limited, the maximum economic return per unit of fertilizer applied plus the application cost is a relevant consideration. The rate that produces this maximum economic return is designated the *minimum rate*.

In Fig. 3–5 the return to fertilizer plus application cost is the ratio of the difference between the response curve OEA and the total fertilizer cost (line BFC) to the total fertilizer cost (line BFC). The relative return rises rapidly to a maximum (frequently at the first or second increment of fertilizer) and then declines to zero at the point where the variable fertilizer cost line crosses the response curve.

Application of this concept to a fertilizer response curve for a crop that may be produced profitably without any fertilizer follows. The fertilizer response curve is written

$$\Delta Y = b_1 X + b_{11} X^2, \quad [26]$$

and the variable fertilizer cost line is:

$$C = m + rX, \quad [27]$$

where m is fixed cost of fertilizer per unit area and r is the cost of the fertilizer in units of response.

The quantity to maximize is the difference between these two equations divided by the latter. This maximum is reached at the level of X given in the following equation:

$$X' = \frac{-mb_{11} \pm [(mb_{11})^2 - mr\,b_1 b_{11}]^{1/2}}{rb_{11}}. \quad [28]$$

In many cases, some fertilizer is required to produce a crop profitably. The total yield curve and the total cost of production line including the variable fertilizer cost must be considered and the quantity to be maximized is the difference between the yield curve

$$\hat{Y} = b_0 + b_1 X + b_{11} X^2 \quad [29]$$

and the total cost line

$$n = p + rX, \quad [30]$$

where b_0 is the yield without fertilizer and p is the sum of the total cost of production and of fertilizer per unit area.

The minimum rate of fertilizer which should be applied is given by the solution for X' in the following equation:

$$X' = \frac{pb_{11} \pm [(pb_{11})^2 - rb_{11}(pb_1 - rh_0)]^{1/2}}{rb_{11}}. \quad [31]$$

This is the rate that maximizes return to investment per unit area in crop production.

The interpretation of the value for X' in Eq. [28] is that all similar units of area should receive this minimum rate of fertilizer if resources permit. If there is not enough fertilizer to cover all units at this level, fertilizer should be applied at this level as far as it will reach, and the other units should be left unfertilized. In the case of Eq. [31], X' is interpreted as the rate to apply to all similar units, but if there is not enough to treat all units at this level, these excess units should be left out of production. Any other action would lead to anticipated reduction in revenue. Applications of these concepts were made by Pesek and Heady (1958) and more recently by Voss (1975).

E. Minimizing Losses

In some cases, projected yields, costs, and returns may strongly indicate a loss regardless of any action. If, in spite of these projections, a crop is to be produced, the fertilizer rate that will minimize losses is the economic optimum rate, which does not consider costs other than the fertilizer. The correctness of this rate may be visualized from Fig. 3–5. If the fixed cost line BHD were elevated above the response curve OEA, the shortest distance (minimum loss) between BHD and OEA would be a tangent to OEA parallel to BFC, which is the economic optimum.

F. The Opportunity Cost Principle

If there are some enterprises that compete for the farmer's capital, one should compare the marginal returns from the various investments before carrying the rates of fertilizer to the point of most profit per hectare. Perhaps putting part of the money intended for fertilizer purchase into a poultry (chicken, *Gallus gallus domesticus*) operation would give a higher incremental return/incremental cost ratio. There are some difficult questions that need to be asked. For example, if there is a cattle (beef, *Bos tourus*) feeding enterprise, would it pay to apply less N fertilizer to the corn produced on the farm and then buy corn from off of the farm to offset the loss in potential corn production due to limited fertilizer? These decisions require the determination of marginal returns for each course of action and equating them. This is possible only if all production functions are known. Thus, one must always consider the cost of lost opportunities in not placing capital in some alternative enterprise when making decisions about optimal fertilizer strategies. If it were not for risk and uncertainty, the choice of an "optimum mix" of input allocations to various enterprises on the farm would be a rather straightforward procedure that could be handled by linear programming methods. Unfortunately, there are many factors that have an influence on the relative profitability of various enterprises on the farm, and it is impossible to predict how important they will be during a particular growing season. Some of these will be discussed in the following section.

G. Dealing With Risk and Uncertainty

The rational economic range within which the choices of fertilizer rates must be made has been described. This range is conditioned by the expected (or average) response, the expected price structure, availability of resources for fertilizer purchase, and alternative uses for capital that might be expended for fertilizer. We considered the solutions as if the yield functions were known and the expected outcome was accurately predicted, or that the average of all functions over time is known and the producer will operate long enough to realize this outcome. Hence, matters were viewed in a *risk* setting.

Actual functions are not the same each year, depending on timeliness of operations and natural hazards such as weather, disease, pests, etc., none of which are known at the beginning of production. Expressed in another way, the variability of yields that will be produced from what is considered to be an optimum rate of fertilizer will be great from farm to farm or from year to year on the same farm. This implies that the calculations used in linear programming for choosing an optimal enterprise mix will not necessarily be reliable for a given farm in a particular season. Even though the results may be viewed from the point of view of *risk* over a period of years, each year's outcome is uncertain. Therefore, we must consider decision making under uncertainty. Decision making under uncertainty is quite different from decision making under risk. The behavior of an individual regarding *uncertainty* is probably affected by age, tenure, family situation, equity, and psychological traits. Concepts for dealing with uncertainty follow in the next section.

H. Game Models in Decision Making

Fertilizer recommendation strategies need to be tailored to various probabilities of natural disasters occurring. The crop producer can choose recommendations accordingly to his aversion to risk.

Walker et al. (1960, 1964) applied game theory models to fertilizer and other crop-producing decisions. There are at least four models that were explored in these studies; those of Laplace, Wald, Savage, and Hurwicz.

The Laplace model is a special case of a simple risk model. In it, it is assumed that each previously observed outcome has an equal chance of occurring the next season, or that the average condition will occur. The alternative chosen is the one that is expected to give the highest return. Because the "best" is always expected, the Laplace solution is an optimistic one and provides no protection against the occurrence of low returns because of the alternative chosen.

In contrast, the Wald model lends to conservative action. It is set up as a game against nature with states of nature and alternatives making up the choice matrix. The decision maker playing against nature assumes that nature "tries to do its worst," and therefore, always selects an alternative that maximizes returns under the worst that nature presents. The fallacy is that nature, unlike a living opponent, does not take conscious action.

A Savage minimum regret model is less conservative than the Wald. Its operation starts with the Wald "states-alternatives" choice matrix, and a new regret matrix is generated by subtracting the highest outcome alternative within each "state" from each other alternative. This new matrix is one of ex post facto opportunity losses. The strategy is to minimize the opportunity loss for a given season or crop.

The Hurwicz criterion depends on the assignment of an optimism-pessimism index a to an operator. This index lies between 0, optimistic, and 1, pessimistic; under the latter, the strategy in a given game is the same as the Wald solution gives. If m is the minimum and M is the maximum outcome under one alternative over all "states," and $1-a$ is the individual's belief that M will occur, then $am + (1-a)M$ is the a index for that alternative. The alternative with the highest a is the preferred one.

Even more favorable outcomes may be determined by using more than one alternative, e.g., part of the crop may be fertilized at one level and another part at a higher or lower level. Simulation models can be used to produce weather data assuming a specified underlying model. Simulated weather data for a number of years are available almost instantly. These data permit the testing of various fertilization strategies used in connection with a given weather outcome and evaluation of the resulting outcome. Research results are lacking at the present time to develop fertilization strategies for individual farm situations.

REFERENCES

Abraham, T. P., and Rao, V. Y. 1965. An investigation on functional models for fertilizer response surfaces. Indian Soc. Agric. Stat. J. 18:45–61.

Anderson, R. L., and L. A. Nelson. 1971. Some problems in the estimation of single nutrient functions. p. 203–222. *In* Bulletin of the International Statistical Institute, 44, Part 1. International Statistical Institute, The Hague, Netherlands.

Anderson, R. L., and L. A. Nelson. 1975. A family of models involving intersecting straight lines and concomitant experimental designs useful in evaluating response to fertilizer nutrients. Biometrics 31:303–318.

Baird, B. L., and J. W. Fitts. 1957. An agronomic procedure involving the use of a central composite design for determining fertilizer response surfaces. p. 135–143. *In* E. L. Baum et al. (ed.) Fertilizer innovations and resource use. Iowa State College Press, Ames, IA.

Balmukand, B. H. 1928. Studies in crop variation. V. The relation between yield and soil nutrients. J. Agric. Sci. 18:602–627.

Bartlett, M. S. 1937. Properties of sufficiency and statistical tests. Proc. R. Soc. (London) A160:268–282.

Baule, B. 1918. Zu Mitscherlichs gesetz der physiologischen beziehungen. Landwirtsch. Jahrb. 51:363–385.

Baum, E. L., E. O. Heady, and J. Blackmore (ed.). 1956. Methodological procedures in the use of fertilizer use data. Iowa State University Press, Ames, IA.

Baum, E. L., E. O. Heady, J. T. Pesek, and C. G. Hildreth (ed.). 1957. Economic and technical analysis of fertilizer innovations and resource use. Iowa State University Press, Ames, IA.

Box, G. E. P. 1954. Exploration and exploitation of response surfaces. Biometrics 10:16–60.

Box, G. E. P., and J. S. Hunter. 1957. Multifactor experimental designs for exploring response surfaces. Ann. Math. Stat. 28:195–241.

Boyd, D. A. 1972. Some recent ideas on fertilizer response curves. Potassium Symp. 9:461–473.

Boyd, D. A., L. T. K. Yuen, and P. Needham. 1976. Nitrogen requirement of cereals. I. Response curves. J. Agric. Sci. 87:149–162.

Cady, F. B., and D. M. Allen. 1972. Combining experiments to predict future yield data. Agron. J. 64:211–214.

Cady, F. B., and R. J. Laird. 1973. Treatment design for fertilizer use experimentation. CIMMYT Res. Bull. 26. International Maize and Wheat Improvement Center, Mexico City, Mexico.

Cochran, W. G., and G. M. Cox. 1957. Experimental designs, 2nd ed. John Wiley & Sons, Inc., New York.

Colwell, J. D. 1967. Calibration and assessment of soil test for estimating fertilizer requirements. I. Statistical models and tests of significance. Aust. J. Soil Res. 5:275–293.

Colwell, J. D. 1968. Calibration and assessment of soil test for estimating fertilizer requirements. II. Fertilizer requirements and an evaluation of soil testing. Aust. J. Soil Res. 6:93–103.

Colwell, J. D. 1970. A comparison of soil test calibrations for the estimation of phosphorus fertilizer requirements. Aust. J. Exp. Agric. Anim. Husb. 10:774–782.

Colwell, J. D. 1978. Computations for studies of soil fertility and fertilizer requirement. Commonwealth Agricultural Bureaux, Farnham House, Farnham Royal, Slough, U.K.

Colwell, J. D. 1979. National soil fertility project, Vol. 2. Commonwealth Scientific and Industrial Research Organization, E. Melbourne, Victoria, Australia.

Denmead, O. T., and R. H. Shaw. 1962. Availability of soil water to plants as affected by soil moisture content and meteorological conditions. Agron. J. 54:385–390.

Draper, N. R., and H. Smith. 1981. Applied regression analysis, 2nd ed. John Wiley & Sons, Inc., New York.

Escobar, G. J. A. 1967. Consideraciones sobre la comparición de disenós de tratamientos, M. S. Thesis. Escuela Nacional de Agricultura, S. A. G. Chapingo, México.

Federer, W. T. 1955. Experimentai design. Macmillan Publishing Co., New York.

Fisher, R. A. 1924. The influence of rainfall on yields of wheat at Rothamsted. Philos. Trans. R. Soc. London Ser. B:213:89–142.

Gomez, K. A. 1972. Techniques for field experiments with rice. The International Rice Research Institute, Los Baños, Philippines.

Hader, R. J., M. E. Harward, D. D. Mason, and D. P. Moore. 1957. An investigation of some relationships between copper, iron and molybdenum in the growth and nutrition of lettuce: I. Experimental design and statistical methods for characterizing the response surface. Soil Sci. Soc. Am. Proc. 21:59–74.

Hartley, H. O. 1950. The maximum F-ratio as a short-cut test for heterogeneity of variance. Biometrika 37:308–312.

Hatheway, W. H., and E. J. Williams. 1958. Efficient estimation of the relationship between plot size and the variability of crop yields. Biometrics 14:207–222.

Heady, E. O., and J. L. Dillon. 1961. Agricultural production functions. Iowa State University Press, Ames, IA.

Heady, E. O., and J. T. Pesek. 1957. Some methodological considerations in the Iowa–TVA Research Project of Fertilizer Use. p. 144–167. *In* E. L. Baum et al. (ed.) Economic and technical analysis of fertilizer innovations and resource use. Iowa State University Press, Ames, IA.

Heady, E. O., J. T. Pesek, and W. G. Brown. 1955. Crop response surfaces and economic optima in fertilizer use. Iowa Agric. Exp. Stn. Res. Bull. 424.

Heady, E. O., J. T. Pesek, W. G. Brown, and J. P. Doll. 1961. Crop response surfaces and economic optima in fertilizer use. p. 475–525. *In* E. O. Heady and J. L. Dillon (eds.) Agricultural production functions. Iowa State University Press, Ames, IA.

Hendricks, W. A., and J. C. Scholl. 1943. Joint effects of temperature and rainfall on corn yields. North Carolina Agric. Exp. Stn. Tech. Bull. 74.

Hildreth, C. G. 1957. Possible models for agro-economic research. p. 176–186. *In* E. L. Baum et al. (eds.) Economic and technical analysis of fertilizer innovations and resource use. Iowa State University Press, Ames, IA.

Jónsson, L. 1974. On the choice of a production function model for nitrogen fertilizer on small grains in Sweden. Swed. J. Agric. Res. 4:87–97.

Laird, R. J., and F. B. Cady. 1969. Combined analysis of yield data from fertilizer experiments. Agron. J. 61:829–834.

Mason, D. D., and D. Cooper. 1958. Characterization and utilization of climatic data in a multiple covariance model for studying the relationships between yield response to fertilizers and soil nutrients. *In* Determining yield response surfaces and economically optimum fertilizer rates for corn under various soil and climatic conditions in North Carolina. TVA Rep. T58-2AE. Tennessee Valley Authority, Muscle Shoals, AL.

Mead, R., and D. J. Pike. 1975. A review of response surface methodology from a biometric viewpoint. Biometrics 31:803–851.

Mitscherlich, E. A. 1909. Das gesetz des minimums and das gesetz des abnehmenden bodenertrages. Landwirtsch Jahrb. 38:537–552.

Mitscherlich, E. A. 1930. Die bestimmung des dungerbedurfnisses des bodens, 3rd ed. West Berlin.

Mombiela, F. A., and L. A. Nelson. 1980. Evaluation of fertilizer recommendation methodologies using simulated corn P responses in North Carolina Norfolk-like soils with varying soil P fertility. North Carolina Agric. Res. Service Tech. Bull. 268.

Mombiela, F., J. J. Nicholaides III, and L. A. Nelson. 1981. Method to determine the appropriate mathematical form for incorporating soil test levels in fertilizer response models for recommendation purposes. Agron. J. 73:937–941.

National Academy of Sciences–National Research Council, Committee on Economics of Fertilizer Use of the Agricultural Board. 1961. Status and methods of research in economic and agronomic aspects of fertilizer response and use. NAS–NRC Pub. 918. National Academy of Sciences–National Research Council, Washington, DC.

Nelder, J. A. 1966. Inverse polynomials, a useful group of multifactor response functions. Biometrics 22:128–141.

Nelson, L. A., and R. J. McCracken. 1962. Properties of Norfolk and Portsmouth soils: Statistical summarization and influence on corn yields. Soil Sci. Soc. of Am. Proc. 26:497–502.

Niklas, H., and M. Miller. 1927. Beiträge zur mathematischen Formulierung des Ertraggesetzes. Z. Pflanzenernaehr. Dueng. Bodenkd. 8A:289–297.

Pesek, J. T., and E. O. Heady. 1958. Derivation and application of a method for determining minimum recommended rates of fertilization. Soil Sci. Soc. Am. Proc. 22:419–423.

Pesek, J. T., E. O. Heady, and L. C. Dumenil. 1961. Influence of residual fertilizer effects and discounting on optimum fertilizer rates. Trans. Int. Congr. Soil Sci. 7th 3:220–227.

Runge, E. C. A. 1968. Effects of rainfall and temperature interactions during the growing season on corn yield. Agron. J. 60:503–507.

Runge, E. C. A., and R. T. Odell. 1958. The relation between precipitation, temperature and the yield of corn on the Agronomy South Farm, Urbana, Illinois. Agron. J. 50:448–454.

Ryan, J. G., and R. K. Perrin. 1973. The estimation and use of a generalized response function for potatoes in the Sierra of Peru. North Carolina Agric. Exp. Stn. Tech. Bull. 214.

Schultz, E. F., Jr. 1955. Rules of thumb for determining expectations of mean squares in analysis of variance. Biometrics 11:123–135.

Smith, H. F. 1938. An empirical law describing heterogeneity in the yields of agricultural crops. J. Agric. Sci. 28:1–23.

Sopher, C. D., and R. J. McCracken. 1973. Relationships between individual soil properties, management practices and corn yields on selected South Atlantic Coastal Plain soils. Agron. J. 65:595–599.

Sopher, C. C., R. J. McCracken, and D. D. Mason. 1973. Relationships between drought and corn yields on selected South Atlantic Coastal Plain Soils. Agron. J. 65:351–354.

SAS Institute, Inc. 1982. SAS user's guide: Statistics, 1982 ed. SAS Institute, Inc., Cary, NC.

Steel, R. G. D., and J. H. Torrie. 1980. Principles and procedures of statistics. 2nd ed. McGraw-Hill, Inc., New York.

van Bavel, C. H. M. 1953. A drought criterion and its application in evaluating drought incidence and hazard. Agron. J. 45:167–172.

von Boguslawski, E., and B. Schneider. 1962. Die dritte annäherung des ertragsgesetzes. 2. Mitteilung. Z. Acker Pflanzenbau 116:113–128.

Voss, R., and J. Pesek. 1967. Yield of corn grain as affected by fertilizer rates and environmental factors. Agron. J. 59:567–572.

Voss, R. D. 1975. Fertilizer N: The key to profitable corn production with changing prices and production costs. p. 215–229. *In* Proc. 30th Annu. Corn and Sorghum Res. Conf., Chicago, IL. December, 1975. American Seed Trade Association, Washington, DC.

Voss, R. E., J. J. Hanway, and W. A. Fuller. 1970. Influence of soil, management, and climatic factors on the yield response by corn (*Zea mays* L.) to N, P, and K fertilizer. Agron. J. 62:736–740.

Walker, O. L., E. O. Heady, and J. T. Pesek. 1964. Application of game theoretic models to agricultural decision making. Agron. J. 56:170–173.

Walker, O. L., E. O. Heady, L. G. Tweeten, and J. T. Pesek. 1960. Application of game theory models to decisions on farm practices and resource use. Iowa Agric. Exp. Stn. Res. Bull. 488.

Wood, C. L., and F. B. Cady. 1981. Intersite transfer of estimated response surfaces. Biometrics 37:1–10.

4

Lime-Fertilizer-Plant Interactions in Acid Soils[1]

Eugene J. Kamprath
North Carolina State University
Raleigh, North Carolina

Charles D. Foy
Agricultural Research Service, USDA
Plant Stress Laboratory
Beltsville, Maryland

Soils of the humid regions have developed under conditions in which rainfall exceeds evapotranspiration during most of the year. Under this condition there has been a gradual depletion of soil bases and the development of soil acidity. The soil clays often contain coatings of Fe and hydroxy Al. These materials significantly affect the retention and availability of fertilizer cations and anions in acid soils. The capacities of these soil materials to fix P, Mo, S, and B are influenced by liming.

Acid mineral soils at PH $\leq$ 5 often contain appreciable amounts of Al and Mn, which are detrimental to plant growth, in the soil solution. Optimum growth and efficient use of fertilizer nutrients in acid soils require the addition of lime to eliminate the toxic effects of Al, H, and Mn. This chapter describes some of the effects of soil acidity on plant growth and fertilizer nutrient availability and shows how these effects are modified by liming.

I. LIMING IN RELATION TO CATION EXCHANGE REACTIONS AND PLANT GROWTH

A. Cation Saturations and Soil Solution Contents of Acid Soils

Since 1950, certain concepts about acid soils have changed. Trivalent Al^{3+} rather than exchangeable H^+ is now recognized as an important exchangeable cation in acid soils. Coleman and Thomas (1967) have given an excellent discussion of the basic chemistry of acid soils and the ion-exchange chemistry of Al.

[1]Joint contribution from the Dep. of Soil Science, North Carolina State Univ., Raleigh, NC, and the Plant Stress Laboratory, USDA–SEA–ARS, Beltsville, MD.

Table 4-1. Exchangeable and percent saturation of Al at various pH levels in mineral soils.

Soil	pH	Exchangeable		Effective	Al
		Al	Bases	CEC	saturation
		meq/100 g			%
Paleudult†	4.5	0.91	0.20	1.11	82
	5.9	0.10	1.60	1.70	6
Paleudult‡	4.7	1.08	0.68	1.76	61
	5.7	0.09	2.14	2.25	4

† Kamprath, 1970. ‡ Richburg and Adams, 1970.

1. Mineral Soils

a. Cation Saturation

i. Exchangeable Aluminum and Basic Cations

Coleman and Thomas (1967) defined exchangeable cations in acid soils as those cations extracted with a neutral unbuffered salt. The sum of these cations was termed the *effective cation exchange capacity*. A neutral unbuffered salt solution will extract only cations that are held at active exchange sites at the particular pH of the soil. The exchangeable acidity extracted from soils with a neutral unbuffered salt is Al^{3+} (Lin & Coleman, 1960; Bhumbla & McLean, 1965).

Examples of the exchangeable Al^{3+} saturation of the effective cation exchange capacity (CEC) (sum of exchangeable Al^{3+} and bases) at different pH values is given in Table 4–1. Below pH 5 Al^{3+} generally accounted for over half of the exchangeable cations, while at pH 6 the effective CEC was essentially base-saturated. The highly leached soils of Brazil, primarily Ultisols and Oxisols, also have exchangeable Al^{3+} as the dominant cation below pH 5 (Pratt & Alvahydo, 1966).

Acid mineral soils at pH $\leq$ 5 have their active exchange sites occupied by Al^{3+} and at pH 6 these sites are countered by exchangeable bases. As will be discussed later, the relative cation saturations have an important effect on the cation composition of the soil solution and in turn on plant growth.

ii. Exchangeable Manganese

The exchangeable Mn^{2+} content of soils is related to soil pH. Highest levels of exchangeable Mn^{2+} are present when pH is $\leq$ 5 and decrease sharply when soils are limed to pH 6 (Foy, 1964; Jones & Fox, 1978).

b. Soil Solution Composition

i. Aluminum

The concentration of Al in the soil solution is related to pH of the soil, the Al saturation of effective cation exchange capacity, and the salt concentration of the system. The solubility of Al in displaced soil solution at various pH values is the same as the solubility of Al in water at the same pH values (Magistad, 1925; Pierre et al., 1932). At a pH of 5.5 the concentration of Al in the soil solution is quite low. However, as the pH drops from 5 to 4.5, the Al concentration increases markedly.

The Al concentration of the soil solution is related to the Al saturation of the effective CEC of the soil. The concentration of Al in the soil solution is quite low until the exchangeable Al^{3+} saturation exceeds 60% and then increases rapidly (Nye et al., 1961; Evans & Kamprath, 1970). When the Al saturation is $> 60\%$, the soil solution concentration of Al is > 1 mg/kg and may be as high as 5 or 6 mg/kg (Evans & Kamprath, 1970; Helyar & Anderson, 1974; Pearson et al., 1977).

As the salt concentration in the soil increases, the concentration of Al in the soil solution increases. This is due to the cation of the salt replacing some of the exchangeable Al^{3+}, which in solution undergoes hydrolysis, thereby lowering the pH and resulting in a greater solubility of Al (Black, 1968). The water-soluble Al content of a soil at pH 5.1 was doubled with heavy fertilization (MacLeod & Jackson, 1967).

ii. Manganese

Water-soluble Mn content of acid soils is closely related to the soil pH, being high below pH 5 but decreasing rapidly as pH increases to 6 (Fried & Peech, 1946; Berger & Gerloff, 1948; Morris, 1948). The addition of a soluble salt that causes a decrease in soil pH results in an increase in water-soluble Mn.

iii. Calcium

The predominant cation in the soil solution of most acid soils ranging from 0.75 to 19 meq/L is Ca, with most concentrations > 3 meq/L. Concentrations of soil solution Ca are increased considerably when acid soils are limed (Adams & Lund, 1966; Evans & Kamprath, 1970; Pearson et al., 1977).

2. Organic Soils

a. Cation Saturation—The exchangeable cation status of organic soils is difficult to describe, since different cations are complexed to various degrees by organic matter. Trivalent cations, such as Al^{3+}, are retained more strongly than are divalent cations, such as Ca^{2+} and Mg^{2+}. Extraction with a neutral unbuffered salt, however, can give an estimate of the readily exchangeable cations in equilibrium with the soil solution.

The negative charges of organic matter in acid soils are generally countered by Al. Less than 50% of the Al bound by organic soils was extractable with KCl. This indicates a large part of the Al is held very strongly by the organic matter (Hargrove & Thomas, 1981; Bloom et al., 1979).

The exchangeable acidity extracted from acid organic soils with a neutral unbuffered salt is primarily Al^{3+} with relatively little H^+ (Table 4–2).

Table 4-2. Content of exchangeable bases, H, and Al in an acid organic soil, 52% organic matter.†

Soil pH	Exchangeable		
	Bases	Al	H
	meq/100 g		
4.1	13.8	7.6	0.8
4.3	19.0	4.6	0.4
4.5	21.0	2.8	0.2
4.7	23.8	2.0	0.2

† E. J. Kamprath, 1967, unpublished data. Exchangeable cations extracted with a neutral unbuffered salt.

b. Soil Solution Composition—The soil solution of acid organic soils contains very little Al at pH $\geq$ 5 (Evans & Kamprath, 1970). Complexing of Al by organic matter reduces the amount of Al in soil solution. The Ca concentration of organic soils limed to pH 4.7 to 5.0 is generally relatively high and more than adequate for plant growth (Evans & Kamprath, 1970).

B. Plant Growth in Relation to Cation Saturation and Solution Concentration

Plant growth in acid soils as influenced by Al, Mn, and Ca has been discussed thoroughly in the monograph *Soil Acidity and Liming* (Adams, 1984b). Therefore, this discussion is limited to some of the general aspects of the problem.

1. Aluminum

a. Relative Tolerance of Plants—Plant species differ in their tolerance to Al. The relative tolerance to Al in decreasing order was corn (*Zea mays* L.), cotton (*Gossypium hirsutum* L.), and barley (*Hordeum vulgare* L.) (Foy & Brown, 1964). Soybeans [*Glycine max* (L.) Merr.] and cotton were considerably less tolerant to Al than corn when grown in solution culture (Rios & Pearson, 1964). Earlier work had indicated that lettuce (*Lactuca sativa* L.), beets (*Beta vulgaris* L.), timothy (*Phleum pratense* L.), and barley were very sensitive to Al; Radishes (*Raphonus sativus* L.), sorghum [*Sorghum bicolor* (L.) Moench], cabbage (*Brassica oleracea* L. var. *capitata*), oats (*Avena sativa* L.), and rye (*Secale cereale* L.) were medium sensitive; and corn was relatively tolerant (McLean & Gilbert, 1927; Ligon & Pierre, 1932). However, within a given species, there may be a wide range in tolerance to Al (Foy, 1974).

b. Aluminum Saturation and Solution Concentration—Growth of plants has been found to be related to the Al saturation of the effective CEC (Kamprath, 1970; Evans & Kamprath, 1970; Fox, 1979; Farina et al., 1980; Alley, 1981). When Al saturations were > 50 to 60%, growth of corn was drastically decreased on mineral soils (Kamprath, 1970; Fox, 1979; Alley, 1981). Yields of corn under field conditions were close to 90% of maximum when Al saturations were $\leq 30\%$. Root growth of corn was not decreased until the Al saturation was 60% (Brenes & Pearson, 1973). Growth of more sensitive crops such as cotton, soybeans, and alfalfa (*Medicago sativa* L.) was optimum when the Al saturation was close to zero (Kamprath, 1970; Alley, 1981). Maximum growth of six Stylosanthes species was obtained at an Al saturation of $< 5\%$ (de Carvalho et al., 1980).

The addition of a high rate of KCl to an unlimed soil increased the concentration of Al in the soil solution and markedly reduced the growth of sweet corn (Ragland & Coleman, 1959) roots. Therefore, where high rates of fertilization are used, it is desirable that the CEC have a relatively low Al saturation to avoid the detrimental effects of Al.

Root growth of cotton in acid subsoils was related to the activity of Al in the soil solution. When the molar activity of Al^{3+} was $> 0.5 \times 10^{-5}$, cotton root decreased (Adams & Lund, 1966). As the concentration of Ca and other basic cations increased, Al toxicity was less severe. Foy et al. (1969b) also reported that, at lower levels of Ca in the nutrient solution, differences in Al tolerances between two varieties of soybeans became larger. Wallace et al. (1971) have suggested that Ca acts as a detoxifier of other ions present in excess.

The pH for optimum growth of plants is lower for organic soils than for mineral soils. Growth of soybeans and corn on organic soils was optimum in the pH range of 4.6 to 5.0 (Welch & Nelson, 1950; Evans & Kamprath, 1970; Mengel & Kamprath, 1978). Aluminum is complexed by the organic matter and at pH 5 the soil solution Al is very low. The addition of humic acid to a water culture prevented the occurrence of Al toxicity in alfalfa (Brogan, 1964).

2. Manganese

There is a close correlation between the water-soluble Mn content of soils and the Mn toxicity symptoms (Berger & Gerloff, 1948; Morris, 1948; Adams & Wear, 1957; Parker et al., 1969). Liming to a pH of around 5.5 generally decreased the water-soluble Mn to a low level and eliminated the toxicity symptoms of cotton, soybeans, and Korean lespedeza (*Lespedza stipulacea* Maxim.).

Manganese toxicity of alfalfa occurred on several acid soils when the exchangeable Mn^{2+} content of the air-dried soils was 50 mg/kg Mn or more. Alfalfa yields were generally highest when the exchangeable Mn^{2+} content was reduced to ≤ 20 mg/kg by liming (Foy, 1964).

Toxicities are less likely to occur on soils with a high CEC because more Mn^{2+} is held on exchange sites, and the concentration of Mn in the soil solution is decreased (Adams, 1984a). Addition of peat to soils on which alfalfa had shown Mn toxicity alleviated the problem (Foy, 1964). Apparently the peat complexed the Mn, thereby reducing its solubility. Manganese toxicity of plants grown in a nutrient solution was reduced when the Ca concentration of the solution was increased (Williams & Vlamis, 1957).

3. Hydrogen Concentration

Within the pH range generally found in acid soils, exchangeable H^+ is not found in any appreciable amounts (Coleman & Thomas, 1967). However, the hydrolysis of Al^{3+} results in H^+ being present in the soil solution. Hydrogen, if present in sufficient amounts in nutrient solutions, will reduce plant growth (Islam et al., 1980). The reduction in growth at low pH is due to damage of root membranes and interference in uptake of nutrients. At pH 5 in culture solutions, the depression in growth is due to the effect of H competition on uptake of basic cations. The competition effect of H^+ can be overcome by supplying more Ca. When the Ca concentration was increased, growth of lettuce at pH 5 was the same as that at pH 6 (Arnon & Johnson, 1942). Thus, when Al and Mn are not present in toxic concentrations, the growth of plants at pH 4.8 to 5.0 is related to the amount of Ca in the soil solution, provided other elements are present in adequate amounts.

4. Calcium

Most acid soils contain sufficient Ca for plant growth; Ca deficiencies of plants are rarely seen in the field. A calcium concentration in the soil solution of 1.5 meq/L was shown to be adequate for high corn yields (Barber et al., 1963). Even at a low pH the Ca concentration in the soil solution of most soils is this high or higher. Therefore, in most cases if the toxic factors such as Al and Mn can be eliminated, the Ca supply is adequate.

There does, however, seem to be a lower limit as to the amount of Ca required in the soil for adequate plant growth. For maximum penetration of cotton roots into a soil, Ca had to comprise 10 to 15% of the total cations expressed in terms of chemical equivalents (Howard & Adams, 1965).

Peanut (*Arachis hypogea* L.) yields decreased when the exchangeable Ca^{2+} level was < 0.75 meq/100 g (Hartzog & Adams, 1973). Large increases in sugarcane (*Saccharum officinarum* L.) yields were obtained from Ca applications when the exchangeable Ca^{2+} content was < 500 kg/ha to a depth of 30 cm, equivalent to about 0.7 meq Ca^{2+}/100 g (Ayres, 1963). In both instances yields were closely related to the level of exchangeable Ca and not to changes in soil pH.

Calcium supply is also important in obtaining high yields on acid organic soils. As discussed previously, organic soils at pH 5 have a very low concentration of Al in the soil solution. An adequate supply of Ca in such soils will offset the competitive effects of H^+ ions.

C. Effective Cation Exchange Capacity and Cation Retention as Related to Liming

1. Increase in Effective Cation Exchange Capacity

Coleman and Thomas (1967) state that the pH-dependent CEC of acid soils is associated with the ionization of H^+ from Fe and Al hydrous oxides and the hydrolysis of trivalent metal ions bound by organic matter. Thus, when acid soils are limed, the effective CEC increases (Bhumbla & McLean, 1965; Kamprath, 1970).

The effective CEC of an Ultisol was increased 60% when the soil pH was increased from 4.9 to 5.9 (Juo & Ballaux, 1977).

2. Effect of Lime on Retention of Fertilizer Cations

The effect of lime on the retention of fertilizer cations is twofold. First, it displaces and precipitates Al, which is held very strongly in acid soils. Where a large portion of effective CEC is occupied by Al, it is difficult for other cations to replace Al for the exchange sites. Greater leaching of fertilizer cations will occur under these conditions, especially K^+ and NH_4^+. Secondly, when a soil is limed, more exchange sites become active.

Munson and Nelson (1963), in a review of work relating to movement of K fertilizers, stated that liming of acid soils generally reduced the amount of fertilizer K that leached from the surface soils. Leaching losses of fertilizer K over a 1-yr period from a sandy loam soil at a pH of 4.3 and 2.5 to 3.5 times greater than those of pH 6.4 (Krause, 1965). Similar results were obtained with acid volcanic ash soils (Mahilum et al., 1970). Leaching losses of Mg^{2+} from acid sandy soils also were decreased when the Ca saturation of the soils was increased by liming (Mehlich & Reed, 1945).

II. EFFECTS OF LIMING ON AVAILABILITY OF SOIL AND FERTILIZER NUTRIENTS

A. Factors Affecting Availabilities of Calcium, Magnesium, and Potassium

1. Factors Affecting Calcium Availability

The availability of Ca to plants growing in soils is influenced by the percent Ca saturation of the soil colloids, the total Ca supply, H ion concentration, the Ca concentration in the soil solution and the presence of toxic ions such as Al and Mn (Adams, 1984a).

a. Total Calcium Supply in Soils—Calcium deficiency, per se, apparently is not a primary growth-limiting factor except in very sandy, acid soils having low CEC (Adams, 1984a; Adams & Lund, 1966; Howard & Adams, 1965; Evans & Kamprath, 1970; Adams et al., 1967). If toxic factors were absent, most acid soils could probably supply adequate Ca for most crops. The addition of soluble salts of Ca to acid soils generally does not increase yields (McCart & Kamprath, 1965) and often decreases plant growth (Fried & Peech, 1946). This detrimental effect results from the displacement of Al from cation exchange sites and a lowering of soil solution pH, which, in turn, increases the solubility and toxicity of Al and/or Mn. Even in those soils in which plants develop Ca-deficiency symptoms, the reduced Ca uptake and utilization observed are more likely due to Al–Ca or other ionic antagonisms than to a low absolute Ca level in the soil (Foy et al., 1969a; Soileau et al., 1969; Long & Foy, 1970).

Table 4–3. Effect of Ca concentration and pH in subsurface nutrient solution on soybean taproot elongation in the nutrient solution (Lund, 1970).

pH	Ca concentration	Taproot† Elongation rate	Taproot Harvest length
	mg/kg	mm/h	mm
5.6	0.05	2.66	461
	0.50	2.87	453
	2.50	2.70	455
4.5	0.05	0.04	24
	0.50	1.36	270
	2.50	2.38	422
LSD at 5%		0.35	62
1%		0.54	97

† Elongation rate during first 4 h in solution. Harvested 7.5 d after entering the solution.

b. Soil pH or Hydrogen Ion—Direct effects of H^+ in soils on the uptake of other ions and on plant growth are difficult to determine, because at the pH levels where the H^+ ion is harmful, Al, Mn, and perhaps other elements are also soluble in toxic concentrations. When plants are grown in nutrient solutions, the detrimental effect of H^+ ions can be clearly shown (Islam et al., 1980). Calcium uptake is considerably reduced at pH $\leqslant 5$. The effect of H^+ ions on growth and Ca uptake can be overcome by addition of soluble Ca. Lund (1970) found that at pH 4.5 the same growth rate of soybean taproots was obtained as at pH 5.6 by increasing the Ca concentration of the solution considerably (Table 4–3). The nodulation and growth of legumes generally required a higher Ca concentration in the nutrient solution at pH 5 than at pH 6 (Andrew, 1976).

c. Type of Soil Colloid and Degree of Calcium Saturation—Allaway (1945) found that the availability of Ca to soybeans was in the order peat > kaolin > illite > Wyoming bentonite > Mississippi bentonite. Mehlich and Colwell (1944) concluded that in acid mineral soils Ca was not readily available to plants at low base saturation and indicated that percent base saturation was more important than the total Ca present. Adams et al. (1967) reported that in acid Norfolk (Typic Paleudults), Magnolia (Rhodic Paleudults), and Greenville (Rhodic Paleudults) subsoils cotton yields were reduced at Ca saturations in the range of 30 to 40% and at Al saturations of 40 to 60%. Henderson (1969) reported Ca deficiency (< 0.58% Ca in tops) in soybeans on Norfolk soil at Ca saturations of < 20% and Al saturations of > 68%. Mahilum et al. (1970) reported that 12% Ca saturation supported near-normal growth of sugarcane in volcanic soils of Hawaii. Maximum growth of cotton roots in subsoils occurred when exchangeable Ca^{2+} was $\geqslant 12\%$ of the total exchangeable cations (Howard & Adams, 1965).

d. Ratios of Calcium to Aluminum and Other Cations in Soil Solutions—Perhaps the single most important factor affecting the availability of Ca in acid soils is the level of exchangeable or soluble Al^{3+} in relation to that of Ca^{2+}. Lance and Pearson (1969) found that reduced Ca uptake was one of the first symptoms of Al damage in cotton seedling roots; the effect was noted after a 1-h exposure of 0.3 mg/kg Al. Inhibition of Ca uptake was avoided and the Ca concentration of the nutrient solution increased. Similar reductions in Ca uptake by Al have been reported by others (Johnson & Jackson, 1964; Evans & Kamprath, 1970; Soileau et al., 1969).

Ratios between Ca^{2+} and other cations are also important in determining Ca uptake. For example, Howard and Adams (1965) found that in nutrient solutions or displaced solutions from Norfolk and Dickson (Glossic Fragindults) soils, a Ca/total cation ratio of 0.10 to 0.15 was required for optimum growth of cotton roots. Lund (1970) reported that in nutrient cultures cotton root elongation rate was highest when the Ca/(Ca + Mg + K) ratios were between 0.10 and 0.20. Low ratios of Ca/total cations were less detrimental when K was substituted for one-half of the Mg. Increasing the Ca level from 10 to 40 mg/kg decreased the injury caused by 0.5, 1.0, or 2.0 mg/kg Al. Geraldson (1957) found that the ratio of Ca/soil solution salts must be maintained in the range of 0.16 to 0.20 to prevent blossom end rot (Ca deficiency) in tomatoes (*Lycopersicon esculentum* Mill.).

2. *Factors Affecting Magnesium Availability*

a. General—Conditions that may lead to the development of Mg deficiency in plants are: acid, sandy, highly leached soils having low CEC values (Kamprath, 1967); acid soils low in available Mg limed with high rates of calcitic lime to neutrality (Grove et al., 1981); and the application of high rates of K fertilizers (Jones & Haghiri, 1963).

The lack of Mg deficiencies on many acid soils has been suggested as being due to the release of Mg from vermiculite and montmorillonite clay structures, which are less stable under acid conditions (Adams, 1984a). Rice and Kamprath (1968) found that in Coastal Plain soils only 10% of the total Mg was in the exchangeable form and that a considerable amount of nonexchangeable Mg was released by cropping in soils with expansive clay minerals.

b. Soil pH—Magnesium uptake by plants from nutrient solutions is influenced by the H^+ ion concentration. Uptake of Mg by plants increased with increasing pH and reached an optimum at approximately pH 5.5 (Leggett & Gilbert, 1969; Islam et al., 1980).

Magnesium deficiencies of plants are generally associated with soil pH values of ≤ 5 (Ferrari & Sluijmans, 1955; Rice, 1966; Kamprath, 1967). Liming of acid soils with calcitic lime often eliminates the Mg deficiencies (Milam & Mehlich, 1954; Rice, 1966). Correction of Mg deficiencies on acid soils is difficult unless the soil pH is increased (Jones & Haghiri, 1963; McCart & Kamprath, 1965).

The effect of soil pH on Mg availability is probably an antagonism of both Al^{3+} and H^+ on Mg uptake, particularly when the percent Al saturation is high. Neutralization of Al^{3+} and H^+ is necessary for optimum Mg availability. When the soils are acid and low in available Mg, the use of dolomitic lime is the best approach to correct Mg deficiencies.

c. Magnesium Saturation—The Mg saturation of the CEC is a better measure of Mg availability than the actual amount of exchangeable Mg^{2+} (Adams & Henderson, 1962). Critical levels of Mg saturation of the CEC as determined with NH_4OAc was found to be 5% (Adams & Henderson, 1962; McLean & Carbonell, 1972).

3. Potassium Availability as Affected by Liming

The effect of liming on K availability has been debated for many years. Liming has been reported to increase, decrease, or have no effect on the K^+ concentration in the soil solution, depending on the initial degree of base saturation and pH (Peech & Bradfield, 1934, 1943; Thomas & Hipp, 1968).

Liming certain strongly acid soils reduces the leaching of K from the plow layer into deeper soil horizons, thereby preventing decreased levels of exchangeable K^+ in surface soils (Krause, 1965; Mahilum et al., 1970). Liming has also been reported to decrease levels of exchangeable K^+ in certain soils. For example, MacLeod et al. (1964) found that liming a strongly acid soil to pH ≥ 5.4 decreased exchangeable K^+. Powell and Hutcheson (1965) concluded that liming, or wetting and drying cycles following liming, reduced exchangeable K^+ levels in soils containing micaceous minerals. However, this reduction in exchangeable K^+ did not reduce plant growth or K uptake. Liming increased the release of nonexchangeable K to the crop and reduced K fixation. Pearson et al. (1970) reported that the K^+ concentration in displaced soil solution of a Norfolk B_2 horizon was reduced sharply by liming the soil from pH 5.2 to 5.8. Bartlett and McIntosh (1969) found that liming a spodosol that was high in extractable Al and in pH-dependent CEC induced a K deficiency in tomato seedlings.

Liming acid soils that have a relatively large Al saturation increased the buffering capacity more than twice when the exchangeable Al^{3+} was neutralized (Goedert et al., 1975). One benefit of liming acid soils, therefore, is the reduction in leaching losses of fertilizer K. However, high rates of lime, which raise pH levels above 6, may decrease the available K and increase the need for K fertilization on certain soils.

B. Transformations and Availability of Inorganic Nitrogen

1. Conversion of Ammonium to Nitrate

The oxidation of NH_4^+ to NO_3^- in soils by *Nitrosomonas* and *Nitrobacter* is inhibited by a low pH (Waksman, 1923). Although solution culture studies have shown that the optimum pH for these organisms is > 7 (Jackson, 1967; Alexander, 1964), Weber and Gainey (1962) reported NO_3^- production in soils with a pH as low as 4.

Soil pH has been found to be the best indicator for the rate at which

NH_4^+ is converted to NO_3^- in mineral soils (Morrill & Dawson, 1967; Dancer et al., 1973). Nitrification was very limited at pH ≤ 5. Maximum nitrification occurred when soils were limed to pH 6.3. The source of NH_4^+ added to acid soils has an important effect on rate of oxidation. Anhydrous ammonia added to sandy soils at pH 5 was oxidized faster than ammonium sulfate [$(NH_4)_2SO_4$ (Eno & Blue, 1957). Application of ammonia causes an initial increase in pH, whereas ammonium sulfate does not.

2. Plant Utilization of Ammonium and Nitrate in Relation to Soil Acidity

Ammonium is used as efficiently as NO_3^- by many plant species when the pH is kept near neutrality. Between pH 5.5 and 6.5 there was generally little difference in the growth of plants supplied with NH_4^+ and those supplied with NO_3^- (Nightingale, 1937). Excised tomato roots utilized NH_4^+ very well when the pH of the solution was maintained at 7, but utilization was markedly reduced at pH 4.5 (Sheat et al., 1959). The pH appeared to affect utilization rather than uptake of NH_4^+. Studies with whole plants in sand cultures indicated that control of ambient acidity around roots supplied with NH_4^+ enhanced its conversion to organic N compounds in the roots and protected the shoots from excess ammonia (Barker et al., 1966).

Absorption of NH_4^+ by plants resulted in a decrease in the pH of the ambient solution. When several species were supplied with NH_4^+, the final pH of the sand leachate varied from 2.8 to 4.2 (Maynard & Barker, 1969). The decrease in pH resulted from the release of H^+ from the root when the NH_4^+ was taken up by the root. The changes in pH may be quite apparent close to the root as compared with the bulk solution. For example, the pH at the surface of apple tree roots (*Malus sylvestris* Mill.) receiving NH_4^+ was in the range of 4 to 4.5, while that of the bulk solution was 6 (Nightingale, 1934). The rhizosphere pH of 3-week-old wheat plants grown in a fine sandy loam soil was 5.8 with no N, 5.7 with calcium nitrate [$Ca(NO_3)_2$], and 4.7 with ammonium sulfate (Smiley, 1974).

Therefore, under relatively acid soil conditions the use of NH_4^+ fertilizers may have some significant consequences on plant growth. Where the soil is sufficiently acid so that NO_3^- production is drastically reduced, the NH_4^+ fertilizer may not be very effectively utilized within the plant. Also, on poorly buffered soils, the absorption of NH_4^+ may result in a marked decrease in the pH around the roots. A decrease in pH of mineral soils to pH < 5 may result in a large increase in the concentrations of Al and Mn in the soil solution, which could impair plant growth. However, where soils are properly limed, NH_4^+ has been shown to be as effective a source of N as NO_3^-.

C. Availability of Soil and Fertilizer Anions—Phosphorus, Sulfur, Molybdenum, and Boron—as Influenced by Liming

1. Phosphorus Availability

a. Phosphorus Adsorption and Availability as Influenced by Liming—The effect of liming on P adsorption and availability needs to be considered from the standpoint of neutralization of exchangeable Al and the reactivity of hydrous oxides of Fe and Al as affected by pH. The reader is referred to the monograph *The Role of Phosphorus in Agriculture* (Khasawneh et al., 1980) for a detailed discussion of the chemistry of P reactions in soils.

i. Phosphorus Adsorption

The adsorption of added P by amorphous aluminum hydroxide decreased as the pH was increased (Hsu & Rennie, 1962). Conversely, there was a greater amount of P in solution. This can be explained as a consequence of a higher OH^- concentration as the pH is increased and a replacement of adsorbed P by the OH^- ions.

The effect of liming acid soils on P adsorption and soil solution P has not been consistent. Liming had no effect on P adsorption of an Oxisol (Jones & Fox, 1978); increased P adsorption on Ultisols (White & Taylor, 1977); and decreased P adsorption on Ultisols at equilibrium solution concentrations of 0.2 μg P/ML (Friesen et al., 1980). Solubility of adsorbed P in limed soils, as measured in dilute $CaCl_2$ solutions, has not shown a consistent effect of liming. Amarasiri and Olsen (1973) found that P solubility initially decreased with liming and then increased. Similar results were obtained by Friesen et al. (1980), who found that soil solution P initially decreased but increased after the pH had been increased to the point where all the exchangeable Al^{3+} was neutralized.

Robarge and Corey (1979) observed that with an Al-saturated cation exchange resin the concentration of solution P was affected by pH, the salt concentration, and the neutral salt cation. The concentration of solution P decreased until the percent Al saturation was zero. The pH observed for zero Al saturation was 4.7 in the presence of 10^{-3} *M* Ca. As the pH increased, the concentration of solution P also increased due to a higher content of OH^- tons.

ii. Phosphorus Availability

Neutralization of exchangeable Al^{3+} by liming has a marked effect on the response of plants to additions of fertilizer P. Much lower rates of fertilizer P were required for optimum growth when exchangeable Al^{3+} was neutralized (Table 4–4). Four times as much superphosphate was required on a soil with pH 5.1 and 49% Al saturation to give the same growth of alfalfa as at pH 6.1 and no exchangeable Al^{3+} (Munns, 1965). Only half as much fertilizer P was required for optimum growth of millet (*Setaria* spp.) on three acid Ultisols when exchangeable Al^{3+} was neutralized (Woodruff & Kamprath, 1965). Lime rates sufficient to neutralize exchangeable Al^{3+} provide conditions for maximum uptake of fertilizer P. However, high rates of lime on acid Oxisols drastically reduced P uptake at pH 7. This apparently was the result of formation of insoluble Ca phosphates (Fox et al., 1964).

The increased availability of P when exchangeable Al^{3+} is neutralized may be due to increased root growth throughout the amended soil volume.

Table 4-4. Phosphorus response of tall fescue as influenced by liming (data from Shoop et al., 1961).

Fertilizer P	Relative yield	
	No lime, 79% Al sat.	Lime, 0% Al sat.
kg/ha	%	
0	5	14
20	7	82
40	27	89
80	66	84
160	82	98
320	84	100

Friesen et al. (1980) found a good relationship between increased surface area of roots and P uptake by corn when an acid soil with a high percent Al saturation was limed.

2. *Sulfur Availability*

a. Effect of pH on Sulfate Adsorption by Soils—Mattson (1927) noted that as the pH of soil colloids was changed from acid to neutral, the amount of sulfate adsorbed decreased markedly. Kamprath et al. (1956) observed that there was a marked decrease in the amount of SO_4^{2-} adsorbed when the pH of an Ultisol was increased from 5 to 6. At pH 6.5 SO_4^{2-} adsorption by the Ap horizon of an Acrohumox was very small (Couto et al., 1979). The effects of pH on SO_4^{2-} adsorption were much more pronounced on soils that contained appreciable amounts of hydrous Fe and Al oxides (Chao et al., 1964).

It is suggested that SO_4^{2-} reacts with the OH^- ions associated with hydrated oxides of Fe and Al (Chao et al., 1964; Chang & Thomas, 1963). Chang and Thomas (1963) proposed that SO_4^{2-} adsorption increases when pH is lowered because the replaced OH^- ions are more effectively neutralized by H^+ ions resulting from hydrolysis of Al replaced by the cation added with the SO_4^{2-}.

The amount of SO_4^{2-} retained is affected by the cation associated with the sulfate and also by the exchangeable cation (Chao et al., 1963). Aluminum-saturated soils adsorbed more SO_4^{2-} than Ca-saturated soils when pH was the same. Adsorption of SO_4^{2-} was also greater from a solution of $CaSO_4$ than from K_2SO_4. However, soil pH had a greater effect on SO_4^{2-} adsorption than did the nature of the cation. Liming of a subsoil from pH 5 to pH 6.1 increased soil solution SO_4^{2-} threefold (Elkins & Ensminger, 1971).

Liming acid soils decreases the amount of SO_4^{2-} adsorbed and increases the amount of SO_4^{2-} in the soil solution. This increases the potential leaching of SO_4^{2-}.

b. Effect of pH on the Oxidation of Elemental Sulfur—The oxidation of elemental S (flour grade and 0.1-mm mesh size) was lower with unlimed soil than with limed soil (Fox et al., 1964). The effect of pH was greater with the 0.1-mm mesh material than with the flour grade. After 8 weeks of incubation, the pH was 5.3 in the unlimed soil and 6.3 in the limed soil. Liming of a loamy sand from pH 5.2 to pH $\geq$ 6.3 increased the amount of S oxidized by 10 to 20% (Attoe & Olson, 1966).

Liming of acid soils, particularly sandy soils, should enhance the oxidation of very fine particle-size elemental S. This would increase the effectiveness of elemental S when used to correct S deficiencies.

3. Molybdenum Availability

a. Effect of pH on Molybdenum Retention and Solubility—Soil pH is one of the most important factors affecting the availability of Mo and its uptake by plants. Liming acid soils increases availability of soil Mo (Gupta & Lipsett, 1981). Stephens and Oertel (1943) suggested that this might be due to OH^- ions replacing adsorbed MoO_4^{2-}. The amount of Mo sorbed by soils and hydrous oxides increased as pH was decreased (Reisenauer et al., 1962). Molybdate ions replaced surface OH^- ions of hydrous oxides of Fe and Al. As pH decreases, surface OH^- ions are held less tightly and can be replaced by MoO_4^{2-} ions.

Water-soluble Mo increased sixfold as pH increased from 4.7 to 7.5 (Barshad, 1951). Replacement of adsorbed Mo by OH^- ions is responsible for the increase in water-soluble Mo as pH increased.

b. Effect of Lime on Response of Plants to Molybdenum—Liming above pH 5.5 increased alfalfa yields very little when Mo was applied, but liming to pH 6.5 was beneficial when no Mo was applied (Evans et al., 1951). The response of soybeans to pH and Mo on a soil with an initial pH of 5.6 is given in Table 4–5. No response to lime was obtained when Mo was applied. However, without Mo application the soil had to be limed to pH 6.4 before the yield was the same as that with Mo (Parker & Harris, 1962).

The application of Mo to well-limed soils may result in quite large concentrations of Mo in the plant because, as the pH is increased, Mo is held less tightly by soil materials. The Mo concentration of alfalfa grown on an adequately limed soil was increased from 3.6 to 28.5 mg/kg (James et al., 1968). High concentrations of Mo in forages can have detrimental effects on animal nutrition.

The response of legumes to liming without Mo fertilization clearly delineates the effects of lime on neutralization of exchangeable Al^{3+} and the availability of adsorbed Mo. At a pH of about 5.6, practically all of the exchangeable Al^{3+} is neutralized and the yield response of legumes to higher

Table 4–5. Response of soybeans to soil pH and molybdenum (Parker & Harris, 1962).

Soil pH	– Mo	+ Mo
	grain, kg/ha	
5.6	1698	2213
5.7	1795	2230
6.0	1800	2148
6.2	2121	2229
6.4	2234	2213

lime rates is primarily due to increased availability of adsorbed Mo as the pH is increased. The extent to which lime applications correct Mo deficiencies depends on the amount of adsorbed Mo in the soil.

4. Boron Availability

a. Effect of pH on Boron Fixation and Availability—Liming of acid soils to pH $\geq$ 7 has often caused B deficiencies (Naftel, 1937). Uptake of B by tall fescue (*Festuca arundinacea* Schreb.) was relatively constant from pH 4.7 to 6.3 but decreased 2.5-fold at pH 7.4 (Peterson & Newman, 1976). The B content of alfalfa plants was considerably less at pH 7.4 than at pH 6.3, even though the water-soluble soil B content was the same (Wear & Patterson, 1962).

Fixation of applied B was much greater at pH $\geq$ 7 than at pH $<$ 7 (Midgley & Dunklee, 1939; Parks & Shaw, 1941; Olson & Berger, 1946). The work of Sims and Bringham (1968b) indicates that hydroxy Al and Fe materials are responsible for B fixation when acid soils are limed. Retention of B by hydroxy Al and Fe compounds was pH-dependent. Maximum retention of B was at pH 7 with hydroxy Al compounds and at pH 8.5 with hydroxy Fe compounds (Sims & Bingham, 1968a). They postulated that the retention of B is due to anion exchange reactions in which borate ions replace OH^- ions. As pH increases, a greater amount of B is also present as the borate ion.

Soil pH also had an effect on the availability of water-soluble B. As pH increased from pH 5.2 to 5.5 to pH 6.3 to 7.4, the concentration of B in the plant decreased, even though the level of water-soluble B was the same (Wear & Patterson, 1962).

b. Effect of Calcium on the Boron Requirement of Plants—Boron absorption by plants decreased much more when both pH and Ca concentration increased than when only one of these factors increased (Cook & Millar, 1939). Uptake of B has been shown to decrease as Ca uptake increased (Jones & Scarseth, 1944). Indications are that with low Ca uptake, a larger proportion of the B in the plant accumulates as soluble B, whereas, with higher concentrations of Ca in the plant the soluble becomes insoluble.

A high pH and high Ca concentration of the nutrient solution decreased B uptake by cotton 50% (Fox, 1968). Neither high Ca nor high pH alone had any effect on the absorption rate of B. It was suggested that the presence of high concentration of Ca^{2+} and OH^- ions affected the B adsorption mechanism. Thus, besides its effect on B fixation, pH appears to have a physiological effect on B absorption by plants when the supply of Ca is high.

D. Availability of Micronutrient Cations in Relation to Liming

The availabilities of Mn, Cu, Zn, and Fe to plants generally decrease as the soil pH increases. Liming soils that are inherently low in these elements may induce micronutrient deficiencies of short or long duration

(Price & Moschler, 1970). Sharp yield depressions, often observed when strongly acid soils are limed to pH $\geq$ 6.0, probably result from deficiencies or unavailabilities of one or more micronutrients (Foy et al., 1965).

1. Copper

Total Cu content of the soil, soil pH, organic matter content, and the presence of other cations are important factors affecting the availability of Cu to plants in soils (Berger & Pratt, 1963; Sauchelli, 1969). Copper is strongly complexed by organic matter. Thus, the higher the level of soil organic matter, the greater the probability of Cu deficiency (Bloomfield & Sanders, 1977; Mathur et al., 1979). According to Hodgson et al. (1966), 98% of the Cu in soil solution may be complexed with organic matter. Mercer and Richmond (1970) found that Cu complexes with molecular weights $<$ 1000 were much more available to plants than those with molecular weights $>$ 5000.

Copper is also bound tightly to inorganic cation exchange sites in soils (Grimme, 1968). Hydrogen ions can displace Cu adsorbed to clay minerals, and, conversely, increasing the soil pH increases the strength of binding and decreases the concentration of Cu in the soil solution. According to Berger and Pratt (1963), added Cu not chelated by soil organic matter is probably precipitated as $Cu(OH)_2$ at soil solution pH values $>$ 4.7. However, Lindsay (1979) noted that soil Cu is much less soluble than $Cu(OH)_2$ and other Cu compounds and, hence, concluded that the minerals governing the solubility of Cu^{2+} in soils is unknown. Increasing the pH of organic soils is also reported to increase the binding of Cu by phenolic and carboxyl groups (Younts & Patterson, 1964). Thus, in general, increasing the soil pH by liming would be expected to decrease the solubility and availability of Cu to plants. However, the pH at which Cu availability is highest appears to vary with soil organic matter content and the presence of other ions (Lucas & Davis, 1961).

Liming to pH $>$ 5.1 lowered the Cu concentration in wheat plants (*Triticum aestivum* L.), grown in a soil with 22.5% organic matter (Younts & Patterson, 1964). Wheat yields were increased by both lime and Cu, but yield responses to Cu were better without lime than with lime. It was suggested that the lower Cu response with lime was due to stimulated root growth, which enabled plants to extract Cu from a larger zone of soil. The yield response to Cu on the unlimed soil may be explained, at least in part, by data of Brown and Foy (1964). They found that on an acid Bladen soil (Typic Albaquult) (pH 4.5) containing 2.5% organic matter, liming to a soil pH range of 5.2 to 5.5 induced Cu deficiency symptoms in wheat. The deficiency was characterized by a rolling of the youngest leaves and reduced translocation of Cu to these plant parts. Older leaves, located immediately below the affected leaves, contained higher than normal Cu concentrations. Adding 2.5 mg/kg $CuSO_4 \cdot 7H_20$ to the soil prevented the symptoms, increased Ca concentrations in young leaves to normal levels, and doubled vegetative yields in a growth chamber experiment. Thus, it appears that Cu is essential for the normal transport of Ca to the growing point of wheat. In an acid soil that is low in Ca and in which Al–Ca antagonism is present, some of the growth benefits of adding Cu may be due to improved utilization of Ca.

Copper toxicity sometimes occurs in soils receiving repeated treatments with Cu fungicides that contain Cu. In Florida, Cu toxicity in citrus has been associated with 150 to 250 mg/kg total Cu in the topsoil and a soil pH of ≤ 5.0. The toxicity is generally corrected by liming the soil to a pH of 6.5 to 7.0 (Younts & Fiskell, 1963).

2. Manganese

Factors affecting the availability of Mn to plants in soils include the level of total and easily reducible Mn (related to soil parent material), pH, presence of other salts, organic matter content, aeration, and microbial activity (Foy, 1973). Small changes in these factors can determine the degree of Mn toxicity or deficiency produced in a given crop. Plants apparently absorb Mn primarily in the divalent state. Lowering soil pH below about 5.5 increases the level of Mn^{2+} in soil solution and increases the likelihood of toxicity. Air drying may greatly increase the level of exchangeable Mn^{2+} in acid soils having high levels of the easily reducible form. Flooding and compaction of soils reduce soil aeration and favor the reduction of Mn to the divalent form in which it may reach toxic levels. Liming and soil drainage can favor the oxidation of Mn to the tetravelent form and correction of toxicity or induction of Mn deficiency in plants.

Solubility and plant availability of Mn in the soil is closely related to activities of microorganisms. Some bacteria oxidize the soluble, divalent Mn to the less-soluble tetravalent state and may thereby induce Mn deficiency in plants (Kamura & Nishitani, 1977). This oxidation process is increased by liming (Bromfield, 1978) and is reported to reach a maximum at about pH 6.5 (Bromfield & David, 1976). Godo and Reisenauer (1980) reported that root exudates, such as hydroxycarboxylases, increased plant uptake of Mn by reducing MnO_2 and complexing the divalent Mn in a more available form. This process was particularly marked below pH 5.5.

According to Berger and Pratt (1963), Mn deficiencies generally occur in organic soils and neutral or alkaline of mineral soils. Hanna and Hutcheson (1968) reported that Mn deficiency is most common between pH 6.5 and 8.0. In the sandy soils of the southeastern Coastal Plain of the USA, liming to pH near neutrality often produces Mn deficiency in certain crops. Examples are soybeans in Georgia (Boswell et al., 1981), peanuts in Virginia (Hallock, 1979), and bermudagrass [*Cynodum dactylon* (L.) Pers.] in Florida (Snyder et al., 1979). Palmer (1981) reported that soybeans develop Mn deficiency at pH ≥ 6.2 in the Coastal Plains soils of South Carolina and adjacent states. Manganese deficiency is also prevalent in soybeans in Ohio (Kroetz et al., 1977) and has been found in wheat in Tanzania (Hoyt & Myovella, 1979).

3. Zinc

Zinc deficiency is prevalent in acid, leached sandy soils with a low Zn content and in neutral and alkaline soils with high levels of P and organic matter (Berger & Pratt, 1963; Chapman, 1966). It is a widespread problem in wetland rice (*Oriza sativa* L.) (Agarwala & Mehrotra, 1979; Animesh & Shahjahan, 1980; Alijakhro, 1980; Wells, 1980), and it is apparently more severe in poorly drained areas of rice fields (Van Breeman & Castro, 1980) and at high pH (IRRI, 1979). Zinc deficiency has also been reported in many other crops, including slash pine (*Pinus elliottii* Engelm.) (McKee, 1976), flax (*Linum* spp.) (Moraghan, 1978), soybeans (Graves et al., 1980), and corn (Clark, 1982).

Berger and Pratt (1963) stated that Zn deficiency generally occurs within the pH range of 6.0 and 8.0; however, Sauchelli (1969) indicated that the availability of Zn in soils may become critical at soil pH values as low as 5.3. The disorder is reported to occur in wetland rice at pH 5.4 to 8.7 (IRRI, 1977). According to Sedberry et al (1978), the optimum pH for Zn availability to rice is about 5.5.

Massey (1957) found that Zn uptake by corn was significantly correlated with soil pH between 4.3 and 7.5. Wear and Hagler (1963) reported that liming sandy Alabama soils to pH near 6.5 produced Zn deficiency in corn. In earlier work, Wear (1956) found that $CaCO_3$ decreased the Zn content of sorghum at soil pH levels of 5.7 to 6.6. Sodium carbonate ($NaCO_3$) also reduced Zn uptake but $CaSO_4$ did not. The reduction in Zn uptake induced by $CaCO_3$ was attributed to the increased soil pH and not the Ca added. Lime reduced Zn uptake by red clover (*Trifolium pratense* L.), bromegrass (*Bromus inermis* Leyss.), and timothy in studies by Percival et al. (1955). Miller et al., 1964 found that the Zn content of 'Coastal' bermudagrass was inversely correlated with soil pH within the range of 5.3 to 7.2 on a Cecil sandy loam (Typic Hapludult) in the Piedmont region. Zinc content increased with increased N fertilization rates. Viets et al. (1957) also found that N fertilizers increased the availability of Zn and this effect was attributable to pH decreases. Similar effects have been reported by Giordano et al (1966). Phosphate fertilization has induced Zn deficiency in corn (Takkar et al., 1976).

Within the soil pH range of 5.0 to 7.0, simple precipitation does not seem to be involved in Zn retention. Most of the Zn in soils is believed to occur as simple ions of Zn^{2+}, $Zn(OH)^{+}$, or $Zn(Cl)^{+}$ absorbed on fine granular constituents of soils, with perhaps some Zn entering holes in the clay crystal lattice (Grimme, 1968; Krauskopf, 1972). Fixation of Zn in nonexchangeable form at pH 5.0 to 7.0 could be due to adsorption on permanent charge sites of clays and complexing with organic matter (McBride & Blasiak, 1979). The intensity of Zn adsorption by goethite increases with increasing pH; therefore, Zn mobility and availability to plants are greatly decreased in neutral and alkaline soils (McBride & Blasiak, 1979).

Zinc also forms both soluble and insoluble complexes with organic matter in soils. It is estimated that as much as 60% of the soluble Zn in soils occurs as soluble organic complexes (Hodgson et al., 1966). Stevenson and Ardakani (1972) concluded that the soluble Zn organic complexes were mainly associated with amino, organic, and fluvic acids and the insoluble complexes with humic acids. Presad et al. (1976) concluded that chelating agents (natural or synthetic) play an important role in overcoming Zn availability problems in alkaline and calcareous soils.

Zinc toxicity has been reported in cotton grown on old peach [*Prunis persica* (L.) Batsch] soils (Faceville sand, Typic Paleudults) at pH 5.3 and was corrected by liming to pH $\geq$ 6.0 (Lee & Pae, 1967). Similar findings were later reported for soybeans (Lee & Craddock, 1969). Zinc toxicity also occurs in areas of Zn minespoils (Rosen et al., 1978; Johnson et al., 1971).

III. LIMING EFFECTS ON MICROBIAL REACTIONS IMPORTANT IN NUTRIENT AVAILABILITY

A. Mineralization of Organic Matter

1. Types and Activities of Microorganisms

Soil pH is an important factor in determining the kinds, amounts, and activities of microorganisms involved in organic matter transformations (Sommers & Biederbeck, 1973; Alexander, 1980). By regulating microbial activity, pH affects the mineralization of organic matter and the subsequent availabilities of N, P, S, and micronutrients to higher plants. In general, organic matter, whether natural or added, decomposes more rapidly in neutral soils than in acid soils. At soil pH values below about 5.5 the fungi are most active in this decomposition, primarily because of lack of competition from other organisms that are more sensitive to acid soil factors. Campbell and Lees (1967) suggested that fungi may carry out nitrification in strongly acid forest soils. At pH values > 6.0 the actinomycetes and bacteria are more prominently involved in organic matter decomposition. Different bacteria and actinomycetes appear to have different optimum pH ranges for reproduction. For example, Corke and Chase (1964) noted differential pH requirements of certain actinomycetes. Other evidence indicates that in certain soils acid-tolerant strains of nitrifying bacteria have developed (Jackson, 1967; Brar & Giddens, 1968).

In some strongly acid soils, Al toxicity, as well as H^+ ion toxicity, may limit microbial breakdown of organic matter. For example, Mutatker and Pritchett (1966) found that CO_2 production was greater in a Ca muck than in a Ca–Al much when both were maintained at pH 4.0. In a latosol and in mixtures of Ca and Al-saturated mucks, CO_2 production was significantly decreased with increasing levels of Al during incubation at pH $\leq$ 4.0. However, on a modified Arredondo fine sand (Grossarenic Paleudult), added Al had very little effect on relative numbers of fungi or bacteria or on CO_2 and NO_3^- production. Denison (1922) reported that Al actually stimulated ammonifying organisms but adversely affected nitrifying bacteria. Whitney (1923) concluded that Al was not important as a toxic factor to soil microorganisms when optimum levels of P, K, and Ca were available. However, according to Sommers and Biederbeck (1973), the soil is generally a suboptimal growth medium for microorganisms; Gray (1970) stated that "microorganisms live in the soil under starvation conditions." Aluminum may limit microbial growth through interference in the uptake and use of essential elements like Ca, Mg, and P.

2. *Release of Nutrients*

a. Organic Nitrogen, Phosphorus and Sulfur—Two important microbially induced reactions affecting N availability in soils are: (i) ammonification, the production of NH_4^+ from the decomposition of organic matter; and (ii) nitrification, the oxidation of NH_4^+ to NO_2^- and NO_3^- brought about primarily by bacteria of the *Nitrosomonas* and *Nitrobacter* genera, respectively (Sommers & Biederbeck, 1973; Alexander, 1980). Ammonification can apparently occur over a wide range of soil pH, but nitrification is markedly reduced at pH values < 6.0 and > 8.0 (Chase et al., 1968; Alexander, 1980).

In most mineral soils, from one-half to two-thirds of the total P is in organic forms (Cosgrove, 1967). A large part of this organic P is of microbial origin and must be mineralized to become available to plants. In acid soils the very insoluble Al and Fe phytates are believed to be the most abundant organic P compounds. Oliveira and Drozdowicz (1980) demonstrated the presence of bacteria and actinomycetes that can mineralize P from deoxyribonucleic acid (DNA), lecithin, and calcium phytate in the Cerrado soils of Brazil. Increasing the soil pH generally causes mineralization of phytate P and thus increases its availability to plants (Cosgrove, 1967). Thompson et al. (1954) found that the mineralization of P was positively correlated with soil pH. The amounts of P, C, and N mineralized in soils are generally roughly correlated with their proportions in the organic matter being decomposed, but the release of organic P is sometimes quite different from that of C and N. For example, Thompson et al. (1954) found that in 50 unlimed soils the mineralization of organic P increased markedly with increase in pH but that of organic C did not. On the basis of such evidence, Cosgrove (1967) suggested that alkaline soils should have higher C/P and N/P ratios than acid soils.

The plant availability of S contained in organic matter appears to be limited by the rate of organic matter breakdown, rather than by the oxidation of the S compounds released (Burns, 1967). This oxidation is brought about largely through biological processes and can take place between pH 2.0 and 9.0. Nelson (1964) found that liming soil from pH 4.0 to 5.0 increased SO_4^{2-} production threefold, but increasing the pH from 5.0 to 7.0 had very little additional effect. Lime additions to soils increases the SO_4^{2-} levels, but this can result from SO_4^{2-} added in the lime, SO_4^{2-} released from organic matter by hydrolysis at higher pH, or SO_4^{2-} released from soil organic matter by bacteria growing more profusely at a favorable pH (Freney, 1967). For additional information regarding S transformations in soils see McLachlan (1975).

b. Micronutrients—Soil organisms may influence the availabilities of nutrients indirectly by changing the pH of the medium or by changing the oxidation state of an element. These reactions are important in determining mineral element solubilities, and hence, toxicities and deficiencies. There is considerable evidence that soil microorganisms alter the solubility or oxidation states of Mn, Zn, Cu, Al, and Mo (Alexander, 1961, 1980).

The availabilities of Mn, Cu, Zn, and Fe are closely related to the transformations of organic matter in soils (Alexander, 1961). For example, Zn deficiency of fruit trees does not occur in sterilized soils. Sterilizing the soil releases Zn from microbial bodies and makes it available to plants. Zinc deficiency in flax has also been attributed to microbial activities.

Copper availability is generally decreased by microbial metabolism during the decomposition of certain crop residues (Alexander, 1961); however, the decay of the more readily decomposable carbonaceous materials, like glucose or asparagine, may increase the level of water-soluble and available Cu.

The availability of Fe may be increased by microbial action through solubilization of organic products that chelate Fe or by reducing conditions (Alexander, 1961). Ferrous Fe^{2+} is likely to be more abundant at pH values < 5.0 and ferric Fe^{3+} is favored at pH > 6.0. Iron can also be precipitated by Fe bacteria in soils and made less available to plants.

Manganese deficiencies and toxicities are closely related to the activities of soil microorganisms in altering the pH and bringing about oxidation-reduction reaction changes, which determine the solubility of Mn (Foy, 1973).

Jones and Woltz (1960) reported an interesting microbial-micronutrient relationship. Liming a Leon soil (Aeric Haplaquod) from 5.8 to 8.0 reduced the incidence and severity of Fusarium wilt of tomatoes, caused by *Fusarium oxysporium, F., sp. lycopersicon,* race 2. Soil additions of Mn plus Zn, or Fe plus Zn lignosulfates, which are plant-available at pH 9.0, reversed the disease control given by lime and increased the severity of the disease. They suggested that liming the soil reduced the severity of tomato wilt by creating an imbalance of micronutrients in the pathogen.

B. Dinitrogen Fixation (Symbiotic)

Legumes in general have been much more responsive to liming than other plants. As discussed earlier, one reason for this is the increased availability of soil Mo and its role in N_2 fixation. Of more importance, perhaps, is the effect of soil acidity and Ca concentration on N_2 fixation. Acid soil factors can reduce N_2 fixation by injuring the host plant directly, by reducing or inhibiting the survival of rhizobia, or by interfering with various stages of the nodulation and N_2 fixation processes. Current thinking regarding the effects of acid soil factors (excess Al and Mn, low pH, and low levels of Ca and P) follows.

The rhizobia of some legume species appear to be more sensitive to Al than do their host plants. For example, Pieri (1974) reported that the nodulation of groundnut (*Arachis hypogaea* L.) was reduced when the Al saturation of the soil CEC reached 30% in sandy soils of Senegal. Higher levels of Al saturation were required for toxicity of the host plant. de Carvalho et al. (1980) concluded that Al toxicity decreased the growth of *Stylosanthes* spp. more severely when plants were dependent on symbiotic N_2 fixation than when fertilizer N was applied. The Al tolerance of six *Stylosanthes* species appeared to be dependent on the ability to nodulate and develop an efficient symbiosis in the presence of Al and on the inherent sensitivity of the host plant to Al. Mengel and Kamprath (1978) found that the shoot growth of soybeans on eight organic, North Carolina soils was significantly related to soil pH between 4 and 5. As pH increased, nodule numbers and weights, N concentration, and total N uptake by plants increased markedly. The critical pH for shoot, root, and nodule growth was generally in the range of 4.6 to 4.8 (salt pH). The growth response to lime was attributed to decreased exchangeable and water-soluble Al, increased water-soluble Ca, and provided favorable pH for rhizobia.

In studies by Keyser and Munns (1979a), Al toxicity and soil acidity, per se, appeared more important than Mn toxicity and Ca deficiency in limiting the rhizobia of cowpea [*Vigna unguiculata* (L.) Walp.] and soybean. Aluminum concentrations of 0.68 and 1.35 mg/kg produced more severe stress than did high Mn (10.8 mg/kg) or low Ca (2 mg/kg). Furthermore, Ca added at 0.2 to 40 mg/kg did not protect rhizobia (two strains studied) against Al toxicity. In general, cowpea rhizobia tended to be more tolerant to Al than *Rhizobium japonicum*. In another study, Keyser and Munns (1979b) found that 1.35 mg Al/kg was more harmful to rhizobia than low pH (4.5) or low P (0.3 mg/kg). Keyser et al. (1979) concluded that cowpea rhizobia contained a large, and perhaps continuous, variation in symbiotic tolerance to soil acidity at pH 4.6. Sixty-five percent of the acid–sensitive strains were identified in nutrient solutions containing 1.35 mg Al^{3+}/kg. No strain that was highly tolerant to the acid soil was identified as Al-sensitive in solution.

The importance of N deficiency in limiting legume growth on acid soils is a debatable question. Munns et al. (1981) have suggested that, unlike other legumes, soybeans may be limited by factors other than nodulation failure. In their studies, involving 2 soybean cultivars and 13 single strain inoculants, liming the soil from pH 4.4 to 6.0 (aqueous paste) doubled growth, regardless of N source, cultivar, rhizobial strain, or numbers. The inoculated plants were nodulated, green, and high in N, even when growth was severely reduced by the acid soil. Plant symptoms indicated that soybean growth on acid soils was limited by Al toxicity to the host plant. This conclusion was supported by nutrient solution studies in which the pH, Al, and Ca were controlled at levels similar to those in soil solution extracts. In the solution experiments, soybean growth was unaffected by low Ca (8 mg/kg) or low pH (4.5) but was depressed by 0.8 to 1.6 mg Al/kg. From soil and nutrient culture studies, Munns et al. (1981) concluded that efforts to increase the acid soil tolerance of soybeans should center on the plant and not the rhizobia. This agrees with our unpublished findings (C. D. Foy and T. E. Devine, 1980, unpublished data) showing that the relative acid soil (Al) tolerances of 'Perry' and 'Chief' soybeans were the same with symbiotic or fertilizer N. For additional information on rhizobial growth in relation to acid soil factors (excess Al and Mn and low availabilities of Ca, Mo, and P), see Andrew (1978), Munns (1980), and Cregan (1980).

Dobereiner (1966) suggested that the sensitivity of many legumes to acid soils could be explained by a specific Mn toxicity on the N_2-fixing process. When bean plants (*Phaseolus vulgaris* L.) were dependent on symbiotic N_2 fixation, 25 mg Mn/kg in sand culture-solutions decreased the total N contents of plants by 30 to 60%. However, when plants were adquately supplied with mineral N, the corresponding reduction was only 13%. In plants that were totally dependent on symbiotic N_2 fixation, the N contents of bean plants decreased linearly with the logarithm of the Mn concentrations in plant tops; however, this was not observed in plants supplied with mineral N. The effects of Mn toxicity on N_2 fixation varied with inoculated strains of *Rhizobium phaseoli*. Souto and Dobreiner (1971) found that excess Mn markedly decreased nodulation in tropical legumes. Kliewer (1961) reported that 20 mg Mn/kg in nutrient solutions reduced the number of nodulated alfalfa plants by 50% but did not affect the nodulation of birdsfoot trefoil (*Lotus corniculatus* L.). Dobreiner (1966) postulated that contradictions in the literature regarding the response of legumes to lime might be attributed to variable concentrations of soluble Mn in acid soils and differences in tolerance to Mn toxicity among both host plants and rhizobial strains.

Freire (1975) concluded that soil pH values of 4.0 to 4.5 were not harmful to the soybean nodulation process in the absence of excess Al and Mn. However, toxic Al and Mn, together with P and Ca deficiencies, were regarded as highly detrimental to nodule formation and N_2 formation. In later work, Freire (1976) found that soybean cultivars with low tolerance to Mn toxicity showed higher response to lime and had higher Ca uptake than Mn-sensitive varieties.

Recent studies of Keyser and Munns (1979a,b) with cowpea and soybean rhizobia indicate that in acid soils, Al toxicity and soil acidity, per se, are probably more important than Mn toxicity and Ca deficiency in restricting N_2 fixation. Aluminum at 0.68 to 1.35 mg/kg in solution produced a greater stress than high Mn (10.8 mg/kg) or low Ca (2 mg/kg). All rhizobial strains that were tolerant to Al were also tolerant to Mn and low Ca. The high Mn (10.8 mg/kg) and low Ca (2 mg/kg) reduced the growth rates of some strains but did not stop the growth of any strain.

IV. TOXICITIES OF PLANTS IN ACID SOILS

A. Aluminum Toxicity

Aluminum and Mn toxicities are the most prominent growth-limiting factors in many acid soils. Much of the poor root development (and drought susceptibility) seen in soils with acid ($pH < 5.0$) subsoil layers is probably due primarily to Al toxicity, which limits both rooting depth and degree of branching.

1. Plant Symptoms

Symptoms of Al toxicity are not always easily identified (Foy et al., 1978; Alam & Adams, 1979; Clark, 1982; McFee et al., 1984). In some plants the foliar symptoms resemble those of P deficiency—overall stunting; small, dark green leaves and late maturity; purpling of stems, leaves, and leaf veins; and yellowing and death of leaf tips. In others, Al toxicity appears as an induced Ca deficiency or reduced Ca transport problem within the plant (curling or rolling of young leaves and a collapse of growing points or petioles). Aluminum injured roots are characteristically stubby and brittle. Root tips and lateral roots become thickened and turn brown. The root system as a whole is corraloid in appearance, with many stubby lateral roots, but lacking in fine branching. Such roots are inefficient in absorbing nutrients and water.

2. Physiological and Biochemical Effects of Aluminum

For plants in general, excess Al has been variously reported to interfere with cell division in root tips and lateral roots, increase cell wall rigidity by crosslinking pectins, reduce DNA replication by increasing the rigidity of the DNA double helix, fix P in less available forms in soils and on plant root surfaces, decrease root respiration, interfere with enzymes governing sugar phosphorylation and the deposition of cell wall polysaccharides, and interfere with the uptake, transport, and use of several essential elements, including Ca, Mg, K, P, and Fe. For details on the older literature, see review papers (Foy, 1974; Foy & Fleming, 1978; Foy et al., 1978; McFee et al., 184).

Metal ions, such as Al^{3+}, are known to form strong complexes with nucleic acid; in fact, they are used to complex and isolate polynucleotides from leaves (Trim, 1959). Matsumoto et al. (1979) found that Al was bound to the P in DNA and not to the histone proteins associated with the DNA double helix. Ulmer (1979) reported that Al had two effects on susceptible wheat cultivars—a rapid inhibition of root elongation, followed by inhibition of DNA synthesis. Naidoo (1976) found the highest Al concentrations in the nuclei of Al-injured snap beans and suggested that Al is bound to esteric P in nucleic acids and membrane lipids; such binding was believed to inhibit cell division by interfering with nucleic acid replication. McLean (1980) found that Al caused an abnormal distribution of ribosomes on the endoplasmic reticulum of barley root cells; interference in protein synthesis was postulated. Schmandke et al. (1979) observed that $AlCl_3$ increased the firmness and decreased the solubility of protein casein fibers in fava bean (*Vicia faba* L.). Trivalent Al was believed to form coordination complexes with carboxyl and sulfhydryl groups of the protein, which results in a cross-linking.

Many of the effects of Al on plants are probably associated with alteration of root membrane structure and function. Plant membranes are visualized as semifluid arrangements of proteins and lipids. Aluminum could bind to either, depending on pH and other conditions. Vierstra and Haug (1978) found that Al decreased lipid fluidity in membranes of *Thermo-*

plasma acidophilium. In their study, 270 mg Al/kg in solution produced dramatic changes, but detectable effects occurred with only 0.27 mg Al/kg at pH 4.0. Gomez-Lepe et al. (1979) reported that Al binds to cell membrane proteins on the inner epidermal cells of onion (*Allium cepa* L.).

Aluminum also affects water use by plants. For example, Horton and Edwards (1976) found that increasing concentrations of Al in nutrient solutions increased the diffusive resistance of peach seedlings. Beradze (1977) reported that Al disrupted the water status of corn. Kaufman and Gardner (1978) concluded that water potential of wheat plants increased when root growth was reduced by soil acidity (probably Al toxicity).

Excess Al may decrease the uptake of certain essential elements and increase that of others (Lau, 1979; Ben et al., 9176; Gurrier, 1979; Alam & Adams, 1979, 1980; Duncan et al., 1980). However, such variations in gross plant composition are difficult to interpret in terms of Al-toxicity mechanisms. No one pattern of elemental accumulation can apply to all cases of Al injury.

Aluminum accumulates on or in the roots of Al-injured plants, often in association with P (McCormick & Borden, 1972), but does not generally accumulate in tops of Al-sensitive plants (Foy, 1974). Even in cases where Al injury in a given plant is correlated with increased Al concentrations in plant tops (James et al., 1978; Alam & Adams, 1979; Duncan et al., 1980), the Al found in the tops may not be directly harmful to the plant tops themselves. It may simply be the result of passive accumulation caused by prior injury to the roots. In fact, the entire array of elements in the tops of Al-injured plants probably represents the accumulated, systematic effects of the initial root injury by Al. Such effects are too far removed from the initial injury to reveal Al toxicity mechanisms.

Aluminum toxicity continues to be associated with P nutrition of plants. For example, James et al. (1978) concluded that Al-induced P deficiency reduced the growth of Sitka spruce [*Picea sitchensis* (Bong.) Carr.] in Scotland. Negative correlations were found between growth, beaded roots (Al-injured), and foliar Al concentrations. Santana and Braga (1977) found that P concentrations in rice tops decreased with increasing Al saturation in the soil. Total Ca and Mg uptake were also decreased over the same range of Al saturation. Helyar (1978) concluded that Al toxicity effects were largely associated with Al interference in P metabolism and with Al binding to root cell wall pectins, which stops root elongation.

Aluminum–iron interactions are frequently mentioned in the literature. For example, Alam and Adams (1980) found that Al induced Fe deficiency in oats and postulated that Al interfered with the reduction of Fe^{3+} to Fe^{2+} within the plant, a process essential for normal Fe metabolism.

Aluminum–calcium interactions have also been reported in recent papers. For example, Al toxicity in peach seedlings has been related to reduced Ca uptake but not transport (Edwards & Horton, 1977). Simpson et al. (1977) attributed poor root growth of alfalfa to Al–Ca interactions. Similar relationships have been found in Kikuyu grass (*Pennisetum clandestinium* Hochst.) grown in soil at pH 4.36 (Awad et al., 1976).

Aluminum has recently been implicated in animal health. Allen and

Robinson (1980) found very high concentrations of Al (2000 to 8000 mg/kg in plants from ryegrass (*Lolium* sp.) pastures, on which grass tetany was prevalent. Rumen content samples of dead tetany animals contained an average of 2373 mg Al/kg compared with 405 mg/kg from nontetany animals. The addition of Al to rumen-fluid-ryegrass buffer solutions decreased Mg solubility by 56% and Ca solubility by 74% within 48 h. It was concluded that high Al levels in forage and in rumen contents are actively involved in the etiology of grass tetany. However, subsequent work at the same location indicated that the very high Al levels reported previously in forage and rumen fluid samples collected at grass tetany sites were due to soil contamination of the forage and ingestion by the grazing animal rather than high Al uptake by plants (Cherney et al., 1983). More recent work (D. L. Robinson, 1983, personal communication) has also shown that the administration of either $Al_2(SO_4)_3$ or Na_2SO_4 to cows *(Bos taurus)* depressed Mg solubility in rumen fluid. Hence, the decrease in Mg solubility induced by $Al_2(SO_4)_3$ and previously attributed to Al appears to be associated instead with the SO_4^{2-} anion. Grass tetany is still consistently associated with high Al contents of rumen fluid, but the role of soil ingestion in the disease has not been determined. It does appear, however, that the Al ion, per se, is not directly involved in the etiology of grass tetany as previously suggested.

B. Manganese Toxicity

Manganese toxicity generally occurs in soils with pH values of ≤ 5.5, if the soil parent materials contain sufficient total Mn (Foy, 1973; McFee et al., 1984). However, it can also occur at higher pH values (≥ 6.0) in poorly drained or compacted soils where reducing conditions favor the production of divalent Mn^{2+}, which is most available for plant uptake. Thus, Mn toxicity can occur at pH values that are too high for Al toxicity. Soils of the Atlantic Coastal Plain of the USA are lower in total Mn than those of the Gulf Coastal Plain. Hence, at a given low pH, Mn toxicity is less likely in the former than in the latter.

1. Plant Symptoms

Unlike Al, excess Mn seems to affect plant tops more directly than roots (Foy, 1973; McFee et al., 1984). In addition, Mn produces more definitive symptoms in plant tops than does Al, and for a given plant, Mn accumulates somewhat in proportion to plant injury (Osawa & Ikeda, 1980). Plant symptoms of Mn toxicity include marginal chlorosis and necrosis of leaves (alfalfa, rape [*Brassica napus* L.], kale [*Brassica oleracea* var. *acephala* Dc.], lettuce), leaf puckering (snap bean, soybean, cotton), chlorosis of young leaves (resembling Fe deficiency), and necrotic spots on leaves (barley, lettuce, soybeans). Necrotic spots in barley and chlorotic leaf margins of lettuce injured by Mn toxicity contain much higher Mn concentrations than do adjacent leaf tissues (Vlamis & Williams, 1973). In severe cases of Mn toxicity, plant roots turn brown, usually after the tops have been severely injured.

Specific plant physiological disorders associated with excess Mn are "crinkle leaf" of cotton (Foy, 1973), "stem streak necrosis" in potato (*Solanum tuberosum* L.) (Lee & McDonald, 1978), "internal bark necrosis" of apple trees (Miller & Schubert, 1977; Schibisz & Sadowski, 1979), growth retardation and "leaf tip burn" in carnation (*Dianthus carophyllus* L.) (Ishida & Masui, 1976), and "fruit cracking" at the blossom end of muskmelon (*Cucumis melo* L.) (Masui et al., 1976).

In several tropical pasture grasses, Mn toxicity has been associated with small, black or dark brown flecks on the lower leaves, and sometimes with an induced Fe deficiency. Buffelgrass (*Cenchrus cilliaris* L.) had these symptoms, plus white bands 1 to 2 cm wide, extending transversely across the leaves at regular intervals along the blades. No other species had this latter symptom (Smith, 1979). Manganese toxicity has been characterized by interveinal chlorosis on upper leaves in bean, tomato, and spinach (*Spinacea oleracea* L.), and by marginal leaf chlorosis on lower leaves of cabbage, lettuce and celery (*Apium graveolens* Mill.). In Welsh onion (*Allium fistulosum* L.), excess Mn induced chlorosis of the older leaves (Osawa & Ikeda, 1977, 1980).

Andrew and Pieters (1976) described Mn toxicity symptoms in *Glycine wightii* (Grah. ex Wight & Arn.) Verdc. as follows: interveinal chlorosis of young expanding leaves, similar to Fe deficiency; mottling and irregularly shaped brown spots in vicinity of midrib and main veins of expanded leaves; puckering of leaf surfaces; and in severe cases, a browning of the midribs and main veins (particularly on the upper leaf surfaces), which sometimes extends to petioles and stems. Jones and Clay (1976) found that Mn injury in another tropical legume, Townsville stylo [*Stylosanthes humilis* (H. B. K.)], was characterized by an overall yellowing, particularly in the youngest leaves and stems, with some interveinal chlorosis and leaf puckering. Under severe Mn toxicity, the tips of emerging leaves became necrotic before the leaf was fully expanded and the necrosis proceeded from the leaf tip downward.

Manganese toxicity causes slow growth of tobacco (*Nicotiana* spp.) seedlings and toxicity symptoms appear 4 to 5 weeks after transplanting in the field (Link, 1975, 1979; Smiley, 1974). The symptoms include a light green or yellowish mottling of leaves between the large veins, dark green leaf veins, and a gradual development of necrotic spots, especially in the older leaves. Leaves may also have a hard, semiglossy appearance. Plants may recover, but overall growth is reduced.

Plant symptoms of Mn toxicity are often detectable at stress levels that produce little or no reduction in vegetative growth. In contrast, Al toxicity can reduce yields greatly (by root damage) without producing clearly identifiable symptoms in plant tops. Hence, in acid soils that contain high levels of both Al and Mn in solution, the plant growth reductions observed may be largely or wholly attributed to Mn toxicity when Al toxicity is the more important of the two factors (Foy, 1976).

2. Physiological and Biochemical Effects of Excess Manganese

In general, Mn toxicity has been associated with the following effects on plants: destruction of auxin (indoleacetic acid IAA) by increasing the activity of IAA oxidase of cotton; a possible amino acid imbalance in potato; increased replication of ribonucleic acid (RNA) in potato spindle tuber virus; increased activities of peroxidase and polyphenol oxidase, decreased activities of catalase, ascorbic acid oxidase, glutathione oxidase and cytochrome *c* oxidase, lower adenosine triphosphate (ATP) contents and lower respiration rates in cotton; reversal of growth inhibition of sorghum roots caused by gibberellic acid, which enhances auxin production; and induction of erythromycin-resistant mitochondrial mutations in yeast. Manganese interacts with many other mineral elements in plant nutrition, and under certain conditions, the addition of these elements (Si, Fe, Ca, and P) may alleviate Mn toxicity. For specific references to the older literature, see review chapters (Foy, 1973; Foy et al., 1978; Foy, 1984).

Recent literature provides additional evidence that excess Mn alters the activities of enzymes and hormones in plants. For example, Sirkar and Amin (1979) found that Mn toxicity reduced respiration in young cotton seedlings and that this effect was reversed by adding IAA. (The addition of IAA did not affect the respiration of normal tissues.)

Horst and Marschner (1978a) have proposed a Mn toxicity hypothesis that involves auxin and Ca. They observed that excess Mn inhibited Ca transport into the leaves of beans and induced typical Ca deficiency symptoms (crinkling of leaves in shoot apices and younger leaves). Manganese toxicity also reduced the cation exchange capacity of leaf tissues and reduced Ca movement into the apparent free space of isolated leaf segments. From these observations, Horst and Marschner (1978a) concluded that Mn toxicity increased the activity of IAA oxidase, which decreased the auxin level (proposed earlier in cotton by Morgan et al., 1966, 1976). It was reasoned that the lower auxin level then reduced cell wall expansion and that the subsequent reduction in formation of new, negative sites decreased Ca translocation into the tissues.

Memon et al. (1980) reported that, in the Mn accumulator *(Acanthopax sciadophylloides)*, Mn accumulated markedly in peripheral cells of leaf petiolule and petiole tissues. High Mn concentrations were found in the epidermis, palisade, and spongy parenchyma cells. Low Mn was found in vascular bundle cells. Such compartmentalization of Mn may act as a protective mechanism against toxicity by keeping excess Mn from key metabolic centers.

Manganese toxicity is often associated with a decrease in the Ca concentrations of plants (Smith, 1979; Horst & Marschner, 1978a; Takayanagi, 1976). Supplying additional Ca to the growth medium sometimes reduces Mn accumulation, decreases the severity of Mn-induced chlorosis, and alleviates growth reductions (Osawa & Ikeda, 1977). Masui et al. (1976) reported that Ca decreased Mn concentrations in muskmelon plants but did not prevent yield decreases by excess Mn.

The addition of Si to nutrient solutions reduced Mn toxicity in barley (Williams & Vlamis, 1957) and bean (Horst & Marschner, 1978b,d). Apparently, this detoxification is not due to reduced Mn uptake, but to the prevention of localized accumulations of Mn associated with necrotic spots on the leaves. For additional information on Mn–Si interactions, see Foy et al. (1978) and Foy (1984).

Manganese and Fe are closely related in plant nutrition, and some investigators have used the Fe/Mn ratio in plant tops as one indicator of toxicity. For example, Hati et al. (1979) found that Mn toxicity occurred when Fe/Mn ratios were 1.36, 1.40, and 0.8 for cotton, wheat, and soybean, respectively, grown on an acid soils at pH 4.1. Iron and Mn contents were negatively correlated. Excess Mn often produces plant symptoms resembling Fe deficiency (Horst & Marschner, 1978c), and increasing the Fe supply may decrease Mn uptake (Masui et al., 1976; Moraghan, 1979). However, not all Mn toxicity is Fe-related. Gupta and Rao (1977) concluded that Mn toxicity in sugarcane is not Mn-induced Fe deficiency. Van der Vorm and Van Diest (1979) stated that under acid, aerobic conditions, the absorption of Fe is affected little by large quantities of Mn. However, excess Mn uptake can interfere with normal Fe metabolism without reducing Fe uptake.

Temperature × Mn interactions affect the severity of Mn toxicity in a given plant. For example, researchers at North Dakota State University reported that Mn toxicity in 'Bragg' soybean was more severe at a lower temperature (Anon., 1980). Rufty et al. (1979) found that in tobacco, the lower the temperature, the lower level of leaf Mn required for toxicity. Leaf Mn concentrations associated with toxicity at different day/night temperatures were as follows: 22/28°C, 700 to 1200 mg/kg; 26/22°C, 2000 to 3500 mg/kg; and 30/26°C, 5000 to 8000 mg/kg. Manganese toxicity was more severe in young leaves, and the severity decreased with age, even at the same internal Mn concentrations. It was concluded that the increased tolerance of tobacco to high leaf Mn contents in warm temperatures was associated with a more rapid rate of leaf expansion, accompanied by increased vacuolar capacity for disposal of accumulated Mn. Such temperature × Mn interactions increase the difficulties of determining critical Mn concentrations in soils and plants.

Another factor affecting Mn toxicity is the form of N used in the growth medium. Osawa and Ikeda (1980) concluded that this had a marked effect on Mn uptake and toxicity in vegetable crops. Manganese toxicity symptoms (interveinal chlorosis on upper leaves of tomato and spinach and marginal chlorosis on lower leaves of cabbage) were both less severe with $NO_3^- + NH_4^+$ than with NO_3^- alone.

C. Other Metal Toxicities

Toxic levels of metals in soils may be caused by natural soil properties or by agricultural, manufacturing, mining, and waste disposal practices (Foy et al., 1978). Any of the metals can be toxic if soluble in sufficient quantities. In near-neutral soils, the heavy metals occur as inorganic compounds or in bound forms with organic matter, clays, or hydrous oxides of Fe, Mn, and Al. This precipitation and absorption of metals by soils protects plants against toxicity. However, decreasing the soil pH can quickly create metal toxicity problems. Iron toxicity can occur under flooded conditions where Fe occurs as the reduced soluble Fe^{2+} form (Foy et al., 1978). Zinc, Cu, and Ni toxicities have occurred rather frequently. Toxicities of Pb, Co, Be, As, and Cd occur only under very unusual conditions. Lead and Cd are of particular interest because they move into the food chain and affect human and animal health. For further details, see Foy et al. (1978).

V. EFFECT OF PLANT SPECIES AND VARIETY ON LIME–FERTILIZER INTERACTIONS

A. Differences in Tolerance to Toxic Factors

Plant species and varieties within species differ widely in tolerance to mineral stresses of toxicity or deficiency. The individual publications that document these findings are too numerous to cite here, but the review papers listed will introduce the reader to the literature—Clark, 1982a; McFee et al., 1984; Rorison, 1980; Vose, 1981, 1984; and Devine, 1982.

1. Aluminum Tolerance

a. Species and Varietal Differences—Plant species classified as Al-tolerant included azalea (*Rhododendron* spp.), Datura spp., rye, cranberry (*Vaccinium* spp.), tea *(Thea sinesis)*, weeping lovegrass [*Eragrostis curvula* (Schrad.) Nees], bermudagrass, stargrass [*Cynodon plectostachyus* (K. Schum.) Pilg.], buckwheat (*Fagopyrum esculentum* Moench), and peanut (McFee et al., 1984). Other species reported to tolerate strongly acid soils having high degrees of Al saturation are pangolagrass (*Digitaria decumen* Stent) (Blue & Rodriguez-Gomes, 1975); rubber [*Hevea brasiliensis* (Willd. ex A. Juss.) Meull.-Arg.] (Santana & Braga, 1977); blueberries (*Vaccinium* spp.) (Cummings, 1978); and Norway spruce (*Picea abies* (L.) Karst.] (Ogner & Tiegen, 1980). In a study of five species, Tanaka and Hayakawa (1975) concluded that the Gramineae and Leguminosea were more tolerant to Al than the Crucifereae and Chenopodiaceae. McCormick and Steiner (1978) reported differential Al tolerance among 11 species of trees. Differential acid soil tolerances found among species of tropical legumes (Munns & Fox, 1977) and ornamental species (Foy & Wheeler, 1979) are probably also due in large part to differential Al tolerance.

Differential Al tolerances of cultivars within species are often greater than species differences based on random cultivar comparisons. Such differences have been found among cultivars or strains of barley, wheat, triticale (X *Triticosecale* Wittmack), rice alfalfa, tomato, soybean, ryegrass, snap bean, cotton, corn, rye sunflower (*Helianthus annuus* L.), pea (*Pisum sativum* L.), sweet potato [*Ipomoea batatas* (L.) Lam.], green algae, and even soil-borne pathogens. Specific references to the older literature are available in review chapters (Foy, 1974; McFee et al., 1984).

Lopez et al. (1976) reported that the diploid and tetraploid wheats tested were highly sensitive to Al. Hexaploid wheats differed widely in tolerance, but all were injured by 25 mg Al/kg in solution. Rye varieties varied from sensitivity to 1 mg Al/kg to tolerance to 30 mg Al/kg in solution. Additional recent references to differential Al tolerances of plant genotypes within species are as follows: wheat (Magalhaes, 1979; Muzilli et al., 1978; Aniol & Kaczkowski, 1979; Sousa et al., 1977; Mugwira et al., 1981b); barley and wheat (Graves et al., 1980); triticale (Aniol & Kaczkowski, 1979; Mugwira et al., 1981b); rye (Aniol & Kaczkowski, 1979); ryegrass (Nelson & Kiesling, 1980); rice (Fageria & Zimmerman, 1979); cotton (Foy et al., 1980); sorghum (Brown & Jones, 1977a); turfgrasses (Murray & Foy, 1978); bermudagrass (Lundberg et al., 1978); sweet potato (Munn & McCollum, 1976); apple rootstocks (Kotze, 1976); soybean (Muzilli et al., 1978; Sartain & Kamprath, 1975; Devine et al., 1979); cassava (*Manihot esculenta* Crantz) (Edwards & Kang, 1978); and cacao (*Theoboroma cacao* L. subsp. *cacao*) (Garcia & Leon, 1978).

b. Genetic Control of Aluminum Tolerance—In certain barley populations, Al tolerance is controlled by one major, dominant gene (Reid, 1971). In wheat, two and possibly three major, dominant genes, plus modifiers, are believed to be involved (Lafever & Campbell, 1978; Campbell & Lefever, 1979). In corn, Al tolerance appears to be qualitatively inherited. A single gene locus with a multiple allelic series has been postulated (Rhue, 1979). Lopez-Benitez (1977) indicated that the DD genome from some hexaploid wheats and the RR genome from diploid rye are the most promising source of Al tolerance for triticale.

Hanson and Kamprath (1979) found no clear-cut genetic effects imparting Al tolerance in soybean populations.

c. Physiology of Differential Aluminum Tolerance—The exact physiological mechanisms of Al tolerance are still debated. Tolerance may be controlled by different genes (through different biochemical pathways) in different plants. Epstein (1969) pointed out that an element present in excess can interfere with metabolism through competition for uptake, inactivation of enzymes, displacement of elements from functional sites, or alteration of the structure of water. Many of these effects probably involve modification of membrane structure and function. Obviously, Al-tolerant plants must be able to prevent the absorption of excess Al or detoxify the Al after it has been absorbed. Some of the plant physiological and biochemical factors associated with differential Al tolerance are discussed below.

Aluminum tolerance has been associated with pH changes in root zones; Al trapping in nonmetabolic sites within plants, P efficiency, Ca and Mg uptake and transport, root cation exchange capacity, root phosphatase activity, internal concentrations of Si, NH_4^+ vs. NO_3^- tolerance or preference, organic acid contents, Fe efficiency, and resistance to drought. For citations from the earlier literature, see review chapters (Foy, 1974; Foy et al., 1978; Foy & Fleming, 1978; Foy, 1984; McFee et al., 1984). The present review will refer only briefly to the older literature and emphasize recent developments in understanding the differential Al tolerance of plants.

i. pH Changes in Root Zones

Certain Al-tolerant cultivars of wheat, barley, rice, peas, and corn (inbreds) increase the pH of their nutrient solutions and thus decrease the solubility and toxicity of Al (Foy & Fleming, 1978; Foy et al., 1978). In contrast, Al-sensitive cultivars of the same species decrease or have no effect on the pH of their nutrient cultures and are thus exposed to higher concentrations of Al for longer periods. In certain wheat genotypes these pH differences have been demonstrated in thin layers of soil removed directly from plant roots. In wheat there is evidence that these differential pH changing abilities are associated with differential anion-cation uptake (Foy & Fleming, 1978).

ii. Ammonium vs. Nitrate Nutrition

The ratio between NO_3^- and NH_4^+ in the nutrient solution determines the rate and direction of plant-induced pH changes in the presence or absence of Al. Superior Al tolerance in certain wheat cultivars is characterized by the ability to use NO_3^- efficiently in the presence of NH_4^+ and to increase the pH of the growth medium. For original references see Foy and Fleming (1978) and Foy et al. (1978).

iii. Aluminum Uptake and Distribution

With respect to Al accumulation, Al-tolerant plants may be divided into at least three groups. In the first group, Al concentrations in the tops are not consistently different from those of Al-sensitive plants, but the roots of tolerant plants often contain less Al than those of sensitive plants. Plants in this category include certain cultivars of wheat, barley, soybean, snap bean, triticale, and pea (Foy, 1974; Mugwira et al., 1981b; Ulmer, 1979; Klimashevski et al., 1976). In such cases, Al tolerance is apparently related to an exclusion mechanism.

In second group of plants, Al tolerance is associated with lower levels of Al in plant tops and/or entrapment of excess Al in roots. Examples reported are azalea, cranberry, rice, triticale, rye, alfalfa clones, and certain cultivars of ryegrass, wheat, barley, and potato (Foy et al., 1978; Foy, 1984). Other examples of Al-tolerant plants having lower levels of Al in their tops are: tomato genotypes (Baumgartner et al., 1976); populations of paper birch (*Betula nigra* L.) (Steiner et al., 1980); and kikuyugrass (*Pennisetum clandestinum* Hochst, ex Chiov.) compared with Al-sensitive cabbage and lettuce (Huett & Menary, 1980).

In a third group of plants, Al tolerance is directly associated with Al accumulation by the tops; hence, such plants must have a high internal tolerance to Al. Examples of Al accumulators are tea, certain Hawaiian grasses, pine trees, and mangrove (*Rhizophora mangle* L.) (Foy et al., 1978). Memon et al. (1981) concluded that in the Al-tolerant tea plant, Al is bound to cell walls of epidermal and mesophyll cells and thereby prevented from reaching critical metabolic sites within the cell. Naidoo (1976) attributed the superior Al tolerance of 'Dade' snap bean (compared with

'Romano') to a greater ability to tolerate Al within the root cell nuclei, where Al was presumably associated with esteric P and membrane lipids.

iv. Calcium Nutrition

Aluminum tolerance in certain cultivars of wheat, barley, soybean, and snap bean has been associated with the ability to resist Al-induced Ca deficiency (Foy et al., 1978). Baumgartner et al. (1976) found that the Al-tolerant 'Santa Cruz Kada' tomato had a lower Ca requirement than the less tolerant 'Ronita' tomato. Aluminum-tolerant varieties of soybeans accumulated more Ca, P, and Mg in their tops per unit of root length than did the more sensitive cultivars (Sartain, 1974).

v. Phosphorus Nutrition

In many plants, Al tolerance appears to be closely related to P use efficiency. For example, in certain cultivars of wheat and tomato and inbred lines of corn, Al tolerance coincides with the ability to tolerate low P levels in nutrient solutions, either in the presence or absence of Al. For original citations, see recent reviews (Foy et al., 1978; Foy, 1984).

Cambrai and Calbo (1980) reported that pretreatment with Al decreased P uptake by 42% in the roots of an Al-sensitive sorghum cultivar ('CMS × S-903') but did not affect that of the Al-tolerant 'CMS × S-106' cultivar. In both cultivars, Al reduced the accumulation of cations in the tops and roots of both cultivars, and because ATPase was regarded as being involved in the active transport of cations, it was suggested that Al could be affecting plants primarily by inhibiting the ion carriers. Lowe and Bortner (1973) suggested that excess Al uptake from an acid soil accentuate a genetically controlled P metabolism problem in tobacco, manifested as leaf spotting.

vi. Aluminum Related to Iron, Magnesium Silicon, and Potassium

Acid soil sensitivity in certain wheat and barley cultivars has been associated with Al-induced Fe deficiency chlorosis in nutrient cultures at pH 4.1 (Otsuka, 1969). Brown and Jones (1977a) found that certain sorghum lines grew poorly and developed chlorosis and purple pigmentation in Al-toxic Bladen soil at pH 4.3. Chlorotic lines contained significantly less Fe and P than those that remained green. Aluminum tolerance has been associated with greater uptake of K and Mg in Irish potato cultivars and with greater Mg uptake in corn inbred lines. Certain Al-tolerant rice cultivars accumulated higher levels of Si in the epidermal cells of their leaves than did Al-sensitive cultivars. Silicon is known to reduce the toxicity of Mn in barley leaves and could play a similar role in detoxifying Al. For original references, see Foy et al. (1978).

vii. Organic Aluminum Complexes in Plants

Jones (1961) suggested that naturally occurring organic acids in Al-tolerant species chelate Al and thereby reduce the Al–P precipitation expected at normal pH levels in plant sap. Small (1946) noted that acid-tolerant plants generally have strong organic acid buffer systems in their cells, but alka-tolerant plants have phosphate buffers dominating the system. The phosphate systems would be more susceptible to disruption by Al–P interaction than the organic acid systems. Aluminum tolerance in certain calcifuge species has been attributed to a chelating mechanism that also has an affinity for Fe (Grime and Hodgson, 1969). Lee (1977) noted that hydroxy Al was toxic to corn seedlings, but Al citrate was not. Klimashevski and Chernysheva (1980) found that Al-tolerant varieties of pea, corn, and barley contained substantially higher concentrations of citric acid than did Al-sensitive varieties of the same species.

viii. Aluminum-Water Relationships

Excess Al can reduce water use by restricting plant root penetration and proliferation in acid subsoils. In addition, Al toxicities can damage roots to the extent that they cannot absorb adequate water, even in moist soils. In preliminary studies, Krizek and Foy (1981) found that the Al-sensitive 'Kearney' barley (developed in the Kansas-Nebraska region) was more tolerant to soil water stress, per se, than was the Al-tolerant 'Dayton' variety (developed in Ohio). However, Kearney was much more sensitive to Al-induced water stress than was Dayton. When plants are grown over acid, Al-toxic subsoils, both types of water stress tolerance are needed to prevent yield reductions from drought.

2. Manganese Tolerance

a. Species and Varietal Differences—Plant species and cultivars differ widely in their tolerances to excess soluble or exchangeable Mn. Corn and rice are reported to be more tolerant than lespedeza, soybean, or barley. Alsike clover (*Trifolium hybridum* L.) and oats are more tolerant than cowpea, lespedeza, and sweet clover (*Melilotus alba* Medik.). The ornamental plants calendula (*Calendula officinalis* L.), snapdragon (*Antirrhinum majous* L.), and *Chrysanthemum* sp. are sensitive to excess Mn and have been suggested as indicator plants. Carnation, poinsettia (*Euphorbia pulcherrima* Willd. ex Kl.), and rose (*Rosa* sp.) are classified as Mn-tolerant. Other species rankings according to Mn tolerance are: tomato > lettuce > barley and bean > clover > potato. For specific references see Foy, 1973; Foy et al., 1978; and McFee et al., 1984. Berg and Vogel (1976) observed that several plant species differed in tolerance to Mn toxicity in acid coal mine spoils. Common lespedeza [*Lespedeza bicolor* Turcz.) and Korean lespedeza developed Mn toxicity symptoms at pH < 5.0 and sometimes at pH 5.0 and 5.4. Hutton et al. (1978) reported differences in Mn tolerance among species of tropical legumes.

As with Al tolerance, any ranking of plant species according to Mn tolerance depends on the cultivars selected to represent the species. Differential Mn tolerance has been reported within several species, including wheat (Andrade et al., 1976), apple (Miller & Schubert, 1977), triticale (Mugwira et al., 1981a), soybean (Brown & Jones, 1977b; Jones & Nelson,

1979; Heenan & Campbell, 1980; Ohki et al., 1980), cotton (Foy et al., 1981), and flax (Moraghan & Ralowicz, 1979).

b. Genetic Control of Manganese Tolerance—Manganese tolerance in alfalfa has been attributed to additive genes having little or no dominance (Dessureaux, 1959). In lettuce, Mn tolerance is reportedly controlled by one to four genes, depending on the particular species of lettuce involved (Eenink & Garretsen, 1977). In soybean, the control of tolerance to excess Mn appeared to be multigeneic, rather than controlled by a one-gene locus (Brown & Devine, 1980). Reciprocal differences in progeny suggested that cytoplasmic inheritance influenced Mn tolerance.

c. Physiology of Differential Manganese Tolerance—Manganese tolerance in plants has been associated with oxidizing powers of plant roots, Mn absorption and translocation rate, Mn entrapment in nonmetabolic centers, high internal tolerance to excess Mn, and the uptake and distribution of Si and Fe. For specific references, refer to previous review chapters (Foy et al., 1978; Foy, 1973, 1984; McFee et al., 1984). The present review will refer only briefly to the older work and emphasize any new findings concerning the mechanism of differential Mn tolerance.

i. Manganese Uptake and Transport

Plant tissue Mn concentrations required to produce Mn toxicity symptoms vary with plant species and genotype within species and with environmental conditions, particularly temperature (Foy et al., 1978). Hence, *critical* Mn concentrations in plants are useful only when applied within a limited set of circumstances. Internal Mn concentrations associated with toxicity symptoms are as follows: apple leaves, 120 to 600 mg/kg (Fisher et al., 1977; Scibisz & Sadowski, 1979); Bragg soybean, 171 to 181 mg/kg (Mask & Wilson, 1978); young leaves of tomato, 450 to 500 mg/kg and old leaves, 900 to 1000 mg/kg (Ward, 1977); sweet sorghum, 445 mg/kg in upper leaves and 1440 mg/kg in lower leaves (Malavolta et al., 1979); flax genotypes, 500 mg/kg (Moraghan, 1979); cotton genotypes, 500 to 1959 mg/kg (Foy et al., 1981; Hati et al., 1979); Easter lily *(Lilium longiflorum* var. *eximium)*, 2000 mg/kg (Holmes & Coorts, 1980); 'Coral' carnation, 2600 mg/kg (Ishida et al., 1977); and corn, 2500 to 6500 mg/kg (Tanner, 1977).

The difficulty of assigning critical internal Mn concentrations for toxicity is seen in the following example: Mask and Wilson (1978) reported that Bragg soybean had toxicity symptoms when leaves contained 171 to 181 mg Mn/kg; however, Jones and Nelson (1978) found that Bragg showed no toxicity symptoms when the leaves contained 320 mg Mn/kg.

Hutton et al. (1978) concluded that Mn tolerance in tropical legumes is associated with reduced Mn uptake in some species and greater internal tolerance in others. Bromfield (1978) noted that rape was more efficient than oats in absorbing Mn from MnO_2. Rape is also more sensitive to excess Mn than oats, and this may be related to greater Mn uptake. Superior Mn tolerance in corn (compared with peanut) has been attributed to reduced transport of Mn from roots and stems to leaves; Mn-sensitive peanut accumulates high levels of Mn in its leaves (Benac, 1976).

In some cases, differential Mn tolerances of plant genotypes within species have also been associated with differential Mn uptake. For example, Holmes and Coorts (1980) found that 'Ace' Easter lily was much more tolerant to excess Mn and absorbed less Mn than the 'Nellie White' cultivar. Gammon and McFadden (1979) reported that Mn accumulation by rose plants on *Rosa fortumana* root stocks was five times that of plants on *R. odorata* rootstocks. However, Mn was not related to flower production.

ii. Internal Tolerance to Excess Manganese

Some cultivars of wheat, cotton, soybean, and lettuce tolerate higher levels of Mn within their tops than do others (Foy et al., 1978). For example, the differential Mn tolerance of Bragg (sensitive) and 'Lee 68' (tolerant) soybean cultivars was not reflected in Mn concentrations of plant tops (Jones & Nelson, 1978). Heenan and Campbell (1981) found that Mn-tolerant Lee and Mn-sensitive Bragg soybean cultivars did not differ in Mn distribution to plant parts or in Mn concentrations in actively growing tissues. Brown and Devine (1980) reported that Mn-tolerant and Mn-sensitive soybean cultivars and F_2 progeny from their reciprocal crosses all contained about the same Mn concentration in their tops (500 mg/kg). Similar findings have been made in cultivars of wheat (Foy, 1973; Brauner, 1979), cotton (Foy et al., 1981) and flax (Moraghan & Ralowicz, 1979).

iii. Interactions of Manganese With Other Mineral Elements

Manganese has been reported to interact with Fe, Mo, P, Ca, and Si in affecting toxicity symptoms and growth. In some plants, excess Mn produces symptoms that resemble Fe deficiency and these can be prevented by adding soluble Fe to the plant or soil. In others, Mn-induced symptoms are quite different (see Foy et al., 1978 and preceding sections of this chapter) and are not corrected by adding Fe. Increased Ca in the growth medium may decrease Mn uptake and toxicity. Phosphorus additions can reduce the toxicity of Mn by rendering it inactive within the plant. Silicon is known to reduce Mn toxicity by preventing localized accumulations of Mn in leaves. From the foregoing evidence, it appears that the toxicity of a given level of soluble Mn in the growth medium, or even within the plant, depends on interactions between Mn and several other mineral elements, particularly Fe and Si. Such interactions may account for the wide variety of plant symptoms and different growth reductions produced by excess Mn in different plant species and cultivars. For details on the older literature see review chapters (Foy, 1973; Foy et al., 1978; McFee et al., 1984).

iv. Organic Manganese Complexes in Plants

There has been little effort to relate organic Mn complexes to Mn tolerance in plants, but this appears to be a promising area for research. Sutcliffe (1962) suggested that organic complexes can reduce the transport of

Mn to plant tops. Tiffin (1967) found that Mn in tomato xylem exudate showed no binding to organic compounds (moved as a cation under electrophoresis), even when the exudate Mn content was enriched by 18-fold. However, Hofner (1970) found a close relationship between Mn and an amino acid and carbohydrate fraction of sunflower stem exudate. White et al. (1981) stated that Mn was bound primarily by citric and malic acids in tomato stem exudates. The overall evidence, based on widely different plants, indicates that Mn occurs as a free cation or bound in soluble form with low molecular weight complexes in xylem fluid. However, the significance of this in relation to Mn tolerance is unknown.

3. Tolerance to Other Metal Toxicities

Some of the mechanisms by which plants may avoid metal toxicity are: exclusion of the metal by precipitation in the growth medium or on plant root surfaces; entrapment of excesses in cell walls or vacuoles; chelation and detoxification by organic acids or other complexes in plant sap; and alteration of enzyme structure, which may reduce metabolic susceptibility. For original references see Foy et al. (1978) and Foy (1984).

B. Differences in Requirement for Essential Elements

1. Calcium Requirement

Loneragan et al. (1970) defined Ca requirement in three parts as follows:

a. Solution Calcium Requirement—The minimal Ca concentration permitted in solution for maximum growth rate.

b. Functional Calcium Requirement—The minimum Ca concentration required at functional sites within plant tissues to sustain maximum growth rate.

c. Critical Calcium Concentration—The Ca concentration actually present in the plant or its organs at the time Ca becomes limiting for growth.

These investigators pointed out that because Ca has low mobility between the organs of plants, the critical concentration can vary greatly with the conditions under which the deficiency is produced and may have little relation to the *functional requirement.* They regarded growth rate and Ca uptake rate as important factors determining Ca requirement in solution and critical Ca requirement in plants. For further details regarding the concept of *nutrient element requirement,* see review chapters (Foy, 1974; Clark, 1982).

Differential Ca uptake and Ca deficiency symptoms observed in plants grown on acid soils may be the results of differential Al–Ca, Mn–Ca, or other cationic antagonisms, rather than an absolute shortage of Ca in the soil or in the plants. Nevertheless, certain plant species and varieties also appear to differ in absolute Ca requirement, even in the absence of excess

Al or other antagonistic cations. For example, Snaydon and Bradshaw (1962) found that crimson clover (*Trifolium incarnatum* L.) required more Ca than alsike clover and that red and white clover (*T. repens* L.) were intermediate in Ca requirement. Loneragan and Snowball (1969a,b) showed that tomato depleted standard nutrient solutions of Ca 10 to 15 times as rapidly as wheat, even when both had the same root weights. Legumes and herbs absorbed Ca faster and appeared to have a higher functional requirement for Ca in their tissues than did grasses and cereals. The lupine species were the only legumes having functional Ca requirements as low as those for grasses and cereals.

Andrew (1976) reported that *Macroptilium lathyroides* (L.) Urb. and *Lotononis bainesii* Miles were more tolerant to low Ca than Spanish tickclover [*Desmodium uncinatum* (Jacq.) DC.] and Kenya clover (*T. semipilosium* Fresen.). European yellow lupine (*Lupinus luteus* L.) is reported to accumulate more Ca than horse bean (*Vicia faba* L.) (Rossignol & Grignon, 1980). Algae require less Ca than angiosperms (Gerloff & Fishbeck, 1969). When grown on a Ca-deficient soil, orchardgrass (*Dactylis glomerata* L.) accumulated more Ca than perennial ryegrass (*Lolium perenne* L.) (Forbes & Gelman, 1981).

Plant varieties (inbred lines, genotypes, etc.) within species also differ in Ca requirement, Ca uptake, and yield response to added Ca. Examples of species showing such differences are as follows: sheep fescue (*Festuca ovina* L.) (Snaydon & Bradshaw, 1961, 1962); perennial ryegrass (Forbes & Gleman, 1981); white clover (Snaydon, 1962); red clover (Hunt et al., 1976); tobacco (McEvoy, 1963; Peedin & McCants (1977); barley (Young & Rasmussen, 1966); cabbage (Maynard et al., 1965); cauliflower [*Brassica oleracea* (Betrytis Group)] (Maynard et al., 1981); corn hybrids (Bradford et al., 1966; Andrade et al., 1975; Furlani et al., 1977); corn inbreds (Porter & Moraghan, 1975; Clark, 1978); tomato (Greenleaf & Adams, 1969; De Brito Giordano, 1980; English & Maynard, 1981); snap bean (Shannon et al., 1967); peanut (Beringer & Taha, 1976; Bathla et al., 1978; Walker & Kiesling, 1978); apple root stocks (Poling and Oberly, 1979); potato (Dyson & Digby, 1975); pearl millet (Oza et al., 1979); and strawberry [*Fragaria chiloensis* (L.) Duch.] (Cuttridge et al., 1978).

Calcium is rather immobile in both soils and plants. Species and varietal differences in Ca requirement have been associated with differences in both uptake and translocation within the plant. Snaydon and Bradshaw (1961) reported that differences in Ca response among populations of sheep fescue were due to differences in Ca-absorbing ability, rather than differences in Ca metabolism at low internal Ca levels in plants. Greenleaf and Adams (1969) found that tomato lines that were more resistant to blossom end rot (Ca deficiency) absorbed and accumulated Ca more effectively in the fruit than did the more susceptible lines. De Brito Giordano (1980) reported that some Ca-efficient tomato lines were more effective than others in depleting nutrient solutions of Ca; others made better use of absorbed Ca, as shown by higher yields from similar Ca uptake. The excellent performance of one line appeared to involve both factors, superior Ca uptake and Ca use efficiency. With low Ca, the efficient line produced twice as much dry weight as inefficient lines. At adequate Ca levels, the efficient and inefficient lines produced comparable dry weights.

Susceptibility to internal tipburn (Ca deficiency) in cabbage varieties has been related to a lower efficiency in Ca uptake and transport from basal and wrapper leaves to head leaves (Maynard et al., 1965). Resistance to tipburn in cauliflower cultivars was related to higher Ca levels in leaves (Maynard et al., 1981). Differential susceptibility of tobacco varieties was associated with differential distribution of Ca to apical regions (Peedin & McCants, 1977). In peanut, resistance to Ca deficiency was related to greater translocation of Ca from shells to nuts (Beringer & Taha, 1976). Clark (1978) found that the Ca-efficient corn inbred 'Oh 43' had fewer Ca deficiency symptoms and produced more dry matter per unit of Ca absorbed than did the Ca-inefficient 'A251' when both were grown at low Ca levels. Both functional and critical Ca concentrations were lower in Oh 42 than in A251. Kawaski and Moritsugu (1979) concluded that differential Ca-deficiency symptoms in corn and sorghum cultivars were due to differential Ca-absorbing abilities of roots.

Barrelclover (*Medicato truncatula* Gaertn.), a tropical legume, appears to require a higher internal Ca concentration for maximum yield than does subterranean clover (Loneragan et al., 1970). The higher Ca requirement of the former was associated with less effective distribution of Ca in apical tissue, plus a higher growth rate. Rossignol and Grignon (1980) found that higher Ca absorption by the calcifuge, yellow lupine, (compared with the calcicole, horse bean) was associated with higher contents of acidic phospholipids and saturated fatty acids.

2. Magnesium Requirement

Plant species and varieties within species differ significantly in their uptake and use of Mg. Two species reported to require rather high levels of Mg in their growth media are tobacco and citrus (Adams & Pearson, 1967). Adams (1969) found that on Norfolk and Magnolia soils corn was more tolerant to both low soil pH and low soil Mg level than was cotton. 'Lotar' orchardgrass and 'Fawn' tall fescue contained 0.20% Mg, whereas wheatgrass forage (*Agropyron* spp.) contained only 0.11% Mg (Fairbourn & Batchelder, 1980). Walker and Graffis (1979) also found that grass species and cultivars differed in Mg uptake. Weeping lovegrass is more effective than tall fescue in absorbing Mg in the presence of Al (Fleming et al., 1974).

Differential Mg uptake and/or use has been reported for cultivars, inbred lines, genotypes, or populations within many plant species. Examples are as follows: corn varieties (Andrade et al., 1975; Broersma & Van Ryswyk, 1979; Agarwala & Mehrotra, 1979; Gallagher et al., 1981); corn inbreds (Foy & Barber, 1958; Clark, 1976, 1982); hops (*Humulus lupulus* L.) (Marocke et al., 1979); wheat (C. D. Foy, 1980, unpublished data; Karlen et al., 1978); perennial ryegrass (Lane & Lamp, 1980); tall fescue (Brown & Sleper, 1980; Nguyen & Sleper, 1981; Currier et al., 1981); and orchardgrass (Stratton & Sleper, 1979). For further information regarding Mg contents of forages in relation to grass tetany see Mayland and Grunes

(1979). Celery varieties that are susceptible to Mg deficiency accumulate as much Mg in root and heart tissue as the more resistant varieties, but less Mg in the stem plate, petiole, and blade tissue (Pope & Munger, 1953). Susceptibility to Mg deficiency in corn inbred lines has been associated with reduced Mg transport to the leaves (Foy & Barber, 1958; Schauble & Barber, 1958; Clark, 1976, 1982). Gallagher et al. (1981) found that corn hybrids that were most efficient in accumulating Mg in their leaves were least efficient in using Mg to produce dry matter yield.

3. Phosphorus Requirement

Plant species differ widely in their abilities to absorb and use P, but species rankings depend on the definition of P requirement or P efficiency. Blair and Cordero (1978) used four criteria to compare the P efficiencies of three species grown in soil and nutrient cultures. The criteria were: (i) top dry weight/unit P applied; (ii) top dry weight per unit of P absorbed; (ii) top dry weight at a constant P level; and (iv) P uptake per unit of root weight. The order of P efficiency among the species varied with the criteria used. These investigators emphasized that the term *P efficient species* must be clearly defined when used.

One measure of *P requirement* or *P efficiency* is the ability to survive and produce a reasonable amount of growth at low levels of soluble P in soils or other media. Cassman et al. (1981) reported that cowpea is much more tolerant to low P stress than soybean, especially when dependent on symbiotic N_2 fixation. As a group, grasses are more tolerant to low-P soils than are legumes. Grasses also appear more effective in translocating absorbed P to plant tops. Legumes retain more P in their roots when grown in low-P soils (Caradus, 1980). Oats are more tolerant to low P levels than barley (Oza et al., 1976).

A review by Salinas and Sanchez (1976) shows that plant species differ in P concentrations required in soil solution and within the plant to achieve high yields. Annual crops classified as most tolerant to low levels of P included rice, sweet potato, and corn. Among tropical pasture species, Townsville stylo, *Centrosema pubescens* Benth., molasses grass (*Melinis minutiflora* Beauf.), and guineagrass *Panicum maximum* Jacq.) were more tolerant to low P than *Glycine wightii* (Grah, ex Wight & Arn.) Verdc., alfalfa, rhodesgrass (*Chloris gayana* Kunth), and dallisgrass (*Paspalum dilatatum* Poir.). Species and varieties tolerant to low P generally produce maximum yields at lower levels of applied P. In some species, tolerance to low P coincides with tolerance to low P generally produce maximum yields at lower levels of applied P. In some species, tolerance to low P coincides with tolerance to Al in acid soils. Fox (1979, 1981) reported that tomato, head lettuce, and soybean require higher P levels in the soil solution (for 95% of maximum yield) than do cassava, peanut, cabbage, corn, or sorghum. In other comparisons, lettuce was much more sensitive than corn. External P requirements in solution were 0.3, 0.03, and 0.003 mg/kg for lettuce, corn, and cassava, respectively. Loneragan (1978) found that at low P levels the roots of sub clover (*Trifolium subterraneum* L.) absorbed

P nearly twice as rapidly as those of lupine (*Lupinus* spp.), even when equal root weights were used. However, at high P levels the lupine absorbed P twice as rapidly as subclover.

Picha and Minotti (1977) reported that tomato had a higher P requirement for optimum growth and a higher P concentration in its tissues than did lambsquarters (*Chenopodium album* L.). Vander Zaag et al. (1980) found that greater yam (*Dioscorea alata* L.) had a higher P requirement and produced higher yields than lesser yam [*D. esculenta* (Lour.) Burk] and white yam (*D. roundata* Poir.). All three species were effective in using low P concentrations in soil solutions. *Andropogon scoparius* (Michx.) *(Schizachyrium scoparium)* was much more tolerant to a low P soil than Kentucky bluegrass (*Poa pratensis* L.), because it absorbed more P and used absorbed P more effectively (Wuenscher & Gerloff, 1971).

Plant genotypes within species also differ widely in P uptake and use. Species in which such differences have been reported include the following: sorghum (Brown & Jones, 1975; Brown et al., 1977; Salako, 1980; Maranville et al., 1980; Clark, 1982); bean (Whiteaker et al., 1976; Lindgren et al., 1977; Gomes & Braga, 1980); cowpea (Safaya & Singh, 1977); orchardgrass (Davis, 1978); white clover (Caradus & Dunlop, 1978); rice (IRRI, 1979; Mahadevappa et al., 1979); pearl millet [*Pennisetum americanium* (L.) Leeke] (Oza et al., 1979); wheat (Palmer & Jessop, 1977); tomato (Hochmuth et al., 1978); barley (Oza et al., 1976); and corn (Clark & Brown, 1974; Fox, 1981). Gerloff (1976) classified snap bean lines into four categories with respect to P response: (i) inefficient responders—those that had low yield at low P and high yield at high P, (ii) efficient nonresponders—those that had the highest yield at low P and not much higher yield at high P, (iii) efficient responders—those that produced among the highest yields at low P and also at high P levels, and (iv) inefficient nonresponders—those that had poor yield with both low and high P levels.

Some plants are so efficient in absorbing P that they accumulate toxic concentrations in their tops. Examples are: several species of the Proteaceae family (Nichols & Beardsell, 1981); certain eucalyptus populations (*Eucalyptus* spp.) (Ladiges, 1977), some soybean varieties (Lee et al., 1970) and wheat (Bhatti & Loneragan, 1970).

The mechanisms involved in differential uptake and use of P by plants may vary in different species and genotypes. Loneragan (1978) associated differential P uptake with three root characteristics: (i) ability to absorb P from dilute solutions, (ii) metabolic activities that may increase the solubility of soil P, and (iii) ability to explore the soil volume. He also noted that all of these root properties can be altered by root-mycorrhizal associations. In this connection, Fox (1979) mentioned that onion apparently needs mycorrhiza for P efficiency and that cowpea plants are more P efficient with mycorrhiza than without. Yam plants also appear to depend on mycorrhiza for adequate P uptake (Vander Zaag et al., 1980). In contrast, Crucifereae tolerate low P levels and do not have mycorrhiza.

Salinas and Sanchez (1976) found that differential tolerance to low P may involve differences in root extension, root exudation, action of mycorrhizal fungi, and P absorption and translocation in relation to growth rate. Rapid translocation of absorbed P appears to be an important part of P efficiency. In this regard, Lonergan (1978) stated that the transport of P from roots to shoots is more sensitive to several environmental factors than is P absorption by roots. Loughman (1977) also emphasized the role of P transport in P efficiency. The ability to absorb P apparently is present within a few hours after the radicle emerges, but the ability to translocate P to plant tops may be delayed for several weeks in some species. For example, Loughman (1977) found that pea and field bean plants absorbed P effectively during the early weeks of growth, but the rate of transport to plant tops was only about 30% of that for oats, barley, mung bean [*Vicia radiata* (L.) Wilczek var. *radiata*], or sunflower.

Semidwarf wheat absorbed more P per unit of fresh root weight and gave a greater grain yield response to superphosphate than did standard height wheat (Palmer & Jessop, 1977). It was suggested that the superior P response of the semidwarf variety was due to a greater root number, root area, and lateral root volume per unit of root fresh weight.

Gerloff (1976) reported that 15 snap bean varieties differed by 100% in P uptake per unit root weight. The most efficient P absorber had the smallest roots (total lowest weight). However, some strains having equal root weights showed wide differences in P uptake per unit root weight.

Bieleski (1971, 1973) suggested that the induction of phosphatase enzymes in plant roots by low P levels in soils may enable a particular plant genotype to survive where P availability depends on the breakdown of organic P forms. Woolhouse (1969) cited evidence that even in the absence of microorganisms in the root zone, higher plants can use phytates and other organic P compounds probably by the action fo phosphatase enzymes on root surfaces. The same investigator presented evidence that P metabolism is involved, at least in part, in the adaptive mechanism of colonial bentgrass (*Agrostis tenuis* Sibth.) populations in different soil types. It was suggested that increased P-feeding powers could result from alterations in the structure and function of phosphatase enzymes, which are induced by low P levels or other conditions in the growth medium. Clark and Brown (1974) found that P stress increased the phosphatase activities in two corn inbreds and that this increase was greater in the P-efficient than in the P-inefficient inbred. The P-efficient line also lowered the pH of nutrient solutions more rapidly than did the P-inefficient line. Both factors could have increased P availability.

'Lincoln' and 'Clark' soybean varieties absorb more P from solutions and are more sensitive to P toxicity than Chief and L-9 varieties. Sensitivity to excess P was associated with lower rates of sugar phosphorylation and protein synthesis (Lee et al., 1976).

C. Fitting the Plant to the Soil

Mineral stress in acid and other problem soils is not always economically correctable with current technology. An alternative or supplemental approach is to fit plants more precisely to soils. This approach does not propose the elimination of liming and fertilization nor the depletion of soil fertility to low levels. Instead, it advocates a more effective use of plant genetic diversity in solving some of the more difficult problems of soil fertility. The results will be plant genotype-soil-lime-fertilizer combinations that may be superior to medical "prescriptions" for people. Within recent years, teams of plant and soil scientists have made considerable progress in the development and release of stress-tolerant germplasm. Examples are described in the following paragraphs.

Certain wheat varieties of Brazil have very high tolerance to Al as a result of being selected on strongly acid, Al-toxic soils over a period of about 50 yr (Silva, 1976). Moderately Al-tolerant 'Titan' wheat has been released for use in eastern Ohio where Al toxicity in acid subsoils is a problem (Lafever, 1978, 1979). Titan wheat is not as tolerant to Al as the Brazilian wheats (like 'BH 1146'), but it is much more tolerant than varieties like 'Sonora 63'. Camargo et al. (1980) reported a source of Al tolerance in short-strawed wheats ('Tordo' and 'Siete Cerros') and suggested genetic approaches for using this trait. Hence, Al-tolerant varieties of the future may not need to be tall-strawed and subject to lodging with high N fertilization.

Two Al-tolerant, drought-tolerant sorghum cultivars have recently been released in Brazil (R. Schaffert et al., 1981, Sete Lagoas, M. G., Brazil, personal communication). Duncan (1981) developed an acid-tolerant, presumably Al-tolerant, sorghum population, 'GPIR'.

Reid et al. (1980) released an Al-tolerant barley population for experimental purposes. This composite cross (XXXIV) contains sources of Al tolerance from a world collection of germplasm.

In alfalfa, recurrent phenotypic selection was effective in increasing tolerance to an acid, Al-toxic Bladen soil. Aluminum tolerances of the resulting tolerant and sensitive populations were confirmed in acid, Al-toxic Tatum soil (Typic Hapludults) (Devine et al., 1976). Bouton and Sumner (1981) found that an alfalfa population developed by recurrent selection on an acid (pH 4.4), Al-toxic soil produced significantly higher yields on an acid soil at pH 4.8 than did a population selected on limed soil at pH 6.5 or 'Apollo' alfalfa used as a check.

A cold-tolerant strain ('PI 364344') of limpo grass [*Hemarthria altissima* (Poir.) Stapf & C. E. Hubb.] has exceptional tolerance to acid soils and Al; hence, it has potential for use on acid, high-altitude minespoils (Oakes & Foy, 1984).

REFERENCES

Adams, F. 1969. Response of corn to lime in field experiments in Alabama soils. Auburn Univ. Exp. Stn. Bull. 391.

Adams, F. (ed.). 1984a. Crop response to lime in the southern United States. *In* Soil acidity and liming, 2nd ed. Agronomy 12:211–165.

Adams, F. (ed.) 1984b. Soil acidity and liming, 2nd ed. Agronomy 12.

Adams, F., and J. B. Henderson. 1962. Magnesium availability as affected by deficient and adequate levels of potassium and lime. Soil Sci. Soc. Am. Proc. 26:65–68.

Adams, F., and Z. F. Lund. 1966. Effect of chemical activity of soil solution aluminum on cotton root penetration of acid subsoils. Soil Sci. 101:193–198.

Adams, F., R. W. Pearson, and B. D. Doss. 1967. Relative effects of acid subsoils on cotton yields in field experiments and on cotton roots in growth chamber experiments. Agron. J. 59:453–456.

Adams, F. and J. I. Wear. 1957. Manganese toxicity and soil acidity in relation to crinkle leaf of cotton. Soil Sci. Soc. Am. Proc. 21:305–308.

Agarwala, S. C., and N. K. Mehrotra. 1979. Relative susceptibility of some maize varieties to magnesium deficiency in sand culture. Indian J. Plant Physiol. 22:9–13.

Alam, S. M., and W. A. Adams. 1979. Effects of aluminum on nutrient composition and yield of oats. J. Plant Nutr. 1:365–375.

Alam, S. M., and W. A. Adams. 1980. Effects of aluminum upon the growth and nutrient composition of oats. Pak. J. Sci. Ind. Res. 23(3–4):130–135.

Alexander, M. 1961. Introduction to soil microbiology. John Wiley & Sons, Inc., New York.

Alexander, M. 1964. Nitrification. *In* M. V. Bartholomew and F. E. Clark (ed.) Soil nitrogen. Agronomy 10:307–343.

Alexander, M. 1980. Effects of acidity in microorganisms and microbial processes in soils. p. 363–374. *In* T. Hutchinson and M. Havas (ed.) Plenum Publishing Corp., New York.

Alijakhro, A. 1980. Zinc deficiency: A problem growing fast. Pakistan Agriculture. 3 October, p. 35–37.

Allaway, W. H. 1945. Availability of replaceable Ca from different types of colloids as affected by degree of Ca saturation. Soil Sci. 59:207–217.

Allen, V. G., and D. L. Robinson. 1980. Occurrence of aluminum and manganese in grass tetany cases and their effects on the solubility of calcium and magnesium in vitro. Agron. J. 72:957–960.

Alley, M. M. 1981. Short-term soil chemical and crop yield response to limestone applications. Agron. J. 73:687–689.

Amarasiri, S. L., and S. R. Olsen. 1973. Liming as related to solubility of P and plant growth in an acid tropical soil. Soil Sci. Soc. Am. Proc. 37:716–721.

Andrade, A. G., H. P. De Haag, G. D. Oliveira, and J. R. De Sarruge. 1975. The differential accumulation of nutrients by five cultivars of maize (*Zea mays* L.). I. Accumulation of macroonutrients. An. Esc. Super. Agric. "Luiz de Queiroz" Univ. Sao Paulo 32:115–149.

Andrade, J. M. V., F. I. F. Carvalho, and J. Mielmczuk. 1976. Identification and selection in the greenhouse of genotypes of wheat (*Triticum aestivum* L.) tolerant of aluminum and manganese with modifications of the chemical characteristics of the soils. Cienc. Cult. (Maracaibo) 28:760–761.

Andrew, C. 1976. Effect of calcium, pH and nitrogen on the growth and chemical composition of some tropical and temperature pasture legumes. I. Nodulation and growth. Aust. J. Agric. Res. 26:611–623.

Andrew, C. S. 1978. Mineral characterization of tropical forage legumes. p. 93–112. *In* C. S. Andrew and E. J. Kamprath (eds.) Mineral Nutrition of legumes in tropical and subtropical soils. Commonwealth Scientific and Industrial Research Organization (CSIRO), East Melbourne, Victoria, Australia.

Andrew, C. S., and W. H. J. Pieters. 1976. Foliar symptoms of mineral disorders in *Glycine wightii*. CSIRO Aust. Div. Trop. Agron. Tech. Paper 18. Commonwealth Scientific and Industrial Research Organization (CSIRO), East Melbourne, Victoria, Australia.

Animesh, C. R., and A. K. M. Shahjahan. 1980. Zinc deficiency—a new problem for irrigated rice in Bangladesh. International Rice Research Newsletter. 3:8–9 International Rice Research Institute, Los Baños, Philippines.

Aniol, A., and J. Kaczkowski. 1979. Wheat tolerance to low pH and aluminum. Comparative aspects. Cereal Res. Commun. 7:113–122.

Anonymous. 1980. Bragg soybeans develop abnormality. Crop Soils 32(5)18–19.

Arnon, D. I., and C. M. Johnson. 1942. Influence of hydrogen ion concentration on the growth of higher plants under controlled conditions. Plant Physiol. 17:525–539.

Attoe, O. J., and R. A. Olson. 1966. Factors affecting rate of oxidation in soils of elemental sulfur and that added in rock phosphate–sulfur fusions. Soil Sci. 101:317–325.

Awad, A. S., D. G. Edwards, and P. J. Mulhain. 1976. Effect of pH and phosphate on soluble soil aluminum and on growth and composition of kikuyu grass. Plant Soil 45:531–542.

Ayers, A. S. 1963. The utility of soil analysis in determining the need for applying calcium to sugar cane. Proc. Int. Soc. Sugar Cane Technol. 11:162–170.

Barber, S. A., J. M. Walker, and E. H. Vasey. 1963. Mechanisms for the movement of plant nutrients from the soil and fertilizer to the plant root. J. Agric. Food Chem. 11:204–207.

Barker, A. V., R. J. Volk, and W. A. Jackson. 1966. Root environment acidity as a regulatory factor in ammonium assimilation by the bean plant. Plant Physiol. 41:1193–9.

Barshad, I. 1951. Factors affecting the molybdenum content of pasture plants. I. Nature of soil molybdenum, growth of plants, and soil pH. Soil Sci. 71:297–313.

Bartlett, R. J., and J. L. McIntosh. 1969. pH Dependent bonding of potassium by a Spodosol. Soil Sci. Soc. Am. Proc. 33:535–539.

Bathla, R. N., S. M. Virmani, and M. L. Chaudhury. 1978. Pattern of calcium and magnesium accumulation in some varieties of groundnut (*Arachis hypogaea* L.) J. Indian Soc. Soil Sci. 26:290–296.

Baumgartner, J. G., H. P. Haag, G. D. Oliveira, and D. Perecin. 1976. Tolerance of tomato (*Lycopersicon esculentum,* Mill) cultivars to aluminum and manganese. An. Esc. Super. Agric. "Luiz de Queiroz" Univ. Sao Paulo 33:513–541.

Ben, J. R., M. Morelli, and V. Estefanel. 1976. Rev. Cent. Cienc. Rurais Univ. Fed. St. Maria 6:117–189.

Benac, R. 1976. Response of a sensitive *(Arachis hypogaea)* and a tolerant *(Zea mays)* species to different concentrations of manganese in the environment. Cah. ORSTON Ser. Biol. 11:43–51. (In French with English summary.)

Beradze, E. N. V. 1977. Some aspects of water status of plants on subtropical soil in relation to effect of mineral elements. Rudz Pedagogicheskikh Institutev Grunzinskoi SSR, Estestvoz nainza: Reperativnyi Zhurnal 6.55.153. 2:3–13.

Berg, W. A., and W. G. Vogel. 1976. Toxicity of acid coal mine spoils to plants p. 57–68. *In* Northeastern For. Exp. Stn, U.S. Forest Service, Berea, KY.

Berger, K. C., and G. C. Gerloff. 1948. Manganese toxicity of potatoes in relation to strong soil acidity. Soil Sci. Soc. Am. Proc. 12:310–314.

Berger, K. C., and P. F. Pratt. 1963. Advances in secondary and micro-nutrient fertilization. p. 287–329. *In* M. H. McVickar et al. (ed.) Fertilizer technology and usage. Soil Science Society of America, Madison, WI.

Beringer, H., and M. A. Taha. 1976. Calcium absorption by two cultivars of groundnut *(Arachis hypogaea)*. Exp. Agric. 12:1–7.

Bhatti, A. S., and J. F. Loneragan. 1970. Phosphorus concentrations in wheat leaves in relation to phosphorus toxicity. Agron. J. 62:288–290.

Bhumbla, D. R., and E. O. McLean. 1965. Aluminum in soils: VI. Changes in pH dependent acidity, cation exchange capacity, and extractable aluminum with additions of lime to acid surface soils. Soil Sci. Soc. Am. Proc. 29:370–374.

Bieleski, R. J. 1971. Recent advances in plant nutrition. p. 143–153. *In* R. M. Samish (ed.) Proc. 6th Int. Colloq. on Plant Analysis and Fert. Problems, Vol. 1. Gordon and Breach Science Publishers, New York.

Bieleski, R. J. 1973. Phosphate pools, phosphate transfer and phosphate availability. Annu. Rev. Plant Physiol. 24:225–252.

Black, C. A. 1968. Soil plant relationships, 2nd ed. John Wiley & Sons, Inc., New York.

Blair, G. J., and S. Cordero. 1978. The phosphorus efficiency of three annual legumes. Plant Soil 50:387–398.

Bloom, P. R., M. B. McBride, and R. M. Weaver. 1979. Aluminum organic matter in acid soils: Salt extractable aluminum. Soil Sci. Soc. Am. J. 43:813–815.

Bloomfield, C., and J. R. Sanders. 1977. The complexing of copper by humidified organic matter from laboratory preparations, soil and peat. J. Soil Sci. 28:435–444.

Blue, W. G., and M. Rodriguez-Gomez. 1975. Effect of sodium and lime on pangola digitgrass growth on Leon fine sand. Soil Crop Sci. Soc. Fla. Proc. 34:66–68.

Boswell, F. C., K. Ohki, M. B. Parker, L. M. Shuman, and D. O. Wilson. 1981. Methods and rates of applied manganese for soybeans. Agron. J. 73:909–912.

Bouton, J. H., and M. E. Sumner. 1981. Selection of alfalfa for yield in acid soil. Agron. Abstr. American Society of Agronomy, Madison, WI, p. 173.

Bradford, R. R., D. E. Baker, and W. I. Thomas. 1966. Effect of soil treatments on chemical element accumulation of four corn inbred lines. Agron. J. 58:614–617.

Brar, S. S., and J. Giddens. 1968. Inhibition of nitrification in Bladen grassland soil. Soil Sci. Soc. Am. Proc. 32:821–823.

Brauner, J. L. 1979. Tolerencia de cultivares de trigo (*Triticum aestivum,* L.) so aluminio e ao manganes: Suh determinacao, influencia na concentracao de nutrientes e absorcao de calcio e fosforo. Ph.D. thesis. University of Sao Paulo, Brazil.

Brenes, E., and R. W. Pearson. 1973. Root response of three gramineae species to soil acidity in an Oxisol and Ultisol. Soil Sci. 116:295–302.

Broersma, K., and A. L. Van Ryswky. 1979. Magnesium deficiencies observed in forage corn varieties. Can J. Plant Sci. 59:2:541–544.

Brogan, J. C. 1964. The effect of humic acid on aluminum toxicity. Trans. Int. Congr. Soil Sci. 8th 3:227–234.

Bromfield, S. M. 1978. The effect of manganese oxidizing bacteria and pH on the availability of manganous ions and manganese oxide to oats in nutrient solutions. Plant Soil 49:23–39.

Bromfield, S. M., and D. J. David. 1976. Sorption and oxidation of manganous ions and reduction of manganese oxide by cell suspensions of a manganese oxidizing bacterium. Soil Biol. Biochem. 8:37–43.

Brown, J. C., R. B. Clark, and W. E. Jones. 1977. Efficient and inefficient use of phosphorus by sorghum. Soil Sci. Soc. Am. J. 41:747–750.

Brown, J. C., and T. E. Devine. 1980. Inheritance of tolerance or resistance to manganese toxicity in soybeans. Agron. J. 72:898–904.

Brown, J. C., and C. D. Foy. 1964. Effect of Cu on the distribution of P, Ca, and Fe in barley plants. Soil Sci. 98:362–370.

Brown, J. C., and W. E. Jones. 1975. Phosphorus efficiency as related to iron inefficiency in sorghum. Agron. J. 67:468–472.

Brown, J. C., and W. E. Jones. 1977a. Fitting plants nutritionally to soils. III. Sorghum. Agron. J. 69:410–414.

Brown, J. C., and W. E. Jones. 1977b. Manganese and iron toxicities dependent on soybean variety. Commun. Soil Sci. Plant Anal. 8:1–15.

Brown, J. R., and D. A. Sleper. 1980. Mineral concentration in two tall fescue genotypes grown under variable soil nutrient levels. Agron. J. 72:742–745.

Burns, G. R. 1967. Oxidation of sulphur in soils. Sulphur Inst. Tech. Bull. 13. The Sulphur Institute, Washington, DC.

Camargo, C. E. O., W. E. Kronstad, and R. J. Metzger. 1980. Parent-progeny regression estimates and association of height level with aluminum toxicity and grain yield in wheat. Crop Sci. 20:355–358.

Cambrai, J., and A. G. Calbo. 1980. Efeito do aluminio sobre a absorcao e sobre o transporte de fosforo em dois cultivares de sorgho (*Sorghum bicolor* L. Moench). Rev. Ceres 27(154):615–625.

Campbell, L. G., and H. N. Lafever. 1979. Heritable and gene effects for aluminum tolerance in wheat. p. 963–977. *In* S. Ramanujam (ed.) Proc. 5th Int. Wheat Genet. Symp., New Delhi, India. 23–28 Feb. 1978.

Campbell, N. E. R., and H. Lees. 1967. The nitrogen cycle. p. 194–215. *In* A. D. McLaren and G. H. Peterson (ed.) Soil biochemistry. Marcel Dekker, Inc., New York.

Caradus, J. R. 1980. Distinguishing between grass and legume species for efficiency of phosphorus use. N. Z. J. Agric. Res. 23:75–82.

Caradus, J. R., and J. Dunlop. 1978. Screening white clover plants for efficient phosphorus use. p. 75–82. *In* A. R. Ferguson et al. (ed.) Plant nutrition. Proc. 8th Int. Colloq. on Plant Analysis and Fert. Problems, Auckland, N. Z. 28 Aug.–1 Sept. 1978. 2.3 DSIR Inf. Series 134. New Zealand Government, Wellington, New Zealand.

Cassman, K. G., A. S. Whitney, and R. L. Fox. 1981. Phosphorus requirements of soybean and cowpea as affected by mode of N nutrition. Agron. J. 73:17–22.

Chang, M. L., and G. W. Thomas. 1963. A suggested mechanism for sulfate adsorption by soils. Soil Sci. Soc. Am. Proc. 27:281–283.

Chao, T. T., M. E. Harward, and S. C. Chang. 1963. Cationic effects on sulfate adsorption by soil. Soil Sci. Soc. Am. Proc. 27:35–38.

Chao, T. T., M. E. Harward, and S. C. Fang. 1964. Iron or aluminum coatings in relation to adsorption characteristics of soils. Soil Sci. Soc. Am. Proc. 28:632–635.

Chapman, H. D. 1966. Zinc. p. 482–499. *In* H. D. Chapman (ed.) Diagnostic criteria for plants and soils. University of California Division of Agricultural Science, Riverside, CA.

Chase, F. E., C. T. Corke, and J. B. Robinson. 1968. Nitrifying bacteria in soil. p. 593–611. *In* T. R. G. Gray and D. Parkinson (ed.) The ecology of soil bacteria. University of Toronto Press, Ontario, Canada.

Cherney, J. H., D. L. Robinson, L. C. Kappel, F. G. Hembry, and R. H. Ingraham. 1983. Soil contamination and elemental concentration sof forages in relation to grass tetany. Agron. J. 75:447–451.

Clark, R. B. 1976. Plant efficiencies in the use of calcium, magnesium and molybdenum. p. 175–191. *In* M. J. Wright and S. A. Ferrarri (ed.) Plant adaptation to mineral stress in problem soils. Cornell Univ. Agric. Exp. Stn.

Clark, R. B. 1978. Differential response of corn inbreds to calcium. Commun. Soil Sci. Plant Anal. 9:729–744.

Clark, R. B. 1982. Plant response to mineral element toxicity and deficiency. p. 71–142. *In* M. N. Christiansen and C. F. Lewis (ed.) Breeding plants for less favorable environments. John Wiley & Sons, Inc., New York.

Clark, R. B., and J. C. Brown. 1974. Differential phosphorus uptake by phosphorus stressed corn inbreds. Crop Sci. 14:505–508.

Coleman, N. T., and G. W. Thomas. 1967. The basic chemistry of soil acidity. *In* R. W. Pearson and F. Adams (ed.) Soil acidity and liming. Agronomy 12:1–41.

Cook, R. L., and C. E. Millar. 1939. Some soil factors affecting boron availability. Soil Sci. Soc. Am. Proc. 4:297–301.

Corke, C. T., and F. E. Chase. 1964. Comparative studies of actinomycete populations in acid podzolic and neutral mull forest soils. Soil Sci. Soc. Am. Proc. 28:65–67.

Cosgrove, D. J. 1967. Metabolism of organic phosphates in soil. p. 216–226. *In* A. D. McLaren and G. H. Peterson (ed.) Soil biochemistry. Marcel Dekker, Inc., New York.

Couto, W., D. J. Lathwell, and D. R. Bouldin. 1979. Sulfate adsorption by two Oxisols and an Alfisol of the tropics. Soil Sci. 127:108–116.

Cregan, P. D. 1980. Soil acidity and associated problems. Guidelines for farmer recommendations. A. G. Bulletin 7, Oct. 1980. Wagga Wagga, Department of Agriculture, New South Wales, Australia.

Cummings, G. A. 1978. Plant and soil effects of fertilizer and lime applied to highbush blueberries. J. Am. Soc. Hortic. Sci. 103:302–305.

Currier, C. G., C. B. Williams, and C. S. Hoveland. 1981. The relationship of grass tetany potential with root growth in tall fescue. Agron. Abstr. American Society of Agronomy, Madison, WI. p. 136.

Cuttridge, C. G., R. F. Hughes, C. R. Sturchcombe, and W. W. George. 1978. Nutrient effects of tipburn of strawberry. Exp. Hortic. 30:36–41.

Dancer, W. S., L. A. Peterson, and G. Chesters. 1973. Ammonification and nitrification of N as influenced by soil pH and previous N treatments. Soil Sci. Soc. Am. Proc. 37:67–69.

Davis, M. R. 1978. Phosphate responses of grasses and legumes growing in subsoil. p. 21–38. *In* J. Orwin (ed.) Revegetation in the rehabilitation of mountain lands. New Zealand Forest Service Research Institute, Rotorua, New Zealand.

De Brito Giordano, L. 1980. Inheritance and physiology of calcium utilization in tomatoes (*Lycopersicon esculentum* Mill.) grown under low Ca stress. Ph.D. thesis. University of Wisconsin, Madison (Diss. Abstr. Int. Vol. 42, no. 01, July 1981).

de Carvalho, M. M., C. S. Andrew, D. G. Edwards, and C. J. Asher. 1980. Comparative performance of six Stylosantheses species in three acid soils. Aust. J. Agric. Res. 31:61–76.

Denison, I. A. 1922. The nature of certain Al salts in the soil and their influences on ammonification and nitrification. Soil Sci. 13:81–106.

Dessureaux, L. 1959. Heritability of tolerance to manganese toxicity in lucerne. Euphytica 8:260–265.

Devine, T. E. 1982. Genetic fitting of crops to problem soils. p. 143–173. *In* M. N. Christiansen and C. F. Lewis (ed.) Breeding plants for less favorable environments. John Wiley & Sons, Inc., New York.

Devine, T. E., C. D. Foy, A. L. Fleming, C. H. Hanson, T. A. Campbell, J. E. McMurtrey III, and J. W. Schwartz. 1976. Development of alfalfa strains with differential tolerance to aluminum toxicity. Plant Soil 44:73–79.

Devine, T. E., C. D. Foy, D. L. Mason, and A. L. Fleming. 1979. Aluminum tolerance in soybean germplasm. Soybean Genetics Newsletter, April, Vol. 6, p. 24–27. USDA and Iowa State University, Ames, IA.

Dobereiner, J. 1966. Manganese toxicity effects on nodulation and N fixation in beans (*Phaseolus vulgaris* L.) in acid soils. Plant Soil 24:153–166.

Duncan, R. R. 1981. Variability among sorghum genotypes for uptake of elements under acid soil field conditions. J. Plant Nutr. 4:21–2.

Duncan, R. R., J. W. Dobson, Jr., and C. D. Fisher. 1980. Leaf elemental concentrations and grain yield of sorghum grown on acid soil. Commun. Soil Sci. Plant Anal. 11:699–707.

Dyson, P. W., and J. Digby. 1975. Effects of calcium on sprout growth of ten potato cultivars. Potato Res. 18:363–377.

Edwards, D. G., and B. T. Kang. 1978. Tolerance of cassava (*Manihot esculenta* Crantz) to high soil acidity. Field Crops Res. 1:337–346.

Edwards, J. H., and B. D. Horton. 1977. Aluminum induced calcium deficiency in peach seedlings. J. Am. Soc. Hortic. Sci. 102:459–461.

Eenink, A. H., and G. Garretsen. 1977. Inheritance of insensitivity of lettuce to a surplus of exchangeable manganese in steam sterilized soils. Euphytica 26:47–53.

Elkins, D. M., and L. E. Ensminger. 1971. Effect of soil pH on the availability of adsorbed sulfate. Soil Sci. Soc. Am. Proc. 35:931–934.

English, J. E., and D. N. Maynard. 1981. Calcium efficiency among tomato strains. J. Am. Soc. Hortic. Sci. 106:552–557.

Eno, C. F., and W. G. Blue. 1957. The comparative rate of nitrification of anhydrous ammonia, urea, and ammonium sulfate in sandy soils. Soil Sci. Soc. Am. Proc. 21:392–396.

Epstein, E. 1969. Mineral metabolism of halophytes. p. 381–397. *In* I. H. Rorison (ed.) Ecological aspects of the mineral nutrition of plants. Blackwell Scientific Publications, Inc., Boston, MA.

Evans, C. E., and E. J. Kamprath. 1970. Lime response as related to percent Al saturation, solution Al, and organic matter content. Soil Sci. Soc. Am. Proc. 34:893–896.

Evans, H. J., E. R. Purvis, and F. E. Bear. 1951. Effect of soil reaction on availability of molybdenum. Soil Sci. 71:117–124.

Fageria, N. K., and F. J. P. Zimmerman. 1979. Screening rice varieties for resistance to aluminum toxicity. Pesquis. Agropecu. Bras. 14:141–148.

Fairbourn, M. L., and A. R. Batchelder. 1980. Factors influencing magnesium in high plains forage. J. Range Manage. 33:435–438.

Farina, M. P. W., M. E. Sumner, C. O. Plank, and W. S. Letzsch. 1980. Exchangeable aluminum and pH as indicators of lime requirement for corn. Soil Sci. Soc. Am. J. 44:1036–1041.

Ferrari, T. J., and C. M. J. Sluijsmans. 1955. Mottling and magnesium deficiency in oats and their dependence on various factors. Plant Soil 6:262–299.

Fisher, A. G., G. W. Eaton, and S. W. Porritt. 1977. Internal bark necrosis of Delicious apple in relation to soil pH and leaf manganese. Can. J. Plant Sci. 57:297–299.

Fleming, A. L., J. W. Schwartz, and C. D. Foy. 1974. Chemical factors controlling the adaptation of plant species to acid mine spoils. Agron. J. 66:715–718.

Forbes, J. C., and A. L. Gelman. 1981. Copper and other minerals in herbage species and varieties on copper deficient soils. Grass Forage Sci. 36:25–30.

Fox, R. H. 1968. The effect of calcium and pH on boron uptake from high concentrations of boron by cotton and alfalfa. Soil Sci. 106:435–439.

Fox, R. H. 1979. Soil pH, aluminum saturation, and corn grain yield. Soil Sci. 127:330–334.

Fox, R. L. 1979. Comparative responses of field grown crops to phosphate concentrations in soil solutions. p. 81–106. *In* H. Mussell and R. C. Staples (ed.) Stress physiology in crop plants. John Wiley & Sons, Inc., New York.

Fox, R. L. 1981. External phosphorus requirements of crops. p. 223–239. (ed.) Chemistry in the soil environment *In* R. H. Dowdy et al. Spec. Pub. 40. American Society of Agronomy, Madison, WI.

Fox, R. L., A. D. Flowerday, F. W. Hosterman, H. F. Rhoades, and R. A. Olson. 1964. Sulfur fertilizers for alfalfa production in Nebraska. Res. Bull. 214. Nebraska Agric. Exp. Stn.

Foy, C. D. 1964. Toxic factors in acid soils of the southeastern United States as related to the response of alfalfa to lime. USDA Prod. Res. Rep. 80. U.S. Government Printing Office, Washington, DC.

Foy, C. D. 1973. Manganese and plants. p. 51–76. *In* Manganese. National Academy of Sciences–National Research Council, Washington, DC.

Foy, C. D. 1974. Effects of aluminum on plant growth. p. 601–642. *In* E. W. Carson (ed.) The plant root and its environment. University Press of Virginia, Charlottesville, VA.

Foy, C. D. 1976. General principles involved in screening plants for aluminum and manganese tolerance. p. 255–267. *In* M. J. Wright and S. A. Ferrari (ed.) Plant adaptation to mineral stress in problem soils. Cornell Univ. Agric. Exp. Stn.

Foy, C. D. 1984. Physiological effects of hydrogen, aluminum, and manganese toxicities in acid soil. *In* F. Adams (ed.) Soil acidity and liming. Agronomy 12:57–97.

Foy, C. D., and S. A. Barber. 1958. Magnesium absorption and utilization by two inbred lines of corn. Soil Sci. Soc. Am. Proc. 22:57–62.

Foy, C. D., and J. C. Brown. 1964. Toxic factors in acid soils. II. Differential aluminum tolerance of plant species. Soil Sci. Soc. Am. Proc. 28:27–32.

Foy, C. D., G. R. Burns, J. C. Brown, and A. L. Fleming. 1965. Differential aluminum tolerance of two wheat varieties associated with plant induced pH changes around their roots. Soil Sci. Soc. Am. Proc. 29:64–67.

Foy, C. D., R. L. Chaney, and M. C. White. 1978. The physiology of metal toxicity in plants. Annu. Rev. Plant Physiol. 29:511–566.

Foy, C. D., and A. L. Fleming. 1978. The physiology of plant tolerance to excess available aluminum and manganese in acid soils. p. 301–328. *In* G. A. Jung (ed.) Crop tolerance to suboptimal land conditions. Spec. Pub. 32. American Society of Agronomy, Madison, WI.

Foy, C. D., A. L. Fleming, and W. H. Armiger. 1969a. Aluminum tolerance of soybean varieties in relation to calcium nutrition. Agron. J. 61:505–511.

Foy, C. D., A. L. Fleming, and W. H. Armiger. 1969b. Differential tolerance of cotton varieties to excess manganese. Agron. J. 61:690–694.

Foy, C. D., J. E. Jones, and H. W. Webb. 1980. Adaptation of cotton genotypes to an acid, Al-toxic soil. Agron. J. 72:833–839.

Foy, C. D., H. W. Webb, and J. E. Jones. 1981. Adaptation of cotton genotypes to an acid, manganese toxic soil. Agron. J. 73:107–111.

Foy, C. D., and N. C. Wheeler. 1979. Adaptation of ornamental species to an acid soil high in exchangeable aluminum. J. Am. Soc. Hortic. Sci. 104:762–767.

Freire, J. R. J. 1975. Compartamento da soja e do seu rizobio ao Al e Mn nos solos do Rio Grande do Sul. Cien. Cult. (Maracaibo) 28:169–170.

Freire, J. R. J. 1976. Inoculation of soybeans. p. 335–380. *In* J. M. Vincent et al. (ed.) Exploiting the legume-rhizobium symbiosis in tropical agriculture. College of Tropic. Agric. Misc. Pub. 145. Department of Agronomy and Soil Science, University of Hawaii, Honolulu.

Freney, T. R. 1967. Sulfur containing organics. p. 229–253. *In* A. D. McLaren and G. H. Peterson (ed.) Soil biochemistry. Marcel Dekker, Inc., New York.

Fried, M., and M. Peech. 1946. The comparative effect of lime and gypsum upon plants grown in acid soils. J. Am. Soc. Agron. 38:614–623.

Friesen, D. K., A. S. R. Juo, and M. H. Miller. 1980. Liming and lime-phosphorus-zinc interactions in two Nigerian Ultisols. I. Interactions in the soil. Soil Sci. Soc. Am. J. 44:1221–1226.

Friesen, D. K., A. S. R. Juo, and M. H. Miller. 1980. Liming and lime-phosphorus-zinc interactions in two Nigerian Ultisols. II. Effects on maize root and shoot growth. Soil Sci. Soc. Am. J. 44:1227–1232.

Furlani, P. R., R. Hiroce, O. C. Bataglia, and W. J. Silva. 1977. Accumulation of micronutrients, silicon and dry matter in two simple maize hybrids. Bragantia 36:223–229.

Gallagher, R. N., M. D. Jellum, and J. B. Jones, Jr. 1981. Leaf magnesium concentration efficiency versus yield efficiency of corn hybrids. Commun. Soil Sci. Plant Anal. 12:345–354.

Gammon, N., Jr., and S. E. McFadden. 1979. Effect of rootstocks on greenhouse rose flower yield and leaf nutrient levels. Commun. Soil Sci. Plant Anal. 10:1171–1184.

Garcia, O. A., and S. A. Leon. 1978. Reaction of 5 cocoa hybrids *(Theobroma cacao)* to toxicity produced by aluminum in nutritive solution and on the plains of Colombia. Rev. Inst. Colomb. Agropecu. 13:219–228.

Geraldson, C. M. 1957. Control of blossom end rot in tomatoes. Proc. Am. Soc. Hortic. Sci. 69:309–317.

Gerloff, G. C. 1976. Plant efficiencies in the use of nitrogen, phosphorus and potassium. p. 161–173. *In* M. J. Wright and S. A. Ferrari (ed.) Plant adaptation to mineral stress in problem soils. Cornell Univ. Agric. Exp. Stn.

Gerloff, G. C., and K. A. Fishbeck. 1969. Quantitative cation requirements of several green and blue-green algae. J. Phycol. 5:109–114.

Giordano, P. M., J. J. Mortvedt, and R. I. Papendick. 1966. Response of corn (*Zea mays* L.) to zinc as affected by placement and nitrogen source. Soil Sci. Soc. Am. Proc. 31:767–770.

Godo, G. H., and H. M. Reisenauer. 1980. Plant effects on soil manganese availability. Soil Sci. Soc. Am. J. 44:993–995.

Goedert, W. J., R. B. Corey, and J. K. Syers. 1975. Lime effects on potassium equilibria in soils of Rio Grande do Sul, Brazil. Soil Sci. 120:107–111.

Gomes, J. C., and J. M. Braga. 1980. Relcao entre a capcidade tampao de fosforo de tres latossolos de Minas. Gerais E A Absorcao differencial de fasforo em tres cultivares de feizao (*Phaseolus vulgaris,* L.) Rev. Ceres 27(150):134–144.

Gomez-Lepe, B. E., O. Y. Lee-Stadelman, J. A. Patta, and E. J. Stadelman. 1979. Effects of actylguanidine on cell permeability and other protmplasmic properties of *Allium cepa* epidermal cells. Plant Physiol. 64:131–138.

Graves, C. R., K. B. Sanders, J. Odom, and M. Smith. 1980. Barley and wheat varieties and soil acidity. Tennessee Farm and Home Science. January–March, p. 15–17.

Gray, T. R. G. 1970. Microbial growth in soil. p. 36–41. *In* Pesticides in the soil: Ecology, degradation and movement. Pestic. Soil: Ecol. Degrad. Mon. Int. Symp. 1970. Michigan State University East Lansing, MI.

Greenleaf, W. H., and F. Adams. 1969. Genetic control of blossom and rot disease in tomatoes. J. Am. Soc. Hortic. Sci. 94:248–250.

Grime, J. P., and J. S. Hodgson. 1969. An investigation of the significance of lime chlorosis by means of large scale comparative experiments. p. 381–397. *In* I. H. Rorison (ed.) Ecological aspects of the mineral nutrition of plants. Blackwell Scientific Publications, Inc., Boston, MA.

Grimme, H. 1968. Adsorption of Mn, Co, Cu and Zn to goethite in dilute solutions. Z. Pflanzenernaehr. Bodenkd. 121:58–65.

Grove, J. H., M. E. Sumner, and J. K. Syers. 1981. Effect of lime on exchangeable magnesium in variable surface charge soils. Soil Sci. Soc. Am. J. 45:497–500.

Gupta, A. P., and G. S. G. Rao. 1977. Physiological studies on manganese nutrition. Indian Sugar 27:547–552.

Gupta, U. C., and J. Lipsett. 1981. Molybdenum in soils, plants and animals. Adv. Agron. 34:73–115. Academic Press, Inc., New York.

Gurrier, G. 1979. Absorption of mineral elements in the presence of aluminum Plant Soil 51:275–278.

Hallock, D. L. 1979. Relative effectiveness of several Mn sources on Virginia type peanuts. Agron. J. 71:685–688.

Hanna, W. J., and J. B. Hutcheson, Jr. 1968. Soil-plant relationships. p. 141–161. *In* L. B. Nelson (ed.) Changing patterns in fertilizer use. Soil Science Society of America, Madison, WI.

Hanson, W. D., and E. J. Kamprath. 1979. Selection for aluminum tolerances in soybeans based on seedling root growth. Agron. J. 71:581–586.

Hargrove, W. L., and G. W. Thomas. 1981. Extraction of aluminum from aluminum-organic matter complexes. Soil Sci. Soc. Am. J. 45:151–153.

Hartzog, D., and F. Adams. 1973. Fertilizer, gypsum, lime experiments with peanuts in Alabama. Auburn Univ. Agric. Exp. Stn. Bull. 448.

Hati, N., T. R. Fisher, and W. J. Upchurch. 1979. Liming of soil. I. Effect on plant available aluminum. J. Indian Soc. Soil Sci. 27:277–281.

Heenan, D. P., and L. C. Campbell. 1981. Soybean *(Glycine max)* nitrate reductase activity influenced by manganese nutrition. Plant Cell Physiol. 21:731–736.

Helyar, K. R. 1978. Effects of aluminum and manganese toxicity on legume growth. p. 207–231. *In* C. S. Andrew and E. J. Kamprath (ed.) Mineral nutrition of legumes in tropical and subtropical soils. Commonwealth Scientific and Industrial Research Organizaiton (CSIRO), East Melbourne, Victoria, Australia.

Helyar, K. R., and A. J. Anderson. 1974. Effects of calcium carbonate on the availability of nutrients in an acid soil. Soil Sci. Soc. Am. Proc. 38:341–346.

Henderson, J. B. 1969. The calcium to magnesium ratio of liming materials for efficient magnesium retention in soils and utilization by plants. Ph.D. thesis. North Carolina state Univ., Raleigh (Diss. Abstr. B. 1970, 30:4871–4872).

Hochmuth, G. J., G. C. Gerloff, and W. H. Gabelman. 1978. Differential P uptake among strains of tomato. Hortic. Sci. 13:54.

Hodgson, J. F., W. L. Lindsay, and J. F. Trierweiler. 1966. Micro-nutrient cation complexing in soil solution. II. Complexing of zinc and copper in displacing solution from calcareous soils. Soil Sci. Soc. Am. Proc. 30:723–726.

Hofner, W. 1970. Elsen and manganhatlige verbindungen in blutungssaft von *Helianthus annuus*. Physiol. Plant. 23:673–677.

Holmes, R. L., and G. D. Coorts. 1980. Effects of excess manganese nutrition on the growth of Easter lily. Hortic. Sci. 15:497–498.

Horst, W. J., and H. Marschner. 1978a. Effect of excessive manganese supply on uptake and translocation of calcium in bean plants. Pflanzenphysiol. 87:137–148.

Horst, W. J., and J. Marschner. 1978b. Effect of silicon on manganese tolerance of bean plants *(Phaseolus vulgaris)*. Plant Soil 50:287–304.

Horst, W. J., and H. Marschner. 1978c. Symptoms of manganese toxicity in *Phaseolus vulgaris*. Z. Pflanzenernaehr. Bodenkd. 141:129–142.

Horst, W. J., and H. Marschner. 1978d. Effect of silicon on the chemical state of manganese in bean leaf tissues. Z. Pflanzenernaehr. Bodenkd. 141:487–498.

Horton, B. D., and J. H. Edwards. 1976. Diffusive resistance rates and stomatal aperatures of peach seedlings as affected by Al concentrations. Hortic. Sci. 11:591–593.

Howard, D. D., and F. Adams. 1965. Calcium requirement for penetration of subsoils by primary cotton roots. Soil Sci. Soc. Am. Proc. 29:558–561.

Hoyt, P. B., and G. G. S. Myovella. 1979. Correction of severe manganese deficiency in wheat with chemical fertilizers. Plant Soil 52:437–444.

Hsu, P. H., and D. A. Rennie. 1962. Reactions of phosphate in aluminum systems. I. Adsorption of phosphate by x-ray amorphous aluminum hydroxide. Can. J. Soil Sci. 42:197–209.

Huett, D. O., and R. C. Menary. 1980. Effect of aluminum on growth and nutrient uptake of cabbage, *Brassiea oleracea,* var-capitata, Cult. Baldhead, lettuce, *Latuca sativa,* Cult. Pennlake and Kikuyu grass *Pennisetum clandestinium,* Cult Whittet, in nutrient solution. Aust. J. Agric. Res. 31:749–762.

Hunt, I. V., J. Frame, and R. D. Harkess. 1976. Removal of mineral nutrients by red clover varieties. J. Br. Grassl. Soc. 31:171–179.

Hutton, E. M., W. T. Williams, and C. S. Andrew. 1978. Differential tolerance to manganese in introduced and bred lines of *Macroptilium atropupureum.* Aust. J. Agric. Res. 29:67–79.

International Rice Research Institute. 1977. Tolerance for adverse soils. Zinc deficiency. IRRI Annu. Rep. for 1976. International Rice Research Institute, Los Baños, Philippines.

International Rice Research Institute. 1979. Genetic evaluation and utilization (GEU) program. Adverse soil tolerance. Screening zinc deficiency tolerance. p. 109–115. IRRI Annu. Rep. for 1978. International Rice Research Institute, Los Baños, Philippines.

Ishida, A., and M. Masui. 1976. Studies on the manganese excess of carnation. II. Manganese excess in relation to certain soils, soil pH and two nitrogen forms. J. Jpn. Soc. Hortic. Sci. 45:283–288.

Ishida, A., M. Masui, T. Ogura, and A. Mikaya. 1977. Studies on the manganese excess of carnation. III. Effects of manganese concentration in nutrient solution on growth and flowering. J. Jpn. Soc. Hortic. Sci. 45:383–388.

Islam, A. K. M. S., D. G. Edwards, and C. J. Asher. 1980. pH Optima for crop growth. Results of a flowing culture experiment with six species. Plant Soil 54:339–357.

Jackson, W. A. 1967. Physiological effects of soil acidity. *In* R. W. Pearson and F. Adams (ed.) Soil Acidity and liming. Agronomy 12:43–124.

Jones, D. W., T. L. Jackson, and M. E. Harward. 1968. Effect of molybdenum and lime on the growth and molbydenum content of alfalfa grown on acid soils. Soil Sci. 105:397–402.

James, H., M. N. Court, D. A. MacLeod, and J. W. Parsons. 1978. Soil factors and mycorrhizal factors activity on basaltic soils in Western Scotland. Forestry 51:105–119.

Johnson, M. S., T. McNeilly, and P. D. Putwain. 1971. Revegetation of metalliferous mine spoil contaminated by lead and zinc. Environ. Pollut. 12:261–277.

Johnson, R. E., and W. A. Jackson. 1964. Calcium uptake and transport by wheat seedlings as affected by aluminum. Soil Sci. Soc. Am. Proc. 28:381–386.

Jones, H. E., and G. D. Scarseth. 1944. The calcium-boron balance in plants as related to boron needs. Soil Sci. 57:15–24.

Jones, J. B., Jr., and F. Haghiri. 1963. Magnesium deficiency on Columbia county soils. Ohio Agric. Exp. Stn. Res. Center 116:18.

Jones, J. P., and R. L. Fox. 1978. Phosphorus nutrition of plants influenced by manganese and aluminum uptake from an Oxisol. Soil Sci. 126:230–236.

Jones, J. P., and S. S. Woltz. 1960. Fusarium wilt of tomato: Interaction of soil liming and micronutrient amendments on disease development. Phytopathology 60:812–813.

Jones, L. H. 1961. Aluminum uptake and toxicity in plants. Plant Soil 13:297–310.

Jones, R. K., and H. J. Clay. 1976. Foliar symptoms of nutrient disorders in Townsville stylo *(Stylosanthes humilis).* Aust. Div. Trop. Agron. Tech. Paper 19:1–11. Commonwealth Scientific and Industrial Research Organization (CSIRO), East Melbourne, Victoria, Australia.

Jones, W. F., and L. E. Nelson. 1978. Response of field grown soybeans to lime. Commun. Soil Sci. Plant Anal. 9:607–614.

Jones, W. F., and L. E. Nelson. 1979. Manganese concentration in soybean leaves as affected by cultivars and soil properties. Commun. Soil Sci. Plant Anal. 10:821–830.

Juo, A. S. R., and J. C. Ballaux. 1977. Retention and leaching of nutrients in a limed Ultisol under cropping. Soil Sci. Soc. Am. J. 41:757–761.

Kamprath, E. J. 1967. Soil acidity and response to liming. International Soil Testing Series Tech. Bull. 4. North Carolina State Univ. Agric. Exp. Stn.

Kamprath, E. J. 1970. Exchangeable aluminum as a criterion for liming leached mineral soils. Soil Sci. Soc. Am. Proc. 34:252–254.

Kamprath, E. J., W. L. Nelson, and J. W. Fitts. 1956. The effect of pH, sulfate and phosphate concentrations of the adsorption of sulfate by soils. Soil Sci. Soc. Am. Proc. 20:463–466.

Kamura, T., and K. Nishitani. 1977. Oxidation mechanisms of manganese in soils. 2. Effects of soil reaction on manganese oxidation. Soil Sci. Plant Nutr. 23:391.

Karlen, D. L., R. Ellis, Jr., D. A. Whitney, and D. L. Grunes. 1978. Influence of soil moisture and plant cultivar on cation uptake by wheat with respect to grass tetany. Agron. J. 70:918–921.

Kaufman, M. D., and E. H. Gardner. 1978. Segmental liming of soil and its effect on the growth of wheat. Agron. J. 70:331–336.

Kawaski, T., and M. Moritsugu. 1979. A characteristic symptom of calcium deficiency in maize and sorghum. Commun. Soil Sci. Plant Anal. 10:41–56.

Keyser, H. H., and D. N. Munns. 1979a. Effects of calcium, manganese and aluminum on growth of rhizobia in acid media. Soil Sci. Soc. Am. J. 43:500–503.

Keyser, H. H., and D. N. Munns. 1979b. Tolerance of rhizobia to acidity, aluminum and phosphate. Soil Sci. Soc. Am. J. 43:519–523.

Keyser, H. H., D. N. Munns, and J. S. Hohenberg. 1979. Acid tolerance of rhizobia in culture and in symbiosis with cowpea. Soil Sci. Soc. Am. J. 43:719–722.

Khasawneh, F. E., E. C. Sample, and E. J. Kamprath (ed.). 1980. The role of phosphorus in agriculture. American Society of Agronomy, Madison, WI.

Kliewer, W. M. 1961. The effects of varying combinations of molybdenum, aluminum, manganese, phosphorus, nitrogen, calcium, hydrogen concentrations, lime and rhizobium strain on growth composition and nodulation of several legumes. Ph.D. thesis. Cornell University, Ithaca NY.

Klimashevski, E. L., and N. F. Chernysheva. 1980. Content of organic acids and physiologically active compounds in plants differing in their susceptibility to the toxicity of Al^{3+}. (In Russian.) Sov. Agric. Sci. 1980(2):5–8.

Klimashevski, E. L., Y. A. Markova, S. M. Zyabkina, G. K. Zirenka, T. E., Zolotuklhn, and S. E. Pavolova. 1976. Aluminum absorption and localization in root tissues of different pea varieties. (In Russian with English summary.) Fizol. Biokhim. Kult. Rast 8:396–401.

Kotze, W. A. G. 1976. Growth and nutrition of apple rootstocks and seedlings in the presence of aluminum. Agrochemophysica 8(3):39–43.

Krause, H. H. 1965. Effect of pH on leaching losses of potassium applied to forest nursery soils. Soil Sci. Soc. Am. Proc. 29:613–615.

Krauskopf, K. B. 1972. Geochemistry of micronutrients. p. 7–40. *In* J. J. Mortvedt, et al. (ed.) Micronutrients in agriculture. Soil Science Society of America, Madison, WI.

Krizek, D. T., and C. D. Foy. 1981. Water stress: Role in differential aluminum tolerance of barley genotypes. Agron. Abstr. American Society of Agronomy, Madison, WI. p. 181–182.

Kroetz, M. E., W. H. Schmidt, J. E. Beuerlein, and G. L. Ryder. 1977. Correcting manganese deficiency increases soybean yields. Ohio Rep. 62:51–53.

Ladiges, P. Y. 1977. Differential susceptibility of two populations of *Eucalyptus viminalis,* Labill. to iron chloroses. Plant Soil 48:581–597.

Lefever, H. N. 1978. Tital—a new high yielding variety of soft red winter wheat. Ohio Rep. 63:44–45.

Lafever, H. N. 1979. Registration of Titan wheat. Crop Sci. 19:749.

Lafever, H. N., and L. G. Campbell. 1978. Inheritance of aluminum tolerance in wheat. Can. J. Genet. Cytol. 20:335–364.

Lance, J. C., and R. W. Pearson. 1969. Effect of low concentration of aluminum on growth and water and nutrient uptake by cotton roots. Soil Sci. Soc. Am. Proc. 33:95–98.

Lane, D. W. A., and C. A. Lamp. 1980. Preliminary selection of ryegrass plants for Mg accumulating ability. Proc. Int. Grassl. Congr. 13th. 385–387.

Lau, C. H. 1979. Effect of potassium and aluminum treatments on growth and nutrient uptake of rubber *(Hevea)* seedlings and on soils. J. Rubber Res. Inst. Malays. 27:92–103.

Lee, C. R., and G. R. Craddock. 1969. Factors affecting growth in high zinc medium. II. Influence of soil treatment on growth of soybeans on strongly acid soil containing Zn from peach sprays. Agron. J. 61:565–567.

Lee, C. R., and M. L. McDonald. 1978. Influence of soil amendments on potato growth, mineral nutrition, and tuber yield and quality in very strongly acid soils. Soil Sci. Soc. Am. J. 41:573–577.

Lee, C. R., and N. R. Page. 1967. Soil factors influencing the growth of cotton following peach orchards. Agron. J. 59:237–240.

Lee, K. W., C. E. Clapp, and A. C. Caldwell. 1970. P^{32} distribution in phosphorus fractions of phosphorus sensitive and tolerant soybeans. Plant Soil 33:707–712.

Lee, Y. S. 1977. Aluminum toxicity in corn seedlings. J. Korean Soc. Soil Sci. Fert. 10:75–78.

Leggett, J. E., and W. A. Gilbert. 1969. Magnesium uptake by soybeans. Plant Physiol. 44:1182–1186.

Ligon, W. A., and W. H. Pierre. 1932. Soluble aluminum studies. II. Minimum concentrations of aluminum found to be toxic to corn, sorghum and barley in culture solutions. Soil Sci. 34:307–317.

Lin, C., and N. T. Coleman. 1960. The measurement of exchangeable aluminum in soils and clays. Soil Sci. Soc. Am. Proc. 24:444–446.

Lindgren, D. T., W. H. Gabelman, and G. C. Gerloff. 1977. Variability of phosphorus uptake and translocation in *Phaseolus vulgaris* L. under phosphorus stress. J. Am. Soc. Hortic. Sci. 102:674–677.

Lindsay, W. L. 1979. Chemical equilibria in soils. John Wiley & Sons, Inc., New York. p. 222.

Link, L. 1975. Barley faces problems from manganese toxicity. Southeast Farm Press. June 25, p. 7.

Link, L. 1979. Critical pH for the expression of manganese toxicity in burley tobacco and the effect of liming on growth. Tobacco Int. 181:18:50–52.

Loneragan, J. F. 1978. The physiology of plant tolerance to low phosphorus availability. p. 329–343. *In* G. A. Jung (ed.) Crop tolerance to suboptimal land conditions, Spec. Pub. 32. American Society of Agronomy, Madison, WI.

Loneragan, J. F., I. C. Rowland, A. D. Robson, and K. Snowball. 1970. The calcium nutrition of plants. Proc. Int. Grassl. Congr., 11th. p. 358–367.

Loneragan, J. F., and K. Snowball. 1969a. Calcium requirements of plants. Aust. J. Agric. Res. 20:465–478.

Loneragan, J. F., and K. Snowball. 1969b. Rate of calcium absorption by plant roots and its relation to growth. Aust. J. Agric. Res. 20:479–490.

Long, F. L., and C. D. Foy. 1970. Plant varieties as indicators of aluminum toxicity in the A_2 horizon of a Norfolk soil. Agron. J. 62:679–681.

Lopez, A., et al. 1976. Reaction to different levels of aluminum concentrations in wheat and rye and their contributions to tolerance in triticale. West. Branch, CSSA Abstr. Western Branch, Crop Science Society of America, p. 11.

Lopez-Benitez, A. 1977. Influence of aluminum toxicity in intergeneric crosses of wheat and rye. Ph.D. thesis. Oregon State University, Corvallis, OR.

Loughman, B. C. 1977. Metabolic factors and the utilization of phosphorus by plants. p. 155–169. *In* R. Porter and D. W. Fitzsimons (ed.) Phosphorus and the environment: Its chemistry and biochemistry. Elsevier Science Publishing Co., Inc., New York.

Lowe, P. H., and C. E. Bortner. 1973. Effect of phosphorus nutrition and soil pH on physiologic spotting of L8 burley tobacco. Agron. J. 65:263–265.

Lucas, R. E., and F. J. Davis. 1961. Relationships between pH values or organic soils and availabilities of 12 plant nutrients. Soil Sci. 92:177–182.

Lund, Z. F. 1970. The effect of calcium and its relation to some other cations on soybean root growth. Soil Sci. Soc. Am. Proc. 34:456–459.

Lundberg, P. E., O. L. Bennett, and E. L. Mathias. 1978. Tolerance of bermuda grass selections to acidity. I. Effects of lime on plant growth and mine spoil material. Agron. J. 69:913–916.

MacLeod, L. B., R. F. Bishop, and F. W. Calder. 1964. Effect of various rates of liming and fertilization on certain chemical properties of a strongly acid soil on the establishment, yield, botanical and chemical composition of a forage mixture. Can. J. Soil Sci. 44:237–247.

MacLeod, L. B., and L. P. Jackson. 1967. Water soluble and exchangeable aluminum in acid soils as affected by liming and fertilization. Can. J. Soil Sci. 47:203–210.

Magalhaes, J. C. A. J. 1979. Effect of liming and P fertility rates on yield of two wheat varieties. Rev. Bras. Cienc. Solo 3:24–29.

Magistad, O. C. 1925. The aluminum content of the soil solution and its relation to soil reaction and plant growth. Soil Sci. 20:181–226.

Mahadevappa, M., H. Ikehashi, and F. N. Ponnamperuma. 1979. Research on varietal tolerance for phosphorus deficient rice soils. International Rice Research Newsletter 4:9–10. International Rice Research Institute, Los Baños, Philippines.

Mahilum, B. C., R. L. Fox, and J. A. Silva. 1970. Residual effects of liming volcanic ash soils in the humid tropics. Soil Sci. 109:102–109.

Malavolta, E., E. L. M. Coutinho, G. C. Vitti, N. V. Alejo, N. J. Novaes, and V. L. Furlani Netto. 1979. Studies on the mineral nutrition of sweet sorghum. 1. Deficiency of macro and micronutrients and toxicity of aluminum, chlorine and manganese. An. Esc. Super. Agric. "Luis de Quiroz" Univ. Sao Paulo 36:173–202.

Maranville, J. W., J. C. Larson, and W. M. Ross. 1980. Uptake and distribution of nitrogen and phosphorus in grain sorghum hybrids and their parents. J. Plant Nutr. 2:267–281.

Marocke, R. G. Correge, and R. Trendel. 1979. Varietal differences in the response of hops to magnesium deficiency. C. R. Seances Acad. Sci., Ser. D. 289:97–100.

Mask, P. L., and D. O. Wilson. 1978. Effect of Mn on growth, nodulation and nitrogen fixation by soybeans grown in the greenhouse. Commun. Soil Sci. Plant Anal. 9:653–666.

Massey, H. F. 1957. Relation between dithizone-extractable zinc in the soil and zinc uptake by corn plants. Soil Sci. 83:123–129.

Masui, M., A. Nukaya, and A. Ishida. 1976. Studies on the manganese excess of muskmelon. V. The manganese, calcium and iron concentrations in nutrient solution. J. Jpn. Soc. Hortic. Sci. 45:267–274.

Mathur, S. P., H. A. Hamilton, and C. M. Preston. 1979. The influence of variation in copper content of an organic soil in the mineral nutrition of oats grown in situ. Commun. Soil Sci. Plant Anal. 10:1399–1409.

Matsumoto, H., S. Morimura, and E. Hirasawa. 1979. Localization of absorbed aluminum in plant tissues and its toxicity studies in the inhibition of pea root elongation. *In* T. Kudrev, et al. (ed.) Mineral nutrition of plants. Vol. I., Proc. 1st Int. Symp. on Plant Nutrition, Varja, Bulgaria, 24–29 Sept. Bulgarian Academic Books, Sofia, Bulgaria. p. 171–194.

Mattson, S. 1927. Anionic and cationic adsorption by soil colloidal materials of varying $SiO_2/Al_2O_3 + Fe_2O_3$ ratios. Trans. Int. Congr. Soil Sci., 1st (Washington) 2:199–211.

Mayland, H. F., and D. L. Grunes. 1979. Soil-climate-plant relationships in the etiology of grass tetany. p. 123–175. *In* V. V. Rendig and D. L. Grunes (ed.). Grass tetany. Spec. Pub. 35. American Society of Agronomy, Madison, WI.

Maynard, D. N., and A. V. Barker. 1969. Studies on the tolerance of plants to ammonium nutrition. J. Am. Soc. Hortic. Sci. 94:235–239.

Maynard, D. N., B. Gersten, and H. F. Vernell. 1965. The distribution of calcium as related to internal tipburn, variety and calcium nutrition of cabbage. Proc. Am. Soc. Hortic. Sci. 86:392–396.

Maynard, D. N., D. C. Warner, and J. C. Howell. 1981. Cauliflower *(Brassica-oleracea)* leaf tipburn, a calcium deficiency disorder. Hortic. Sci. 16:193–195.

McBride, M. B., and J. J. Blasiak. 1979. Zinc and copper solubility as a function of pH in an acid soil. Soil Sci. Soc. Am. J. 43:666–870.

McCart, G. D., and E. J. Kamprath. 1965. Supplying Ca and Mg for cotton on sandy low cation exchange capacity soils. Agron. J. 57:404–406.

McCormick, L. H., and F. Y. Borden. 1972. Phosphate fixation by aluminum in plant roots. Soil Sci. Soc. Am. Proc. 36:779–802.

McCormick, L. H., and K. C. Steiner. 1978. Variation in aluminum tolerance among six genera of trees. For. Sci. 24:565–568.

McEvoy, E. T. 1963. Varietal differences in calcium level in leaves of flue cured tobacco. Can. J. Plant Sci. 43:141–145.

McFee, W. W., F. T. Adams, C. S. Cronan, M. K. Firestone, C. D. Foy, R. D. Harter, and D. W. Johnson. 1984. The acidic deposition phenomenon and its effects. p. 21–27. *In* Effects on soil systems. U.S. Environmental Protection Agency and North Carolina State University, Raleigh, NC.

McKee, W. H. 1976. Response of potted slash pine seedlings on imperfectly drained coastal plain soil to additions of zinc. Soil Sci. Soc. Am. J. 40:586-588.

McLachlan, K. D. 1975. Sulphur in Australian agriculture. University of Sydney, Sydney, Australia.

McLean, E. E., and M. D. Carbonell. 1972. Calcium, magnesium and potassium saturation ratios in two soils and their effects upon yield and nutrient contents of german millet and alfalfa. Soil Sci. Soc. Am. Proc. 36:927–930.

McLean, F. T., and B. E. Gilbert. 1927. The relative aluminum tolerance of crop plants. Soil Sci. 24:163–175.

McLean, I. B. 1980. The toxic aluminum reaction in corn and barley roots: An ultrastructural and morphological study. Michigan State University, East Lansing. Univ. Microfilm 1314642, Ann Arbor, MI.

Mehlich, A., and W. E. Colwell. 1944. Influence of the nature of soil colloids and degree of base saturation on growth and nutrient uptake by cotton and soybeans. Soil Sci. Soc. Am. Proc. 8:179–184.

Mehlich, A., and J. F. Reed. 1945. The influence of degree of saturation, potassium level, and calcium additions on removal of calcium and potassium. Soil Sci. Soc. Am. Proc. 10:87–83.

Memon, A. R., M. Chino, Y. Takeoka, K. Hora, and M. Yatazawa. 1980. Distribution of manganese in leaf tissues of manganese accumulator: *Acanthopanax sciadophylloides* as revealed by electronprobe x-ray microanalyzer. J. Plant Nutr. 2:457–476.

Memon, A. R., M. Chino, and Y. Yatazawa. 1981. Microdistribution of aluminum and manganese in tea leaf tissues as revealed by x-ray analyses. Commun. Soil Sci. Plant Anal. 12:441–452.

Mengel, D. B., and E. J. Kamprath. 1978. Effect of soil pH and liming on growth and nodulation of soybeans in Histosols. Agron. J. 70:959–963.

Mercer, E. R., and J. L. Richmond. 1970. Fate of nutrients in soil: Copper. p. 9. *In* Letcombe Laboratory Annual Report. Letcombe Laboratory, Agricultural and Food Research Council, Wantage, Oxfordshire, U.K.

Midgley, A. R., and D. E. Dunklee. 1939. The effect of lime on the fixation of borates in soil. Soil Sci. Soc. Am. Proc. 4:302–307.

Milam. F. M., and A. Mehlich. 1954. Effect of soil-root ionic environment on growth and mineral content of *Crotaloria straiata*. Soil Sci. 77:227–236.

Miller, S. S., and O. E. Schubert. 1977. Plant manganese and soil pH associated with internal bark necrosis in apple. Proc. W. V. Acad. Sci. 49:97–102.

Miller, W. J., W. E. Adams. R. Nusspaumer, R. A. McCreery, and H. F. Perkins. 1964. Zinc content of Coastal bermudagrass as influenced by frequency and season of harvest, location, and level of N and lime. Agron. J. 56:198–201.

Moraghan, J. T. 1978. Chlorotic dieback in flax *(Linum usitatissimum)*. Agron. J. 70:501–505.

Moraghan, J. T. 1979. Manganese toxicity in flax grown on certain calcareous soils low in available iron. Soil Sci. Soc. Am. J. 43:1177–1180.

Moraghan, J. T., and P. Ralowicz. 1979. Relative responses of four flax differential lines to FeEDDHA. Crop Sci. 19:9–11.

Morgan, P. W., H. E. Hoham, and J. V. Amrin. 1966. Effect of manganese toxicity on the indole lacetic acid oxidase system of cotton. Plant Physiol. 41:718–724.

Morgan, P. W., D. M. Taylor, and H. E. Hoham. 1976. Manipulations of IAA-oxidase activity and auxin deficiency symptoms in intact cotton plants with manganese nutrition. Physiol. Plant. 37:149–156.

Morrill, L. G., and J. E. Dawson. 1967. Patterns observed for the oxidation of ammonium to nitrate by soil organisms. Soil Sci. Soc. Am. Proc. 31:757–760.

Morris, H. D. 1948. The soluble manganese content of acid soils and its relation to the growth and manganese content of sweet clover and lespedeza. Soil Sci. Soc. Am. Proc. 13:361–362.

Mugwira, L. M., M. Floyd, and S. V. Patel. 1981a. Tolerances of triticale lines to manganese in soil and nutrient solution. Agron. J. 73:319–322.

Mugwira, L. M., V. T. Sapra, S. V. Patel, and M. A. Choudhury. 1981b. Aluminum tolerance of tricale and wheat cultivars developed in different regions. Agron. J. 73:470–475.

Munn, D. A., and R. E. McCollum. 1976. Solution culture evaluation of sweet potato cultivar tolerance to aluminum. Agron. J. 68:989–991.

Munns, D. N. 1965. Soil acidity and growth of a legume. I. Interactions of lime with nitrogen and phosphorus on growth of *Medicago sativa* L. and *Trifolium subterraneum* L. Aust. J. Agric. Res. 16:733–741.

Munns, D. N. 1980. Mineral nutrition and nodulation. Proc. World Soybean Res. Conf., 2nd. Westview Press, Inc., Boulder, CO.

Munns, D. N., and R. L. Fox. 1977. Comparative lime requirements of tropical and temperate legumes. Plant Soil 46:533–548.

Munns, D. N., J. S. Hohenberg, T. L. Righetti, and D. T. Lauter. 1981. Soil acidity tolerance of symbiotic and nitrogen fertilizer soybeans. Agron. J. 73:407–410.

Munson, R. D., and W. L. Nelson. 1963. Movement of applied potassium in soils. J. Agric. Food Chem. 11:193–201.

Murray, J. J., and C. D. Foy. 1978. Differential tolerances of turfgrass cultivars to an acid soil high in exchangeable aluminum. Agron. J. 70:769–774.

Mutatker, V. K., and W. L. Pritchett. 1966. Influence of added aluminum on carbon dioxide production in tropical soils. Soil Sci. Soc. Am. Proc. 30:343–346.

Muzilli, O., D. Santos, J. B. Palhano, J. Manetti F. O., A. F. Lantmann, A. Garcia, and A. Cataneo. 1978. Soil-acidity tolerance in soybean and wheat cultivars. Rev. Bras. Cienc. Solo 2:34–40.

Naftel, J. A. 1937. Soil liming investigations. V. The relation of boron deficiency to overliming injury. J. Am. Soc. Agron. 29:761–771.

Naidoo, G. 1976. Aluminum toxicity in two snapbean varieties. Ph.D. thesis. University Tennessee, Knoxville (Diss. Abstr. Int. B, 1977, 27:54788. Eng. 77:791).

Nelson, L. E. 1964. The effect of pH on the acetate-soluble sulfur content of a Mayhew soil in Mississippi before and after incubation. Soil Sci. Soc. Am. Proc. 28:290–291.

Nelson, L. R., and T. C. Kiesling. 1980. Selection for ryegrass genotypes with tolerance to aluminum toxicity. Commun. Soil Sci. Plant Anal. 11:451–458.

Nguyen, H. T., and D. A. Sleper. 1981. Genetic variability of mineral concentrations in *Festuca arumdinacea* Schreb. Theor. Appl. Genet. 59:57–63.

Nichols, D. G., and D. V. Beardsell. 1981. The response of phosphorus-sensitive plants to slow-release fertilizers in soil-less potting mixtures. Sci. Hortic. 15:301–109.

Nightingale, G. T. 1934. Ammonium and nitrate nutrition of dormant delicious apple trees at 48°F. Bot. Gaz. (Chicago) 95:437–452.

Nightingale, G. T. 1937. The nitrogen nutrition of green plants. Bot. Rev. 3:85–174.

Nye, P., D. Craig, N. T. Coleman, and J. L. Ragland. 1961. Ion exchange equilibria involving aluminum. Soil Sci. Soc. Am. Proc. 25:14–17.

Oakes, A. J., and C. D. Foy. 1984. A winter hardy, aluminum tolerant perennial pasture grass for reclamation of acid mine spoils. J. Plant Nutr. 7:929–951.

Oyner, G., and O. Tiegen. 1980. Effects of acid irrigation at different temperatures on seven clones of Norway spruce. Norwegian For. Res. Inst. Rep. 36.3. Norwegian Forest Research Institute, Oslo, Norway.

Ohnki, K., D. O. Wilson, and O. E. Anderson. 1980. Manganese deficiency and toxicity sensitivities of soybean cultivars. Agron J. 72:713–716.

Oliveira, R. G. B., and A. Drozdowicz. 1980. Occurrence of microorganisms capable of decomposing organic phosphates in Cerrado soil in Brazil. Zentralbl. Bakteriol. Parasitenkd. Infektionskr. Hyg. Abt. 2 Naturwiss. Mikrobiol. Landwirtsch. Technol. Umweltschutzes 135:467–476.

Olson, R. V., and K. C. Berger. 1946. Boron fixation as influenced by pH, organic matter content and other factors. Soil Sci. Soc. Am. Proc. 11:216–220.

Osawa, T., and H. Ikeda. 1977. Heavy metal toxicities in vegetable crops. VI. The effect of potassium and calcium concentration in the nutrient solution on manganese toxicity in vegetable crops. J. Jpn. Hortic. Sci. 46:181–188.

Osawa, T., and H. Ikeda. 1980. Heavy metal toxicities in vegetable crops. B. Effect of nitrogen form supplied and pH levels of the nutrient solution on manganese toxicities in vegetable crops. J. Jpn. Soc. Hortic. Sci. 49:197–202.

Otsuka, K. 1969. Aluminium induced Fe chlorosis. Aluminum and manganese toxicities for plants. (Part 4) J. Sci. Soil Manure Jpn. 40:177–220. (Abstr. in Soil Sci. Plant Nutr. 16:140, 1970.)

Oza, A. M., S. Sarjeet, and M. B. Kamath. 1979. Differential accumulation of phosphorus and calcium by pearl millet cultivars from applied sources. J. Nucl. Agric. Biol. 8:54–57.

Oza, A. M., M. Singh, N. S. Varma, and P. K. Khanna. 1976. Varietal differences in the uptake of phosphorus from soil and applied source by barley *(Hordeum vulgare)*. J. Nucl. Agric. Biol. 5:79–81.

Palmer, B., and R. S. Jessop. 1977. Some aspects of wheat cultivar response to applied phosphate. Plant Soil 47:63–73.

Palmer, J. H. 1981. Manganese, critical element for soybeans. Southeast Farm Press. 24 June, p. 17.

Parker, M. B., and H. B. Harris. 1962. Soybean response to molybdenum and lime and relationship between yield and chemical composition. Agron. J. 54:480–483.

Parker, M. B., H. B. Harris, H. D. Morris, and H. F. Perkins. 1969. Manganese toxicity of soybeans as related to soil and fertility treatments. Agron. J. 61:515–518.

Pakrs, R. Q., and B. T. Shaw. 1941. Possible mechanisms of boron fixation in soil: I. Chemical. Soil Sci. Soc. Am. Proc. 6:219–223.

Pearson, R. W., R. Perez-Escolar, F. Abruna, Z. F. Lund, and E. J. Brenes. 1977. Comparative responses of three crop species to liming several soils of the southeastern United States and of Puerto Rico. J. Agric. Univ. P. R. 61:361–382.

Pearson, R. W., L. F. Ratliff, and H. M. Taylor. 1970. Effect of soil temperature, strength and pH on cotton seedling root elongation. Agron. J. 62:243–246.

Peech, M., and R. Bradfield. 1934. The effect of lime and neutral calcium salts upon the solubility of soil potassium. Am. Soil Surv. Assoc. Bull. 15:104–106.

Peech, M., and R. Bradfield. 1943. The effect of lime and magnesia on the soil potassium and on the absorption of potassium by plants. Soil Sci. 55:37–48.

Peedin, G. F., and C. B. McCants. 1977. Influence of variety and soil applications of Ca on development of Ca deficiency in tobacco. Agron. J. 69:71–76.

Percival, G. P., D. Josselyn, and K. C. Beeson. 1955. Factors affecting the micronutrient element content of some forages in New Hampshire. New Hampshire Agric. Exp. Stn. Tech. Bull 93.

Peterson, L. A., and R. C. Newman. 1976. Influence of soil pH on the availability of added boron. Soil Sci. Soc. Am. J. 40:280–282.

Picha, D. H., and P. L. Minotti. 1977. Differential response of tomato and lambsquarter to P levels and N source. Proc. Annu. Meet. Northeast. Weed Sci. Soc. 31:140.

Pieri, C. 1974. First experimental results on the susceptibility of groundnut to aluminum toxicity. Agron. Trop. (Paris) 29:685.

Pierre, W. H., G. G. Pohlman, and T. C. McIlvaine. 1932. Soluble aluminum studies. I. The concentration of aluminum in the displaced soil solution of naturally acid soils. Soil Sci. 34:145–160.

Poling, E. B., and G. H. Oberly. 1979. Effect of rootstock on mineral composition of apple leaves. J. Am. Soc. Hortic. Sci. 104:799–801.

Pope, D. T., and H. M. Munger. 1953. Heredity and nutrition in relation to magnesium deficiency chlorosis in celery. Proc. Am. Hortic. Sci. Soc. 61:472–480.

Porter, O. A., and Moraghan. 1975. Differential response of two corn inbreds to varying root temperature. Agron. J. 67:515–518.

Powell, A. J., and T. B. Hutcheson. 1965. Effect of lime and potassium additions on soil potassium reactions and plant response. Soil Sci. Soc. Am. Proc. 29:76–78.

Pratt, P. F., and R. Alvahydo. 1966. Cation-exchange characteristics of soil of Sao Paulo, Brazil. Bull. 31 IRI Research Institute, New York.

Presad, B., M. K. Sinha, and N. S. Randhawa. 1976. Effect of mobile chelating agents on diffusion of zinc in soils. Soil Sci. 122:260–266.

Price, N. O., and W. W. Moschler. 1970. Residual lime effect in soils on certain mineral elements in barley. J. Agric. Food Chem. 18:5–8.

Ragland, J. L., and N. T. Coleman. 1959. The effect of soil solution aluminum and calcium on root growth. Soil Sci. Soc. Am. Proc. 23:335–357.

Reid, D. A. 1971. Genetic control of reaction to aluminum in winter barley. *In* R. A. Nilan (ed.) Barley Genet. Proc. Int. Barley Genet. Symp., 2nd. p. 409–413. Washington State University Press, Pullman, WA.

Reid, D. A., L. A. J. Slootmaker, and J. C. Craddock. 1980. Registration of barley composite cross XXXIV. Crop Sci. 20:416–417.

Reisenauer, H. M., A. A. Tabilh, and P. R. Stout. 1962. Molybdenum reactions with soils and the hydrous oxides of iron, aluminum and titanium. Soil Sci. Soc. Am. Proc. 26:23–27.

Rhue, R. D. 1979. Differential aluminum tolerance in crop plants. p. 61–80. *In* H. Mussell and R. C. Staple (ed.) Stress physiology in crop plants. John Wiley & Sons, Inc., New York.

Rice, H. B. 1966. Magnesium availability in selected Coastal Plain soils of North Carolina. Ph.D. thesis. North Carolina State University, Raleigh (Diss. Abstr. B, 1967 27:4213).

Rice, H. B., and E. J. Kamprath. 1968. Availability of exchangeable and nonexchangeable Mg in sandy Coastal Plain soils. Soil Sci. Soc. Am. Proc. 32:386–388.

Richburg, J. S., and F. Adams. 1970. Solubility and hydrolysis of aluminum in soil solutions and saturated paste extracts. Soil Sci. Soc. Am. Proc. 34:728–734.

Rios, M. A., and R. W. Pearson. 1964. The effect of some chemical environmental factors on cotton root behavior. Soil Sci. Soc. Am. Proc. 28:232–235.

Robarge, W. P., and R. B. Corey. 1979. Adsorption of phosphate by hydroxy-aluminum species on a cation exchange resin. Soil Sci. Soc. Am. J. 43:481–487.

Rorison, I. H. 1980. The effect of soil acidity on nutrient availability and plant response. p. 283–304. *In* T. C. Hutchinson and M. Havas (ed.) Effects of acid precipitation on terrestrial ecosystems. Plenum Publishing Corp., New York.

Rosen, J. A., C. S. Pike, M. L. Golden, and J. Freedman. 1978. Zinc toxicity in corn as a result of a geochemical anomaly. Plant Soil 50:151–159.

Rossignol, M., and C. Grignon. 1980. Root phospholipid composition as a factor of the differential Ca-sensitivity of plants. p. 0–0. *In* P. Mazliak et al. (ed.) Biogenesis and function of plant lipids. Elsevier Science Publishing Co., Inc., New York.

Rufty, T., G. S. Miner, and C. D. Raper. 1979. Temperature effects on growth and manganese tolerance in tobacco. Agron. J. 71:638–644.

Safaya, N. M., and B. Singh. 1977. Differential susceptibility of two varieties of cowpea (*Vigna unguiculata* (L.) Walp) to phosphorus-induced zinc deficiency. Plant Soil 48:279–290.

Salako, E. A. 1980. Evaluation of phosphorus uptake efficiency of sorghum genotypes. Ph.D. thesis. Kansas State University, Manhattan (Diss. Abstr. Int. B, 1981. 41:2422).

Salinas, J. G., and P. A. Sanchez. 1976. Soil plant relationships affecting varietal and species differences in tolerance to low available soil phosphorus. Cienc. Cult. 28:156–168.

Santana, M. B., and J. M. Braga. 1977. Aluminum-phosphorus interactions of acidic soils in southern Bahia. Rev. Ceres 24(132):200–211.

Sartain, J. B. 1974. Differential effects of aluminum on top and root growth, nutrient accumulation and nodulation of several soybeans varieties. Ph.D. diss. North Carolina State University, Raleigh, NC. Diss. Abstr. 35:641.

Sartain, J. B., and E. J. Kamprath. 1975. Aluminum tolerance of soybean cultivars based on root elongation in solution culture compared with growth in acid soil. Agron. J. 70:17–20.

Sauchelli, V. 1969. Trace elements in agriculture. Van Nostrand Reinhold Company, Inc., New York.

Schauble, C. E., and S. A. Barber. 1958. Magnesium immobility in the nodes of certain corn inbreds. Agron. J. 50:651–653.

Schmandke, H., G. Muschiolik, M. Schultz, G. Schmidt, and H. D. quade. 1979. The effect of aluminum ions on chemical and functional properties of spun protein fibers. Nahrung 23:229–236.

Scibisz, K., and A. Sadowski. 1979. Internal bark necrosis in Stark Crimson Delicious apple trees in relation to mineral nutrition. Gartenbauwissenschaft 44:177–181.

Sedberry, J. E., Jr., P. G. Schilling, F. E. Wilson, and F. J. Peterson. 1978. Diagnosis and correction of zinc problems in rice production. Louisiana State Univ. Agric. Exp. Stn. Bull. 708.

Shannon, S. J., J. Nath, and J. D. Atkin. 1967. Relation of calcium nutrition to hypocotyl necrosis of snapbean (*Phaseolus vulgaris,* L.) Proc. Am. Soc. Hortic. Sci. 90:180–190.

Sheat, D. E. G., B. H. Fletcher, and H. E. Street. 1959. Studies on the growth of excised root. VIII. The growth of excised tomato roots supplied with various inorganic sources of nitrogen. New Phytol. 58:128–141.

Shoop, G. J., C. R. Brooks, R. E. Blaiser, and C. W. Thomas. 1961. Differential responses of grasses and legumes to liming and phosphorus fertilization. Agron. J. 53:111–115.

Silva, A. R. 1976. Application of the plant genetic approach to wheat culture in Brazil. p. 223–231.. *In* M. J. Wright and Sheila A. Ferrari (ed.) Plant adaptation to mineral stress in problem soils. Spec. Pub., Cornell Univ. Agric. Exp. Stn.

Simpson, J. R., A. Pinkerton, and J. Lazdovokis. 1977. Effects of subsoil calcium on the root growth of some lucerne genotypes (*Medicago sativa,* L.) Aust. J. Agric. Res. 29:629–638.

Sims, J. R., and F. T. Bingham. 1968a. Retention of boron by layer silicates, sesquioxides, and soil materials. II. Sesquioides. Soil Sci. Soc. Am. Proc. 32:364–396.

Sims, J. R., and F. T. Bingham. 1968b. Retention of boron by layer silicates, sesquioxides and soil materials. III. Iron and aluminum-coated layer silicates and soil materials. Soil Sci. Soc. Am. Proc. 32:369–373.

Sirkar, S., and J. V. Amin. 1979. Influence of auxins on respiration of manganese toxic cotton plants. Indian J. Exp. Biol. 17:618–619.

Small, J. 1946. pH and Plants. Van Norstrand Reinhold Company, Inc., New York.

Smiley, R. W. 1974. Rhizospere pH as influenced by plants, soils and nitrogen fertilizer. Soil Sci. Soc. Am. Proc. 38:795–799.

Smirnov, Y. S., T. A. Krupnikova, and M. Y. Shkol'nik. 1977. Sov. Plant Physiol. 24:270–276.

Smith, F. W. 1979. Tolerance of seven tropical pasture grasses to excess manganese. Commun. Soil Sci. Plant Anal. 10:853–867.

Snaydon, R. W. 1962. The growth and competitive ability of contrasting populations of *Trifolium repens* on calcareous and acid soils. J. Ecol. 50:439.

Snaydon, R. W., and A. D. Bradshaw. 1961. Differential response to calcium within the species *Festuca ovina,* L. New Phytol. 60:219–234.

Snaydon, R. W., and A. D. Bradshaw. 1962. Differences between natural populations of *Trifolium repens* (white clover) in response to mineral nutrients. I. Phosphate. J. Exp. Bot. 13:422–434.

Snyder, G. H., E. O. Burt, and G. J. Gascho. 1979. Correcting pH induced manganese deficiency in bermudagrass turf. Agron. J. 71:603–608.

Soileau, J. M., O. P. Englestad, and J. B. Martin, Jr. 1969. Cotton growth in an acid fragipan subsoil. II. Effects of soluble calcium, magnesium and Al on roots and tops. Soil Sci. Soc. Am. Proc. 33:919–924.

Sommers, L. E., and V. O. Biederbeck. 1973. Tillage management principles: Soil microorganisms in conservation tillage. Proc. Annu. Meet. Soil Conserv. Soc. Am. 28:87–108.

Sousa, C. N. A., S. R. Dotto, A. C. Baier, and J. C. Moreira. 1977. Reaction of wheat cultivars of varying origin to scorch. Abstr. 29th Annu. meet. Brazilian Society for Scientific Progress, Passo Fundo, Brazil.

Souto, S. M., and J. Dobereiner. 1971. Manganese toxicity in tropical forage legumes. Pesquis. Agropecu. Bras. 4:128–138.

Steiner, K., D. H. McCormick, and D. S. Canavera. 1980. Differential response of paper birch *(Betula papyrifera)* provenances to aluminum in solution cultures. Can. J. For. Res. 10:25–29.

Stephens, C. G., and A. C. Oertel. 1943. Responses of plants to molybdenum in pot experiments on Cressy shaley clay loam. Aust. J. Counc. Sci. Ind. Res. 16:69–73.

Stevenson, F. J., and M. S. Ardakani. 1972. Organic matter reactions involving micronutrients in soils. p. 79–114. *In* J. J. Mortvedt et al. (ed.) Micronutrients in agriculture. Soil Science Society of America, Madison, WI.

Stratton, S. D., and D. A. Sleper. 1979. Genetic variation and interrelationships of several minerals in orchardgrass herbage. Crop Sci. 19:477–481.

Sutcliffe, J. F. 1962. Mineral salts absorption in plants. Pergamon Press, Inc., Elsmford, NY.

Tahtinen, H. 1978. Determining the sensitivity of cereal varieties to copper deficiency in a pot experiment. Ann. Agric. Fenn. 17:147–151.

Takayanagi, H. 1976. Effects of EDTA on the absorption of manganese by tea plants and solubilization of manganese from tea soils. Chagy Gijutsu Kenkyu 51:44–52.

Takkar, P. N., M. S. Mann, R. L. Bansal, N. S. Randhawa, and Harder Singh. 1976. Yield and uptake response of corn to zinc as influenced by phosphorus fertilization. Agron. J. 68:942–946.

Tanaka, A., and Y. Hayakawa. 1975. Comparison of tolerance to soil acidity among crop plants. 2. Tolerance to high levels of aluminum and manganese—studies on comparative plant nutrition. J. Sci. Manage. 46(2):19–5.

Tanner, P. D. 1977. Toxic effects of manganese on maize as affected by calcium and molybdenum application in sand culture. Rhod. J. Agric. Res. 15:25–32.

Thomas, G. W., and B. Hipp. 1968. Soil factors affecting potassium availability. p. 269–280. *In* V. J. Kilmer et al., (ed.) The role of potassium in agriculture. American Society of Agronomy, Madison, WI.

Thompson, L. M., C. A. Black, and J. A. Zoellner. 1954. Occurrence and mineralization of organic phosphorus in soils, with particular reference to associates of nitrogen, carbon and pH. Soil Sci. 77:185–196.

Tiffin, L. O. 1967. Translocation of manganese, iron cobalt, and zinc in tomato. Plant Physiol. 42:1427–1432.

Trim, R. R. 1959. Metal ions as precipitants for nucleic acids and their use in the isolation of polynucleotides from leaves. Biochem. J. 72:298–304.

Ulmer, S. E. 1979. Aluminum toxicity and root DNA synthesis in wheat. Ph.D. thesis. Iowa State University, Ames, IA.

Van Breeman, N., and R. V. Castro. 1980. Zinc deficiency in wetland rice along a toposequence of hydromorphic soils in the Philippines. II. Cropping experiment. Plant Soil 57:215–221.

Van der Vorm, P. D. J., and A. Van Diest. 1979. Aspects of the iron and manganese nutrition of rice—*Oryza-sativa* plants. 1. Iron and manganese uptake by rice plants grown under aerobic and anaerobic conditions. Plant Soil 51:233–246.

Vander Zaag, R. L. Fox, P. K. Kwakye, and G. O. Obigbesan. 1980. The phosphorus requirements of yams (*Dioscorea* spp.) Trop. Agric. (Trinidad) 47:97–106.

Viestra, R., and A. Haug. 1978. The effect of Al^{3+} on the physical properties of membrane lipids in *thermoplasma acidophilium*. Biochem. Biophys. Res. Commun. 84:138–143.

Viets, F. G., T. C. Boawn, and C. L. Crawford. 1957. The effect of nitrogen and types of nitrogen carrier on plant uptake of indigenous and applied zinc. Soil Sci. Soc. Am. Proc. 21:197–201.

Vlamis, J., and D. E. Williams. 1973. Manganese toxicity and marginal chlorosis of lettuce. Plant Soil 39:245–251.

Vose, P. B. 1981. Crops for all conditions. New Scientist. 12 March, p. 688–690.

Vose, P. B. 1984. Effects of genetic factors on nutritional requirements of plants. *In* P. B. Vose and S. Blixt (ed.) Crop breeding—a contemporary basis. Pergammon Press, Inc., Elmsford, NY.

Waksman, S. A. 1923. Microbial analysis of soils as an index of soil fertility. V. Methods for the study of nitrification. Soil Sci. 15:241–260.

Walker, M. E., and T. C. Keisling. 1978. Response of five peanut cultivars to gypsum fertilization on soils varying in calcium content. Peanut Sci. 5:57–60.

Walker, W. M., and D. W. Graffis. 1979. Grass species and cultivar differences in magnesium concentration. Commun. Soil Sci. Plant Anal. 10:1239–1248.

Wallace, A., S. M. Sufi, and E. M. Romney. 1971. Regulation of heavy metal uptake and response in plants. p. 547–558. *In* R. M. Samish (ed.) Proc. 6th Int. Colloq. on Plant Analysis and Fert. Problems, Vol. 2. Gordon and Breach Science Publishers, New York.

Ward, G. M. 1977. Manganese deficiency and toxicity in greenhouse tomatoes. Can. J. Plant Sci. 57:107–115.

Wear, J. I. 1956. Effect of soil pH and calcium on uptake of zinc by plants. Soil Sci. 81:311–315.

Wear, J. I., and T. B. Hagler. 1963. Zinc status and needs of the southern regions. Plant Food Rev. 9:2–5.

Wear, J. I., and R. M. Patterson. 1962. Effect of soil pH and texture on the availability of water-soluble boron in the soil. Soil Sci. Soc. Am. Proc. 26:344–345.

Weber, D. F., and P. L. Gainey. 1962. Relative sensitivity of nitrifying organisms to hydrogen ions in soils and in solutions. Soil Sci. 94:138–145.

Welch, C. D., and W. L. Nelson. 1950. Calcium and magnesium requirements of soybeans as related to the degree of base saturation of the soil. Agron. J. 42:9–13.

Wells, B. R. 1980. Zinc nutrition of rice growing on Arkansas soils. Arkansas Agric. Exp. Stn. Bull. 848, p. 3–16.

White, M. C., F. D. Baker, R. L. Chaney, and M. A. Decker. 1981. Metal complexion in xylem fluid. II. Theoretical equilibrium model and computational computer program. Plant Physiol. 67:301–310.

White, R. E., and A. W. Taylor. 1977. Effect of pH on phosphate adsorption and isotopic exchange in acid soils at low and high additions of soluble phosphate. J. Soil Sci. 28:48–61.

Whiteaker, G., G. C. Gerloff, W. H. Gabelman, and D. Lindgren. 1976. Interspecific differences in growth of beans at stress levels of phosphorus. J. Am. Soc. Hortic. Sci. 101:472–475.

Whitney, A. L. 1923. Inorganic substances, especially Al, in relation to the activities of soil organisms. Agron. J. 15:277–289.

Williams, D. E., and J. Vlamis. 1957. The effect of silicon on yield and ^{54}Mn uptake and distribution in the leaves of barley plants grown in culture solution. Plant Physiol. 32:404–409.

Woodruff, J. R., and E. J. Kamprath. 1965. The phosphorus adsorption maximum as measured by the Langmuir isotherm and its relationship to phosphorus availability. Soil Sci. Soc. Am. Proc. 29:148–150.

Woolhouse, H. W. 1969. Differences in the properties of the acid phosphatases of plant roots and the significance in the evolution of edaphic ecotypes. p. 357–380. *In* I. Rorison (ed.) Ecological aspects of the mineral nutrition of plants. Blackwell Scientific Publications, Inc., Boston, MA.

Wuenscher, M. L., and G. C. Gerloff. 1971. Growth of *Andropogon scoparius* (little bluestem) in phosphorus deficient soils. New Phytol. 70:1035–1042.

Young, W. I., and B. C. Rasmussen. 1966. Variety differences in strontium and calcium accumulation in seedlings of barley. Agron. J. 58:481–483.

Younts, S. E., and J. G. Fiskell. 1963. Copper status and needs in the southern region. Plant Food Rev. 9:6–10.

Younts, S. E., and R. W. Patterson. 1964. Copper-lime interactions in field experiments with wheat. Yield and chemical composition data. Agron. J. 56:229–232.

5

Fertilizer-Plant Interactions in Alkaline Soils

David E. Kissel
Kansas State University
Manhattan, Kansas

D. H. Sander
University of Nebraska
Lincoln, Nebraska

R. Ellis, Jr.[1]
Kansas State University
Manhattan, Kansas

Alkaline soils are soils with a pH > 7. Soil chemical reactions involving a number of essential plant nutrients depend on soil pH. Soil pH influences both the availability of soil nutrients to the crop as well as chemical reactions that may affect losses of certain nutrients from the soil system.

The chemistry of alkaline soils is often dominated by calcium carbonate ($CaCO_3$) found in the soil. When a soil's $CaCO_3$ content is > 2 to 3% by weight, its pH will generally be in the range of 7.6 to 8.3.

The amounts of $CaCO_3$ in the soil may exceed 50% of the soil weight in some highly calcareous soils. Nevertheless, the pH of such soils will still generally be in the range of 7.6 to 8.3. Within this pH range, the carbon dioxide (CO_2) content of the soil atmosphere will have a major influence on the pH of a calcareous soil. A rise in the CO_2 content of the soil solution (due to an increase in microbial and/or root respiration) will generally result in a decrease in the soil pH because of the following reactions described by Lindsay (1979) and Turner (1958):

1. $CO_2(g) + H_2O \rightleftharpoons H_2CO_3$ pK = 1.46
2. $H_2CO_3 \rightleftharpoons H^+ + HCO_3^-$ pK = 6.36
3. $CaCO_3(s) + H_2O \rightleftharpoons Ca^{2+} + CO_3^{2-} + H_2O$ pK = 8.35
4. $CO_3^{2-} + H^+ \rightleftharpoons HCO_3^-$ pK = −10.33

The result of increased CO_2 in the soil is an increase in carbonic acid (H_2CO_3), which in turn increases H^+ in the soil. The final result of increased CO_2 is an increase in Ca^{2+} and HCO_3^- in the soil solution as well as the drop in pH.

[1]Deceased 9 Sept. 1982.

If substantial quantities of Na are present in a calcareous soil, the soil pH may often exceed 8.3 and can be as high as 10 in extreme cases of sodic soils (U.S. Salinity Lab. Staff, 1954, p. 160). In these cases, the presence of Na allows soluble sodium carbonate (Na_2CO_3) in the soil solution to raise the soil pH above those normally found in calcareous soils.

The presence of $CaCO_3$ in soils has both a direct and sometimes an indirect influence on availability of various plant nutrients essential to crops. In particular, availability of P, Fe, and Zn are most directly influenced by $CaCO_3$. The application of S is sometimes used to control availability of other nutrients by neutralizing $CaCO_3$ and reducing soil pH. The chemistry of N and K is also influenced by $CaCO_3$ presence in the soil, although they are less directly influenced. Each of these nutrients will be discussed in more detail.

I. NITROGEN TRANSFORMATIONS

Soil pH affects the rates of several soil chemical and biochemical reactions involving N and, as a result, influences the efficiency of N fertilizer use by crops. Because pH values of calcareous soils are generally buffered within the pH range of 7.6 to 8.3 (or higher in the case of sodic-calcareous soils), the reaction rates will be influenced accordingly. Some N-cycle reactions are more sensitive to pH than others. In particular, the processes of nitrification and volatilization of ammonia (NH_3) are especially sensitive to soil pH. Other N-cycle processes such as urea hydrolysis and denitrification, which influence applied N fertilizer, are less sensitive to soil pH. Each of these processes will be discussed separately.

A. Nitrification

1. Process Description

Nitrification, in general terms, means the conversion of ammonium (NH_4^+) to nitrate (NO_3^-). Autotrophic nitrifiers are primarily responsible for the two-step reaction, given by Eq. [1] and [2].

$$2\,NH_4^+ + 3O_2 \longrightarrow 2\,NO_2^- + 2\,H_2O + 4\,H^+ \qquad [1]$$

$$2\,NO_2^- + O_2 \longrightarrow 2\,NO_3^- \qquad [2]$$

Equation [1] is accomplished by *Nitrosomonas* spp. and Eq. [2] by *Nitrobacter* spp. The details of the nitrification process are well known and the reader is referred to Focht and Verstraete (1977) for more detail on the biochemical ecology of nitrification and to Keeney (1980) for a discussion of nitrification inhibition in agronomic systems.

2. Effect of Soil pH on Nitrification

Nitrification by autotrophic bacteria proceeds most rapidly in soils of neutral to slightly alkaline pH (pH 7 to 8). In acid soils, the rate of autotrophic nitrification is slowed and approaches zero at around pH 5 (Morrill & Dawson, 1967). Nitrification can occur at soil pH 5, but is due primarily to heterotrophic nitrifiers (Focht & Verstraete, 1977).

The influence of soil pH on the activity of autotrophic nitrifiers can manifest itself in two ways. First, the numbers of autotrophic nitrifiers that live in soil are quite sensitive to soil pH. For example, Pang et al. (1975) measured the number of nitrifiers in three soils with pH varying widely from very acid (pH 5.4) to calcareous (pH 8.2). The numbers of *Nitrosomonas* spp. were around 1000 counts/g of dry soil on the neutral and calcareous soils; the number dropped to 7 counts/g of acid dry soil. Liming the acid soil did not improve its nitrifying capacity appreciably until the soil was inoculated with nitrifiers. Others have also shown low numbers of autotrophic nitrifiers in acid soil (Herlichy, 1973; Ayanaba & Omayuili, 1975; Cooper, 1975).

The second influence of soil pH on autotrophic nitrification is related to the rate of increase of nitrifier populations and/or their activity per unit of population. Nitrifier populations increase in response to the addition of NH_4^+ substrate to the soil. Several studies noted that the optimum pH for autotrophic nitrification is in the range of 7 to 8 (Myers, 1974; Dancer et al., 1973; Frederick & Broadbent, 1966). More specifically, however, the soil pH range of 7 to 8 is least likely to allow buildup of the two inorganic N compounds that inhibit the metabolism of the nitrifying organisms. Inhibition at soil pH above the optimum for nitrification is caused by ammonia toxicity, while inhibition at soil pH less than the optimum is due to nitrous acid (HNO_2) toxicity (Anthonisen et al., 1976). A graph showing the nitrifiers tolerance to NO_2^-–N and NH_4^+–N as a function of pH is given by Focht and Verstraete (1977).

3. *Acidity Produced During Nitrification*

Another consideration of nitrification relative to a soil's pH and buffering properties is the acidity produced during the first step of nitrification, as shown in Eq. [1]. When nitrification proceeds on poorly buffered, noncalcareous soils, the production of H^+ in the first step of nitrification can slow the nitrification rate considerably by lowering soil pH. On the other hand, a calcareous soil would be buffered against a large drop in pH, so that nitrification would proceed normally after application of ammoniacal N fertilizer, as demonstrated by the work of Jones and Hedlin (1970) and Pang et al. (1975). Jones and Hedlin (1970) also demonstrated that on acid soils those N fertilizer sources with an initial alkaline reaction after soil application such as urea and anhydrous ammonia would have a more rapid nitrification rate than acid fertilizers like ammonium sulfate [$(NH_4)_2SO_4$]. On calcareous soils, however, all ammoniacal N sources would nitrify at a faster and possibly equal rate due to the pH buffering of $CaCO_3$. At high rates of application to a calcareous soil, N sources such as anhydrous ammonia or urea may result in toxic levels of ammonia at the site of application, which would temporarily slow nitrification; an acid N source such as ammonium nitrate (NH_4NO_3;AN) may nitrify more rapidly immediately after application.

4. Effect of Salinity on Nitrification

Alkaline soils may often be saline in arid and sometimes in semiarid regions. Soil salinity generally reduces the rate of nitrification. Johnson and Guenzi (1963) found a linear decrease in nitrification as soil solution osmotic potential increased. Nitrification generally ceased at an osmotic potential of −30 bar. They also noted that the organisms in calcareous soils were slightly more tolerant of salinity than those in the noncalcareous soil. Johnson and Guenzi (1963) also noted a slight influence of salt type on microbial activity. Their work was expanded by Westerman and Tucker (1974), who also showed decreased nitrification with higher salt concentrations in the soil solution. They found that high concentrations of $CuCl_2$ and $CaCl_2$ salts inhibited nitrification more than Na salts.

B. Ammonia Volatilization

1. General

Those aspects of ammonia volatilization related to the influence of $CaCO_3$ and fertilizer technology will be covered here. More detail on ammonia volatilization can be obtained from recent reviews on the subject (Terman, 1979; Freney et al., 1981; Nelson, 1982).

Ammonia volatilization from the application of ammoniacal N fertilizer will generally be of practical significance only if the pH is > 7 at the site of fertilizer application (Martin & Chapman, 1951). For the soil to exceed pH 7 significantly, the soil must be calcareous or the reaction of the fertilizer material must be sufficiently alkaline to raise the soil pH into the alkaline range. This latter condition may be met by application of anhydrous ammonia or N fertilizers that contain urea. Also, in order for ammonia loss from a soil to be significant, the pH of the soil must be sufficiently alkaline between the site of placement and the soil surface. Losses by ammonia volatilization are most severe when fertilizer is applied to the soil surface without incorporation into the soil (Ernst & Massey, 1960; Fenn & Kissel, 1976).

2. Ammonia–Ammonium Equilibrium

After fertilizer application, loss of ammoniacal N by volatilization can only occur as ammonia. The amount of ammonia present in the soil solution at any time depends strongly on the soil pH from the expression

$$NH_4^+ \rightleftharpoons NH_3 + H^+. \qquad [3]$$

The pKa for Eq. [3] is 9.28. From Eq. [3], the ammonia/ammoniacal N ratio in the mixture can be calculated. After examination, the proportion of N in the ammonia form would be approximately 50% at pH 9.3, 10% at pH 8.3, and 1% at pH 7.3.

When the ammoniacal N in a solution has a pH < 7, the work of Winter et al (1981) indicated volatilization of ammonia occurred at a low rate

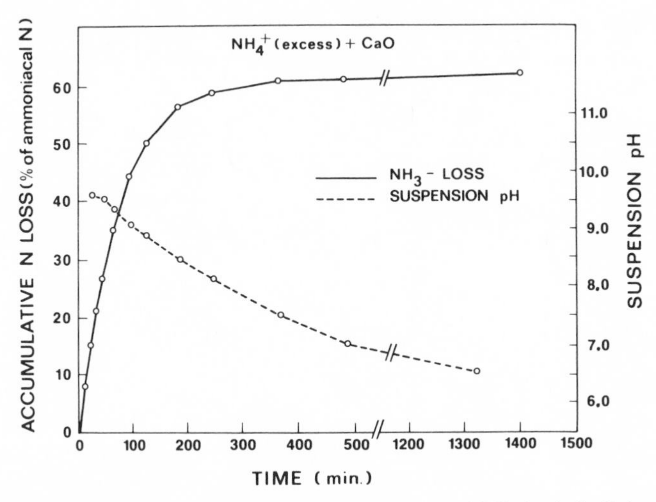

Fig. 5–1. Loss of ammonia from a solution of ammoniacal N with limited alkalinity and excess ammonium (NH_4^+) (Winter et al., 1981).

since most of the ammoniacal N was ionized NH_4^+. In this research (Fig. 5–1.), excess ammoniacal N was stirred with a limited quantity of calcium hydroxide [$Ca(OH)_2$] and the pH dropped as ammonia was volatilized. It can be noted in Fig. 5–1 that ammonia volatilization losses from the vigorously stirred solution of excess ammoniacal N with limited OH^- ions slowed considerably as the pH dropped to around 7 due to loss of ammonia (and loss of alkalinity; see Vlek & Stumpe, 1978) and the loss rate approached zero at pH 6.5.

3. Soil and Environmental Influences

The soil pH that results after fertilizer application is determined by the soil's initial pH, its buffering capacity, and the chemical reactions taking place in the soil immediately after application (Avnimelech & Laher, 1977; Fenn & Kissel, 1973). It is also quite apparent that a host of dynamic soil physical properties interact with the chemistry at the site of fertilizer application to determine the rate of chemical reactions and loss of ammonia from the soil. Clearly, soil temperature and soil water content influence ammonia volatilization (Ernst & Massey, 1960; Fenn & Kissel, 1974; Fenn & Escarzaga, 1976), and these are most dynamic at the soil surface, often due to rapidly changing environmental conditions of air temperature, solar radiation, air velocity, and rainfall.

4. Ammonia Losses from Ammonium Fertilizers on Calcareous Soils

a. Mechanisms Involved—Ammonia losses should be greatest from surface application of ammonium fertilizers to calcareous soils. Studies of such situations indicate that substantial loss of ammonia can occur (Jewitt, 1942; Steenbjerg, 1947; Martin & Chapman, 1951).

The effect of $CaCO_3$ is more than that of a buffer at the soil surface maintaining sufficiently high pH values to cause ammonia loss. Rather, a chemical reaction of ammonium compounds with $CaCO_3$ can occur to form an unstable solution with NH_4^+, HCO_3^- and CO_3^{2-} ions.

The degree to which this reaction proceeds appears to depend on the anion associated with the NH_4^+ fertilizer (Terman & Hunt, 1964; Fenn & Kissel, 1973). Those fertilizers that react to form Ca-reaction products of low solubility [ammonium sulfate will produce gypsum ($CaSO_4$) of low solubility] will lose considerably more ammonia than those ammonium fertilizers that produce Ca-reaction products of relatively higher solubility; AN will produce highly soluble calcium dinitrate [$Ca(NO_3)_2$]. The less-soluble Ca reaction product will force the reaction to form more CO_3^{2-} and especially bicarbonate (HCO_3^-) ions (Feagley & Hossner, 1978) in solution with a resulting general rise in pH, often above the initial pH of the calcareous soil. This rise in pH will, of course, increase the rate of ammonia loss. The reaction described above can be written for the case of ammonium sulfate as follows, assuming a bicarbonate intermediate as proposed by Feagley and Hossner (1978):

$$2\,CaCO_3 + (NH_4)_2SO_4 \rightleftharpoons Ca^{2+} + 2\,OH^- + CaSO_4 + 2\,NH_4HCO_3. \quad [4]$$

In solution, ammonium bicarbonate (NH_4HCO_3) is unstable and decomposes to allow loss of ammonia and carbon dioxide, as follows:

$$NH_4HCO_3 \rightleftharpoons NH_3(g) + CO_2(g) + H_2O. \quad [5]$$

As can be seen from Eq. [4] and [5], the reaction uses $CaCO_3$ in sustaining the loss of ammonia. To give an example using these equations, the application of 314 kg N/ha to the soil surface would use 0.75% $CaCO_3$ in the top 1 cm of soil, assuming that the reaction was restricted to the top 1 cm of soil. The reactions also suggest that the amount of $CaCO_3$ in the soil may influence the rate and amount of ammonia loss.

b. Effect of Calcium Carbonate Content—Early research in the Netherlands suggested that the amount of $CaCO_3$ and its particle size influenced ammonia volatilization in reclaimed polders (Harmsen & Kolenbrander, 1965). More recent laboratory research (Fenn & Kissel, 1975) has shown an increase in ammonia loss from surface-applied ammonium sulfate as percent $CaCO_3$ in the soil increased to approximately 10%. With further increases in $CaCO_3$, the ammonia loss did not increase. The $CaCO_3$ used in the latter study was finely divided reagent-grade $CaCO_3$. Later research by Ryan et al. (1981) has shown that the particle size of $CaCO_3$ also influences the rate of ammonia loss. In the case of very fine $CaCO_3$, their results agreed closely with those of Fenn and Kissel (1975). Ryan et al. (1981) described the soil $CaCO_3$ in a range of soils according to their percent reactive $CaCO_3$, which was a function of $CaCO_3$ particle size as well as the percent $CaCO_3$ in the soils. A graph of ammonia lost as a function of soil-reactive $CaCO_3$ taken from their work is shown in Fig. 5–2.

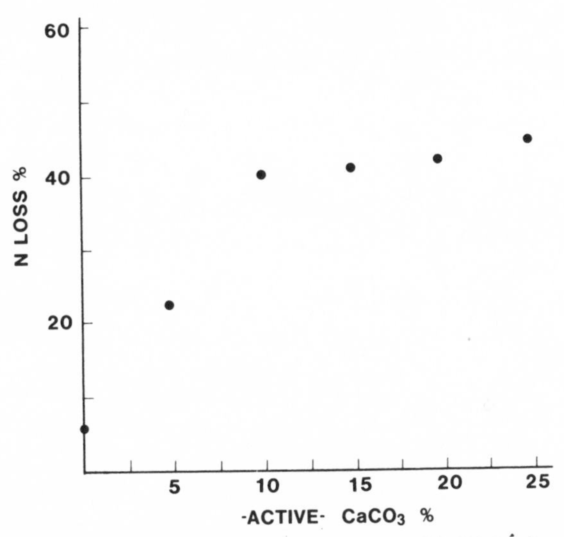

Fig. 5–2. Volatile loss of ammonia from ammonium sulfate [$(NH_4)_2SO_4$] surface-applied to soil with variable active calcium carbonate ($CaCO_3$) (Ryan et al., 1981).

c. Ammonia Losses from Limed Acid Soils—Liming the surface of an acid soil without incorporation by tillage (as in the case of grass sods or no-tillage cropping) will temporarily make the surface layer of an acid soil calcareous. For example, if 2 metric tons of $CaCO_3$ is applied per hectare and it mixes with the top 6 mm of soil, the soil would contain approximately 6% $CaCO_3$ in that layer. Subsequent applications of ammonium fertilizers could then be subject to substantial loss of N by ammonia volatilization. Evidence of such losses from a limed turf were presented by Volk (1961). He found losses of approximately 3% from surface-applied AN but nearly 20% from ammonium sulfate. More recent work by Winter et al. (1981) also showed increased ammonia volatilization when lime was applied with N solution.

d. Ammonia Loss from Ammonium Fertilizers and Fertilizer Mixtures—Fenn and Kissel (1973, 1974) showed that ammonium fertilizers, such as ammonium sulfate, that react with $CaCO_3$ to form Ca-reaction products of low solubility lose more ammonia and at a faster rate than ammonium fertilizers that do not form precipitates with Ca^{2+} [AN and ammonium chloride (NH_4Cl) are examples]. Formation of Ca-precipitates apparently leads to formation of ammonium bicarbonate in the soil solution, which promotes a high pH at the soil surface conducive to ammonia volatilization. The addition of more highly buffered acidic compounds with materials such as ammonium sulfate can reduce volatilization losses. Fenn (1975) found that the addition of monoammonium phosphate (MAP) to either ammonium sulfate or ammonium fluoride (NH_4F), with MAP constituting 30% of the N mixture, markedly reduced ammonia loss. Although Fenn (1975) explained the reduced loss in terms of less precipitation with mixtures containing MAP, it seems more likely that the lower buffered pH of MAP (around 4.5) created more resistance to a rise in pH as the ammonium sulfate and ammonium fluoride reacted with $CaCO_3$, thus reducing ammonia losses. The pH explanation seems more plausable since the reaction products of MAP and diammonium phosphate (DAP) should be about the same, yet loss of ammonia is considerably greater from DAP than from MAP. Bremner and Douglas (1971) found a similar reduction in soil pH when urea was applied as urea-phosphate ($NH_2CONH_2 \cdot H_3PO_4$). As a result, their average ammonia loss of 21% from urea was reduced to only 1% loss from urea-phosphate. The reduced volatilization loss could not be ascribed entirely to the pH reduction of phosphoric acid (H_3PO_4) since the hydrolysis rate with urea-phosphate was slower compared with urea alone. Nevertheless, pH reduction by the phosphoric acid was no doubt a substantial factor in reducing ammonia loss.

5. *Ammonia Volatilization from Urea Fertilizers*

Fertilizers that contain urea as part of all of their N source can lose substantial amounts of ammonia when surface-applied to acid soils. Ammonia loss is possible on acid soils, since the hydrolysis of urea causes an increase in alkalinity at the site of fertilizer placement and may easily raise the soil surface pH above 7 in some instances. With a calcareous soil, the addition of alkalinity by urea hydrolysis may raise the pH even higher and result in more loss than from an acid soil. This principle was illustrated well by the work of Ernst and Massey (1960), who found increasing ammonia loss from surface-applied urea as the pH of an acid Dickson silt loam (Glossic Fragiudults) was increased by liming pretreatments.

An additional influence of calcareous soils on ammonia loss is the generally high amount of exchangeable Ca^{2+} (and Mg^{2+}) that can influence ammonia loss. The amount of Ca^{2+} and Mg^{2+} depends on the soil cation exchange capacity (CEC), which would largely be saturated with Ca^{2+} and, to a lesser extent, with Mg^{2+} in a calcareous soil. Additionally, some Ca^{2+} and Mg^{2+} exist in soil solution as well. Fenn et al. (1981a,b) have described the influence of Ca^{2+} and Mg^{2+} in reducing ammonia loss from surface-applied urea. The influence can be described by the following reactions for a Ca^{2+} saturated soil:

$$\text{Urea} + 3\,H_2O \xrightarrow[\text{enzyme}]{\text{urease}} (NH_4)_2CO_3 \qquad [6]$$

$$(NH_4)_2CO_3 + Ca^{2+}\text{–Soil} \rightleftharpoons CaCO_3 + 2\,NH_4^+\text{–Soil}. \qquad [7]$$

Since the ammonium carbonate [$(NH_4)_2CO_3$] is removed from solution by the second reaction and exchangeable NH_4^+ results (which is nonvolatile), exchangeable or soil solution Ca^{2+} helps to reduce ammonia loss. Exchangeable Mg^{2+} or soil solution Mg^{2+} acts in much the same way. Fenn and co-workers also showed that the addition of Ca^{2+} salts, such as calcium chloride ($CaCl_2$) with the urea or ammonium fertilizers, can further reduce ammonia volatilization. In addition to Eq. [6] and [7], the addition of soluble Ca^{2+} and Mg^{2+} salts also reduces soil pH slightly (irrespective of Eq. [6] and [7]), which in turn can slow losses of ammonia.

C. Urea Hydrolysis

Hydrolysis of urea in soils has received considerable research attention since the early 1940s. Recent reviews (Bremner & Mulvaney, 1978; Mulvaney & Bremner, 1981; Ladd & Jackson, 1982) provide an up-to-date analysis of our knowledge on urea hydrolysis and urease activity in soils.

The activity of urease in calcareous soils is similar to its activity in noncalcareous soils. Urease activity is moderately sensitive to pH, as shown by the work of Pettit et al. (1976) and Nannipieri et al. (1978). The optimum pH given by these authors varies somewhat with the source of urease, but is generally in the range of 7 to 8.5. Although this is in the range of most calcareous soils, it is doubtful that actual hydrolysis rates of applied urea fertilizer would be much faster in calcareous soils since (i) the site of urea application will become alkaline rather quickly as hydrolysis begins, even in moderately acid soils, and (ii) the hydrolysis rate is only moderately slower at pH 6 than at pH 7.5 (Pettit et al., 1976).

D. Denitrification

Denitrification is the biological process by which NO_3^- is reduced to nitrogen gases, usually N_2 and N_2O. Under agronomic practices, denitrification can be an important mechanism of NO_3^- loss from the soil, particularly under very wet soil conditions (Ralston et al., 1978). Other factors also influence denitrification and a recent review by Firestone (1982) provides a comprehensive analysis of our knowledge on denitrification.

Denitrification is relatively insensitive to pH. In the pH range of most agronomic soils (approximately 6 to 8), there is little effect of pH on denitrification rates (Burford & Bremner, 1975; Stanford et al., 1975). Hence, denitrification losses from calcareous soils would probably differ little from those of noncalcareous soils.

E. Influence of Ammonia on Germination and Seedling Growth in Alkaline Soils

From the earlier discussion of ammonia volatilization, it follows that higher levels of free ammonia may exist in alkaline soils than in acid soils after application of some sources of N fertilizer. The effect of free ammonia on corn (*Zea mays* L.) germination and seedling growth was well demonstrated by the work of Allred and Ohlrogge (1965), Cummins and Parks (1961), and Colliver and Welch (1970). Allred and Ohlrogge (1965) found that continuous exposure of germinating corn for 7 d to vapor pressures of ammonia as low as 0.063 mm of Hg severely retarded the germination process. In addition, they showed that DAP retarded emergence of corn seedlings in sand at much lower concentrations than MAP. The addition of 3% $CaCO_3$ to the sand had an additional and strong negative influence on seedling emergence, especially for DAP, apparently due to more ammonia being formed. Their results agree well with the above discussions about the amounts of ammonia loss resulting from various ammonium fertilizers and the influence of soil $CaCO_3$ content (sections I B 4 a, b, and d). It seems likely that the most severe germination and seedling damage from seed placement of fertilizer would occur from sources of fertilizer and soils that form the most ammonia soon after application.

II. PHOSPHORUS

Phosphorus is one of the essential elements for plant nutrition that may have limited availability in calcareous soils. Calcium carbonate dominates the chemistry of these soils. The amount of surface area of the $CaCO_3$ has an influence on the chemical reactions that P undergoes, particularly the rates of the chemical reactions. Free $CaCO_3$ controls the activity of Ca in the soil solution, which also influences the reactions of P in soils.

The chemistry of the native soil P is useful in supplying information on the forms that added P will eventually reach in soils. This topic will not be covered in any detail in this chapter. Primary emphasis will be on chemical reactions that take place when fertilizer P is added to calcareous soils and the influence these reactions have on P availability to plants. The book *The Role of Phosphorus in Agriculture,* edited by Khasawneh et al. (1980), provides an excellent discussion on all phases of P in agriculture. Reactions of native soil P have been discussed in detail by Larsen (1967), Olsen and Flowerday (1971), and Lindsay and Vlek (1977). Readers are referred to these sources for a more thorough discussion of P chemistry in calcareous soils.

A. Factors Determining Availability of Phosphorus to Plants in Calcareous Soils

1. General Discussion

Concentration of P in the soil solution is the factor most closely related to the P availability to plants. This has been designated the intensity factor by most research workers. Khasawneh (1971) defined an intensity factor and a relative factor. The *intensity factor* is a measure of the gradient in the electrochemical potential across the absorbing surfaces of plant roots and the relative factor is the effect of other ions in solution on P uptake. The *relative factor* is nearly one for P because other ions do not compete to any extent with P uptake.

The intensity factor alone is inadequate to describe P availability to plants. Since the amount of P in the soil solution at any time is small compared with crop requirements for P, it is also necessary to describe the ability of the soil system to renew the P in the soil solution (quantity factor) as crop uptake depletes it. The terms quantity factor, capacity factor, and labile P have been used to describe this factor. *Quantity factor* is probably the best term. Olsen and Khasawneh (1980) stated that the term *capacity* has been used to describe gradients that relate quantity and intensity factors. *Labile P* is normally considered to be the fraction of soil P that is isotopically exchangeable with ^{32}P within a specified time period. The amount of labile P is related to P availability to plants, but may not completely describe the quantity of any particular form of P present in the soil.

A useful concept might be to consider a quantity factor and a capacity factor for each form of P in the soil system. The capacity factor would describe the ability of a particular form of P to furnish P to the soil solution. There would be a quantity factor for adsorbed P, diacalcium phosphate dihydrate (DCPD), octacalcium phosphate (OCP), hydroxyapatite (HA), etc. Each quantity factor would have a capacity to furnish a certain amount of P to the soil solution. If a certain form of P is depleted, that quantity factor would become zero. The concept of having a series of quantity and capacity factors associated with different forms of P in soils is complex, but the chemistry of P in soils is also complex. The methods used to date probably measure an average quantity factor and, although useful in describing the amount of P available to plants, do not adequately describe the complex chemistry of soil P.

2. Chemistry of Phosphorus in Calcareous Soils

Unfortunately, in most cases research workers have studied the chemistry of P added to calcareous soils as either an adsorption or a precipitation process. In general, low concentrations of added P have been studied as an adsorption process and higher concentrations of added P have been studied as precipitation reactions (Cole et al., 1953). The Langmuir adsorption isotherm has been applied to the adsorption of P by calcite and calcareous soils (Cole et al., 1953; Kuo & Lotse, 1972). The fact that P precipitates and forms crystalline P compounds in soils has been shown by Lindsay et al., 1962; Bell and Black, 1970a,b,c; and Subbarao and Ellis, 1977.

In reality, adsorption is likely the initial step in the precipitation process. Griffin and Jurinak (1973, 1974) concluded that two reactions occurred when P was added to calcite. The first reaction was the adsorption of P at the calcite surface. In the second reaction, amorphous calcium phosphate formed at the calcite surface and was subsequently transformed into crystalline calcium phosphates. Sample et al. (1980) stated, "P retention should be regarded as a continuum embodying precipitation, chemisorption and adsorption." It is regretable that this concept has not been pursued by more research workers studying the chemistry of P in soils. The concept that the chemistry of P added to calcareous soils is a continuum of adsorption, formation of amorphous calcium phosphate compounds, followed by a gradual change to crystalline phosphate compounds will be adopted in this discussion. From a practical standpoint, the availability of P added to a crop in one season is not determined by the final equilibrium form of added P, but rather is determined by the rates of reaction involved in going through the various stages. Therefore, kinetics control the availability of added P to crops during a growing season.

In high-pH soils, the initial P compound formed and the rates of formation of subsequent compounds will be controlled by the ratios of Ca^{2+} and P activities in the soil system. As a first approximation, the activity of Ca^{2+} in solution can be regarded as a constant controlled either by equilibrium with $CaCO_3$ or by dissociation of Ca^{2+} from soil colloids. Therefore, the rate of application of P will determine the ratio of Ca^{2+} to P activities and the calcium phosphate compound formed. The influence of the rate of P application on the calcium phosphate compound formed and the rates of reaction has been shown by Subbarao and Ellis (1977) and by Amer and Mostafa (1981). If sufficient soluble P is added to a calcareous soil, DCPD will be formed. As the concentration of P is lowered in the solution with time, the DCPD will dissolve and OCP will form. When all the DCPD has dissolved, the amount of P in the soil solution *(intensity factor)* will be controlled by OCP. When the amount of P in the soil solution will no longer support OCP, it will dissolve and HA will form. If HA is the stable form in a particular soil, this will be the final calcium phosphate compound formed and will represent the lower limit of P in the soil solution and the lower limit of P available to plants. Consequently, the availability of P to plants will be controlled by the rate of application of soluble P, which controls the calcium phosphate compounds formed and the rates of transformation from one compound to another. Enfield et al. (1981) proposed a model to predict the rates of reaction for different soils based on measurement of certain parameters. This study, based on addition of P in waste waters, provides some useful information, but needs refinement as a general model for application of soluble P materials to soils.

There is agreement among researchers that DCPD will maintain more P in the soil solution than is necessary for maximum plant growth, and that the amount of P in the soil solution that is required for maximum plant growth is adequately supplied by OCP (Olsen & Khasawneh, 1980). If these reaction products can be maintained throughout a growing season, sufficient P should be present for maximum growth of plants. However, if HA is the compound controlling the amount of P in the soil solution, P will be limiting for maximum growth of most agronomic crops.

3. *Method of Application*

It has been common knowledge for many years that banding phosphate fertilizer on some soils results in greater availability of P and higher yields of crops than when the same quantity of P is applied broadcast and worked into the soil. The difference in efficiency between banded and broadcast applications of P materials is usually greater on soils testing low in available P and during seasons in which adverse climatic conditions occur. Differences are especially noticeable when cold and wet conditions limit the uptake of P by crops. In some cases a rule of thumb has been used to recommend twice as much P if the fertilizer is broadcast rather than banded. There seems to be little scientific information to support this concept. The relative efficiency (band/broadcast) may vary from 3:1 for soils low in available P to 1:1 for soils testing medium to high in available P when fertilizer P is being applied to maintain a given test level.

The relative efficiency of broadcast P compared with banded P is determined by the concentration of P per volume of soil, which in turn determines which P compounds are formed in calcareous soils. If the same P compounds are formed when P is applied either broadcast or banded, similar crop availability of P should be expected, provided the band can be intercepted by the plant roots. However, if a higher concentration of P results from banding the P in a small volume of soil, more soluble P compounds will be formed in the band and cause greater uptake of the fertilizer P by plants. Heterogeneity exists in the microenvironment around each fertilizer granule with broadcast application of P, but concentration of the P in a smaller volume of soil with band applications increases the heterogeneity and results in drastic alterations in the chemical properties of the soil in the band. The pH may either increase or decrease, depending on the pH of the saturated solution of the fertilizer materials. Soil cations will be displaced and soil minerals and organic matter may be dissolved in the band. Sample et al. (1980) pointed out that research is needed to determine the concentration of ions and the reaction products formed at various distances from the point of application in the band. Changes that take place over time should also be evaluated.

Research is needed to determine the influence of subsequent tillage operations at various time periods on the changes in reaction products with time compared with leaving the bands intact in the soil. This information would be very useful in comparing no-tillage, reduced-tillage, and conventional-tillage systems. Engelstad and Terman (1980) report that several research workers have found uptake of P from surface application was equal in effectiveness to P incorporated into the soil in minimum-tillage systems. Most of the reported work was done in the more humid regions in the eastern USA, but there may be problems with broadcast-surface application of P in no-tillage cropping systems in the drier regions where surface soils are often dry for periods during the growing season. On the other hand, deep placement of P and N in minimum-tillage systems could have the advantages of banding plus the advantage of having the fertilizer materials in moist soil.

Advantages for deep placement of N and P in the same band have been reported by Murphy et al. (1976, 1978) and Leikam et al. (1983), but this method needs to be evaluated for different cropping and tillage systems. Research in this area is needed because reduced-tillage systems are being adopted to conserve energy and reduce soil loss by erosion. With increasing prices for fertilizer materials, it is important to obtain information on how to maximize efficiency of fertilizer use with different tillage and cropping systems.

B. Phosphate Fertilizer Materials

1. Orthophosphates

The relatively insoluble ground phosphate rock (PR) materials are of no value in supplying P for plants in calcareous soils. This should be obvious since most calcareous soils contain considerable amounts of calcium phosphate compounds with equal or greater solubility. When soluble orthophosphate fertilizers are added to calcareous soils, differences between P sources are likely to be small if the fertilizers are broadcast and distributed throughout the soil. Heterogeneity does exist in the microenvironment around granules, but the effects will probably not last very long. If the fertilizers are concentrated in a band, there may be important differences in reaction products formed from different fertilizers. Saturated solutions of monocalcium phosphate (MCP) and MAP are acid, while a saturated solution of DAP is alkaline. Sample et al. (1980) and Olsen and Flowerday (1971) report that DCPD is the most likely reaction product formed when soluble orthophosphate fertilizers are added to calcareous soils. However, Bell and Black (1970b) reported that some OCP was formed when DAP was added to calcareous soils. Research is needed to find if these differences are significant in determining the P availability to plants.

2. Polyphosphates

The use of fertilizers containing polyphosphates adds to the complexity of P chemistry in soils. Polyphosphates are higher-analysis fertilizers that result in lower transportation costs, although some additional cost is involved in producing them. Polyphosphates work well in solutions, and they can sequester some of the micronutrients that can aid in handling clear liquid fertilizers.

Polyphosphates must be hydrolyzed to orthophosphate before the P can be utilized by plants. After polyphosphates have been hydrolyzed to orthophosphate, the chemistry should be the same as that of orthophosphates in soils. Availability of P from polyphosphates depends on the solubility of precipitates of polyphosphates in soils and the rate of hydrolysis of the soluble polyphosphate to orthophosphate.

Polyphosphates can undergo enzymatic hydrolysis and chemical hydrolysis. Chemical hydrolysis is very slow at normal soil temperatures so the rate of enzymatic hydrolysis is the factor controlling the rate of hydrolysis to orthophosphate in most cases. Lindsay (1979) points out that precipitates of polyphosphates are not stable in soils and that $B\text{–}Ca_2P_2O_7$ is more soluble than $CaHPO_4 \cdot 2H_2O$. Therefore, solubility of the polyphosphate precipitates should not limit the availability of P from polyphosphates. The rate of hydrolysis of poly- to orthophosphate could limit the availability of P to plants if enzymatic activity is low. On the other hand, if the rate of hydrolysis could be controlled, it might result in keeping P available to plants longer, even though polyphosphates will eventually be converted to orthophosphate in soils. Research is needed in this area, because most field experiments show the two sources are equally effective from an agronomic standpoint. In a few isolated cases, there have been reports where one of the sources is superior. Reasons for the superiority of either ortho- or polyphosphate in these isolated cases needs to be determined.

C. Water Solubility of Phosphate Fertilizers

The importance of water solubility of phosphate fertilizers is a question that has persisted for many years. Webb and Pesek (1958) conducted a series of field experiments and concluded that for hill or row fertilization, at least 60% of the P should be water-soluble. Lawton et al. (1956) reported that for phosphate fertilizer applied in a row or band, not less than 40% of the P should be water-soluble. Webb and Pesek (1959) found that water solubility was not important when P was broadcast and incorporated for corn grown on acid to neutral soils; in another study, Webb et al. (1961) found that water solubility of P did increase yields of corn when P was broadcast and incorporated on calcareous soils in Iowa. Norland et al. (1958) concluded that P applied in solution and P applied in the form of very soluble coarse granules were about equally effective on the soils used in their experiments. Therefore, the decision whether to use materials in liquid or solid form needs to be made on the basis of price, equipment available, and ease of application.

Uptake of P decreases with decrease in soil temperature. Considering this fact, one would expect water solubility of P to be a greater factor in colder areas with shorter growing seasons for crops. The combination of high water solubility of P and banding of the fertilizer material should be advantageous on low P testing soils when soils are cold during the early part of the growing season.

D. Determining Availability of Phosphorus in Soils

1. Soil Tests

Soil tests have been the most common method used to determine availability of P for plants. Correlation of soil tests for available P with actual response by crops to applications of P has been very useful in making P fertilizer recommendations. An ideal P-extracting solution should remove from soils the forms of P that furnish significant amounts of P for use by plants without removing forms that do not contribute to P use by plants. This is difficult to achieve with any P-extracting solution. Olsen et al. (1954) showed the bicarbonate test to be the most consistently correlated with P uptake by plants in calcareous soils. The extracting solution used with the Bray and Kurtz P-1 test has given good correlations with uptake of P by plants grown in calcareous soils, particularly when the soil/solution ratio was widened from the usual 1:10 to 1:50 (Smith et al., 1957). The soils used in that study all contained $< 8\%$ $CaCO_3$, however; the 1:50 soil/solution ratio may not work as well on soils with greater $CaCO_3$ contents.

2. Phosphate Potentials

For this discussion, *phosphate potentials* are defined as a function of phosphate activities and are described further by Olsen and Khasawneh

(1980). Although phosphate potentials have limitations, they are useful in identifying P compounds in soils that cannot be isolated and identified by x-ray diffraction, optical microscopy, or infrared spectroscopy methods. Since the original work with phosphate potentials by Aslyng (1954), this technique has been used by such researchers as Lindsay and Moreno (1960), Taylor and Gurney (1962), Weir and Soper (1963), Withee and Ellis (1965), and Subbarao and Ellis (1977) to identify P compounds in soils and to determine changes in phosphate compounds with time after P has been applied to soils. Use of phosphate potentials has provided useful information and is a good technique if its limitations are understood.

Phosphate potentials are useful in identifying P compounds in soils and in following changes in P compounds with time after P application to soils. This can be done by comparing activity products and the solubilities of known compounds or as shown graphically by Lindsay and Moreno (1960). Sample et al. (1980) point out some limitations in determining phosphate potentials. One disadvantage in interpreting phosphate potentials after they have been measured is the fact that the potentials only identify the most soluble source furnishing P to solution. Also, phosphate potentials do not measure the quantity of the P compound that furnishes the P to solution. Phosphate potentials are known for phosphate compounds that are crystallized well. However, P likely precipitates as amorphous compounds that then crystallize with further reaction time. This probably explains why points for phosphate potentials determined for soils seldom fall exactly on the line in phosphate potential diagrams. However, this limitation should not present a serious obstacle in interpreting phosphate potentials.

3. Isotopic Exchange Methods

There is evidence that labile P measured by isotopic exchange is related to P available to plants (White, 1976). The rate of isotopic exchange should give a measure of the capacity of a soil to renew P to solution. Also, some estimate of the quantity factor may be obtained. McAuliffe et al. (1948) pointed out that isotopic exchange of P consisted of a fast reaction essentially complete in 24 h and a slow reaction that could continue for days. This was also discussed by White (1976). A variable rate of reaction suggests that different forms of soil P are involved in isotopic exchange. If the isotopic exchange reaction is heterogeneous, this certainly makes interpretation of the data more difficult.

4. Desorption of Phosphorus by Anion-Exchange Resins

Anion-exchange resins provide a method for determining the rate of release of P from soils by providing a "sink" for the released P that could simulate P uptake by plants. How closely adsorption of P by anion-exchange resins simulates uptake of P by plants could depend on bonding characteristics of the resin and other factors. This technique has been used by Amer et al. (1955), Cooke and Hislop (1963), Cooke (1966), and

Gunary and Sutton (1967). Use of this method for determining available P for plants would appear to have merit. However, it has not been adopted to any extent as a routine soil test method for determining available P, probably due to the extra effort and time required for the analytical determinations.

E. Fractionation of Soil Phosphorus

The fractionation method of Chang and Jackson (1957), with several modifications by other investigators, has been used to partition forms of P in soils by means of extractions with different chemical solutions. It is very difficult to completely extract one form of P from soils without extracting small amounts of other forms. Regardless of its limitations, the method has provided useful information on the forms of P present in different soils. Lindsay et al. (1968) pointed out that the method is capable of separating stable forms of P in soils such as variscite, strengite, and the apatites, but the method would not separate the metastable forms such as DCPD and OCP or amorphous Fe and Al phosphates. Therefore, the method would have more application for determining the stable forms of native soil P rather than for determining reaction products that develop after adding P to soils. However, knowing the stable forms of native soil P in a soil gives some indication of the compounds that added P should eventually form at equilibrium.

F. Determining Availability of Residual Fertilizer Phosphorus in Soils

Soil tests for available P have been used mainly as a basis for deciding if a crop is likely to respond to applications of P on a particular soil. It is now recognized that tests for available P may be even more beneficial in determining how soil test P may change over time with continued fertilization and crop removal of P.

As the cost of P fertilizers increases, it is becoming more important to determine the availability of residual P from fertilizer applications. Barrow (1980) presents a thorough discussion on evaluation and use of residual P in soils. A precise method for determining the residual value of applied P would be to compare its influence on yield and uptake of P with fresh applications of P in field trials or pot tests. The main disadvantages of this method are the time and cost involved with the trials and the questionable extrapolation of the results to different soils.

Soil tests can be used as a measure of availability of residual P for plants. Gwin (1977) found a good correlation between the residual P measured by the Bray P-1 test and the response of irrigated corn to residual P in west central Kansas. Other soil tests would be expected to give similar results. Barrow (1980) stated that such correlations are not likely to be very precise. This is probably correct, but there is a need for simple measurements that will give at least approximate values for the availability of residual P. Combining data from phosphate-potential studies with data from desorption studies might provide useful information on the availability of residual P. It is evident that additional research is needed on this subject.

III. POTASSIUM

The behavior of K in calcareous soils is similar in most respects to its behavior in noncalcareous soils. The availability of K to plants is related to pH in some general ways, but is not directly influenced by the presence of $CaCO_3$ in calcareous soils. Some of the general influences of H^+, Ca^{2+}, and Mg^{2+} ions will be discussed here, but these influences are only indirectly related to the presence or absence of $CaCO_3$ in the soil.

More detailed discussions of K chemistry and fertility in soils can be found elsewhere and the reader is referred to several excellent reviews in *Potassium in Agriculture* (Munson, 1985).

In general, calcareous soils contain more total K and cation-exchangeable K than noncalcareous soils, primarily due to less weathering in calcareous soils. The K-bearing minerals in soils are primarily the K micas and their derivatives and K feldspars.

A. Effect of Hydrogen Ions on Weathering

When weathering or slow decomposition of K minerals takes place, the K–O bond in the K-bearing 2:1 layer clays is realtively easily broken due to the large size of the K^+ ion and the large number of oxygens surrounding the ion. Therefore, as weathering of these minerals proceeds, K is among the first decomposition products. This K exists within the hexagonal ring formed by the silica tetrahedra and is not cation exchangeable prior to decomposition. Removal of K^+ from the clay crystal could be viewed as K^+–H^+ exchange in the interlayer wedge positions (Rich, 1968).

Weathering (or decomposition) is accomplished to a large degree by H^+ ions present in the soil solution. As a result, weathering of K^+ from clay minerals will occur at a faster rate in acid soils than in calcareous soils. When viewed over geological time, less weathering of K^+ and subsequent leaching has occurred in the calcareous soils of areas with drier climates.

B. Potassium Fixation

1. General

The reverse of weathering of K from the hexagonal ring structure formed by the silica tetrahedra is referred to as *K fixation*. A thorough description of the main forces causing K fixation was presented by Rich (1968). Briefly, electrostatic forces of attraction between the clay platelets and K tend to pull the clay platelets together, fixing K in the hexagonal ring. Opposite forces are hydration of interlayer cations (Dennis & Ellis, 1962) and hydroxy–Al interlayers (Rich & Obenshain, 1955) that tend to prop layers apart.

Potassium fixation most likely occurs when the equilibria between exchangeable and nonexchangeable K is disturbed as proposed by Bray and DeTurk (1939). One example of such a disturbance is the application of K fertilizer. Fertilizer application will put an amount of K on the exchange sites that is in disequilibrium with K in nonexchangeable positions of the clay. As a result, the clay will tend to fix some of the applied K. Fortunately, this attempt to reach a new equilibrium is slow in most soils; moderate rates of K fertilization are usually effective in supplying K to crops.

2. Effect of Soil pH on Potassium Fixation

Martin et al. (1946) studied the effect of a wide range in soil pH on K fixation. They found essentially no fixation around pH 2; the greatest fixation occurred above pH 7. A later interpretation of their data by Thomas and Hipp (1968) suggested that H_3^+O ions at a low pH tended to aid K exchange and reduce fixation, as shown by Rich (1964); at a higher pH, a decrease in hydroxy–Al polymers would tend to allow more K fixation. These studies all suggest more potential for K fixation in calcareous soils.

C. Crop Availability of Potassium in Calcareous Soils

1. Soil Reserves of Potassium

While the potential for K fixation in calcareous soils is substantial, it is also true that these soils are generally less weathered than noncalcareous soils and have much greater reserves of nonexchangeable K. This nonexchangeable K is in equilibrium with K on the exchange sites, and as a result, exchangeable K is usually quite high in calcareous soils.

High levels of exchangeable K depend on sufficient quantities of unweathered K-bearing minerals, but as pointed out by Thomas and Hipp (1968), soils of the Great Plains (which may be calcareous) generally have considerable quantities of dioctahedral mica, while those soils in the western and southwestern USA often have substantial quantities of biotite mica to supply K. There are some soils that are exceptions, however, and with continued K removal by crops, K reserves in calcareous soils will be reduced.

2. Cation Exchange of Potassium Ion by Other Cations

The dominant cation competing with K^+ for exchange sites in slightly acid to calcareous soils is Ca^{2+}. It is for this reason that considerable research effort has been expended to study the cation exchange of K^+ by Ca^{2+} and its subsequent influence on plant uptake of K^+.

In very acid soils, cation exchange of K^+ by Al^{3+} can occur to a greater degree than with Ca^{2+}. Because much of the Al in very acid soils is trivalent and cation-exchangeable, it can exchange with K^+ more easily than divalent Ca^{2+}. As a result, more K^+ will occur in the soil solution of very acid soils than in neutral and calcareous soils, provided all other things (such as the amount of exchangeable K^+) are equal (Thomas & Coleman, 1959).

3. Soil Solution Composition of Potassium

Schofield (1947) was the first to show that the ratio of K^+ to $(Ca^{2+})^{1/2}$ was constant for a given soil across a relatively wide range of concentrations in the soil solution, as would be predicted from the Donnan Equilibrium Theory. The ratio law was later expanded to include Mg so, for a given soil, $K^+/(Ca^{2+} + Mg^{2+})^{1/2}$ was constant across a wide range of soil solution concentrations.

Between soils, wide variations occur in the soil solution ratio of $K^+/(Ca^{2+} + Mg^{2+})^{1/2}$ due to differences in types and amounts of clay and soil organic matter content and exchangeable cation composition of the soil. There has been considerable interest in using the ratio $K^+/(Ca^{2+} + Mg^{2+})^{1/2}$ in soil solution as an indicator of K availability (first suggested by Woodruff, 1955). This ratio would indicate the availability of K^+ in the soil solution at any instant (intensity factor). Beckett (1964) later developed a procedure for describing a soil's ability to maintain its original ratio in the soil solution as plant uptake proceeds (the capacity factor).

Considerable research has been done on the $K^+/(Ca^{2+} + Mg^{2+})^{1/2}$ ratio as an indicator of plant availability. However, some research does not support the use of this relationship in estimating K availability to plants. For example, Collander (1941) found little effect on K^+ uptake when Ca^{2+} was increased from 0.1 to 1.0 m*M* in culture solution. He also found that similar changes in K levels had little influence on Ca^{2+} uptake. Later research with soils also casts some doubt on the use of this ratio for K^+ availability (Barber, 1968).

4. Percent Potassium Saturation

More recent research on K^+ availability as related to percent K^+ saturation of the CEC has shown adequate K^+ availability for optimum crop production across a wide range of exchangeable $K^+/(Ca^{2+} + Mg^{2+})$ ratios in the soil (Eckert & McLean, 1981; Liebhardt, 1981). Knudsen (1970) also found the exchangeable K^+ level in a soil was a better predictor of K availability than the $K^+/(Ca^{2+} + Mg^{2+})$ ratio in the soil or in the soil solution. He also found that exchangeable Ca^{2+} and Mg^{2+} levels had little influence on K^+ uptake, but abnormally high soil K^+ levels did result in low Mg^{2+} uptake.

In summary, the results of a number of studies indicate that increasing amounts of K^+ in soils can reduce uptake of Ca^{2+} and Mg^{2+}, but uptake of K^+ does not seem to be reduced by increases in Ca^{2+} and Mg^{2+} uptake by the plant. Potassium uptake appears to be related more to the quantity of K^+ available in the soil than to the ratios described above.

While generally good correlations exist between cation-exchangeable K^+ and response to applied K fertilizers, there may be some exceptions. For example, Skogley and Haby (1981) have reported increases in wheat yields when K fertilizer was applied to soils with high exchangeable K^+ levels. Their studies were conducted under climatic conditions involving cool soil temperatures and semiarid moisture regimes.

IV. SULFUR

Sulfur deficiencies have become more widespread in recent years. In 1947, S was reported to be deficient for crop production in 12 states and three Canadian provinces (Bledsoe & Blaser, 1947); by 1971, the number of reported S deficiencies increased to 29 states and five of the Canadian provinces (Beaton, 1970); and in 1982, S deficiencies were reported in 35 states and 10 Canadian provinces. While much of the increased recognition of S deficiencies is undoubtedly due to more experimentation and cropping of less desirable marginal land, there are other reasons for more prevalent S deficiencies. Farmers continue to use more high-analysis fertilizers that contain less S, crop yields are increasingly causing greater S removal, soil erosion is depleting surface organic horizons, and air-quality standards have reduced atmospheric S.

Since soil organic matter is recognized as the major contributor of plant-available S, declining soil organic matter content is often given as a contributing factor. However, since most cultivated soils in the USA either have reached or are near equilibrium with their management environment, organic matter levels have stabilized except for marginal soils recently brought under cultivation. While organic matter may be stabilized, the equilibrium level of these soils may not provide adequate S for high crop yields.

Sulfur-deficient soils are generally low in organic matter, coarse-textured, and well-drained. Beaton (1971) has an excellent review of known or suspected S-deficient soil areas in the USA, which shows both acid and alkaline soils.

Sulfates are quite mobile in soils and can be leached quite easily. Because alkaline soils generally are found in areas of low precipitation, they are not usually S deficient. These soils often contain sulfate salts, especially gypsum, somewhere in the root zone and even at the soil surface. However, because of losses of about 50% of the organic matter after 50 to 70 yr or cropping, S deficiencies are becoming more prevalent, even on the chernozemic soils of Canada. Sulfur deficiencies in Canada were at one time confined to the leached Luvisolic soils (Bettany et al., 1980).

There have been many reviews of the reactions that occur in soils, as well as responses to applied S and plant needs for S (Eaton, 1966; Anderson, 1952; Jordan & Reisenauer, 1957; Jordan & Ensminger, 1953; Terman, 1978; Freney et al., 1962; McLachlan, 1974; Burns, 1967; Harward & Reisenauer, 1966). None of these reviews are specific for alkaline soils. Since most S deficiencies are associated with acid soils, relatively few investigators have studied alkaline soils to any great extent. More general reviews of S, S fertilizers, S for acidification, and predicting S needs are found in chapter 11 of this book.

A. Factors Affecting Native Soil–Sulfur Availability and Fertilizer Need in Alkaline Soils

1. Inorganic Sulfur

The availability of S in alkaline soils depends largely on the amount of sulfate salts, mainly gypsum ($CaSO_4$), found in the surface horizons and the entire root zone. Some investigators have reported that two-thirds of the total S found in calcareous soils was inorganic (Freney et al., 1962). The available sulfate (SO_4^{2-}) found throughout or in certain layers of the soil profile often accounts for a crop's recovery from an earlier deficiency. While the soil may contain large amounts of sulfate in the root zone, the surface soil may be leached of available sulfate, resulting in early deficiencies, especially on coarse-textured soils.

Sulfate adsorption in soils is primarily a function of soil pH and the amount of Fe and Al hydrous oxides present in the soil (Chao et al., 1962; Ensminger, 1954; Beaton, 1968) or as constituents of the clay minerals (Kingston et al., 1972). Kamprath et al. (1956) found that kaolinitic clays retain more sulfate than montmorillonitic clays and that adsorption decreases markedly as the soil pH increases from 4.0 to 6.0. Williams and Steinbergs (1962) found sulfate adsorption to be negligible above pH 6.5 in Australian soils. Adsorption on alkaline soils generally is considered negligible unless organic matter contents are high. Chao et al. (1962) have shown that soil organic matter can adsorb significant amounts of sulfate. These authors indicate that the amphoteric nature of organic matter and the resulting anion exchange capacity is probably responsible. While sulfate can be adsorbed in soil, it generally is considered weakly held, since phosphates will displace or reduce sorption of sulfates (Harward & Reisenauer, 1966).

The gypsum found in alkaline and calcareous soils apparently is quite available to plants because it dissolves freely, replenishing the soil solution. Reitemeier (1946) found that in arid soils containing gypsum, the sulfate concentration of the soil solution increases with increasing soil moisture. In one soil containing small amounts of gypsum, all gypsum was dissolved with high dilutions of water. This indicates that S, like most other nutrients, is more available when the soil contains an adequate amount of water. As a result of this solubility, drainage water often moves dissolved sulfates from surface soil horizons to deeper soil horizons.

While gypsum found in alkaline and calcareous soils generally is quite available as a S source, Williams and Steinbergs (1962) and Williams et al. (1960) have reported insoluble sulfate associated with the $CaCO_3$ in calcareous soils in Australia and Scotland. Most of this insoluble S is soluble in hydrochloric acid (HCl). One soil in Australia contained as much as 93% of its S in this insoluble form. Since fine grinding did not affect solubility, sulfate was assumed to be evenly distributed throughout the mass of calcareous material and not adsorbed to the $CaCO_3$ surfaces. Williams and Steinbergs (1962) suggest that this sulfate occurs as a coprecipitated or co-crystallized impurity in the $CaCO_3$. In further research, Williams and Steinbergs (1964) reported that S found as an impurity in $CaCO_3$ has limited availability for plants. No reports were found as to the stability of calcium sulfates associated with calcareous soils in the midwestern and western USA.

Relatively large amounts of sulfate are found in many irrigation waters in the west. Often the quantity added by normal irrigation supplies the crop needs. Water analysis is necessary to evaluate the S needs of crops grown where irrigation water S content is high.

2. *Organic Sulfur*

Organic transformation of S has not been studied specifically in alkaline soils; at least no research was found, which suggests that alkaline and acid soils react similarly. However, the kinds of microorganisms and the rate of organic matter decomposition and the resultant S reaction products are probably affected by soil pH.

While many alkaline soils probably contain sulfates somewhere in the soil profile, organic S is the primary source if sulfates are not present. Neptune et al. (1975) found that three alkaline soils in Iowa contained 98% of their total S in the organic form. Therefore, plants may have to depend, especially at early growth stages, on S mineralized from the organic matter.

Mineralization of S from organic matter depends on essentially the same factors that affect the mineralization of N, such as soil temperature, soil water content, and oxygen supply. As one might expect, incorporation of plant residues low in S content can cause S deficiencies during decomposition of the plant material, as it does with N. Stewart et al. (1966) found that when wheat straw (*Triticum aestivum* L.) containing $< 0.15\%$ S was added to the soil, depressed wheat growth occurred because of S deficiency.

Recent research on soil organic S has involved organic fractionation in an attempt to learn what specific organic compounds are affecting S availability and to improve prediction of S fertilizer needs. Freney et al. (1975) used a detailed fractionation of organic S and found that 60% of the S taken up by plants was derived from the C-bonded S compounds, such as cystine and methionine amino acids. While the C-bonded S fraction was most important, other fractions contributed significantly. No one fraction of any great value was found for predicting S needs. Other investigators have noted that organic S is found primarily with the clay fraction in Canadian soils. Anderson et al. (1974, 1981) showed that 70% of the total soil S was associated with the fine and coarse clay fractions. This indicates that a significant portion of the soil organic matter and organic S is actually associated intimately with the inorganic constituents. Ultimate understanding of organic matter and its role in supplying S, as well as other nutrients, may lie in our understanding of what happens in these clay–organic matter relationships.

V. IRON

Iron chlorosis, sometimes referred to as Fe deficiency or Fe stress, is very common in alkaline soils, especially calcareous soils. In fact, susceptible plants grown in calcareous soils are most likely to be chlorotic. Many use the terms *iron chlorosis* and *iron deficiency* interchangeably. Technically, a deficiency would mean that the plant Fe content is low; however, very often Fe-chlorotic plants may be higher in Fe content than normal plants. Some investigators have avoided this problem by referring to *iron stress,* since chlorosis may be caused either by an Fe deficiency or a condition called lime-induced Fe deficiency or chlorosis. In calcareous soils, we rarely have true Fe deficiencies, but rather the lime-induced chlorosis.

The problem of Fe chlorosis is widespread in the western USA. Mortvedt (1975) reported a survey showing that $>$ 5 million ha or 5% of the total cultivated area in the USA has alkaline and calcareous soils on which susceptible crops either cannot be grown or require some Fe fertilization for profitable production. About 4 million of the 5 million ha of alkaline and calcareous soils are found in the Great Plains states.

Highly susceptible crops include most fruit trees and citrus, many ornamentals and turf grasses, field beans (*Phaseolus vulgaris* L.), soybeans [*Glycine max* (L.) Merr.], sorghums [*Sorghum bicolor* (L.) Moench], and many vegetables. While chlorosis may develop on any calcareous soil, chlorosis is more likely during wet, cold weather, following irrigation, on poorly drained soils, on compacted soils, or where root growth is restricted. Chlorosis has been related to excess bicarbonate, high available P and Ca, and other heavy metals, especially Cu and Zn.

Because of the various interactions of Fe with other nutrients, the environmental effects associated with the solubility and plant availability of these nutrients, and widely varying plant adaptability, the chlorosis problem has been very complex and perplexing for many investigators. The literature is plentiful, but the problem still is not well understood. This review will provide insight into the Fe problem and what corrective measures can be taken. Excellent reviews on Fe have been published by Brown (1961), Wallace and Lunt (1960), Murphy and Walsh (1972), Hodgson (1963), and Follett et al. (1981).

A. Soil Iron

Lindsay (1972) discussed equilibrium constants and showed the solubility of various Fe compounds in soil. While soil contains very large amounts of total Fe, considerably more than any other micronutrient, its solubility is very low. The amount of Fe in solution is controlled primarily by the solubility of Fe^{3+} oxides, which is pH dependent. Ferric ion activity decreases 1000 times for each increase of one pH unit; while Fe^{2+} activity decreases 100 times for each increase of one pH unit. At a pH of 7.8, the concentration of Fe^{3+} and Fe^{2+} are equal at 10^{-21} *M*. By contrast, P, which is often considered insoluble in soils, may have a soil-solution concentration of 3.3×10^{-3} M, expressed as the dihydrogen phosphate ion ($H_2PO_4^-$), or approximately 3×10^9 times more concentrated than Fe in the soil solution (Follett et al., 1981). Under reduced conditions, Fe^{2+} concentrations increase drastically, but Fe availability to plants may remain low because of very low root activity in situations of low O_2.

While the Fe requirement of plants is much lower than that for P, it is still apparent that the inorganic Fe phase in the soil may contribute relatively little to the Fe nutrition of plants in calcareous soils. While no data were found for Fe, investigators showed that 75% of the Zn and 98 to 99% of the Cu found in soil solutions are in the form of organic complexes (Hodgson et al., 1966). It is probable that a large amount of the Fe in the soil solution is also of organic origin. Organic-complexing agents are known to be released from plant roots as well as derived from organic matter decomposition and microbiological activity (Rovira, 1962). Elgawhary et al. (1970) concluded from simulated root studies with ceramic tubes, that organic complexes released by roots can contribute greatly to the availability of micronutrients in the soil. It is generally accepted that plants take up Fe^{3+} ions, but they must be reduced to Fe^{2+} at the root surface, at least when associated with organic complexes (Chaney et al., 1972).

B. Induced Iron Chlorosis

The chlorosis that commonly occurs in plants growing on calcareous soils often has been referred to as *lime-induced chlorosis*. This term has been used because plants suffering from lime-induced chlorosis may or may not be low in Fe. Immobilization of Fe in the tissues is an accepted major factor. Iron content of chlorotic plants often is higher than that of healthy plants (DeKock, 1955).

The primary factor associated with chlorosis on calcareous soils seems to be the effect of the bicarbonate ion on Fe uptake and/or translocation in the plant. As early as 1945, it was observed that chlorosis of apple (*Malus Sylvestris* Mill.) and pear trees (*Pyrus communis* L.) was related to the bicarbonate content of irrigation water. However, the actual effect of bicarbonate on Fe nutrition of plants has been more difficult to pinpoint. Brown et al. (1959b) showed with split-root studies that when chlorosis-susceptible soybeans were grown through a soil mix containing Fe into a solution containing no Fe or P, adding bicarbonate did not cause chlorosis or reduce Fe adsorption. If similar soybeans were grown into a complete nutrient solution, adding bicarbonate induced chlorosis. Brown et al. (1959a) attributed the chlorosis to the bicarbonate treatment, which maintained about twice as much P in solution as treatments containing no bicarbonate. They also found that the increased availability of P affected Fe nutrition in soybean plants. The main effect of bicarbonate in calcareous soils is one of decreasing Ca^{2+} activity, which increases solubility of P, which in turn affects Fe availability. Other investigators have indicated that bicarbonate inactivates Fe in plants, retards Fe movement in plants, and causes reduced Fe content in some plants (Brown, 1960).

Since chlorotic plants may be higher in Fe than healthy plants, DeKock (1955) used the P/Fe ratio as an index of the Fe status in plants. The P/Fe ratio is usually larger in chlorotic plants than in healthy plants.

In addition to the bicarbonate ion effect on Fe availability, there are several other nutrient interactions that are known to affect plant utilization and uptake of Fe. Investigators have observed that as Fe supplies become low because of the high pH in alkaline and calcareous soils, other micronutrients will induce Fe deficiencies. The effect of other micronutrients is often more serious on alkaline soils. These soils are often low in organic matter, which affects both micronutrient supply and buffering capacity. Brown et al. (1959a) found that soybean chlorosis was affected by both Cu and Mn, which were influenced by P and Ca. Brown and Tiffin (1962) studied 14 plant species for Fe and Zn deficiencies. They found that added P accentuated both Zn and Fe deficiencies and that added Zn induced Fe deficiencies in corn and millet (*Panicum miliaceum* L.).

Kersch et al. (1960) observed that with tomato (*Lycopersicon esculentum* Mill.), Mo accentuated Fe deficiencies. Since Mo availability is high on alkaline soils and Fe availability is low, Olsen and Watanabe (1979) studied the Mo–Fe interaction on alkaline soils. Authors used SO_4^{2-} to depress the MoO_4^{2-} ion uptake according to the findings of Gerloff et al. (1959). Results confirmed the interaction on six soils, showing that gypsum decreased Mo from 2.33 to 1.26 mg/kg and increased Fe from 56 to 64 mg/kg in sorghum.

Susceptibility to Fe chlorosis depends on a plant's ability to respond to Fe stress. Investigators found that this ability is controlled genetically and is related specifically to the plant's ability to release H^+ ions and reductase from its roots (Brown et al., 1972). Brown and Jones (1975) studied four cultivars of sorghum and found that the Fe-efficient cultivars released more reductant to the nutrient solution and took up less P than the Fe-inefficient cultivars. They concluded that the accumulation of P, as well as the ability to produce reductant, are causative factors of Fe chlorosis. Esty et al. (1980) found in studies with grain sorghum that the ability to transport Fe from the roots to plant tops is controlled genetically. These authors also observed that the more efficient genotypes accumulate less P in the roots and shoots. Phosphorus appeared to interfere with Fe translocation.

Phosphorus appears to be a primary factor affecting Fe chlorosis. A number of studies have been conducted on soils or nutrient solutions under controlled conditions to evaluate this problem. Usually alkaline, and especially calcareous, soils are naturally low in available P. However, P-induced Fe chlorosis in the field has been documented (Mathers, 1970). While the nature of the interaction is not well understood, excessive P fertilization of alkaline soils certainly should be avoided.

C. Correcting Iron Chlorosis

Iron chlorosis of susceptible crops on calcareous soils is not corrected easily. In fact, for field crops, the most common recommendation is not to grow chlorosis-sensitive crops but to switch to less sensitive crops. It is not that Fe cannot be effectively applied to crops. Several Fe fertilizers are available that can correct Fe chlorosis. However, the application rates either are too high or the number of applications needed with severe Fe chlorosis is too many to be economically feasible on all but the highest value cash crops.

The most common inorganic Fe sources are ferrous sulfate ($FeSO_4$), ferric sulfate [$Fe_2(SO_4)_3$], ferrous carbonate ($FeCO_3$), and ferrous ammonium sulfate [$Fe(NH_4)_2(SO_4)_2 \cdot 6\ H_2O$]. Organic chelates include the Fe salts of EDTA (ethylenediaminetetraacetic acid), EDDHA [ethylenediamine di(*o*-hydroxy-phenylacetic acid)], DTPA (diethylenetriaminepentaacetic acid), and HEDTA (hydroxyethylethylenediaminetriacetic acid). The effectiveness of these Fe fertilizers varies greatly, depending on soil pH. To correct chlorosis with inorganic Fe sources by soil applications, very high rates generally have been required. Olson (1950) found that none of the naturally occurring or synthetic Fe oxides, such as hematite, magnetite, or limonite, were of any value for correcting chlorosis in grain sorghum grown on a calcareous Colby soil (Ustic Torriorthent). Even ferrous sulfate at 227 kg/ha failed to affect growth. Acidification with sulfuric acid (H_2SO_4) lowered soil pH, increased yields, and decreased chlorosis. Mathers (1970) also found that it required as much as 227 kg/ha of ferrous sulfate to correct chlorosis in sorghum in both the field and greenhouse.

High rates of inorganic Fe have some carryover effect on subsequent crops. Hodgson et al. (1972) observed that when 170 kg/ha of ferric sulfate was applied to a calcareous soil, there was no appreciable decrease in relative growth response of corn after three harvests. A 90-min extraction of the soil with EDDHA recovered only 1 to 2% of the Fe added. This was, however, three times the amount of Fe removed by the three corn crops.

The possibility of neutralizing only a small amount of the soil to increase Fe availability has been studied. It requires 43 960 kg of sulfuric acid/ha to neutralize each 1% of $CaCO_3$ in the soil to a depth of 30 cm. While it has been shown that only a small fraction of the root is needed to absorb adequate Fe (Gile & Carrero, 1920), it appears that the amount of soil that needs to be neutralized may vary. Wallace and Mueller (1978) showed that only small volumes of soil acidified with sulfuric acid equivalent to as little as 0.4% of the free $CaCO_3$ in the soil, were required to correct soybean chlorosis. Mathers (1970) found that both banded sulfuric acid and ferrous sulfate were effective in correcting chlorosis (Table 5–1). A rate of application of 560 kg/ha of either or both of these amendments was required to maximize sorghum yields, although chlorosis was corrected with about half this rate. Ryan et al. (1974) noted that they had to neutralize nearly all of the free $CaCO_3$ in Arizona soils before there was a significant increase in DTPA–extractable Fe. Surface-applied sulfuric acid was not as effective as either ferrous sulfate or Fe–EDDHA on bermudagrass [*Cynodon dactylon* (L.) Pers.] (Ryan et al., 1975). However, in the same study, an application of 3% sulfuric acid in irrigation water was as effective as either of the two above materials.

Table 5-1. Effects of ferrous sulfate ($FeSO_4$) and sulfuric acid (H_2SO_4) on grain sorghum yields (Mathers, 1970).

Applied sulfuric acid	Applied ferrous sulfate, kg Fe/ha		
	0	112	560
	kg/ha		
0	434 d*	1460 bc	2275 ab
112	605 cd	1538 ab	2474 a
560	2169 ab	2429 ab	2230 ab
5600	1885 ab	1971 ab	1810 ab

* Yield values followed by the same letter are not significantly different at the 5% level. Standard error of the mean was 304.

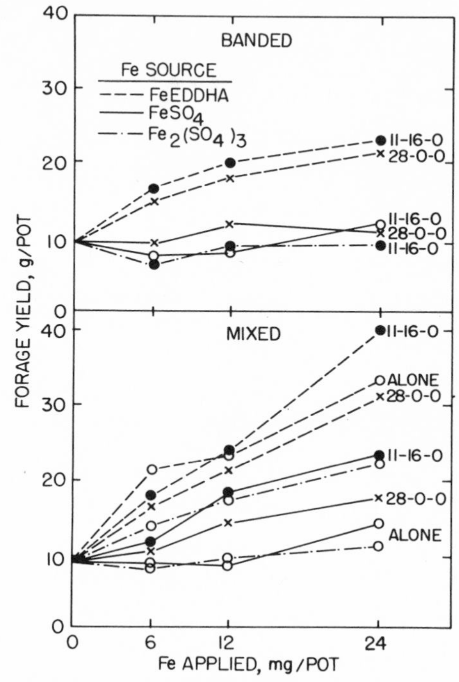

Fig. 5–3. Yields of sorghum forage, as affected by placement of several Fe sources alone or with fluid fertilizers (Mortvedt & Giordano, 1971).

In an effort to increase the efficiency of inorganic Fe sources, Mortvedt and Giordano (1971) studied the effect of placing Fe fertilizers together with ammonium polyphosphate (APP) fertilizers on a calcareous soil in the greenhouse (Fig. 5–3). The authors found that grain sorghum yields were higher with Fe–EDDHA than with ferrous sulfate or ferric sulfate. Inorganic sources were more effective when applied with polyphosphate (11–16–0) (11–36.64–0). Mixed placement was better than band placement, but this may have been due to limited root volume under greenhouse conditions.

Since ammonium thiosulfate [$(NH_4)_2S_2O_3$] is a widely used S source and is a good reducing agent, Mortvedt and Giordano (1973) determined its value for increasing Fe availability. On a calcareous soil from western Nebraska, grain sorghum yields and Fe uptake were doubled with a band application of ferrous sulfate in APP. When ammonium thiosulfate was included, crop response was increased another 15 to 25%. Application of ammonium thiosulfate alone or with APP increased the levels of extractable Fe in the soil.

Chelates have been shown to be superior sources of Fe for plants because lower rates are required than inorganic Fe sources. However, the cost of chelates generally has been too high to permit economic use. Holmes and Brown (1955) studied the effectiveness of the five chelates EDTA, HEDTA, DTPA, CDTA (cylcohexanediaminetetraacetic acid), and EDDHA for correcting soybean chlorosis when grown on 17 calcareous soils in the greenhouse; only two of the chelates, DTPA and EDDHA, were effective. DTPA was required at a rate of 120 mg/kg to correct soybean chlorosis compared with only 30 mg/kg of EDDHA. The study also showed that the chelate alone, without Fe, made soil Fe available to the plant.

The effectiveness of chelates in alkaline soils is related to their stability at various pH levels (Norvell, 1972). For example, EDDHA is selective for Fe only over a pH range of 4 to 9. This selectivity makes EDDHA a much better Fe source for calcareous soils than EDTA, which chelates Fe primarily at a pH < 6.3; above pH 6.8, EDTA reacts with Ca^{2+}, making it an ineffective chelate on high-pH soils. DTPA chelates Fe up to a pH of about 7.5; above a pH of 7.5, Ca^{2+} occupies this chelate.

It is apparent that in an alkaline soil system where Fe solubility and availability to plants is extremely low, adding soluble Fe cannot be very effective, unless the soluble Fe is in a form that does not react with the soil system and yet can be released effectively to the plant. Even if all of the free $CaCO_3$ in calcareous soils is neutralized by acid, the pH is about 7.3 to 7.5, where Fe solubility is still low.

Because of the high cost of chelates and the generally ineffective nature of inorganic Fe sources, foliar applications have been the most effective method of correcting Fe chlorosis. However, the number of applications required to correct chlorosis and to increase yield have often limited use to high-value cash crops, turf, or ornamentals. While studies have generally shown that several spray applications are required to correct chlorosis with ferrous sulfate, Follett et al. (1981) showed data by Randall in Minnesota that indicated only 0.17 kg/ha of FeEDDHA increased the yield from chlorotic soybeans from 1035 to 3152 kg/ha. It was important to apply the Fe treatment early. One application applied at the second trifoliate stage was adequate to maximize yield at several Fe-deficient locations. Follett et al. (1981) also showed the recommended rates and sources of Fe for several crops.

VI. ZINC

Zinc is probably the most commonly deficient nutrient on alkaline soils, next to N and P. While Zn was recognized as essential for the production of peaches and citrus in the 1930s, it was not until the 1950s that Zn became widely known as a needed nutrient for crop production, especially corn (Viets, 1951; Viets et al., 1954). Zinc deficiencies became prominent at this time, largely because of increased irrigation development. Land leveling, which necessarily removed or moved the topsoil, resulted in widespread Zn deficiencies for corn production in Nebraska and Kansas (Pumphrey et al., 1963). Leveling often completely removed the high–organic matter A horizon where Zn had accumulated during soil formation. Lower soil horizons are often calcareous and low in available Zn, as well as P, which results in severe Zn and P deficiencies. Zinc deficiencies were often of such severity that corn plants produced essentially no yield and even died if corrective Zn fertilizers were not applied. Soil compaction in these situations was also a contributing factor that impeded root growth and reduced soil Zn availability.

Since the mid-1950s, much Zn fertilizer has been applied on old, leveled land areas so that deficiency symptoms are rarely seen, except in areas where topsoil has been recently removed for terraces or land has been shaped for center pivot irrigation. However, accelerated soil erosion in recent years, with its inherent loss of topsoil, is cause for concern. Erosion of alkaline soils certainly increases the likelihood of encountering more deficiencies in the future; the lower soil horizons generally have a higher pH, which limits Zn availability. Deficiencies may or may not be visible and are best detected by soil and plant analysis. Producer experience with severe Zn deficiencies has often resulted in indiscriminate recommendations and Zn use, whether it was needed or not (Olson et al., 1982).

There have been many good reviews on Zn, some quite recent, to which the reader is referred for more detailed information than can be covered here (Viets, 1966; Hodgson, 1963; Swaine, 1955; Thorne, 1957; Mortvedt et al., 1972; Murphy & Walsh, 1972).

A. Factors Affecting Zinc Availability

1. Soil pH

Soil pH is the most important single factor regulating the Zn supply available for plants in alkaline soils. As early as 1941, Peach (1941) concluded that as the pH of the soil was increased with $CaCO_3$, the amount of extractable Zn decreased; at pH 9.0 virtually all of the Zn was fixed. Wear (1956) showed that by mixing various rates of $CaCO_3$, sodium carbonate, and gypsum with soil, Zn uptake by soybeans was affected only by the pH changes caused by $CaCO_3$ and sodium carbonate and not directly by Ca. Ninety-two percent of the variation in Zn uptake from applied zinc sulfate ($ZnSO_4$) was accounted for by changes in pH (Fig. 5–4).

More recently, the relationship of pH to the solubility of soil Zn has been shown by Lindsay (1972) and Lindsay and Norvell (1969). These authors show in Fig. 5–5 that the Zn-soil complex from five acid and calcareous soils produces a concentration of Zn^{2+} of only 10^{-12} *M* in soil solution at pH 9. Figure 5–5 also indicates that the solubility of the relatively insolu-

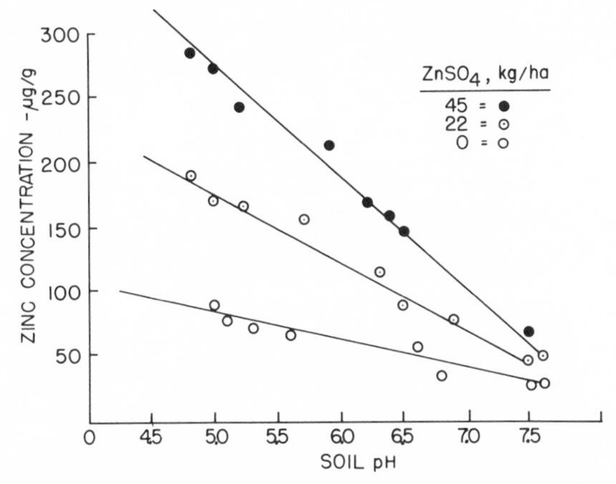

Fig. 5–4. Zinc concentration of sorghum as affected by soil pH (Wear, 1956).

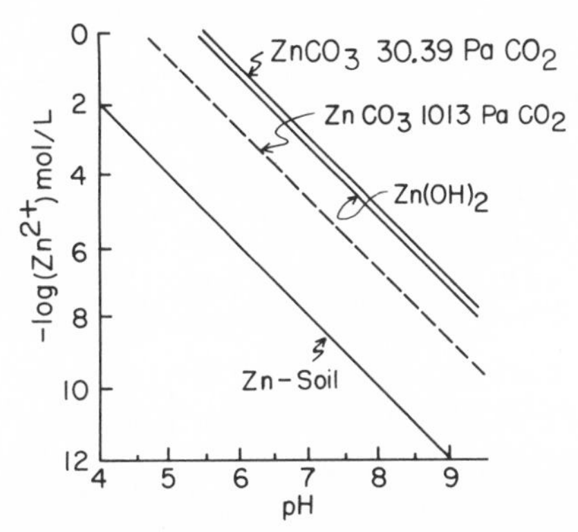

Fig. 5–5. The solubility of various Zn minerals compared with the solubility of Zn in soils (Lindsay, 1972).

ble zinc carbonate ($ZnCO_3$) and zinc hydroxide [$Zn(OH)_2$] compounds is considerably greater than the Zn-soil complex and is affected somewhat by CO_2 pressure. The critical Zn^{2+} level in solution for plants, according to Halvorson and Lindsay (1977), is $10^{-10.6}$ M and, contrary to earlier investigations, the form of Zn absorbed by the plant is the Zn^{2+} and not the chelate (Wallace, 1963). Therefore, the soil pH would need to be < 8 for Zn to be available to plants in sufficient amounts. This explains why Zn deficiencies commonly occur on calcareous soils, even though other Zn species may be present (Fig. 5–6), and why exchangeable Zn is not a factor affecting Zn availability in alkaline soils (Tiller et al., 1972).

The drastic effect of pH on soil-Zn availability reemphasizes the importance of not removing the neutral to slightly acid topsoil of soils in the western USA, where subsoils are often calcareous or alkaline at best. If topsoil removal is required, then stockpiling and replacement of the topsoil can eliminate the continuing micronutrient problems that may be associated with these cut-soil areas.

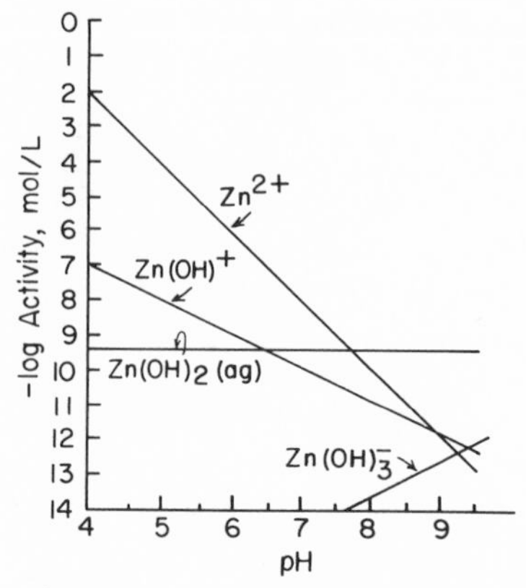

Fig. 5–6. Soluble Zn species in solution in equilibrium with soil Zn (Lindsay, 1972).

It should be mentioned that although Zn availability is much reduced by pH in alkaline and calcareous soils, the area around the plant root always has a pH much lower than 8. Therefore, all alkaline or even calcareous soils with a pH of 8 are not necessarily Zn deficient. In addition, pH effects on soil-Zn availability discussed above do not provide for the Zn that is in organic complexes. Hodgson et al. (1966) showed in 20 calcareous soils in Colorado that an average of 60% of the Zn in solution was in the form of organic complexes. Halvorson and Lindsay (1977) indicated that while only Zn^{2+} is actually absorbed by the plant root, organic chelation aids the transport and movement of metal ions to the root.

2. *Temperature, Moisture, and Light*

Zinc deficiencies are often more prevalent during cold, wet springs than during warm ones. Sensitive crops such as corn sometimes show Zn deficiencies on old, previously leveled soils where low Zn levels have been corrected. Such borderline deficiencies are often only temporary. As soil temperatures increase and roots explore more soil volume, the deficiency symptoms often disappear and yields are not affected. Several investigators have shown that increasing soil temperature has a positive effect on Zn uptake by plants (Ellis et al., 1964; Martin et al., 1965; Edwards & Kamprath, 1974).

Giordano and Mortvedt (1978) studied the effects of soil moisture and temperature on P and Zn uptake by corn in the greenhouse. They found that, while low soil temperatures reduced Zn and P uptake as well as concentration, soil moisture had little effect on Zn and P uptake. Therefore, wet soils in the spring contribute to increased Zn deficiencies, probably because wet soils warm up in the spring much more slowly than dry soils. Giordano and Mortvedt (1978) stated that climatic effects in early spring contribute to Zn deficiency in a very complex manner; light is probably also a contributing factor. Edwards and Kamprath (1974) observed that under low-light conditions, corn-kernel Zn supply accounted for nearly 100% of the total Zn found in young corn seedlings, even when exposed to external Zn.

B. Factors Affecting Zinc Availability in Fertilizer

1. Zinc Fertilizers

There are many Zn fertilizers available, but the most common inorganic compounds are zinc sulfate and zinc oxide (ZnO). Other inorganic Zn sources are zinc carbonate, zinc sulfide, (ZnS), Zn frits, and zinc phosphate [$Zn_3(PO_4)_2$]. Organic Zn fertilizer sources are the chelates, EDTA, NTA (nitriloacetate), and HEDTA. Other organic sources include Zn ligninsulfonates and Zn polyflavonoid, by-products of the wood pulp industry.

Many researchers compared the various Zn fertilizers on many different crops; often the results conflicted. No particular Zn fertilizer has been consistently better than others for either alkaline or acid soils. Most investigators reported that the different inorganic Zn sources perform similarly. Boawn et al. (1957), in a greenhouse study, showed that insoluble compounds, such as zinc oxide, zinc phosphate, and zinc carbonate, are just as effective as water-soluble zinc sulfate for grain sorghum if finely divided and well mixed with the soil. These authors indicate that only those Zn compounds, such as Zn frits and basic slag, which are not soluble in 0.1 *M* HCl are less available to plants. Shaw et al. (1954) also found that corn utilized as much Zn from zinc carbonate as from zinc sulfate if both sources were powders mixed with the soil.

Giordano and Mortvedt (1969) found that while Zn may be applied in the water-soluble form, it remains soluble only long enough to affect Zn uptake when combined with different macronutrient fertilizers. In greenhouse studies on an alkaline sandy clay loam soil, corn yields and Zn uptake were increased by water solubility of various combinations of zinc sulfate and zinc oxide granulated with MAP and to some degree AN, but were unaffected when granulated with ammonium polyphosphate (APP). Water solubility of the Zn source might be expected to be more important for noncalcareous soils, since the concentration of plant-available Zn^{2+} in the soil solution is regulated by pH (Lindsay, 1972).

The organic-chelated Zn sources generally, but not always, have been reported as superior to the inorganic Zn sources. Rehm et al. (1980) found that Zn uptake by corn plants was higher for Zn–EDTA on a calcareous silt loam than either zinc oxide or zinc sulfate; there were no differences on an acid sandy soil. Zinc fertilizers were banded in this study with poly- or ortho-phosphate suspensions. Shukla and Morris (1967) compared zinc sulfate, zinc oxide, Zn polyflavonoid, and Zn–EDTA in a greenhouse study on acid soils from Georgia and found the zinc sulfate and zinc oxide to be equal or superior to either organic Zn source. Anderson also found that chelates were much more effective at low rates than zinc sulfate on calcareous soils in eastern Colorado (Anderson, 1964).

There have been conflicting reports from various investigators on the effectiveness of various Zn fertilizers; however, it appears that, in general, most Zn compounds soluble in dilute HCl are suitable Zn fertilizers. On calcareous soils the chelates are probably more effective carriers of Zn in terms of plant uptake, since they prevent the Zn from being adsorbed to the soil complex and rendered unavailable (Jackson et al., 1962). Because the plant requirement for Zn is so low, differences in yield may not be apparent. The difference between chelates and inorganic sources often becomes largely academic. There is no definitive research evidence to indicate a particular ratio of effectiveness comparing chelates and Zn sources, such as zinc sulfate or zinc oxide. While effectiveness ratios between chelates and inorganic sources have been reported as high as 10:1, the actual efficiency ratio is probably closer to 2:1 or 3:1 (Follett et al., 1981). Any ratio of effectiveness will depend on soil properties and especially the degree of deficiency. Varying reported ratios of effectiveness between chelates and inorganic carriers often have been confounded because of differences in soils, especially texture and pH, fertilizer form, degree of granulation and methods of application.

2. Method of Application

Methods of application for Zn fertilizers affect fertilizer performance by changing soil-fertilizer contact and/or probability of root-fertilizer contact. Brown and Krantz (1966) found that when zinc sulfate, and Zn–EDTA were mixed into the root zone of a calcareous loam soil, the two sources performed similarly. However, the Zn–EDTA was superior to the inorganic Zn sources when banded. They found that granulation greatly decreased the effectiveness of zinc sulfate, zinc ammonium phosphate, and Zn polyflavonoid. Brown and Krantz (1966) concluded that granulation, band placement, or both greatly reduced the effectiveness of the relatively immobile Zn source, but had no effect on the mobile Zn-EDTA source.

The effect of Zn fertilizer distribution in the soil, as a factor affecting Zn availability, also was studied by Sorensen et al. (1970). They found in leaching studies that inorganic Zn sources do not move appreciably. Therefore, low-analysis materials and small fertilizer particle size increase probability of root-fertilizer contact and thereby increase availability. Mixing with soil, therefore, would be a more effective method of application than banding. However, banding a mobile chelated Zn source should result in a vertical distribution in the soil that would increase effectiveness over other less-mobile Zn sources. If one considers mobility, soil texture is an important factor influencing Zn fertilizer effectiveness because of the greater Zn mobility in sandy soils compared with fine-textured soils.

Another factor that affects Zn fertilizer effectiveness is the time of application. Pumphrey et al. (1963) studied several methods of applying zinc sulfate to corn on calcareous soils in western Nebraska. Methods of application in order of decreasing effectiveness were: (i) broadcasting and incorporating prior to planting, (ii) banding Zn with N at planting, (iii) side-dressing Zn with N when corn was 15- to 30-cm tall, (iv) banding Zn only at planting, (v) side-dressing Zn only, and (vi) spraying with 1% $ZnSO_4$ solution when corn was 15- to 30 cm tall.

Foliar applications have been used extensively for treatment of Zn deficiencies on fruit trees (Childers, 1954); however, in general, foliar applications of Zn are not recommended for field crops, except as an emergency treatment. Soil applications are used in spite of the fact that foliar applications may be much more efficient (McNall, 1967). Foliar sprays are usually not recommended for the following reasons: (i) by the time a deficiency is obvious, much of the yield loss probably has already occurred; (ii) more than one foliar application is often required, which increases cost; and (iii) soil applications last for several years. The low rates used for foliar applications are effective only on the crop being sprayed.

It is recognized that efficiency is low when Zn fertilizer is applied to the soil. This results in considerable residual value (Brown et al., 1964). Boawn et al. (1960) showed that on an alkaline soil in Washington, a 4-yr sequence of crops (grain sorghum–potato–sugarbeets–grain sorghum) recovered $< 2\%$ of one 14 kg/ha application of Zn as zinc sulfate. They calculated that 65% of the applied Zn could not be accounted for by either Zn in the crops or Zn in the soil when extracted by 0.1 *M* HCl. The 35% soluble portion in HCl was probably available to plants.

In later studies on a calcareous and noncalcareous soil, Boawn (1976) showed that after applying zinc sulfate, soil DTPA–extractable Zn reached an equilibrium after about 3 yr. This equilibrium was at a stable level greater than the original soil content. One application of 22.4 kg/ha of Zn maintained soil Zn levels for 7 yr and appeared to supply an adequate level of extractable Zn for an indefinite period.

3. *Other Nutrients*

Zinc deficiencies became prominent, especially in the western areas of the USA in the early 1950s. Since then, many studies have attempted to determine the cause of P-induced Zn deficiencies. There have been several good reviews of this interaction. Two of the most complete and recent are by Olsen (1972) and Adams (1980).

Of all the nutrient interactions that occur in plants, the P–Zn interaction has probably been studied the most. While high P soils may be Zn deficient, the interaction is more commonly a P fertilizer–induced Zn deficiency and occurs primarily on alkaline and especially calcareous soils where Zn availability is low (Langen et al., 1962). Pauli et al. (1968) studied the effect of $CaCO_3$ on the P–Zn interaction. They found that when $CaCO_3$ was added, less Zn and more P was translocated to the leaves of bean plants. Therefore, $CaCO_3$ had a differential effect on the translocation of P and Zn from the roots to the leaves.

The presence of $CaCO_3$ is not necessarily essential for the interaction to occur. Boawn and Brown (1968) showed that when P was supplied from a nutrient solution, P induced a Zn deficiency without decreasing Zn content in the plant. Adams et al. (1982) adjusted the pH of two Alabama soils from 5.0 to 7.3 and found that high rates of P (392 kg P/ha) induced Zn deficiencies, even on soil with a pH of 5.2.

Although many investigators have reported on various aspects of the P–Zn interaction, the basic casual relationships are still not clear. Investigators agree that it is not a reaction between P and Zn in the soil (Adams,

1980). Rather, it appears to be related to some complex relationships within the plant, perhaps involving the translocation of Zn from the root to the leaves (Stukenholz et al., 1966).

One aspect of the P–Zn interaction that has intrigued investigators is the inducement of severe Zn-deficiency symptoms in the leaves without a reduction in Zn concentration (Boawn & Brown, 1968; Christensen & Jackson, 1981). Investigating this phenomenon, Loneragan et al. (1979, 1982) concluded from studies with nutrient solutions that Zn deficiency interferes with P metabolism, thereby increasing the amounts of P absorbed by roots and transported to the plant tops. Under high available-P conditions, P may accumulate to toxic levels, causing symptoms similar to Zn deficiency.

Any factor that increases P fertilizer uptake also will aggravate the P–Zn interaction. For example, banding generally increases P fertilizer uptake; choice of P sources such as polyphosphates may affect Zn availability (Adriano & Murphy, 1970; Giordano et al., 1971), as does presence or absence of mycorrhiazae, which affect P uptake (Lambert et al., 1979).

Unlike P, N fertilizers generally increase Zn uptake by plants (Pumphrey et al., 1963). Viets et al. (1957) showed that N application increased Zn uptake of both native and applied Zn. The effect depended on the change in soil pH brought about by the N fertilizer. Part of the N effect may be related to an increased ability of the root system to absorb Zn, simply due to increased root volume.

REFERENCES

Adams, F. 1980. Interactions of phosphorus with other elements in soils and plants. p. 655–680. *In* F. E. Khasawneh et al. (ed.) The role of phosphorus in agriculture. American Society of Agronomy, Madison, WI.

Adams, J. F., F. Adams, and J. W. Odom. 1982. Interaction of phosphorus rates and soil pH on soybean yield and soil solution composition of two phosphorus-sufficient ultisols. Soil Sci. Soc. Am. J. 46:323–328.

Adriano, D. C., and L. S. Murphy. 1970. Effects of ammonium polyphosphates on yield and chemical composition of irrigated corn. Agron. J. 62:561–567.

Allred, S. E., and A. J. Ohlrogge. 1964. Principles of nutrient uptake from fertilizer bands: VI. Germination and emergence of corn as affected by ammonia and ammonium phosphates. Agron. J. 56:309–313.

Amer, F., and H. A. Mostafa. 1981. Effect of pyrophosphate on orthophosphate reactions in calcareous soils. Soil Sci. Soc. Am. J. 45:842–847.

Amer. R., D. R. Bouldin, C. A. Black, and F. R. Duke. 1955. Characterization of soil phosphorus by anion exchange resin adsorption and ^{32}P-equilibration. Plant Soil 6:391–408.

Anderson, A. J. 1952. The significance of sulfur deficiency in Australian soils. J. Aust. Inst. Agric. Sci. 18:135–139.

Anderson, D. W., E. A. Paul, and R. J. St. Arnaud. 1974. Extraction and characterization of humus with reference to clay associated humus. Can. J. Soil Sci. 54:317–323.

Anderson, D. W., S. Sagger, J. R. Bettany, and J. W. B. Stewart. 1981. Particle size fractions and their use in studies of soil organic matter: I. The nature and distribution of forms of carbon, nitrogen and sulfur. Soil Sci. Soc. Am. J. 45:767–772.

Anderson, W. B. 1964. Effect of synthetic chelating agents and sources of Zn for calcareous soils. Ph.D. diss. Colorado State University, Fort Collins, CO.

Anthonisen, A. C., R. E. Loehr, T. B. S. Prakasan, and E. G. Srinath. 1976. Inhibition of nitrification by ammonia and nitrous acid. J. Water Pollut. Control Fed. 48:835.

Aslyng, H. C. 1954. The lime and phosphate potentials of soils: The solubility and availability of phosphate, Royal Agricultural College, Copenhagen, Denmark.

Avnimelech, Y., and M. Laher. 1977. Ammonia volatilization from soils: Equilibrium considerations. Soil Sci. Soc. Am. J. 41:1080–1084.

Ayanaba, A., and A. P. O. Omayuili. 1975. Microbial ecology of acid tropical soils. A preliminary report. Plant Soil 43:519.

Barber, S. A. 1968. Mechanism of potassium absorption by plants. p. 293–310. *In* J. J. Kilmer et al. (ed.) The role of potassium in agriculture. American Society of Agronomy, Madison, WI.

Barrow, N. J. 1980. Evaluation and utilization of residual phosphorus in soils. p. 332–359. *In* F. E. Khasawneh et al. (ed.) The role of phosphorus in agriculture. American Society of Agronomy, Madison, WI.

Beaton, J. D. 1968. Determination of sulfur in soils and plant material. Sulfur Inst. Tech. Bull. 14. Sulfur Institute, Washington, DC.

Beaton, J. D. 1971. Crop responses to sulfur in North America. Sulfur Inst. Tech. Bull. 18. Sulfur Institute, Washington, D.C.

Beckett, P. H. T. 1964. Studies on soil potassium. I. Confirmation of the ratio law: Measurement of potassium potential. J. Soil Sci. 15:1–8.

Bell, L. C., and C. A. Black. 1970a. Comparison of methods for identifying crystalline products produced by interaction of orthophosphate fertilizers with soils. Soil Sci. Soc. Am. Proc. 34:579–582.

Bell, L. C., and C. A. Black. 1970b. Crystalline phosphates produced by interaction of orthophosphate fertilizers with slightly acid and alkaline soils. Soil Sci. Soc. Am. Proc. 34:735–740.

Bell, L. C., and C. A. Black. 1970c. Transformation of dibasic calcium phosphate dihydrate and octacalcium phosphate in slightly acid and alkaline soils. Soil Sci. Soc. Am. Proc. 34:583–587.

Bettany, J. R., S. Saggar, and J. W. B. Stewart. 1980. Comparison of the amounts and forms of sulfur in soil organic matter fractions after 65 years of cultivation. Soil Sci. Soc. Am. J. 44:70–75.

Bledsoe, R. W., and R. E. Blaser. 1947. The influence of sulfur on the yield and composition of clovers fertilized with different sources of phosphorus. J. Am. Soc. Agron. 39:146–152.

Boawn, L. C. 1976. Sequel to residual availability of fertilizer zinc. Soil Sci. Soc. Am. Proc. 40:467.

Boawn, L. C., and J. C. Brown. 1968. Further evidence for a P–Zn imbalance in plants. Soil Sci. Soc. Am. Proc. 32:94–97.

Boawn, L. C., F. G. Viets, Jr., and C. L. Crawford. 1957. Plant utilization of zinc from various types of zinc compounds and fertilizer materials. Soil Sci. 83:219–227.

Boawn, L. C., F. G. Viets, Jr. C. L. Crawford, and J. L. Nelson. 1960. Effect of nitrogen carrier, nitrogen rate, zinc rate, and soil pH on zinc uptake by sorghum, potatoes, and sugar beets. Soil Sci. 90:329–337.

Bray, R. H., and E. E. DeTurk. 1939. The release of potassium from non-replaceable forms in Illinois soils. Soil Sci. Soc. Am. Proc. 3:101–106.

Bremner, J. M., and L. A. Douglas. 1971. Decomposition of urea phosphate in soils. Soil Sci. Soc. Am. Proc. 35:575–578.

Bremner, J. M., and R. L. Mulvaney. 1978. Urease activity in soils. p. 149–196. *In* R. G. Burns (ed.) Soil enzymes. Academic Press, Inc., New York.

Brown, A. L., and B. A. Krantz. 1966. Source and placement of zinc and phosphorus for corn (*Zea mays* L.). Soil Sci. Soc. Am. Proc. 30:86–89.

Brown, A. L., B. A. Krantz, and P. E. Martin. 1964. Effect of zinc applied to soils. Soil Sci. Soc. Am. Proc. 28:236–238.

Brown, J. C. 1960. An evaluation of bicarbonate induced iron chlorosis. Soil Sci. 89:246–247.

Brown, J. C. 1961. Iron chlorosis in plants. Adv. Agron. 13:329–369.

Brown, J. C., J. E. Ambler, R. L. Chaney, and C. D. Foy. 1972. Differential responses of plant genotypes to micronutrients. p. 389–418. *In* J. J. Mortvedt et al. (ed.) Micronutrients in agriculture. Soil Science Society of America, Madison, WI.

Brown, J. C., R. S. Holmes, and L. O. Tiffin. 1959a. Hypotheses concerning from chlorosis. Soil Sci. Soc. Am. Proc. 23:231–234.

Brown, J. C., and W. E. Jones. 1975. Phosphorus efficiency as related to iron deficiency in sorghum. Agron. J. 67:468–472.

Brown, J. C., O. R. Lunt, R. S. Holmes, and L. O. Tiffon. 1959b. The bicarbonate ion as indirect cause of iron chlorosis. Soil Sci. 88:260–266.

Brown, J. C., and L. O. Tiffin. 1962. Zinc deficiency and iron chlorosis dependent on the plant species and nutrient-element balance in Tulane clay. Agron. J. 54:356–358.

Burford, J. R., and J. M. Bremner. 1975. Relationships between the denitrification capacities of soils and total, water soluble and readily decomposable soil organic matter. Soil Biol. Biochem. 7:389–394.

Burns, G. R. 1967. Oxidation of sulfur in soils. Sulfur Inst. Tech. Bull. 13. Sulfur Institute, Washington, DC.

Chaney, R. L., J. C. Brown, and L. O. Tiffin. 1972. Obligatory reduction of ferric chelates in iron uptake by soybeans. Plant Physiol. 50:208–210.

Chang, S. C., and M. L. Jackson. 1957. Fractionation of soil phosphorus. Soil Sci. 84:133–144.

Chao, T. T., M. E. Harward, and S. C. Fang. 1962. Soil constituents and properties in the adsorption of sulfate ions. Soil Sci. 94:275–283.

Childers, N. F. (ed.). 1954. Fruit nutrition. Horticultural Publications, Rutgers University, New Brunswick, NJ.

Christensen, N. W., and T. L. Jackson. 1981. Potential for phosphorus toxicity in zinc-stressed corn and potato. Soil Sci. Soc. Am. J. 45:904–909.

Cole, C. V., S. R. Olsen, and C. O. Scott. 1953. The nature of phosphate sorption by calcium carbonate. Soil Sci. Soc. Am. Proc. 17:352–356.

Collander, R. 1941. Selective absorption of cations by higher plants. Plant Physiol. 16:691–720.

Colliver, G. W., and L. F. Welch. 1970. Toxicity of preplant anhydrous ammonia to germination and early growth of corn: II. Laboratory studies. Agron. J. 62:346–348.

Cooke, I. J. 1966. A kinetic approach to the description of soil phosphate status. J. Soil Sci. 17:56–64.

Cooke, I. J., and J. Hislop. 1963. Use of an anion-exchange resin for the assessment of available soil phosphate. Soil Sci. 96:308–312.

Cooper, J. E. 1975. Nitrification in soils incubated with pig slurry. Soil Biol. Biochem. 7:119.

Cummins, D. G., and W. L. Parks. 1961. The germination of corn and wheat as affected by various fertilizer salts at different soil temperatures. Soil Sci. Soc. Am. Proc. 25:47–49.

Dancer, W. S., L. A. Peterson, and G. Chesters. 1973. Ammonification and nitrification of N as influenced by soil pH and previous N treatments. Soil Sci. Soc. Am. Proc. 37:67–69.

DeKock, P. C. 1955. Iron nutrition of plants at high pH. Soil Sci. 79:167–175.

Dennis, E. J., and R. Ellis, Jr. 1962. Potassium ion fixation, equilibria, and lattice changes in vermiculite. Soil Sci. Soc. Am. Proc. 26:230–233.

Eaton, F. M. 1966. Sulfur. *In* H. D. Chapman (ed.) Diagnostic criteria for plants and soils. University of California, Division of Agricultural Science, Riverside, CA.

Eckert, D. J., and E. O. McLean. 1981. Basic cation saturation ratios as a basis for fertilizing and liming agronomic crops: I. Growth chamber studies. Agron. J. 73:795–798.

Edwards, J. H., and E. J. Kamprath. 1974. Zinc accumulation by corn as influenced by phosphorus, temperature and light intensity. Agron. J. 66:479–482.

Elgawhary, S. M., W. L. Lindsay, and W. D. Kemper. 1970. Effect of complexing agents and acids on the diffusion of zinc to a simulated root. Soil Sci. Soc. Am. Proc. 34:211–214.

Ellis, R., Jr., J. F. Davis, and D. L. Thurlow. 1964. Zinc availability in calcareous Michigan soils as influenced by phosphorus level and temperature. Soil Sci. Soc. Am. Proc. 28:83–87.

Enfield, C. G., T. Kahn, D. M. Walters, and R. Ellis, Jr. 1981. Kinetic model for phosphate transport and transformation in calcareous soils: I. Kinetics of transformation. Soil Sci. Soc. Am. J. 45:1059–1064.

Engelstad, O. P., and G. L. Terman. 1980. Agronomic effectiveness of phosphate fertilizers. p. 311–332. *In* F. E. Khasawneh et al. (ed.) The role of phosphorus in agriculture. American Society of Agronomy, Madison, WI.

Ensminger, L. E. 1954. Some factors affecting the adsorption of sulfate by Alabama soils. Soil Sci. Soc. Am. Proc. 18:259–264.

Ernst, J. W., and H. F. Massey. 1960. The effects of several factors on volatilization of ammonia formed from urea in the soil. Soil Sci. Soc. Am. Proc. 24:87–90.

Esty, J. C., A. B. Onken, L. R. Hossner, and R. Matheson. 1980. Iron use efficiency in grain sorghum hybrids and parental lines. Agron. J. 72:589–592.

Feagley, S. E., and L. R. Hossner. 1978. Ammonia volatilization reaction mechanism between ammonium sulfate and carbonate systems. Soil Sci. Soc. Am. J. 42:364–367.

Fenn, L. B. 1975. Ammonia volatilization from surface applications of ammonium compounds on calcareous soils: III. Effects of blending a non-precipitate forming with a precipitate forming ammonium compound. Soil Sci. Soc. Am. Proc. 39:366–368.

Fenn, L. B., and R. Escarzaga. 1976. Ammonia volatilization from surface applications of ammonium compounds on calcareous sells: V. Soil water content and method of nitrogen application. Soil Sci. Soc. Am. J. 40:537–541.

Fenn, L. B., and D. E. Kissel. 1973. Ammonia volatilization from surface applications of ammonium compounds on calcareous soils: I. General theory. Soil Sci. Soc. Am. Proc. 37:855–859.

Fenn, L. B., and D. E. Kissel. 1974. Ammonia volatilization from surface applications of ammonium compounds on calcareous soils: II. Effects of temperature and rate of ammonium nitrogen applicaton. Soil Sci. Soc. Am. Proc. 38:606–610.

Fenn, L. B., and D. E. Kissel. 1975. Ammonia volatilization from surface applications of ammonium compounds on calcareous soils: IV. Effect of calcium carbonate content. Soil Sci. Soc. Am. Proc. 39:631–633.

Fenn, L. B., and D. E. Kissel. 1976. The influence of cation exchange capacity and depth of incorporation on ammonia volatilization from ammonium compounds applied to calcareous soils. Soil Sci. Soc. Am. J. 40:394–398.

Fenn, L. B., J. E. Matocha, and E. Wu. 1981a. Ammonia losses from surface-applied urea and ammonium fertilizers as influenced by rate of soluble calcium. Soil Sci. Soc. Am. J. 45:883–886.

Fenn, L. B., R. M. Taylor, and J. E. Matocha. 1981b. Ammonia losses from surface-applied nitrogen fertilizer as controlled by soluble calcium and magnesium: General theory. Soil Sci. Soc. Am. J. 45:777–781.

Firestone, M. K. 1982. Biological denitrification. *In* F. J. Stevenson (ed.) Nitrogen in agricultural soils. Agronomy 22:289–326.

Focht, D. D., and W. Verstraete. 1977. Biochemical ecology of nitrification and denitrification. Adv. Microb. Ecol. 1:135–214.

Follett, R. H., L. S. Murphy, and R. L. Donahue. 1981. Fertilizers and soil amendments. Prentice-Hall, Inc., Englewood Cliffs, NJ.

Frederick, L. R., and F. E. Broadbent. 1966. Biological interactions. p. 198–212. *In* M. H. McVickar et al. (ed.) Agricultural anhydrous ammonia technology and use. American Society of Agronomy, Madison, WI.

Freney, J. R., N. J. Barrow, and K. Spencer. 1962. A review of certain aspects of sulfur as a soil constituent and plant nutrient. Plant Soil 17:295–308.

Freney, J. R., G. E. Melville, and C. H. Williams. 1975. Soil organic fractions as sources of plant available sulfur. Soil Biol. Biochem. 7:217–221.

Freney, J. R., J. R. Simpson, and O. T. Denmead. 1981. Ammonia volatilization. *In* F. E. Clark, and T. Rosswall (ed.) Terrestrial nitrogen cycles. Ecol. Bull. 33:291–302.

Gerloff, G. C., P. R. Stout, and L. H. P. Jones. 1959. Molybdenum-manganese-iron antagonisms in the nutrition of tomato plants. Plant Physiol. 34:608–613.

Gile, P. L., and J. O. Carrero. 1920. Cause of lime induced chlorosis and availability of iron in the soil. J. Agric. Res. 20:33–62.

Giordano, P. M., and J. J. Mortvedt. 1969. Response of several corn hybrids to level of water-soluble zinc in fertilizers. Soil Sci. Soc. Am. Proc. 33:145–148.

Giordano, P. M., and J. J. Mortvedt. 1978. Response of corn to Zn in ortho- and pyrophosphate fertilizers, as affected by soil temperature and moisture. Agron. J. 70:531–534.

Giordano, P. M., E. C. Sample, and J. J. Mortvedt. 1971. Effect of ammonium ortho- and pyrophosphate on Zn and P in soil solution. Soil Sci. 111:101–106.

Griffin, R. A., and J. J. Jurinak. 1973. The interaction of phosphate with calcite. Soil Sci. Soc. Am. Proc. 37:847–850.

Griffin, R. A., and J. J. Jurinak. 1974. Kinetics of the phosphate interaction with calcite. Soil Sci. Soc. Am. Proc. 38:75–79.

Gunary, D., and C. D. Sutton. 1967. Soil factors affecting plant uptake of phosphorus. J. Soil Sci. 18:167–173.

Gwin, R. E. 1977. Phosphorus increases yield of irrigated corn in west-central Kansas. Kansas Agric. Exp. Stn. Bull. 602.

Halvorson, A. D., and W. L. Lindsay. 1977. The critical Zn^{2+} concentration for corn and the non-absorption of chelated zinc. Soil Sci. Soc. Am. J. 41:531–534.

Harmsen, G. W., and G. J. Kolenbrander. 1965. Soil inorganic nitrogen. *In* W. V. Bartholomew, and F. E. Clark (ed.) Soil nitrogen. Agronomy 10: 43–92.

Harward, M. E., and H. M. Reisenauer. 1966. Reactions and movement of inorganic soil sulfur. Soil Sci. 101:326–335.

Herlichy, M. 1973. Distribution of nitrifying and heterotrophic microorganisms in cutover peats. Soil Biol. Biochem. 5:621.

Hodgson, J. F. 1963. Chemistry of the micronutrient elements in soils. Adv. Agron. 15:119–159.

Hodgson, J. F., W. L. Lindsay, and J. F. Trierweiler. 1966. Micronutrient cations complexing in soil solution: II. Complexing of zinc and copper in displaced solution from calcareous soils. Soil Sci. Soc. Am. Proc. 30:723–726.

Hodgson, J. F., K. L. Neeley, and J. C. Pusher. 1972. Iron fertilization of calcareous soils in the greenhouse and laboratory. Soil Sci. Soc. Am. Proc. 36:320–356.

Holmes, R. S., and J. C. Brown. 1955. Chelates as corrections for chlorosis. Soil Sci. 80:167–179.

Jackson, W. A., K. A. Heinly, and J. H. Caro. 1962. Solubility status of zinc carriers intermixed with N–P–K fertilizers. J. Agric. Food Chem. 10:361–364.

Jewitt, T. N. 1942. Loss of ammonia from ammonium sulfate applied to alkaline soils. Soil Sci. 54:401–409.

Johnson, D. D., and W. D. Guenzi. 1963. Influence of salts on ammonium oxidation and carbon dioxide evolution from soil. Soil Sci. Soc. Am. Proc. 27:663–666.

Jones, R. W., and R. A. Hedlin. 1970. Ammonium, nitrite, and nitrate accumulation in three Manitoba soils as influenced by added ammonium sulfate and urea. Can. J. Soil Sci. 50:331–338.

Jordan, J. V., and L. E. Ensminger. 1953. The role of sulfur in soil fertility. *In* A. G. Norman (ed.) Adv. Agron. 10:407–434.

Jordan, J. V., and H. M. Reisenauer. 1957. Sulfur and soil fertility. p. 107–111. *In* Soil. U.S. Department of Agriculture, Washington, DC.

Kamprath, E. J., W. L. Nelson, and J. W. Fitts. 1956. The effect of pH sulfate and phosphate concentrations on the adsorption of sulfate by soils. Soil Sci. Soc. Am. Proc. 20:463–466.

Keeney, D. R. 1980. Factors affecting the persistence and bioactivity of nitrification inhibitors. p. 33–46. *In* Nitrification inhibitors—potentials and limitations. Spec. Pub. 38, American Society of Agronomy, Madison, WI.

Kersch, R. K., M. E. Harward, and R. G. Peterson. 1960. Interrelationships among iron, manganese, and molybdenum in the growth and nutrition of tomatoes grown in culture solution. Plant Soil 12: 259–275.

Khasawneh, F. E. 1971. Solution ion activity and plant growth. Soil Sci. Soc. Am. Proc. 35:426–436.

Khasawneh, F. E., E. C. Sample, and E. J. Kamprath. 1980. The role of phosphorus in agriculture. American Society of Agronomy, Madison, WI.

Kingston, F. J., A. M. Posner, and J. D. Quick. 1972. Anion adsorption by geothite and gibbsite: I. The role of the proton in determining adsorption envelopes. J. Soil Sci. 23:177–192.

Knudsen, D. 1970. The interrelations of exchangeable bases in the soil with plant availability of potassium, magnesium and calcium. Ph.D. diss. Kansas State University, Manhattan, KS.

Kuo, S., and E. C. Lotse. 1972. Kinetics of phosphate adsorption by calcium carbonate and Ca-kaolinite. Soil Sci. Soc. Am. Proc. 36:725–729.

Ladd, J. N., and R. B. Jackson. 1982. Biochemistry of ammonification. *In* F. J. Stevenson (ed.) Nitrogen in agricultural soils. Agronomy 22:173–228.

Lambert, D. H., D. E. Baker, and H. Cole, Jr. 1979. The role of mycorrhizae in the interactions of phosphorus with zinc, copper and other elements. Soil Sci. Soc. Am. J. 43:976–980.

Langen, E. J., R. C. Ward, R. A. Olson, and H. F. Rhoades. 1962. Factors responsible for poor response of corn and grain sorghum to phosphorus fertilization: II. Lime and P placement effects on P–Zn relations. Soil Sci. Soc. Am. Proc. 26:574–578.

Larsen, D. 1967. Soil phosphorus. Adv. Agron. 19:151–210.

Lawton, K., C. Apostolakis, R. L. Cook, and W. L. Hill. 1956. Influence of particle size, water solubility, and placement of fertilizer on the nutrient value of phosphorus in mixed fertilizers. Soil Sci. 82:465–476.

Liebhardt, W. C. 1981. The basic cation saturation ratio concept and lime and potassium recommendations on Delaware's coastal plain soils. Soil Sci. Soc. Am. J. 45:544–549.

Leikam, D. F., L. S. Murphy, D. E. Kissel, D. A. Whitney, and H. C. Moser. 1983. Effects of nitrogen and phosphorus application method and nitrogen source on winter wheat grain yield and leaf tissue phosphorus. Soil Sci. Soc. Am. J. 47:530–535.

Lindsay, W. L. 1972. Inorganic phase equilibria of micronutrients in soils. p. 41–57. *In* J. J. Mortvedt et al. (ed.) Micronutrients in agriculture. Soil Science Society of America, Madison, WI.

Lindsay, W. L. 1979. Chemical Equilibria in Soils. John Wiley & Sons, Inc., New York.

Lindsay, W. L., A. W. Frazier, and H. F. Stephenson. 1962. Identification of reaction products from phosphate fertilizers in soils. Soil Sci. Soc. Am. Proc. 26:446–452.

Lindsay, W. L., and E. C. Moreno. 1960. Phosphate equilibria in soils. Soil Sci. Soc. Am. Proc. 24:177–182.

Lindsay, W. L., and W. A. Norvell. 1969. Equilibrium relationships of Zn^{2+}, Fe^{3+}, Ca^{2+}, and H^{+} with EDTA and DTPA in soils. Soil Sci. Soc. Am. Proc. 33:62–63.

Lindsay, W. L., P. F. Pratt, F. L. Blair, and A. W. Frazier. 1968. Effectiveness of the Chang and Jackson procedure for extracting well characterized phosphorus compounds from soils. Agron. Abstr. American Society of Agronomy, Madison, WI., p. 84.

Lindsay, W. L., and P. L. G. Vlek. 1977. Phosphate minerals. p. 639–672. In J. B. Dixon and S. B. Weed (ed.) Minerals in soil environments. Soil Science Society of America, Madison, WI.

Loneragan, J. F., T. S. Grove, A. D. Robson, and K. Snowball. 1979. Phosphorus toxicity as a factor in zinc-phosphorus interactions in plants. Soil Sci. Soc. Am. J. 43:966–972.

Loneragan, J. F., D. L. Grunes, R. M. Walch, E. A. Aduayi, A. Tengah, V. A. Lazar, and E. E. Gary. 1982. Phosphorus accumulation and toxicity in leaves in relation to zinc supply. Soil Sci. Soc. Am. J. 46:345–352.

Martin, J. C., C. R. Overstreet, and D. R. Hoagland. 1946. Potassium fixation in soils in replaceable and non-replaceable forms in relation to chemical reactions in the soil. Soil Sci. Soc. Am. Proc. 10:94–101.

Martin, J. P., and H. D. Chapman. 1951. Volatilization of ammonia from surface-fertilized soils. Soil Sci. 71:25–30.

Martin, W. E., J. G. McClean, and J. Quick. 1965. Effect of temperatures on occurrence of phosphorus induced zinc deficiency. Soil Sci. Soc. Am. Proc. 29:411–413.

Mathers, A. C. 1970. Effect of ferrous sulfate and sulfuric acid on grain sorghum yields. Agron. J. 62:555–556.

McAuliffe, C. D., N. S. Hall, L. A. Dean, and S. B. Hendricks. 1948. Exchange reactions between phosphates and soils: Hydroxylic surface of soil minerals. Soil Sci. Soc. Am. Proc. 12:119–123.

McLachlan, K. D. (ed.). 1974. Handbook on sulphur in Australian agriculture. Commonwealth Scientific and Industrial Research Organization (CSIRO), East Melbourne, Victoria, Australia.

McNall, L. R. 1967. Foliar application of micronutrients. Fert. Solutions 11(6):8, 10–11, 13.

Morrill, L. G., and J. E. Dawson. 1967. Patterns observed for the oxidation of ammonium to nitrate by soil organisms. Soil Sci. Soc. Am. Proc. 31:757–760.

Mortvedt, J. J. 1975. Iron chlorosis. Crops Soils 27(9):10–12.

Mortvedt, J. J., and P. M. Giordano. 1971. Response of grain sorghum to iron sources applied alone or with fertilizers. Agron. J. 63:758–761.

Mortvedt, J. J., and P. M. Giordano. 1973. Grain sorghum response to iron in a ferrous sulfate-ammonium thiosulfate-ammonium polyphosphate suspension. Soil Sci. Soc. Am. Proc. 37:951–955.

Mortvedt, J. J., P. M. Giordano, and W. L. Lindsay (ed.). 1972. Micronutrients in agriculture. Soil Science Society of America, Madison, WI.

Mulvaney, R. L., and J. M. Bremner. 1981. Use of urease and nitrification inhibitors for control of urea transformations in soils. p. 153–196. *In* E. A. Paul and J. N. Ladd (ed.) Soil biochemistry, Vol. 5. Marcel Dekker, Inc., New York.

Munson, R. D. (ed.). 1985. Potassium in agriculture. American Society of Agronomy, Crop Science Society of America, and Soil Science Society of America, Madison, WI.

Murphy, L. S., K. W. Kelley, P. J. Gallagher, and C. W. Swallow. 1976. Tillage implement applications of amhydrous ammonia and liquid ammonium polyphosphate. p. 293–305. Proc. 8th Int. Fertilizer Congr., Moscow, USSR. 21–26 June. Centre International Des Engrais Chimiques, Vienna.

Murphy, L. S., D. R. Leikam, R. E. Lamond, and P. J. Gallagher. 1978. Dual applications of N and P—better agronomics and economics? Fert. Solutions 22(4):8–20.

Murphy, L. S., and L. M. Walsh. 1972. Correction of micronutrient deficiencies with fertilizers. p. 347–387. *In* J. J. Mortvedt et al. (ed.) Micronutrients in agriculture. Soil Science Society of America, Madison, WI.

Myers, R. J. K. 1974. Soil processes affecting nitrogenous fertilizers. p. 13–26. *In* D. R. Leece (ed.) Fertilizers and the environment. K. T. R. Printing, Croydon, New South Wales, Australia.

Nannipieri, P., B. Ceccanti, S. Cervelli, and P. Sequi. 1978. Stability and kinetic properties of humus-urease complexes. Soil Biol. Biochem. 10:143–147.

Nelson, D. W. 1982. Gaseous losses of nitrogen other than through denitrification. *In* F. J. Stevenson (ed.) Nitrogen in agricultural soils. Agronomy 22:327–363.

Neptune, A. M. L., M. A. Tabatabai, and J. J. Hanway. 1975. Sulfur fractions and carbon nitrogen-phosphorus-sulfur relationships in some Brazilian and Iowa soils. Soil Sci. Soc. Am. Proc. 39:51–55.

Norland, M. A., R. W. Starostka, and W. L. Hill. 1958. Crop resonse to phosphate fertilizers as influenced by level of phosphorus solubility and by time of placement prior to planting. Soil Sci. Soc. Am. Proc. 22:529–533.

Norvell, W. A. 1972. Equilibria of metal chelates in soil solution. p. 115–138. *In* J. J. Mortvedt et al. (ed.) Micronutrients in agriculture. Soil Science Society of America, Madison, WI.

Olsen, S. R. 1972. Micronutrient Interactions. p. 243–264. *In* J. J. Mortvedt et al. (ed.) Micronutrients in agriculture. Soil science Society of America, Madison, WI.

Olsen, S. R., C. V. Cole, F. S. Watanabe, and L. A. Dean. 1954. Estimation of available phosphorus in soils by extraction with sodium bicarbonate. USDA Circ. 939, U.S. Government Printing Office, Washington, DC.

Olson, S. R., and A. D. Flowerday. 1971. Fertilizer phosphorus interactions in alkaline soils. p. 153–185. *In* R. A. Olson et al. (ed.) Fertilizer technology and use, 2nd ed. Soil Science Society of America, Madison, WI.

Olsen, S. R., and F. E. Khasawneh. 1980. Use and limitations of physical-chemical criteria for assessing the status of phosphorus in soils. p. 361–410. *In* F. E. Khasawneh et al. (ed.) The role of phosphorus in agriculture. American Society of Agronomy, Madison, WI.

Olsen, S. R., and F. S. Watanabe. 1979. Interaction of added gypsum in alkaline soils with uptake of iron, molybdenum, manganese, and zinc by sorghum. Soil Sci. Soc. Am. J. 43:125–130.

Olson, R. A., K. D. Frank, P. H. Grabouski, and G. W. Rehm. 1982. Economic and agronomic impacts of varied philosophies of soil testing. Agron. J. 74:492–499.

Olson, R. V. 1950. Effects of acidification, iron oxide addition, and other soil treatments on sorghum chlorosis and iron adsorption. Soil Sci. Soc. Am. Proc. 15:97–101.

Pang, P. C., C. M. Cho, and R. A. Hedlin. 1975. Effects of pH and nitrifier population on nitrification of band-applied and homogeneously mixed urea nitrogen in soils. Can. J. Soil Sci. 55:15–21.

Pauli, A. W., R. Ellis, Jr., and H. C. Moser. 1968. Zinc uptake and translocation as influenced by phosphorus and calcium carbonate. Agron. J. 60:394–396.

Peach, M. 1941. Availability of ions in light sandy soils as affected by soil reaction. Soil Sci. 51:473–486.

Pettit, N. M., A. R. J. Smith, R. B. Freedman, and R. G. Burns. 1976. Soil urease: Activity, stability and kinetic properties. Soil Biol. Biochem. 8:479–484.

Pumphrey, F. V., F. E. Koehler, R. R. Allmaras, and S. Roberts. 1963. Method and rate of applying zinc sulfate for corn on zinc-deficient soils in western Nebraska. Agron. J. 55:235–238.

Ralston, D. E., D. L. Hoffman, and D. W. Toy. 1978. Field measurement of denitrification: I. Flux of N_2 and N_2O. Soil Sci. Soc. Am. J. 42:863–869.

Rehm, G. W., R. A. Wiese, and G. W. Hergert. 1980. Response of corn to zinc source and rate of zinc band applied with either ortho- or polyphosphate. Soil Sci. 129:36–44.

Reitemeier, R. F. 1946. Effect of moisture content on the dissolved and exchangeable ions of soils in arid regions. Soil Sci. 61:195–214.

Rich, C. I. 1964. Effect of cation size and pH on potassium exchange in Nason soil. Soil Sci. 98:100–106.

Rich, C. I. 1968. Mineralogy of soil potassium. p. 79–108. *In* V. J. Kilmer et al. (ed.) The role of potassium in agriculture. American Society of Agronomy, Madison, WI.

Rich, C. I., and S. S. Obenshain. 1955. Chemical and clay mineral properties of a red-yellow podzolic soil derived from mica schist. Soil Sci. Soc. Am. Proc. 19:334–339.

Rovira, A. D. 1962. Plant root exudates in relation to the rhizosphere microflora. Soils Fert. 25:167–172.

Ryan, J. S., D. Curtin, and I. Safi. 1981. Ammonia volatilization as influenced by calcium carbonate particle size and iron oxides. Soil Sci. Soc. Am. J. 45:338–341.

Ryan, J. S., S. Miyamoto, and H. L. Bohn. 1974. Solubility of manganese, iron, and zinc as affected by application of sulfuric acid to calcareous soils. Plant Soil 40:421–427.

Ryan, J. S., L. Stroehlein, and S. Miyamoto. 1975. Sulfuric acid application to calcareous soils: Effects on growth and chlorophyll content of common bermudagrass in the greenhouse. Agron. J. 67:633–637.

Sample, E. C., R. J. Soper, and G. J. Racz. 1980. Reaction of phosphate fertilizers in soils. p. 263–310. *In* F. E. Khasawneh et al. (ed.) The role of phosphorus in agriculture. American Society of Agronomy, Madison, WI.

Schofield, R. K. 1947. A ratio law governing the equilibrium of cations in the soil solution. Int. Congr. Pure Appl. Chem. Proc. 11th 3:257–261.

Shaw, E., R. G. Menzel, and L. A. Dean. 1954. Plant uptake of Zinc-65 from soils and fertilizers in the greenhouse. Soil Sci. 77:205–207.

Shukla, V. C., and H. D. Morris. 1967. Relative efficiency of several zinc sources for corn (*Zea mays* L.). Agron. J. 59:200–202.

Skogley, E. O. and V. A. Haby. 1981. Predicting crop responses on high-potassium soils of frigid temperature and ustic moisture regimes. Soil Sci. Soc. Am. J. 45:533–536.

Smith, F. W., B. G. Ellis, and J. Grava. 1957. Use of acid-fluoride solutions for the extraction of available phosphorus in calcareous soils and in soils to which rock phosphate has been added. Soil Sci. Soc. Am. Proc. 21:400–404.

Sorensen, R. C., E. J. Penas, and A. D. Flowerday. 1970. How to select and apply zinc fertilizer. Nebraska Quarterly. Spring 1970, p. 9–11.

Stanford, G., R. A. Vander Pol, and S. Dzienia. 1975. Denitrification rates in relation to total extractable soil carbon. Soil Sci. Soc. Am. Proc. 39:284–289.

Steenbjerg, F. 1947. Ammonia loss from nitrogen-containing commercial fertilizers when applied to topsoil. Chem. Abstr. 41:4878.

Stewart, B. A., L. K. Porter, and F. G. Viets, Jr. 1966. Effect of sulfur content of straws on rates of decomposition and plant growth. Soil Sci. Soc. Am. Proc. 30:355–358.

Stukenholtz, D. D., R. J. Olson, G. Gogan, and R. A. Olson. 1966. On the mechanism of phosphorus–zinc interaction in corn nutrition. Soil Sci. Soc. Am. Proc. 30:759–763.

Subbarao, Y. V., and R. Ellis, Jr. 1977. Determination of kinetics of phosphorus mineralization in soil under oxidizing conditions. USEPA Rep. 6001/2-77-180. U.S. Government Printing Office, Washington, DC.

Swaine, D. J. 1955. The trace element content of soils. Tech. Commun. Commonw. Bur. Soil Sci. 48:157.

Taylor, A. W., and E. L. Gurney. 1962. Phosphate equilibria in an acid soil. J. Soil Sci. 13:188–197.

Terman, G. L. 1978. Atmospheric sulfur—the agronomic aspects. Sulfur Inst. Tech. Bull. 23. Sulfur Instutute, Washington, DC.

Terman, G. L. 1979. Volatilization losses of nitrogen as ammonia from surface-applied fertilizers, organic amendments, and crop residues. Adv. Agron. 31:189–223.

Terman, G. L., and C. M. Hunt. 1964. Volatilization losses of nitrogen from surface applied fertilizers as measured by crop response. Soil Sci. Soc. Am. Proc. 28:667–672.

Thomas, G. W., and N. T. Coleman. 1959. A chromatographic approach to the leaching of fertilizer salts in soils. Soil Sci. Soc. Am. Proc. 23:113–116.

Thomas, G. W., and B. W. Hipp. 1968. Soil factors affecting potassium availability. p. 269–291. *In* V. J. Kilmer et al. (ed.) The role of potassium in agriculture. American Society of Agronomy, Madison, WI.

Thorne, W. 1957. Zinc deficiency and its control. Adv. Agron. 9:31–65.

Tiller, K. G., J. L. Honeysett, and M. P. C. DeVries. 1972. Soil zinc and its uptake by plants: II. Soil chemistry in relation to prediction of availability. Aust. J. Soil Res. 10:165–182.

Turner, R. C. 1958. A theoretical treatment of the pH of calcareous soils. Soil Sci. 86:32–34.

U.S. Salinity Laboratory Staff. 1954. Diagnosis and improvement of saline and alkali soils. USDA Handb. 60. U.S. Government Printing Office, Washington, DC.

Viets, F. G., Jr. 1951. Zinc deficiency of corn and beans on newly irrigated soils in central Washington. Agron. J. 43:150.

Viets, F. G., Jr. 1966. Zinc deficiency in the soil plant system. p. 90–128. *In* A. S. Prasad (ed.) Zinc metabolism. Charles C. Thomas, Publisher, Springfield, IL.

Viets, F. G., Jr., L. C. Boawn, and C. L. Crawford. 1954. Zinc contents and deficiency symptoms of 26 crops grown in a zinc deficient soil. Soil Sci. 78:305–316.

Viets, F. G., Jr., L. C. Boawn, and C. L. Crawford. 1957. The effect of nitrogen and types of nitrogen carrier on plant uptake of indigenous and applied zinc. Soil Sci. Soc. Am. Proc. 21:197–201.

Vlek, P. L. G., and J. M. Stumpe. 1978. Effect of solution chemistry and environmental conditions on ammonia volatilization losses from aqueous systems. Soil Sci. Soc. Am. J. 42:416–421.

Volk, G. M. 1961. Gaseous loss of ammonia from surface applied nitrogenous fertilizers. J. Agric. Food Chem. 9:280–283.

Wallace, A. 1963. Review of chelation on plant nutrition. J. Agric. Food Chem. 11:103–107.

Wallace, A., and O. R. Lunt. 1960. Iron chlorosis in horticultural plants: A review. Proc. Am. Soc. Hortic. Sci. 75:819–841.

Wallace, A., and R. T. Mueller. 1978. Complete neutralization of a portion of calcareous soil as a means of preventing iron chlorosis. Agron. J. 70:888–890.

Wear, J. I. 1956. Effect of soil pH and calcium on uptake of zinc by plants. Soil Sci. 81:311–315.

Webb, J. R., K. Eik, and J. T. Pesek. 1961. An evaluation of phosphorus fertilizers applied broadcast on calcareous soils for corn. Soil Sci. Soc. Am. Proc. 25:232–236.

Webb, J. R., and J. T. Pesek. 1958. An evaluation of phosphorus fertilizers varying in water solubility: I. Hill applications for corn. Soil Sci. Soc. Am. Proc. 22:533–538.

Webb, J. R., and J. T. Pesek. 1959. An evaluation of phosphorus fertilizers varying in water solubility: II. Broadcast applications for corn. Soil Sci. Soc. Am. Proc. 23:381–384.

Weir, C. C., and R. J. Soper. 1963. Solubility studies of phosphorus in some calcareous Manitoba soils. J. Soil Sci. 14:256–261.

Westerman, R. L., and T. C. Tucker. 1974. Effect of salts and salts plus nitrogen-15-labeled ammonium chloride on mineralization of soil nitrogen, nitrification, and immobilization. Soil Sci. Soc. Am. Proc. 38:602–605.

White, R. E. 1976. Concepts and methods in the measurement of isotopically exchangeable phosphate in soil. p. 9–16. *In* J. R. Ansiaux et al. (ed.) Phosphorus in agriculture, no. 67. World Phosphate Industry Association, Paris.

Williams, C. H., and A. Steinbergs. 1962. The evaluation of plant-available sulfur in soils: I. The chemical nature of sulphate in some Australian soils. Plant Soil 17:279–294.

Williams, C. H., and A. Steinbergs. 1964. The evaluation of plant-available sulfur in soils: II. The availability of adsorbed and insoluble sulfates. Plant Soil 21:50–62.

Williams, C. H., E. G. Williams, and N. M. Scott. 1960. Carbon nitrogen sulfur and phosphorus in some Scottish soils. J. Soil Sci. 11:334–346.

Winter, K. T., D. A. Whitney, D. E. Kissel, and R. B. Ferguson. 1981. Ammonia volatilization from lime urea ammonium nitrate suspensions before and after soil application. Soil Sci. Soc. Am. J. 45:1224–1228.

Withee, L. V., and R. Ellis, Jr. 1965. Change of phosphate potential of calcareous soils on adding phosphorus. Soil Sci. Soc. Am. Proc. 29:511–514.

Woodruff, C. M. 1955. Ionic equilibrium between clay and dilute salt solutions. Soil Sci. Soc. Am. Proc. 19:36–40.

6

Plant Nutrient Behavior in Flooded Soil

W. H. Patrick, Jr.
Louisiana State University
Baton Rouge, Louisiana

Duane S. Mikkelsen
University of California
Davis, California

B. R. Wells
University of Arkansas
Fayetteville, Arkansas

Flooding the soil has a marked effect on the behavior of several important plant nutrients and on the growth and yield of rice (*Oryza sativa* L.), the crop most widely grown under flooded conditions. After flooding, the availability of some nutrients to the crop is increased, while others are subjected to greater fixation or loss from the soil as a result of flooding. Alternate flooding and drying of the soil is especially detrimental to the availability of both fertilizer and native soil sources of N and phosphate (PO_4^{3-}).

The changes in plant nutrient availability resulting from flooding are largely due to biological oxidation-reduction processes brought into play by the exclusion of oxygen from the flooded soil. The layer of water over the soil surface restricts the movement of oxygen to the soil and causes microorganisms that only grow in the absence of oxygen, and some of those that ordinarily use oxygen, to metabolize other reducible substances to carry on respiration. These biological reduction reactions and the chemical changes that accompany them critically influence nutrient behavior in flooded soils. This chapter deals with the biological and physiochemical reactions taking place in a flooded soil. Their effect on plant nutrient availability and the mineral nutrition of rice is also covered.

I. EFFECT OF FLOODING ON THE SOIL

A. Physical Characteristics of Flooded Soils

1. Gaseous Exchange

The immediate consequence of flooding is an interruption of the normal processes of gaseous exchange between air and soil. Water covering the soil or filling soil pores prevents the entry of gaseous oxygen and decreases oxygen diffusion by a factor of more than 10 000. When a soil is flooded, the oxygen level begins to decline and drops to zero within about a day. Wetting or flooding of a dry soil usually stimulates the activity of the soil microflora, which rapidly decomposes the available oxidizable substrate, creating a large demand for oxygen and liberating large amounts of respired CO_2.

The rapid disappearance of oxygen from the soil is accompanied by the release of other gases, produced largely through processes of microbial respiration. Carbon dioxide, N_2, methane (CH_4), and H_2 accumulate in flooded soils. The gases vary in composition from about 1 to 20% CO_2, 10 to 95% N_2, 15 to 75% CH_4, and 0 to 10% H_2. The composition of gases varies with time after flooding and is dependent on soil and other environmental conditions, the composition of the soil microflora, and the nature of the organic and inorganic substances available as substrate.

The restriction of oxygen diffusion does not mean that the entire profile of a flooded soil is uniformly low in oxygen. Oxygen is usually present in a layer extending from a few millimeters to 1 cm below the soil-water interface. Floodwater usually contains some oxygen in a dissolved state from the atmosphere or from various hydrophytes that carry on photosynthesis in the water. This oxygen is available for various biological processes. In the flooded soil immediately beneath this thin oxygenated layer, the oxygen content drops off sharply and with increased depth becomes virtually zero. The depth of oxygen penetration into the soil is an equilibrium value determined by the supply and the biological demand. A schematic cross-section of a flooded soil profile, indicating the concentration of dissolved oxygen at different depths, is shown in Fig. 6–1.

The thin oxygen-containing surface layer of soil immediately below the water layer supports aerobic microorganisms that carry on biological processes similar to those in aerated soils. Various mineral constituents exist here in typical oxidized forms such as nitrates, sulfates, and ferric and manganic compounds. The soil may appear reddish-brown due to the presence of oxidized Fe so that the color resembles that of unflooded soil, which is in sharp contrast to underlying soil that is considerably darker. In the soil beneath the thin oxidized layer, oxygen is virtually absent; anaerobic respiration predominates in this layer. Biological reduction processes result in the accumulation of reduced chemical ions and products of anaerobic respiration.

Another aerobic-anaerobic interface exists in the root rhizosphere of lowland rice and other wetland plants. Oxygen diffusing in the gaseous phase through specialized aerenchyma tissue from the aboveground portions of the plant to the roots creates a thin aerobic layer at the root-soil interface. A number of important reactions take place at these two aerobic-anaerobic interfaces.

2. Other Physical Factors Influenced by Flooding

Flooding creates physical changes aside from decreasing soil aeration, but their effects on nutrient behavior are mostly indirect. It has been ob-

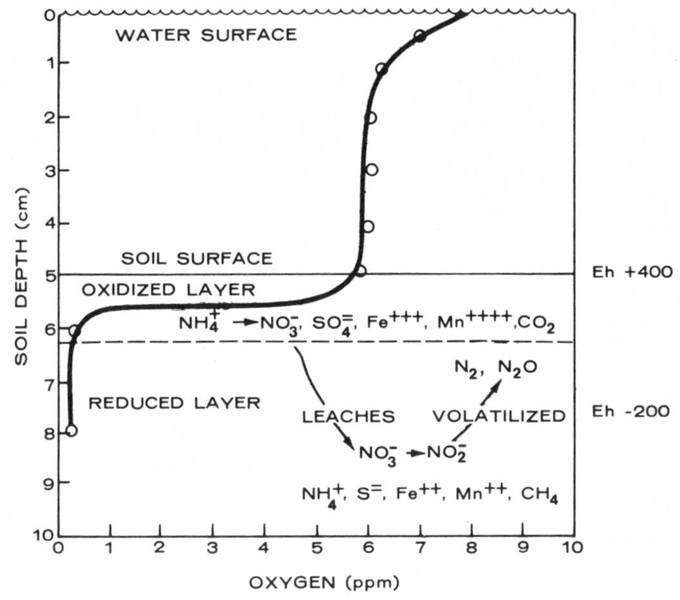

Fig. 6–1. Oxygen profile in flooded soil. The stable forms of inorganic nitrogen, carbon, sulfur, iron, and manganese ions and compounds in the surface oxidized layer and in the underlying reduced layer are shown.

served, for example, that flooding may improve the mechanical properties of the soil by removing impedance to roots of aquatic plants. Soil structure may disintegrate if the agents that cement soil particles and affect cohesive forces are disrupted. In areas where mechanical power is not available for seedbed preparation, advantage is taken of the softening effect of excess water to reduce the mechanical power required for soil manipulation. Permeability to water will be reduced if chemical, physical, or biological changes allow clogging of soil pores—heavy silt loads in floodwater, deflocculating agents, and various salts in water may influence soil permeability. In many rice-growing areas in Asia, the permeability is reduced by puddling in the soil. Flooding also alters soil temperature through the cooling effect caused by evaporation of floodwater, by a layering effect related to heat conductance, and by insulating the soil against rapid diurnal temperature changes.

B. Biological Transformations

Within a short time after waterlogging, aerobic organisms exhaust the soil oxygen supply. In this oxygen-deficient environment, the aerobic organisms become latent or die and new inhabitants, largely facultative and obligate anaerobic bacteria, take over (Fig. 6–2). The rates of many microbial transformations decline, and some processes may be eliminated and replaced by new ones. New microbiological processes, involving oxidized soil constituents rather than oxygen as electron acceptors in respiration, develop with the production of a wide variety of reduction products.

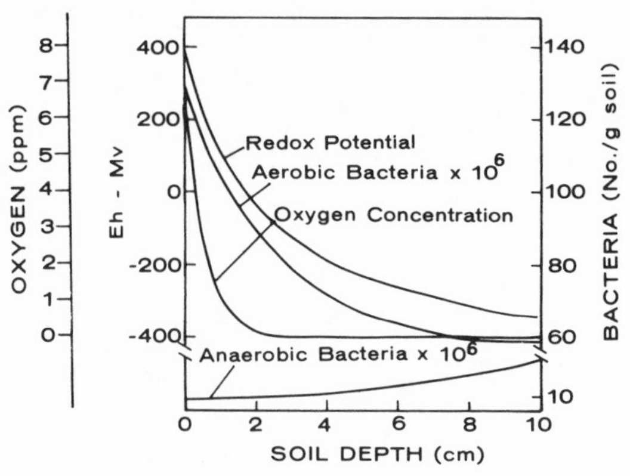

Fig. 6–2. Aerobic and anaerobic organisms, oxygen concentration, and redox potential (Eh) distribution in a flooded soil profile.

1. Changes in Metabolic Processes

All microorganisms must have substrate materials to build cellular constituents, provide energy necessary for vital processes, and function as electron acceptors in reactions that yield energy. The reactions by which most of the soil microflora secure their metabolic requirements involve biological oxidation. Organic and inorganic compounds, to be used as an energy source, must give up electrons and thus become oxidized. Since each oxidation reaction must be accompanied by reduction, there must be electron acceptors to take up electrons from the energy source. Oxygen is the most commonly occurring electron acceptor and it is utilized in aerated and waterlogged soils by aerobes and facultative anaerobes as long as it is available.

In the absence of oxygen, many facultative and obligate anaerobes oxidize organic compounds with the release of energy in a process called *anaerobic fermentation*. Only a partial oxidation of the organic matter occurs and only a small percentage of the potential energy is released. Anaerobic fermentation usually produces lactic acid as a first product; this is subsequently converted to acetic, formic, and butyric acids. These products as well as alcohols, CO_2, CH_4, and H_2 are derived from carbohydrates; NH_3, amines, mercaptans, and hydrogen sulfide (H_2S) are produced from protein sources. The end products of some reactions may accumulate and increase in concentration until they become toxic to crops such as rice. The partially oxidized materials become both electron donors and electron acceptors, which means that some of the same molecular types become oxidized while others are reduced. A key component in such oxidation-reduction reactions is nicotinamide adenine dinucleotide (NAD). Coupled with the oxidation-reduction and the nicotinamide adenine dinucleotide phosphate (NADP) reaction is the synthesis of high-energy phosphate bonds, such as adenosine triphosphate (ATP). The energy of ATP is used by organisms in the synthesis of new cellular materials.

In addition to the striking difference in the end products of aerobic and anaerobic decomposition, there is also a large disparity in the amount of energy released; this greater energy release allows a more efficient synthesis of cellular material per unit of organic nutrient. Under aerobic conditions, utilization of substrate C is relatively high, ranging from 20 to 40%, depending on the microbial population. Anaerobic bacteria typically realize a C assimilation rate of only 2 to 5%. Consequently, organic matter decomposition is retarded in flooded soils.

C. Physiochemical Changes

The three most important physiochemical properties of the soil that are affected by flooding are pH, oxidation-reduction potential (Eh or redox potential), and ionic strength.

1. pH

The pH of most soils tends to change toward the neutral point after flooding, with acid soils increasing and alkaline soils decreasing in pH. Increases as great as 3 pH units have been measured in some acid soils. The equilibrium pH for waterlogged soils is usually between pH 6.5 and 7.5. The tendency for soils of low pH to decrease in acidity and for soils of high pH to increase in acidity when submerged indicates that the pH of a submerged soil is buffered around neutrality by substances produced as a result of reduction reactions (Fig. 6–3). Among the more likely compounds involved in buffering the pH of waterlogged soils are Fe and Mn compounds in the form of hydroxides and carbonates, and carbonic acid.

2. Ionic Strength

Flooding the soil causes an increase in the concentration of ions in the soil solution, although the increase may not persist throughout the growing season. In slightly acid and acid soils, the reduction of insoluble Fe, and possibly of Mn compounds, to more soluble forms accounts for much of the increase in cations. In neutral to slightly alkaline soils, Ca^{2+} and Mg^{2+} in the soil solution make significant contributions to the ionic strength. Ferrous and manganous ions (Fe^{2+} and Mn^{2+}, respectively) produced through reduction reactions displace other cations from the exchange complex to the soil solution.

3. Oxidation-Reduction Potential

The physicochemical measurement that best differentiates a flooded soil from a well-drained soil is the oxidation-reduction, or redox potential. Well-aerated soils are characterized by Eh $\geqslant +400$ mV, whereas waterlogged soils have redox potentials as low as -300 mV if the reduction processes are sufficiently intense. The redox potential of the soil is determined by the degree of oxidation or reduction of the various redox systems in the soil. These include the oxygen, nitrate, nitrite, manganese, iron, and sulfur systems, as well as various organic compounds. Free oxygen functions both chemically and biologically to maintain these systems in the oxidized form, hence, the positive redox potential of oxygenated soils. The redox potential of a flooded soil gives a fairly good indication of the intensity of oxida-

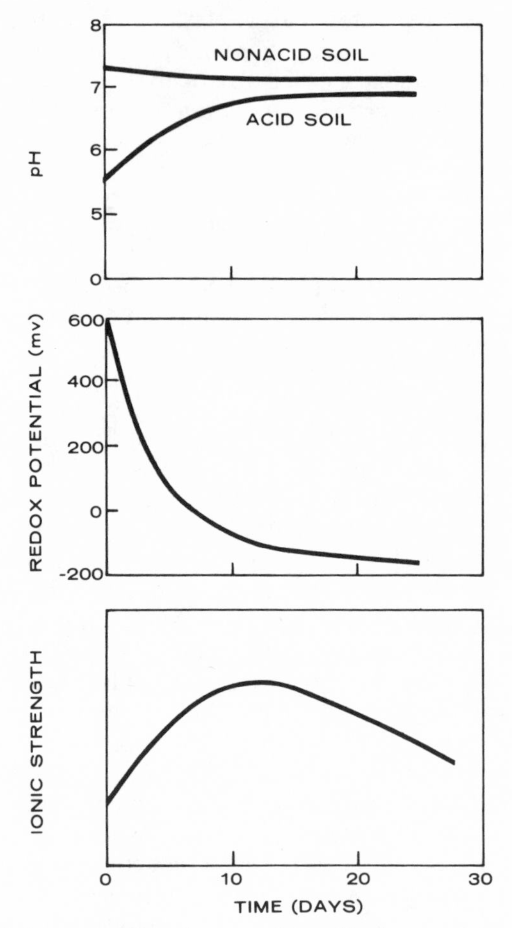

Fig. 6–3. Typical changes in pH, redox potential, and ionic strength in soils after flooding.

tion or reduction in the soil. The redox potential of a submerged soil is a mixed potential from a number of oxidation-reduction systems and is difficult to interpret in terms of any single system. A change in the potential signals a shift in the oxidation-reduction equilibrium and usually involves changes in more than one system. Although its field measurement presents problems, the redox potential is the best single indicator of the degree of anaerobiosis in a flooded soil. It allows reasonable predictions concerning the behavior of several important nutrients and toxins.

4. Changes in Oxidation-Reduction System in Flooded Soils

When oxygen disappears from the soil the requirement of facultative anaerobic and true anaerobic organisms for electron acceptors results in the reduction of several oxidized compounds in the soil. Oxygen, (NO_3^-), the higher oxides of Mn, hydrated ferric oxide (Fe_2O_3) and other ferric (Fe^{3+}) compounds, sulfate (SO_4^{2-}), and CO_2 will be reduced if an energy

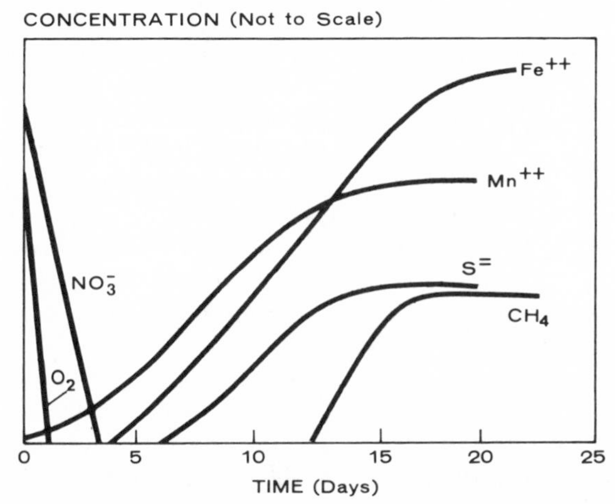

Fig. 6–4. The sequential reduction of several soil redox systems in rice soils after flooding as indicated by the disappearance of oxygen and nitrate and the appearance of reduced forms of manganese, iron, sulfur, and carbon.

source is available to the microorganisms. Nitrate and manganese dioxide (MnO_2) are reduced at fairly high redox potentials, whereas SO_4^{2-} and CO_2 are reduced only under the strictly anaerobic conditions associated with Eh $\leqslant -150$ mV. Hydrated ferric oxide is intermediate between the easily reduced components and the hard-to-reduce components. The reduction of these components is somewhat sequential; oxygen, NO_3^-, and Mn^{4+} compounds are reduced first, followed by the reduction of Fe^{3+} compounds, and then the reduction of SO_4^{2-} and CO_2 (Fig. 6–4). Usually one component is not completely reduced before the reduction of the next one begins, although there are some exceptions. Studies have shown that oxygen and NO_3^- are depleted from the soil before Fe^{3+} is reduced. Sulfate and CO_2 are not reduced to sulfide (S^{2-}) and CH_4 in the presence of oxygen or NO_3^-.

Conditions may exist in which the intensity of reduction is great enough for Fe^{3+} to be reduced to the Fe^{2+}, but not great enough for SO_4^{2-} to be reduced to sulfide. In such instances, SO_4^{2-} can be found in waterlogged soils coexisting with Fe^{2+} and Mn^{2+}. An important consequence of the sequential reduction of Fe^{3+} compounds and SO_4^{2-} is that in soils containing reducible Fe compounds there will always be Fe^{2+} to precipitate sulfide as ferrous sulfide (FeS) a the time the reduction processes are intense enough for H_2S to be produced. Otherwise, toxic levels of H_2S will build-up. Toxicity from H_2S has been reported in soils low in reducible Fe.

Soil reduction does not appear to limit rice growth, except possibly at the extremely low redox potentials in low-Fe soils where free H_2S may be formed, or in soils that have a high concentration of Fe^{2+} and a low pH. The oxygen present in the root rhizosphere as a result of internal transfer through the aerenchyma plays an important role in neutralizing toxic reduced substances. There is usually some benefit in maintaining the redox potential of a flooded rice soil at an intermediate value. If such is the case, then all of the oxidation-reduction systems that function between the oxygen-water and sulfate-sulfide systems will serve to buffer the potential at intermediate values. Because of its high content in most rice soils, hydrated ferric oxide is important in preventing the redox potential from reaching extremely low values. Redox potentials low enough for SO_4^{2-} to be reduced do not generally occur until most of the reducible Fe is in the Fe^{2+} form. Manganese has an effect in stabilizing the redox potential of a flooded soil that is greater than would be anticipated in view of its usual low content in the soil. A rapid decrease in the redox potential after flooding is characteristic of soils with low contents of reducible Fe and Mn. Oxygen, NO_3^-, Mn^{4+} compounds, and Fe^{3+} compounds are all important in slowing or preventing drastic reduction in a flooded soil. However, both oxygen and NO_3^- are usually present in such small amounts in soils that they are rapidly reduced and, therefore, retard reduction for only a few hours, or a few days, after waterlogging. The Fe and Mn systems tend to buffer the redox potential of the soil at about +100 to +300 mV.

II. REACTIONS OF PLANT NUTRIENTS IN FLOODED SOILS

A. Nitrogen Behavior in Flooded Soils

The behavior of N in flooded soils is markedly different from its behavior in drained soils that receive atmospheric oxygen. Flooding the soil results in the accumulation of (NH_4^+), the instability of NO_3^-, and a lowered N requirement for organic matter decomposition as a result of incomplete decomposition of plant residues by anaerobic bacteria. Flooding the soil affects the behavior of added fertilizer N as well as native soil N, and must be taken into consideration in the N fertilization of lowland rice (Fig. 6–5). The unique reactions undergone by N in flooded soils result in considerable loss of applied N fertilizer. Even with the best management currently available, the utilization of added N is generally poorer in flooded soils than in well-drained soils.

1. Forms of Nitrogen in Flooded Soils

Most of the N in soils is organically bound, with only a very small amount present in the inorganic form. In most soils, inorganic N is present as both NO_3^- and NH_4^+ (with a trace amount of nitrite [NO_2^-]); but in flooded soils, the inorganic N is present almost exclusively as NH_4^+. The only part of the flooded soils in which NO_3^- is stable is the thin surface aerobic layer, but NO_3^- can readily diffuse and leach into the underlying anaerobic layer where it is denitrified. Nitrate present in the soil when it is flooded may be removed by leaching, plant uptake, or, more often, by denitrification. If reducing conditions are slow to set in, some of the NO_3^- may be assimilated into microbial tissue, although this is generally not the case. It has been suggested that some of the NO_3^- present in the soil at the time of flooding could be reduced to NH_4^+; however, experiments utilizing ^{15}N show the amount reduced to be insignificant, except where reducing conditions are

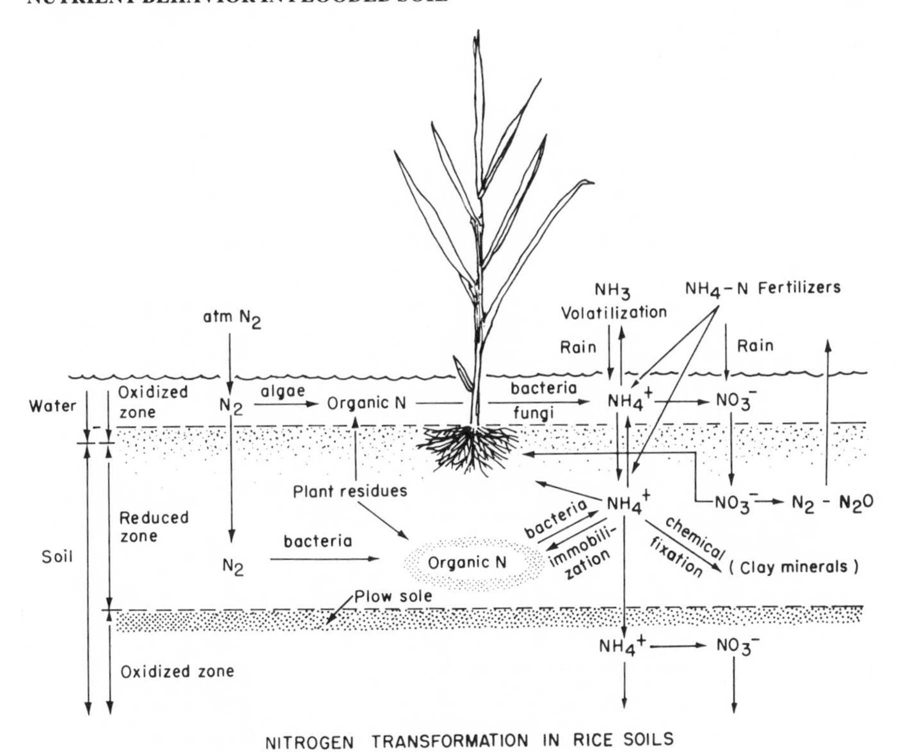

Fig. 6–5. A diagram showing various N transformations in the plant-soil-water system.

very intense. Although NO_3^- can be detected in waterlogged soils under certain conditions, the amount present is too low, usually ≤ 3 mg/kg, to be of any significance as a nutrient or as a toxic substance.

Ammonium-N in waterlogged soils is largely adsorbed on the cation exchange complex, with only a small amount present in the soil solution. A small percentage of the NH_4^+–N may be fixed in a nonexchangeable form between the lattices of silicate minerals.

Although representing only a small fraction of the total soil N, the inorganic NH_4^+–N is the most dynamic part of the N system in waterlogged soils, and is simultaneously being released to and withdrawn from the soil solution and exchange complex. Absorption by the plant, incorporation into microbial tissue, oxidation by *Nitrosomonas,* loss by volatilization, and fixation in the crystal lattice are fairly rapid processes by which NH_4^+ is removed from the exchange complex and soil solution.

2. Reaction of Ammonium Nitrogen in Flooded Soils

a. Chemical Reactions—Ammonium-N added to waterlogged soils, or that formed as a result of organic matter breakdown or urea hydrolysis, moves into the soil solution, where it is rapidly adsorbed onto the cation exchange complex of the soil. Little of the NH_4^+ is left in the soil solution, although the soluble and adsorbed forms are in equilibrium. In reduced soils that have large amounts of Fe^{2+} and Mn^{2+} adsorbed on the exchange complex, a larger percentage of NH_4^+ will be present in the soil solution. This condition will accentuate the loss of NH_4^+ through leaching if downward percolation of the floodwater occurs. The distribution of cations between the exchange complex and the soil solution is also pH dependent, with a larger fraction in the solution phase at low pH values.

b. Biological Mineralization-Immobilization Reactions—In flooded soils, organic N is mineralized to the NH_4^+-form. Immobilization is the opposite process—the incorporation of inorganic N into microbial tissue. Both processes go on at the same time, although one or the other usually predominates, resulting in a net increase or decrease in NH_4^+ content. Except at the aerobic soil surface and at the oxygenated root rhizosphere, NO_3^- is not biologically stable in flooded soils; and NH_4^+ is the only inorganic form involved to any extent in mineralization-immobilization reactions. Additionally, NH_4^+−N enters the organic N pool much more readily than does NO_3^-. Most investigators have found that the heterotrophic microorganisms responsible for N immobilization prefer NH_4^+ to NO_3^-.

Flooded soils differ greatly from drained soils in mineralization and immobilization reactions of N. One of the major differences is in the rate of organic matter decomposition; organic matter breakdown proceeds at a slower rate in a submerged soil as compared with a drained soil. In the presence of free oxygen, organic matter is degraded by heterotrophic bacteria, fungi, and actinomycetes, with fungi probably playing the biggest role. In a waterlogged soil, on the other hand, the absence of oxygen results in a less efficient and more restricted group of bacteria that do not depend on free oxygen as a terminal electron acceptor for decomposition. In spite of the slower rate of decomposition taking place in flooded soils, which might be expected to retard the release of N, the low requirement of the anaerobic organisms for N causes organically bound N to be released to a rice crop earlier in the growing season than would be the case for an upland crop. This ready release of NH_4^+ from anaerobically decomposing organic matter is partially responsible for the good response of rice to added organic matter, although anaerobic decomposition of organic matter can create toxic substances. Perhaps more important in some cases than the rapid release of N, the low N requirement of anaerobic bacteria means that added fertilizer N will not be subject to as great a degree of immobilization in a rice soil as it would be in a well-drained soil. In California, a field study showed that the N requirement for decomposition of rice straw in a flooded soil was 0.54% N, compared with 1.5% N for aerobic decomposition. The generally poor utilization of applied N by rice is apparently not due as much to the immobilization of the N in the flooded soil as to the loss of N through denitrification, NH_3 volatilization, and runoff.

c. Denitrification—Denitrification, or the microbial conversion of NO_3^- to N_2 or nitrous oxide gas (N_2O), is one of the major mechanisms of N loss from flooded soil. Denitrification and NH_3 volatilization account for most of the loss of fertilizer N from flooded rice soils. Nitrate is almost never added to a flooded soil as a N fertilizer, but it is always being produced from NH_4^+ in the aerobic portion of the soil-water-plant system. Denitrification occurs when the NO_3^- diffuses into the anaerobic zone of the soil. Denitrification losses can be reduced if urea and ammonium fertilizers are placed several centimeters deep in the soil immediately before flooding. For transplanted rice, deep placement can be achieved by point placement of large granules of urea (supergranules) in the anaerobic zone of the puddled soil. This zone of the soil becomes anaerobic after flooding, which renders the NH_4^+–N safe from nitrification. Other practices are also used to minimize denitrification losses. These include (i) delayed applications until the plants are large enough to rapidly absorb most of the added N, and (ii) split topdress applications to increase N utilization, and (iii) nitrification inhibitors to prevent the oxidation of NH_4^+ to NO_3^-—a necessary first step for denitrification. Although several nitrification inhibitors have been found that are specific for the NH_4^+ oxidizers, they are not widely used. Slow-release materials have also shown promising results, but because of cost and other considerations, their development beyond the experimental stage has been slow.

d. Ammonia Volatilization—Nitrogen can be lost from the soil and water by volatilization of NH_3 under certain conditions. High NH_3 concentration in the floodwater, high floodwater pH, high temperature, and high wind speed all increase the likelihood of N loss by this mechanism. With the proper combination of these conditions, NH_3 volatilization losses can be high. Urea topdressed into the floodwater is more likely to be lost than ammonium sulfate [$(NH_4)_2SO_4$] because ammonium carbonate mixtures produced by hydrolysis of urea produces alkalinity, which sustains a high floodwater pH. Photosynthesis in the water column also raises the pH as a result of CO_2 (carbonic acid) depletion. For this reason, daytime pH values can exceed 9, which makes dissolved NH_4^+ unstable and increases its volatilization, especially under hot, windy conditions. Volatilization losses are expected to be most pronounced for seedling rice because of the low uptake rate of applied N by the small plants. Considerable research is currently being directed to determining the extent of NH_3 volatilization and the effect of the above-listed variables on the loss. Losses are generally severe enough that farmers are encouraged to add all or a part of the N as a preplant or pretransplant subsurface dressing. Placement several centimeters below the soil surface instead of on the soil surface decreases the amount of NH_4^+ in the water column. For direct-seeded rice with delayed flooding, urea can be topdressed onto dry soil prior to flooding and will be moved down into the soil by the floodwater, provided urea hydrolysis has not occurred prior to flooding.

e. Clay Fixation—Ammonium can be fixed in a plant-unavailable form in the interlayer space of 2:1 type clays. The NH_4^+ ion occupies space usually filled by K^+, and the NH_4^+ is rendered unavailable when the layered clays contract, usually as a result of drying. The slow reversibility of this fixation reaction limits the availability of this NH_4^+ to the current crop. Usually, only a relatively small amount, $\leq 10\%$, of added NH_4^+–N (or urea that hydrolyses to NH_4^+) undergoes fixation. Because NH_4^+ competes with

K^+ for the fixation sites, fixation is usually low in soils with high available K^+.

f. Losses in Runoff Water and Deep Percolation—Inorganic forms of N (some of the NH_4^+ and all of the NO_2^- and NO_3^-) are soluble in water and will be removed in the irrigation water that is lost by runoff or by deep percolation. Although downward percolation below the root zone is ordinarily not severe since most rice soils are very slowly permeable; medium-textured or coarse-textured soils that have high permeability can have significant deep percolation. Under proper management, neither runoff nor deep percolation is likely to be a very significant mechanism of N loss under most field conditions.

g. Nitrogen Fixation—Not all N reactions in flooded soils result in N loss. Significant amounts of atmospheric N_2 can be fixed in a flooded soil. Dinitrogen fixation by blue-green algae in the water column and on the surface of the soil and on the stems and leaves of aquatic plants has been recognized for a long time. Fixation by anaerobic bacteria in the flooded soils can also take place. In recent years, N_2 fixation by nonsymbiotic bacteria living in association with rice roots (associative N_2 fixation) has been recognized as an improtant mechanism of fixation in flooded rice fields. These mechanisms of N_2 fixation are probably partly responsible for sustained yields of rice being relatively higher than yields of upland grain crops in subsistence farming systems. Another source of biologically fixed N_2 is the *Azolla–Anabaena* association, which, when properly managed, can supply a considerable amount of the crop's needs.

B. Phosphate Behavior in Flooded Soils

The requirement of lowland rice for phosphate fertilization is not as universal as is the need for N. This section will describe the various forms of P existing in soils and the reactions undergone in flooded soils by both native forms of phosphate and added phosphate fertilizers.

1. Forms of Phosphorus in Soils

Phosphorus occurs in both organic and inorganic combinations. Inorganic P usually predominates and 15 to 70% of the total P is normally in adsorbed or insoluble inorganic forms. The balance of the soil P is present as organic P, which when mineralized serves as a major source of P for soil microorganisms and plants. Organic forms of P are not utilized directly by crops, but become available as orthophosphate following organic matter mineralization. Because of the small amounts released (2 to 4% per year) organic P is usually not as important to the nutrition of rice as is inorganic P.

Inorganic soil P has been classified for convenience into four main groups: Ca phosphates, Al phosphates, Fe phosphates, and reductant-soluble phosphates, all of which coexist in soils. This classification is oversimplified, but serves to characterize the major inorganic forms. Calcium phosphates tend to predominate in young, relatively unweathered soils with a neutral to alkaline reaction and are converted during soil weathering to the less-soluble Al and Fe phosphate fractions in medium to strongly acid soils. Aluminum and Fe phosphates occur in the soil as insoluble crystalline and amorphous compounds with sesquioxides and clays, with P occluded on their surfaces.

Iron phosphates, including occluded Fe phosphates are dominant in acid soils, while Al phosphates are dominant in soils developed from volcanic tuff. The reductant-soluble P is too insoluble to benefit crops in well-drained soils, but is important to lowland rice.

2. Phosphorus Fixation in Soils

Phosphorus fixation and release in the soil is influenced by a number of soil factors, among them the kind and amount of clay, the amounts and activities of Fe, Al, Ca, and Mg, pH of the soil, and the oxidation-reduction status of the soil. Phosphorus reacts with clay in several ways. One reaction is the replacement of hydroxyl ions (OH^-) in the surface layer of clay minerals. This type of anion exchange reaction is most common with the 1:1 type clays, since these clays have more exposed-hydroxyl ions than do the 2:1 type clays. There is also evidence of P exchange with silicate ions in clay. Digestion of kaolinite in a strong phosphate solution has been shown to release silica in a quantity proportional to the amount of P fixed. Phosphorus also reacts with Fe and Al ions released from the surface decomposition of clay to form insoluble Al and Fe phosphates. Phosphorus fixation in calcareous soils is largely the result of reaction with Ca ions and possibly with solid-phase calcium carbonate ($CaCO_3$). Complex Ca phosphate compounds of varying solubility result from these reactions. Under such conditions, the P concentration in the soil solution of calcareous soils is a function of Ca^{2+} activity. The simplest Ca phosphate compounds in soils—monocalcium, diacalcium, and tricalcium phosphates—decrease in solubility with an increase in Ca content. Insoluble tricalcium phosphate is more likely to be found at a high pH. In alkaline soils containing free calcium carbonate, phosphate ions are precipitated on the surface of the calcium carbonate particles. The amount of exposed surface will determine the amount of P precipitated. In a Ca-saturated clay or a Na-saturated clay soil containing Ca^{2+} ions, calcium phosphate will precipitate as a separate phase above pH 6.5.

The extent of fixation of P fertilizer in the soil will depend on the chemical properties of the soil, the characteristics of the added P material, and the method and rate of application. In acid soils, P is usually precipitated as Al phosphate or absorbed onto the clay surfaces. With time, the phosphate is converted to the more insoluble ferric phosphate ($FePO_4$). Thoroughly mixing the P fertilizer with the soil results in faster, more complete fixation than occurs with banding the fertilizer in the soil.

3. Effects of Soil Submergence on Phosphorus Fixation

Saturation of the soil with water usually increases the availability of soil P to the rice crop. Soils that require the addition of phosphate fertilizer to correct deficiencies for upland crops will, when submerged, usually supply enough P from native sources to meet the requirements of rice. The reduced demand for phosphate fertilizer is not due to a lower plant requirement, but results from the release of P from the soil as a result of reactions brought on by soil submergence and by increased diffusion at a high moisture content. In some rice-growing areas, the soil remains flooded throughout the growing season; in others, the supply of water is intermittent and the soil is alternately dry and flooded. Phosphorus behavior is not the same under these two conditions.

a. Phosphorus Behavior in Continuously Flooded Soil—The increase in soluble and extractable P as a result of soil flooding is well documented. The increase in availability involves a reduction of ferric phosphate to the more soluble ferrous form [$Fe_3(PO_4)_2$], the hydrolysis of Fe and Al phosphates as the pH is increased, and dissolution of Ca phosphates because of higher CO_2 pressures in the soil solution. Among the reactions cited, the reduction of ferric phosphate to ferrous phosphate appears to be the most important. Experiments have shown that uptake of soluble fertilizer phosphate is about the same in flooded as in nonflooded soils, but that uptake of P from Fe phosphates is much better in flooded soils. In acid soils containing free Fe oxides, phosphate also occurs in an occluded form. Flooding usually causes reduction of the hydrated ferric oxide to ferrous hydroxides and releases a part of the occluded phosphate. This is the phosphate fraction most difficult to solubilize and it is almost completely unavailable to upland crops. The occluded phosphate is a part of the soil reductant–soluble phosphate fraction.

b. Phosphorus Reactions under Alternate Submergence and Drying—Alternating anaerobic and aerobic conditions in the soil brought on by flooding and drying cause marked fluctuations in soluble and extractable phosphate. Phosphate solubility, which increases during the flooding phase, usually decreases as the chemical reactions are reversed when the soil begins to dry.

Drying of the soil results in soil oxidation and a decrease in soil pH facilitates conversion of Fe^{2+} precipitates or hydrated magnetite (Fe_3O_4) to amorphous ferric hydroxides. Because of the high chemical reactivity and specific area of these Fe compounds, soluble phosphates are rapidly adsorbed and their availability decreases sharply. Manganese reduction and oxidation, carbonate (CO_3^{2-}) and sulfide formation during flooding may also play a role in this phosphate reversion.

Fertilizer phosphate added to the soil is rapidly converted to an unavailable form when the flooded soil undergoes drying. Soluble P is rapidly fixed on the soil solid phase, particularly on the amorphous ferric oxyhydroxide (FeOOH) formations. The fixation of P is more extensive and less reversible under alternating flooding-draining than under either continuously flooded or continuously moist soil conditions. Alternate flooding and drying increases the amount of P in the Fe phosphate and reductant-soluble Fe phosphate fraction at the expense of the soluble and Al phosphate fractions. The large need for phosphate fertilizer by crops grown in rotation with rice is due in a large part to fixation of P as a result of alternate flooding and drying. Because P availability is usually very low in drained rice soils, banded phosphate fertilizers are usually superior to broadcast applications for upland crops.

C. Potassium Behavior in Flooded Soils

Soil K is less affected by flooding than are N and P. There is little evidence that K fertilization of rice differs significantly from other cereal crops, except possibly for an increased diffusion rate at higher moisture contents. In worldwide rice production, K is usually less limiting than N or P but need increases with cropping intensity. Potassium responses occur most frequently on light-textured soils developed from parent materials low in K or on soils where poor drainage and toxic substances occur and where soil and irrigation sources are high in Ca and dissolved CO_2.

Soil K is classified in four forms—soluble, exchangeable, nonexchangeable, and mineral—all in a state of dynamic equilibrium. Flooding enhances the release of exchangeable K^+ to the soluble form by stimulating the reduction of Fe^{3+} and Mn^{4+}. Soluble K^+ reaches maximum values at the peak of soil reduction and the release is correlated with production of soluble Fe and Mn. Anaerobic production of NH_4^+ also contributes to the displacement of K^+ from the exchange complex. While flooding increases soluble K^+, alternate flooding and drying may also affect increased K^+ availability on soils rich in 2:1 types of clay minerals. While the increased concentration of soluble K^+ is beneficial to rice, it may result in greater leaching losses from permeable soils. Some of the K added as a fertilizer can be fixed in an unavailable form by the clay, but there is little evidence that K availability in flooded soils is seriously limited by fixation.

Exchangeable soil K^+ values correlate well with availability for upland cereals, but in flooded soils the inclusion of nonexchangeable K gives a better indication of the soil K status. Where soluble or exchangeable Na^+ occurs, it can substitute for K^+ to a limited extent and must be considered in criteria used for estimating K needs.

D. Sulfur Behavior in Flooded Soils

1. Forms and Behavior

Most of the S in flooded and unflooded soils is present as a constituent of soil organic matter. A small but very active portion of soil S occurs in inorganic form as SO_4^{2-} and is the immediate source of S for higher plants. The inorganic S is produced from the decomposition of S-containing organic compounds and from S-bearing fertilizers. In oxygen-deficient soils sulfate-sulfur ($SO_4^{2-}-S$) may be reduced to sulfide, which can occur either as H_2S dissolved in the soil solution, as insoluble metal sulfides such as Fe^{2+}, Mn^{2+}, Cu or Zn sulfide, or as H_2S gas, which forms under conditions of high sulfide production, low pH, and low mineral content of the soil. Hydrogen sulfide may escape from the soil to the overlying water and to the atmosphere. If it comes in contact with oxygen or NO_3^-, the sulfide released from organic matter decomposition is utilized by chemoautrophs as an energy source and converted to SO_4^{2-}.

Sulfate formed in rice soils, either in the root rhizosphere or in the surface aerobic-anaerobic zone, or SO_4^{2-} that moves into the soil from irrigation water is subject to reduction to sulfide by sulfate reducers in the anaerobic soil. The soil must be very reduced for this reaction to occur and the presence of even trace amounts of oxygen will inhibit the reaction. Thus, in aerobic soils, the stable inorganic form of S is SO_4^{2-}, while in anaerobic soils the meta-stable inorganic form of S is sulfide. In addition to transformation of inorganic S to and from the oxidized and reduced forms, there is also a dynamic exchange between the organic and inorganic soil fractions. Sulfur is mineralized from organic to inorganic forms as a result of decomposition of high S-containing organic matter and is immobilized from inorganic to organic forms when plant residues with a low S content are decomposed by soil microorganisms.

E. Effect of Submergence on Secondary and Micronutrient Elements

Submergence of the soil has an effect on the availability of plant nutrients other than N, P, and K. Excess water can act in several ways to influence the availability of nutrients. The effect may be through (i) increased solubility of relatively insoluble compounds due to the dilution effect of the excess water, (ii) pH changes associated with changes in the oxidation-reduction status of the soil, (iii) increased availability due to a greater mobility of nutrients in a saturated soil, (iv) changes in the oxidation-reduction equilibria in the soil as a result of exclusion of oxygen, and (v) precipitation as insoluble complexes with hydroxide, carbonate bicarbonate (CO_3^{2-}–HCO_3^-), organic acids, or sulfides, depending on equilibria and environmental conditions. The excess water may also act directly to increase leaching losses of any soluble nutrients in permeable soils. Permeability is so low in most rice soils, however, that leaching may not be a significant factor in removing plant nutrients from the root zone. The results of research carried out in California on micronutrient deficiencies in rice as affected by various environmental factors are shown in Table 6–1.

Soil submergence has different effects on the various secondary and micronutrient elements. Some nutrients such as Ca and Mg are altered only to a limited extent by chemical changes associated with flooding; others such as Fe, Mn, and S undergo marked chemical changes following waterlogging. As mentioned above, one way in which excess water can influence the supply of plant nutrients is to increase the solubility of sparingly soluble compounds. The solubility and availability of B and Mo, and possibly of Co, Cu, and Zn, should be increased slightly as a result of the increased solubility of sparingly soluble compounds. As a possible result of

Table 6-1. Factors contributing to micronutrient deficiencies in rice in California.

Factor affecting deficiencies	Boron	Copper	Iron	Manganese	Molybdenum	Zinc
Low micronutrient composition	X	X	X	X	X	X
High soil pH	X	X	X	X		X
Low soil pH	X				X	
Calcareous soil			X	?		X
High bicarbonate			X			X
Low temperature			X	X		X
Low organic matter			X			X
High organic matter		X	X			?
Oxidation-reduction changes		?	X	X		
Micronutrient interactions		X	X	X	X	X
Low light intensity				X		X
High light intensity	X			X		X
Varietal differences			X			X

higher solubility resulting from excess water, plants grown on poorly drained soils have been observed to contain higher amounts of some micronutrients than plants grown on well-drained soils.

In addition to its effect on solubility, soil submergence changes the availability of several of the micronutrient elements by initiating reduction reactions. Oxidation-reduction reactions can change the solubility of micronutrients in several ways. The reduction of insoluble Fe and Mn compounds solubilizes appreciable quantities of these two micronutrients. Several Fe^{3+} compounds, such as hydrated ferric oxide and ferric phosphate, are reduced under anaerobic conditions to forms that are much more soluble than the oxidized forms. Other heavy metal micronutrients, such as Co and Mo, may be made more available under reducing conditions through chelation of the metal ions by organic molecules produced by anaerobic decomposition of plant material. Organic matter decomposition is not as complete under anaerobic as under aerobic conditions, and the long-chained organic molecules produced under anaerobiosis have been found to increase the solubility of some heavy metal ions.

While most nutrient elements increase in availability in submerged soils, Zn is subject to various chemical reactions that both increase and decrease solubility. Zinc availability is quite pH dependent, decreasing in solubility about 100 times when the pH is increased one pH unit. The pH of flooded acid soils increases following submergence, which may reduce Zn availability. In calcareous soils, flooding should enhance Zn availability, but this does not always occur. Free carbonates rapidly absorb Zn, reducing its availability, and the presence of bicarbonate ions will antagonize Zn uptake. In submerged soils, insoluble Zn compounds such as sulfide, silicate, and phosphate compounds may form.

Changes in the soil pH as a consequence of flooding can affect the availability of other micronutrients. Although the specific effect of pH change on micronutrient solubility varies with each element, Fe, Mn, Cu, and Zn solubility decreases about 100-fold for each unit pH increase. A pH increase induced by flooding would thus decrease the availability with increasing soil pH. Soil reduction brought about by flooding can increase the solubility of Fe in acid soils to cause both a direct and/or indirect toxicity to rice. However, soluble Fe has an ameliorating effect by preventing Mn toxicity in some submerged soils.

F. Formation of Toxic Substances in Flooded Soils

While flooded soils often constitute an adverse and sometimes toxic medium for all plants, rice and other swamp plants have evolved strategies to avoid or compensate for the toxic conditions brought about by oxygen deficiency. The major mechanism in survival is avoidance, by which plants develop structures that transport internal atmospheric oxygen to the roots to provide the oxygen necessary to carry on aerobic root respiration in an anaerobic substrate. Many wetland plants, such as rice, also have the capacity to carry on anaerobic respiration, and although this metabolic process is much less efficient than aerobic respiration, it enables the plant to function in an environment that would be lethal to upland plants.

Reduced soils exhibit two kinds of toxicity to plants. One form of toxicity is caused by the oxygen demand of the reduced chemical constituents of the flooded soil; unless enough oxygen is provided by internal diffusion to the roots to create an oxidizing environment for the root, respiration is impaired and plant growth decreases. A large supply of readily decomposible organic matter such as fresh plant residues may create a strong soil oxygen demand so that the aerobic root functions cannot be maintained and the plant will die.

In addition to the toxicity resulting from a high oxygen demand, some of the reduced components of a flooded soil exhibit specific toxicities. The most important toxins in flooded soils are hydrogen sulfide, Fe^{2+}, organic acids, and, in cases of rapid decomposition of plant residue, CO_2. Except for CO_2, all of these toxins are the reduced products of anaerobic decomposition processes and have two adverse effects on plants: (i) an increase in the requirement for oxygen in the root and the root rhizosphere, and (ii) a specific toxicity to root cell functions.

Hydrogen sulfide is an important toxin in some high organic matter soils where there is not enough Fe to neutralize the sulfide by precipitation as ferrous sulfide. Organic S compounds that are toxic to plants, such as methyl mercaptan (CH_4S), are also produced in anaerobic soils. A physiological plant disorder associated with the occurrence of soluble sulfides, hydrosulfides, and metal sulfides occurs in some areas where lowland rice is grown. Sulfides produce by sulfate reduction are directly toxic to rice by destroying the oxidative power of the rice roots, impairing nutrient uptake and adversely affecting normal physiological processes.

Organic acids such as acetic, butyric, and propionic acids are also formed as a result of microbial fermentation and can sometimes accumulate in concentrations that are injurious to the plant. This condition is most likely to occur soon after crop residues are incorporated and the soil is flooded.

Ferrous iron has the potential for toxicity in most flooded soils because of the large amount of Fe^{3+} compounds reduced to Fe^{2+}. Fortunately, the rise in pH of acid soils after flooding causes the precipitation of oxyhydroxides of Fe^{2+}, which decrease the activity of Fe^{2+} to the point where the internally supplied oxygen in the root can cope with the Fe^{2+} at the root surface. In some very acid soils, where flooding does not result in a pH increase large enough to decrease Fe^{2+} activity, Fe toxicity can result. Under such low pH conditions Al^{3+} toxicity is also likely to occur.

III. FERTILIZER USE FOR LOWLAND RICE

Fertilization of rice, especially with N and to a lesser extent with P and K, is necessary on most soils if optimum grain yields are desired. Fertilizers must be used efficiently to obtain the highest possible yields with a minimum input of fertilizers and with proper concerns for environmental quality. Efficient fertilizer use necessitates consideration of the following factors (i) proper selection of the form and source of fertilizer; (ii) timely applications to provide for optimum growth, grain yield, and crop quality; (iii) suitable placement to provide the most efficient crop utilization; and (iv) the economics of fertilizer use.

A. Effective Water Management and Fertilizer Use

Much of the world rice area is planted with flooded or *lowland* rice. Flood control, however, varies from none to excellent. Flooding as such is not an absolute requirement for rice, but a saturated soil in a chemically reduced state is usually required for optimum grain production. Flooding soils has long been known to enhance the growth and yield of rice. Characteristics of flooded soils that produce high yields are (i) elimination of moisture as a crop factor; (ii) greater availability of nutrients, such as P, Fe, and Mn; (iii) suppression of weed competition; and (iv) a microclimate favorable to plant development.

The preference of a flooded soil environment for the production of rice imposes special considerations on fertilizer use, especially N. Factors to consider for obtaining high fertilizer use efficiency are (i) prevention of losses attributable to leaching and/or surface runoff in water; (ii) prevention of losses of various gaseous forms (NH_3 volatilization and denitrification); (iii) control of nutrient mobilization-immobilization influenced by physical, chemical, and biological processes (immobilization of N in organic matter, clay-fixed NH_4, etc.); and (iv) minimizing of chemical reactions between fertilizers, water, and various soil constituents that would be detrimental to crop growth (e.g., sulfide toxicity).

Since rice benefits from an environment of continuous soil submergence, fertilizer applications are usually designed to optimize fertilizer use efficiency in a chemically reduced soil system. Fertilizer losses can be appreciable (by various factors already indicated) if soils are drained and oxidized and then returned to a reducing condition. Fertilizer materials placed in the reduced soil layer undergo a variety of biological and chemical processes that generally enhance plant nutrient availability if reducing conditions are maintained.

Under normal conditions it is neither essential nor desirable to remove the water once a permanent flood has been applied. Draining the water allows the soil to become aerated, thus enlarging the oxidized soil layer and making possible increased rates of aerobic microbial activity and chemical oxidation of soil constituents. Aerating the soil is detrimental to fertilizer N conservation since nitrates are formed and are subsequently either lost through leaching or denitrified. Drying the soil may lead to plant-water stress and also cause reversion of nutrients to less available forms through oxidation and immobilization.

In some areas where large amounts of crop residues are incorporated and decomposed anaerobically at relatively high temperatures or where soil degradation has occurred, timely drainage is an acceptable practice. Under these conditions, drainage is practiced to allow soil aeration to increase the redox potential, and to dissipate toxic gases, organic acids, and reduced inorganic constituents that may damage the rice crop. Examples where drainage might be beneficial to rice occur when physiological diseases such as "straighthead" or sulfide toxicity occur in association with certain reduced soils. On these soils, the damage can be prevented or minimized by draining and drying the soil during the vegetative stage of plant development.

B. Fertilizer: Sources, Timing, and Placement

1. Nitrogen

Nitrogen fertilization is needed almost everywhere that rice is grown. Adequate N is needed to realize the full yield potential of improved rice varieties. The only general exceptions occur where rice is planted on new fertile soil or where considerable amounts of N-rich residues or green manure crops have been incorporated into the soil. No other plant nutrient has such a direct bearing on rice grain yields or is subjected to such a wide variety of biological, chemical, and physical reactions. Some of the major reactions undergone by fertilizer N after addition to the soil are shown in Fig. 6–6. Careful N management is extremely important since both deficiencies influence plant growth, grain yield, and quality.

a. Forms of Nitrogen for Rice—Considerable attention has been given to determining which form of N, NH_4^+ or NO_3^-, is more effective in the culture of rice. Evidence indicates that rice grown on flooded soils consistently produces better growth and higher grain yield when fertilized with NH_4^+ rather than NO_3^-. Actually, rice uses either form of N about equally well after uptake, but it is not possible to maintain NO_3^-–N in a flooded soil. Losses of NO_3^- through leaching and denitrification account for 20 to 50%, and occasionally as high as 70%, of the total applied. In contrast, NH_4^+–N properly placed effectively supports the plant as long as the supply

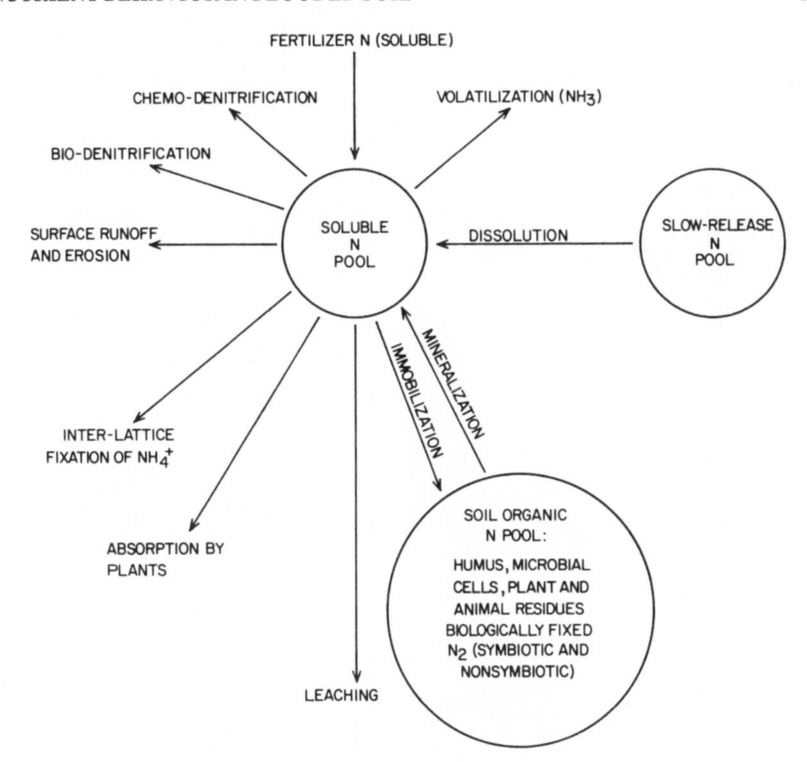

Fig. 6–6. Various reactions of soluble fertilizer N after addition to soil.

lasts. Appreciable losses of NH_4^+ are found only in soils with a high permeability and a low cation exchange capacity.

b. Sources of Nitrogen—Many experiments have been conducted to determine the effect of source of N on yields of rice. The commonly used NH_4^+ or NH_4^+-producing sources appear to be about equally effective.

Various workers have compared the major NH_4^+ sources of N— i.e., ammonium sulfate, urea, ammonium chloride (NH_4Cl), anhydrous NH_3, and aqua NH_3. Most results show that there are no significant differences for increasing grain yields. Urea sometimes gives the most inconsistent results and, if evaluated in comparison with ammonium sulfate, is usually slightly inferior in performance. The reason for the lower production response from urea is not well established, but it is likely due to volatilization of NH_3 following hydrolysis of urea.

Ratings of N sources as they influence rice yields have been made by various investigators. Relative ratings of materials applied preplant, using ammonium sulfate as a standard of 100, range as follows: ammonium chloride, 97 to 100; urea, 92 to 100; aqua and anhydrous NH_3, 85 to 100; and ammonium nitrate (NH_4NO_3) 57 to 70 (Table 6–2). Differences in efficiency are related to the properties of fertilizer sources, application methods, and, in the case of aqua and anhydrous NH_3, to soil conditions during application.

Table 6-2. Relative effectiveness of various N sources for flooded rice by state and method of application (data from O. P. Engelstad, Tennessee Valley Authority).

Nitrogen source	California	Louisiana		Mississippi	Texas			Sri Lanka	India				Japan			Taiwan	
	Basal	Basal	Top-dressed	Basal	Basal		Top-dressed						Yamageta	Shiga	Hiroshima		
Ammonium sulfate	100	100	100	100	100	100	100	100	100	100	100	100	100	100	100	100	100
Urea	90	95	98	94	90	93	94	32	57	82	69	57	96	100	98	84	99
Ammonium chloride	97	--	--	--	--	--	--	--	87	--	99	88	--	--	--	92	99
Ammonium phosphate	--	--	--	--	--	--	--	--	84	86	105	--	--	--	--	--	--
Ammonium nitrate	57	--	--	76	52	81	58	40	44	92	--	44	--	--	--	57	83
Sodium nitrate	--	--	--	60	42	39	46	--	13	40	--	--	--	--	--	--	--
Application technique	All N drilled in before planting	Planting treatment incorporated		Incorporated	Planting treatment incorporated			33% of N added basally remainder as topdressing	Probably incorporated at planting Data from 44.8 kg rate only				All N plowed in with immediate flooding			50% Added at transplanting time; remainder as topdressing	

Urea has gained considerable favor among growers who desire a cheap material with high analysis. Urea has generally replaced ammonium sulfate as the major N source for rice in many parts of the world. Urea or ammonium chloride are used in preference to ammonium sulfate to avoid crop injury from sulfate reduction. Anhydrous NH_3, aqua NH_3, and N solutions are also gaining favor in recent years because of cost advantages. Nitrogen solutions with no free NH_3 are commercially available for topdressing and can be used without causing serious foliar burn to the rice.

Various ammonium phosphate combinations are also effective for rice requiring both N and P. Such materials as mono- and diammonium phosphate, ammoniated superphosphate, and ammonium sulfate–phosphate combinations have been proven effective and have virtually the same characteristics of combining simple fertilizers. Ammonium bicarbonate is used extensively for rice in the People's Republic of China.

Controlled-release N materials, including sulfur-coated urea (SCU), urea formaldehyde, synthetic organic polymers, isobutylidene diurea (IBDU), have been evaluated in rice culture. Sulfur-coated urea has been the most widely tested of these compounds, but is not presently available in commercial form. Under conditions of good water management crop response to SCU has been generally equal to or superior to single preplant applications of urea and is often superior to split applications of urea when applied at the same rate of N. Under conditions of delayed or intermittent flooding, SCU has been shown to be consistently superior to split applications of urea.

c. Placement of Nitrogen—Losses of N can be minimized and availability to rice enhanced by placement of ammonium forms of N into the reduced soil zone. Methods of achieving this placement vary with the cultural methods utilized to produce the rice. For a water-seeded, permanently flooded cultural system, placement of the ammonium form of N 5 to 10 cm into the soil shortly prior to flooding has been shown to be superior to split-topdress applications. For transplanted rice, effective placement of N into the saturated puddled soil presents special problems. The hand or machine placement of large particles of urea (or ammonium bicarbonate [NH_4HCO_3] in the case of the People's Republic of China) has shown promise for increasing N efficiency but is still at the experimental stage. For a drill or broadcast dry-seeded culture with flooding delayed until the seedling rice is 10 to 15 cm in height, preplant placement of NH_4^+−N often results in excessive nitrification during the 3- to 5-week interval between seeding and flooding. With this type of rice culture, research in the southern USA has shown that split-topdress applications timed to meet the demands of the rice plant result in highest N use efficiency. Under this system, the first N application is applied onto a dry soil prior to flooding. Urea N topdressed onto dry soil is moved by the floodwater through several centimeters of soil if flooding occurs before the urea is hydrolyzed, whereas urea N topdressed into the floodwater tends to remain in the top centimeter of soil, where loss due to volatilization and denitrification are likely to be higher.

Nitrogen is frequently topdressed into the floodwater as late as the time of panicle initiation. By this growth stage, rice has developed an extensive root system at the soil-water interface. Once this root system has developed, topdressed N can be absorbed by the plant within 5 d. Also, by the last part of the vegetative growth period, a complete leaf canopy has formed, restricting air movement and thereby decreasing the likelihood of volatilization loss.

d. Nitrification Inhibitors—Considerable attention has been focused on the use of chemical nitrification inhibitors to retard the conversion of NH_4^+ to NO_3^-. The products include such compounds as nitapyrin [2-chloro-6-(trichloromethyl)pyridine], etradiazole [5-ethoxy-3(trichloromethyl)-1,2,4-thiadiazole], 2-amino-4-chloro-6-methyl pyrimidine, and dicyandiamide. Nitrapyrin and etradiazole have been extensively tested for rice under water-sown, drill-seeded, and postemergence flooding cultural conditions in the USA. Properly applied, these compounds have been shown to delay nitrification for periods of 3 to 6 weeks after application. Some of the materials are volatile so they must be immediately and thoroughly incorporated into the soil. Also, temperature and soil pH effect their performance; increases in temperature and pH stimulate nitrification. In general, these compounds have not given consistent economic yield increases when applied with fertilizer N just prior to flooding or as a topdressing during vegetative development.

e. Timing of Nitrogen Applications for Rice—Nitrogen losses from flooded soils can be large if placement and/or proper timing are not obtained. Under cultural conditions that preclude preplant placement, good timing requires that fertilizers be applied to coincide with plant demand. The effects of timing, N rate, and water management on rice yields in an Arkansas experiment are shown in Table 6–3. It is apparent that preplant N coupled with intermittent flooding resulted in lower rice yields.

Rice development is normally divided into three phases: vegetative,

Table 6–3. Grain yields of 'Starbonnet' rice in Arkansas as influenced by water management, N timing, and N rate (1976).

	Preplant N		Topdress N	
Nitrogen rate	Continuous flooding	Intermittent flooding	Continuous flooding	Intermittent flooding
	kg/ha			
0	5690 g*	3700 h		
67	7840 abcd	6250 fg	7600 bcd	6230 fg
101	7620 bcd	6870 ef	8090 ab	7620 bcd
135	8290 ab	6840 ef	8160 ab	8240 ab
168	8280 ab	7140 de	8570 a	7720 bcd
202	7980 abc	7300 cde	8010 abc	8460 a
Avg†	7620 a	6350 c	7690 a	6990 b

* Means within the table not followed by a common letter differ significantly at the 5% level.

† Averages within the line not followed by a common letter differ significantly at the 5% level.

reproductive, and grain filling. The plant has two peak demand periods for N, the first being during the tillering portion of the vegetative phase and the second during the early reproductive phase. During the tillering phase, the number of panicles is determined. The reproductive phase is important to grain yield since the number of grains per panicle and grain weight are determined during this period. The correct timing of N applications consists of matching application time to the correct development stage for a given variety. Additional considerations include the nutrient-supplying capacity of the soil, climatic factors, and various crop management practices.

In U.S. rice-growing areas, the following general fertilizer practices have been shown from research and grower experience to meet most requirements for appropriate timing of N applications. The appropriate choice will depend on the cultural system in use by the grower.

1. A single application of N applied preflood or at the time of seeding or transplanting. The N should be placed 5 to 10 cm deep or applied to a dry soil so that the fertilizer will move down in the soil after flooding or rainfall. This system is most applicable with a water-seeded, continuous flood culture, where flooding is delayed until seedling establishment, or when a controlled-release N source or a nitrification inhibitor is used to reduce the rate of nitrification. It is important to prevent nitrification of the applied N.
2. A single topdressing applied after seedling emergence or at the end of active tillering. This system is normally followed when only small amounts of N are added.
3. Split applications of N, one basal or pretillering application consisting of about two-thirds of the N and one topdress application at or near panicle initiation supplying the remainder.
4. Multiple applications of N with approximately one-half to two-thirds applied at basal or pretillering and the remainder as topdressing at panicle initiation and panicle differentiation development stages.

Topdress N applications made prior to tillering should be placed on a dry soil surface and followed by flooding.

f. Fate of Fertilizer Nitrogen Applied to Rice and Rotation Crops—Nitrogen recovery by the rice plant is consistently low, rarely exceeding more than 60 to 70% of applied N (Fig. 6–7). Numerous loss pathways have been discussed in this chapter, including denitrification, NH_3 volatilization, and leaching. In addition, a portion of the applied N is incorporated into the organic fraction of the soil. Studies utilizing ^{15}N have shown that this immobilized N is only sparingly available to subsequent crops.

2. Phosphorus

The response of rice to P fertilization is not as general, nor the effect on yield as large, as is generally obtained with N. However, if P is limiting, plant growth is inhibited, tillering is reduced, root development is restricted, maturity is delayed, and utilization of other nutrients is impaired.

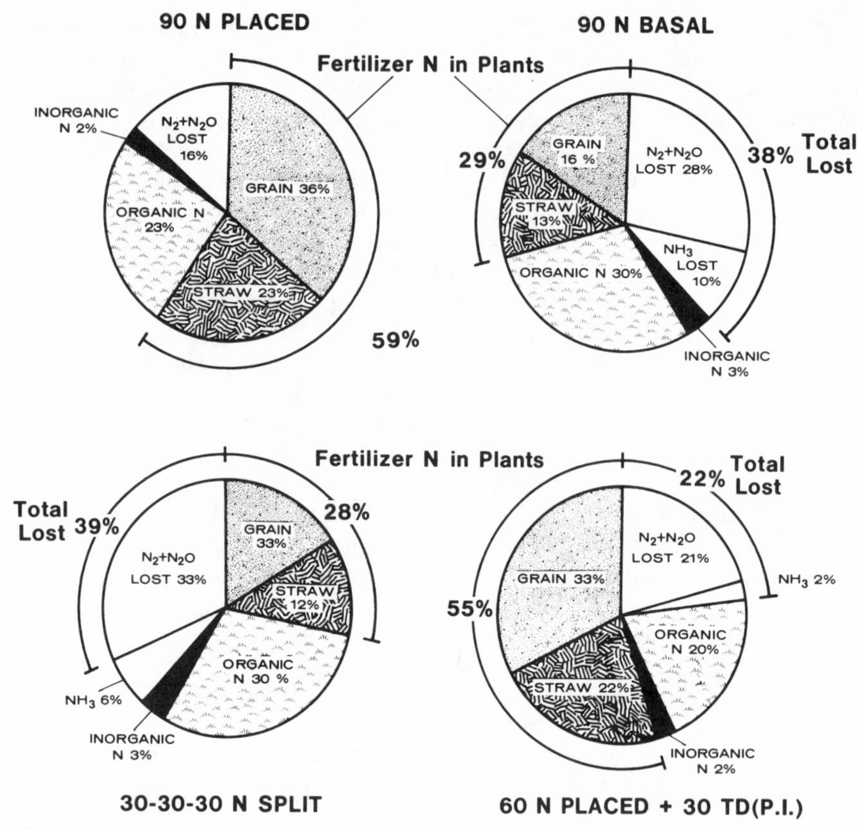

Fig. 6–7. Typical N recovery following application by different methods under California conditions. The most efficient recovery is from preflood placement in the anaerobic soil zone.

Rice often does not respond to P fertilization on the same soils where upland crops do poorly or fail without addition of phosphate. This is not necessarily related to a difference in P requirement, but more directly to the fact that flooding enhances P availability. Phosphorus deficiency, however, does occur on extensive areas of Ultisols, Oxisols, acid sulfate soils, Vertisol, and certain Inceptisols. Many of these soils are low in available P, fix fertilizer P in a highly insoluble form, and show only a slight increase in availability of P with flooding.

The source of P and time and method of application influence plant growth response and yield increase. Environmental factors such as soil pH, organic matter status, and soil temperature may greatly influence results and should be considered in determining the need for P fertilizers.

a. Source of Phosphorus—Numerous yield trials provide evidence that except on extremely acid or alkaline soils, the common P sources are about equally effective. The water-soluble P sources, such as superphosphate, are effective on all soil types except those that are extremely acid. In extremely acid soils, soluble P reacts readily with Fe and Al, which tends to reduce its availability. On these soils, rock phosphates of high citrate solubility are usually as effective as a more soluble source.

b. Placement of Phosphorus—On most soils where direct applications of P are made for the rice crop, there is no advantage to split applications. Broadcast-incorporated P or banded P have been shown to be superior to surface application in increasing rice yields. Surface application has the further disadvantage of stimulating grassy weeds and algae scums.

c. Time of Phosphorus Application—The rice plant requires a continuous supply of P throughout its life cycle, although maximum absorption occurs during the flowering stage. However, for P to be utilized effectively, it should be applied preplant and incorporated into the soil. Phosphorus application delayed until the early tillering stage will still provide a significant benefit to plant growth and yield, but its value decreases with time when applied after the seedling stage. In soils where a positive response to P fertilization is obtained, the major source of plant P comes from fertilizer during the first 2 months of growth. Soil P makes its major contributions during later periods of uptake after chemical soil reduction has enhanced the availability of native soil P.

Where multiple deficiencies of plant nutrients occur, it is usually advantageous to apply nutrients simultaneously, unless the fertilizers are incompatible. As a general rule, rice soils requiring P also require N. The requirement for placement of N so that it ultimately occurs in the reduced soil zone also improves the utilization of P and contributes to added efficiency in fertilizer applications. Also, in areas where rice is rotated with upland crops, the P and K requirements for rice may be met by applying phosphate–potassium fertilizers to the upland crop.

d. Fate of Fertilizer Phosphorus Applied to Rice and Rotation Crops—Fertilizer P reacts rapidly with the soil, resulting in greatly reduced solubility and mobility of the P. Plant recovery of added P is low, rarely exceeding 10% by the rice crop during the year applied; thus, most of the added P remains in the soil with varying degrees of availability to subsequent crops.

Flooding of a soil influences its sorption properties for P. Previously flooded soils tend to show greater P sorption and require more fertilizer phosphate for subsequent arable crops as compared with soils that have not been subjected to flooding.

3. Potassium

Potassium deficiencies in flooded rice occur less frequently than N and P deficiencies. The young and alluvial nature of many rice soils and their high clay and silt content, as well as enhanced K availability after flooding, accounts in part for the generally favorable K status. In the highly weathered soils of the tropics and semitropics, where the K minerals have decomposed and high rates of leaching occur, response of rice to K additions may be significant.

Potassium deficiency is frequently associated with the occurrence of physiological diseases in rice and additions of K may aid in ameliorating these problems.

a. Potassium Sources—The two major K sources, potassium chloride (KCl) and potassium sulfate (K_2SO_4), are about equally effective in terms of plant utilization. Potassium chloride is normally the preferred source as its cost per unit of K is lower. Where K is required on soils subject to reduction of sulfate to sulfide, potassium sulfate use should be avoided. Where salinity is a problem and K is needed, the sulfate form tends to show less salinity damage per unit of K as compared with the chloride form.

b. Time and Method of Application—Rice has a rather continuous requirement for K, extending throughout its entire growth period. It is essential for the transport of carbohydrates between plant parts, regulation of certain organic acids, the adjustment of stomatal movement, and water relations.

Potassium fertilizers are usually added as a single application or as part of a mixed fertilizer during seedbed preparation. Rice is seldom grown on highly permeable soils, but where excessive leaching occurs, rice will respond to split applications of K. Where rice is rotated with other crops it may be more beneficial to apply the K to the rotation crop in sufficient quantity to supply the needs for the entire rotation. One example would be the rice-soybean [*Glycine max* (L.) Merr.] rotation practiced in the southern USA. Potassium fertilizer is applied to the soybean plants since they have the higher K requirement and are more likely to show a yield response to K addition.

c. Fate of Fertilizer Potassium Added to Rice—Potassium losses from the soil are through the mechanism of plant uptake, leaching, and runoff. Soil reduction may increase the K concentration of the soil solution by increasing competition for certain exchange sites by Fe and Mn. This may result in increased losses by either leaching or runoff.

The magnitude of K removal by plant uptake is related to that portion of the crop removed from the field. If the grain is removed, only moderate amounts of K are depleted. If the grain and straw are both removed, the result is loss of large amounts of K from the soil.

4. Sulfur

Sulfur deficiency is not widespread in rice and S fertilization is not as common as N, P, K, and Zn fertilization. There are areas where native soil sources are inadequate for production of high rice yields and S additions are required. Well-documented cases of S deficiency occur throughout rice-growing areas; the greatest deficiency occurs on rain-fed upland soils. Irrigation water that contains S can be an important source. Sulfur deficiency also occurs in some alluvial areas, with the floodplain of the Amazon River being the most striking example.

The trend toward use of N and P sources that do not contain S and the withdrawal of larger amounts of soil S as a result of higher yields is contributing to increased S deficiency. On many S-deficiency soils, S is added as a constituent of fertilizer compounds that are added to correct other deficiencies. Thus, sufficient S is often supplied as ammonium sulfate or as single superphosphate to supply the crop's needs. Another source of S, especially in industrial countries, is atmospheric S produced from combustion of high-S coal. Since most rice is grown in nonindustrial countries, this source of S is not as important for rice worldwide as for the upland crops grown in industrial countries. Biogenic atmospheric S from coastal and marine environments is also a source for rice, although the significance of this source is not known.

Rice removes about 2 kg of S/metric ton of grain; however, field studies indicate that 15 to 20 kg S/ha is required annually to prevent S deficiency since some of the S may be lost through various processes. These include immobilization by microorganism, reduction to sulfides to form less-available compounds, or by direct volatilization. Among the common fertilizers containing S, ammonium sulfate and single superphosphate are considered excellent sources for rice. Gypsum, elemental S, and various other S-containing materials may also be used. Broadcast applications are usually effective where soluble sulfate forms are used and should ideally be applied preplant, although topdressings are effective. Experience reveals that rates of 15 to 20 kg S/ha results in residual effects and often supplies the need of at least two consecutive rice crops, even though losses occur through leaching, volatilization, and immobilization.

5. Calcium, Magnesium, and Silicon Requirements

Calcium, Mg, and Si deficiencies seldom occur in flooded rice because wetland soils and irrigation water are usually well-supplied with these elements. Calcium and Mg are displaced from cation exchange sites during flooding, which increases their concentration in the soil solution. Exchangeable Mg is replaced more readily than Ca and it is not unusual that they occur in the soil solution in about equal proportions. Nutritional problems with Ca, Mg, and Si occur more frequently in upland rice and under special conditions such as in acid sulfate soils.

The essentiality of Si for rice is not well established, even though it appears to play a role in maintaining erect leaves, resistance to disease and insects, and in osmotic regulation. Dissolved Si occurs as a condensation polymer of orthosilicic acid, which generally increases in solubility after flooding. Increases in soluble Si are associated with decreased pH, and the reduction of amorphous Al and ferric hydroxide forms in flooded soils. The average Si content of rice straw is about 11% and lesser values are thought to justify applications of silicate slags ($CaSiO_3$). Large areas of rice land in Japan and Korea are fertilized annually with slag.

6. Micronutrient Elements

Micronutrient disorders in rice were relatively unknown prior to 1936; in recent years the identification of Zn, Fe, Mn, and Cu deficiencies have been more common and have stimulated wide interest. Greater incidences of micronutrient deficiencies, especially in developing countries, may be attributed to higher yields from improved rice varieties and increased use of major nutrients. Rice has a relatively low requirement for micronutrients and the irrigation water used for flooding often supplies a part of the crop needs. A large number of factors may contribute to micronutrient deficiencies in rice and are referred to in Table 6–1.

a. Zinc—Zinc is the most widespread micronutrient deficiency recognized in worldwide rice production. Deficiencies have been reported from most of the major rice-growing areas of the world. Deficiencies are most frequently observed in seedling rice grown on calcareous and high-pH soils since availability is pH dependent. Deficiencies often occur where topsoil has been removed and also in some soils with high organic matter. Evidence suggests that carbonate–bicarbonate (CO_3^{2-}–HCO_3^-) relationships, organic fermentation products, sulfide formation, interactions, and antagonisms such as Zn–N, Zn–Fe, Zn–P, and Zn–Cu are factors in Zn deficiency.

Zinc deficiency has been corrected with soil applications of various Zn salts. Soluble salts such as Zn sulfate, Zn nitrate, and Zn chloride have been equally effective when applied at rates of 2 to 4 kg Zn/ha. Even sparingly soluble Zn oxide has given good correction when applied as finely ground material or when incorporated in an acidic macronutrient carrier. Various synthetic and natural chelates applied to the soil may satisfy Zn requirements, but the greater agronomic efficiency usually obtained with these sources for upland crops has not been consistent for flooded rice. In water-seeded rice, surface-placement of Zn is superior, so less-mobile Zn sources are desired. Drilled rice usually responds better to ethylenediaminetetraacetate (EDTA)-chelated Zn than inorganic zinc sources. The material should be incorporated at least 2 to 5 cm deep prior to planting. In transplanted rice, deficiencies are often controlled in the nurseries, by soil application in the field, and by dipping rice seedlings in a 1% Zn oxide suspension prior to transplanting. Coating seeds with Zn oxide and foliar sprays have also been effective. Zinc deficiency occurs more frequently in soils subject to prolonged waterlogging. In areas of southeastern Asia, the incidence is increasing as cropping intensity increases in irrigated areas. Drainage and soil aeration are reported to alleviate the chlorosis and allow resumption of growth.

b. Iron—Iron deficiencies are not common in flooded rice, because flooding enhances the reduction and solubility of soil Fe. Availability is very pH dependent, however; calcareous and sodic soils have the most frequent problems. With the same soils, upland rice is much more susceptible to Fe deficiency. Iron-deficient plants grown on alkaline soils accumulate Fe in the roots, but it is not translocated to the shoots. Immobilization of Fe in the roots has been associated with carbonate–bicarbonate buffering effects on conductive tissue and Fe precipitation in the root.

Correction of Fe deficiency is often difficult because of rapid reversion of soluble Fe materials such as ferric and ferrous sulfate, and natural and synthetic chelates to less soluble forms; high rates of Fe oxide are only partially effective. Acidification of the soil and/or the irrigation water with sulfuric acid or acid-forming salts has alleviated the problem in several rice areas. Foliar Fe applications have provided partial correction, but the small leaves of seedling plants do not provide for effective absorption. The incorporation of organic residues or wastes and its subsequent anaerobic decomposition helps alleviate Fe deficiency.

Iron toxicity occurs when excessive soluble Fe accumulates in highly reduced soils. Concentrations of soluble Fe as high as 300 mg/kg decreases the absorption of other nutrients and causes direct Fe toxicity. Poorly drained acidic latosols, acid sulfate, and peaty soils are most susceptible to toxic Fe problems. Liming and drainage are the most effective ways to deal with Fe toxicity.

c. Manganese—Manganese is seldom deficient in medium or fine-textured alluvial soils. Reduction of Mn following flooding greatly increases its solubility. Only highly weathered soils with low levels of total Mn or calcareous soils planted to upland rice are likely to be deficient. Soluble Mn salts, such as manganese sulfate ($MnSO_4$), are usually applied to correct this deficiency. The use of Mn–EDTA has not usually been effective. Manganese toxicity may be a problem after flooding manganiferrous and acid soils, but it seldom affects rice directly. Manganese toxicity may sometimes affect upland crops grown in rotation with rice.

d. Copper—The Cu requirement of rice is low, but deficiencies have been reported on sandy soils and highly organic soils such as peats and mucks. The range between Cu adequacy and toxicity is very narrow and care must be exercised in fertilization. Yield increases and correction of a physiological disorder called *straight-head* are attributed to the alleviation of Cu deficiency in Japan, Portugal, India, and Indonesia. This is apparently not the same disorder called straight-head in the USA since it does not respond to Cu additions.

e. Molybdenum, Boron, and Chlorine—In contrast to some other micronutrients, Mo deficiency occurs mostly on acid, sandy soils, and acid organic soils. Availability usually increases in flooded soils since flooding increases soil pH. Rice has a low requirement for Mo with toxicity occurring at tissue concentrations of < 2 mg/kg. Sodium and ammonium molybdate are the most commonly used sources.

Although B and Cl are essential elements, their concentration in rice soils and irrigation water have usually been sufficient for rice. Boron and Cl toxicity are more common and are most often associated with saline conditions or areas where alkaline well water is used for irrigation, but are not necessarily limited to these soils.

7. Soil Acidity and Liming

Rice grows well over a pH range extending from about pH 5.0 to 7.5. Most flooded soils attain an equilibrium pH between 6.5 and 7.5 when flooded. In extremely acid soils Al toxicity is a potential problem, especially on upland soils or during the early stages of flooding. Acid sulfate soils require special management. Acid sulfate soils not only have active acidity that must be dealt with, but have reserves of pyrite that can generate large amounts of acid under drained conditions. In alkaline soils, micronutrient availability, especially Zn and Fe, are frequent problems.

Lime may be applied to excessively acid soils to improve fertilizer use efficiency, supply Ca and/or Mg, enhance the mineralization of organic matter and release of NH_4^+–N, to favor N_2 fixation by blue-green algae, reduce metal toxicity, and to favor a rotation crop. Excessive liming should be avoided, since it reduces the availability of most micronutrients, except Mo. Liming alone is usually not a feasible treatment for acid sulfate soils because the very high potential acidity can neutralize very large amounts of lime.

GENERAL REFERENCES

De Datta, S. K. 1981. Chemical changes in submerged rice soils. p. 89–145. *In* Principles and practices of rice production. John Wiley & Sons, Inc., New York.

Institute of Soil Science, Academia Sinica. (ed.). 1981. Proceedings of symposium on paddy soils. Springer-Verlag, Berlin.

International Rice Research Institute. 1978. Soils and rice. International Rice Research Institute, Los Baños, Philippines.

Patrick, W. H., Jr., and I. C. Mahapatra. 1968. Transformations and availability to rice of nitrogen and phosphorus in waterlogged soils. Adv. Agron. 20:323–359.

Ponnamperuma, F. N. 1972. The chemistry of submerged soils. Adv. Agron. 24:29–96.

7

Production, Marketing, and Use of Nitrogen Fertilizers

Fred C. Boswell
University of Georgia
Experiment, Georgia

J. J. Meisinger
Agricultural Research Service, USDA
Beltsville, Maryland

Ned L. Case
Phillips Petroleum Company
Pavillion, Wyoming

I. NITROGEN IN NATURE

A. The Nitrogen Resource

Large quantities of free N are found in the atmosphere, which is about 80% N by volume. Although this abundance of N exists, it cannot be used by higher plants until it is combined with hydrogen (H_2) or oxygen (O_2). If inexpensive methods were available to produce these combined N compounds, the problem of supplying adequate N for crop production would be simplified. However, much energy and expense are required to fix atmospheric N.

Substantial quantities of N also occur in chemically combined forms in soils, geologic formations, and oceans. Nitrogenous compounds may be either organic or inorganic. Some nitrogenous compounds, like the elemental dinitrogen (N_2) are gaseous but the most important compounds in agriculture are nitrate (NO_3^-) and ammonium (NH_4^+).

Of the six known isotopes of N, only those having mass numbers of 14 and 15 are stable and occur naturally. The isotope of mass 13 is the longest lived of the four radionuclides, with a half-life of 10.05 min. This relatively brief half-life is too short for practical use in most research on biological systems. The stable N isotopes (^{14}N and ^{15}N) have been used almost exclusively as tracers for biological and related types of research, because the two occur naturally in an almost constant ratio (^{14}N to ^{15}N is about 272:1).

The ^{15}N tracer technique is more accurate than the older methods of N recovery, which consider the difference between N removed from control soils by the crop and that removed from fertilized soils. The tracer method is based on actual recovery of added ^{15}N in the crop plus that remaining in the soil. These techniques are expanding our understanding of the fate of N, especially as related to its use for crop production. However, there remains many unanswered questions relative to optima N production, marketing, and use.

B. Role of Nitrogen in Plants

Established as an essential mineral element for rooted plants in the 1800s, N has been recognized to be responsible for lush vegetative growth and a dark green leaf color. Its deficiency is usually recognized first by the pale green or yellowish-green color, especially in the grasses, and the premature necrosis of the older leaves beginning from the tip and progressing along the midrib toward the collar and the edges. This association with green coloration is likely due to the fact that N, along with Mg, is one of two soil-derived constituents of chlorophyll ($C_{33}H_{72}O_5N_4Mg$). Adequate N for plants promotes aerial vegetative growth, increases the top/root ratio, and is essential for fruit and seed formation. Being an essential constituent of amino acids, N is required in protein synthesis, making up 12 to 19% of various proteins and averaging about 16% by weight. Because grain formation depends on certain threshold levels of protein, grain production is significantly related to N supply, especially in cereal crops. Abundance of N in the growth medium also is reflected in the crude protein levels of grain and in forages. Among the essential mineral elements for the growth and reproduction of higher green plants, there are more atoms (approximately threefold) of N in dry organic matter than of any other soil-derived element (not from water or the atmosphere). In terms of mass, N in plant material is often found in larger quantities than any other element. Although the concentration of K may be higher in some plant materials, N exceeds the combined total of all other soil-derived essential mineral elements in the seeds of commonly grown agricultural crops.

Considering the abundance of N in plants, its central role in plant functions and its reactivity in the biosphere, it is not surprising that this element is the most universally deficient for optimum crop production.

C. Dinitrogen Fixation in Nature

Rooted green plants may absorb N in several combined forms and through different organs, but the preponderance is usually absorbed through the roots in ionic form either as NH_4^+ or NO_3^-. Some plants can absorb and utilize only NH_4^+, but most do absorb both ions; the ion absorbed depends more on the environment than on plant capability. Hence, rice (*Oryza sativa* L.) grown under flooded conditions is most likely to absorb its N as NH_4^+, and maize (*Zea mays* L.) grown in well-aerated soils will absorb mostly NO_3^-. Also, NO_3^- is likely to predominate as the absorbed form in late spring and summer when aerated soils are warm, resulting in N transformation to the NO_3^- form.

Because rooted plants cannot utilize gaseous N, they must have a supply of combined N for growth. There are a number of processes in nature that convert N_2 to forms useful for plants; some involve living organisms, while others are independent.

1. Lightning

Electrical discharges related to meteorological events fix a certain amount of N in the atmosphere. The products are most likely to be nitrogenous oxides. This direct oxidation of N by electrical discharge, demonstrated by Cavendish, was the basis for early attempts at commercial processes for producing combined N. The reactions may be shown as follows:

$$N_2 + O_2 \longrightarrow 2NO$$

$$2NO + O_2 \longrightarrow 2NO_2$$

$$3NO_2 + H_2O \longrightarrow 2HNO_3 + NO.$$

Precipitation washes these compounds from the air, along with various other organic and inorganic forms of N originating as atmospheric intrusions from activities near the earth's surface. Thus, it is not possible to determine the importance of direct oxidation by lightning. The amount of N deposited annually in precipitation varies among location, with 10 kg/ha being about the average value. This is not nearly enough N for high yields of most crops.

2. Fixation by Organisms

Organisms that can utilize free N directly (i.e., fix it) may be divided into two classes—symbiotic organisms and free-living organisms. Symbiotic organisms can function only in symbiotic relationships with a rooted green plant; free-living organisms function independently, usually depending only on a fixed energy source. The blue-green algae, however, are even independent of fixed energy because of photosynthesis in these lower green plants. There are at least 12 kinds of blue-green algae that have this capability. This is emphasized because algae are often implicated in eutrophication of lakes and streams due to nutrient enrichment of the water. It is evident that the blue-green algae can multiply in waters without N enrichment.

Besides the blue-green algae, *Rhodopseudomonas, Thodomiusbium, Chlorobium,* and *Chromatium* also are autotropic and fix N_2. These organisms need certain essential elements, light, CO_2, and N_2, for growth and N_2 fixation.

The other free-living organisms that can fix N_2 are saprophytes, i.e., they require an external energy source to grow and to fix N_2. Like the others, these require essential elements for growth but tend not to fix atmospheric N_2 when N sources, especially NH_4^+, are readily available in the growth medium. Some of these such as *Clostridium* sp. and *Aerobacter* sp. are anaerobic, while others like *Azotobacter* sp., *Achromobacter* sp., and *Beijerinckia* sp. are aerobic.

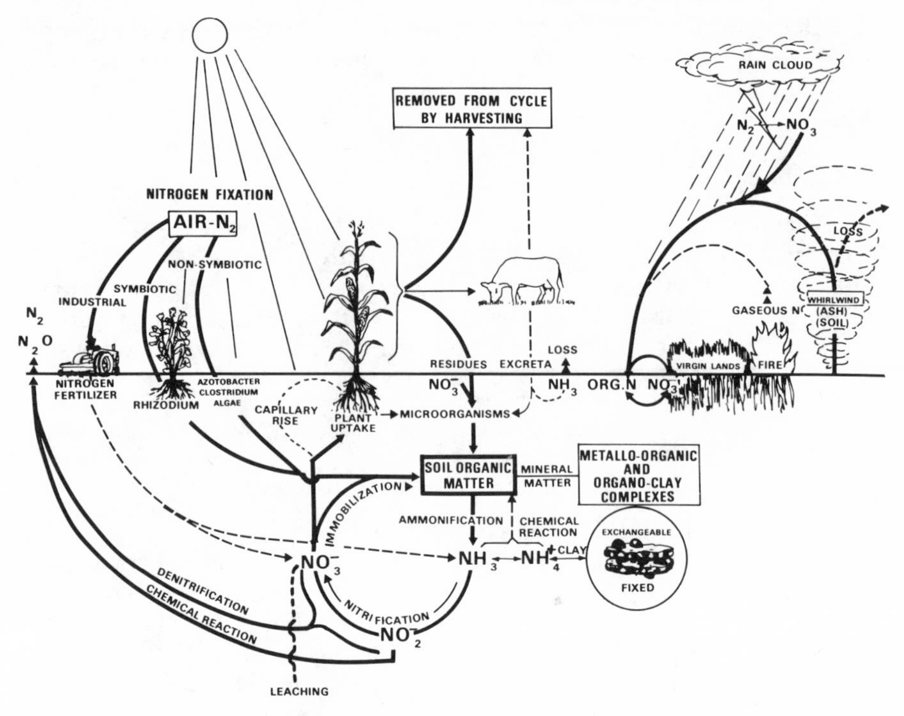

Fig. 7–1. The nitrogen cycle (courtesy of Stevenson, 1964).

D. The Nitrogen Cycle

A graphic portrayal of the relationship of the various forms of N in nature is given in Fig. 7–1. It shows the two larger reservoirs of N mostly as N_2 in the atmosphere and as combined N in the soil and other parts of the earth's crust.

The use of the latter of the two reservoirs is conditioned by its location, time, and usually by the presence of higher green plants; a very small portion, however, is found in deposits that are exploited for fertilizer and other uses. Atmospheric N, on the other hand, is quite uniformly distributed over the earth and it can be fixed favorably, either by the physiochemical processes described later in this chapter or by growing leguminous species of higher green plants. Dinitrogen fixation by organisms nonsymbiotically and naturally by lightning is not yet controlled adequately for general use; however, this N is significant for producing food, feed, and fiber under native conditions.

Regardless of various ways that N may be added to the soil, it often undergoes many transformations before it is removed. Organically combined N is subjected to complex changes such as protein formation and decomposition. Nitrogen also appears in forms that can be appropriated by microorganisms and higher plants, immobilized and slowly released by mineralization, lost by volatilization, leached in the NO_3^- form, and/or removed by erosion. These processes serve to utilize, recycle, or lose N from the soil-crop system.

II. COMMERCIAL DINITROGEN FIXATION PROCESSES

Nitrogen fertilizer production capacity and use have increased tremendously throughout the world since about 1975. A number of obsolete ammonia (NH_3) plants have been replaced with new modern plants and other old, high–production cost plants have been shut down.

The basic processes have changed little from the original means of synthetically producing anhydrous NH_3, the starting product for nearly all N fertilizers. The reactions are the same and only variations in pressure, temperature, catalyst, mechanical equipment, and gas purification distinguish the different commercial processes available or in operation (Axelrod & O'Hare, 1964; Sauchelli, 1960, 1964). Significant economic improvements have been realized in modern plants by the use of heat interchange throughout the various process steps and by the use of centrifugal compressors. Advances in technology, metallurgy, mechanical equipment design, and instrumentation have made possible the modern single-train manufacturing facility. The application of centrifugal compressors has been one of the chief factors in the development of today's single-train plant; it is not uncommon to find one centrifugal compressor doing the work formerly done by 30 to 40 reciprocating compressors. Advances in metallurgy have brought about much higher pressure steam-hydrocarbon reformers, which have greatly reduced the investment in new single-train plants.

A. Ammonia

Anhydrous NH_3 is the basic building block for almost all synthetic N fertilizers. Formula, molecular weight, and physical properties are as follows:

Formula	NH_3
Percent N (by weight)	82
Molecular weight	17.03
Critical temperature	132.4°C
Critical pressure	11.26 MPa
Heat of vaporization (−33.33°C)	1061 cal/g
Liquid density (O°C)	637.8 g/L
Vapor density (0.101 MPa, O°C)	0.7708 g/L
Boiling temperature (0.101 MPa)	−33.3°C
Freezing temperature (0.101 MPa)	−77.7°C

Gaseous NH_3 is lighter than air, but on compression and cooling it becomes a liquid about 60% as heavy as water. It is readily absorbed in water up to concentrations of 30 to 40% by weight, with the resulting liquid of low vapor pressure. Due to the high vapor pressure of anhydrous NH_3 at normal temperatures, it is transported in pressure containers. Gaseous NH_3 is extremely irritating to the eyes and respiratory system in concentrations up to 0.07% by volume. A concentration of 0.17% causes convulsive coughing, and concentrations of 0.5 to 1.0% are rapidly fatal after short exposure. While NH_3 does not sustain combustion, it will burn in the presence of air or oxygen if a source of ignition is present, but combustion will stop when the source of ignition is removed. The ignition temperature is around 650°C. Flammable and explosive mixtures are formed with air in concentrations of 16 to 25% NH_3 by volume.

1. Synthetic Ammonia Processes

All processes used are basically modifications of the original Haber-Bosch process in that the NH_3 is synthesized from a 3:1 volume mixture of H and N at elevated temperature and pressure in the presence of an iron catalyst. The H may be obtained from water, natural gas, oil, or coal, and all the N used is obtained from air. Other than the source of H, the major differences in the various processes used are the manner in which N is obtained from the air, the method employed to purify intermediate gas streams, and the synthesis converter operating pressures. With one or two exceptions, the arc process, the cyanide process, the nitridic process, and the cyanamide process have all faded into obscurity. The manufacture of NH_3 can be broken down into three steps: synthesis gas production, purification, and NH_3 synthesis. Synthesis gas preparation involves the production of a raw hydrogen gas and the addition of the stoichiometric amount of N. Purification comprises the removal of CO_2 and CO, which are synthesis catalyst poisons. Synthesis constitutes the catalytic reaction of N and H at elevated temperature and pressure. The manufacture of NH_3 involves the handling of three simple materials: natural gas, steam, and air. In those parts of the world where natural gas is not yet available, it is replaced by naphtha or other hydrocarbons and the process is only slightly more involved. Refinery gas and cokeoven gas may also be used as the raw H source for NH_3 synthesis. A simplified flow diagram for a modern single-train ammonia plant is shown in Fig. 7–2.

2. Synthesis Gas Production

a. Hydrogen Manufacture—The steam-hydrocarbon reforming process is now the most widely used method of generating H, and natural gas is the most popular and the preferred hydrocarbon feedstock. All traces of S must be removed from any hydrocarbon feedstock before the reforming step can take place, because S is a reforming catalyst poison.

After desulfurization, the feed gas is mixed with superheated steam, preheated to 315 to 425°C, and passed through tubes filled with nickel catalyst in a fired furnace. The reactions that take place between the steam and hydrocarbon are shown below for methane (CH_4), the principal component in a natural gas feedstock (the first reaction shown is by far the most predominant and is highly endothermic):

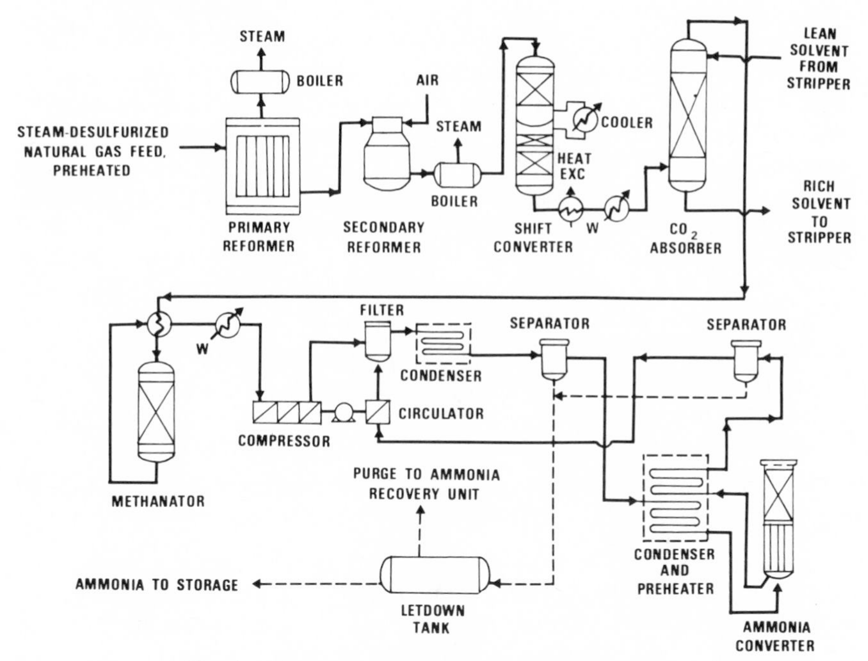

Fig. 7–2. Flow diagram for ammonia synthesis plant.

$$CH_4 + H_2O \longrightarrow CO + 3H_2$$

$$CO + H_2O \longrightarrow CO_2 + H_2.$$

The effluent gas stream from the primary reformer will usually contain enough unconverted hydrocarbon to justify further reforming. This is accomplished by passing the gas over another bed of nickel catalyst. Compressed air is injected into the gas stream entering the secondary reformer. The combustion of the oxygen in this air with a part of the H in the process stream provides the heat required for the reforming reaction, and the noncombustible N component of this air gives the N required to be catalytically fixed with the remaining H to form NH_3.

Partial oxidation is probably the second most popular process for the manufacture of a raw gas for NH_3 synthesis. In this process hydrocarbons are oxidized under pressure to produce a gas that contains H and oxides of C. The reaction takes place in a refractory-lined combustion chamber at temperatures up to 1480°C. Oxygen or oxygen-enriched air is required for this process. The reaction products are cooled and scrubbed to remove C and are then ready for the next step of synthesis gas production.

b. Shift Conversion—The water–gas shift reaction presents a convenient and economic way in which to react or remove CO, which is a synthesis catalyst poison. The raw H stream is cooled to about 370°C, the steam content is adjusted if necessary, and the gas is then passed over an iron-chrome catalyst where the following reaction takes place:

$$CO + H_2O \longrightarrow CO_2 + H_2.$$

The economic advantage to this step is the fact that an additional mole of H_2 is produced for each mole of CO shifted. The removal of the additional mole of CO_2 that is formed is a relatively simple step. In modern plants CO conversion usually takes place in two steps. The first step, which is described above, is exothermic. The exit gas is cooled or quenched to about 200°C and is then passed over a copper–zinc catalyst. The first catalyst bed reduces the CO content of the gas to 2 to 3% by volume and the second catalyst bed further reduces the CO content to the range 0.2 to 0.5% by volume. These two reactions are commonly referred to as high and low temperature shift conversions. The low temperature shift catalysts have been developed since about 1975.

3. Purification

Oxygen will poison the synthesis catalyst and cause it to decrease in activity. For this reason, CO and CO_2 must be removed.

a. Carbon Dioxide Removal—Scrubbing with monethanolamine (MEA) and with hot potassium carbonate (K_2CO_3) are two processes used for CO_2 removal. The Giammarco-Vitro coke and the catacarb processes have improved the hot carbonate removal system by including a catalyst in the circulating scrubbing solution. The use of sulfinol solution, a mixture of diisopropylamine and sulfolane in water, can also be used to scrub the raw H for CO_2 removal.

b. Carbon Monoxide Removal—Carbon monoxide can be removed by scrubbing with a solution of cuprous ammonium acetate, or cuprous ammonium formate. Both solutions are regenerated with heat. These two systems have the disadvantages of high chemical cost, a rather complicated chemistry, requirement of very close chemical control of the scrubbing solution, and high maintenance costs.

The use of liquid N scrubbing and catalytic methanation have all but replaced copper liquor scrubbing. Liquid N scrubbing removes both CO and CH_4, and this followed by methanation gives the purest H_2 for NH_3 synthesis. Catalytic methanation employs a nickel catalyst at about 315° C to give the following reactions:

$$CO + 3H_2 \longrightarrow CH_4 + H_2O$$

$$CO_2 + 4H_2 \longrightarrow CH_4 + 2H_2O.$$

4. Gas Compression—Synthesis

Gas reforming and purification processes operate at 0.1 to 3.4 MPa [20 to 500 pounds/inch^2 gauge (psig)]. Compression of the process gases up to eventual synthesis pressure of 13.6 to 38.4 MPa (2000 to 5000 psig) may take place at any point in the process; however, most purification processes are operated at $\leqslant$ 13.6 MPa (2000 psig) and there is a final compression step between purification and synthesis. Most modern plants employ centrifugal compressors and the synthesis loops operate at 13.6 to 24.2 MPa (2000 to 3500 psig). Some new low-tonnage plants still employ reciprocating compressors and operate at pressures of 101 MPa (15 000 psig).

The catalytic reaction of H and N to form NH_3 is exothermic and it is favored by high pressure because there is a decrease in volume as the reaction takes place. The conversion of the reactants to NH_3 will vary from 10 to 30% for each pass of the synthesis gas mixture over the catalyst bed. In the generally used process, the purified synthesis makeup gas is compressed and added to the recycle gas within the synthesis loop. The mixed gases then pass through the water and refrigerant-cooled exchangers in the loop where the NH_3 formed in the last pass through the catalyst bed is condensed. This condensed liquid NH_3 scrubs the fresh makeup gas and removes the last traces of water and CO_2. The recycle gas, essentially free of all oxygen-containing compounds, is then reheated by exchange with the hot gas from the converter and flows to the catalyst bed in the converter at about 275 to 300°C. The reduced iron catalyst is normally contained in a vertical vessel in one or more distinct catalyst beds. The reaction is exothermic and temperatures at the exit of the catalyst beds range from 500 to 540°C. Circulation of synthesis gas through the catalyst bed normally runs from 5 to 10 times the volume of fresh makeup. Internal heat exchangers in the converters are used to control catalyst temperatures by transferring the heat of reaction to the incoming recycle gas. Some converters also add fresh makeup or quench at various points within the catalyst bed to control the bed temperatures.

Leaving the converter, the hot gas is cooled by water or by exchange with gas going back to the converter. It is then mixed with fresh makeup gases, compressed, and recycled back to the refrigeration exchangers, product separator, and converter.

Ammonia synthesis catalyst when charged consists of magnetite (Fe_3O_4) usually promoted with alumina, KCl, and Ca. The catalyst comes in the form of crushed granules nominally sized from 3 to 12 mm. Converters had a capacity of 50 to 100 metric tons/d of production about 30 yr ago, whereas today's modern plants can have converters with capacities up to 750 metric tons/d and higher.

Gases other than H_2 and N_2 are nonreacting and will concentrate in the recycle gas. These gases are unreacted methane, methane formed in the methanator, rare atmospheric gases admitted with the air injected into the secondary reformer, and H and other inert gases present in the feed gas. These must be continuously purged from the synthesis loop.

5. Ammonia by Gasification of Coal

The rapidly escalating cost of natural gas and petroleum derivatives as feedstocks for H used in production of NH_3 has focused interest on coal, a more abundant resource in the USA and other countries. The basic coal gasification technology was developed and used in Germany during World War II. South Africa has about 30 yr of further experience, and India has two large ammonia plants based on coal. Tennessee Valley Authority was operating a demonstration plant gasifier and purification system retrofitted to an existing ammonia plant in 1983.

Investment for a coal-based ammonia plant was estimated at more than double that for a plant based on natural gas in 1982. Natural gas prices by 1990–1995 may become high enough to make the more complex and difficult coal gasification technology cost competitive.

B. Ammonia Derivatives

Anhydrous NH_3, a liquefied gas, is the most concentrated N fertilizer (82% N) available and is the raw material for all other synthetic N fertilizers. Special storage, transportation, and application equipment are required. The advantages of storing, transporting, and applying solid fertilizers and nonpressure solutions containing N have brought about the development of numerous solid- or dry-type fertilizers. With NH_3 and by-products produced by an ammonia plant, it is possible to manufacture ammonium nitrate (NH_4NO_3) and urea without additional raw materials. Other N-base fertilizers may be manufactured with NH_3 and various other raw materials. (See Table 7–1 for physical properties of these).

1. Nitric Acid

Nitric acid (HNO_3), as such, finds limited use in the mixed fertilizer industry or for agricultural purposes. But nitric acid, which is made from NH_3 and air, is reacted with additional NH_3 to manufacture ammonium nitrate.

Nitric acid is produced by oxidizing NH_3 with air at a temperature near 950° C in the presence of a platinum catalyst. The resultant gas is absorbed in water to form nitric acid. The chemical reactions for NH_3 oxidation and absorption steps are:

Oxidation	1. $4NH_3 + 5O_2 \longrightarrow 4NO + 6H_2O$
	2. $2NO + O_2 \longrightarrow 2NO_2$
Absorption	3. $3NO_2 + H_2O \longrightarrow 2HNO_3 + NO$
	4. $2NO + O_2 \longrightarrow 2NO_2$
	5. Reactions 3 and 4 repeat through absorber

There are three types of nitric acid plants: (i) oxidation and absorption at atmospheric pressure, (ii) oxidation at atmospheric pressure followed by compression of the oxidation product gases and absorption under pressure, and (iii) oxidation and absorption under pressure.

The oxidation catalyst used in all three types of plants is Pt alloyed with 5 to 10% of Pd or Rh or both. This alloy is drawn into wire about 0.076 cm in diam and then woven into a mesh with approximately 120 openings per centimeter. The usual catalyst charge is 62 g (2 Troy ounces)/metric ton per day of 100% acid manufacturing capacity.

Table 7-1. Physical properties of solid nitrogen fertilizers.†

Fertilizer Fertilizer material	Formula	Molecular weight	Density	Melting point	Boiling point	Solubility in water			
						Cold		Hot	
				°C	°C	g/kg	°C	g/kg	°C
Ammonium chloride	NH_4Cl	53	1.53	340 D	520 S	2.97	0	7.58	100
Ammonium nitrate	NH_4NO_3	80	1.73	170	210	11.83	0	87.1	100
Ammonium sulfate	$(NH_4)_2SO_4$	132	1.77	235 D	--	7.06	0	10.38	100
Calcium cyanamide	$CaCN_2$	80	--	1300 S	--	D		D	
Calcium nitrate	$Ca(NO_3)_2$	164	2.50	561	--	10.2	0	36.4	100
Oxamide	$NH_2(CO)_2NH_2$	88	1.67	256 S	290 D	0.004	7.5	--	
Sodium nitrate	$NaNO_3$	85	2.26	307	380 D	7.3	0	18.0	100
Thiourea	NH_2CSNH_2	76	1.41	180	D	0.98	12	--	
Urea	NH_2CONH_2	60	1.33	133	D	7.8	5	--	

† D = decomposes; S = sublimes.

a. Oxidation—Because pressure oxidation and absorption is the most commonly used process today, this is the process that will be described. Ammonia is vaporized and superheated to about 120° C. The superheated NH_3 vapor is mixed with hot, filtered, compressed air to give a mixture about 10.5% NH_3, by volume. This mixed gas is then preheated to about 230° C and allowed to flow over the sandwich of wire meshes that constitute the catalyst. Once lighted with either an electric filament heater or a hydrogen torch, the reaction between NH_3 and the oxygen in air is exothermic and selfsustaining. Approximately 95% of the NH_3 is converted to N oxides. The hot gases from the catalyst flow through a heat exchanger train and the heat is used for the following: (i) generate steam, (ii) preheat air, (iii) preheat air–ammonia mix, (iv) preheat tail gas from absorber, (v) preheat boiler feed water, and (vi) filter for Pt metal recovery.

b. Absorption—Processed gas from the end of the heat exchange train flows to a water-cooled dilute acid condenser. Water formed in reaction 1 is condensed here as dilute acid at 20 to 23% by weight. This dilute acid is fed to the absorber trays that operate at this acid strength. Nitric acid absorbers usually have between 30 to 35 trays with cooling of the liquid on the trays. Cooling with water is necessary because reactions 3 and 4 are exothermic and it is necessary to remove this reaction heat in order that both reactions will alternately proceed to completion to get maximum conversion of NO to NO_2 and absorption of NO_2 to make nitric acid. When properly sized and designed, an acid absorber can give about 97 to 98% efficiency.

Catalytic combustors are now included as a part of almost every new nitric process. The kinetics of reactions 3 and 4 make it impossible to convert and recover all of the N oxides. About 0.2 to 0.4% N oxides by volume will be present in the absorber tail gas. This gas, which contains 2 to 3% oxygen by volume, is reheated, mixed with a hydrogen or hydrocarbon fuel, and is then passed over the catalytic combustor catalyst. Active catalysts can reduce most of the N oxides to N. This reaction is also exothermic and thus will increase the tail gas temperature for additional power recovery when let down through a hot gas expander.

2. *Ammonium Nitrate*

Ammonium nitrate (NH_4NO_3) is a logical solid fertilizer for manufacture in a synthetic ammonia plant. Nitric acid is manufactured, which, in turn, is neutralized with additional NH_3 to form ammonium nitrate. Ammonium nitrate is a white crystalline solid containing 350 g N/kg (35% N). (The fertilizer grade material after coating with a conditioning agent contains 33.5 to 34.0% N). About 75% of the ammonium nitrate produced is used to manufacture dynamite, high explosives, and solid rocket propellants; it can be mixed with about 6% fuel oil by weight as a substitute for dynamite.

The reaction between ammonia and nitric acid to produce ammonium nitrate is a simple acid-base neutralization reaction:

$$HNO_3 + NH_3 \longrightarrow NH_4NO_3.$$

Vaporized and superheated ammonia and nitric acid are sparged or injected below the liquid level in the neutralizer. The reaction is exothermic and the heat of reaction is sufficient to concentrate the neutralized solution to about 83%. It is then further concentrated to 95 to 96% and pumped to the top of a prilling tower. From here it is sprayed and, in falling to the bottom of the tower through a counter-current air stream, the drops of solution chill and solidify. The solid prills are collected at the bottom of the tower, screened to remove oversized material, and then conveyed to one or more dryers where heated air is employed to reduce the moisture content to $\leq 0.5\%$. The dried prills are cooled and again screened to remove over- and undersized material. The dry, cooled ammonium nitrate is then coated with about 3% clay or parting agent to minimize moisture pickup from the atmosphere, prevent caking, and keep the product free-flowing.

An alternate to this process is the use of a shorter tower and the prilling of an ammonium nitrate melt at a concentration of about 99.8%. The solid prills will contain $\leq 0.2\%$ moisture; therefore, it requires no additional drying. Prills produced by this process are slightly more dense than prills made from prilling a 95 to 96% solution.

The Stengel process also produces a high-density ammonium nitrate, but the product is angular as opposed to the smooth, round prill produced by the prilling process. Nitric acid is preheated in this process to further increase the reaction heat and water vaporization from the neutralizer. Ammonium nitrate solution from the neutralizer is concentrated to a melt of near 99.8% and is flowed onto a cooled continuous stainless steel belt. The melt solidifies on this belt and is then chipped off, ground, and screened to give the desired particle size. The product is cooled and coated with clay.

Ammonium nitrate for fertilizer use is also produced in granular form by *spray-drum* granulation (spherodizing) and by granulation in revolving inclined pans. (Reed & Reynolds, 1973; TVA, 1980). The granules are harder and more durable and can readily be produced in larger size.

Ammonium nitrate in all its forms is a strong oxidizer, but it is not an explosive in its pure form or when mixed with other major nutrients. Before mixing with metals or carbonaceous materials, safety precautions should be observed.

3. Ammonium Sulfate

Most of the ammonia used in agricultural ammonium sulfate [$(NH_4)_2SO_4$] is produced as a by-product of coal coke ovens and in caprolactum manufacture; however, it may also be made synthetically. The reaction is the same for both sources of NH_3

$$2NH_3 + H_2SO_4 \longrightarrow (NH_4)_2SO_4.$$

Ammonium sulfate is a white crystalline solid, slightly less soluble in water than urea, with a N content of 21%. It is an excellent source of the secondary nutrient S. The shift to higher-analysis fertilizers, such as ammonium nitrate and urea, has reduced the ammonium sulfate market to the point that by-product ammonium sulfate alone will satisfy the demand. By-product ammonium sulfate generally enjoys a cost advantage over the synthetic product; by virtue of this and the oversupply situation, there is very little synthetic material being produced in the USA today.

The process takes place in a reactor or saturator where sulfuric acid (H_2SO_4) and superheated NH_3 gas are sparged into a supersaturated solution of ammonium sulfate. The reaction is exothermic and water is added to vaporize as steam and thus dissipate the heat of reaction. The crystals that form are maintained in suspension by agitation or forced circulation. The crystal growth and size can be controlled by the length of time the crystal remains in the saturator, by reactor temperature, by the percent free acid in the saturator, and by the liquor recycle rate. When crystals reach the desired size, a continuous drawoff of solution and suspended crystals is started. This drawoff stream is filtered or centrifuged to recover the crystals, and the filtrate (mother liquor) is returned to the saturator. The crystals are then dried with heated air, cooled, and coated.

4. Urea and Urea Derivatives

a. Urea—Like ammonium nitrate, urea is a well-integrated product for a synthetic ammonia plant (Scaglione, 1963; Anon., 1963). The two feed streams required for the manufacture of urea are ammonia and carbon dioxide, both of which are products or by-products of an ammonia plant. The general reaction for producing urea is:

$$CO_2 + 2NH_3 \longrightarrow CO(NH_2)_2 + H_2O$$

Urea is a white crystalline solid, very soluble in water, and the clay-coated agricultural grade is 45% N.

There are many variations of urea manufacturing processes; most of the variations are in the methods used to recover, separate, and recycle unreacted NH_3 and CO_2. The various processes may also be subdivided into: once-through, partial recycle, and total recycle.

The urea reactor is a large continuous autoclave in which compressed CO_2 gas and heated liquid NH_3 are reacted at 19.2 to 27.3 MPa (2800 to 4000 psig) and 180 to 195° C. The reactants combine to form urea and ammonium carbonate $[(NH_4)_2CO_3]$. Conversion and eventual recovery of ammonia N as urea N will run about 30% for the once-through process, 50 to 80% for the partial recycle processes, and up to 98% for the total recycle process. Most processes, if not all, operate with excess NH_3 in the reactor. The solution from the reactor is heated to dehydrate the carbonate to form urea and water, and then the different methods of separating, recovering, and recycling the unreacted NH_3 and CO_2 and the carbonate that is not decomposed come into use. These are so numerous and complicated that no attempt will be made to discuss them in this book.

Either prills, granules, or crystals may be produced for the end product. Urea solution from the reactor at 75 to 87% concentration is vacuum-concentrated to 99.7% melt that is either prilled or granulated (Reed & Reynolds, 1973; TVA, 1980, 1982). For some uses, the urea is first crystallized and then melted and prilled or granulated. An additive in small proportion provides good storage and handling properties without a surface coating. Grade of product urea is 46% N.

Urea became the world's leading N fertilizer in the mid-1970s.

b. Sulfur-Coated Urea (SCU)—Granular or prilled urea may be coated with S and a sealant to give a controlled-release N fertilizer (TVA, 1966, 1968, 1979). The solid urea is preheated to 65 to 70° C and molten S at 143° C is then atomized and sprayed onto the urea in a rotary drum. This is followed by a spray of the molten wax-like sealant (polyethylene-brightstock mixture). The urea is preheated so that the molten S does not chill too quickly and gives a uniform coating. The material is then cooled and coated with a conditioner. Typical finished product contains 36% N and analyzes about 78% urea, 17% S, 2% sealant, and 2.5% diatomaceous earth conditioner.

c. Urea-Formaldehyde—Urea-formaldehyde or ureaform also is a slow-release N fertilizer, but unlike SCU, it is slowly water soluble. Ureaform is a slightly hygroscopic, odorless, white, granular solid containing about 38% N. Actually, the term *ureaform* covers many products that are mixtures of methylene ureas. They range from short-chain water-soluble molecules to long-chain water-insoluble molecules.

Ureaform encompasses urea-formaldehyde products from the urea-formaldehyde resins or plastics with almost completely unavailable N to products with some unreacted urea along with methylene ureas, which give both readily available and slowly available N.

d. Thiourea—Thiourea is a crystalline, colorless solid (prisms or needles) having the formula

$$H_2N-\overset{\overset{\displaystyle S}{\|}}{C}-NH_2$$

and a melting point of 180 to 182° C. It is relatively insoluble in water and only slightly soluble in ether. The three most common methods of preparation are (i) heating ammonium thiocyanate (NH_4CNS) to about 180° C, where equilibrium with thiourea is established; (ii) the reaction of hydrogen sulfide at about 180° C with calcium cyanamide; and (iii) the reaction of dicyandiamide and ammonium sulfide at 60 to 70° C.

e. Crotonylidene Diurea—Crotonylidene diurea is currently used in nonfarm applications as a slow-release N fertilizer. It is a high-cost product manufactured by the reaction between cronotonaldehyde and urea. The chemical name for this material is 2-oxo-4-methyl-6-ureidohexahydropyrimide.

f. New Urea Forms—Other urea fertilizers in the slow-release group are being investigated for their advantages in economics and manufacture and for their agronomic value. Some examples of these are propylidene diurea, isobutylidene diurea, trifurfurylidene triurea, and urea pyrolzates. Information on the soil reactions of the slow-release materials is included in another chapter.

5. Other Ammonium Fertilizers

a. Ammonium Nitrate–Sulfate—This material, which is manufactured by at least one major fertilizer supplier, is produced by granulating ammonium nitrate onto ammonium sulfate nuclei and then coating the finished product with a conditioning agent. It can also be manufactured as a double salt of ammonium sulfate and ammonium nitrate containing 26% N. Its main use is to provide nutrient S that is needed in some areas and cropping situations.

b. Ammonium Chloride—Fertilizer-grade ammonium chloride (NH_4Cl) is generally recovered as a by-product from an ammonium soda operation or Solvay Process plant. In the Solvay Process, a brine solution is ammoniated and then carbonated; sodium bicarbonate and ammonium chloride are formed and the solution still contains considerable unreacted sodium chloride. The sodium bicarbonate, sparingly soluble in this mixture of salts, is separated from the liquor by filtration and then converted to soda ash by heating.

C. Other Nitrogen Fertilizers

1. Calcium Nitrate

Calcium nitrate [$Ca(NO_3)_2$] is manufactured principally in Europe as a coproduct of nitric phosphate production. It is a white crystalline hygroscopic solid containing 15.5% N. It can be produced by reacting nitric acid and crushed limestone as shown in the following equation:

$$CaCO_3 + 2HNO_3 \longrightarrow Ca(NO_3)_2 + CO_2 + H_2O.$$

The co-product calcium nitrate from nitric phosphate production is coated with paraffin as a conditioner for the hygroscopic material. Calcium nitrate is popular in Europe but is not widely used in the USA, because of its low N content and the availability of lime-nitrate-urea fertilizers. Its principal use in the USA is on citrus, for which it is well suited.

2. Sodium Nitrate

Essentially all sodium nitrate ($NaNO_3$) used for fertilizer purposes is a natural product mined and processed in Chile. However, a synthetic sodium-nitrate process has been used in the USA. In this process, salt and a 70% HNO_3 are heated to boiling in a reaction vessel, evolving chlorine and nitrosyl chlorine as shown in the reaction below:

$$3NaCl + 4HNO_3 \longrightarrow 3NaNO_3 + Cl_2 + NOCl + H_2O.$$

The evolved gases are then oxidized with hot 80% HNO_3 in a column to produce nitrogen dioxide and chlorine, as indicated by the following reaction:

$$2NOCl + 4HNO_3 \longrightarrow 6NO_2 + 2H_2O + 2Cl.$$

3. Oxamide

Oxamide is a slow-release N fertilizer that in pure form has a N content of 31.8%. It has the structural chemical formula

$$NH_2{-}\overset{\overset{\displaystyle O}{\|}}{C}{-}\overset{\overset{\displaystyle O}{\|}}{C}{-}NH_2.$$

Recent work on the production of oxamide has been conducted in the USA (TVA), Federal Republic of Germany (Hoechst A. G.), and Japan (Ube Industries). Hoechst has developed a process to pilot-plant stage in which hydrogen cyanide is oxidized in an acetic acid/water mixture with a copper nitrate catalyst (Riemenschneider, 1976). TVA's laboratory studies have centered around the oxidative carbonylation of alcohols with palladium catalysts and quinone oxidants in a single-step process. Ube Industries' work has been similar to TVA's but in a two-step operation.

The slow-release characteristics of oxamide depend to a large extent on its granule size. Some evidence also indicates that certain bacterial action is required before N is released. Cost of production is projected as substantially higher than for N as urea, and for that reason the use of oxamide would be restricted to specialty fertilizer markets.

4. Calcium Cyanamide

One of the oldest methods employed for the fixation of atmospheric N for agricultural purposes is the nitrification of calcium carbide to form calcium cyanamide ($CaCN_2$). The first step in this process is the production of calcium carbide by fusion of lime and coke. This is carried out in an electric arc furnace to provide the high level of temperature needed. The following reaction occurs:

$$CaO + 3C \longrightarrow CaC_2 + CO$$

$$CaC_2 + N_2 \longrightarrow CaCN_2 + C.$$

The second step of the process is initiated at approximately 1000° C and, once started, is self-sustaining.

Pure N must be used to avoid oxidizing the carbide to lime and carbon monoxide. These operations must be blanketed in an inert atmosphere to avoid explosions.

Calcium cyanamide is only produced in Europe. Total production was < 55 000 metric tons of N in 1980.

D. Solutions Containing Nitrogen

Liquid anhydrous NH_3 is by far the leading form of N fertilizer applied in the USA. Table 7–2 shows N consumption by types of fertilizer in 1982.

Water solutions of any solid N fertilizer may be used for direct application or as the N source for blended or granulated fertilizers. Ammonia, ammonium nitrate, and urea are the most commonly used N compounds in fertilizer solutions.

1. Aqua Ammonia

Water solutions of NH_3 are commonly referred to as *aqua* and the most widely used concentration is about 25% NH_3, which corresponds to 20% N. Some aqua is made at the ammonia plant and shipped to the consumer, but the preferred procedure is to ship anhydrous NH_3 to the retail point where it is then converted to aqua. Manufacture in the field is accomplished by metering and mixing the NH_3 and water, followed by cooling to remove the heat of mixing. The reaction or mixing heat is quite high, and the product must be cooled to a temperature of about 32° C to bring the total vapor pressure down below the atmospheric pressure and thus minimize NH_3 loss from the solution.

2. Ammonium Nitrate

Ammonium nitrate–water solutions are used almost exclusively for blending mixed fertilizer solutions and as the N source for granulated fertilizer. The most commonly used solution contains 83% ammonium nitrate, which has a crystallization temperature of 66° C and a density of 1.385.

3. Ammonium Nitrate–Urea–Ammonia

The main constituents used in ammonium nitrate–urea–ammonia solutions are ammonium nitrate and urea with ammonia added in some types. Different combinations of these ingredients are used to obtain the

Table 7-2. Consumption of various types of N fertilizers in 1982.

Fertilizer type	Application
	metric tons of N × 10^6
Anhydrous NH_3	4.7
Nitrogen solutions	2.3
Solid urea	1.1
Solid ammonium nitrate	1.0
Ammonium sulfate	0.16
Aqua NH_3	0.14

desired properties. These solutions may be classified into two types—low pressure and nonpressure.

The low-pressure solutions contain free NH_3 and are used primarily for ammoniating superphosphate. These solutions have a wide range of vapor pressure [0.07 to 3 MPa (1 to 50 psig) at 40° C] and crystallization temperature (−45 to +10° C). Nitrogen analysis varies from 37 to 49% N. For direct application, these solutions have the advantage of lower vapor pressures than anhydrous NH_3 and higher N analyses than aqua.

4. Urea–Ammonium Nitrate Solution

The nonpressure urea–ammonium nitrate solutions contain ammonium nitrate and urea. The high mutual solubility of the two salts gives a relatively high-analysis solution containing 32% N and makes it one of the most popular of the N solutions. This solution normally contains 33 to 35% urea, 45 to 47% ammonium nitrate, and 32% N. The solution has a density of 1.33 kg/L and a crystallization (salting-out) temperature of 0° C. The addition of sufficient water to dilute this solution to 28% N gives a solution with a density of 1.285 kg/L and a crystallization temperature of −18° C.

5. Others

There are also a few special solutions that are used to supply the supplemental elements—Na, Ca, or S solutions—when necessary for special soil or crop requirements. These include sodium nitrate–ammonium nitrate (20% N), calcium nitrate–ammonium nitrate (17% N), ammonium bisulfite (8.5% N, 19.4% S), and ammonium polysulfide (20.6% N, 45% S).

E. Mixed Fertilizers Containing Nitrogen

Mixed fertilizers are important to agriculture but are difficult to cover. Only introductory treatise is given in this chapter with more detailed information in chapters 9, 10, 12, and 13 of this book.

1. Potassium Nitrate, 13.85–0–38.66 (13.85–0–46.39)

In a pure state, potassium nitrate (KNO_3) contains 13.85% N and 38.66% K. It is also known as salt-peter or nitre and is a white, crystalline solid. It has been produced domestically for years by the double decomposition reaction between sodium nitrate and potassium chloride (KCl), which follows:

$$NaNO_3 + KCl \longrightarrow KNO_3 + NaCl.$$

In this process, solid potassium chloride is reacted in a hot solution of sodium nitrate. Sodium chloride is crystallized from solution and is separated from the hot potassium nitrate liquor. Solid potassium nitrate is produced by subsequent crystallization and filtration. More information on this product is given to chapter 10 of this book.

2. Urea-Phosphate, 17.7–19.6–0 (17.7–44.88–0)

Urea-phosphate is an additional compound of urea, a very weak base, and phosphoric acid (H_3PO_4), a relatively strong acid. A white, crystalline, dry powder is formed, which contains 17.7% total N and 19.6% P. Its specific gravity is 1.759 and its melting point is 117.5° C. This product, known as carbamide–phosphoric acid, is very soluble in water (202 g/L at 46° C).

Urea-phosphate is prepared by the reaction of concentrated phosphoric acid and solid urea:

$$CO(NH_2)_2 + H_3PO_4 \longrightarrow CO(NH_2)_2\,H_3PO_4.$$

The product is precipitated by cooling and recovered by filtration.

TVA is conducting pilot-plant studies of the production of urea phosphate from wet-process phosphoric acid and urea. In one process, the urea phosphate crystals are separated by centrifuging as comparatively pure material containing 17% N and 19% P (17–19.36–0) (17–44–0). Only about 15% of the impurities from the wet-process acid are retained in the product. That is a good intermediate for preparation of liquid fertilizers of high quality. The urea phosphate crystals are agglomerated by heat of fusion in a simple system to give the product good handling and shipping properties.

In an alternative process, the reaction mass of urea and wet-process acid is granulated in a specially designed drum reactor–granulator. The product granules contain 16% N and 18% P (16–41–0) (16–18.04–0). The product made by the simpler and lower-cost process contains all of the impurities from the acid, so it would not be suitable for production of liquid fertilizers of low-impurity content. It loses very little N on surface application and shows good promise in decreasing the loss of N from mixtures with urea.

3. Ammonium Phosphates

The term *ammonium phosphates* encompasses a wide variety of fertilizers produced by ammoniation of phosphoric acid. Such fertilizers may contain ammonium sulfate or ammonium nitrate as an additional source of N; the ammonium phosphate may be present as the monoammonium or the diammonium salt or mixtures of the two. More detailed information is presented in chapter 9 of this book.

4. Other Ammonium Phosphates

a. Metal Ammonium Phosphates—Small-scale production of a number of metal ammonium phosphates having the general formula

$$MeNH_4PO_4 \cdot XH_2O$$

have been produced on a pilot-plant basis. These compounds are capable of supplying N, P, and various metals over long periods of time. A magnesium ammonium phosphate containing 8% N, 17.5% P, and 14.5% Mg is an example of these compounds.

b. Ammonium Polyphosphates—Ammonium polyphosphate fertilizers are made by reacting ammonia and superphosphoric acid. In the process, the acid and ammonia are reacted at moderate pressure and at an elevated temperature to produce a fluid melt, which is granulated. This solid product contains about half monoammonium phosphate and half ammonium polyphosphate.

Ammonium polyphosphate can also be produced by ammoniation of orthophosphoric acid, using the heat of reaction to evaporate water and form polyphosphate. This process, utilizing merchant-grade wet-process acid, is currently in operation. For more details see chapter 9 of this book.

Superphosphate, monoammonium phosphate (MAP), or diammonium phosphate (DAP) are used as sources of phosphate to make suspension fertilizers that are less costly than ammonium polyphosphate. A small amount of clay is added to promote suspension of the fine solid nutrients. Suspensions allow higher-analysis triple nutrient fertilizers than liquids (solutions), plus the addition of higher concentrations of secondary and micronutrients, herbicides, and insecticides with manageable physical properties. See chapter 12 of this book for more details.

III. ENERGY REQUIREMENTS

A. Energy for Fertilizer Manufacturing

For comparative purposes, average energy requirements for manufacturing common nitrogen, phosphate, and potash fertilizers are summarized in Table 7–3. These estimates are based on an energy-use survey in North America during 1979 and information from other sources. Consequently, these estimates are representative of actual energy use by fertilizer plants in operation. However, one must be careful in generalizing these estimates for the developing countries because of the differences in technology, processes, management, and efficiency. As a result, the energy requirements in manufacturing fertilizers may be underestimated for developing countries. Furthermore, the energy requirements are based on high heating value and total rather than battery-limits energy estimates.

There is a substantial variation in energy requirements across different fertilizers, ranging from 79.5 GJ/metric ton of nutrient content for prilled urea (highest) to 3.8 GJ/metric ton for nongranular potassium chloride (among the lowest). Both phosphate and potash fertilizers use very little energy; most of it is in the form of fuel, electricity, and steam. Since steam and electricity can be generated from any commercial fuel, the manufacture of phosphate and potash fertilizers presents a fairly wide latitude in choice of basic energy sources. On the other hand, N fertilizers are highly energy-intensive and require energy both as feedstock (source of H for NH_3) and fuel (including steam). Most of the direct and indirect energy uses in the existing fertilizer sector depend primarily on nonrenewable hydrocarbons.

Table 7-3. Average energy use for manufacturing selected nitrogen, phosphate, and potash fertilizers.

Product	Percent nutrients	Average energy input, GJ	
		Per metric ton of product	Per metric ton of nutrient
Nitrogen fertilizers (N)			
Ammonia	82	46.9	57.2
Urea: prilled	46	36.6	79.5
granular	46	35.0	76.1
Ammonium nitrate: prilled	34	24.9	73.4
granular	34	24.4	71.8
Ammonium sulfate: synthetic	21	12.6	60.6
by-product	21	4.7	22.4
Phosphate fertilizers (P)†			
Ground rock‡	13	1.2	9.2
Phosphoric acid	24	5.3	22.1
TSP, granular	20	4.3	21.5
DAP, granular	20	14.3§	20.0
MAP, granular	24	10.8¶	18.8
SSP, nongranular	9	1.0	11.1
SSP, granular	9	1.7	18.9
Potash fertilizers (K)			
Muriate:# granular	50	2.9	5.8
nongranular	50	2.3	4.6
average	50	2.6	5.2
Muriate:†† average	50	4.6	9.2

† TSP = triple superphosphate; DAP = diammonium phosphate; MAP = monoammonium phosphate; SSP = single superphosphate.
‡ For direct application, dried, and finely ground.
§ Contains 18% N. Energy use is 57.2 GJ/metric ton of N.
¶ Contains 11% N. Energy use is 57.2 GJ/metric ton of N.
# For North America.
†† For Europe.

The production of urea requires commercial energy, which is over 8 times that of triple superphosphate (TSP) and 19 times that of potassium chloride. This is precisely the reason why many countries that produce NH_3 based on imported feedstocks find it difficult to continue operating the existing N fertilizer industry or plan to phase it out.

B. Energy For Fertilizer Distribution

After the chemical fertilizer is manufactured, commercial energy is also required to pack it in bags, transport it to the farm level, and apply it to the crop. However, the total amount of energy required for fertilizer distribution and application is rather small relative to its manufacture. The major share of energy required for distribution is used in transportation of fertilizer products and raw materials.

Average energy use for fertilizer packaging, transportation, and application is 8.6 GJ/metric ton of N, 22.4 GJ/metric ton of P, and 8.8 GJ/metric ton of K. Of this amount, transportation accounts for 52, 58, and

63%; packaging accounts for 30, 27, and 24% for N, P, and K, respectively. In the case of P and K, since their manufacture is relatively less energy-intensive, packaging and transportation account for approximately 46% of the total energy needs on a nutrient basis.

Even though it is difficult to make any generalizations, two other features for fertilizer transportation stand out. First, trucks are the most common mode of transporting fertilizer, especially at the secondary level. Second, truck transportation is highly energy-intensive. On the average, trucks use four times more energy than rail and almost nine times more energy than waterways for transporting 1 metric ton of fertilizer for a distance of 1.6 km.

C. Energy Conservation

Work has been intense since the early 1970s in efforts to decrease the energy required to produce ammonia, urea, and other nitrogen fertilizers. Engineering firms who design and construct process plants have developed schemes for substantial decreases in natural gas requirements for process needs and fuel in new ammonia plants. In new ammonia and urea processes, various schemes have been worked out to decrease energy requirements as electricity and steam, and to more efficiently utilize available process heat.

Some of the more effective schemes for energy conservation can be retrofitted to existing plants. Benefit/cost ratios often greatly favor the required investments.

IV. TRANSPORTATION AND MARKETING OF NITROGEN FERTILIZERS

Close to 10.6 million metric tons of actual N in fertilizer materials are transported from manufacturing plants to distribution terminals, dealer storage, and farms in the USA. This volume is increasing and has made it necessary to develop improved systems of marketing and of movement by railcars, trucks, barges, and pipelines.

A. Transportation

Railcars are the primary means of N fertilizer transportation. The economy in rates has improved by increasing the capacity of the cars from 24 to 72 metric tons. Rail rates are reduced further through multiple-car shipments.

Some of the most significant changes have been in the transport of nitrogenous fertilizers. There are now approximately 15 000 ammonia railcars in service. Twenty-four metric ton capacity railcars are used within a radius of 320 to 645 km from manufacturing plants, terminals, and other distribution points. The 72 metric ton capacity (jumbo) cars are used between distances of 645 to 1600 km with the multiple-car shipments used for the longest distances.

River barges are also used to transport 6300 to 9000 metric ton quantities of fertilizer from Gulf Coast production points to terminals in the Corn Belt. Many smaller capacity barges are now obsolete and are no longer used.

Trucks, carrying approximately 18 metric tons, transport fertilizer from 1 to 480 km to retail points and farms with more economy than any other means. One truck, because of its quick turn-around time, can transport as much fertilizer on short hauls as 25 small railcars. The N fertilizer industry in the USA is built around anhydrous NH_3, partially because of its relatively low per-unit cost. Continued competitive advantage will cause consumption to increase. Its physical characteristics, which make pipeline transportation possible, adds considerably to its efficiency and demand.

Pipelines have been built and are operating on the concept of satisfying the foregoing market conditions. The new relationship between the cost of natural gas and freight rates has made the location of production plants a major economic consideration. The majority of new ammonia plant capacity in the USA has been constructed in the gas-producing areas of Texas and the Gulf Coast.

A kind of three-ring circle of service has developed along the pipelines, with the pipeline acting as the long-distance transporter. The inner ring is small trucks of 11 500 to 19 000 L water capacity loading at pipeline terminals and distributing the load directly to farm applicators within a radius of 50 km. The second ring, from 50 to 180 km, is large truck transports with 38 000 to 42 000 L water capacity, loading at pipeline terminals and distributing the load to bulk plants having 45 000 to 67 000 L of storage. The outer ring is tank cars delivering to bulk plants having 76 000 to 227 000 L of storage and served by railroads that can provide prompt service on demand.

Pipeline terminals consist of high-pressure storage tanks and truck loading docks capable of loading 100 transports per day and are located at strategic points along the pipeline. Deliveries of anhydrous NH_3 to the terminals are regulated from a central traffic control center.

Two emergency shut-down stations, one located at the truck dock and the other at the terminal building, can be used to shut down the entire terminal in case of an emergency.

The first anhydrous NH_3 pipeline was MAPCO's 1300-km line (Mid-American Pipeline System) from production at Borger, TX, to distribution points in Kansas, Nebraska, and Iowa. Its capacity is 1170 metric tons/d. The Gulf Central Pipeline Company system is about 2700 km long and has a capacity of 1 million metric tons/yr or 2700 metric tons/d. Ammonia enters this pipeline at Luling, LA, and at Donaldsonville, LA. A 1100-km truck line leads to a junction at Herman, MO. An eastern branch goes to markets in Illinois and Indiana, and a western branch serves Iowa and eastern Nebraska (See Fig. 7–3).

Nonpressure N fertilizer solutions are increasing in usage. Railcars, barges, and trucks are transporting the bulk of this product. Moving and

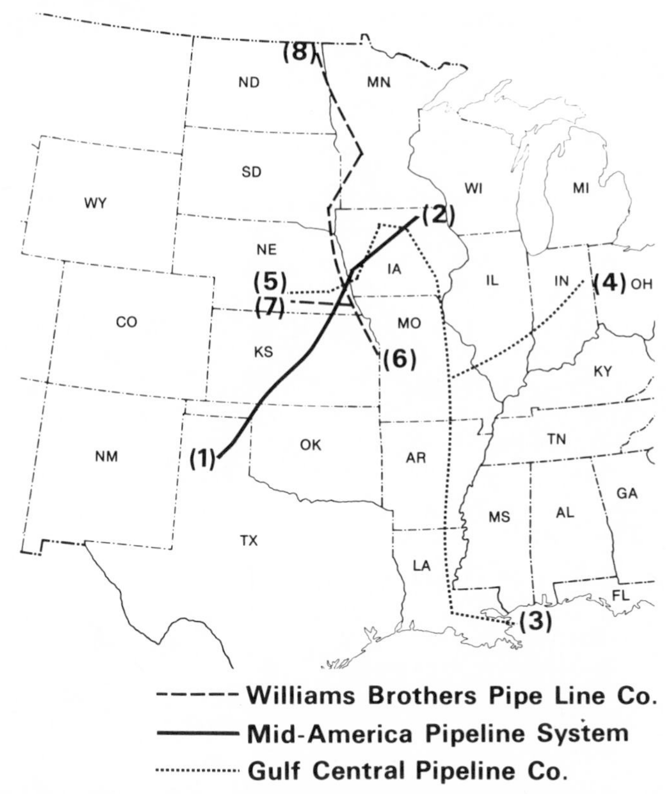

Fig. 7–3. Anhydrous NH_3 and nitrogen solution carrying pipeline. Mid-America starts at Borger (1) and ends at Gamer (2); Gulf Central starts at Luling (3) and runs to Huntington (4) or to Garner (2) and then Aurora (5); Williams Brothers begins at Kansas City (6) and proceeds to Doniphan (7) or Grand Forks (8).

distributing these materials through existing petroleum pipelines is feasible and one pipeline company (Williams Brothers Pipe Line Co., Fig. 7–3) is now moving fertilizer solutions from Kansas City to points in Nebraska, Iowa, Minnesota, and North Dakota. Even though this means of transporting solutions is in its infancy, it is expected to increase with usage of these products.

B. Marketing

Marketing N fertilizers, as we know it today, becomes a set of coordinated forces that makes things happen beyond the point of production. The old marketing concept in which

1. the *focus* was production capacity,
2. the *means* was selling and promotion, and
3. the *end* was profit through volume

has proven itself to have serious limitations. The need for change has been brought about by three factors: first, we now produce and have for sale more fertilizer than the farmers are using; second, the numbers of retailers have increased; and third, the levels from which farmers make the decision to buy have been elevated.

The general marketing policy calls for
1. the *focus* point to be the farm customer,
2. the *means* to be integrated marketing, and
3. the *end* to be profit through customer satisfaction.

This is accomplished through more and more customer services.

Nitrogen fertilizers are marketed through efficiency-oriented channels between the manufacturer and retailer. This does not differ from the marketing of the other fertilizer products. The manufacturers are also known as suppliers and have their own sales organizations who service the retailer. Some brokers exist but are few in numbers and sell limited quantities. The two types of retail organizations are independent retail businesses and supplier-owned retail centers.

1. Independent Retailers

In the independent retailer operation, all degrees of ownership of land, storage, and handling of equipment exist. These may be owned by the retailer, but in many instances the supplier retains ownership to maintain a permanent outlet for the product.

An independent retailer may buy the product as an on-the-spot purchase or contract for resale. Some independent retailers operate as a contracted commission agent and never hold ownership of the product, and the supplier carries the accounts receivable.

Application equipment is usually owned by the retailers because of its maintenance problems.

2. Supplier-Owned Retail Centers

In the supplier-owned retail center operation, all land, storage and handling, and application equipment is owned and managed by the supplier; all personnel and management decisions are made by the supplier. The advantage of this type of outlet is that the supplier controls all aspects of the business and retains ownership of the physical plant to assure a favorable position for the future.

Agricultural colleges have been active in determining the merits of the various types of N fertilizers. They have also ascertained the best time, rates, and types of application. Most manufacturers and retailers base their sales approach and recommendations on research reports and recommendations published by their respective state universities.

V. TRANSFORMATIONS OF NITROGEN SOURCES IN SOILS

About 90% of the fertilizer N used in the USA is in the form of NH_3 or ammonium-producing compounds and about 65% of this amount is anhydrous NH_3. The remainder is chiefly accounted for by nitrate in ammonium nitrate. Improving N fertilizer effectiveness requires an understanding of the factors governing N fertilizer transformations in soils. This section will discuss the short-term chemical and biological transformations of fertilizer N in soils, as well as their long-term behavior, which considers the fate of added fertilizer N, e.g., leaching and denitrification.

A. Immediate Reactions of Mineral Nitrogen Fertilizers

The direct reactions of mineral N fertilizers center on the chemical transformations of ammonia and nitrate immediately after application, as opposed to slower transformations that take place over several days or months.

1. Transformations of Nitrate Ions

Nitrate-N (NO_3^-−N) is relatively unreactive in soils because (i) nitrate compounds are readily soluble and (ii) there is little tendency for them to be adsorbed on the negatively charged surfaces of soil colloids. Nitrate is therefore susceptible to movement through diffusion and mass transport in the soil water.

2. Transformations of Ammonia and Ammonium Ions

If ammonium salts are applied uniformly to a soil, ammonium ions can undergo cation exchange reactions and/or precipitation of insoluble reaction products [e.g., calcium sulfate ($CaSO_4$) from ammonium sulfate]. Since ammonium-N (NH_4^+-N) is held by the negatively charged soil colloids, there is little tendency for it to move with the soil water in most agricultural soils, which have cation exchange capacities (CEC) > 2.

If ammonium fertilizers are applied to a localized zone, however, the area around the application zone becomes dominated by the chemistry of the N source. The scale of this fertilizer-dominated zone may be large (e.g., the 5-cm diam zone around an anhydrous NH_3 injection) or small (e.g., the 5-mm zone around a urea granule) depending on factors such as soil texture, pH, organic matter content, water content, and N source. Initial effects of acidic N sources (ammonium nitrate, ammonium sulfate, monoammonium phosphate) involve mainly cation exchange reactions and precipitation reactions. Initial reactions of basic N sources (anhydrous NH_3, urea, diammonium phosphate) are more complex and include (i) the hydrolysis of urea to ammonium carbonate, (ii) a rise in pH, often to ≥ 9.0, (iii) a very high level of NH_3 and NH_4^+, often > 2000 mg/kg, and (iv) a decline in the number of soil organisms, including nitrifying organisms (Frederick & Broadbent, 1966; Pang et al., 1973; Wetselaar et al., 1972; Creamer & Fox, 1980; Hauck & Stephenson, 1965). The initial sink for NH_3 is the soil solution, because NH_3 is very soluble in water (42 g NH_3−N/100 g of water at 20° C) (Sharp, 1966). The distribution between NH_3 and NH_4^+ is determined by the pH (Fig. 7–4) according to the equation:

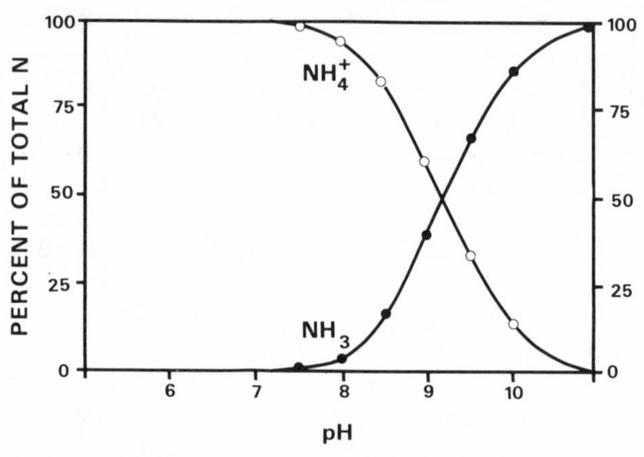

Fig. 7–4. Percent of total N as NH_3 and NH_4^+ at various pH levels.

$$NH_4^+ + OH^- \rightleftharpoons NH_3 + H_2O. \quad [1]$$

Dissolved NH_3 will move in response to diffusion gradients and mass flow of the soil solution. It will be converted to NH_4^+ by reacting with water, clay minerals, and organic matter. Soil inorganic fractions supply hydrogen ions (H^+) for this conversion from exchangeable H^+ on the exchange complex, from hydroxyl groups on the edge of clay minerals, and from the dissociation of water on clay-adsorbed hydrated ions such as aluminum (Al^{3+}) or calcium (Ca^{2+}) (Mortland, 1966). The soil organic fractions supplies H^+ from carboxyl groups (–COOH) and phenolic groups (–OH) (Broadbent & Stevenson, 1966). It is important to note that soils have a large capacity to absorb NH_3 over a broad pH range. The soil CEC is a dominant factor in such retention because it is related to clay content, type of clay, and organic matter content. Under acid conditions, NH_3 retention is probably dominated by reactions with soil minerals; retention under basic conditions is primarily associated with the organic fraction (Parr & Papendick, 1966).

3. Ammonia Fixation in Organic and Mineral Forms

By and large, the above NH_3 transformations produce NH_4^+, which is chemically and physically accessible for plant uptake. Other direct reactions can occur, however, which place the fertilizer N in compounds of low biological availability.

Some of the NH_3 may be chemically fixed by reacting with organic matter. The reaction mechanisms are vague, but are thought to involve NH_3 reacting with oxidized lignin and various quinones accompanied by polimerization, which results in complexes resistant to dissolution and enzyme action (Broadbent & Stevenson, 1966). These chemical fixation reactions are favored by high pH and high NH_3 concentrations; these are the expected conditions in an anhydrous NH_3 injection zone.

The ion NH_4^+ is also subject to chemical fixation within the crystal lattice of expanding 2:1 clay minerals (e.g., vermiculite, illite). It is just the right size to enter into the hexagonal openings between oxygen atoms of the silicon layer of clay minerals. If the crystal lattice contracts, NH_4^+ becomes fixed and is not readily accessible for ion exchange reactions. Potassium interacts with NH_4^+ fixation and release because it is the same size as NH_4^+. Potassium can maintain a clay lattice in a contracted state and thus block the release of fixed NH_4^+. A soil's capacity to fix NH_4^+ increases with its content of expanding 2:1 clays, with pH, and with the NH_4^+ concentration. These conditions will most readily occur in an alkaline high NH_3 environment. The biological availability of fixed NH_4^+ has generally been considered to be small, based on many laboratory and greenhouse studies (e.g., Mortland, 1966). Fixed NH_4^+ is not readily nitrified, but it may be slowly released over the course of a growing season depending on the K status of the soil. Recent field investigations show fixed NH_4^+ to be more readily available to crops than was previously thought, possibly because the plant is removing interfering K^+ (Nommik, 1981; Kowalenko, 1978; Kowalenko & Cameron, 1978; Van Praag et al., 1980).

B. Nitrification of Ammonium Fertilizers and Other Biological Interactions

Nitrification of ammonium salts was investigated for a number of years before anhydrous NH_3 and urea came into prominence. These studies revealed the nature of nitrification and the importance of environmental factors such as soil pH, soil water content, aeration, and temperature (Allison, 1973; Black, 1968). In general, the conversion of NH_4^+ to NO_3^- seldom affects plant growth, particularly if the NH_4^+ is dispersed throughout a large volume of soil. However, the practice of NH_3 injection and localized urea application has imposed a new set of conditions on the nitrification process.

Nitrification is the biological oxidation of NH_4^+ to NO_3^-, which can be represented by the following equation:

$$NH_4^+ \xrightarrow[+\ \text{oxygen}]{\textit{Nitrosomonas}} H^+ + NO_2^- \xrightarrow[+\ \text{oxygen}]{\textit{Nitrobacter}} NO_3^-. \qquad [2]$$

Three important features of nitrification are apparent (i) it is a biological process, which means it is affected by several environmental factors, (ii) it requires oxygen, and (iii) it is an acidifying process. The organisms primarily responsible for these steps are the aerobic autotrophic bacteria *Nitrosomonas* and *Nitrobacter* (Alexander, 1977). In contrast with the heterotrophic soil organisms, whose source of C and energy are plant residues and soil organic matter, the autotrophs use CO_2 for C and obtain energy directly from the oxidation reactions. Since these two groups of bacteria are solely dependent on the supplies of NH_4^+ and NO_2^- for growth, it is apparent that the amount of these substrates governs their number and consequently the oxidation rate. When an ammonium fertilizer is added to a soil an initial lag phase is often observed in the nitrification process. This does not usually occur during nitrification of the low levels of N released by ammonification of the soil organic matter. The delay in oxidation of fertilizer N can occur at the NH_4^+ to NO_2^- step or the NO_2^- to NO_3^- step, although normally the conversion of NO_2^- to NO_3^- is rapid and only traces of NO_2^- occur in soils. The lag is usually attributed to insufficient numbers of nitrifying organisms resulting from specific inhibitors (see chapter 8 of this book), from high levels of NH_3, or from high salt concentrations. The delay in nitrification is not a long-term effect; rather, it is a temporary event during the overall process of oxidizing NH_4^+ to NO_3^-. Nevertheless, these delays do affect the reactions of N fertilizers with soils (see later discussion of NO_2^- reactions in soils).

1. pH and Nitrification

Nitrification can take place over a pH range of about 5 to 10, but it is most rapid near pH 7 and is markedly reduced at a pH < 5.5, although the lower limit varies with the soil texture and the native organisms (Alexander, 1977). Since nitrification produces acids, it is not uncommon to see nitrification rates decline with time, especially in poorly buffered sandy soils fertilized with acidic N sources such as ammonium sulfate. The slower nitrification rate in strongly acid soils is primarily due to slower growth of the bacteria, but some believe it may be due to greater loss of N from chemical decomposition of NO_2^- in acid environments (Allison, 1973).

The ammonium oxidizers, *Nitrosomonas,* have an optimum pH in the alkaline range (ca. pH 7 to 9), while the nitrite oxidizers, *Nitrobacter,* prefer a neutral to slightly alkaline environment (ca. pH 6.5 to 7.5) (Alexander, 1977; Frederick & Broadbent, 1966). Under acid or neutral conditions, NO_2^- does not accumulate; when pH values exceed 8 and NH_3 levels are high, *Nitrobacter* activity may be restricted, which causes a temporary accumulation of NO_2^-. As noted above (section V A 2), these conditions frequently occur in an anhydrous NH_3 injection zone or in the microsite around a urea granule. With time, the NO_2^- will diffuse away from this area and become oxidized or will undergo various reactions associated with HNO_2, except, perhaps, in certain highly calcareous soils where NO_2^- is long-lived (Russell, 1961). Nitrification of a localized NH_3 or urea application begins at the outside edge, where NO_3^- is readily produced, and proceeds toward the center of the zone where NO_2^- may have temporarily accumulated. Eventually, the NH_3 is oxidized to NO_3^- and the accompanying acidity will lower the pH below its initial value, depending on the soil buffering capacity (Frederick & Broadbent, 1966; Pang et al., 1973; Wetselaar et al., 1972).

2. Soil Water, Aeration, and Nitrification

The nitrifying bacteria have an absolute requirement for oxygen, (since they are obligate aerobes) as well as certain demands for water. The precise relationship between nitrification and soil water content/aeration is difficult to assess because of the complex interrelations between soil gas transport, soil water content, and biological activity. However, research indicates that reducing the oxygen concentration from 20 to 11% does not appreciably change the nitrification rate, but a further reduction to about 2% reduces nitrification by 50% and $< 2\%$ oxygen nitrification is negligible (Amer & Bartholomew, 1951). As soil water content increases above field capacity, nitrification declines sharply and with complete water logging nitrification ceases, except for a small amount at the soil surface. Below field capacity, nitrification continues over a wide range of soil moistures extending nearly to the wilting point, with the optimum in the range from 70 to near 100% of field capacity (Allison, 1973; Tisdale & Nelson, 1975).

3. *Temperature and Nitrification*

Nitrification is strongly affected by temperature, with maximum rates occurring between 30 and 35° C (Alexander, 1977). Frederick and Broadbent (1966) summarized the temperature-nitrification relation by noting that at 5, 10, 15 and 20° C nitrification was 10, 20, 50, and 80%, respectively, when compared with the 25° C nitrification rate. However, soils vary considerably in their nitrification rate since some soils have nitrification rates near 100 kg N/ha per week at 5° C (Sabey et al., 1959). Nitrification probably occurs at freezing temperatures and it appears that oxidation of NO_2^- is slowed more by decreasing temperatures than NH_4^+ oxidation (Black, 1968).

Interest in the temperature-nitrification relations has been stimulated by continued questions about the feasibility of fall fertilizer applications. As indicated above, however, nitrification can be substantial, even at cool soil temperatures. Nitrate produced during the autumn, when there is little plant growth, is more likely to be lost by leaching or denitrification than nitrate produced when a crop is actively growing during spring or summer. Field studies often show that fall-applied N is utilized less efficiently than N applied shortly before the growing season or during the growing season (Welch et al., 1971; Stevenson & Baldwin, 1969; Nelson & Uhland, 1955). Nitrification inhibitors offer the prospect of reducing these overwinter losses under certain conditions (Hergert & Wiese, 1980; Nelson & Huber, 1980; Touchton & Boswell, 1980; also see chapter 8 of this book).

4. *Nitrification and Nitrogen Source, Placement, and Particle Size*

The N source, placement, and particle size can also influence nitrification. As noted above, alkaline N sources (anhydrous NH_3, urea, diammonium phosphate) usually have an associated high pH and high NH_3 zone near the fertilizers, which can reduce nitrification rates and may even cause nitrites to temporarily accumulate. Acidic N sources (ammonium sulfate, ammonium nitrate, monoammonium phosphate) can affect nitrification in a similar manner in certain alkaline soils (Broadbent et al., 1957; Bezdicek et al., 1971), but more common occurrences would be delayed nitrification rates from (i) salt effect (Wetselaar et al., 1972; Hauck & Stephenson, 1965) and (ii) the reduced pH associated directly with the N source or from the acidity produced during nitrification (Broadbent et al., 1957; Hauck & Stephenson, 1965; Nommik, 1966). The particle size also affects nitrification somewhat, with nitrification rates generally decreasing as particle size increases (Hauck & Stephenson, 1965; Nommik, 1966). The above effects are enhanced by fertilizer management practices that concentrate the fertilizer in a small volume of soil so that the fertilizer dominates the soil-fertilizer interaction, e.g., large granules or band fertilizer applications. The delayed nitrification rates discussed above are usually of a transitory nature and under normal conditions nitrification is complete within several weeks.

5. *Nitrification and Soil Mineralization-Immobilization*

Nitrification of fertilizer N can also be affected by the microbial mineralization-immobilization process, which is continuously occurring in soils. The balance between mineralization and immobilization is greatly affected by the proportion of C and N (i.e., the C/N ratio) in the substrate undergoing microbial decomposition; other factors such as quantity of organic matter, soil pH, and environmental factors can also affect it. Substrates rich in N emphasize the mineralization phase, which results in a net NO_3^- production, while substrates low in N emphasize the immobilization phase which may readily deplete the soil mineral N pool and therefore delay nitrification (Alexander, 1977; Black, 1968). Commercial fertilizers take part in the mineralization-immobilization cycle, particularly ammonium sources, which are more readily immobilized than nitrate sources (Allison, 1973; Legg & Meisinger, 1982).

If a substrate with a low N content [e.g., corn stalks (*Zea mays* L.) or wheat straw (*Triticum aestivum* L.)] is being decomposed when N fertilizers are added, it is quite likely that a significant quantity of fertilizer N will be incorporated into organic N compounds that are only slowly available (Frederick & Broadbent, 1966). In the absence of recent residue additions, net mineralization is the dominant process. However, it should be realized that some immobilization takes place whenever a N fertilizer is added to soil, particularly an ammonium fertilizer (Frederick & Broadbent, 1966, Legg & Meisinger, 1982).

C. Gaseous Nitrogen Losses

Gaseous N losses constitute important N transformations in virtually all soil-crop systems. These losses primarily involve (i) denitrification, (ii) NH_3 volatilization, and (iii) other losses such as NO_2^- decomposition and nitrous oxide (N_2O) evolution during nitrification. These N transformations involve the interaction of several biological, chemical, and physical factors within the soil-air-water system, which are difficult to describe and control under experimental conditions. Under field conditions, the interrelations are not well understood and our present knowledge of the underlying principles controlling the processes is inadequate. Nevertheless, there is much evidence from field N balance studies, lysimeters, and laboratory studies to indicate the magnitude of such losses and to outline the conditions that promote such losses.

1. Denitrification

Denitrification is microbial respiration that uses nitrate or nitrite as the terminal electron acceptor and thereby reduces N to various gaseous products such as N_2 and/or N_2O (Alexander, 1977; SSSA, 1984). Losses from denitrification range from 0 to > 70%, although most reported values fall between 10 and 30% for customary N rates applied to field crops on well-drained soils (Legg & Meisinger, 1982; Hauck, 1981). The general sequence for denitrification is thought to be the following:

$$NO_3^- \longrightarrow NO_2^- \longrightarrow NO \longrightarrow N_2O \longrightarrow N_2$$

although soil microbiologists are not totally agreed on the precise pathway (Alexander, 1977; Firestone, 1982). Denitrifying bacteria are found in at least 12 different genera and are quite abundant in soils (Firestone, 1982; Knowles, 1981). Denitrifiers are facultative anaerobic bacteria and most obtain energy from oxidizing the soil organic C, although some can also oxidize inorganic chemical compounds such as S. Under aerobic conditions they use oxygen as the terminal electron acceptor, but under anaerobic conditions they use NO_3^- according to the above sequence.

Several factors determine the rate and extent of denitrification in soils. The denitrifying organisms must, of course, be present along with a supply of NO_3^-; but in addition, there are other factors such as the (i) supply of oxidizable energy, (ii) aeration, (iii) soil pH, (iv) soil temperature, and (v) plants.

a. Energy Sources and Denitrification—Many investigators have observed that denitrification generally increases as soil organic C increases; denitrification is often negligible in soils containing < 1% organic C (Bremner & Shaw 1958a, b; McGarity, 1961). This general relation exists because organic C is the chief source of electrons used to reduce nitrate and because C oxidation also decreases the soil oxygen supply. The source of C could be the native soil organic matter or an organic amendment (e.g., manure, which has been shown to markedly increase gaseous loss of fertilizer N [Rolston et al., 1978]). However, the important factor is not necessarily the total C content but rather the biologically available C. Easily decomposed compounds, such as simple sugars, promote denitrification more readily than more lignified sources such as straw or corn stalks. Laboratory studies have shown that the level of water-soluble C is strongly related to denitrification (Bremner & Shaw, 1958b; Burford & Bremner, 1975; Stanford et al., 1975b). If denitrification is intermittent under field conditions, it is quite likely that the level of soluble C will determine the short-term denitrification rate.

b. Aeration and Denitrification—The soil oxygen supply is a major controlling factor in denitrification. Low soil oxygen levels result in an increase in the synthesis and activity of denitrifying enzymes and also affect the supply of NO_3^- by slowing nitrification (see above). Concentrations of dissolved oxygen need to be quite low, approximately < 0.2 ppm, before denitrification begins (Greenwood, 1962; Woldendorp, 1968). Soil water content, as such, has little effect on denitrification, but it greatly influences the oxygen diffusion rate since oxygen diffuses through water 10 000 times slower than air. Because of this, denitrification is not normally a problem in soils drier than about 80% of field capacity. Several complex factors influence denitrification, such as total soil porosity (influenced by texture and aggregation), length of the oxygen diffusion path (influenced by water content and aggregate size), and the oxygen consumption rate of the soil. In nature the situation is further complicated by the fact that soil organic matter is not uniformly mixed throughout the soil but is present as pieces of root or residue debris. It has generally been accepted that soils can simultaneously have well-aerated zones near air-filled pore spaces and poorly aerated zones near the center of large aggregates or near rapidly decomposing organic matter. The proportion of the well-aerated vs. the poorly aerated zones can be affected by seemingly minor changes in the soil water content or oxygen consumption rate. For example, a rain shower or irrigation can readily trigger denitrification losses (Ryden & Lund, 1980).

Aeration also affects the proportion of N_2O and N_2 produced. The dominant product is N_2 under strictly anaerobic conditions, but the proportion of N_2O usually increases as the oxygen stress decreases. It is evident from the foregoing discussion that soil denitrification is favored by poor aeration, but poor oxygen supplies may occur in several ways: (i) through long-term, large-scale flooding, e.g., poorly drained soils or paddy culture; (ii) through a long-term, small-scale process associated with an anaerobic microsite surrounded by an aerobic soil, e.g., the zone surrounding decomposing organic matter; or (iii) through a short-term, large-scale process produced from temporary periods of poor aeration, e.g., gaseous losses associated with rainfall or irrigation. The largest potential denitrification losses occur in very wet soils that have an abundant supply of available C.

c. Soil pH and Denitrification—It is not possible to precisely define the relation between denitrification and pH for all soils due to the diversity of the denitrifying organisms and the potential changes in soil pH under waterlogged conditions. However, it is generally observed that the soil reaction has little effect on denitrification between pH 6 and 8; as the pH drops < 5, denitrification usually decreases markedly (Bremner & Shaw, 1958b; Burford & Bremner, 1975; Firestone, 1982), although soils are frequently encountered with significant denitrifying activity $<$ pH 5 (Ekpete & Cornfield, 1965; Gilliam & Gambrell, 1978). Soil acidity also affects the composition of the gaseous products, i.e., the relative abundance of N_2O vs. N_2. Above pH 6, N_2 is usually the predominant gaseous product with some N_2O given off during the early stages of denitrification. As the pH falls < 6, the proportion of N_2O increases with N_2O often becoming the dominant product when soil pH is ≤ 5 (Alexander, 1977; Firestone, 1982).

d. Temperature and Denitrification—Since denitrification is a microbial process, it is markedly affected by temperature with the optimum usually near 30° C. From 10 to 30° C, denitrification rates increase steadily with the 10° C rate often being about one-half the 20° C rate, which in turn is about one-half the 30° C rate (Bremner & Shaw, 1958b; Cho et al., 1979; Stanford et al., 1975a). Below 10° C, denitrification decreases rapidly and the reported lower limits vary from 2 to 10° C, depending on factors such as incubation technique, oxygen demand, and available C (Bremner & Shaw, 1958b; Cho et al., 1978; Gilliam & Gambrell, 1978; Stanford et al., 1975a). Below 2° C, denitrification is negligible.

e. Plants and Denitrification—Plants may either increase or decrease denitrification, depending on the particular soil factor limiting denitrification such as available C, oxygen supply, or NO_3^- level. Plants increase denitrification by (i) consuming oxygen through root respiration, (ii) releasing readily available C through exudation and sloughing of root tissue, and (iii) by maintaining a high microbial population in the rhizosphere. They can reduce denitrification by (i) removing NO_3^- through crop uptake, (ii) removing NH_4^+, which could readily be converted to NO_3^-, (iii) reducing the soil water content, which improves the oxygen supply, and (iv) directly increasing oxygen availability in the rhizosphere of plants that transport oxygen, such as paddy rice. Research evidence supports both of the above views, i.e., plants may either enhance or hinder gaseous N losses.

Allison (1955) used data from several lysimeter experiments and concluded that plants generally promoted N losses. He estimated the average gaseous loss to be about 20% from 51 cropped lysimeters, while comparable uncropped lysimeters lost about 12%. Stefanson & Greenland (1970) and Stefanson (1972b) grew plants in gas-tight chambers and found that cropping enhanced denitrification losses when the soil water content was 50 to 80% of field capacity. Later work also showed greater gaseous N losses with plants, particularly when nitrate sources were added to soils near field capacity (Stefanson, 1972 a,c) and to soils with low available C (Stefanson, 1976). Similar findings of greater gaseous N losses with crops have been reported by Rolston et al. (1978) and by Volz et al. (1976), who used instrumental field microplots. Woldendorp (1968) and Smith and Tiedje (1979) concluded that denitrification is stimulated in the vicinity of plant roots, particularly when NO_3^- levels are high.

Other researchers have found that plants either decrease or have little effect on gaseous N losses. Carter et al. (1967) used field microplots and recovered 92% of the added ^{15}N fertilizer on fallow plots as compared with a 94% recovery on plots cropped to sudangrass [*Sorghum Sudanense* (Piper) Stapf]. Craswell and Strong (1976) also used ^{15}N in microplots and reported that wheat was an effective N conservation agent because of its direct N uptake and its depletion of soil water. Similar fertilizer N recoveries were observed in fallowed and cropped plots by Kowalenko and Cameron (1978). Smith and Tiedje (1979) showed that plants can effectively reduce denitrification when NO_3^- concentrations are low. It is also noteworthy that although Stefanson usually observed greater N losses with plants, this effect was not observed with undisturbed soil cores (Stefanson, 1972d).

Under field conditions, the above plant effects interact with soil properties that limit denitrification and with the type of denitrification event (e.g., large scale flooding vs. periodic poor aeration) to produce the final gaseous N output. The end result of these complex processes is difficult (some would say impossible) to predict; however, the prevailing opinion seems to be that denitrification rates are increased by plants because denitrification in most fertilized soil-crop systems is C-limited.

2. Ammonia Volatilization

Ammonia volatilization is a major avenue for N loss for topdressed ammonium sources on calcareous soils, for surface-applied urea on acid or alkaline soils, and for anhydrous NH_3. Ammonia volatilization is a complex process affected by several interacting chemical and physical factors as well as soil, fertilizer management, and environmental factors. Under some conditions, losses of 50% may be encountered but more common values would be 5 to 20%, depending on several factors discussed below (Nelson, 1982; Freney & Simpson, 1981).

a. General Principles—Ammonia losses from anhydrous NH_3 result from its rapid conversion from a liquid to a gas during and after injection. These losses can usually be avoided by applying it at a sufficient depth and under appropriate soil physical and moisture conditions that ensure a rapid closure of the injection channel.

Ammonia losses from urea occur readily because urea is enzymatically hydrolyzed to ammonium carbonate [$(NH_4)_2CO_3$] through the ubiquitous enzyme urease. Ammonium carbonate readily decomposes to produce NH_3 and CO_2 according to the following reaction:

$$H_2N\text{-}CO\text{-}NH_2 \xrightarrow[H_2O]{\text{urease}} (NH_4)_2CO_3 \begin{array}{l} \xrightarrow{-H_2O} 2NH_3 + CO_2 \\ \uparrow -2H_2O \\ \xrightarrow{+H_2O} 2NH_4^+ + OH^- + HCO_3^- \end{array} \quad [3]$$

The proportion of NH_3 to NH_4^+ is determined by the local pH (Fig. 7–4). An ammonium carbonate solution has a pH of about 8.6 (Hauck & Stephenson, 1965), which sets the stage for large NH_3 losses from hydrolyzed urea whenever the NH_3 is free to escape to the atmosphere (e.g., surface applications). Such losses can be particularly large for urea topdressed on meadows or small grains and should be expected on either acid or alkaline soils due to the high microsite pH surrounding a urea granule (Freney & Simpson, 1981; Nelson, 1982; Terman, 1979).

When ammonium fertilizers are applied to calcareous soil, the solubility of the calcium salt formed from the anion accompanying NH_4^+ must also be considered as shown below:

$$NH_4X + CaCO_3 \rightarrow CaX + (NH_4)_2CO_3 \begin{array}{l} \xrightarrow{-H_2O} 2NH_3 + CO_2 \\ \uparrow -2H_2O \\ \xrightarrow{+H_2O} 2NH_4^+ + OH^- + HCO_3^- \end{array} \quad [4]$$

If CaX is insoluble (e.g., if X were SO_4^{2-} or HPO_4^{2-}), the reaction favors production of ammonium carbonate and subsequent decomposition to NH_3. If CaX is soluble (e.g., if X were NO_3^- or CL^-), then ammonium carbonate formation is not favored and NH_3 losses are greatly reduced (Fenn & Kissel, 1973; Terman, 1979). Ammonia losses from monoammonium phosphate and ammonium polyphosphate are thought to be low due to the formation of metastable reaction products such as $Ca(NH_4)_2(HPO_4)_2 \cdot H_2O$ and $Ca(NH_4)_2P_2O_7 \cdot H_2O$ (calcium diammonium phosphate and calcium ammonium polyphosphate) (Terman & Hunt 1964; Terman, 1979).

Ammonia losses also occur quite readily if N fertilizers are added to the floodwaters of paddy rice. Reported losses for urea are as high as 20 to 50% while ammonium sulfate losses are often less (Vlek & Craswell, 1979; Mikkelsen et al., 1978). The extent of these losses are governed by the floodwater pH, which, in turn, is related to aquatic photosynthesis and respiration as well as the chemistry of the N source. The pH of the paddy water may reach 9.5 or 10 near midday due to CO_2 assimilation, but it can drop to about neutrality at night due to CO_2 released from respiration.

b. Factors Affecting Ammonia Losses—Ammonia volatilization is a complex process involving chemical reactions within the soil, physical transport out of the soil, and biological interactions within the soil. It is not surprising that NH_3 losses are influenced by a number of factors such as soil pH, soil CEC, calcium carbonate ($CaCO_3$) content, temperature, water loss, and the rate and method of N fertilizer application.

Ammonia losses generally increase as soil pH increases, but significant losses can also occur on acid soils when urea or other non-acid-forming N sources are applied on the surface. The pH immediately surrounding the fertilizer reaction zone is most important. Soil pH values > 7.0 increase NH_3 losses by increasing the NH_3/NH_4^+ ratio (Fig. 7–4). Soil CEC affects NH_3 losses by serving as a temporary sink for NH_4^+, which reduces the aqueous NH_4^+ concentration and the NH_3 concentration. It is commonly found that NH_3 losses are greatest on low CEC soils (coarse texture, low organic matter) and least on high CEC soils (fine texture, high organic matter) (Freney & Simpson, 1981; Nelson, 1982; Terman, 1979).

The soil calcium carbonate content affects NH_3 loss by affecting the soil pH as well as the mechanism of N loss through possible precipitation of the NH_4^+ fertilizer reaction products (see above equations). Woldendorp (1968) presented results from 176 soils throughout the world, which showed that NH_3 losses were more related to calcium carbonate content than the soil pH. High temperatures also favor NH_3 losses by increasing the hydrolysis rate of urea fertilizers, by increasing the NH_3/NH_4^+ ratio, by decreasing the solubility of NH_3 in water, and by increasing gaseous diffusion rates. Equations [3] and [4] above suggest that NH_3 losses should be encouraged by water loss and research has generally demonstrated that NH_3 losses do increase as drying conditions intensify (low humidity or high temperature). Lastly, the rate and method of N application affect losses with high N rates and surface applications favoring losses. Incorporating N sources at least 5 cm into the soil greatly reduces or eliminates losses. Ammonia losses are often greatly decreased by relatively small water additions, which may transport a N source deeper into the soil where it can be adsorbed. This is especially true for urea since it is readily leached before hydrolysis (Freney & Simpson, 1981; Nelson, 1982; Terman, 1979).

In summary, NH_3 losses are greatest with (i) surface-applied fertilizers, (ii) under environmental conditions that favor drying, (iii) high pH, and (iv) in soils with a reduced capacity to absorb NH_3 or NH_4^+.

3. *Other Gaseous Losses*

Other possible avenues of gaseous N loss include the evolution of N_2O during nitrification, the self-decomposition of nitrous acid (HNO_2), and the reaction of nitrous acid with various soil constituents such as organic matter, transition metals, clay minerals, and ammonium and amino groups (Allison, 1973; Nelson, 1982; Woldendorp, 1968). Nelson (1982) recently reviewed these mechanisms, as well as their supporting evidence, and concluded that the most likely mechanisms are the self-decomposition and reactions of nitrous acid with organic matter and the evolution of nitrous oxide during nitrification.

The self-decomposition of nitrous acid produces nitric oxide (NO) and nitrogen dioxide (NO_2), which may be volatilized from the soil or react within the soil to produce NO_3^-. Decomposition of nitrous acid requires an acid environment because the pk_a of the acid is 3.2 and only about 2% of the NO_2^- is present as HNO_2 at pH 5. Acid soils would obviously promote decomposition, but Nelson (1982) has also postulated that the surface of clay minerals would also be a likely environment since the pH there is often 2 pH units lower than the bulk soil solution. Researchers do not agree on the importance of N losses due to HNO_2 decomposition. Some feel that most of the N is retained through oxidation to NO_3^- (Allison, 1973; Broadbent & Clark, 1965) while others feel that such losses are quite important (Nelson, 1982).

The reaction of nitrous acid with soil organic matter involves both liberation of N gases and the fixation of N into slowly available forms. The phenolic substances in soil organic matter have been shown to be largely responsible for the production of N_2 and N_2O with losses being as much as 20% (Bremner & Nelson, 1968; Nelson & Bremner, 1970). Other work has shown that NO_2^- can react with the lignin-derived fraction of soil organic matter in a wide range of soils and form relatively resistant organic compounds (Nelson & Bremner, 1969). There seems little doubt that NO_2^- can react with organic matter to produce N gases or stable organic N complexes, but the extent of these reactions under field conditions remains uncertain.

Small amounts of nitrous oxide can also be evolved during the biological oxidation of NH_4^+ to NO_2^- by *Nitrosomonas* in aerobic soils (Bremner & Blackmer, 1978; Blackmer et al., 1980). These losses increase with an increase in the NH_4^+–N concentration. Fertilizer N losses under field conditions have been as high as 5% with anhydrous NH_3, while losses from ammonium sulfate and urea were < 1% (Bremner et al. 1981). The nitrous oxide losses from anhydrous were greatest between the second and fourth week after application.

Many questions remain about the practical importance of the above NO_2^- losses under field conditions. Since NO_2^- accumulates most easily in alkaline environments, it seems unlikely that the nitrous acid reactions could account for significant N losses. However, it has already been pointed out that NO_2^- frequently accumulates around urea particles and around the anhydrous NH_3 injection channel; this NO_2^- could diffuse into the surrounding acid soil (made more acid by nitrification) and then be lost through one of the mechanisms discussed above. The extent of field nitrous oxide loss during nitrification is also an open question; it appears that such gaseous losses can be significant for anhydrous NH_3. Much more research is needed to determine the extent and magnitude of N losses under field conditions.

D. Nitrogen Losses Through Water Transport

Nitrate-N is readily transported in water due to its high solubility and its nonadsorption on the soil exchange complex. It is subject to loss in the surface runoff and in the water percolating through the soil.

Legg and Meisinger (1982) concluded that soluble N losses through surface runoff are generally very small, except when high rates of fertilizer are surface-applied just before large rainfall events. In many cases, the gain of N in the precipitation is greater than the soluble N lost in the runoff.

There is general agreement that N leaching is often the major avenue of N loss from field soils in humid climates (Allison, 1973; Legg & Meisinger, 1982). Leaching losses occur when (i) the soil contains significant quantities of NO_3^-−N and (ii) water is moving downward through the soil. A number of factors influence these two prerequisites, such as (i) rate, time of applications, and source of N; (ii) crop growth and N uptake; (iii) soil characteristics that affect quantity and type of percolation; and (iv) the quantity, pattern, and time of water inputs.

The major fertilizer management practices that affect leaching are the rate and time of N application. If N rates are at or slightly below the crop assimilation capacity, the crop is usually a good sink for N (Fig. 7–5). However, soil nitrates accumulate quite readily once N rates exceed the crop assimilation capacity (Herron et al. 1968; Broadbent & Carlton, 1978). Time of N application also affects leaching. Nitrogen applied well before the period of crop uptake (e.g., fall-applied N) is subject to greater leaching losses than N applied just before crop uptake (Stevenson & Baldwin, 1969; Welch et al., 1971). The source of N can also influence the short-term leaching losses, since NO_3^- sources are subject to immediate loss and NH_4^+ sources must be nitrified first. However, N source is usually not a major long-term factor, since the NH_4^+−N is usually nitrified in a few weeks and factors such as crop uptake, N rate, and timing will determine the overall leaching losses in upland soils.

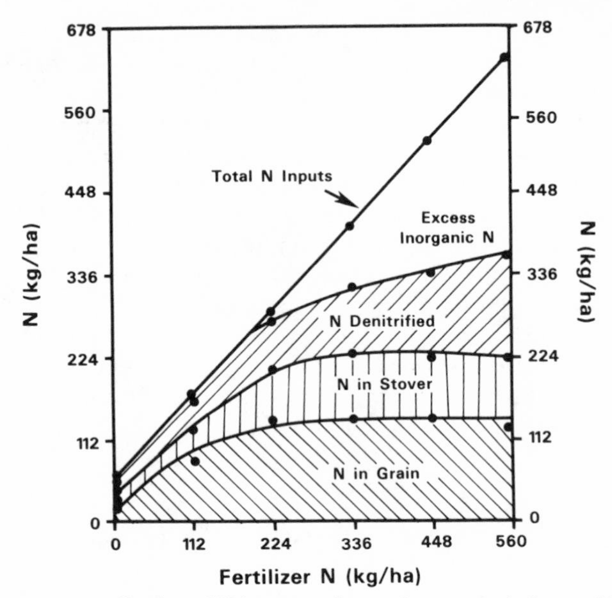

Fig. 7–5. Compartmentalization of N in a corn plant-soil system as influenced by fertilizer N rates (Broadbent & Carlton, 1978; Legg & Meisinger, 1982).

Crop N uptake is indirectly related to N leaching because crop uptake is usually the largest N reservoir and as more N is utilized by the crop, less remains for leaching. Using adapted, high-yielding varieties and maintaining adequate soil fertility will thus decrease leaching by increasing crop utilization. Cover crops grown between cropping seasons will also reduce leaching (Allison, 1955, 1973).

Assuming that NO_3^-–N is present in the soil, the major factor affecting leaching is soil water percolation. Soil nitrate leaching has often been conveniently described as *piston flow* or *complete displacement,* but field observations frequently indicate that this description is inadequate. Water percolation often involves a rapidly moving portion of water that penetrates deeply into the soil through large pores and by-passes much of the soil solution. Interconnected with this large-pore water is the small-pore water, which is also displaced downward during percolation, but at a slower rate. As water and NO_3^--N percolate through soil, a portion will appear relatively deep in the soil, but the greatest concentration of NO_3^--N will usually be much shallower (Boswell & Anderson, 1964; Thomas & Phillips, 1979). Viets and Hageman (1971) suggest the generalization that the maximum NO_3^-–N concentration will advance about 30 cm for each 30 cm of water infiltration. In multipore soil systems, the peak NO_3^-–N concentration advances like a broad wave, which widens and becomes more diffuse as it moves deeper into the soil. Several soil properties affect the type of percolation (multipore vs. piston flow) and the quantity of percolation. For example, soil structure, clay type, soil texture, and organic matter content affect the pore size, pore continuity, and soil texture affects the water holding capacity. Coarse-textured soils are more vulnerable to leaching due to their low water holding capacity and the greater likelihood of piston flow (low clay and organic matter contents, poor structure). Other soil properties that affect N leaching are the water infiltration rate and textural discontinuities in the profile.

Leaching usually increases as the quantity of percolating water increases. Water percolation, in turn, is determined by the balance between water added through precipitation and infiltration and water lost through evaporation and transpiration (ET). During the growing season, water use through ET exceeds water inputs and percolation is negligible, except on soils with low water holding capacities and in high-rainfall or high-irrigation areas. During the winter months, water inputs in humid areas exceed ET and leaching is common (Allison, 1973; Chichester, 1977; Chichester & Smith, 1978). This general leaching pattern is modified by local factors such as soil freezing, irrigation practices, soil water holding capacities, and the type of soil water percolation. Thus, it is generally true that percolation is most likely over winter and will usually increase as the precipitation increases, but how effective this participation is in leaching NO_3^-−N below the crop root zone will depend on local factors such as soil freezing and the extent of large-pore water movement.

E. Acidity and Basicity of N Sources

Nitrogen fertilizers can increase, decrease, or leave soil pH unchanged depending on (i) whether the form of N is NH_4^+ or NO_3^-, (ii) the accompanying anion or cation, (iii) the crop grown, and (iv) the ultimate fate of the fertilizer N. Ammonium sources produce acidity during nitrification (see Eq. [2]) and the theoretical acidity is equivalent to 3.6 kg $CaCO_3$/kg N. The final acidity may be greater if the NH_4^+ is associated with an acidic anion such as sulfate (SO_4^{2-}). Thus, ammonium sulfate is usually the most acidic N source. Nitrate sources accompanied by basic cations such as Na^+ or Ca^{2+} may raise the soil pH. The crop grown can affect pH through its uptake of basic cations relative to acidic ions. Pierre and Banwart (1973) have shown that the aboveground portion of the common field crops take up about 0.5 equivalent of excess base per equivalent of N, but most cereal grains removed < 0.1 equivalent of excess base per equivalent of N. This means that the actual nitrification acidity may only be about 50% of the theoretical value (or 1.8 kg $CaCO_3$/kg N) if the entire crop is removed and $< 10\%$ of the theoretical value if only the grain is harvested. The fate of the added N also affects the acidity. Nitrified N left in the soil or leached out would produce the full theoretical acidity, while N lost in denitrification will nullify the acidity (Pierre & Banwart, 1973; Pierre et al., 1970, 1971).

A procedure was proposed by Pierre (1933) for estimating the potential acid- or base-forming tendencies of a given fertilizer based on its content of acidic anions (SO_4^{2-}, Cl^-), basic cations (Na^+, K^+, Ca^{2+}, Mg^{2+}), and the crop uptake of acid and basic ions. This approach predicts that ammonium nitrate, urea, anhydrous NH_3, or solutions of these materials possess an equivalent acidity of about 1.8 kg $CaCO_3$/kg N, while ammonium sulfate and monoammonium phosphate require about three times as much calcium carbonate. These figures, however, should be used for relative comparisons among sources rather than accurate forecasts, because the above discussion has clearly shown that one cannot accurately predict the degree of acid formation under field conditions. Field research indicates that about 30% of the theoretical acidity developed on plots receiving an average annual N rate of 160 kg N/ha; 60% of the theoretical acidity developed on heavily fertilized plots, probably due to greater leaching losses on the over-fertilized plots (Pierre et al., 1971).

Another facet of N fertilizer acidity occurs with no-till systems. Since N sources are broadcast on the surface and the soil is not mixed by tillage, a highly acidic surface layer often develops (Blevins et al., 1977). The pH may be lowered as much as 1 pH unit in the surface 5 cm, which can inactivate surface-applied triazine herbicides. The acidifying effect of N fertilizers must therefore be carefully considered in modern agriculture.

VI. USING NITROGEN FERTILIZERS

Nitrogen plays a central role in modern agriculture. It is an essential nutrient and it is also the major limiting nutrient in most agricultural soils growing nonleguminous crops. Nitrogen is an energy-intensive input; the energy invested in 7 kg of N is equivalent to the energy in 8 L of diesel fuel (i.e., 1 gallon of diesel fuel equals about 7.3 pounds of N; Lockeretz, 1980; Cervinka, 1980). Nitrogen fertilizer often accounts for about 30% of the energy budget of a modern, nonirrigated corn production system (Mitchell & Teel, 1977). Increasing costs of energy have underscored the need for improving N uptake efficiencies, which are greatly affected by N rate, source, timing, and application method.

A. Determining Nitrogen Needs

Procedures for determining N fertilizer needs rest on N-balance principles, which have recently been reviewed by Meisinger (1984). There are two broad approaches, but, in general, they both reveal that (i) N fertilizer needs are directly related to the crop N requirement, (ii) fertilizer needs are indirectly related to the N use efficiency, and (iii) the N supplied from the soil through mineralization and residual mineral N should also be taken into account in the general case.

1. Determining Crop Nitrogen Requirements

The crop N requirement is involved in determining N fertilizer needs, because the crop is a major N consumer and assimilates anywhere from 30 to 70% of the fertilizer N applied. Nonleguminous crops are known to vary in their N needs. Crops that produce large amounts of dry matter generally need more N than lower producing crops; thus, small grains usually require less N than corn. The likely dry matter production of a given crop is therefore an important input in determining N needs.

The internal crop N requirement can be defined as the N concentration in the total aboveground dry matter at near-maximum yield. The internal crop N requirement seems fairly constant over a range of environmental and cultural conditions (Stanford, 1966, 1973). Therefore, it can serve to forecast the total N needs for a given total dry matter production. The internal N requirements have been estimated for several crops as follows: about 12 g N/kg (1.2%) for corn (Stanford, 1966, 1973), about 12.5 g N/kg (1.25%) for wheat (Stanford & Hunter, 1973), and about 16 g N/kg (1.6%) for the whole plant of sugarbeets (*Beta vulgaris* L.) (Carter et al., 1976; Stanford et al., 1977).

The final estimate of the crop N needs can be made by multiplying the expected total dry matter production by the internal N requirement. For example, corn contains about 50% of its total dry matter in the grain so an expected yield of 10 metric tons/ha at 15.5% moisture would require about 200 kg N/ha. Of course, precise knowledge of yield for the coming year is not possible, but most farmers/advisors can forecast yields accurately enough based on past yields, current management practices, and some measure of available water obtained from direct soil measurements or the likely rainfall of an area. The overall objective is to estimate N needs with sufficient accuracy so that fertilizer N can be added to meet crop needs, but not to fertilize the crop beyond the assimilation capacity, which leads to a marked reduction in N use efficiency.

2. *Estimating the Soil Nitrogen Supply*

Once the crop N needs have been estimated, the soil N supply should be assessed to determine the amount of supplemental N fertilizer needed. The only case where a soil N assessment can be overlooked is when the soil-crop system is at steady-state soil N levels, in which case the fertilizer needs can be predicted from the crop N removals and the total recovery efficiencies (Meisinger, 1984). Steady-state conditions, however, are long-term approaches that neglect annual fluctuations in available soil N. Furthermore, they are difficult to justify experimentally. Soil scientists have therefore had a long-term interest in methods of evaluating available soil N. The available soil N pool can be divided into two components: an inorganic component composed mainly of residual NO_3^-–N and an organic component that is mineralized throughout the growing season.

a. Residual Mineral Nitrogen—The soil mineral N content was formerly considered to be of little value toward estimating available soil N due to sampling difficulties and the transitory nature of the soil NO_3^-–N pool. Many field investigations, however, have indicated that the soil mineral N pool significantly affects crop yield and should be considered when estimating the soil N supply. Spring NO_3^-–N evaluations are especially useful when (i) previous N inputs have exceeded the crop assimilation capacity due to large fertilizer or manure additions, (ii) crop N removals have been low due to droughts, disease, or practices such as summer fallowing, and (iii) N has not been leached below the root zone due to low percolation, large pore water movement, or deeply rooted crops. These conditions oc-

Table 7–4. Summary of N recommendation systems used in the USA for corn in selected states (Keeney, 1982; Meisinger, 1984).

State	Soil N analysis: Mineral	Soil N analysis: Organic	Soil properties requested	Corn N factor†	N credits for: Manure‡	N credits for: Alfalfa§	N credits for: Soybeans§
				kg N/metric ton		—kg N/ha—	
Colo.	NO_3^--N	OM	Texture	38	2.5	55	30
Neb.	NO_3^--N	None	Soil type	25	2	110	55
Iowa	None	None	Soil assoc.	27	2.5	155	45
Ill.	None	None	Soil type	25	2.5	110	45
Ind.	None	None	Soil type	25	None	80	20
Pa.	None	None	Soil type	21	2.5	135	None

† Kilograms N needed to produce 1 metric ton of dry grain.
‡ Kilograms N credit per metric ton of fresh manure.
§ Nitrogen credit if previous crop was a good stand of alfalfa or was soybeans.

cur most frequently in the western portion of the USA; consequently, soil NO_3^-−N tests have been readily adopted in that area (see Table 7–4).

Recommended NO_3^-−N procedures involve intensively sampling the root zone of a relatively uniform area of soil representing similar soil properties and management histories. Since NO_3^-−N is mobile, it is essential to sample below tillage depth and most desirable to sample the entire root zone; suggested depths are usually 120 cm or deeper. Samples are usually air-dried immediately after collection and are later extracted with a salt solution and analyzed for NO_3^-−N by a convenient method suited to large numbers of samples (N.C. Region Soil Test. Comm., 1975; Soltanpour et al., 1979; Weise & Penas, 1979). The soil NO_3^-−N is usually considered to be approximately equivalent to fertilizer N and suggested N rates are adjusted accordingly. Despite the large spatial variability of NO_3^-−N, and its accompanying sampling problems, it has been conclusively shown that a soil mineral N evaluation is an important component of available soil N in modern agricultural systems, particularly in subhumid areas. Its usefulness in humid areas has also been demonstrated under conditions such as low winter rainfall, deep soils, and large fall NO_3^-−N levels (Maples et al., 1977; Vander Paauw, 1962, 1963).

b. Soil Total Nitrogen Content—Soil total N tests are aimed at estimating mineralization by determining the size of the organic N pool. Rather than analyze for total N directly, most laboratories determine soil organic matter, which contains about 5% N and is an easier and quicker determination. Because a total analysis is involved, these procedures are not greatly affected by sample preparation or time of sample collection. Disadvantages of total analysis procedures stem from the fact that only a portion of the organic N becomes available to the crop during a particular growing season. The portion that becomes available is very difficult to predict because environmental factors such as temperature and moisture influence the amount of mineralization. Also, the addition of fresh organic matter provides a greater proportion of readily available N than the more stable soil N compounds. Nevertheless, several states currently use a total-analysis procedure to estimate mineralization, usually by taking a fixed percentage of the organic matter, which amounts to a 1 to 3% annual mineralization rate (Keeney, 1982; Meisinger, 1984). Thus, although total analyses have not proved useful in making precise estimates of mineralization, they can provide a useful first approximation.

c. Biological Incubating—Incubation tests attempt to separate a more active component of the soil total N pool by employing the soil's own biological agents to oxidize the organic matter under controlled temperature and aeration conditions. Proposed tests employ temperatures ranging from 20 to 40° C and a wide range of aeration conditions that vary from well-aerated sand-soil mixtures to waterlogged anaerobic conditions. The suitability of such a wide range of incubation conditions stems from the fact that the most frequent rate-limiting step in mineralization is the breakdown of organic N to form NH_4^+, i.e., ammonification, which is carried out by many different organisms. Nitrification, however, is greatly affected by aeration so it is necessary to measure both NO_3^-–N and NH_4^+–N produced during the incubation.

Correlations between biological incubation procedures and greenhouse N uptake are usually good once soil NO_3^-–N has been taken into account. However, when testing proceeds to field conditions, the correlations are considerably lower, especially if field trials cover a range of climates, soils, or years (Keeney, 1982; Meisinger, 1984; Stanford, 1982). Biological tests are markedly affected by sample pretreatment, especially those involving short-term aerobic incubations. To obtain meaningful comparisons, it is necessary to carefully standardize sample preparation and incubation conditions. Recent investigations have shown that it may be possible to extend laboratory aerobic mineralization data to field temperature and water regimes, at least within a limited area (Stanford et al., 1977; Smith et al., 1977). This approach is less empirical, but is quite time consuming and is tied to average climatic conditions for predictive purposes. It is being further evaluated as a research approach to estimate mineralization.

The principal drawbacks to biological procedures are sample handling, pretreatment effects, and long processing times (at least 1-week incubations). The currently recommended biological index is the NH_4^+–N produced after 1 week of anaerobic incubation at 40° C (Keeney, 1982). Biological indexes were used in several statewide N recommendations systems 25 yr ago, but are not currently used in any state soil testing laboratory (Keeney, 1982; Meisinger, 1984). Nevertheless, it is generally recognized that biological incubations that measure the total mineral N production (NO_3^-–N plus NH_4^+–N) provide a good measure of the soil's mineralization ability.

d. Chemical Extractions—The goal of chemical extraction procedures is to chemically separate a portion of the active soil N pool from the large total N pool. Advantages of such an extractant would be a simple and rapid analysis, which would be compatible with P and K tests. With these advantages in mind, it is not surprising that soil scientists have searched many years for a suitable extractant. Such a chemical extractant, however, has not been identified because soil N transformations are dominated by biological reactions, which strongly interact with the environment over time at any given location.

The range of extractants tested includes water, weak to strong salt solutions, mild acids and bases, and strong acids and bases with or without an oxidant. The extracted material has likewise been analyzed in a number of ways including C content, total N content, distillable NH_3, and absorption of ultraviolet radiation. The results of those strictly empirical procedures are usually influenced by sample preparation, sample drying, extraction time, and soil/extractant ratios.

Interest in basic and acidic extractions developed as a natural extension of the procedures used to fractionate soil organic matter based on differential solubilities in acids or bases. Various reagents are employed, e.g., sulfuric acid (Purvis & Leo, 1961; Richard et al., 1960), calcium hydroxide [$Ca(OH)_2$] (Prasad, 1965), and sodium hydroxide (NaOH) (Cornfield, 1960). Strong acids or bases attack a large portion of the soil organic N and the N removed by such treatments usually bears a close relation to the soil total N content. Such procedures, thus, have the same disadvantages as discussed above for total N analysis. Milder acidic or basic extractions have been evaluated in several studies, but have performed quite inconsistently (Jenkinson, 1968; Keeney & Bremner, 1966; Stanford, 1978a,b). Oxidizing agents (chromate or permanganate) have also been tested in acidic or basic solutions with the hope that they would selectively oxidize a biologically meaningful fraction of the organic matter (Nommik, 1976; Stanford, 1978a,b). The alkaline permanganate method has been widely tested but has given very inconsistent results (Keeney & Bremner, 1966; Stanford, 1978a; Stanford, 1982). More recently, oxidative procedures have been investigated under acid conditions with encouraging preliminary results (Stanford, 1978b), but the general applicability of these latest procedures is not known.

Most recent research has centered on mild extractants such as water or mild salt solutions, e.g., 0.01 *M* calcium chloride ($CaCl_2$). This approach is based on the observations and procedures of Livens (1959a,b) as modified by Keeney and Bremner (1966) and Smith and Stanford (1971). Keeney and Bremner (1966) considered the total N removed by a boiling water extract to be as good as a biological incubation method; furthermore, they found it was not subject to sample drying effects, which influenced the microbial method. The currently recommended method consists of an overnight (16 h) autoclaving in 0.01 *M* calcium chloride and subsequent determination of the NH_4^+−N released (Keeney, 1982). The method is rapid, is not greatly affected by sample pretreatment, and is amenable to automated analysis. But it has not been extensively evaluated under field conditions, although the field results to date are encouraging (Fox & Piekielek, 1978; Roberts et al., 1972). Workers in the USA are not employing a chemical extraction method to estimate mineralizable N, despite the large research effort that has gone into evaluating various methods.

e. Indirect Methods—Indirect methods involve documenting a local variable, which is linked to an increase in mineralization; examples include N credits for previous legume crops and recent manure additions. Legume N credits are used throughout the USA and range from 55 to 155 kg N/ha for previous alfalfa crops (*Medicago sativa* L.) (depending on stand and years since alfalfa) and from 20 to 55 kg N/ha for prior soybean [*Glycine max* (L.) Merr.] crops (Table 7–4). Manure N credits are about 2.5 kg N/metric ton of manure (approximately 5 pounds N/2000 pounds manure). Indirect methods are easy to use and have generally performed well as initial broad-scale adjustments for the soil N supply.

3. Estimating Nitrogen Uptake Efficiency

The final factor to consider in estimating N fertilizer needs is *N utilization efficiency*. This term must be carefully defined since at least two definitions are in general use (Meisinger, 1984). The first is defined through the N uptake in the aboveground portion of the crop, which commonly varies between 40 and 60%; the second is defined through the N recovered within the whole soil-crop-root system and commonly varies between 65 and 85%. The distinction becomes important when considering methods to estimate these terms and in applying these terms to conceptual soil-crop systems.

Fertilizer efficiencies are functions of several interacting N transformations such as leaching, denitrification, and NH_3 volatilization, as well as several management variables such as N source, placement, and timing. It has not been possible to predict these parameters from first principles. Soil scientists have therefore approached this problem empirically by estimating fertilizer efficiencies over a range of soil and climatic conditions. The end result is that efficiencies can usually be estimated in a general sense, but one cannot accurately predict the efficiency for a specific soil-crop system. Obviously, further research is needed to investigate factors affecting fertilizer efficiency to better predict this parameter for specific locations.

4. Recommendations for Nitrogen Use on Crops

Before a N recommendation system can be established, one must carefully define the boundaries of the soil-crop system in time and space and also assess the sources of variation that underlie varying N fertilizer needs (e.g., crop differences, soil NO_3^-–N levels). Nitrogen recommendation systems should systematically combine the above components of crop N requirements, soil N supply, and N utilization into a working N advisory system that supplies N in quantities sufficient to meet the crops needs and in a manner that leads to high crop utilization. Excessive N inputs should be avoided since they inevitably lead to environmental impacts and inefficient N use.

When defining the soil-crop system, scientists have basically chosen either the *aboveground* approach or the *whole-crop* approach. The former considers the N contained in the grain plus stover along with an estimate of N taken up from the soil. It usually considers an annual time frame. The whole-crop approach considers the N contained in the root-soil complex along with the aboveground N uptake. The whole-crop method is a long-term approach, which deals with N mineralization in relation to N residues returned; if these two are equal the system is at a steady-state condition where the net change in organic N is small. A basic question to answer in designing a N recommendation system is: How close is the soil-crop system to steady-state conditions? (See Meisinger, 1984 for further details.)

Current N recommendation systems differ greatly in local details, but most contain the following general features (see Table 7–4): (i) use of a crop N factor (a combination of the crop N requirement and efficiency term), which considers yield goals, (ii) use of broad-scale soil properties such as soil association and soil drainage class, (iii) use of historical data such as legume N credits or manure credits, (iv) use of soil mineral N content for areas where percolation in small (western USA), (v) occasional use of economic factors such as the cost of N in relation to the price of corn to fine-tune the final N recommendation, and (vi) the general nonuse of an estimate of N mineralization, although there are a few exceptions.

For continuous cereal crops, N fertilizer needs are generally based on yield goals (i.e., crop N removals) as modified by soil NO_3^-–N levels (Table 7–4). These features are characteristic of soil-crop systems near steady-state organic N conditions and may explain why soil N mineralization tests are not in general use. In addition to these nearly steady-state systems, there are also instances of nonsteady-state conditions such as manured sites and systems that oscillate between two soil organic N levels such as legume–cereal crop rotation systems. These nonsteady-state systems are usually accounted for by manure and legume N credits, which are used in virtually every state. Current N recommendation systems can therefore be described as using the long-term, steady-state approach with periodic exceptions for manured and rotated sites.

The most cost-effective N recommendation system will depend heavily on economic considerations, particularly on the cost of N in relation to price of crop product. Current recommendation systems were developed in an era of inexpensive energy, and consequently inexpensive N, and have basically resulted in a long-term, steady-state approximation. Present and future needs will no longer reflect these conditions. Cost-effective N recommendations systems for expensive N will most likely require a more detailed evaluation of each sites' crop N needs, a more accurate and complete evaluation of the soil mineral and organic N status, as well as careful N management (source, timing, and placement) to ensure high N utilization within the soil-crop systems.

B. Response of Crops to Nitrogen

The primary purpose of applying N fertilizer to crops is to increase the yield of dry matter. An adequate supply of N is essential for optimum yields and is usually associated with vigorous vegetative growth and dark green color. Excessive quantities of this element will usually prolong the growth period, retard maturity, and may result in lodging and susceptibil-

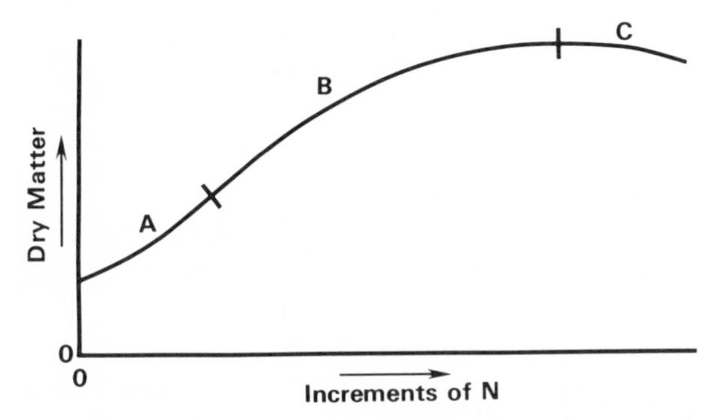

Fig. 7–6. Idealized response in dry matter production of a non-legume to increments of N fertilizer (Viets, 1965).

ity to diseases in certain crops. Crops usually remove from 30 to 70% of the applied N, depending on the crop, yield, and amount of N applied. With good management, uptake efficiency of applied N by corn grain and stover is in the range of 50 to 70% (Stanford, 1973).

An ideal N response curve for nonleguminous crops as developed by Viets (1965) shows: (i) a region where succeeding N fertilizer increments produce successively increased yield (Fig. 7–6, segment A), (ii) a region of diminishing returns that contains the most profitable response to applied N (segment B), and (iii) a region of yield depression if excess N leads to lodging or salt damage (segment C). Many factors influence the crops' N response. Some are not subject to management such as drought, certain diseases, or excess soil moisture. Others are manageable in a long-term sense, such as crop rotations, soil drainage, and grass-legume mixtures for forage production. Still others are subject to short-term management such as N sources, rates and methods of application, and the time of application. These latter factors will be discussed in the following sections.

1. Sources of Nitrogen

Most plant roots in soils absorb N as NO_3^- since that form occurs in higher concentrations than NH_4^+ and is free to move to plant roots, primarily by mass flow. Some NH_4^+ is also present and affects plant growth and metabolism to generally unknown degrees (Reisenaur, 1978). Since NH_4^+ is theoretically more efficiently utilized within the plant and is less subject to leaching and denitrification, it should be the preferred source. However, Arnon (1937) and, more recently, Hageman (1980) concluded that, from physiological considerations, either NO_3^- or NH_4^+ can serve as an adequate source of N for plant growth and productivity, but in general NO_3^- salts are considered the "safer" fertilizer. In addition, studies in Wisconsin (Jung et al., 1972; Schrader et al., 1972) and Indiana (Warncke & Barber, 1973) indicated that corn plants prefer a mixture of NH_4^+ and NO_3^- for maximal growth. Schrader et al. (1972) reported that corn plants absorbed NO_3^- and NH_4^+ at equal rates when both forms were present. These field and greenhouse corn trials showed that a combination of these two forms of N enhances uptake of N, growth, and yield.

The relative effectiveness of N sources for crop production will depend on the manner in which they are applied, the chemistry of the soil, and the moisture regime as related to soil textures.

Numerous workers have shown that ammonium sources that form sparingly soluble reaction products with calcium carbonate are inferior when applied to the surface of calcareous soils (Martin & Chapman, 1951; Fenn & Kissel, 1973; Hargrove & Kissel, 1979; Hargrove et al., 1977; Meyer et al., 1961; and Overrein & Moe, 1967) due to NH_3 volatilization (see section V 2). These losses are of major concern in relation to crop production because the use of ammonium sources has increased greatly since the 1950s or 1960s because of their relative low cost of production (Douglas & Cogswell, 1966).

More recently, urea in granular and N solution forms is rapidly becoming the principal N source for agricultural fertilizers (Bridges, 1980). Urea fertilizers have often been inferior to sources such as ammonium nitrate when applied without incorporation. No-tillage management for crop production, which is gaining in popularity because of its potential for soil and water conservation (Phillips et al., 1980), often precludes the easy incorporation of these N sources. Bandel et al. (1980) reported ammonium nitrate was superior to granular or prilled urea and urea–ammonium nitrate solution for no-till corn on acid soils. However, direct measurements of NH_3 losses from these types of system under field conditions have not been reported.

Under conditions of high rainfall, coarse-textured soils, and significant leaching (percolation) potential, NO_3^- sources are usually inferior. However, nitrification is rapid in good agricultural soils when temperature and moisture are optimum; therefore, any advantage for ammonical N is small. Nitrate sources also are inferior due to denitrification under conditions where soils may be waterlogged for periods of time during the growing season. Inferior crop performance in the first case is the result of NO_3^- being physically removed from the system, especially from sandy areas, while in the second case it results from the chemical reduction of NO_3^- to volatile nitrogenous compounds.

Although not addressed in this section, other sources of N for crop utilization may include animal manures, sewage sludge, crop residues, and other organic forms. These considerations of N source will become more important if N costs increase (due to increasing costs of energy) more than unit value of crop output and as greater attention is focused on the possible adverse effects of N on environmental quality.

2. Rates and Methods of Application

A basic principle related to rates and methods of N fertilization for optimum crop response is that N be present in sufficient quantities at all times to meet the requirements for plant growth. Too little available N limits production and quality; excess N may lower production, reduce quality, cause lodging, increase disease and insect incidents, and may result in toxic levels in the forage relative to animal consumption. Interactions of the above effects complicate the prediction for the proper N fertilizer application rates. Under certain conditions and in some areas, the soil may supply ample N to meet the needs of the crop.

In recent years, N rates and methods of application for crop production have changed. These changes have occurred because of higher yield levels, increased acreage of irrigated areas for crop production, improved cultivars and genotypes, improved application equipment, and modified management systems such as no-tillage and multiple cropping.

Average N rates for corn have increased markedly over the past 20 yr, from about 45 kg N/ha in 1960 to 125 kg N/ha in 1970 and about 145 kg N/ha in 1980. Novoa and Loomis (1981) pointed out that the demand for N is determined by the growth rate and the N composition of the new tissue. Stanford and Hunter (1973) estimated that an average N content of 1.4% was associated with maximum attainable dry matter yields of wheat grain plus straw. They pointed out the remarkable consistency of that value for several wheat varieties. This value for wheat is higher than that previously reported for corn where the uptake of N per unit of dry matter (grain and stover) associated with maximum attainable yield was essentially constant (1.2%) for a wide range of varieties and growing condition (Stanford, 1973). When Stanford and Hunter projected a wheat grain yield of 3360 kg/ha, they calculated the uptake of N required for attaining yields to be 33 g/kg compared with 21 g/kg for corn.

Different crops respond differently to N; therefore, the most profitable rates vary among crops. Generally, the perennial grasses in the humid and subhumid regions of the USA make the most effective use of N. Corn, as it is grown in the USA, efficiently utilizes N up to the flat portion of the N response curve (last part of segment B of Fig. 7–6). With the possible exception of the new short-strawed wheat and rice varieties, which are less susceptible to lodging than the old varieties, the small grains efficiently utilize less N. Part of this is due to the regions where they are grown, but their genetic potential also is involved. Similarly, the relatively short grasses, such as Kentucky bluegrass (*Poa pratensis* L.), make efficient use of lower rates of N than do the tall, cool-season grasses and such relatively new crops as bermudagrass [*Cynodon dactylon* (L.) Pers.] hybrids. Within plant species, improvements in the yield potential have been made in recent years, which permit higher fertilizer rates than were possible previously. For example, corn has been studied quite extensively to show differences in N response among corn genotypes (Beauchamp et al., 1976; Moll & Kamprath 1977; Pollmer et al., 1979; Moll et al., 1982). At low N supply, differences among hybrids for N use efficiency have been shown to be due largely to variation in internal utilization of accumulated N, while at high N levels differences were due largely to variation in uptake efficiency (Moll et al., 1982). They concluded that in breeding for improved N use efficiency, it would seem desirable for both uptake efficiency and utilization efficiency to be improved simultaneously.

Efficiency in uptake and utilization of N in the production of grain requires that those processes associated with absorption, translocation, assimilation, and redistribution of N operate effectively. One management factor that influences these phenomena is placement and/or method of application. Earlier studies have pointed out the importance of proper placement and mixing of ammonium sources of N with the soil to reduce or prevent NH_3 volatilization (Ernst & Massey, 1960; Gasser, 1964). Overrein and Moe (1967) showed a striking decrease in ammonia volatilization when urea was placed 2.5 cm in the soil. They also showed the decrease to be more rapid with depth and with increasing soil moisture.

In addition to reducing or preventing gaseous losses of N, proper placement is necessary to make the element available to plant roots at the desired time. When band placed, the N should be placed in the proper area so the nutrient can be available quickly to plant roots, yet far enough to avoid seedling damage due to high salt or NH_3 levels (Cummins & Parks, 1961). When N is deficient and N fertilizer is placed near the root system, both root and plant growth are improved due to the banding. In a review, Grunes (1959) showed that N placement often increases the availability of P and other nutrients placed in the same band.

On soils that supply little N and those that are subject to significant N losses by leaching or denitrification, the split-application technique is often employed. When N fertilizer costs are high and commodity prices are relatively low, this practice may be desirable to improve crop utilization efficiency.

In experiments from Nebraska, the grain/stover ratio of corn was 1.4 times greater for sidedressed N than fall or spring applications with the same amount of fertilizer N (Olson et al., 1964). Recently, Thomas (1980) concluded from more than 10 yr of studies in Kentucky that delaying at least half of the applied N fertilizer until 4 to 6 weeks after planting ("knee-high" growth states) resulted in improved efficiency for corn production on soils (i) where drainage is less than perfect, (ii) under no-tillage, and (iii) especially with no-tillage on imperfectly drained soils.

The trend in crop production appears to be toward fewer tillage operations for annual crops. With the concern about energy conservation, siltation of our streams, reduced irrigation water usage, and improved moisture regime, this trend should accelerate. Reducing the amount of tillage might have two effects on N supply from the soil. The first is deeper penetration of water and NO_3^- through larger pores in the wetter, mulched soil. The second effect is cooler soil temperatures, lower evaporation, and slower release of N. These conditions can lead to increased N losses through leaching and denitrification (Thomas et al., 1973, 1981; Unger, 1978). Considering all of these factors, it appears that the N requirement for crops grown with reduced tillage will be somewhat greater than for conventional tillage.

In addition to N losses from the soil-plant system, a significant portion of the N applied in no-till or reduced tillage systems can become immobilized in the decaying residue mulch and, thereby, reduce the amount of N available to the growing crop (Doran, 1980). If N fertilizers are broadcast on top of this decaying mulch, it will likely result in greater immobilization than if N is placed into the soil below the mulch. Good fertilizer management practices (rates and methods of application) will therefore be especially important in no-till systems because of the potential N losses and immobilization.

3. Time of Nitrogen Application

Ensuring the existence of an adequate but not excessive supply of available N for crops during the growing season requires good management relative to time of application and is often related to the particular crop-soil system. For soils with sandy texture, where the indigenous N supply is often low and potentials for N losses by leaching are great, the problem is especially apparent. Increased use of irrigation in recent years has influenced the time of N applications for effectiveness in increasing crop yields. Use of irrigation on sandy soils obviously accentuates possible N loss by leaching. However, certain irrigation management systems may provide a convenient vehicle for multiple application of N, resulting in improved efficiency of this element.

In general, reported research suggests that spring application of N is more effective than fall. Spring application is not always superior, but rarely is inferior to fall applications. Olson et al. (1964) reported that summer side-dressing of fertilizer N for row crops, regardless of chemical form, was usually superior to fall or spring applications. Pearson et al. (1961), working with several soil textures in Alabama, Georgia, and Mississippi, found that N broadcast in November or December was only 49% as effective in increasing corn yields as the corresponding spring application. Fall application of N to corn was < 50% as effective as applied at planting time on two wet soils in western Kentucky (Miller et al., 1975). Jung et al. (1972) found N applied to corn 5 to 8 weeks after planting in Wisconsin was the most effective period of application as shown by increased grain and tissue yield. Nitrogen applied after week 8 showed a distinct reduction in N uptake and yields. In Illinois, fall-applied N was 80 to 90% as effective as spring applications for corn at rates of 67 and 134 kg/ha, but about equally effective at rates of 200 and 268 kg/ha (Welch et al., 1971). Boswell (1971) compared the effects of N plowed down (166 kg/ha) in the fall with winter or spring plowed down N and sidedress N on yields of corn and cotton in Piedmont and Coastal Plain soils of Georgia. No differences in yield were obtained on the heavier Piedmont soil, but on light-textured Coastal Plain soils, yields from N side-dressed in the spring were significantly higher than fall plow-down N.

Timing of N applications for wheat is also important. Hamid (1972) showed that the maximum recovery of fertilizer N in the grain occurred when the N was applied at tillering rather than at earlier or later stages. Earlier work by Olson and Rhodes (1953) indicated similar conclusions. Numerous workers found the relative efficiency of N applied in the fall or winter period was inferior to N applied in the spring near the tillering period (Stanford & Hunter, 1973; Welch et al., 1966; Boswell et al., 1976; Doll, 1962).

Most crops require a continuing supply of N throughout the growing season and this need will vary as to stage of crop maturity. For example, the greatest N uptake period by the corn plant is at 40 to 50% maturity. However, the need for N by the corn plant continues until the corn plant approaches maturity.

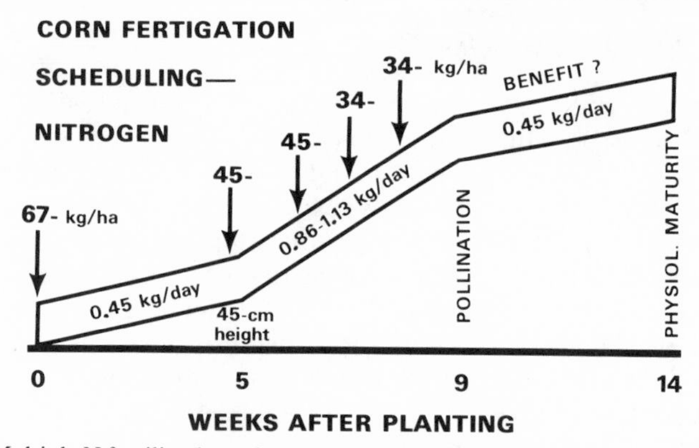

Fig. 7–7. Multiple N fertilization scheme and daily N requirements of corn (Wesley, 1979).

Theoretically, it would be most desirable to add N as closely as possible to the time of maximum uptake by the plant during the growing season. This is especially true if N is applied by fertigation, the practice of applying N through irrigation systems. For example, 225 kg N/ha is recommended for corn yields of 13×10^3 L/ha in Georgia (Wesley, 1979). To increase the N recovery efficiency at this high rate, multiple applications are recommended. If injection equipment is available, four to five applications of N throughout the growing season will increase the N efficiency (Fig. 7–7), based on daily N requirements of corn.

Interest has developed in the timing and splitting of applications of N for forage production. Cool-season grasses normally make a tremendous amount of growth in the spring, and this is stimulated by fertilizing with N. The grass is so efficient in absorbing N that even heavy applications of N in the spring will have a relatively small effect on the growth in the fall. It has been found that two N applications, one early in the spring and one in mid- to late summer, bring about higher yields and better distribution of forage production for grazing. A further subdivision of the first application providing a third application after the first grazing cycle or hay removal further improves the production distribution of the sward.

4. Efficiency of Nitrogen Use by Plants

Response of crops to applied N is an indication of plant efficiency. There are several ways in which efficiency can be considered. The most direct method is to calculate the fertilizer N sorbed by subtracting total N in unfertilized plants from total N in fertilized plants and dividing this value by the N added. Some of the processes affecting efficiency of N use are crop species, soil moisture control, sources of N, rates and time of application, and methods of application. These may, in turn, depend on other factors such as leaching, denitrification, soil N mineralization rate, and past history, which may influence residual N levels.

Grassland has the reputation of being highly efficient in use of N, while efficiencies are relatively low for potatoes (*Solanum tuberosum* L.) and sugarbeets (*Beta vulgaris* L.) (Tinker, 1979). These species may be difficult to compare since, in general, questions of N efficiency revolve around the economically useful parts of the crop. It is quite rare for the amount of N in the roots to be measured; however, if such were practical, estimates of N uptake should therefore be increased. Root functions as related to efficiency may be important where heavy rates of N are necessary for maximum yield.

Pesek et al. (1971) discussed the internal efficiency of plant communities as related to N utilization and pointed out that all individuals of the community of intertilled crops tend to draw on the same environmental resources simultaneously. These efficiencies can be related to plant populations, nature of the root zone as related to moisture, and factors such as sunlight or CO_2 supply.

While a strong effort is made to ensure that plant communities contain only one species of plants, weeds do frequently invade to compete with the crop plants for nutrients and other growth factors. Even modest weed infestations can cause significant reductions in yields with conventional cultivation and no-till systems (Kapusta, 1979; Moomaw & Martin, 1978; Staniforth, 1957). At least part of this competition appears to be for N and other fertilizer elements. Efficiency of N utilization of the crop is, therefore, reduced, and somewhat more N is required to have the same effect on the crop.

5. Legumes as a Source of Nitrogen

Prior to the advent of inexpensive fertilizer N, legumes were often included in rotations to supply N for nonlegume crop production. This practice dates back to ancient times and even early years of the Roman Republic (Pieters & McKee, 1938). Due to increased cost of fertilizer N and diminishing energy resources, renewed interest has developed in legumes as a N source as well as a rotation crop, especially in no-till systems (Ebelhar & Frye, 1981; Mitchell & Teel, 1977; Adams et al., 1970; Hargrove, 1982; Triplett et al., 1979; Touchton et al., 1982). Hairy vetch (*Vicia villosa* Roth) has been reported to supply from 150 to 225 kg N/ha, varying with the amount of top growth. Crimson clover (*Trifolium incarnatum* L.) contains less N than hairy vetch (Mitchell & Teel, 1977). They found that hairy vetch and crimson clover mixtures produced corn grain yields comparable to those obtained by the application of 112 kg N/ha as inorganic fertilizer. They also reported that approximately one-third of the total N from the mulch covers were released to the corn in a single season. The remaining N would be expected to be released at a lower rate during subsequent years.

Although legumes have been utilized over a long period of time to boost N levels for nonlegume plant production, numerous management aspects need additional research. For example, one of the primary problems with winter legumes as a N source is the cost associated with establishment. Development of early maturity species is needed for multiple cropping systems. Information on pest control for the legume and the influence of the legume cover crop on diseases, nematode, and insect damage to subsequent crops is not fully understood. Additional studies are needed to develop cropping systems in which legumes, as the N source, may be used most effectively.

VII. CONCLUSIONS

Since 1955, N consumption in the USA has increased almost linearly. The largest total N use increases occurred in the periods of 1965 to 1970 and 1975 to 1980. During the last 5-yr period, N costs have increased appreciably due to current diminishing energy resources. Increasing energy costs have underscored the need for improving N uptake efficiencies.

Areas of high energy requirement as related to N are manufacturing and distributing. Nitrogen fertilizers are highly energy-intensive and require energy both as feedstock (source of H for NH_3) and fuel (including steam). Most of the direct and indirect energy uses in the fertilizer sector depends primarily on nonrenewable hydrocarbons. Despite the high energy input for N fertilizer, the increased energy in crop output is often greater than used for N production.

After the N fertilizer is manufactured, commercial energy is also required for transport preparation, for transportation to the farm level, and for application to the crop. However, the total amount of energy required for distribution and application is rather small relative to its manufacture. The major share of energy required for distribution is in transportation of fertilizer products and raw materials. This transportation is usually by railcars, river barges, trucks, and/or pipelines. The pipeline is currently acting as the long-distance transporter for anhydrous NH_3 and N solutions. Even though this means of transportation is in its infancy, it is expected to increase with usage of these products.

When N sources were available at lower costs, marketing policies were related to production capacity and, through sales and promotion, profits were attained via volume. Current management policies call for the marketing focus to be toward the farm customer's satisfaction and thus attain positive margins through integrated marketing profits.

About 90% of the fertilizer N used in the USA is in the form of ammonia- or ammonium-producing compounds; about 65% of this amount is anhydrous NH_3. The remainder is chiefly accounted for by nitrate, primarily from ammonium nitrate. Improving N fertilizer effectiveness requires an understanding of the factors governing N transformations in soils. Added fertilizer N becomes intimately involved in the soil N cycle, which is composed of many complex biological, physical, and chemical processes. The fertilizer N may be lost from the soil-crop system as a gas or in percolating water, it may be converted to slowly available organic compounds, or it may be taken up by crops. Ammonium fertilizers are subject to substantial NH_3 volatilization losses if they are surface-applied and are in an alkaline environment. Ammonium fertilizers will usually be readily converted to nitrate within several weeks, unless nitrification is retarded due to chemical inhibitors or environmental conditions. During the ensuing weeks the NO_3^-–N may be utilized by plants, denitrified, or leached, but a vigorously growing crop is generally the major sink for fertilizer N during the growing season. Any excess N remaining in the soil beyond the growing season is subject to leaching and/or denitrification losses in humid climates. In dry climates, the excess N will likely remain in the soil as NO_3^-–N.

Soil scientists have studied the behavior of fertilizer N in soils intensively since the 1950s or 1960s to get more fertilizer into the crop. Crop N utilization can be improved through practices such as liming, soil tillage, water management, and crop rotations, but improvements in N uptake efficiency are most readily accomplished through fertilizer management practices such as source, rate, time, and method of application. High N use efficiencies require a combination of all these factors; a given N source should be applied to minimize gaseous losses and immobilization, at a rate that does not exceed the crop assimilation capacity, and at a time in phase with crop demand.

Although much research effort has gone into estimating N uptake efficiencies by the plant, additional studies are needed to investigate factors affecting efficiency to better predict this parameter for specific locations. Such studies require careful definition of the boundaries of the soil-crop system in time and space and also assess the sources of variation that underlie varying N fertilizer needs (e.g., crop differences and soil NO_3^-–N levels).

The most cost-effective N recommendation system will depend heavily on economic consideration, particularly on the cost of N in relation to product price. Current recommendation systems were developed in an era of inexpensive energy, and consequently inexpensive N, and have basically resulted in a long-term, steady-state approximation. Present and future needs will no longer reflect these conditions. Cost-effective N recommendation systems for expensive N will most likely require a more detailed evaluation of each sites' crop N needs, a more accurate and complete evaluation of the soil mineral and organic N status, as well as careful N management (source, timing, and placement) to ensure high N utilization within the soil-crop system.

Even though urea use almost doubled from 1975 to 1980, it is predicted that its use will continue to increase rapidly. Additional information is needed relative to its use and efficiency. With increased costs, it becomes more important to improve uptake efficiency and provide "prescription-type" recommendations for all N fertilizer sources.

Although legumes have been utilized over a long period of time to boost N levels in the soil for nonleguminous plants, numerous management aspects need additional evaluation. Development of cropping systems in which legumes may be used more effectively as N sources still must be studied.

We should continue to be aware of the importance of N fertilizers in producing adequate food of superior quality to feed the growing population of the USA and the world. Biological scientists will have to be sure that N fertilizer is used efficiently and responsibly.

REFERENCES

Adams, W. E., H. D. Morris, and R. N. Dawson. 1970. Effect of cropping systems and nitrogen levels on corn *(Zea mays)* yields in the southern Piedmont region. Agron. J. 62:655–659.

Alexander, M. 1977. Introduction to soil microbiology, 2nd ed. John Wiley & Sons, Inc., New York.

Allison, F. E. 1955. The enigma of soil nitrogen balance sheets. Adv. Agron. 7:213–250.

Allison, F. E. 1973. Soil organic matter and its role in crop production. Elsevier Science Publishing Co., Inc., New York.

Amer, F. M., and W. V. Bartholomew. 1951. Influence of oxygen concentration in soil air on nitrification. Soil Sci. 71:215–219.

Anonymous. 1963. Toyo Koatsu urea processes with latest progress. Bull. 4. Toyo Koatsu Industries, Inc., Tokyo, Japan.

Arnon, D. I. 1937. Ammonium and nitrate nitrogen nutrition of barley at different seasons in relation to hydrogen-ion concentrations, manganese, copper and oxygen supply. Soil Sci. 44:91–120.

Axelrod, L. C., and T. E. O'Hare. 1964. Production of synthetic ammonia. p. 58–88. *In* V. Sauchelli (ed.) Fertilizer nitrogen—its chemistry and technology. Reinhold Publishing Corporation, New York.

Bandel, V. A., S. Dzienia, and G. Stanford. 1980. Comparison of N fertilizers for no-till corn. Agron. J. 72:337–341.

Beauchamp, E. G., L. W. Kannenberg, and R. B. Hunter. 1976. Nitrogen accumulation and translocation in corn genotypes following silking. Agron. J. 68:418–422.

Bezdicek, D. F., J. M. MacGregor, and W. P. Martin. 1971. The influence of soil-fertilizer geometry on nitrification and nitrite accumulation. Soil Sci. Soc. Am. Proc. 35:997–1002.

Black, C. A. 1968. Soil-Plant Relationships, 2nd ed. John Wiley & Sons, Inc., New York.

Blackmer, A. M., J. M. Bremner, and E. L. Schmidt. 1980. Production of nitrous oxide by ammonium-oxidizing chemoautotrophic microorganisms in soil. Appl. Environ. Microbiol. 40:1060–1066.

Blevins, R. L., G. W. Thomas, and P. L. Cornelius. 1977. Influence of no-tillage and nitrogen fertilization on certain soil properties after 5 years of continuous corn. Agron. J. 69:383–386.

Boswell, F. C. 1971. Comparison of fall, winter or spring N–P–K fertilizer applications for corn and cotton. Agron. J. 63:905–907.

Boswell, F. C., and O. E. Anderson. 1964. Nitrogen movement in undisturbed profiles of fallowed soils. Agron. J. 56:278–281.

Boswell, F. C., L. R. Nelson, and M. J. Bitzer. 1976. Nitrification inhibitor with fall-applied vs split nitrogen applications for winter wheat. Agron. J. 68:737–740.

Bremner, J. M., and A. M. Blackmer. 1978. Nitrous oxide: Emission from soils during nitrification of fertilizer nitrogen. Science (Washington, D.C.) 199:295–296.

Bremner, J. M., G. A. Breitenback, and A. M. Blackmer. 1981. Effect of anhydrous ammonia fertilizer on emission of nitrous oxide from soils. J. Environ. Qual. 10:77–80.

Bremner, J. M., and D. W. Nelson. 1968. Chemical decomposition of nitrite in soils. Trans. Int. Congr. Soil Sci. 9th 1968 2:495–503.

Bremner, J. M., and K. Shaw. 1958a. Denitrification in soil. I. Methods of investigation. J. Agric. Sci. 51:22–39.

Bremner, J. M., and K. Shaw. 1958b. Denitrification in soil. II. Factors affecting denitrification. J. Agric. Sci. 51:40–52.

Bridges, J. D. 1980. Fertilizers trends—1979. Bull. Y-150. National Fertilizer Development Center, Muscle Shoals, AL.

Broadbent. F. E., and A. B. Carlton. 1978. Field trials with isotopically labeled nitrogen fertilizer. p. 1–41. *In* D. R. Nielsen and J. G. MacDonald (ed.) Nitrogen in the environment, Vol. 1. Nitrogen behavior in field soil. Academic Press, Inc., New York.

Broadbent, F. E., and F. E. Clark. 1965. Denitrification. *In* W. V. Bartholomew and F. E. Clark (ed.) Soil nitrogen. Agronomy 10:347–362.

Broadbent, F. E., and F. J. Stevenson. 1966. Organic matter interactions. p. 169–187. *In* M. H. McVicker et al. (ed.) Agricultural anhydrous ammonia technology and use. American Society of Agronomy, Madison, WI.

Broadbent, F. E., K. B. Tyler, and G. N. Hill. 1957. Nitrification of ammonical fertilizers in some California soils. Hilgardia 27:247–267.

Burford, J. R., and J. M. Bremner. 1975. Relationships between the denitrification capacities of soils and total water soluble and readily decomposable soil organic matter. Soil Biol. Biochem. 7:389–394.

Carter, J. N., O. L. Bennett, and R. W. Pearson. 1967. Recovery of fertilizer nitrogen under field conditions using nitrogen-15. Soil Sci. Soc. Am. Proc. 31:50–56.

Carter, J. N., D. T. Westermann, and M. E. Jensen. 1976. Sugarbeet yield and quality as affected by nitrogen level. Agron. J. 68:49–55.

Cervinka, V. 1980. Fuel and energy efficiency. p. 15–21. *In* D. Pimental (ed.) Handbook of energy utilization in agriculture. CRC Press, Inc., Boca Raton, FL.

Chichester, F. W. 1977. Effect of increased fertilizer rates on nitrogen content of runoff and percolate from monolith lysimeters. J. Environ. Qual. 6:211–216.

Chichester, F. W., and S. J. Smith. 1978. Disposition of ^{15}N labeled fertilizer nitrate applied during corn culture in field lysimeters. J. Environ. Qual. 7:227–233.

Cho, C. M., L. Sakdinan, and C. Chang. 1979. Denitrification intensity and capacity of three irrigated Alberta soils. Soil Sci. Soc. Am. J. 43:945–950.

Cornfield, A. H. 1960. Ammonia released on treating soils with *N* sodium hydroxide as a possible means of predicting the nitrogen supplying power of soils. Nature (London) 187:260–261.

Craswell, E. T., and W. M. Strong. 1976. Isotopic studies of the N balance in a cracking clay. III. Nitrogen recovery in plant and soil in relation to the depth of fertilizer addition and rainfall. Aust. J. Soil Res. 14:75–83.

Creamer, F. L., and R. H. Fox. 1980. Toxicity of banded urea or diammonium phosphate to corn as influenced by soil temperature, moisture, and pH. Soil Sci. Soc. Am. J. 44:296–300.

Cummins, D. G., and W. L. Parks. 1961. The germination of corn and wheat as affected by various fertilizer salts at different soil temperatures. Soil Sci. Soc. Am. Proc. 25:47–49.

Doll, E. C. 1962. Effects of fall-applied nitrogen fertilizer and winter rainfall on yield of wheat. Agron. J. 54:471–473.

Doran, J. W. 1980. Soil microbial and biochemical changes associated with reduced tillage. Soil Sci. Soc. Am. J. 44:765–771.

Douglas, J. K., Jr. and S. A. Cogswell. 1966. Past, present, and future production and use of anhydrous ammonia. p. 73–100. *In* M. H. McVicker et al. (ed.) Agricultural anhydrous ammonia technology and use. American Society of Agronomy, Madison, WI.

Ebelhar, S. A., and W. W. Frye. 1981. Legumes boost nitrogen for no-till corn. Crops Soils 34:10–11.

Ekpete, D. M., and A. H. Cornfield. 1965. Effect of pH and addition of organic materials on denitrification losses from soil. Nature (London) 208:1200.

Ernst, J. W., and H. F. Massey. 1960. The effects of several factors on volatilization of ammonia formed from urea in the soil. Soil Sci. Soc. Am. Proc. 24:87–90.

Fenn, L. B., and D. E. Kissel. 1973. Ammonia volatilization from surface applications of ammonium compounds on calcareous soils: I. General theory. Soil Sci. Soc. Am. Proc. 37:855–859.

Firestone, M. K. 1982. Biological denitrification. *In* F. J. Stevenson et al. (ed.) Nitrogen in agricultural soils. Agronomy 22:289–326.

Fox, R. H., and W. P. Piekielek. 1978. Field testing of several nitrogen availability indexes. Soil Sci. Soc. Am. J. 42:747–750.

Frederick, L. R., and F. E. Broadbent. 1966. Biological interactions. p. 198–212. *In* M. H. McVicker et al. (ed.) Agricultural anhydrous ammonia technology and use. American Society of Agronomy, Madison, WI.

Freney, J. R., and J. R. Simpson. 1981. Ammonia volatilization. *In* F. E. Clark and T. Rosswall (ed.) Terrestrial nitrogen cycle processes, ecosystem strategies and management impacts. Ecol. Bull. 33:291–302.

Gaser, J. K. R. 1964. Some factors affecting losses of ammonia from urea and ammonium sulfate applied to soils. J. Soil Sci. 15:258–272.

Gilliam, J. W., and R. P. Gambrell. 1978. Temperature and pH as limiting factors in loss of nitrate from saturated Atlantic Coastal Plain soils. J. Environ. Qual. 7:526–532.

Greenwood, D. J. 1962. Nitrification and nitrate dissimilation in soil. Plant Soil 17:365–391.

Grunes, D. L. 1959. Effect of nitrogen on the availability of soil and fertilizer phosphorus to plants. Adv. Agron. 11:369–396.

Hageman, R. H. 1980. Effect of form of nitrogen on plant growth. p. 47–62. *In* J. J. Meisinger et al. (ed.) Nitrification inhibitors—potentials and limitations. Spec. Pub. 38. Soil Science Society of America, Madison, WI.

Hamid, A. 1972. Efficiency of N uptake by wheat as affected by time and rate of application using ^{15}N labelled ammonium sulphate and sodium nitrate. Plant Soil 37:389–394.

Hargrove, W. L. 1982. Proceedings of the minisymposium on legume cover crops for conservation tillage production systems. Univ. of Georgia College of Agric. Exp. Stn. Spec. Pub. 19. p. 21.

Hargrove, W. L., and D. E. Kissel. 1979. Ammonia volatilization from surface applications of urea in the field and laboratory. Soil Sci. Soc. Am. J. 43:359–363.

Hargrove, W. L., D. E. Kissel, and L. B. Fenn. 1977. Field measurements of ammonia volatilization from surface applications of ammonium salts to a calcareous soil. Agron. J. 69:473–476.

Hauck, R. D. 1981 Nitrogen fertilizer effects on nitrogen cycle processes. *In* F. E. Clark and T. Rosswall (ed.) Terrestrial nitrogen cycles, processes, ecosystem strategies and management impacts. Ecol. Bull. 33:551–562.

Hauck, R. D. 1981. Nitrogen fertilizer effects on nitrogen cycle processes. *In* F. E. Clark and nitrogen source, size and pH of the granule, and concentration. J. Agric. Food Chem. 13:486–492.

Hergert, G. W., and R. A. Wiese. 1980. Performance of nitrification inhibitors in the Midwest (west). p. 89–105. *In* J. J. Meisinger et al. (ed.) Nitrification inhibitors—potentials and limitations. Spec. Pub. 38. American Society of Agronomy, Soil Science Society of America, Madison, WI.

Herron, G. M., G. L. Terman, A. F. Dreier, and R. A. Olson. 1968. Residual nitrate nitrogen in fertilized deep loess-derived soils. Agron. J. 60:477–482.

Jenkinson, D. S. 1968. Chemical tests for potentially available nitrogen in soil. J. Sci. Food Agric. 19:160–168.

Jung, P. E., Jr., L. A. Peterson, and L. E. Schrader. 1972. Response of irrigated corn to time, rate and source of applied N on sandy soils. Agron. J. 64:668–670.

Kapusta, G. 1979. Seedbed tillage and herbicide influence on soybean *(Glycine max)* weed control and yield. Weed Sci. 27:520–526.

Keeney, D. R. 1982. Nitrogen-availability indices. *In* A. L. Page et al. (ed.) Methods of soil analysis, part 2. Agronomy 9:711–733.

Keeney, D. R., and J. M. Bremner. 1966. Comparison and evaluation of laboratory methods of obtaining an index of soil nitrogen availability. Agron. J. 58:498–503.

Kowalenko, C. G. 1978. Nitrogen transformations and transport over 17 months in field fallow microplots using ^{15}N. Can. J. Soil Sci. 58:69–76.

Kowalenko, C. G., and D. R. Cameron. 1978. Nitrogen transformations in soil-plant systems in three years of field experiments using tracer and non-tracer methods on an ammonium-fixing soil. Can. J. Soil Sci. 58:195–208.

Knowles, R. 1981. Denitrification. p. 323–369. *In* E. A. Paul and J. N. Ladd (ed.) Soil biochemistry, Vol. 5. Marcel Dekker, Inc. New York.

Legg, J. O., and J. J. Meisinger. 1982. Soil nitrogen budgets. *In* F. J. Stevenson et al. (ed.) Nitrogen in agricultural soils. Agronomy 22:503–566.

Livens, J. 1959a. Contribution to a study of mineralizable nitrogen in soil. (In French.) Agricultura 7:27–44.

Livens, J. 1959b. Studies concerning ammoniacal and organic soil nitrogen soluble in water. (In French.) Agricultura 7:519–532.

Lockeretz, W. 1980. Energy inputs for nitrogen, phosphorus, and potash fertilizers. p. 23–24. *In* D. Pimental (ed.) Handbook of energy utilization in agriculture. CRC Press, Inc., Boca Raton, FL.

McGarity, J. W. 1961. Denitrification studies on some south Australian soil. Plant Soil 14:1–21.

Maples, R., J. G. Koegh, and W. E. Sabbe. 1977. Nitrate monitoring for cotton production in Loring-Calloway silt loam. Arkansas Agric. Exp. Stn. Bull. 825.

Martin, J. P., and H. D. Chapman. 1951. Volatilization of ammonia from surface fertilized soils. Soil Sci. 71:25–34.

Meisinger, J. J. 1984. Evaluating plant-available nitrogen in soil-crop systems. p. 391–416. *In* R. D. Hauck (ed.) Nitrogen in crop production. American Society of Agronomy, Madison, WI.

Meyer, R. D., R. A. Olson, and H. F. Rhoades. 1961. Ammonia losses from fertilized Nebraska soils. Agron. J. 53:241–244.

Mikkelsen, D. S., S. K. DeDatta, and W. N. Obeemea. 1978. Ammonia volatilization losses from flooded rice soils. Soil Sci. Soc. Am. J. 42:725–730.

Miller, H. F., J. Kavanaugh, and G. W. Thomas. 1975. Time of nitrogen application and yields of corn in wet, alluvial soils. Agron. J. 67:401–404.

Mitchell, W. H., and M. R. Teel. 1977. Winter-annual cover crops for no-tillage corn production. Agron. J. 69:569–573.

Moll, R. H., and E. J. Kamprath. 1977. Effect of population density upon agronomic traits associated with genetic increases in yield of *Zea mays* L. Agron. J. 69:81–84.

Moll, R. H., E. J. Kamprath, and W. A. Jackson. 1982. Analysis and interpretation of factors which contribute to efficiency of nitrogen utilization. Agron. J. 74:562–564.

Moomaw, R. S., and A. R. Martin. 1978. Weed control in reduced tillage corn production systems. Agron. J. 70:91–94.

Mortland, M. M. 1966. Ammonia interactions with soil minerals. p. 188–197. *In* M. H. McVicker et al. (ed.) Agricultural anhydrous ammonia technology and use. American Society of Agronomy, Madison, WI.

Nelson, D. W. 1982. Gaseous losses of nitrogen other than through denitrification. *In* F. J. Stevenson et al. (ed.) Nitrogen in agricultural soils. Agronomy 22:327–363.

Nelson, D. W., and J. M. Bremner. 1969. Factors affecting chemical transformations of nitrite in soils. Soil Biol. Biochem. 1:229–239.

Nelson, D. W., and J. M. Bremner. 1970. Gaseous products of nitrite decomposition in soil. Soil Biol. Biochem. 2:203–215.

Nelson, D. W., and D. M. Huber. 1980. Performance of nitrification inhibitors in the Midwest (east). p. 75–88. *In* J. J. Meisinger et al. (ed.) Nitrification inhibitors—potentials and limitations. Spec. Pub. 38. American Society of Agronomy, Soil Science Society of America, Madison, WI.

Nelson, L. B., and R. E. Uhland. 1955. Factors that influence loss of fall-applied fertilizers and their probable importance in different sections of the United States. Soil Sci. Soc. Am. Proc. 19:492–496.

Nommik, H. 1966. Particle-size effects on the rate of nitrification of nitrogen fertilizer materials, with special reference to ammonium-fixing soils. Plant Soil 24:181–200.

Nommik, H. 1976. Predicting the nitrogen-supplying power of acid forest soils from data on release of CO_2 and NH_3 on partial oxidation. Commun. Soil Sci. Plant Anal. 7:569–584.

Nommik, H. 1981. Fixation and biological availability of ammonium on soil clay minerals. *In* F. E. Clark and T. Rosswall (ed.) Terrestrial nitrogen cycles, processes, ecosystem strategies and management impacts. Ecol. Bull. 33:273–279.

North Central Region-13 Soil Testing Committee. 1975. Recommended chemical soil test procedures for the North Central Region. North Dakota Agric. Exp. Stn. Bull. 499.

Novoa, R., and R. S. Loomis. 1981. Nitrogen and plant production. Plant Soil 58:177–204.

Olson, R. A., A. F. Dreier, C. Thompson, and P. H. Grabousk. 1964. Using fertilizer nitrogen effectively on grain crops. Nebraska Agric. Exp. Stn. Res. Bull. SB 479.

Olson, R. A., and H. F. Rhoades. 1953. Commercial fertilizers for winter wheat in relation to the properties of Nebraska soils. Nebraska Agric. Exp. Stn. Res. Bull. 172.

Overrein, L. N., and P. G. Moe. 1967. Factors affecting urea hydrolysis and ammonia volatilization in soil. Soil Sci. Soc. Am. Proc. 31:57–61.

Pang, P. C., R. A. Hedlin, and C. M. Cho. 1973. Transformation and movement of band-applied urea, ammonium sulfate, and ammonium hydroxide during incubation in several Manitoba soils. Can. J. Soil Sci. 53:331–341.

Parr, J. F., and R. I. Papendick. 1966. Retention of ammonia in soils. p. 213–236. *In* M. H. McVicker et al. (ed.) Agricultural anhydrous ammonia technology and use. American Society of Agronomy, Madison, WI.

Pearson, R. W., H. V. Jordon, O. L. Bennett, C. E. Scarsbrook, W. E. Adams, and A. W. White. 1961. Residual effects of fall- and spring-applied nitrogen fertilizers on crop yields in the southeastern United States. USDA Tech Bull. 1254. U.S. Government Printing Office, Washington, DC.

Pesek, John, G. Stanford, and N. L. Case. 1971. Nitrogen production and use. p. 217–269. *In* R. A. Olson et al. (ed.) Fertilizer technology and use, 2nd ed. Soil Science Society of America, Madison, WI.

Phillips, R. E., R. L. Blevins, G. W. Thomas, W. W. Frye, and S. H. Phillips. 1980. No-tillage agriculture. Science (Washington, D.C.) 208:1108–1113.

Pierre, W. H. 1933. A method for determining the acid- or base-forming property of fertilizer and the production of nonacid-forming fertilizers. Am. Fert. 79:5–8, 24, 26–27.

Pierre, W. H., and W. L. Banwart. 1973. Excess-base and excess base/nitrogen ratio of various crop species and parts of plants. Agron. J. 65:91–96.

Pierre, W. H., J. J. Meisinger, and J. R. Birchett. 1970. Cation-anion balance in crops as a factor in determining the effect of nitrogen fertilizer on soil acidity. Agron. J. 62:106–112.

Pierre, W. H., J. R. Webb, and W. D. Shrader. 1971. Quantitative effect of nitrogen fertilizer on the development and downward movement of soil acidity in relation to level of fertilization and crop removal in a continuous corn cropping system. Agron. J. 63:291–297.

Pieters, A. J., and R. McKee. 1938. The use of cover and green manure crops. Soils and men. p. 431–444. *In* USDA Yearbook of Agriculture U.S. Government Printing Office, Washington, DC.

Pollmer, W. G., D. Eberhard, D. Klein, and B. S. Dhillow. 1979. Genetic control of nitrogen uptake and translocation in maize. Crop Sci. 19:82–86.

Prasad, R. 1965. Determination of potentially available nitrogen in soil—a rapid procedure. Plant Soil 23:261–264.

Purvis, E. R., and M. W. M. Leo. 1961. Rapid procedure for estimating potentially available soil nitrogen under greenhouse conditions. Agric. Food Chem. 9:15–17.

Reed, R. M., and J. C. Reynolds. 1973. Spherodizer granulation process. Chem. Eng. Prog. 69(2):62–66.

Reisenauer, H. M. 1978. Absorption nad utilization of ammonium nitrogen by plants. p. 157–170. *In* D. R. Nielsen and J. G. MacDonald (ed.) Nitrogen in the environment, Vol. 2. Academic Press, Inc., New York.

Richard, R. A., O. J. Attoe, S. Moskal, and E. Truog. 1960. A chemical method for determining available soil nitrogen. Trans. Int. Cong. Soil Sci. 7th 1960 23:261–264.

Riemenschneider, W. 1976. Cyanogen or oxamide from HCN in one step. CHEMTECH 6:658–661.

Roberts, S., A. W. Richards, M. G. Day, and W. H. Weaver. 1972. Predicting sugar content and petiole nitrate of sugarbeets from soil measurements of nitrate and mineralizable nitrogen. J. Am. Soc. Sugar Beet Technol. 17:126–133.

Rolston, D. E., D. L. Hoffman, and D. W. Toy. 1978. Field measurement of denitrification. I. Flux of N_2 and N_2O. Soil Sci. Soc. Am. J. 42:863–869.

Russell, E. W. 1961. Soil conditions and plant growth, 9th ed. John Wiley & Sons, Inc., New York.

Ryden, J. C., and L. J. Lund. 1980. Nature and extent of directly measured denitrification losses from some irrigated vegetable crop production units. Soil Sci. Soc. Am. J. 44:505–511.

Sabey, B. R., L. R. Fredereck, and W. V. Bartholomew. 1959. The formation of nitrate from ammonium nitrogen in soils: III. Influence of temperature and initial population of nitrifying organisms on the maximum rate and delay period. Soil Sci. Soc. Am. Proc. 23:462–465.

Sauchelli, V. (ed.). 1960. Chemistry and technology of fertilizers. ACS Monogr. Ser. 148. Reinhold Publishing Corporation, New York.

Sauchelli, V. (ed.). 1964. Fertilizer nitrogen. ACS Monogr. Ser. 161. Reinhold Publishing Corporation, New York.

Scaglione, P. 1963. Recent developments in the production of urea. Chim. Ind. (Milan) 45:1069–1075.

Schrader, L. E., D. Domska, P. E. Jung, Jr., and L. A. Peterson. 1972. Uptake and assimilation of ammonium-N and nitrate-N and their influence on the growth of corn (*Zea mays* L.). Agron. J. 64:690–695.

Sharp, J. C. 1966. Properties of ammonia. p. 21–31. *In* M. H. McVicker et al. (ed.) Agricultural anhydrous ammonia technology and use. American Society of Agronomy, Madison, WI.

Smith, S. J., and G. Stanford. 1971. Evaluation of a chemical index of soil nitrogen availability. Soil Sci. 111:228–232.

Smith, S. J., L. B. Young, and G. E. Miller. 1977. Evaluation of soil nitrogen mineralization potential under modified field conditions. Soil Sci. Soc. Am. J. 41:74–76.

Smith, S. M., and J. M. Tiedje. 1979. The effect of roots on denitrification. Soil Sci. Soc. Am. J. 43:951–955.

Soil Science Society of America. 1984. Glossary of soil science terms, revised ed. Soil Science Society of America, Madison, WI.

Soltanpour, P. N., A. E. Ludwick, and J. O. Reuss. 1979. Guide to fertilizer recommendations in Colorado—soil analysis and computer process. Colorado State Univ. Coop. Ext. Service.

Stanford, G. 1966. Nitrogen requirements of crops for maximum yield. p. 237–257. *In* W. H. McVicker et al. (ed.) Agricultural anhydrous ammonia technology and use. American Society of Agronomy, Madison, WI.

Stanford, G. 1973. Rationale for optimum nitrogen fertilization in corn production. J. Environ. Qual. 2:159–166.

Stanford, G. 1978a. Evaluation of ammonia released by alkaline permanganate extraction as an index of soil nitrogen availability. Soil Sci. 126:244–253.

Stanford, G. 1978b. Oxidative release of potentially mineralizable soil nitrogen by acid permanganate extraction. Soil Sci. 126:210–218.

Stanford, G. 1982. Assessment of soil nitrogen availability. *In* F. J. Stevenson (ed.) Nitrogen in agricultural soils. Agronomy 22:651–688.

Stanford, G., J. N. Carter, D. T. Westermann, and J. J. Meisinger. 1977. Residual nitrate and mineralizable soil nitrogen in relation to nitrogen uptake by irrigated sugarbeets. Agron. J. 69:303–308.

Stanford, G., S. Dzienia, and R. A. VanderPol. 1975a. Effect of temperature on denitrification rate in soils. Soil Sci. Soc. Am. Proc. 39:867–870.

Stanford, G., and A. S. Hunter. 1973. Nitrogen requirements of winter wheat (*Triticum aestivum* L.) varieties 'Blueboy' and 'Redcoat'. Agron. J. 65:442–447.

Stanford, G., R. A. VanderPol, and S. Dzienia. 1975b. Denitrification rates in relation to total and extractable soil carbon. Soil Sci. Soc. Am. Proc. 39:284–289.

Staniforth, D. W. 1957. Effects of annual grass weeds on yield of corn. Agron. J. 49:551–555.

Stefanson, R. C. 1972a. Effect of plant growth and form of nitrogen fertilizer on denitrification from four South Australian soils. Aust. J. Soil Res. 10:183–195.

Stefanson, R. C. 1972b. Soil denitrification in sealed soil-plant systems. I. Effect of plants, soil water content and soil organic matter content. Plant Soil 37:113–127.

Stefanson, R. C. 1972c. Soil denitrification in sealed soil-plant systems. II. Effect of soil water content and form of applied nitrogen. Plant Soil 37:129–140.

Stefanson, R. C. 1972d. Soil denitrification in sealed soil-plant systems. III. Effect of disturbed and undisturbed soil samples. Plant Soil 37:141–149.

Stefanson, R. C. 1976. Denitrification from nitrogen fertilizers placed at various depths in the soil-plant system. Soil Sci. 121:353–363.

Stefanson, R. C., and D. J. Greenland. 1970. Measurement of nitrogen and nitrous oxide evolution from soil-plant systems using sealed growth chambers. Soil Sci. 109:203–206.

Stevenson, C. K., and C. S. Baldwin. 1969. Effect of time and method of nitrogen application and source of nitrogen on the yield and nitrogen content of corn (*Zea mays* L.). Agron. J. 61:381–384.

Stevenson, F. J. 1964. Soil nitrogen. p. 18–39. *In* V. Sauchelli (ed.) Fertilizer nitrogen, its chemistry and technology. Reinhold Publishing Corporation, New York.

Tennessee Valley Authority. 1966. New developments in fertilizer technology. Sixth demonstration, 4–5 Oct. 1966. Tennessee Valley Authority, Muscle Shoals, AL.

Tennessee Valley Authority. 1968. New developments in fertilizer technology. Seventh demonstration, 1–2 Oct. 1968. Tennessee Valley Authority, Muscle Shoals, AL.

Tennessee Valley Authority. 1979. Sulfur coated urea. TVA Tech. Update Z102. Tennessee Valley Authority. Muscle Shoals, AL.

Tennessee Valley Authority. 1980 TVA development of pan granulation processes for nitrogen fertilizers. TVA Bull. Y-160. Tennessee Valley Authority, Muscle Shoals, AL.

Tennessee Valley Authority. 1982. Certain granulation process. TVA Tech. Update Z-129. Tennessee Valley Authority, Muscle Shoals, AL.

Terman, G. L. 1979. Volatilization losses of nitrogen as ammonia from surface-applied fertilizers, organic amendments, and crop residues. Adv. Agron. 31:189–223.

Terman, G. L., and C. M. Hunt. 1964. Volatilization losses of nitrogen from surface-applied fertilizers, as measured by crop response. Soil Sci. Soc. Am. Proc. 23:667–673.

Thomas, G. W. 1980. Delayed nitrogen applications on corn. Soil Sci. News and Views. 1(2):1–2. Coop. Ext. Service University of Kentucky, Lexington, KY.

Thomas, G. W., R. L. Blevins, R. E. Phillips, and M. A. McMahon. 1973. Effect of a killed sod mulch on nitrate movement and corn yield. Agron. J. 63:736–739.

Thomas, G. W., and R. E. Phillips. 1979. Consequences of water movement in macropores. J. Environ. Qual. 8:149–152.

Thomas, G. W., K. L. Wells, and L. Murdock. 1981. Fertilization and liming. *In* R. E. Phillips et al. (ed.) No-tillage research: Research reports and reviews. Univ. of Kentucky, College of Agric. Exp. Stn.

Tinker, P. B. H. 1979. Uptake and consumption of soil nitrogen in relation to agronomic practices. p. 101–122. *In* E. J. Hewitt and C. V. Cutting (ed.) Nitrogen assimilation of plants. Academic Press Inc., New York.

Tisdale, S. L., and W. L. Nelson. 1975. Soil fertility and fertilizers, 3rd ed. MacMillan Publishing Co., New York.

Touchton, J. T., and F. C. Boswell. 1980. Performance of nitrification inhibitors in the Southeast. p. 63–74. *In* J. J. Meisinger et al. (ed.) Nitrification inhibitors—potentials and limitations. Spec. Publ. 38. American Society of Agronomy, Soil Science Society of America, Madison, WI.

Touchton, J. T., W. A. Gardner, W. L. Hargrove, and R. R. Duncan. 1982. Reseeding crimson clover as a N source for no-tillage grain sorghum production. Agron. J. 74:283–287.

Triplett, G. B. Jr., F. Haghiri, and D. M. Van Doren, Jr. 1979. Plowing effects on corn yield response to N following alfalfa. Agron. J. 71:801–803.

Unger, P. W. 1978. Straw mulch effects on soil temperature and sorghum germination and growth. Agron. J. 70:858–864.

Vander Paauw, F. 1962. Effect of winter rainfall on the amount of nitrogen available to crops. Plant Soil 16:361–380.

Vander Paauw, F. 1963. Residual effect of nitrogen fertilizers on succeeding crops in a moderate marine climate. Plant Soil 19:324–331.

Van Praag, H. J., V. Fischer, and A. Riga. 1980. Fate of fertilizer nitrogen applied to winter wheat as $Na^{15}NO_3$ and $(^{15}NH_4)_2SO_4$ studied in microplots through a four-course rotation. 2. Fixed ammonium turnover and nitrogen reversion. Soil Sci. 130:100–105.

Viets, F. G., Jr. 1965. The plants need for and use of nitrogen. *In* W. V. Bartholomew and F. E. Clark (ed.) Soil nitrogen. Agronomy 10:503–549.

Viets, F. G., and R. H. Hageman. 1971. Factors affecting the accumulation of nitrate in soil, water, and plants. USDA-ARS Agric. Handb. 413. U.S. Government Printing Office, Washington, DC.

Vlek, P. L. G., and E. T. Craswell. 1979. Effect of nitrogen source and management on ammonia volatilization losses from flooded rice-soil systems. Soil Sci. Soc. Am. J. 43:352–358.

Volz, M. G., M. S. Ardakani, R. K. Schulz, L. H. Stolzy, and A. D. McLaren. 1976. Soil nitrate loss during irrigation: Enhancement by plant roots. Agron. J. 68:621–627.

Warncke, D. D., and S. A. Barber. 1973. Ammonium and nitrate uptake by corn (*Zea mays* L.) as influenced by nitrogen concentration and NH_4^+/NO_3^- ratio. Agron. J. 65:950–953.

Weise, R. A., and E. J. Penas. 1979. Fertilizer suggestions for corn. Univ. of Nebraska Coop. Ext. Service Pub. G74-174.

Welch, L. F., P. E. Johnson, J. W. Pendleton, and L. B. Miller. 1966. Efficiency of fall versus spring-applied nitrogen for winter wheat. Agron. J. 58:271–274.

Welch, L. F., D. L. Mulvaney, M. G. Oldham, L. V. Boone, and J. W. Pendleton. 1971. Corn yields with fall, spring, and sidedress nitrogen. Agron. J. 63:119–123.

Wesley, W. K. 1979. Irrigated corn production and moisture management. Univ. of Georgia College of Agric. Ext. Bull. 820.

Wetselaar, R., J. B. Passioura, and B. R. Singh. 1972. Consequences of banding nitrogen fertilizers in soil. I. Effects on nitrificatioin. Plant Soil 36:159–175.

Woldendorp, J. W. 1968. Losses of soil nitrogen. Strikstof (no. 12) p. 32–46. Central Nitrogen Sales Organization, Ltd., The Hague, Netherlands.

8

Slow-Release and Bioinhibitor-Amended Nitrogen Fertilizers

Roland D. Hauck
National Fertilizer Development Center
Tennessee Valley Authority
Muscle Shoals, Alabama

Numerous nitrogen (N) tracer and nontracer studies on crop recovery of applied N indicate that 50 to 60% of fertilizer N added to soil usually is taken up by crop plants during the season of application. The percentage often is lower for flooded rice (*Oryza sativa* L.) or for crops growing in high-rainfall areas. Many of the factors that contribute to incomplete plant recovery of applied N are a result of rapid dissolution of the fertilizer and thereby, release of N at high concentration. These factors affect N movement within and from the plant-soil system, maintenance in plant-available forms, and use by the plant.

One approach to increasing the efficiency of N fertilizer use by plants is to control the rate of N fertilizer dissolution. This can be done by (i) developing compounds with limited water solubility and (ii) modifying water-soluble materials to delay release of their contained N to the soil solution. A second approach is to combine N fertilizers with chemicals that control unwanted N transformations in soil, i.e., developing fertilizers amended with inhibitors of biochemical activity in soil, such as nitrification and urea hydrolysis. Among the reviews on this subject, attention is drawn to those on slow-release N by Army (1963), Hamamoto (1966), Hayase (1967), Hauck and Koshino (1971), Prasad et al. (1971), Hauck (1972), and Allen (1984), and to those on use of nitrification inhibitors by Hauck and Koshino (1971), Prasad et al. (1971), Hauck (1972, 1980), and Slangen and Kerkhoff (1984). These references offer guides to the literature regarding the use of slow-release N and nitrification inhibitors in various cropping systems and their transformations and effects in soils. This chapter focuses on materials that were discussed only briefly or not at all in the earlier reviews and refers to work published before 1970 only where necessary to provide background information or to emphasize an important finding. In this regard, some of the introductory discussion from the author's previous reviews will be repeated.

I. CONCEPTS OF SLOW RELEASE

Efficiency of N fertilizer use can be defined in terms of plant recovery of applied N, plant metabolism and crop quality, and economic return for fertilizer investment.

From the viewpoint of improving N uptake by plants, three main advantages cited for use of slow-release N fertilizers are (i) reduction of N loss via leaching and surface runoff from soil, (ii) reduction of chemical and biological immobilization reactions in soils that decrease the supply of plant-available N, and (iii) reduction of N loss via ammonia (NH_3) evolution or denitrification following nitrification.

From the viewpoint of plant physiology, a slow-release N fertilizer ideally should supply nutrient to the soil solution at rates and concentrations that allow the growing plant to maintain maximum expression of its genetic capability. To achieve this goal, it is not clear whether the nutrient supply should be kept in phase with normal growth and turnover rates, or whether some luxury consumption and accumulation would be advantageous. Often, the ideal specifications of a slow-release fertilizer are given in terms of uptake and use patterns of plants as determined by chemical analyses of plant tissue at intervals during the growth sequence. Such patterns may offer an agronomic prescription for the fertilizer technologist to follow. However, uptake patterns often reflect N supply as well as plant needs. Also, because uptake and use patterns vary considerably among different plant species grown under similar conditions of N supply, it is unlikely that any single pattern of N release from a material will satisfy the N requirements of all cropping situations.

Advantages cited for using slow-release N fertilizers include the reduction of seed and seedling damage from high local concentrations of fertilizer salts, reduction of leaf burn from heavy rates of surface-applied N fertilizers, improved crop quality, reduction of plant disease infestation, reduction of stalk breakage, better season distribution (e.g., for forage), increased residual value of applied N, better economy of use (e.g., from off-season or single vs. multiple applications of materials), improved storage and handling properties, and product differentiation resulting in improved market potential. Although slow-release N fertilizers inevitably cost more than conventional N fertilizers, benefits obtained from their use in some cropping situations may justify the initial higher cost.

II. MODE OF RELEASE

Slow release, slow acting, controlled release, and *metered release* are terms used with materials that release their N to the soil solution in some manner that matches the need of the growing plant. *Delayed release* implies that little or no N is available for an initial time period, followed by either a gradual or rapid release of plant-available N to the soil. Controlled release is considered here to be a generic term that can be applied in its broadest sense either to fast-acting or slow-acting materials. Metered and delayed release are specific examples of slow release, which is the collective term that will usually be used here. A listing of most of the commercially available slow-release materials are given in Table 8–1.

The accuracy by which the rate and pattern of N release can be predicted is determined by the composition of the fertilizer and one's ability to control or predict the effects of the physical, chemical, and biological characteristics of the soil-plant system in which the fertilizer is placed.

Slow-release N fertilizers are of four types: (i) water-soluble materials containing ammonium (NH_4^+) and/or nitrate (NO_3^-) where dissolution is controlled by a physical barrier, e.g., by a coating; (ii) materials of limited water solubility containing plant-available forms of N (e.g., metal ammonium phosphates; (iii) materials of limited water solubility, which, during their chemical and/or microbial decomposition, release plant-available N (e.g., the ureaforms, oxamide); and (iv) water-soluble or relatively water-soluble materials that gradually decompose, thereby releasing plant-available N (e.g., guanylurea salts). The N release rates of all types of slow-release materials can be modified by using chemical additives such as nitrification or urease inhibitors, which affect N transformations in soils.

Coatings, encapsulations, and matrixes have been used to change the rate of N entry from water-soluble materials into the soil solution. Coatings applied to soluble materials usually are of three types: (i) impermeable coatings with tiny holes through which solubilized materials diffuse; (ii) impermeable coatings that must be broken by abrasive, chemical, or biological action before N can be released; and (iii) semipermeable coatings through which water diffuses until the internal osmotic pressure ruptures the coating or distends it to increase its permeability. An example of a simple encapsulation is a perforated polyolefin bag. In a matrix, fertilizer N is entrapped in a mixture so that diffusion and outward flow of fertilizer N are impeded by the tortuosity of the matrix. Coatings and matrixes may function only as physical barriers or be a source of plant nutrient (e.g., sulfur-coated urea, SCU).

Uncoated slow-release N fertilizers usually are sparingly soluble. Once in solution, they decompose to form plant-available N, usually NH_4^+. The rate of N release depends on the rate of dissolution for materials such as the metal ammonium phosphates and on the rate of dissolution and degradation such as for the urea-aldehyde condensation products. Rate of dissolution depends on solubility of the material and rate of removal of solubilized material from the particle surface, either through diffusion, mass flow, or degradation. Dissolution rate and N release can be controlled by the size and compactness of the fertilizer particle; larger and harder particles dissolve at a slower rate than smaller and softer ones. Interrelationships between particle dissolution and decomposition should be kept in mind when comparing different slow-release N fertilizers in different plant-soil systems. Thus, in aerobic, moist soil, oxamide and melamine particles of the same size and compactness go into solution at different rates, even though they have similar solubilities in water. Oxamide, once in solution, is quickly decomposed to release NH_4^+ and therefore dissolves faster than melamine, which even in solution is relatively resistant to chemical and biological degradation.

Table 8–1. Some commercially available slow-release nitrogen fertilizers.

Material	Trade name(s)	Producer or main supplier	Factory (office)† location
		Coated materials	
Sulfur-coated urea (SCU)	--	Canadian Industries Ltd.	Courtright, Canada (Montreal)
Sulfur-coated fertilizers	LESCO® Sulfur-Coated Fertilizers; Fairway Fertilizers	Ag Industries Manufacturing, Corp. (Lakeshore)	Columbia, AL
	SC KASEI®	Mitsui Toatsu Chemicals	Hiroshima, Japan (Tokyo)
Polymer-coated NPK fertilizers	Osmocote®; Agriform®; Sierrablen®	Sierra Chemical Co.; Sierra Chemical Europe B.V.	Mipitas, CA DeMeern, The Netherlands
	CSR®	Showa Denko K.K.	Kawasaki, Japan (Tokyo)
Petroleum-coated urea	LP-COTE®	Chisso-Asahi Fertilizer Co., Ltd. Chisso Corp.	Minamata, Japan (Tokyo)
Petroleum-coated NPK Ca	Nutricote®	Chisso-Asahi Fertilizer Co., Ltd. Chisso Corp.	Minamata, Japan (Tokyo)
Petroleum-coated NPK Ca	Ficote®	Fisons (distributed for Chisso, Ltd.)	(Felixstowe, Suffolk, England)
Polyethylene NPK packets	ROOTCONTACT PAKET®	S&D Products, Inc.	Prairie du Chien, WI
		Uncoated inorganic materials	
NPK stake	Green Pile	Chisso-Asahi Fertilizer Co., Ltd. Chisso Corp.	Minamata, Japan (Tokyo)
	Ross Super Tree Stakes® Ross Boomers®	Ross Daniel, Inc.	West Des Moines, IA
NPK spike (various formulations)	Job's® Spikes	International Spike Co.	Lexington, KY
	HYPONEX® Fertilizer Spikes	Hyponex Co., Inc.	Fort Wayne, IN
	HYPOSTYX® Plant Food Sticks	Hyponex Co., Inc.	Fort Wayne, IN
Magnesium ammonium phosphate	Map Amp®	W. R. Grace & Co.	Charleston, SC
Potassium ammonium phosphate	--		(New York, NY)

		Uncoated organic materials	
Ureaforms (UF)	Nitroform®	Hercules Incorporated	Wilmington, DE
		(marketed by BFC Chemicals, Inc.)	Wilmington, DE
Ureaform + NPK	--	Mitsui Toatsu Chemicals	Hiroshima, Japan (Tokyo)
	--	Sumitomo Chemical Industries	Niihama, Japan (Osaka)
Ureaform concentrates	UF Concentrates	Bordon Incorporated	Louisville, KY (New York, NY)
		Georgia Pacific Corporation	Columbus, OH (Portland, OR)
Ureaform concentrate N mixtures	--	Kaiser Aluminum and Chemical Corp.	Savannah, GA (Oakland, CA)
Ureaform solution	UF Solutions, e.g., FLUF®	W. A. Cleary Corporation	Somerset, NJ
Crotonylidine diurea (CDU)	Crotodur®, Floranid®	Badische Anilin and Sodafabrik (BASF)	Ludwigshafen-am-Rhein, FRG
	CD-Urea	Chisso-Asahi Fertilizer Co., Ltd. Chisso Corp.	Minamata, Japan (Tokyo)
Isobutylidene diurea (IBDU)	--	Mitsubishi Chemical Industries	Kitakyushu, Japan (Tokyo)
		(U.S. supplier, Estech, Inc.)	Winterhaven, FL (Chicago, IL)
		IB Chemical Co. (Virginia Chemicals/Mitsubishi)	Bucks, AL (Portsmouth, VA)
		Hoechst AG	Frankfort, FRG
Melamine (triaminotriazine)	Nitrazine 66™		
Melamine + urea	MCI® Fertilizer	Melamine Chemicals, Inc.	Donaldsonville, LA
Oxamide	--	Allied Corp. Chemicals Co. Division	Morristown, NJ (Claymount, DE)
	--	Guardian Chemical Corp., Eastern Chemical Division	Smithtown, NY
	--	Hummel Chemical Corp., Inc.	S. Plainfeld, NJ
	FOX-RINKAAN®	Ube Industries, Ltd.	Ube, Japan (Tokyo)
Guanylurea phosphate	--	Nitto Chemical Industry Co.	Kushiro, Japan (Tokyo)
Guanylurea sulfate	--	Nitto Chemical Industry Co.	Kushiro, Japan (Tokyo)

(continued on next page)

Table 8–1. Continued.

Material	Trade name(s)	Producer or main supplier	Factory (office)† location
Dicyandiamide	Dicyan, Dd	Showa Denko, K.K.	Kawasaki, Japan (Tokyo)
	Didin®, DCD™	SKW Trostberg AG	Trostberg, FRG
	--	Nippon Carbide Industries Co., Ltd.	Uozu, Japan (Tokyo)
Dicyandiamide-ammonium sulfate	Alzodin®	SKW Trostberg AG	Trostberg, FRG
Dicyandiamide + NPK	22-4-8-DCD Turf Food	Lebanon Chemical, Turf Products Division	Lebanon, PA
	Processed waste products		
Sewage sludge	Chicago Heat-Dried Fertilizer	Metropolitan Sanitary District of Greater Chicago, IL	--
	Hou-Actinite®	City of Houston, TX	--
	Nitrohumus®	Los Angeles County, CA	--
	Milorganite	City of Milwaukee, WI	--
Tankage	Hynite®	Hynite Corporation	Oak Creek, WI
	Mainly Green®	Maine Resources Corp.	Brooks, ME
Tankage/methylene urea polymers	Organiform®	Omnicology, Inc.	Gloversville, NY

† Factory location is followed by corporate headquarters, where applicable, in parentheses.

III. COMMERCIAL AND EXPERIMENTAL MATERIALS

A. Coated Materials

1. Coating Compositions

Various materials have been tried as coatings for solid N fertilizers, including asphalts, tars, gums, latexes, oils, paraffins, waxes, acrylic and epoxy resins, polyolefins, polyvinyl chloride, polystyrene, polyurethane, urea formaldehydes, vinyl acetate, sulfur (S), and others, alone or in combination. The most commercially important of these materials are waxes, polymers, and S.

Description of the many substances used as coatings and methods of applying them may be found in the patent literature (see Powell, 1968; TVA, 1972, 1978). Only those coating materials that are used in commercial fertilizer processes will be discussed here.

Most of the information that is available to the fertilizer technologist is found in the patent literature. It concerns mainly details of manufacture and the results of preliminary tests. Experimental work on the agricultural use of coated N fertilizers generally was concerned with (i) measuring rates of dissolution, nutrient release, and N movement; (ii) measuring nitrification rates; (iii) studying seed and seedling damage; and (iv) evaluating the efficiency of the coated fertilizer in terms of crop yield and nutrient uptake. Little work has been reported on chemical and biochemical reactions of the coating materials in soils.

Some coating materials were considered attractive because of their high nutrient content, as well as their physical properties. Examples of these are dicyandiamide, melamine-formaldehyde, S, and the urea formaldehydes. Of these, only S has been found to be effective. Most of the materials tested as coatings have proven to be either ineffective or too effective in retarding N movement from the coated particle to the soil solution.

Two main problems in producing an effective coating are (i) the coating must be thin or else it occupies a large percentage of the entire particle volume (about 40%, even with a relatively thin shell), resulting in an undesirable lowering of the N content of the particle; and (ii) the relatively porous, rough, and irregular surface of the fertilizer granule or prill makes uniform coating free of imperfections difficult.

2. Sulfur-Coated Urea

Coatings for urea have received much attention because of urea's high N content, widespread use, and problems associated with use of uncoated solid urea, especially when it is applied to the soil surface. During the period 1961–1981, the Tennessee Valley Authority (TVA) was the leading developer of S as a low-cost coating for urea. Sulfur-coated urea (SCU) was first produced commercially in 1972 by Imperial Chemical Industries, Ltd. (ICI) in pilot-plant amounts and marketed under the trade name Gold-N® (no longer on the market). An ICI subsidiary, Canadian Industries, Ltd., began the first fully commercial production of SCU during 1975 in a plant having a production capacity of 30 000 metric tons annually. Refinements of the coating process by TVA continued, eventuating in the construction of a 10 metric tons/h demonstration plant in 1978. During that same year, Ag Industries Manufacturing Corporation (Lakeshore) began producing SCU and other sulfur-coated mixed fertilizers, and marketing them under the trade name LESCO® sulfur-coated fertilizers. Mitsui Toatsu Chemicals, Incorporated (Japan) has manufactured a mixed fertilizer coated with S and wax since 1976, which is marketed under the name SC KASEI®.

All commercial processes for producing SCU are based on the TVA method whereby molten S is sprayed on a falling curtain of preheated urea particles in a rotating drum. A sealant is then applied to close pores in the S coating, followed by a conditioner to improve handling properties. A typical TVA formulation may contain 36% N (as urea granules), 10 to 16% S, and 5% sealant (polyethylene oil/viscous oil mixture) plus conditioiner (diatomaceous earth). The dissolution rate is controlled by varying the average thickness of the S coating. The product produced by Canadian Industries, Ltd. contains 32% N as prilled urea with a coating that is 26% by weight consisting of S, microcrystalline wax, and clay. Lakeshore produces two sizes of SCU granules, the smaller granule designed for use on golf greens and low-cut fairways. A similar product using miniprills is produced by the Canadian firm.

The rate of N release from SCU is characterized by a dissolution test that is a measure mainly of the number of imperfectly coated particles. In this test, a measurement is made of the amount of urea that is released during a 7-d period from 100 g of SCU immersed in 400 mL water at 37.8° C (100° F).

Allen (1984) has summarized some of the factors that modify the dissolution rates of SCU particles: (i) N release rate increases with increase in soil temperature, which suggests some biodegradation of the coating; (ii) the coating ruptures sooner in a relatively dry soil (e.g., 10% water content) than in a wet soil; (iii) particles lying on the soil surface tend to rupture sooner than particles within the soil; (iv) plant roots accelerate the dissolution rate of SCU; (v) dissolution appears to be little affected by soil pH in the range 5 to 8; and (vi) biological oxidation of the S in the coating residue is slow but eventually lowers soil pH because of sulfuric acid formation.

Sulfur-coated urea has been evaluated as a slow-release N fertilizer for a variety of container and ornamental crops and on agricultural crops such as corn (*Zea mays* L.), wheat (*Triticumaestivum* L.), forage, turfgrasses, cotton (*Gossypium hirsutum* L.), rice (*Oryza sativa* L.) sugarcane (*Saccharum officinarum* L.), cranberries (*Vaccinium macroacrpon* Ait.), forest trees, and vegetables (for references, see TVA, 1976, 1980a, 1981a,b,c). They indicate a significant advantage for SCU over soluble N fertilizers when used for turfgrasses, cranberries, and flooded rice, a discussion of which will be included in section IV.

3. Polymer-Coated Fertilizers

Fertilizers coated with Osmocote® are the only polymer-coated fertilizers produced in the USA. First developed by the Archer Daniels Midland Company, Osmocote products are now produced by Sierra Chemicals Company. The coating is applied in several layers of different compositions and has as its main component a copolymer of dicyclopentadiene with a glycerol ester derived from soybean (Powell, 1968). Sierra Chemical Company markets several N-, K-, or NPK-containing fertilizers. Nutrient release is controlled by the thickness and composition of the coating. Trade names are Osmocote® and Sierrablen®, the latter being a product that contains about 20% of soluble, uncoated nutrient. Osmocote is also produced in Holland by Sierra Chemical Europe B.V.

The nutrient release mechanism of products coated with S and Osmocote® are somewhat different. With SCU, water vapor transfers through the S coating, solubilizes the urea within the shell, and builds up sufficient osmotic pressure to disrupt the coating, thereby releasing urea solution. With Osmocote®, water vapor transfer through microscopic pores in the coating causes the internal osmotic pressure to distend the semipermeable and flexible coating, thereby enlarging the pores sufficiently to permit solution to pass through the coating.

Three-components (polyethylene and other hydrocarbons) were used by the Minnesota Mining and Manufacturing Company to coat urea, ammoniated superphosphoric acid, and KCl (Stanford Res. Inst. Int., 1981). The products were marketed under the trade name Precise® but production was discontinued in 1979.

Coated fertilizer was first produced in Japan by Showa Denko K. K. in 1970. Their product, CSR®, consists of ammonium sulfate, ammonium phosphate, and KCl coated with a phenolic resin. In 1975, Chisso-Asahi Fertilizer Company, Ltd. began producing polyoleum-coated urea, LP-COTE®, and a mixed coated fertilizer, Nutricote®, containing ammonium- and nitrate-N, P, K, and Ca. The latter product is distributed in the United Kingdom by Fisons under the trade name Ficote®.

The polymer-coated materials are used mainly on lawns, professional turf, container-grown ornamental and horticultural plants, and, to a limited extent, in vegetable production (to be discussed later).

4. Matrixes and Encapsulations

Coatings usually are applied in intimate contact with the surface of the fertilizer particle. Matrixes and encapsulations are two variations of this procedure. In a matrix, soluble fertilizer is dispersed into asphalts, gels, oils, paraffins, polymers, resins, or waxes. Nitrogen fertilizers of limited water solubility, such as urea formaldehydes, also have been incorporated with or have been caused to permeate clay, glass, frit, perlite, and expanded vermiculite, among other materials (Powell, 1968). Encapsulations, as used here, usually include a group of fertilizer particles rather than single particles. A simple encapsulation is a perforated polyethylene bag containing soluble fertilizer particles that either are relatively uniform in size or may be present as pellets or tablets of varying size.

Although numerous formulations have been patented, commercial production of matrix-incorporated fertilizers is low and restricted to the micronutrients, mainly in the form of glass frit. During the mid-1970s, the Minnesota Mining and Manufacturing Company produced a mixed fertilizer consisting of ammoniated superphosphoric acid, urea, and potassium chloride that was encapsulated with polyethylene and two other hydrocarbons. Nutrient release over a 3- to 4-month period was through tiny pores in the encapsulation. Several formulations under the trade name Precise® were marketed for house and garden use, but manufacture was discontinued in 1979. Sealed polyethylene packets of mixed fertilizer [16–3.5–5 (16–8–6)] are produced by S&D Products, Incorporated. Marketed under the name ROOTCONTACT PAKET®, packets come in several sizes containing from 10 to 114 g of fertilizer. Nutrient release is through micropores in the encapsulation.

Slow-release materials in the physical form of spikes, stakes, sticks, pellets, and tablets, although neither true matrix- nor encapsulation-type materials, are included here with these types. They consist of water-soluble fertilizers mixed with a binder to produce a highly compressed solid form. These products expose a relatively small surface area relative to their volume to the solvent action of soil water, thereby slowing the rate of dissolution and nutrient release.

Large spikes and stakes are produced mainly for use on trees and shrubs; smaller spikes, sticks, and tablets are mainly for home and garden use. A stake measuring about 30 cm in length and 19 mm in thickness, consisting of nitrophosphate and potassium chloride, is manufactured by Chisso-Asahi Fertilizer Company, Inc. (Japan) and is marketed under the trade name Green Pile for use as a tree fertilizer. A variety of products is manufactured in the USA. Six formulations are offered by the International Spike Company for use on fruit and shade trees, shrubs, evergreens, container plants, house plants, and tomatoes. They are marketed under the trade name Jobe's® *X* Spikes (where *X* equals vegetable, flower, fruit and shade trees, shrubs, and evergreens) e.g., Jobe's® Flower Pot Spikes [analysis of 4–5–3(4–12–4)] or Jobe's® Evergreen Spikes [12–3.5–7(12–6–8)]. All formulations except those for house and container plants contain ammonium sulfate, diammonium phosphate, and potassium chloride. Ureaform, triple superphosphate, and potassium sulfate supply the nutrients for pot plants. The Hyponex Company, Inc. produces three formulations that are marketed as HYPONEX® Fertilizer Spikes and Organic Fish Brand Sticks and two formulations as HYPOSTYX® Plant Food Sticks. Ross Daniel Inc. uses ammonium sulfate, diammonium phosphate and/or ammonium polyphosphate, potassium chloride, iron, and zinc in three formulations marketed as Ross Super Tree Stakes® for use on shrubs and fruit, shade, and evergreen trees, and three formulations marketed as Ross Boomers® for use on flowers, shrubs, and tomatoes. Nitrogen is supplied as ammonium sulfate and ureaform in a formulation that is designed for use on azalea and other ornamental shrubs. Ureaform also is an ingredient of Agriform® tablets produced by Sierra Chemical Company, which offers six formulations that are recommended for use on fruit and nut trees, ornamental shrubs and trees, container- and nursery-grown ornamentals, grapes, and garden plants. These tablets are produced in addition to their polymer-coated formulations.

Production and sales data for the materials discussed in this section are not published elsewhere. Annual production for the 1980s is estimated at 4500 metric tons.

B. Uncoated Inorganic Materials

These materials comprise a group of compounds that have the general formula $MeNH_4PO_4 \cdot xH_2O$, where Me is a divalent metal ion (Me^{2+}) Potassium analogues of the ammonium also can be made. All are slightly water-soluble and several have been developed as multinutrient slow-release fertilizers. See Bridger et al. (1962) for a summary of their physical characteristics and Bridger (1968) for a comprehensive review of their potential as slow-release fertilizers.

Magnesium ammonium phosphate, first suggested for use as a fertilizer over a century ago (Murray, 1858), is the most common of these compounds. It has been commercially produced in the Federal Republic of Germany (FRG) since 1928 (Burns & Smith, 1966) and more recently in Japan and the USA. In the USA, the W. R. Grace and Company produces a magnesium ammonium/potassium phosphate [7–40–72–12(Mg), 7–17–6–12(Mg)] marketed under the trade name MagAmp®. The product is offered in two particle sizes, which are said to supply nutrients to the soil solution over 1 and 2 yr. The largest market for MagAmp® is in floriculture, where the products are mixed with potting soil and with bedding soil for greenhouse use. They are used also for addition directly into the planting hole for forest, landscape, and orchard trees.

C. Uncoated Organic Materials

Urea is the major organic form of N used as fertilizer in agricultural soils, but because it is readily soluble in water and quickly decomposed to release NH_4^+-N, it does not act as a slow-release N fertilizer. However, urea forms several chemical reaction products that are useful as slow-release N fertilizers. Most organic N compounds have little or no potential as slow-release fertilizers because of (i) too slow or rapid rate of degradation in soil, (ii) toxicity to higher plants, (iii) excessive leaching before nutrient release, (iv) low nutrient content, and (v) high cost of production (for a compilation of over 300 references on the agronomic value of various organic N compounds, see Isbell and Fleming (1969). The focus of the discussion here will be on organic N compounds now in commercial use as slow-release fertilizers. Little reference will be made to experimental materials that were discussed previously (Hauck & Koshino, 1971; Hauck, 1972).

1. Urea-Adehyde Condensation Products

Urea reacts with any of several aldehydes to form compounds that are sparingly soluble in water. The reaction is technically practical and uses at least one low-cost material, urea. Compounds of interest have N contents about 30% and slowly decompose in soil by chemical and/or biological action. These factors combine to make the urea-aldehyde condensation reaction an attractive approach to the development of a slow-release N fertilizer.

a. Ureaforms—Urea reacts with formaldehyde in the presence of a catalyst to form a mixture of compounds known under the generic name ureaform (UF). Ureaforms are white, odorless solids. They consist mainly of methylene urea polymers that vary in chain length and, with the longer-chain molecules, in degree of cross-linking. The composition of the final product is controlled by the mole ratio of urea to formaldehyde, type of catalyst, and reaction conditions. Increase in mole ratio increases the amount of water-soluble components. Decrease in ratio increases the average molecular weight and proportion of insoluble components. Commercial ureaforms have urea/formaldehyde mole ratios in the range of 1.25:1 to 1.5:1 and a N content of about 36 to 38%, of which about two-thirds is water-insoluble. The main problem of manufacture is obtaining a suitable balance between relatively soluble, biologically labile short-chain and more insoluble, degradation-resistant longer-chain polymers, the extremes of which have N release rates that are too fast or too slow, respectively. Ureaform compounds that are regarded to have desirable N release rates are dimethylenetriurea (Hayase, 1967) and trimethyleneurea.

Solubility has been related to rate of degradation and nitrification in soil through the activity index (Morgan & Kralovec, 1953). This index is derived from an empirical test that measures the percentage of the N insoluble in cold (25°C) water that is soluble in a hot (100°C) phosphate buffer solution (pH 7.5). Modifications of this procedure have been suggested. For example, Yamazoe and Hayase (1966) measured the amount of N that is soluble in cold buffer solution (pH 7) minus that soluble in cold water to distinguish between soluble ureaform components from unreacted urea that may be present in some formulations.

Ureaform was first patented as a chemical in 1924 by the Badische Anilin- and Sodafabrik AG and later as a fertilizer by Rohner and Wood (1947). Ureaforms currently are marketed as solids, UF concentrates, and UF solutions. The E. I. duPont de Nemours and Company (Wilmington, DE) and the Nitroform Corporation began commercial production of solid ureaform around 1955. Production of its material, Uramite®, was discontinued by duPont in 1976. Following a series of corporate changes, Nitroform® now is manufactured by Hercules Incorporated under contract to BFC Chemicals, Inc. (U.K.), which has full marketing rights. Nitroform contains 38% N (29% of which is water-soluble) and is marketed in powder and granular forms. Mitsui Toatsu Chemicals and Sumitomo Chemical Industries (Japan) produce ureaforms for incorporation into mixed fertilizers.

Ureaform concentrates have urea/formaldehyde mole ratios of about 4.5:1 or higher. These partly polymerized concentrates are used as intermediates by fertilizer manufacturers to increase the amount of water-insoluble N in their products. Additional urea is added to the UF concentrate during granulation with other fertilizer materials. Principal suppliers of UF concentrates to other fertilizer companies are Borden Incorported and Georgia Pacific Corporation. The Kaiser Aluminum and Chemical Corporation produces solutions containing urea, formaldehyde, ammonia, and ammonium nitrate, UF components being produced in situ. Productions of the UF concentrate Nuform 30® by Hawkeye Chemical Company and the UF concentrates N-Dure® and HUF® by Allied Corporation was discontinued during 1978.

Commercial production of UF solutions is of recent origin (1977), although such solutions have been of interest since the mid-1960s (e.g., see Hays & Haden, 1966). The solutions contain water-soluble short chain UF derivatives, mainly methylene diurea and methylol ureas, plus unreacted urea. The solutions may be used to produce suspensions containing relatively insoluble UF derivatives, thereby producing a variety of materials having different N release rates. An example of such a fluid mixture is Fluf®, produced by the W. A. Cleary Corporation. For numerous references on the manufacture and properties of ureaform products, see TVA (1973, 1980b).

b. Crotonylidene Diurea—(2-oxo-4-methyl-6-ureido-hexahydro-pyrimidine, CDU). The acid catalysis of either urea-acetaldehyde or urea-crotonaldehyde yields a ring-structured compound, CDU, containing two ring and two side-chain N atoms. Several water-soluble or hydroalyzable compounds also are formed, but manufacturing conditions keep the level of these components $< 3\%$. The total N content is about 32%. Two products are manufactured by the Badische Analin- and Sodafabrik, FRG: Crotodur®, which is commercially pure CDU, and Floranid®, a specialty product containing 28% N, of which 10% is derived from nitrate. Chisso-Asahi Fertilizer Company, Ltd. (Japan) has manufactured CDU (CD-urea) since 1964. Japanese production from 1975–1985 has been about 7000 to 11 000 metric tons annually, as compared with about 2500 to 3500 metric tons of UF products. German and Japanese CDU is marketed uncombined or as a slow-release N component of compound fertilizers containing P, K, and sometimes micronutrients. Details of CDU manufacture are discussed in Powell (1968), Jung and Detmer (1968), and Ushioda (1969). The CDU products are used mainly on grasses, vegetables, and fruit trees and other perennials.

c. Isobutylidene Diurea (IBDU)—The major condensation product of urea and isobutyraldehyde is a relatively water-insoluble ($< 0.1\%$) compound, isobutylidene-diurea (1-diureido isobutane). The acid-catalyzed reaction occurs either in aqueous solution or in liquid aldehyde that is interspersed with solid urea. The commercial product contains about 30 to 31% N (32.18% theoretical). About 35 000 metric tons is manufactured in Japan (Mitsubishi Chemical Industries) and about 7000 metric tons in FRG

(Hoechst AG). The material is marketed for direct application as spherical granules ranging in size from 0.7 to 8mm in diam or is granulated with soluble sources of N, P, and K to provide a range of mixed fertilizer formulations. Swift and Company introduced Japanese IBDU into the U.S. market in 1970 under the names Vigoro Once and Par-X I.B.D.U. Estech, Inc. is the sole supplier of IBDU® for Mitsubishi in the USA and Canada. Two size ranges usually offered are 0.5 to 1.0 mm (fine) and 0.75 to 2.5 mm (coarse). During 1985, IB Chemical Company began manufacturing IBDU at Bucks, AL. Manufacture is a joint venture of Virginia Chemicals, Mitsubishi Chemical Industries America, and Mitsubishi International Corporation.

References to the patent literature and discussion of IBDU properties and early use are given in two comprehensive reviews by Hamamoto (1966, 1968). Although the principal current use of IBDU is probably on professional turf and for container crops, this slow-release N fertilizer has potential for use on high cash crop vegetables and flooded rice, depending on its cost relative to that of conventional, soluble N fertilizers.

d. Other Urea-Aldehyde Products—Urea reacts with acetaldehyde in neutral aqueous solution to form a mixture that consists mostly of ethylenediurea and diethylenetriurea, with some unreacted urea. The mixtures, known as urea-Z, contain 33 to 38% N and were first tested as slow-release N sources in Germany. Urea-Z is not produced commercially. See TVA (1977a) for references to the patent literature and other articles.

Japanese agronomists have reported work with the reaction products of urea with propionaldehyde or furfurylaldehyde to form propylidinediurea of furfurylidendiureide, respectively, but these polymers are too expensive for commercial production (see Hauck, 1972, for references). The reaction product of urea with glyoxal forms glycoluril, a relatively water-insoluble (0.2%) material containing 39.4% N. The U.S. patent holder, Monsanto Chemical Company, has not commercialized this product.

2. Triazines (Urea Pyrolyzates)

Urea heated under ammonia pressure to $\geq 150°$ C yields a six-membered ring containing three C and three N atoms, symetrically disposed (i.e., alternating C and N atoms forming the triazine ring). Hydroxyl groups attached to the ring C atoms can be partly or completely replaced by amino groups as temperature and ammonia pressure are increased. Successive replacement of the three hydroxyls results in the formation of a series of compounds—cyanuric acid (32.4% N), ammelide (43.7% N), ammeline (55.1% N), and melamine (66.5% N). Their high N content, relatively low solubility in water, and ease of production make these materials attractive as potential slow-release N fertilizers. However, they degrade slowly in soils, and greenhouse and field tests conducted in the 1960s gave little promise for their use on agronomic crops (Terman et al., 1964). More recently, there has been renewed interest in melamine for use on flooded rice and melamine-urea mixtures for upland agronomic and tree crops. Melamine Chemicals, Inc. (MCI) has initiated testing with powdered melamine (triamino-triazine, Nitrazine 66™, formerly SUPER 60®) mixed with urea. The company offers Granular MCI Fertilizers, which are produced by physically combining Nitrazine 66™ powder with urea in a granulator to form a durable heterogeneous granule. Granules of different total N contents are made by varying the proportions of urea and melamine. For example, products analyzing 60% N contain 45% melamine-N/15% urea-N. The 55% N formulation contains 27.5% melamine-N/ 27.5% urea-N. Standard- and forestry-grade products have a nominal granule diam of 2.5 and 3.5 mm, respectively.

When added to moist soil, the urea component of the granule dissolves, dispersing powdered melamine. Although degrading slowly in soil, powdered melamine mineralizes faster than when in the form of larger granules (Hauck & Stephenson, 1964). Its slightly faster N release rate in waterlogged soil indicates its possible usefulness for flooded rice (see also section IV A). A second market area of main interest is forest fertilization for pulpwood production. Sufficient field data are not available to justify comment on the value of urea-melamine formulations for any of these market areas.

3. Oxamide

Oxamide, the diamide of oxalic acid, is a white, nonhygroscopic compound containing 31.8% N. Small quantities for industrial use are made by the Chemicals Company Division of Allied Corporation, Eastern Chemical Division of the Guardian Chemical Corporation, and Hummel Chemical Corporation, Inc., but it is not produced commercially in the USA as a fertilizer. Ube Industries, Ltd. (Japan) produces oxamide as a slow-release N fertilizer marketed under the trade name FOX-RINKAAN® An estimated 4000 metric tons of mixed fertilizer containing oxamide is produced annually (compared with about 100 000 metric tons of mixed fertilizer containing IBDU produced by Mitsubishi Chemical Industries).

The value of oxamide as a fertilizer was suggested as early as 1917 (Brigham, 1917). Its value as a slow-release N fertilizer on paddy rice and upland crops was established by Ogata and Hino (1958a,b,c,d). The problem is its high cost of production. Approaches taken by the Tennessee Valley Authority during the period 1960–1982 include: (i) oxidation of hydrocyanic acid to cyanogen with either NO_2 or CuO, followed by acid hydrolysis of the cyanogen to oxamide; (ii) oxidative carboxylation of alcohols using quinone as an oxidant to produce oxalic esters that can act as intermediates in the production of oxamide and other slow-release N fertilizers; and (iii) nitric acid oxidation of paper or other waste carbohydrates to oxalic acid, followed by reaction with ammonia to form oxamide. None of these processes are sufficiently economical for large-scale oxamide production at this time, although the availability of low-cost cyanogen as a byproduct would make process *i* attractive, and Ube Industries, Ltd. produces relatively small amounts of fertilizer-grade oxamide using a variation of process *ii* whereby alcohol (R–OH) is oxidized to oxalic diester with carbon monoxide and oxygen. Because of its attractive agronomic characteristics (section IV), development of a relatively low-cost synthesis process probably will continue to be of interest.

The possible value of oxamide or its derivatives as a plant growth promoter is being investigated on turfgrasses and citrus. Oxamide or oxamic acid and its lower alkyl esters and alakli metal salts sprayed on plant leaves at concentrations in the range between 10^{-1} *M* to 10^{-6} *M* are said to regulate plant growth by effecting ethylene inhibition (Rehberg, 1984). Preliminary results indicate that this possible use for oxamide merits further study.

4. Guanylurea

This compound and its salts were included in agronomic studies conducted in France, Japan, and the USA since about 1930, but its commercial production in small quantities began in Japan during the mid-1970s. It is made by polymerizing cyanamide to dicyandiamide, followed by hydrolysis with phosphoric or sulfuric acid to form guanylurea phosphate (27.8% N) or guanylurea sulfate (37.0% N), respectively. Nitto Chemical Industry Company, Ltd. produces both guanylurea salts; small amounts of guanylurea sulfate also are produced by Nippon Carbide Industries Company, Ltd. Annual production varies between about 500 to 1800 metric tons, mainly of guanylurea phosphate.

5. Other Organic Compounds

Among the many compounds tested as slow-release N fertilizers are biuret, triuret, glycoluril, hexamethylenetetramine, imidazoles, pyrimidines, polyamides, urethanes, methylolurea, and other urea and cyanamide derivatives. Except for dicyandiamide and thiourea, none are in commercial production. For references concerning their evaluation as N sources, see TVA, 1977b. Dicyandiamide and thiourea will be discussed later in connection with their possible dual function as a slow-release N fertilizer and nitrification inhibitor.

6. Processed Waste Products

About 360 to 410 thousand metric tons of processed organic wastes, containing about 10 to 11 thousand metric tons of N, are produced and used annually as fertilizers in the USA. Annual consumption of processed wastes has been declining at the rate of about 4.5 to 5% during the period 1974–1984 (as estimated from the annual and monthly issues during this period of *Commercial Fertilizers,* USDA, 1974–1984). About 70 to 75% of the materials are applied directly; the remainder are used as a component of mixed fertilizers. Organic wastes processed as fertilizers include bagasse, castor pumice, compost, dried blood, hair, hoof, horn, lignins, manures, molasses, sawdust, seaweed, seed hulls and meals, sewage, tankage, waste industrial liquors, whey, and wool and leather scraps. Processing may vary from a single improvement in physical condition to a large change in the material's chemical composition. Materials may be heated, oxidized, digested, ammoniated, nitrated, or purified to raise their nutrient content and lower the content of water-insoluble or refractory components.

Dried blood (about 145 000 metric tons) and sewage sludge (about 160 000 metric tons) make up the bulk of the processed organic wastes that are used as fertilizers. The main products made from sewage sludge are Chicago Heat-Dried Fertilizer (6.4% N) processed by the Metropolitan Sanitary District of Greater Chicago, IL; Hou-Actinite® (about 5% N), processed by the Waste Water Division, Public Work Department of Houston, TX; Nitrohumous® (0.75% N), produced by Los Angeles County, CA; and Milorganite® (6% N), produced by the Milwaukee Sewage Commission, Milwaukee, WI. Other municipalities produce relatively small amounts of processed sewage sludge, mostly for local use. Much of the Chicago product is shipped in bulk to Florida and used as a base for fertilizer applied to citrus. Milorganite® is bagged for use throughout the USA as a direct application on lawns, turfs, and gardens or sold to bulk blend fertilizer manufacturers. Nitrohumous® is blended with other organic wastes or serves as a base for fertilizers used in gardens, landscapes, and nurseries. The Houston product is applied directly or as a fertilizer blend, mainly to Florida citrus and vegetables.

About 55 to 64 thousand metric tons of leather tankage were processed annually during 1981–1983 (about 30% less than a decade earlier). Products include Hymite ® (11% N), Mainly Green® (12% N), and Organiform® (26% N); the latter is a granular product made from Mainly Green® polymerized with methylene ureas. Florida citrus and vegetable growers are the main users of processed leather tankage.

Although processed organic wastes represent the largest tonnage of slow-release N materials used, they supply only about 0.1% of U.S. consumption of fertilizer N. By comparison, agricultural, industrial, and municipal wastes contain enormous amounts of N, e.g., animal manures produces annually in the USA contain an estimated 5.3 to 10.5 million metric tons of N. A small percentage of these unprocessed waste materials is applied directly to soil, a discussion of which can be found in articles by Sommers and Giordano (1984) and Bouldin et al. (1984).

IV. NITROGEN AVAILABILITY AND PLANT RESPONSE

A. Nitrogen

Coated soluble fertilizers can be characterized by the rate at which the soluble component is released to the soil solution. Relatively water-insoluble inorganic materials, such as magnesium ammonium phosphate, can be similarly characterized. Organic slow-release materials can be grouped according to the manner by which their mineralization occurs in soil. Usually, solubility is not a useful criterion for their classification. Most organic slow-release N fertilizers are relatively water-insoluble, notable exceptions being guanylurea and short-chain urea-formaldehyde polymers. As indicated in section II, N release is related to rate of solution in soil, which, in turn, is very dependent on factors that affect the removal of solubilized material from the particle surface. Removal is accomplished either through degradation of the material or its movement from the particle surface by diffusion or mass flow.

The ureaforms, oxamide, the triazines, and CDU (the latter in slightly acidic to alkaline soils) are degraded mainly by microbial action. Of these, oxamide is the most labile so that particle size and hardness (which also affect dissolution rate) greatly affect the rate of oxamide mineralization. Powdered oxamide mineralizes and nitrifies at a rate similar to that of urea. Conversely, the degradation rates of long-chain, cross-linked ureaform polymers and the triazine, melamine, do not keep pace with their respective dissolution rates. For these materials, the effect of particle size on N release is much less pronounced than for oxamide. Except for CDU in acid soils, mineralization rate of CDU, the ureaforms, triazines, and oxamide decreases with decrease in temperature and is negligible in sterile soils.

Mineralization of the methylolureas and IBDU is partly a function of soil chemical activity. Their hydrolysis releases some free urea and the corresponding aldehyde. Unlike formaldehyde (released from methylolureas), isobutyraldehyde (from IBDU) is not toxic to higher plants and apparently does not suppress microbial activity (Hamamoto, 1968). Thus, N release from IBDU, and to a lesser extent from CDU, is accomplished both through chemical and microbial activity in soil, chemical hydrolysis being more pronounced in acid than alkaline soils, microbial degradation being favored in alkaline soils.

Ureaform, CDU, and the triazines degrade slowly, thereby releasing less N than oxamide or IBDU for plant use during the growing season and leaving more unmineralized residue.

Surface application of oxamide can result in N loss as ammonia. Similar to urea, ammonium carbonate is formed at the particle microsite, resulting in a high local concentration of NH_4^+ at high pH. Other organic slow-release N fertilizers and SCU probably can be surface-applied without risk of ammonia loss. Materials that mineralize at a comparatively rapid rate appear to be less susceptible to leaching. For example, solubilized CDU or bioresistant ureaform fractions may leach to a greater extent than solubilized oxamide or IBDU. A limited amount of data is available to support this observation (e.g., see Hamamoto, 1968).

Guanylurea and the triazines (cyanuric acid, ammelide, ammeline, and melamine) degrade faster in waterlogged than upland soils. Nitrogen release from these compounds appears to be related to the maintenance of reducing conditions. The lag period preceding mineralization varies considerably among soils but can be shortened by pretreating the soil with urea. Melamine-urea mixtures (e.g., Nitrazine 66™) are being evaluated as N sources for paddy rice on the assumption that melamine mineralization will be sufficiently rapid to supply the N needs of the rice plant. The marked effects of soil type and redox potential on triazine degradation rates and the observed stimulation by organic matter of triazine degradation may be explained on the basis of microecology, i.e., a particular combination of soil factors may be conducive for growth of microorganisms that can cleave the triazine ring, thereby facilitating mineralization of the triazine residue by other microorganisms.

Oxamide, IBDU, and SCU also release plant-available N under waterlogged conditions at rates adequate to achieve optimum growth of rice. Immobilization may reduce the efficient use by plants of N derived from slow release fertilizers, but definitive data from experiments using ^{15}N which address this point, are not available.

The foregoing information on N release from slow-release fertilizers was summarized in greater detail with appropriate references by Hauck (1972). Since the mid-1970s there has been a marked decline in work published in this area.

B. Crop Response

Grasses, because of their relatively long growing season and need for a rather uniform distribution of N supply over the growing season, have been widely used as test plants for slow-release materials. Sequential harvest experiments generally show that the growth response to slow-release N fertilizers is initially slow but more uniform throughout the growing season than with conventional soluble fertilizers.

For ureaform, N uptake by grasses during the first and second harvests was found to be directly related to increase in mole ratio of urea to formaldehyde (U/F) and corresponding increase in activity index (Knop et al., 1965). Materials with U/F ratios of 1.0 and 1.1 had solubilities too low for optimum growth of ryegrass (*Lolium multiflorum* Lam.) (Ishizuka & Takagishi, 1962). Materials with mole ratios of about 1.25 applied once supported satisfactory growth, but supplemental additions of soluble N fertilizers were needed to support optimum growth of bermudagrass [*Cynodon dactylon* (L.) Pers.] and ryegrass during the first 30 d of growth (Armiger et al., 1948). Generally, N uptake from a single spring application of ureaform is less than for the same amount of N from soluble sources distributed in multiple applications during the growing season. Long-term laboratory incubation tests indicate that about one-half to two-thirds of the N in commercial ureaforms usually is mineralized. Crop recovery of ureaform-N is lower, usually ranging from 20 to 40%. Crop recoveries of N from oxamide and IBDU are markedly higher. For example, in a sequential harvest experiment with field-grown sorghum-sudangrass hybrid [*Sorghum bicolor* (L.) Moench × *S. sudanense* (Piper) Stapf] the amount of N taken up from ^{15}N-labeled urea by the time of first harvest was much greater than from ^{15}N-labeled oxamide (−8 + 10 Tyler mesh, 1.70- to 2.36-mm particles), but by the third harvest, yield response and N uptake from oxamide was greater than from urea (Westerman et al., 1972). At the end of the second year, the total amount of N taken up during the two growing seasons and the amount of N remaining in the soil was not significantly different for the two N sources (Westerman & Kurtz, 1972).

Often in experiments in which slow-release N fertilizers are compared with soluble, fast-acting N sources, yield response to N from either type of fertilizer may be similar, but less N is taken up from the slow-release materials. This would be a desirable feature of slow-release fertilizers if the fertilizer N that remains in the soil would become available for subsequent plant growth. There is evidence also that use of slow-release materials results in a more desirable partitioning of N among plant parts, as compared with fast-acting N sources. Increased growth (Armiger et al., 1948) and N content (Hays, 1964) of roots and crowns of bermudagrass after ureaform addition has been observed, a feature considered desirable for turfgrass production.

The reviews by Hauck and Koshino (1971) and Allen (1984) discuss numerous reports on the use of coated and uncoated slow-release N materials on pasture and turfgrasses, ornamentals, forest and tree crops, vegetables, upland grains and other row crops, and paddy rice. Additional references on the effects of specific slow-release N fertilizers on crops can be found in the following bibliographies: on CDU (TVA, 1977c); IBDU (TVA,1977d, 1982a); urea-Z (TVA, 1977a); oxamide (TVA, 1977e); Osmocote (TVA 1982b); SCU (TVA, 1976, 1980a, 1981a,b,c); and various other materials (TVA, 1977b). These references are representative of the ever increasing literature in this subject area. An examination of them indicates patterns in use of slow-release N that are useful in projecting future use.

V. CURRENT AND PROJECTED USE OF SLOW-RELEASE NITROGEN FERTILIZERS

Accurate statistics on consumption of slow-release N fertilizers are difficult to assemble because such information is scattered among many fertilizer distributors who do not keep accurate records of the end use of the products sold, and where accurate sales records are kept, such information often is proprietary. Data that are available usually are given in terms of total product rather than amount of N or other nutrient element; products vary widely in N content, e.g., from about 5% in some organic wastes to 66% in melamine. The estimates of fertilizer consumption given below are based on information taken from various sources, including the *Fertilizer Reference Manual* (Fertilizer Institute, Washington, DC), Association of American Plant Food Control Officials, Stanford Research Institute marketing research reports, Lawn and Garden, Green Markets, and personal communications.

Processed organic wastes currently comprise about 65 to 70% of the total tonnage of slow-release fertilizers used in the USA. About 25% of these processed wastes are used on agricultural crops, mainly on citrus and strawberries (*Fragaria* × *ananassa* Duch.). Of all commercial slow-release N sources, urea-formaldehyde reaction products account for approximately one-half of the N used (about 50 to 55 thousand metric tons of N equivalent to about 135 000 metric tons of product). Most of the ureaform-N (an estimated 85%) is for nonfarm use (lawns, gardens, landscaping, professional turf, nurseries). Less than 90 metric tons of N from IBDU or polymer-coated urea is used on agricultural crops in the USA. Slightly more than 1800 metric tons of N from these materials is used for nonfarm purposes.

Consumption of SCU appears to be increasing; an estimated 45 000 metric tons of product (containing 31 to 37% N) was used during 1982. Market analysts project its increased use because it is an effective controlled-release N source with an economic advantage over other slow-release N products (e.g., it is about $165/metric ton cheaper than ureaforms). However, as currently manufactured, it cannot be granulated with other fertilizers or used with fluids, a limitation that makes it unattractive to manufacturers of turf fertilizers and excludes it from liquid lawn care services. Sulfur-coated urea shows some promise for use on flooded rice where its higher initial cost for a single preflood application may be offset by the savings made by eliminating postflood aerial applications. However, nurseries, turf management, and lawn care appear to be the most promising areas for its increased use, especially after uniform-size granules that are compatible with mixed fertilizers become available.

Slow-release N fertilizers clearly will not compete with conventional soluble N fertilizers in terms of total N use in agriculture. There may be some encroachment into cropping systems where the advantage of slow release may justify the higher cost. Flooded rice already has been mentioned (see also section VII). Where labor is costly or in short supply, it may be profitable to use slow-release N on sugarcane. Because its growth extends over many months, adding fertilizer after the canopy closes cannot readily be done mechanically. Cane growers would welcome a low-cost, slow-release material that could meet the crop's need for N with a single early growth application. Similarly, N cannot easily be added to strawberries growing through plastic mulch. Accordingly, slow-release fertilizers are used on about 70% of California strawberries. Use of slow-release fertilizers also may increase with high-value cash crops such as peppers (*Capsicum annum* L. var *annum*) and staked tomatoes (*Lycopersicom esculentum* Mill.).

VI. NITRIFICATION INHIBITORS

Nitrification, defined here as the biological oxidation of NH_4^+ to NO_3^-, is a key process in soil that affects the nutrition of higher and lower plants, promotes movement of N within the soil, and may result in N loss from the soil. Nitrate supply and uptake as an important crop production factor are reviewed by Hageman (1984). This section focuses on the disadvantages of NO_3^- formation in soils, which include: (i) promotion of N loss via bio- and chemodenitrification, leaching, and runoff; (ii) seedling damage from accumulated NO_2^-; (iii) NO_3^- movement away from the plant-root zone; (iv) increase in subsoil acidity following repeated application of nitrifiable fertilizers; and (v) nitrate accumulation in ground waters. Where the disadvantages of nitrification outweigh the advantages, control of nitrification should lead to increased efficiency of N use measured in terms of increased yield or improved quality or both.

Many chemicals have been tested for their potential as selective inhibitors of nitrification. The ideal substance should specifically inhibit the oxidation of NH_4^+ to NO_2^-; have the same degree of mobility in soil as NH_4^+; retain its effectiveness for several weeks; be chemically and physically compatible with fertilizers during manufacture, storage, and application; be economical; and be nontoxic to higher plants, beneficial soil microorganisms, and humans. Chemicals used commercially or patented as nitrification inhibitors are discussed in reviews by Gasser (1965), Prasad et al. (1971), and Hauck (1972).

Many laboratory and greenhouse experiments have shown that nitrification inhibitors under certain conditions can reduce N loss and increase crop yields. More importantly, yield responses to nitrification inhibitors are often demonstrated in the field. In general, yield increases are obtained under conditions where formation of NO_3^- would have led to N loss via leaching or denitrification. Such loss can occur in warm, permeable soils in high-rainfall areas; soils that are abnormally wet in the spring; irrigated, aerobic soils; and paddy.

Only two chemicals are registered for use as nitrification inhibitors in the USA. They are 1-chloro-6-(trichloromethyl) pyridine, developed by the Dow Chemical Company and marketed under the trade name N-Serve® Nitrogen Stabilizer, common name, nitrapyrin; and 5-ethoxy-3-(trichloromethyl)-1, 2, 4-thiadiazole, developed by Olin Corporation and marketed under the trade name Dwell®. The active ingredient has been marketed as a fungicide under the trade marks Terrazole®, Pansoil®, Truban®, Koban®, and Aaterra®. Dwell® was registered for use with fertilizers late in 1982, but Olin Corporation has sold its interest in this chemical to Uniroyal Chemical Company, Bethany, CT, and its further market development is open to question. Market testing of a third chemical, dicyandiamide, was begun in the USA during 1984 (to be discussed later).

The first commercial sale of nitripyrin was in 1976. It is used mainly with anhydrous ammonia, although it can be used effectively with liquids incorporated into the soil. Permissible application rates currently are limited to 1.12 kg/ha, which usually is sufficient to retard nitrification in the injection band. Recommended application rates range from 0.28 to 1.12 kg of nitrapyrin/ha, depending on N rate and crop. Terrazole® can be applied with ammonia or N solutions or as a preplant spray incorported into the soil following fertilization. The recommended application rate is about 0.55 kg/ha of active ingredient.

Because nitropyrin and Terrazole® have high vapor pressures, there is need for nitrification inhibitors that can be used during the manufacture of solid N fertilizers. To this end, interest has been renewed in the use of dicyandiamide (1-cyanoguanidine) and thiourea. Both materials have been used commercially in limited quantities as nitrification inhibitors in Japan for almost 20 yr. Dicyandiamide (dicyan, DCD, Dd) was first developed as a nitrification inhibitor by Showa Denko, K. K. (Japan). It is being developed for use in Europe, North America, and elsewhere by SKW Trostberg AG (FRG) for use as a nitrification inhibitor and as a slow-release N fertilizer. As a nitrification inhibitor, dicyandiamide is sold in Europe, under the trade name Didin®, mostly for use in liquid manures. As a slow-release N source, dicyandiamide (66.7% N) is a component of Alzodin®, a product containing 20% N, 90% of which is present as ammonium sulfate and 10% as dicyandiamide for use mainly on lawns and turf. In the USA, Lebanon Chemical (Turf Products Division) and SKW Trostberg AG are sharing technologies to produce and introduce into the U.S. market a mixed fertilizer, Lebanon 22–4–8–DCD® Turf Food; SKW also is distributing dicyandiamide (DCD) through agreements made with Estech, Inc. (Bartow, FL) and Southern Farmers Association (Little Rock, AR). During 1984, DCD was used commercially mainly with urea on about 2700 ha planted to flooded rice (about 2200, 450, and 25 ha in Arkansas, Mississippi, and Louisiana, respectively).

A DCD concentration of 10% DCD–N by weight of the total N in the fertilizer usually inhibits nitrification without being toxic to higher plants. Its degradation products in soil reportedly include guanylurea, guanidine, and urea (Amberger, 1981). Thiourea (36.8% N) also theoretically can serve the dual function of nitrification inhibitor and slow-release N source, but has not been used for this purpose in the USA.

Use of nitrification inhibitors may increase crop yields when nitrification leads to N loss from the soil, thereby making N a limiting growth factor. Farmers usually cannot forecast whether such adverse conditions will occur during any particular growing season. If their experience suggests that N loss via leaching or denitrification sometimes or often has resulted in yield reduction, farmers may use nitrification inhibitors to ensure that substantial N loss will not occur if the weather in the coming growing season should be unfavorable. Under some cropping situations, it is highly probable that use of a nitrification inhibitor will be beneficial, e.g., for the direct-seeded rice system. In this system, at least some fertilizer N usually is added at seeding. While the seedlings establish themselves, soil conditions are conducive to nitrification. After flooding, the soil becomes anaerobic and nitrate formed during the preflood period is almost completely denitrified. Clearly, control of nitrification under these conditions conserves N, thereby maintaining the crop production value of the fertilizer (for appropriate references, see Prasad, 1966 and Prasad et al., 1970).

Voluminous literature is available on the effects of nitrification inhibitors in soil-plant systems. Results of field experiments in the USA, mainly with nitrapyrin, have been summarized recently (Meisinger, 1980). Current thought on their potential for use in U.S. agriculture is discussed by Hoeft (1984). Also available are the proceedings of two technical workshops on the use of dicyandiamide as a nitrification inhibitor (SKW, 1981; Hauck & Behnke, 1981). Numerous references on nitrification inhibitor use are listed in the bibliographies on dicyandiamide (TVA, 1981d), thiourea (TVA, 1981e), and nitrapyrin (TVA, 1982c).

VII. UREASE INHIBITORS

Solid urea and N solutions that contain urea account for almost 30% of the N used by U.S. farmers. Urea is the dominant form of N fertilizer in world agriculture (e.g., about 80% of Asian N consumption). Urea use worldwide has increased about threefold since 1970 (e.g., from about 7.5 million metric tons of urea-N to about 22.7 million metric tons in 1983). However, some problems associated with its use as a fertilizer remain. These problems are the result of the rapid urease-catalyzed hydrolysis of urea to ammonium carbonate. The resultant increase in alkalinity and NH_4^+ concentration can lead to loss of N as NH_3, especially when urea is surface-applied without incorporation into the soil. Damage to germinating seed and to seedlings also may occur from band application of urea. Formerly, farmers were advised not to surface-apply without incorporation and not to place urea too close to seed, but new crop and soil management practices such as minimum tillage make surface application and single-pass application of seed and fertilizer practicable. Therefore, it is sometimes desirable to slow the rate of urea hydrolysis or to delay hydrolysis until urea can be washed into the soil. One approach is to add a substance that will inhibit soil urease activity to urea.

Many compounds have been identified that will interfere with urease activity. Many of these chemicals inhibit urease prepared from jackbean (*Canavalia ensiformis* (L.) DC.] but do not inhibit soil urease. Some may be somewhat effective in light-textured soils but not in heavier ones. Also, there appears to be no clear relationship between the molecular structure of a chemical and its potency as a urease inhibitor. This may be because the structure of the urease complex, particularly in soil, is largely unknown. For comprehensive reviews of soil urease activity, see Bremner and Mulvaney (1978), Mulvaney and Bremner (1981), Ladd and Jackson (1982), and Gorelik and Gritsevich (1984).

Among the chemicals that inhibit urease to some degree are (i) substituted ureas such as methylurea, dimethylurea, thiourea, and phenylurea (Sor et al., 1966; Geissler et al., 1970); (ii) dithiocarbamates (Hyson, 1963; Tomlinson, 1967); (iii) mercaptans (Gould et al., 1978); (iv) quinones and polyhydric phenols (Anderson, 1969, 1970); (v) heavy metal salts (Sor, 1969); and (vi) various halogens, cyanides, boron- or fluorine-containing compounds, copper or nickel chelates, and urea complexes. Antimetabolites toward urease-producing microorganisms have been suggested for use with urea, such as γ-benzenehexachloride and pyridine-3-sulfuric acid (Peterson & Walter, 1970).

The ideal urease inhibitor for use with urea fertilizer should effectively retard urea hydrolysis in soil for about 10 to 20 d, possess chemical and physical characteristics that will permit its ready addition to urea (preferably at the site of urea manufacture), remain stable in storage after incorporation into fertilizer, and be nonphytotoxic to crop plants, specific toward urease (innocuous toward other beneficial soil enzymes), safe to use, and relatively inexpensive (Voss, 1984). Chemicals that show promise do not, however, meet all of the above specifications. For example, phenylmercuric acetate inhibits soil urease at relatively low concentration (about 5 mg/L but is unacceptable as a fertilizer amendment because of its mercury content. Para-benzoquinone and 2, 5-dimethyl benzoquinone, effective in some soils as urease inhibitors, are carcinogens. The most promising chemicals are phenylphosphorodiamidate (PPDA) and phosphorotriamidate (PTA). The experimental materials are expensive, but an economical commercial process for their production probably could be developed using phosphenous chloride (POCl) as the main intermediate. Phenylphosphorodiamidate has been found to decrease ammonia volatilization from urea applied to floodwater (Vlek et al., 1980); in greenhouse pots cropped to rice, yields were higher from urea amended with PPDA than with urea not amended with urease inhibitor (Byrnes et al., 1983). However, unpublished studies at TVA and elsewhere in the greenhouse under aerobic soil conditions suggest that PPDA is rapidly hydrolyzed and/or degraded in soil, thereby reducing its effectiveness within a few days after application. Several phosphorodiamidates and phosphorotriamidates were found less potent than PPDA as inhibitors of soil urease activity (Martens & Bremner, 1984). Current TVA work indicates that certain phosphoronitriles, of low potency initially, degrade in soils to form products that show promise as inhibitors of soil urease activity. The bibliography compiled by TVA (1981f) attests to growing interest in PPDA and similar compounds as a urease inhibitor for use in ruminants and with urea fertilizers.

VIII. POTENTIALS AND RESEARCH NEEDED

The rationale for using slow-release N fertilizers and N fertilizer amendments is essentially the same as discussed earlier (Hauck & Koshino, 1971; Hauck, 1972). In the USA, slow-release N fertilizers are used mostly on ornamentals, lawns, and professional turfgrasses, but there is a growing trend toward using them for high-cash agricultural crops under special situations of use. These situations include crops growing under mulch, in highly permeable soils in high rainfall areas, and under conditions where nitrification/denitrification is almost certain to occur, e.g., in direct-seeded, flooded rice culture. During the mid-1970s, some new products have entered the market while a few have been withdrawn. There appears to be a growing trend to combine slow-release N fertilizers with soluble sources of P, K, and other nutrients. This should stimulate needed studies on the effect that soluble components of a mixture have on N release from the less soluble component.

Commercial use of a nitrification inhibitor (nitrapyrin) in the USA began in 1976. A second product (Terrazole®) has been registered but its commercial availability is uncertain. However, dicyandiamide, an effective nitrification inhibitor for use on agronomic crops, is being marketed in small quantities as a slow-release N fertilizer (mixed with other nutrient sources) for professional turf and garden use, and will be marketed as a nitrification inhibitor for use on agronomic crops.

The potential market for nitrification inhibitors is large; probably $< 5\%$ is being exploited. There is little question that chemicals already available will effectively inhibit nitrification and that N loss and loss of crop yield can accordingly be decreased. Expanding use of nitrification inhibitors will depend on (i) farmer understanding and acceptance of the concept of inhibitor use as a N management asset, (iii) increasing costs of fertilizer N relative to profit obtained from fertilizer use, (iii) farmer experience in determining the cost effectiveness of using a nitrification inhibitor, and (iv) delineation of cropping situations under which economic return for use of a nitrification inhibitor is highly probable. As with slow-release N fertilizers, convenience and labor savings may be important factors that will prompt some farmers to use a nitrification inhibitor. The desire to fall-apply N on midwestern corn or to delay flooding on wetland rice often makes use of a nitrification inhibitor attractive. In some situations, such as direct-seeded, flooded rice, use of either slow-release N or conventional N amended with a nitrification inhibitor will decrease N loss.

Cost-benefit ratios for use of slow-release N fertilizers often remain difficult to determine because their accurate determination must consider convenience factors and crop quality as well as yield and fertilizer costs. A large knowledge gap still exists on the relationship between rate of N uptake and use by plants. Not known, for example, is why guanylurea is more effective in increasing rice growth in the presence of NH_4^+-N or why ureaform produces better growth of long-season, tall rice varieties than of short-season, stiff-strawed ones. Nitrogen uptake and use by rice from IBDU and oxamide appears to be a complex function of N source and rate, particle size (N release rate), and method of rice culture or watering regime (unpublished data). The physiological reasons for these observations have not been determined.

Although increased knowledge of the physiology of N use by crop plants does not necessarily lead directly to expanded use of slow-release N fertilizers, it may be useful in delineating cropping situations where these materials have special value.

Considerable information is available from laboratory and greenhouse experiments concerning use of slow-release N fertilizers and nitrification inhibitors. Attention is being focused on their use in productive ecosystems. Studies with nitrapyrin have progressed beyond the question of whether the chemical inhibits nitrification. In one listing of more than 300 studies with nitrapyrin, 11 were concerned with plant diseases such as root rot (e.g., *Fusarium graminearum*), stalk rot (e.g., *Diploidia zeae* Lev), and potato scap *(Streptomyces scabies)*; 36 with nutrient uptake, concentration within the plant, and use; 14 with effects on soil microbial populations; 11 with toxicity to higher plants; and 19 with inhibitor degradation and/or movement in soil. Crops used in these and other studies include corn, rice and other cereals, cotton, forage and other grasses, sugar crops, fruits and vegetables, and ornamentals. Also, many agricultural chemicals including herbicides, fungicides, and insecticides have been tested for their effect on nitrification, urease activity, and other N transformations in soils (see Goring & Laskowski, 1982 for a comprehensive listing). The recent literature suggests that practical questions concerning nitrification inhibitor use are being answered. Continued evaluation of experimental and commrcial use of nitrification inhibitors is needed for more accurate determination of cost/benefit ratios for specific cropping situations. More information is needed on the residual effects of slow-release N using N tracer techniques. There is evidence that nitrification inhibitors increase heterotrophic immobilization of applied NH_4^+ (e.g., see Juma & Paul, 1983). Will such immobilization under reduced tillage be increased where N fertilizers amended with nitrification inhibitors are used? If so, is the immobilized N remineralized during the fall and early spring when the land is in fallow?

Considerable effort is being directed toward increasing the efficiency of N use by agricultural and nonagricultural plants. An effective urease inhibitor has not yet been developed, but a variety of slow-release N fertilizers and nitrification inhibitors are available to achieve this objective. Each type of material will be used in increasing amounts, according to need for them in particular soil-plant systems and situations.

Plant physiologists and pathologists are invited to join soil chemists and microbiologists to answer questions about the effects of slow-release N and fertilizer N amendments in soil. Fertilizer chemists and engineers can respond to the information obtained by developing new materials and modifying old ones. Agronomists can delineate new areas of need, eventually leading to increased production and more favorable economics of use.

REFERENCES

Allen, S. E. 1984. Slow-release nitrogen fertilizers. p. 195–206. *In* R. D. Hauck (ed.) Nitrogen in crop production. American Society of Agronomy, Crop Science Society of America, and Soil Science Society of America, Madison, WI.

Amberger, A. 1981. Dicyandiamide as a nitrification inhibitor. p. 3–17. *In* R. D. Hauck and H. Behnke (ed.) Proc. Technical Workshop on Dicyandiamide. Muscle Shoals, AL. December. SKW Trostberg AG, Trostberg, FRG.

Anderson, J. R. 1969. Fertilizing process and composition. British Patent 1 142 245. Date issued: 5 February Abstr. in Chem. Abstr. 70:95902.

Anderson, J. R. 1970. Fertilizer compositions. U.S. Patent 3 515 532. Date issued: 2 June. Abstr. in Chem. Abstr. 73:65535.

Armiger, W. H., I. Forbes, R. E. Wagner, and F. O. Lundstrom. 1948. Urea-form—a nitrogenous fertilizer of controlled availability: Experiments with turf grasses. J. Am. Soc. Agron. 40:342–356.

Army, T. J. 1963. Coated fertilizers for the controlled release of plant nutrients. Agric. Chem. 18(8):26–28, 81–82.

Bouldin, D. R., S. D. Klausner, and W. S. Reid. 1984. Use of nitrogen from manure. p. 221–245. *In* R. D. Hauck (ed.) Nitrogen in crop production. American Society of Agronomy, Crop Science Society of America, and Soil Science Society of America, Madison, WI.

Bremner, J. M., and R. L. Mulvaney. 1978. Urease activity in soils. p. 149–196. *In* R. G. Burns (ed.) Soil enzymes. Academic Press, Inc., New York.

Bridger, G. L. 1968. Magnesium ammonium phosphate and related compounds. p. 256–284. *In* New fertilizer materials. Noyes Development Corp., Park Ridge, NJ.

Bridger, G. L., M. L. Salutsky, and R. W. Staroska. 1962. Metal ammonium phosphates as fertilizers. J. Agric. Food Chem. 10:181–188.

Brigham, R. O. 1917. Assimilation of organic nitrogen by *Zea mays* and the influence of *Bacillus subtilis* on such assimilation. Soil Sci. 3:155–195.

Burns, A. F., and A. M. Smith. 1966. Available magnesium in mixed fertilizers. Agric. Chem. 21:45–47.

Byrnes, B. H., N. K. Savant, and E. T. Craswell. 1983. Effect of a urease inhibitor, phenylphosphorodiamidate, on the efficiency of urea applied to rice. Soil Sci. Soc. Am. J. 47:270–274.

Gassner, J. K. R. 1965. Soils. Rep. Prog. Appl. Chem. 50:201–212.

Geissler, P. R., K. Sor, and T. M. Rosenblatt. 1970. Fertilizer compositions containing borax and (or) copper sulfate as urease inhibitors. U.S. Patent 3 523 018. Date issued: 4 August. Abstr. in Chem. Abstr. 73:98051.

Goring, C. A. I., and D. A. Laskowski. 1982. The effects of pesticides on nitrogen transformations in soils. *In* F. J. Stevenson (ed.) Nitrogen in agricultural soils. Agronomy 22:689–720.

Gorelik, L. A. and Yu. G. Gritsevich. 1984. Inhibitors of soil urease activity. Agrokhimiya 1984(2):103–126.

Gould, W. D., F. D. Cook, and J. A. Bulat. 1978. Inhibition of urease activity by heterocyclic sulfur compounds. Soil Sci. Soc. Am. J. 42:66–72.

Hageman, R. H. 1984. Ammonium versus nitrate nutrition of higher plants. p. 67–85. *In* R. D. Hauck (ed.) Nitrogen in crop production. American Society of Agronomy, Crop Science Society of Agronomy, and Soil Science Society of Agronomy, Madison, WI.

Hamamoto, M. 1966. Isobutylidene diurea as a slow acting nitrogen fertilizer and the studies in this field in Japan. Proc. Fert. Soc. 90:1–64.

Hamamoto, M. 1968. Isobutylidene diurea = IBDU (a slow acting nitrogen fertilizer). p. 28–37. *In* New fertilizer materials. Noyes Development Corp. Park Ridge, NJ.

Hauck, R. D. 1972. Synthetic slow-release fertilizers and fertilizer amendments. p. 633–690. *In* C. A. I. Goring and J. W. Hamaker (ed.) Chemicals in the soil environment, Vol. 2. Marcel Dekker, Inc., New York.

Hauck, R. D. 1980. Mode of action of nitrification inhibitors. p. 19–32. *In* J. J. Meisinger (ed.) Nitrification inhibitors—potentials and limitations. Spec. Pub. 38. American Society of Agronomy, Crop Science Society of America, and Soil Science Society of America, Madison, WI.

Hauck, R. D., and H. Behnke (ed.) 1981. Proc. of the Technical Workshop on Dicyanidiamide, Muscle Shoals, AL. December. SKW Trostberg AG, Trostberg, FRG.

Hauck, R. D., and M. Koshino. 1971. Slow-release and amended fertilizers. p. 455–494. *In* R. A. Olson et al. (ed.) Fertilizer technology and use, 2nd ed. Soil Science Society of America, Madison, WI.

Hauck, R. D., and H. F. Stephenson. 1964. Nitrification of triazine nitrogen. J. Agric. Food Chem. 12:147–151.

Hayase, T. 1967. On the slowly available nitrogen fertilizers, Part 1. Nogyo Gijutsu Kenkyusho Shiryo B 18:129–303. (In Japanese with English summary.)

Hays, J. T. 1964. Fertilizers for controlled release of nitrogen. Hercules Chem. 49:1–7.

Hays, J. T., and W. W. Haden. 1966. Soluble fractions of ureaforms—nitrification, leaching, and burning properties. J. Agric. Food Chem. 14:339–341.

Hoeft, R. G. 1984. Current status of nitrification inhibitor use in U.S. agriculture. p. 561–570. *In* R. D. Hauck (ed.) Nitrogen in crop production. American Society of Agronomy, Crop Science Society of America, and Soil Science Society of America, Madison, WI.

Hyson, A. M. 1963. Fertilizer composition comprising urea and dithiocarbamates. U.S. Patent 3 073 694. Date issued: 15 January. Abstr. in Chem. Abstr. 58:9590h.

Isabell, R. E., and J. D. Fleming. 1969. Agronomic value of organic nitrogen compounds. Literature survey. Spec. rep. no. S-438. Tennessee Valley Authority, Muscle Shoals, AL.

Ishizuka, Y., and H. Takagishi. 1962. Urea formaldehyde, "Ureaform," as nitrogenous fertilizer: III. Fertilizer value of various ureaforms. Nippon Dojo Hiryogaku Zasshi 33:83–87. (In Japanese with English summary.)

Juma, N. G., and E. A. Paul. 1983. Effect of a nitrification inhibitor on N immobilization and release of ^{15}N from nonexchangeable ammonium and microbial biomass. Can. J. Soil Sci. 63:167–175.

Jung, J., and O. H. Detmer. 1968. Crotonylidenediurea = CDU (a slow-acting nitrogen compound). p. 16–27. *In* New fertilizer materials. Noyes Development Corp., Park Ridge, NJ.

Knop, K., V. Vanek, M. Zazvorka, and J. Neuberg. 1965. Some characteristics of ureaform fertilizers. Agricultural College, Prague, p. 159–168.

Ladd, J. N., and R. B. Jackson. 1992. Biochemistry of ammonification. *In* F. J. Stevenson (ed.) Nitrogen in agricultural soils. Agronomy 22:173–228.

Martens, D. A., and J. M. Bremner. 1984. Effectiveness of phosphorodiamidates for retardation of urea hydrolysis in soils. Soil Sci. Soc. Am. J. 48:302–305.

Meisinger, J. J. (ed.) 1980. Nitrification inhibitors—potentials and limitations. ASA Spec. Pub. 38. American Society of Agronomy, Crop Science Society of America, and Soil Science Society of America, Madison, WI.

Morgan, W. A., and R. D. Kralovec. 1953. Chemical method for available fertilizer nitrogen in urea-formaldehyde compositions. J. Assoc. Off. Agric. Chem. 36:907–914.

Mulvaney, R. L., and J. M. Bremner. 1981. Control of urea transformations in soil. p. 153–196. *In* E. A. Paul and J. M. Ladd. (ed.) Soil biochemistry, Vol. 5. Marcel Dekker, Inc., New York.

Murray, J. 1958. Notices and abstracts. Br. Assoc. Adv. Sci. Rep. 28:54–55.

Ogata, T., and K. Hino. 1958a. On the properties of oxamide as gradually acting nitrogenous fertilizer: II. Effect of oxamide on paddy rice (1). Nippon Dojo Hiryogaku Zasshi 29:383–388. (In Japanese with English summary.)

Ogata, T., and K. Hino. 1958b. On the properties of oxamide as gradually acting nitrogeneous fertilizer: III. Effect of oxamide on paddy rice (2). Nippon Dojo Hiryogaku Zasshi 29:441–443 (In Japanese with English summary.)

Ogata, T., and K. Hino. 1958c. On the properties of oxamide as gradually acting nitrogenous fertilizer: IV. Effect of oxamide on upland crops. Nippon Dojo Hiryogaku Zasshi 29:501–504. (In Japanese with English summary.)

Ogata, T., and K. Hino. 1958d. On the properties of oxamide as gradually acting nitrogenous fertilizer: V. Effect of oxamide on paddy rice (3). Nippon Dojo Hiryogaku Zasshi 29:535–540. (In Japanese with English summary.)

Peterson, A. F., and C. R. Walter, Jr. 1970. Regulation of urea hydrolysis in soil. U.S. Patent 3 547 614. Date issued: 15 December. Abstr. in Chem. Abstr. 74:63585.

Powell, R. 1968. Controlled release fertilizers. Noyes Development Corp., Park Ridge, NJ.

Prasad, R. 1966. Slow-acting nitrogenous fertilizers and nitrification inhibitors. Fert. News 11:27–32.

Prasad, R., G. B. Rajale, and B. A. Lakhdive. 1970. Effect of time and method of application of urea and its treatment with nitrification inhibitors on the yield and nitrogen uptake by irrigated upland rice. Indian J. Agric. Sci. 40:1118–27.

Prasad, R., G. B. Rajale, and B. A. Lakhdive. 1971. Nitrification retarders and slow-release nitrogen fertilizers. Adv. Agron. 23:337–383.

Rehberg, B. E. 1984. Ethylene inhibition in plants. U.S. Patent 4 441 918. Date issued: 10 April. Abstr. in Official Gazette, 10 April, p. 718.

Rohner, L. V., and A. P. Wood. 1947. Fertilizers containing insoluble nitrogen. U.S. Patent 2 415 705. Date issued: 11 February. Abstr. in Chem. Abstr. 41:2837e.

SKW. 1981. SKWDIDIN® nitrogen stabilizer for liquid manure. Bayer. Landwirtsch. Jahrb. 58(7).

Slangen, J. H. G., and P. Kerkhoff. 1984. Nitrification inhibitors in agriculture and horticulture. Fert. Res. 5:1–76.

Sommers, L. E., and P. M. Giordano. 1984. Use of nitrogen from agricultural, industrial, and municipal wastes. p. 207–220. *In* R. D. Hauck (ed.) Nitrogen in crop production. American Society of Agronomy, Crop Science Society of America, and Soil Science Society of Agronomy, Madison, WI.

Sor, K. M. 1969. Fertilizer Composition consisting of urea, a urease inhibitor and a hydrocarbon binder. British Patent 1 157 400. Date issued: 9 July. Abstr. in Chem. Abstr. 71:69813.

Sor, K. M., R. L. Stansbury, and J. D. DeMent. 1966. Agricultural nutrient containing urea. U.S. Patent 3 232 740. Date issued: 1 February. Abstr. in Chem. Abstr. 64:14918c.

Stanford Research Institute. 1981. Fertilizers: Controlled release and nutrient efficiency. p. 535.8000A–535.8004N. *In* Chemical economics handbook. Stanford Research Institute, Palo Alto, CA.

Tennessee Valley Authority. 1972. Coatings for slow-release fertilizers. TVA Bibliography no. 1011. Technical Library, Tennessee Valley Authority, Muscle Shoals, AL.

Tennessee Valley Authority. 1973. Ureaform-manufacture and properties. TVA Bibliography no. 1345. Technical Library, Tennessee Valley Authority, Muscle Shoals, AL.

Tennessee Valley Authority. 1976. Sulfur coated urea, production and use (1947–1976). TVA Bibliography no. 1325. Technical Library, Tennessee Valley Authority, Muscle Shoals, AL.

Tennessee Valley Authority. 1977a. Urea-Z (urea acetaldehyde). TVA Bibliography no. 1349. Technical Library, Tennessee Valley Authority, Muscle Shoals, AL.

Tennessee Valley Authority. 1977b. Miscellaneous slow-release fertilizers. TVA Bibliography no. 1350. Technical Library, Tennessee Valley Authority, Muscle Shoals, AL.

Tennessee Valley Authority. 1977c. CDU and floranid. TVA Bibliography no. 1348. Technical Library, Tennessee Valley Authority, Muscle Shoals, AL.

Tennessee Valley Authority, 1977d. IBDU (isobutylenediurea). TVA Bibliography no. 1351. Technical Library, Tennessee Valley Authority, Muscle Shoals, AL.

Tennessee Valley Authority. 1977e. Oxamide (1958–1976). TVA Bibliography no. 1347. Technical Library, Tennessee Valley Authority, Muscle Shoals, AL.

Tennessee Valley Authority. 1978. Coatings for slow-release fertilizers (1972–1978). TVA Bibliography no. 1011 S (Suppl.). Technical Library, Tennessee Valley Authority, Muscle Shoals, AL.

Tennessee Valley Authority. 1980a. Sulfur-coated urea, production and use (1976–1979). TVA Bibliography no. 1325 S (Suppl.). Technical Library, Tennessee Valley Authority, Muscle Shoals, AL.

Tennessee Valley Authority. 1980b. Ureaform—manufacture and properties. TVA Bibliography no. 1345 S (Suppl.). Technical Library, Tennessee Valley Authority, Muscle Shoals, AL.

Tennessee Valley Authority. 1981a. Sulfur-coated urea, production and use (1978–1980). TVA Bibliography no. 1325 S(2) (Suppl. no. 2). Technical Library, Tennessee Valley Authority, Muscle Shoals, AL.

Tennessee Valley Authority. 1981b. Sulfur-coated urea—use on corn (1971–1980). TVA Bibliography no. 1325-A. Technical Library, Tennessee Valley Authority, Muscle Shoals, AL.

Tennessee Valley Authority. 1981c. Sulfur-coated urea—use on turf grass (1970–1980). TVA Bibliography no. 1325-B. Technical Library, Tennessee Valley Authority, Muscle Shoals, AL.

Tennessee Valley Authority. 1981d. Dicyandiamide (1907–1981). TVA Bibliography no. 1671. Technical Library, Tennessee Valley Authority, Muscle Shoals, AL.

Tennessee Valley Authority. 1981e. Thiourea (1932–1980). TVA Bibliography no. 1675. Technical Library, Tennessee Valley Authority, Muscle Shoals, AL.

Tennessee Valley Authority. 1981f. Phenylphosphorodiamidate (1942–1981). TVA Bibliography no. 1672. Technical Library, Tennessee Valley Authority, Muscle Shoals, AL.

Tennessee Valley Authority. 1982a. IBDU (isobutylenediurea) (1977–1982). TVA Bibliography no. 1351 S (Suppl.). Technical Library, Tennessee Valley Authority, Muscle Shoals, AL.

Tennessee Valley Authority. 1982b. Osmocote (1977–1982). TVA Bibliography no. 1694. Technical Library, Tennessee Valley Authority, Muscle Shoals, AL.

Tennessee Valley Authority. 1982c. Nitrapyrin (1969–1982). TVA Bibliography no. 1689. Technical Library, Tennessee Valley Authority, Muscle Shoals, AL.

Terman, G. L., J. D. DeMent, C. M. Hunt, J. T. Cope, Jr., and L. E. Ensminger. 1964. Crop response to urea and urea pyrolysis products. J. Agric. Food Chem. 12:151–154.

Tomlinson, T. E. 1967. Controlling urea hydrolysis in soils. British Patent 1 094 802. Date issued: 13 December. Abstr. in Chem. Abstr. 68:48637.

U.S. Department of Agriculture. 1974–1984. Commercial fertilizers. USDA, Econ. and Stat. Service Board, Crop Rep. Board Publ. SpCr 7. U.S. Department of Agriculture, Washington, DC.

Ushioda, T. 1969. Development and commercial scale of CDU (slow-acting fertilizer) technology. Jpn. Chem. Q. 5(4):27–32.

Vlek, P. L. G., J. M. Stumpe, and B. H. Byrnes. 1980. Urease activity and inhibition in flooded soil systems. Fert. Res. 1:191–202.

Voss, R. D. 1984. Potential for use of urease inhibitors. p. 571–577. *In* R. D. Hauck (ed.) Nitrogen in crop production. American Society of Agronomy, Crop Science Society of Agronomy, and Soil Science Society of Agronomy, Madison, WI.

Westerrman, R. L., and L. T. Kurtz. 1972. Residual effects of nitrogen-15 labeled fertilizers in a field study. Soil Sci. Soc. Am. Proc. 36:91–94.

Westerman, R. L., L. T. Kurtz, and R. D. Hauck. 1972. Recovery of nitrogen-15 labeled fertilizers in field experiments. Soil Sci. Soc. Am. Proc. 36:82–86.

Yamazoe, F., and T. Hayase. 1966. Chemical method for urea-formaldehyde components. Nippon Dojo Hiryogaku Zasshi 37:69–73. (In Japan with English summary.)

9

Production, Marketing, and Use of Phosphorus Fertilizers

Ronald D. Young
National Fertilizer Development Center
Tennessee Valley Authority
Muscle Shoals, Alabama

D. G. Westfall
Colorado State University
Fort Collins, Colorado

Gary W. Colliver
Farmland Industries
Kansas City, Missouri

Phosphorus fertilizer production primarily involves the mining, upgrading, and conversion of phosphate rock to more soluble P compounds that can be effectively utilized by plants. Certain chemical processes, such as the solubilization of phosphate rock with acids, have been basic to the phosphate industry for many years. However, continued technological improvements have resulted in more economical production of products of higher analysis and better quality by these processes. New P fertilizers have also been produced through the efforts of research in phosphate chemistry and accompanying developments in technology. Work continues in efforts to develop new products that will be useful as fertilizers. Growth in the use of liquid and suspension fertilizers has resulted in some P compounds being used as fertilizers for the first time and has also influenced marketing, distribution, and usage practices.

Starting in the 1970s, there have been major environmental improvements in mining and processing of phosphate rock and in production of P fertilizers. Major efforts have also been directed toward improving the energy efficiency of all steps of processing of phosphate rock to fertilizers.

The marketing of P fertilizers, like all types, has changed markedly since about 1950 to adapt to the availability of economical fertilizers of much higher analysis that became available. The evolution in marketing concepts and practices was affected even more by the sharp increase in fertilizer nutrients use (almost fivefold between 1950 and 1982) that was spurred by research, extension, and industry educational efforts. Large basic producers of fertilizer intermediates expanded to take over wholesale distributing facilities and retail outlets. Bulk blending developed as a logical pattern for this system and grew to represent 40% of NPK fertilizers in the USA in 1982. Bulk handling and distribution became highly predominant.

Fluid mixed fertilizers experienced consistent growth in the USA from the mid-1950s to account for about 20% of the total mixed fertilizers in 1982. Both bulk blending and fluid fertilizer systems required radical innovations in marketing concepts and techniques to reach the status they have attained. Phosphorus fertilizer intermediates were developed or adapted to fit these systems. Some of the major fertilizer intermediates are granular diammonium phosphate (DAP), granular triple superphosphate (TSP), and wet-process superphosphoric acid.

Continuing research by agronomists and soils scientists has shed much more light on the complex systems of P fertilizer reactions and movement in the soil. Factors that affect uptake and utilization of P by plants have been defined more clearly. The physical and chemical characteristics of the soil largely control reactions that take place and the availability of P to plants. Phosphorus fertilization practices have evolved in efforts to obtain maximum benefits, and the growing practice of conservation tillage or reduced tillage poses additional factors to be considered.

The purpose of this chapter is to review some of the main principles and practices involved in the production, marketing, and use of P fertilizers. Emphasis is on developments since about 1960.

I. PHOSPHATE ROCK—THE STARTING POINT

Phosphate rock is the starting point for all P fertilizers. With few exceptions, the larger deposits of the world are either directly or indirectly of sedimentary origin, having been laid down in beds under the ocean and subsequently elevated to land masses, or redeposited from surface water that percolated through such beds. The phosphate usually is in the form of small pellets cemented together by $CaCO_3$. In some cases the cementing material is leached away, leaving more or less loose pebbles; in others, the whole mass has been compressed to form hard rock. The principal phosphate mineral form in most deposits is *francolite,* a carbonate fluorapatite represented by the formula $Ca_{10}F_2(PO_4)_6 \cdot XCaCO_3$. Principal impurities of concern are Fe, Al, and Mg. Fluorine is recovered as saleable by-products.

The world is richly endowed with phosphate deposits. Estimates of reserves range from 30 to 50 billion metric tons, and these include only those deposits that could be mined economically within the limitations of present technology. World production of phosphate rock in 1981 was 141.7 million metric tons. The leading phosphate-producing countries (Anon., 1983b) are shown below.

During the 1970s, world production of phosphate rock increased by 87%, and USA production increased by 47%. Historically, new world re-

Country	Million metric tons	% of world total
USA	53.6	38
USSR	25.9	18
Morocco	19.7	15
People's Republic of China	12.0	8
Tunisia	4.6	3
Jordan	4.2	3
Republic of South Africa	3.0	2
Brazil	2.8	2
Togo	2.2	2
Israel	2.4	2
Nauru	1.5	1
Senegal	2.2	2
Syria	1.3	1

serves of phosphate rock have been discovered at a much higher rate than mining has increased.

Almost 40% of world phosphate rock production in 1980 entered world trade. Morocco was the leading exporter, with 16 million metric tons (32% of the world total). The USA was second with 14 million metric tons (28% of the total), and the USSR was third with 5 million metric tons (9% of the total). The USA led in total exports of rock and processed phosphate fertilizers with 33% of the world total in 1980.

Phosphate rock reserves in the USA are located principally in Florida; North Carolina; the western states of Idaho, Utah, Wyoming, and Montana; and Tennessee. The greatest reserves are in the western states, but Florida leads in production as the data from 1981 show below.

State	Million metric tons	% of total
Florida	41	76
North Carolina	5	10
Utah, Wyoming, Idaho, and Montana	5	10
Tennessee	3	4

North Carolina deposits were the latest to be exploited; mining was started there by one large producer in the mid-1960s. A second producer was operating in 1981 with the implementation of a large mining, beneficiation, and phosphate fertilizer production complex.

II. MINING AND BENEFICIATION

Almost all the mining of phosphate rock in the USA and throughout the world is done by strip mining. Only a small portion of the production in the western USA and a few other locations is by underground mining. In the Florida pebble fields, overburdens averaging about 4.5 m thick must be removed to expose the phosphate matrix, which averages about 7 m deep. In North Carolina the mining is a pumping and dredging operation since the deposits are far below the water table. Stripping of the overburden and mining are done by very large mobile dragline excavators; the largest of these have capacities of 1500 metric tons or more h^{-1}. Generally the draglines deposit the matrix in pits, where it is broken up by hydraulic jets before being pumped as a slurry to the beneficiation plant. Alternatively, transport of the matrix is by very large trucks.

Florida phosphate matrix typically has a P content of only 6 to 7% and must be beneficiated to make it suitable for processing into fertilizers. The beneficiation process is a relatively complex procedure that involves wet screening, hydroseparation, and concentration by flotation using chemical flotation agents. The process results in several size fractions that include a pebble fraction > 1.19 mm (+ 16-mesh screen) containing from 13 to 15% P (30–35% P_2O_5) and smaller size fractions averaging somewhat higher in grade, ranging up to 15.5% P. These products are dried and sometimes calcined or ground before shipment.

The North Carolina deposit has heavy overburden (12 m) and rests on a major aquifer. Mining is 43 m below sea level, and 227 million L of fresh water must be pumped from the aquifer daily. All handling of mined rock is by pumping a slurry. Calcination is a necessary step in beneficiation, but organics in the rock provide a substantial part of the fuel.

Mining in the Tennessee fields is carried out by strip mining similar to that in Florida, but beneficiation is simpler because the rock is used largely in the production of elemental P, in which relatively low-grade material can be used. The higher-grade matrix is used without beneficiation and the lower-grade material is upgraded by screening and washing.

Some western rock analyzes up to 15% P as mined and can be used without treatment. However, lower-grade ore comprises the bulk of western production and must be handled like Florida ore unless it is to be used in P furnaces.

In the course of beneficiation of phosphate ore of flotation methods, the impurities removed are discarded as slimes and tailings. This waste material contains a considerable amount of phosphate, ranging from one-third to one-half of that in the original ore, that defies recovery because of its extremely fine particle size. It is generally disposed of by allowing it to dewater in mud ponds, which sometimes are the pits in mined-out areas. The tailings are potentially a source of pollution of streams and lakes if the ponds are allowed to leak. Also, the mining operation itself can be detrimental to the environment because it leaves the land in a disturbed state. Environmental pressure continued to increase on phosphate mining and beneficiation operations in the 1970s. Regulations and compliance procedures increased the cost and time required to obtain mining permits and bring a mine into production. Safe containment of slimes from beneficiation was very rigidly enforced after earlier incidents of major spills into streams.

Substantial progress has been made in all aspects of environmental controls. Phosphate mining companies have put much effort into rehabilitating mined-out areas, converting them into highly desirable residential or commercial property in several cases.

During the 1970s the impurity content of Florida phosphate rock from the main source in the central part of the state increased gradually as the better grade material became depleted. The grade of product rock became slightly lower, and modifications in processes for its use were necessary because of impurities, particularly Mg. Mining is expected to gradually shift farther south during the 1980s to material that lies deeper and will be more costly to mine. The impurities content also will be higher.

A few estimates in the late 1970s of U.S. phosphate rock reserves gave a pessimistic outlook for self-sufficiency beyond the turn of the century. A review and assessment of a number of estimates by government, industry, and engineering consulting organizations indicate that U.S. phosphate rock reserves total 3.5 to 4 billion metric tons. This is equivalent to $>$ 60 yr of production at the 1981 rate of 53.6 million tons/yr. It is certain, however, that the "bread-and-butter" Florida rock will decline in quality and be more expensive to mine and process.

III. PHOSPHATE MANUFACTURE

After phosphate rock is mined and beneficiated, the task of converting the fluorapatite structure to a more soluble form that can be effectively utilized by crops remains. The performance of this task is the basic purpose of all phosphate manufacturing processes. Phosphate manufacturing started in the early 1800s when it was recognized that pulverized bones could be made more available for plant nutrition by treatment with sulfuric acid (H_2SO_4). In 1842 the first commercial production of superphosphate from phosphate rock and H_2SO_4 was started in England by Lawes. This product prevailed as the world's leading phosphate fertilizer for $>$ 100 yr.

The principal routes used by the fertilizer industry in going from phosphate rock to intermediates and finished fertilizers are shown in Fig. 9–1.

It can be seen from Fig. 9–1 that H_2SO_4 is the main intermediate, along with phosphate rock, for production of phosphate fertilizers. Thus, phosphate rock and elemental S (for use in production of H_2SO_4) are the key raw materials. Their availability establishes the location of major phosphate fertilizer production centers. In the USA these production complexes are located at the phosphate rock mining sites in Florida, North Carolina, western states, and Tennessee. Sulfur moves via the Gulf of Mexico from Louisiana and Texas to the Florida production centers and further by ocean to North Carolina. Sulfur recovered from smelting operations provides a large part of that used in the western states.

Several North African countries and the Republic of South Africa, as pointed out earlier, have become major producers of phosphate rock, and are moving to production of large quantities of phosphate fertilizers for domestic use and export. Phosphate fertilizer production centers in Europe,

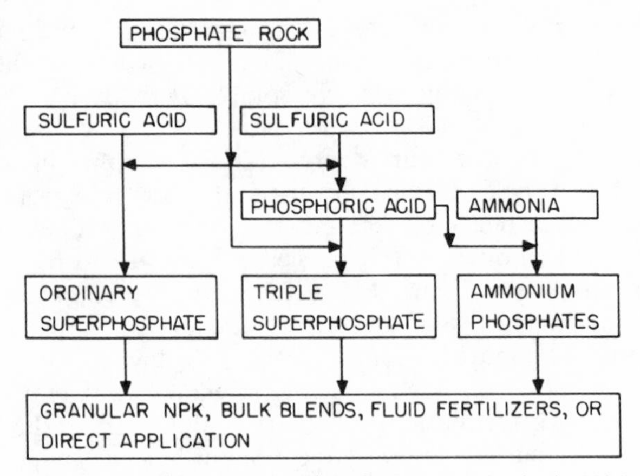

Fig. 9–1. Principal routes of phosphate processing in the USA.

Asia, and Latin America generally import both phosphate rock and S. A few Asian and South American countries are beginning to utilize their domestic sources of phosphate rock.

Wherever practical, the phosphate production centers are located for receipt of raw materials and shipment of products economically by ocean or river barge.

The production of ordinary superphosphate led in the USA until 1964 when TSP became the leading phosphate intermediate. The very popular and dependable Tennessee Valley Authority (TVA) process for granular DAP was introduced in 1961 and led quickly to ammonium phosphate becoming dominant in the USA by 1967 and in the world by 1974. Production of the main phosphate fertilizers in the USA in 1981 (TVA, 1982, unpublished data) is shown below.

Fertilizer	Million metric tons of P
Ammonium phosphates	2.4 (5.4 P_2O_5)
Triple superphosphate	0.7 (1.6 P_2O_5)
Ordinary superphosphate	0.13 (0.3 P_2O_5)

World production of phosphates was about 13.5 million metric tons of P (31 metric tons of P_2O_5) in 1980. It is estimated that ammonium phosphates (monoammonium phosphate [MAP] and DAP) accounted for about 60% of the total production (Anon., 1981).

A. Phosphoric Acid

Phosphoric acid (H_3PO_4) is the most important intermediate in the production of phosphate fertilizers. In addition to its use in the production of ammonium phosphates and concentrated superphosphate, it is an intermediate for mixed fertilizers, both liquid and solid.

Two basic methods are used commercially for the production of H_3PO_4: the wet process, which is a chemical method; and the thermal process, which requires elemental P as an intermediate. In 1950 production of H_3PO_4 by the two methods was about equal in the USA, but by 1981 production by the wet process was about 15 times that by the thermal process—4.1 million metric tons of P by the wet process and 0.3 million metric tons by the thermal route. Because of the much higher cost, essentially no thermal H_3PO_4 has been used in the production of fertilizers in the USA since the mid-1970s. A very high percentage of the wet-process acid goes into fertilizer production, with small amounts treated specially for use in animal feed–grade materials, detergents, and industrial phosphates.

1. Wet Process

Wet-process H_2PO_4 is produced by the action of H_2SO_4 on pulverized, beneficiated phosphate rock. The principal chemical reaction involved may be represented by the equation

$$\underset{\text{Fluorapatite}}{Ca_{10}F_2(PO_4)_6} + 10H_2SO_4 + 20H_2O \longrightarrow \underset{\text{Gypsum}}{10CaSO_4 \cdot 2H_2O} + 2HF + 6H_3PO_4.$$

A simplified flow diagram of the operation is shown in Fig. 9–2. The process consists of three steps: (i) the reaction of phosphate rock and H_2SO_4 to form H_3PO_4 and $CaSO_4 \cdot 2H_2O$, (ii) separation and washing of the acid from the sulfate, and (iii) concentration of the acid. Variables important to successful operation of the process are the type of phosphate rock used, the amount of H_2SO_4, the temperature of the reaction, and the slurry density controlled by recycling weak acid to the digestion step. Im-

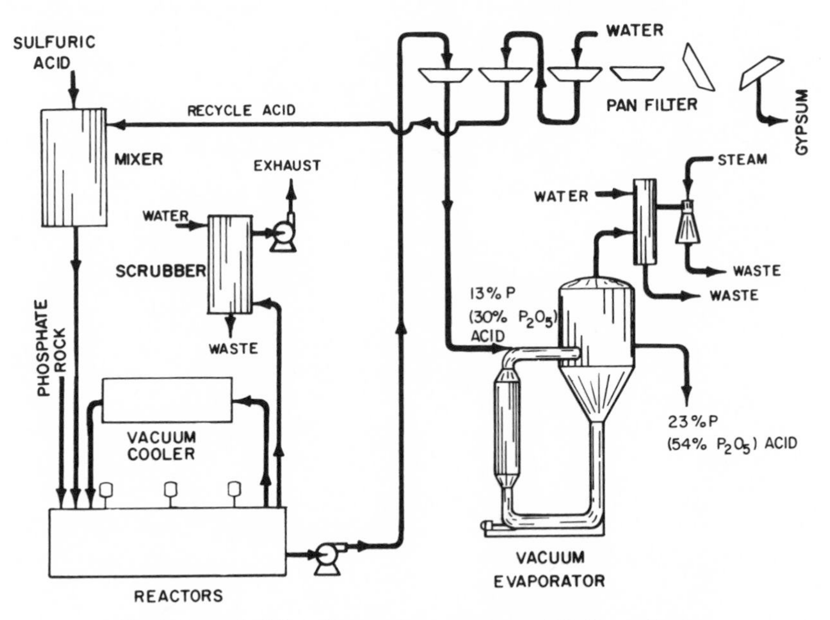

Fig. 9–2. Typical flow diagram for wet-process H_3PO_4 production.

proper control of these variables results in poor filtration, incomplete dissolution of the rock, or both. The level of impurities in the phosphate rock, particularly Fe, Al, and Mg, affect H_3PO_4 production and subsequent use of the product acid.

There are three variations of the basic process. Each has the same general flow arrangement but is differentiated by the forms in which the calcium sulfate is crystallized: as the anhydrite ($CaSO_4$), the hemihydrate ($CaSO_4 \cdot 1/2H_2O$), or the dihydrate ($CaSO_4 \cdot 2H_2O$). The state of hydration of the calcium sulfate is determined by the temperature and concentration during digestion. The anhydrite and hemihydrate processes have the advantage of producing acid of higher concentration, but the dihydrate process does not impose as many operating problems and is the process used almost exclusively in the USA.

During the 1970s, several hemihydrate or combination hemihydrate-dihydrate plants were built throughout the world.

Among the individual installations of the dihydrate process, the number and size of the reaction vessels and the type of filtration equipment used vary considerably. There has been a strong trend since the early 1960s toward the use of a single vessel system for extraction and larger rotary tilting pan filters for removal of gypsum. This trend has led to the development of very large single-train plants since the mid-1970s, with much saving in investment and operating costs. Modern plants can be built with single production trains having a capacity of 400 metric tons of P (900 metric tons of P_2O_5) per day.

Acid produced by the dihydrate process contains about 13% P (30% P_2O_5) and generally must be concentrated to 17 to 24% P (40–54% P_2O_5) before use. Concentration usually is accomplished in continuous steam-heated vacuum evaporators.

Wet-process acid is an impure product containing a complex system of impurities that include compounds of Fe, Al, Ca, Mg, and F, as well as free H_2SO_4 and organic matter. A portion of the impurities is present as solids due to imperfect filtration and precipitation during and after concentration. They cause difficulties in storage and handling of the acid. These problems were not so difficult to cope with when the acid was used promptly at the point of production, but in recent years the acid has become an important article of commerce, shipped widely in tank cars and ocean vessels as a fertilizer intermediate. Consequently, considerable effort has been made to decrease the solids content of the product. This is done usually by settling, although the settling rate is slow and postprecipitation occurs for a long while. The use of calcined rock in the production of the acid greatly decreases the tendency toward postprecipitation and simplifies the clarification process.

A more sophisticated treatment for purification involves extraction of a portion of the acid from the impure product by an organic solvent, usually an aliphatic alcohol. This process is not widely used, however, because of its cost.

Several new approaches have been studied and proposed, but the only radical departure in the production of wet-process acid that has been commercialized even to a minor extent is one that involves extraction of the rock with hydrochloric acid (HCl) rather than H_2SO_4. In the version developed by Israel Mining Industries, H_3PO_4 is separated from calcium chloride ($CaCl_2$) solution by extraction with butyl or isoamyl alcohol. This acid is in turn separated from the solvent with water and is concentrated by evaporation. The resulting acid is relatively pure, but the cost is considerably higher than the H_2SO_4 approach.

Pollution control is a substantial problem in the manufacture of wet-process acid because a large portion of the reactants become by-products. The principal by-product of the dihydrate process is impure gypsum of little practical use. It is generally discarded, although space for this often becomes costly and difficult to find since the gypsum comprises about two-thirds of the phosphate rock used and almost all of the H_2SO_4. In areas of high alkalinity, by-product gypsum can sometimes be sold as a soil amendment. In Japan and a few other areas of the world where the more pure natural gypsum is unavailable, the by-product is used in plaster and wallboard production. Conversion of gypsum to H_2SO_4 is technically feasible but quite unattractive economically. Environmental concerns and controls continued to intensify during the 1970s; disposal of gypsum and control of aqueous effluents from its handling became more rigid.

The only other by-product, F, is easier to cope with. It is a constituent of the gases evolved during digestion of phosphate rock and can be recovered with water by a relatively simple scrubbing procedure. The water is either recycled through a pond where the F reacts with gypsum to form insoluble calcium fluoride (CaF_2), or a multistage scrubbing system is used to recover saleable fluosilicic acid of about 25% concentration used for fluoridation of water and for other purposes. There is limited commercial practice in recovery of F as compounds used as intermediates in production of synthetic cryolite. Most of the water utilized in H_3PO_4 production is recycled, but the portion that must be discharged to the environment is first treated with lime to neutralize acids and remove phosphates and F by precipitation. The much more rigid environmental regulations on F during the 1970s increased the difficulty and expense of controlling F emissions in the working areas from plant stacks and in aqueous effluents.

The economic recovery of U from H_3PO_4 became established commercially during the 1970s. Uranium is present in several commercial phosphate rocks in the proportion of 100 to 200 mg/kg. Most of it is extracted and dissolved in the product H_3PO_4 during its production. A high percentage of the U can be recovered by solvent extraction and precipitation.

2. Thermal Phosphoric Acid

Very little thermal H_3PO_4 has been used in production of fertilizer since the early 1970s because of its much higher cost. Therefore, this acid is touched on only briefly here. Elemental P is produced by the reduction of phosphate rock with coke in an electric arc furnace. The phosphate, along with coke and silica charged into the furnace crucible, is heated by an electric arc from three electrodes to the temperature at which reduction takes place. By-product calcium silicate slag and ferrophosphorus are drawn off intermittently at the bottom of the furnace; P is volatilzed continuously from the top and collected in a water-cooled condenser.

The production of H_2PO_4 from elemental P requires a combustion chamber in which the P is oxidized with air to form P_2O_5, a hydrator for absorbing the oxide in weak acid to form H_3PO_4, and means for recovering as weak acid the P_2O_5 and acid mist not retained in the hydrator. The combustion chamber and hydrator are usually vertical, water-cooled towers of unlined stainless steel. Recovery of residual phosphorus pentoxide from the gas leaving the hydrator is accomplished most commonly by use of a high-energy scrubber (Striplin, 1969; Allgood et al., 1967).

The acid can be readily produced in concentrations up to superphosphoric acid of 33 to 37% P (76–85% P_2O_5; 105–115% H_3PO_4). TVA developments in use of thermal superphosphoric acid in liquid fertilizers led to preparation of this type of acid by evaporation of wet-process acid.

3. *Wet-Process Superphosphoric Acid*

Developments by TVA have led to commercial production of a highly concentrated acid, usually referred to as *superphosphoric acid.* More than 50% of its P is in the pyro and more highly condensed forms. Wet-process superphosphoric acid is made by evaporation of water to concentrate orthophosphoric acid beyond the equivalent of 100% H_3PO_4, at which point molecular dehydration causes the formation of molecules of the general formula $H_{x+2}P_xO_{3_{x+1}}$. Because of the impurities in wet-process acid, the practical concentration attained is 30 to 31% P (68–72% P_2O_5). Higher concentrations cause the product to be unmanageable due to its viscosity, even though the normal practice is to handle it at elevated temperature to decrease the viscosity.

TVA development of a simple pipe reactor process for preparation of liquid fertilizers of high polyphosphate content from wet-process superphosphoric acid of only 30 to 31% P (68–70% P_2O_5) was a boon to commercial production, handling, and use of this acid. This low-conversion superphosphoric acid became established in the early 1970s as the standard commercial product. Preparation of acid of this concentration is easier and more economical, and its much lower viscosity makes handling, transport, and use easier (Meline et al., 1972).

Most of the commercial systems for preparation of wet-process superphosphoric acid use vacuum evaporators of alloy steel construction. Limiting the concentration of product acid to 30 to 31% P (68 to 70% P_2O_5 with 25 to 35% as polyphosphate) requires only steam for heating. The low-conversion acid has a viscosity of only 1 to 3 kPa/s (100–300 cP) at 65° C (Whitehurst, 1974).

Superphosphoric acid finds its use principally in liquid fertilizers, where it makes possible the production of higher analysis products. The polyphosphates also sequester certain impurities that would otherwise cause trouble in the handling of liquid fertilizers. This acid is produced from North Carolina, Florida, and western phosphate rocks and is shipped routinely in specially designed railroad tank cars. Ocean shipment of large quantities of low-conversion superphosphoric acid to the USSR in specially designed tankers was started in the late 1970s by a major producer in Florida. Other world acid exporters were planning in 1982 for future shipment of this more concentrated acid to take advantage of substantial savings in transport cost. Slack (1968) describes in detail the production of H_3PO_4.

B. Superphosphates

Superphosphate is a term first applied to the fertilizer produced by the action of H_2SO_4 on phosphate rock. Later, when H_2PO_4 became available to the fertilizer industry, the more concentrated product made with this acid was called triple, double, or concentrated superphosphate, and the earlier product became ordinary, or normal, superphosphate. Both hold a declining but important place in the fertilizer industry primarily because they are relatively cheap to produce. The chief fertilizer constituent of each is monocalcium phosphate monohydrate, formed by the reaction of the acid with the fluorapatite in the rock according to the following reactions:

$$Ca_{10}PO_4)_6F_2 + 7H_2SO_4 + 3H_2O \longrightarrow 3Ca(H_2PO_4)_2 \cdot H_2O + 7CaSO_4 + 2HF$$

or

$$Ca_{10}(PO_4)_6F_2 + 14H_3PO_4 + 10H_2O \longrightarrow 10Ca(H_2PO_4)_{2P14} \cdot H_2O + 2HF.$$

Other reactions occur that involve impurities in the rock: a part of the hydrofluoric acid (HF) reacts with $CaCO_3$ to yield CO_2. Evolution of these gaseous products gives the superphosphates their characteristic porous structure.

Ordinary superphosphate generally contains 8.7% available P (20% P_2O_5); triple superphosphate, more concentrated by virtue of the absence of $CaSO_4$, usually contains 20% P (46% P_2O_5). Both types of superphosphate are used extensively for direct application to the soil and in the production of mixed fertilizers. In the latter use, particular importance is attached to their ability to react with, or fix, anhydrous ammonia, the most economical source of N available to the industry.

1. Ordinary Superphosphate

In 1842 Lawes in England followed up on the pioneering work of Liebig and received a patent on the use of H_2SO_4 and organic phosphate material (instead of bones) to produce superphosphate. An industry grew slowly, and in 1862 about 0.15 million metric tons of what later became referred to as *normal,* or *ordinary superphosphate* (OSP), was produced in England. By 1870 there were 70 ordinary superphosphate plants in the

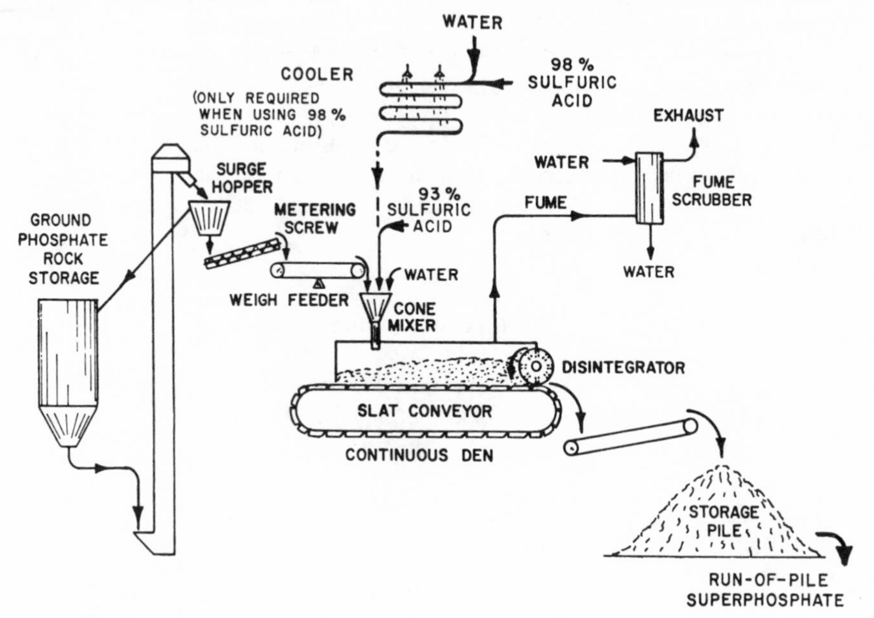

Fig. 9–3. Continuous process for the manufacture of normal superphosphate.

United Kingdom and seven in the Charleston, SC area where phosphate rock (nodules) had been discovered.

In 1888 the commercial shipment of phosphate rock from Florida was initiated, and a major industry in that state followed. Ordinary superphosphate was later produced in local plants from phosphate rock shipped from that area (Young & Davis, 1980).

Ordinary superphosphate is the simplest and oldest (since 1842) of manufactured phosphate fertilizer. Pulverized phosphate rock is treated with H_2SO_4 in a comparatively simple plant to produce a product containing about 9% P (20% P_2O_5). Products in the earlier years were of lower analysis, ranging from 7 to 8% P (16–18% P_2O_5). Ordinary superphosphate can be used for direct application, for bulk blending, or in production of granular NPK fertilizers. If the OSP is to be used for direct application or bulk blending, it should be granulated. The main equipment consists of a mixer, a den, and, if granulation is used, some suitable type of equipment for granulation with steam or water. The simple TVA cone mixer is used widely in a continuous process for mixing of phosphate rock and H_2SO_4. The dens can be either the batch or continuous type. The slat-conveyor continuous den, commonly referred to as a *Broadfield den,* is used in many continuous systems (Young & Davis, 1980). A diagram of a typical cone mixer and den of this type is shown in Fig. 9–3.

Various types of continuous mixers are available, but the lowest cost and simplest is the TVA cone mixer that has no moving parts, as shown in Fig. 9–4. Mixing is accomplished by the swirling action of the acid. Short, single-shaft or double-shaft pug mills are also used for mixing.

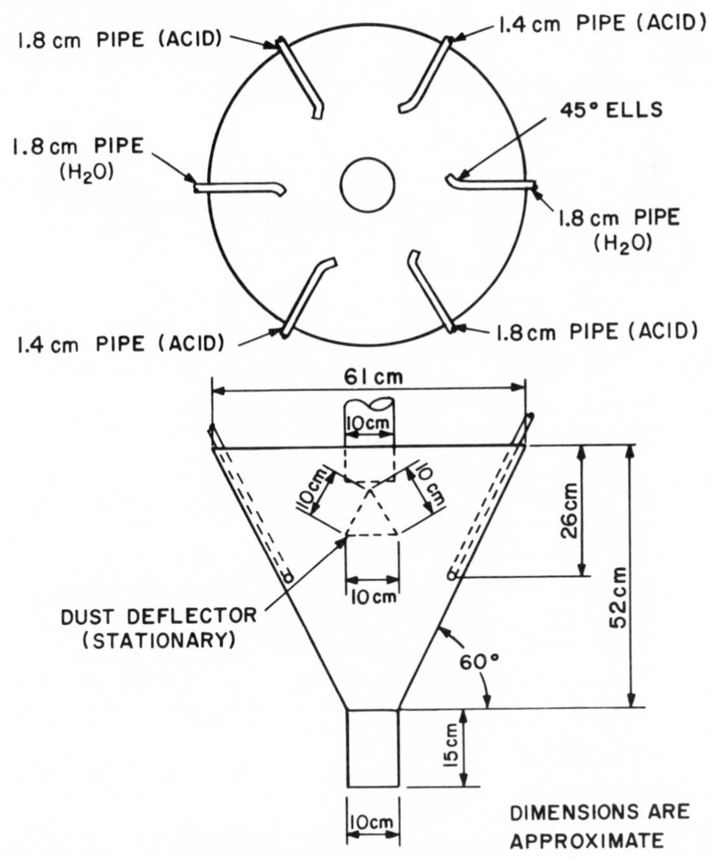

Fig. 9–4. Typical cone mixer for production of 23 to 27 metric tons/h of ordinary superphosphate.

The proportioning of phosphate rock with H_2SO_4 to produce OSP is usually based on a simple relationship of about 0.6 kg of H_2SO_4 (100% H_2SO_4 basis)/kg of phosphate rock (13–14% P; 30–32% P_2O_5). The rock is pulverized to about 90% −100-mesh size and 70% −200-mesh size.

Gases released as the superphosphate is solidifying (setting) results in a porous, friable mass in the den. Ordinary superphosphate will set in 40 to 50 min in a continuous den and can be cut and handled satisfactorily.

The superphosphate usually is held in storage piles (cured) for 4 to 6 weeks to obtain better handling properties and to allow the chemical reaction to continue. The usual grade of OSP made from Florida rock is 9% P (20% available P_2O_5); an analysis is shown below.

P								
Total	Available	Water soluble	Ca	Free acid	S	F	R_2O_5	Mg
			% by weight					
8.8	8.6	7.9	20.1	3.7	11.9	1.6	1.6	0.09

Because of the low analysis of OSP, economics favor shipping the phosphate rock (14% P, or 32% P_2O_5) to local plants where the superphosphate is produced and usually used in formulations for granular NPK fertilizers.

In Australia and New Zealand, OSP is granulated in large quantities as it is discharged from the den. Pan granulators and rotary drums are used. Smaller amounts are granulated in Western Europe by use of steam and water in rotary drums.

Ordinary superphosphate production in the USA in 1981 was only about 1.4 million metric tons of product (0.12 million metric tons of P).

2. Triple Superphosphate

Triple superphosphate (TSP) is made by acidulation of phosphate rock with H_3PO_4 with equipment and process similar to that for OSP. Triple superphosphate containing 20% P (45–46% P_2O_5) did not appear on the scene until wet-process H_3PO_4 (see section IIIA3) was produced commercially. TVA initiated production of similar concentrated superphosphate in the late 1930s by use of electric furnace H_3PO_4. Widespread agronomic testing and market development through TVA programs with universities and industry led to rapid acceptance of this much higher analysis phosphate intermediate. Producers of phosphate rock in the USA and other countries moved into production of wet-process acid and TSP. Logistics favored production of the higher analysis TSP—20% P (46% P_2O_5) vs. 14% P (32% P_2O_5) for phosphate rock—in large plants near the source of rock and shipping this intermediate to mixed fertilizer plants near the markets.

The TVA cone mixer has been almost universally used in production of nongranular TSP. Since the set time for TSP is only 14 to 20 min compared with 40 to 50 min for OSP, a simpler, cupped conveyor belt is used to hold the acidulate until the TSP solidifies. The production rate averages 40 to 50 metric tons/h with a belt 1.5 to 1.8 m wide and 31 m long. A diagram of the cone mixer and wet-belt system is shown in Fig. 9–5 (Young & Davis, 1980).

Proportioning for TSP is typically 1.1 kg of P from acid for each 0.45 kg of P from rock. The TSP usually is cured 4 to 6 weeks before shipment or use at the site. Typical chemical analysis of TSP made with Florida rock is shown below.

P									
Total	Available	Water soluble	Free acid	Ca	R_2O_3	Mg	F	Water
			% by weight					
20.5	20.2	18.3	3.4	13.8	3.1	0.3	2.7	4.5

Use of TSP in granular NPK fertilizer formulations, together with or in place of OSP, allowed production of higher-analysis grades of finished

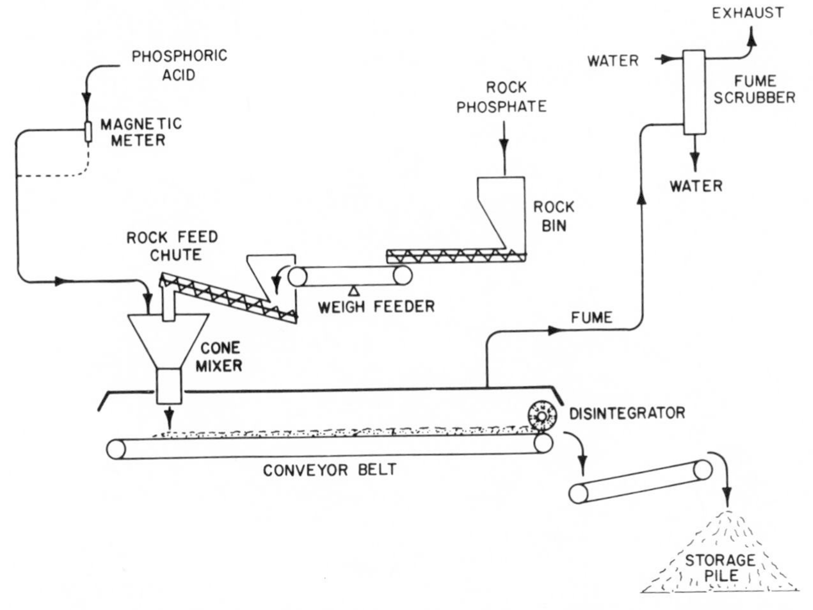

Fig. 9–5. Continuous system for triple superphosphate production.

granular fertilizers such as 14–6–12 (14–14–14) instead of 10–4.4–8.3(10–10–10). Production of TSP in the USA in 1981 was about 3.2 million metric tons of product (0.7 million metric tons of P) (TVA, 1982, unpublished data).

Triple superphosphate is produced in large quantities in granular form for use in direct application and bulk blends. In some processes, cured TSP is granulated in a rotary drum or pan granulator using steam and water to promote granulation. In Australia, New Zealand, and the United Kingdom, the superphosphate is granulated in a pan or drum as it comes from the den.

A slurry-type granulation process as shown in Fig. 9–6 is used widely throughout the world. Pulverized phosphate rock is treated with wet-process H_3PO_4 in a two-stage reaction system, and the slurry is sprayed into a pug mill or rotary drum for layering on recycle at a ratio of 10 to 12 kg of recycle/kg of product. Product granules with 19.5% P (45% P_2O_5) are quite spherical and dense with excellent physical properties. The lower-grade results from the need to decrease the acidulation ratio slightly to minimize prolonged stickiness. For production rates higher than about 25 metric tons/h, a rotary drum is used instead of a pug mill (IFDC/UNIDO, 1979a).

Most of the granular TSP in the USA is used in bulk blending.

C. Ammonium Orthophosphates

Although ammonium phosphates did not come on the scene in significant quantities until the early 1960s, they have become the leading form of phosphate in the USA and in the world. Almost all new phosphate fertil-

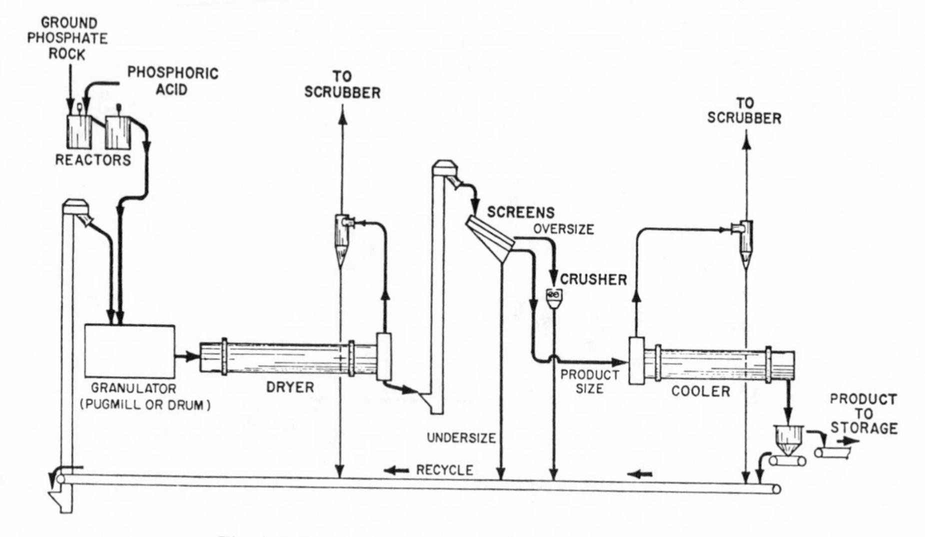

Fig. 9–6. Process for granular triple superphosphate.

izer complexes built since the 1970s and those planned and being built in the 1980s are for production of ammonium phosphate as the major product. Production of ammonium phosphates in the USA in 1981 totaled about 11.7 million metric tons of product (2.4 million metric tons of P).

Crystalline diammonium phosphate was produced by TVA starting in the late 1940s, and demonstration programs showed it to be a very good high-analysis fertilizer. Smaller amounts were produced by others as a by-product. When the simple and dependable TVA process for granular DAP of 18–20–0 grade (18–46–0) was developed in 1960 to 1961, it was rapidly adopted by the industry (Young, et al., 1962). Many granular DAP plants have production capacities of about 45 metric tons/h with a few as high as 65 to 90 metric tons/h.

1. Granular Diammonium Phosphate

A flow diagram of a typical granular DAP production unit of the TVA type that has become standard in the industry is shown in Fig. 9–7. Wet-process H_3PO_4 of about 17.5% P (40% P_2O_5) content is fed to a preneutralizer where anhydrous ammonia is sparged to neutralize the acid to an NH_3–H_3PO_4 mole ratio of about 1.4. This is in a range of maximum solubility of ammonium phosphate as shown in the solubility curve of Fig. 9–8. The heat of reaction evaporates considerable water; the water content of the slurry is only 16 to 20%, but it is quite fluid.

The slurry is pumped at a metered rate and distributed on the bed in a rotary drum, TVA-type ammoniator-granulator. Ammonia is sparged beneath the bed to ammoniate the slurry further to near DAP; the usual finishing NH_3–H_3PO_4 mole ratio is 1.85 to 1.95. Ammonia evolved from the preneutralizer and granulator is recovered in acid of about 13% (30% of P_2O_5) content in a scrubbing circuit. Recycled material at a rate of 4 to 6 kg/kg of product is the primary control of granulation.

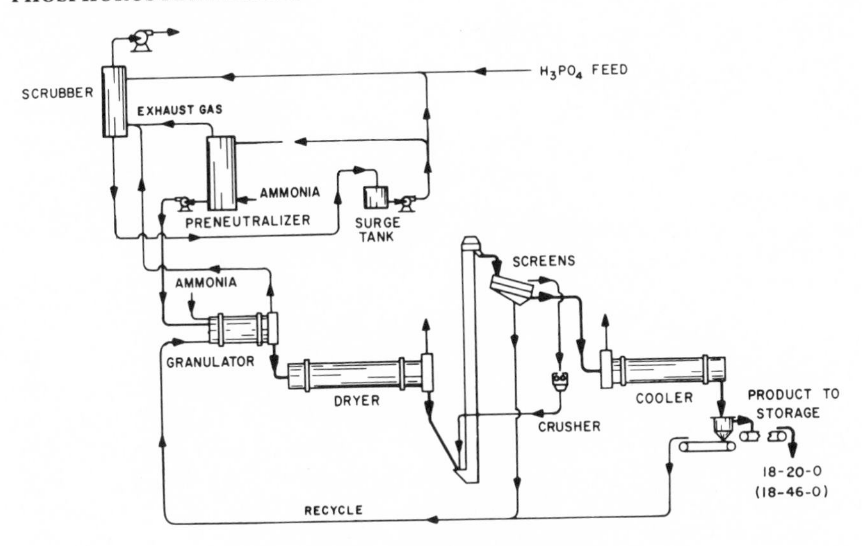

Fig. 9–7. TVA process for granular DAP.

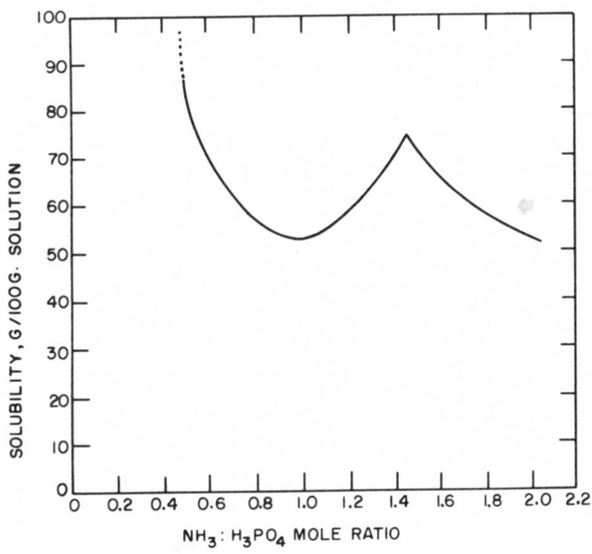

Fig. 9–8. Effect of NH_3–H_3PO_4 mole ratio on solubility in the ammonium phosphate system.

Discharge from the granulator is dried with moderate heat to 82 to 88° C product temperature. Most plants screen hot, and cool only the product fraction since the material is not very sticky. Rotary coolers or the compact and very efficient fluidized bed type of coolers are used. The product has excellent storage and handling properties in bags or in bulk. Millions of tons are shipped in bulk and in bags throughout the world.

2. *Granular Monammonium Phosphate*

Granular DAP (and the grade 18–46–0) has become a household term in the world fertilizer industry, but substantial interest in granular MAP has developed, particularly where soils are mainly alkaline, as in Canada and Pakistan. Also, where the primary interest is in producing and shipping phosphate, the practical 11–23–0 to 10–24–0 (11–52–0 to 10–54–0) grades are attractive to provide higher phosphate payload.

TVA developed two comparatively minor modifications of the granular DAP process to allow production of MAP in existing plants (Young & Hicks, 1967). In one method, the acid is ammoniated to only an NH_3–H_3PO_4 mole ratio of about 0.6 in the preneutralizer and then to about 1.0 in the granulator drum. In the other procedure, which has been adopted by industry, acid in the preneutralizer is ammoniated to about 1.4 as with DAP, and additional wet-process acid is distributed onto the bed to the granulator to adjust back to the MAP mole ratio of about 1.0 (see Fig. 9–8). The remainder of the process with either modification is the same as for DAP, but higher drying temperature can be used to increase the production rate for MAP.

3. *Nongranular Monoammonium Phosphate*

Starting about 1968, simple processes have been developed for production of nongranular, sometimes called powdered, MAP of grade 10–24–0 (10–54–0). The main processes were developed in England, the USA, Scotland, and Japan. A number of these comparatively low-cost units have been built commercially, including plants in the United Kingdom, the Netherlands, Japan, Australia, Spain, the USA, Brazil, and Iran. This intermediate usually is shipped to other plants where it is used in production of NPK fertilizers. The nongranular MAP has not attained the popularity that had been predicted in the mid-1970s (Young & Davis, 1980).

D. Ammonium Polyphosphates

Polyphosphates are formed by the linking of two or more PO_4 tetrahedrons (orthophosphate) through the process of removal or exclusion of part of the chemically combined water as shown in Fig. 9–9. The linking of two molecules produces pyrophosphate, the linking of three produces tripolyphosphate, and the linking of four produces tetrapolyphosphate. This process can continue to form chains (polymers) containing hundreds of PO_4 groups (Fleming, 1969).

Ammonium polyphosphates are very important in liquid fertilizers, since higher analysis grades can be produced and impurities can be held in solution (sequestered) to ensure prolonged quality of the products. Polyphosphates are properly referred to as the backbone of liquid (solution) fertilizers. Granular ammonium polyphosphate fertilizers have not been established in commercial production, although TVA has developed processes and produced these fertilizers for several years in demonstration plants (Siegel & Young, 1969).

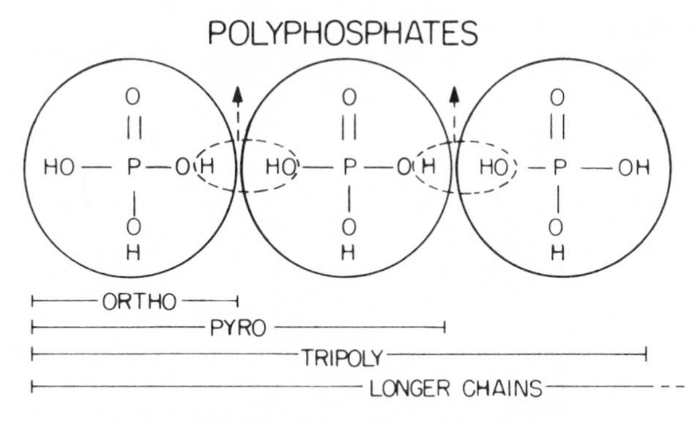

Fig. 9–9. Formation of polyphosphates.

1. Ammonium Polyphosphate Liquids

TVA has done most of the research and development work and demonstration of ammonium polyphosphate fertilizers. The first work was in production of 11–16–0 (11–37–0) liquid by ammoniation of electric furnace superphosphoric acid in a tank-type reactor. This base liquid, from TVA and others, became the foundation of a growing and dynamic liquid fertilizer industry by 1960. Polyphosphate content of the 11–16–0 (11–37–0) was 50 to 60% of the total P (Siegel & Young, 1969).

A sound commercial base for ammonium polyphosphate liquid fertilizers followed TVA's development of wet-process superphosphoric acid, made by concentration of regular merchant-grade H_3PO_4. A quantum jump forward came in the early 1970s when TVA developed a very simple pipe reactor process for preparing 10–15–0 (10–34–0) of 60 to 70% polyphosphate content by ammoniation of wet-process acid that contains only 20 to 30% polyphosphate (see section IIIA3; Meline et al., 1972).

Use of the pipe reactor greatly simplified and decreased the cost of preparation of wet-process superphosphoric acid. This acid (30% P, or 68% P_2O_5) also has lower viscosity and is much easier to handle and ship than acid of high polyphosphate content. It is used $>$ 130 regional plants in the USA. The 10–15–0 (10–34–0) ammonium polyphosphate liquid is shipped to hundreds of satellite plants for preparation of NPK liquid grades by blending (Achorn et al., 1974).

2. Ammonium Phosphate Suspension Fertilizers

TVA developed and introduced 12–17.5–0 (12–40–0) ammonium polyphosphate suspension fertilizer to industry in 1963. This suspension was made by ammoniation of furnace superphosphoric acid and shipped to a large number of commercial plants where it was blended with other materials to produce NPK suspension grades. A 13–18–0 (13–41–0) grade, containing finer crystals of DAP that gave it better storage properties, was introduced in 1969. Suspending-type clay is added in proportion of 2 to 2.5% in preparing suspensions (Young et al., 1968). An orthophosphate suspension with grade of 13–17–0 (13–38–0) was developed by TVA and introduced in demonstration production in 1974. This process and product.

made by ammoniation of merchant-grade wet-process acid (24% P; 54% P_2O_5), has been adopted by industry (Achorn & Salladay, 1978).

TVA's latest suspension that was introduced in demonstration production in 1980 contains about 25% of its phosphate as polyphosphate. This level of polyphosphate is formed easily and economically by ammoniation of merchant-grade wet-process H_3PO_4 (53–54% P_2O_5) in a simple pipe reactor. The grade of the product is 9–14–0 (9–32–0). The main advantage is that the polyphosphate content ensures fluidity for satisfactory handling and storage of the suspension at temperatures as low as −17° C. Industry is showing great interest in this economical product and process (Mann et al., 1981).

Suspensions grew at a higher rate than any other type of fertilizer in the USA throughout the 1970s. They accounted for about 45% of all fluid mixed fertilizers in 1981.

3. Granular Ammonium Polyphosphate

TVA developed a process and operated a demonstration plant for production of granular ammonium polyphosphate by ammoniation of furnace superphosphoric acid in a pressure reactor and granulation of the melt in a pug mill. Grade of the product was 15–27–0 (15–61–0). It was used mainly in preparation of liquid NPK fertilizers (Kelso et al., 1968).

Because of adverse economics of furnace acid after about 1970, TVA shifted emphasis to polyphosphate products based on merchant-grade wet-process acid. A demonstration plant, using a pipe reactor for ammoniation and a pug mill for granulation, was put in operation in 1973. The granular 11–24–0 (11–55–0) product contains about 25% polyphosphate and is used mainly by industry in preparation of NPK suspensions (Lee et al., 1974).

E. Other Phosphates

1. Urea–Ammonium Phosphate

TVA has developed processes for production of granular urea–ammonium phosphate fertilizers that are physical cogranulated mixtures of urea and ammonium phosphate. The most promising process involves granulation of urea melt (99.7%) with ammonium polyphosphate melt from a pipe reactor in the demonstration plant described above for production of 11–24–0 (11–55–0) granular ammonium polyphosphate. Grades produced are 28–12–0 (28–28–0) and 35–7.5–0 (35–17–0) used mainly for bulk blending (Lee et al., 1974).

Urea–ammonium phosphate (NPK) grades are produced commercially in India, Korea, Japan, Norway, and the Netherlands (McCamy et al., 1979).

2. Nitric Phosphates

Fertilizers referred to as nitric phosphate or nitrophosphate are produced by acidulation of phosphate rock with nitric acid (HNO_3) or with mixtures of nitric and sulfuric or phosphoric acids. The reaction of phosphate rock with HNO_3 alone is shown below.

$$Ca_{10}F_2(PO_4)_6 + 20HNO_3 \longrightarrow 6H_3PO_4 + 10Ca(NO_3)_2 + 2HF.$$

This reaction is carried out in a slurry-type operation. If the resulting solution is simply ammoniated, dried and granulated, essentially all of the phosphate would be present as water-insoluble dicalcium phosphate and would contain enough hygroscopic calcium nitrate to seriously impair its handling characteristics. There are three main process variations designed to cope with these problems.

In the most widely used variation, known as the *Odda process,* a portion of the calcium nitrate is removed before ammoniation by cooling, crystallization, and centrifuging. The calcium nitrate is prilled and sold as a by-product (15% N) or is treated with NH_3 and CO_2 to form ammonium nitrate coproduct and calcium carbonate. The remaining solution is ammoniated and granulated or prilled to produce fertilizers containing ammonium nitrate, dicalcium phosphate, and ammonium phosphate. The ratio of ammonium to dicalcium phosphate depends on the proportion of Ca that was removed as nitrate and determines the water solubility of the P in the product. Commercial products range between 25 and 60% water solubility, although 80% is feasible by refrigerated cooling.

The second basic type of nitric phosphate process is one in which either H_3PO_4 or H_2SO_4 is added along with HNO_3 to the extraction step to react with enough of the Ca to prevent the presence of calcium nitrate in the product. When H_3PO_4 is added in sufficient quantity, the ammoniation reaction yields ammonium phosphate, and a part of the P in the product is water soluble. The usual practice is to formulate for 50 to 60% water solubility. Several different process designs are used, but the main variations are the PEC process, which uses a special granulator-dryer (Spherodizer, C&I–Girdler, Inc., Louisville, KY) to convert the ammoniated extract to solid form, and the TVA process, which uses a rotary drum for the final stage of ammoniation and for granulation. These mixed-acid processes are relatively simple, there is no coproduct, and the N–P ratio of products can be varied in the range of 1:0.9 to 1:0.2 (N–P_2O_5 ratio from 1:2 to 2:1) (Young, 1965).

Another modification, which has not become important commercially, copes with the excess Ca through the addition of soluble sulfate salt to precipitate calcium sulfate. The addition may be made before or during ammoniation of the extraction slurry. The calcium sulfate may be left in the product, but this produces low-analysis grades. Alternatively, it may be removed by filtration, producing higher analysis grades with nearly complete water solubility of the phosphate. Several different sulfate salts have been used in this type of process, but the most attractive is ammonium sulfate, since it is a widely available by-product. In another version, the calcium sulfate may be reacted with NH_3 and CO_2 to regenerate ammonium sulfate for recycling, and precipitate calcium carbonate, which is removed as a by-product (Meline et al., 1971).

A variety of processes and equipment have been used in Europe since the late 1930s. (An in-depth coverage of nitric phosphates is given in the IFDC/UNIDO *Fertilizer Manual,* 1979b.) Central and South America and Asia also have several plants. Popular grades include 14–6–12 (14–14–14), 22–5–9 (22–11–11), 20–9–0 (20–20–0), and 26–6–0 (26–13–0). In 1982 there were no nitric phosphate plants operating in the USA, since the remaining three have been modified substantially to use much higher proportions of phosphoric acid and other materials.

The primary advantage of nitric phosphate processes is that no S of less S is required compared with ammonium phosphates; this is particularly important during a shortage of S or where S must be shipped long distances over land. Interest in nitric phosphates has decreased with the price and availability of S but there has been little added production in the world since the late 1960s. World production capacity for nitric phosphates in 1980 was about 21 million metric tons of product (1.4 million metric tons of P). About 90% of this capacity is in Western and Eastern Europe. Production in 1980 was estimated at 1.0 million metric tons of P.

F. Minor Phosphate Fertilizers

1. Phosphate Rock for Direct Application

Direct application of finely pulverized phosphate rock has been utilized from near the beginning of fertilization practice. This unprocessed phosphate is, of course, the lowest cost material, but its agronomic effectiveness depends on the chemical and mineralogical nature of the rock, pH, and other characteristics of the soil, as well as the crop produced. TVA researchers have characterized phosphate rocks from a large number of sources as to their reactivity (Lehr & McClellan, 1974). Phosphate rocks from North Carolina (USA) and Gafsa (Tunisia) are at the top and about equal in reactivity and suitability for direct application. Other phosphate rocks that are not so reactive but are reasonably effective and marketed for this purpose include some types from Morocco, Israel, and a few other locations.

In 1980 a world total of about 4.5 million metric tons of phosphate rock was used for direct application. The USA used only about 0.1 million metric tons. Interest in use of this lowest cost form of phosphate appears to be growing primarily in developing countries that have indigenous deposits of phosphate rock.

2. Partially Processed Phosphates

Aluminum phosphates from Christmas Island are calcined at 499 to 593° C and sold under the trade name Calciphos (Australian Phosphates, Ltd., Adelaide, South Australia, Australia) material from Senegal is sold under the name Phosphal (Senegalese Phosphates, Ltd., Dakar, Senegal). TVA has granulated finely pulverized North Carolina Phosphate rock with concentrated urea solution and partial (25%) acidulation with H_2SO_4 in a pilot-plant pan granulator. The same types of products also were prepared with Christmas Island Calciphos. These fertilizers are being tested agronomically. The premise is that the soluble binding material would allow the granules to disintegrate in the soil moisture and yield the original finely divided phosphate. The partially acidulated rock would provide a portion of the phosphate in a readily soluble form to promote early plant growth. The International Fertilizer Development Center (IFDC) at Muscle Shoals, AL is continuing development work on, and evaluation of, partially processed phosphate rock directed toward utilization of these simpler fertilizers in developing countries.

Partially acidulated phosphate rock is produced and marketed in substantial quantities in the Federal Republic of Germany (FRG), Finland, and Brazil. Finely pulverized (90% −300-mesh size) phosphate rock is granulated commercially in France by use of moisture and a binder.

3. Rhenania Phosphate

Kalichemie in the FRG has produced a calcium sodium phosphate ($CaNaPO_4$) fertilizer for several years, and a substantial number of tons have been used in the FRG and exported. The general formula for the reaction is

$$\underset{\text{Phosphate rock}}{Ca_{10}F_2(PO_4)_6} + 2SiO_2 + 4Na_2CO_3 = \underset{\text{Rhenania phosphate}}{6CaNaPO_4} + 2Ca_2SiO_4 + 2NaF + 4CO_2.$$

The feed materials—Kola (Russian) apatite, silica sand, and sodium carbonate—are proportioned, mixed, and then calcined in a rotary kiln at maximum temperature of about 1260°C. The discharged clinker is cooled, pulverized finely, and granulated in a drum by addition of water and potato starch as a binder. The granules are dried and screened, and Mg, B, or both are added to some products. Grade of the straight Rhenania phosphate is 12 to 13% P (28–30% P_2O_5). Granular mixed grades include 0–6.5–20 (0–15–25) and 0–5–16.6 (0–12–20) + 3Mg + 0.15B (Young & Davis, 1980).

4. Basic Slag

Basic slag, a by-product of the steel industry, has been widely used as a fertilizer in Europe and South America and to a lesser extent in Asia and North America. This material, which also is referred to as *Thomas slag*, has a P content of about 7%. World production of basic slag totaled about 0.6 million metric tons of P in 1962. It stayed at that level through 1971 but decreased to about 0.4 million metric tons of P in 1980.

5. Fused Calcium Magnesium Phosphate

The production and use of fused calcium magnesium phosphate made by an electric furnace-type operation is substantial. About 0.82 million metric tons were produced in Japan and Taiwan in 1977. It is produced by addition of a Mg source (usually olivine or serpentine) along with phosphate rock to the furnace. The glassy product is pulverized to improve its availability to plants; it contains about 9% P (20% P_2O_5) and 9% Mg.

6. Potassium Phosphates

Potassium phosphates and proceses for their production have been studied for a long time, but commercial production has been very limited and primarily experimental. Some early TVA studies involved reaction of wet-process H_3PO_4 with KCl and heating the mixture to produce potassium metaphosphate with a grade of about 0–23–30 (0–54–37). Major problems were corrosion due to HCl evolved and poor market prospects for the by-product HCl.

Pilot-plant studies of several schemes for production of potassium phosphate fertilizers were conducted during the 1970s by major organizations in the USA, Ireland, Scotland, Israel, and Spain. Product grades included 9–10–13 (9–48–16), 0–26–16 (0–60–20), 8–24–15 (8–56–18), and 0–21–25 (0–47–31). One process was tested in a retrofitted system on semicommercial scale, but work was discontinued after very limited operation. Potassium phosphates appear to have potential but have never approached actual commercialization because of economic and marketing restraints and technological problems (Young & Davis, 1980).

G. New Phosphate Products

1. Urea Phosphate

Urea phosphate, made by reaction of wet-process H_3PO_4 and urea, has been studied in pilot-plant work by TVA since the mid-1970s. The product crystallizes and is separated by centrifuging to prepare nongranular material with a grade of 17–19–0 (17–44–0). Only about 15% of the impurities in the feed acid are retained in the precipitated urea phosphate. This is the basis for TVA's main interest in urea phosphate as an intermediate for preparation of liquid fertilizers with low Fe, Al, and Mg content.

The crystalline urea phosphate is of largely fine particle size not well suited for shipment and handling. The product is agglomerated by application of heat to the moist crystals to prepare material more suitable for shipment. The urea phosphate can be polymerized through heat of ammoniation and added heat to prepare a melt with 50 to 80% of the phosphate as polyphosphate. Water and ammonia are added to prepare a clear base liquid fertilizer of 15–12–0 (15–28–0) grade (TVA, 1978; 1980).

TVA is evaluating the urea phosphate route as an attractive alternative to purification of wet-process H_3PO_4 by solvent extraction. It may prove attractive in some situations as an alternative to wet-process superphosphoric acid. Urea phosphate mixed with urea and granulated has given very promising results in suppressing the evolution of NH_3 from urea, particularly when the fertilizer is surface applied. The urea phosphate has also given promising results in minimizing precipitation and plugging problems when added to spray-type irrigation systems. This new type of phosphate material shows promise because of its versatility.

Pilot-plant studies were underway in 1982 by TVA on direct production of a lower-purity urea phosphate by simultaneous reaction-granulation in a rotary drum. Grade of the product that contains all of the impurities in the wet-process acid is about 16–18–0 (16–41–0). Urea phosphate made in this way would be more economical but would not be suitable for use in production of liquid (solution) fertilizers with low impurities content.

H. Problems of the Phosphate Industry

The phosphate industry carried over some major problems of the 1970s into the 1980s. The main problems are (i) escalating of investments for production facilities and equipment, (ii) continuing rigid controls on all aspects of the environment in and around mining and production facilities, and (iii) coping with ever increasing energy costs from phosphate rock mining through production and transport of intermediates and products.

1. Higher Investment

Costs for equipment and complete production facilities doubled in the USA between 1970 and 1980. This increase, along with additional investment for environmental controls, caused interest, depreciation, and other capital costs to account for a much higher percentage of production costs. An estimate for a phosphate fertilizer complex in 1982 indicated that 56% of the total production cost was directly related to investment. Escalation in cost of equipment and facilities eased somewhat to about 5%/yr between 1982 and 1984.

2. Environmental Controls

As mentioned earlier, the phosphate industry made great strides in the 1970s in meeting environmental standards from phosphate rock mining through production of finished products. These include good control and confinement of slimes from phosphate rock beneficiation, reclamation of mined-out areas, control of gypsum and F in effluent water in H_3PO_4 production, and control of F emissions in working areas and from plant stacks. A large part of the F evolved during H_3PO_4 and phosphate fertilizer production is recovered as salable products. The control and recovery of dust in phosphate plants, along with closer approach to zero discharge of materials and less heat in effluent water, are main areas to be tightened in the 1980s. Bag filters were beginning to be used in the later 1970s for very effective recovery of dust. Considerable effort and expense also will be required to maintain environmental controls that have been achieved.

3. Energy Costs

Energy costs soared during the later half of the 1970s, and the phosphate industry was hit in all areas from mining to finished fertilizer production and transport. Listed below are energy requirements for production of wet-process H_3PO_4 and main phosphate fertilizers.

	BTU × 10^6/metric ton P
Wet-process acid (24% P)	4.3
Diammonium phosphate (in addition to H_3PO_4 production)	0.5
Granular triple superphosphate (in addition to H_3PO_4 production)	0.4

The energy required for production of phosphate fertilizers now represents a significant part of the total production cost. However, the total energy consumption for the phosphate industry represents only about 0.1% of the total U.S. consumption.

Better plant maintenance, good housekeeping measures, and some process innovations have resulted in substantial energy conservation. More efficient utilization of the heat of chemical reaction of ammonia and acids by use of TVA-developed pipe reactors has paid good dividends. This simple but very effective technology continued to be adopted more widely in the early 1980s in NPK fertilizer granulation plants and in granular ammonium phosphate plants. Substantial savings in energy costs and in the investment for new facilities are being achieved (Fig. 9–10). Wet grinding of phosphate rock for use in production of wet-process H_3PO_4 and more effective utilization of recovered heat from H_2SO_4 production are other energy conservation practices (Young & Davis, 1980; Achorn & Salladay, 1976).

The early 1980s saw several phosphate production complexes moving to cogeneration of electricity using steam recovered in production of H_2SO_4. In large plants, cogeneration usually is quite attractive economically, with capacity as high as 40 to 60 MW. Estimates indicate that savings of about 10% in the cost of P fertilizer production can be realized (Anon., 1983a).

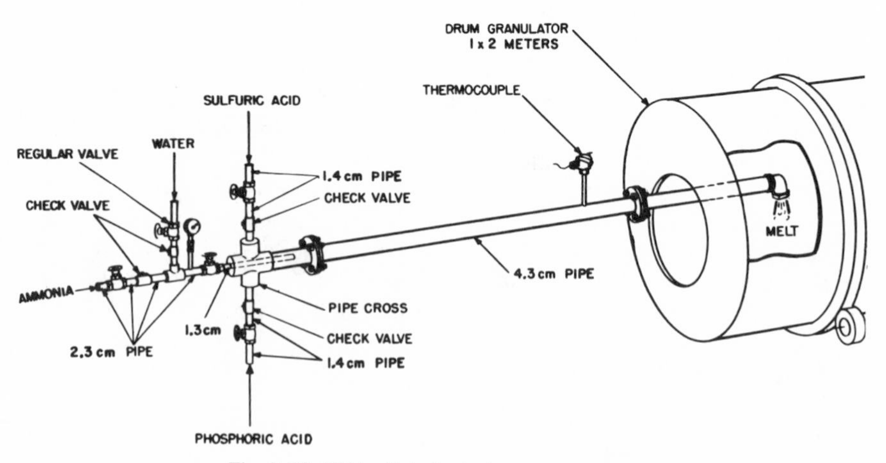

Fig. 9–10. TVA pilot plant pipe-cross reactor.

IV. MARKETING OF PHOSPHORUS FERTILIZERS

The marketing of P fertilizers, as well as marketing of all fertilizers, has undergone marked changes since the mid-1950s. Two significant factors that led to these developments were improvements in the manufacture of fertilizer that permitted more economical production of higher analysis fertilizers and a sharp expansion in fertilizer use enhanced by research, extension, and industry educational efforts to show the value of fertilizers to farmers, along with an increased worldwide demand for more food production.

In the early marketing of phosphate fertilizers in the USA, individual firms produced phosphate rock or superphosphate and sold it to manufacturers who bought N and other materials from other sources. From these materials the manufacturer-wholesaler produced mixed fertilizers that were sold to retail dealers who in turn sold them to farmers. This pattern changed rapidly during the 1960s. As fertilizer demand increased and higher analysis materials became available, larger plants were built, and the basic producers began to do more of the processing themselves. Expansions and integrations took place in which larger companies became basic producers of all three plant nutrients—N, P, and K. The fertilizer industry also was integrated with, or acquired by, other industries, particularly the petroleum industry.

With larger capacity plants and the necessary capital for expansion in both production and marketing, the basic producers increasingly made efforts to ensure their continuing markets. Thus, they began to acquire wholesale distributing facilities and in many cases retail fertilizer outlets. Many of these retail outlets became bulk-blend plants as emphasis shifted from manufactured mixed grades to prescription mixed complete dry bulk blends to meet farmers' fertilizer needs. The advent of rapid growth in dry bulk-blend plants was brought about largely because of improvements in fertilizer product quality. Manufacturing advances made it possible to make consistently uniform products that blended well and spread well as complete blend mixes. In particular, the improvements in the production of concentrated superphosphates and DAPs contributed to this growth in bulk blends. These products now were appropriately size-matched with potash and granular or prilled N materials, leading to excellent blends with good spreading qualities.

This is not to say that all of the growth has been in dry bulk blends. There have also been large increases, especially on a percentage basis, in the manufacturing and consumption of fluid fertilizers. Rapid growth has occurred in the consumption of the fluid ammonium polyphosphates, primarily 10–14.8–10 (10–34–0), 11–16.2–0 (11–37–0), and the dry products DAP and MAP.

The marketing of P fertilizers at the retail level is usually accomplished with one of two areas of potential emphasis. The first is the use of phosphate as one component of the complete fertilizer program for overall crop needs. The second is with emphasis on the use of phosphate in a banded application with the planter or drill as a starter-type application. In the first case, the marketing of P fertilizer is based, in most instances, on the use of a soil testing program in which needs are determined based on soil test level of available P. There are different recommendation philosophies for the amount of fertilizer needed, with the two major philosophies being one as an application of both buildup and maintenance amounts of P so that soil levels are increased; and, the second one with meeting crop needs with minimal rates of P application. In either case, soil testing helps define the amount of fertilizer needed, usually in conjunction with a yield goal and perhaps a P removal factor according to the crop to be grown.

Many local retail outlets then further complete their marketing program by offering custom application services, along with custom blending of either dry or fluid fertilizers to meet specific crop and soil needs, sometimes known as prescription fertilizer application. In the second case, considering band application of P at planting time, total crop needs are not necessarily emphasized. The emphasis may be placed on the need for P as a necessary nutrient for stimulation of plant growth, particularly in more northern regions with cold soils. Also, in the case of many specialty crops, especially vegetables, the timing of marketing may be influenced by early planting and quick maturity so that P may serve to stimulate early growth to the advantage of the grower. Phosphate materials for this application may be either dry or fluid, but there is a strong tendency for starter applications to be as fluids, particularly the clear solutions such as 10–14.8–0 (10–34–0) with or without the addition of additional N or potash, and in some cases micronutrients and secondary nutrients. Aggressive fertilizer retailers have been able to innovate in this market to the point of providing farmers either special application equipment or to help them rig their planters or drills with special equipment; their marketing program includes not only the product but also the services needed to support the use of that product.

Reduction in tillage for most crops is the by-word of the 1980s. This, of course, is being brought on by the increasing emphasis on reduction in soil erosion as well as energy conservation by decreasing trips across the field. The emphasis on reduced tillage (referred to as *minimum tillage, conservation tillage,* or *no-till*) has brought about the concern of whether or not fertilizer needs to be injected in the soil or can be applied on the surface without incorporation. This is particularly of concern with P, which is known to move very little in most soils. Injection of fertilizer is a new marketing technique for both fluid and dry fertilizers. There are, as mentioned above, application techniques for fluids, but in addition, there are application techniques for dry products such as MAP or DAP that can be utilized in reduced tillage and soil injection. These subjects are discussed in more detail in a later section.

Those involved in all phases of marketing of fertilizers will need to be even more informed than ever concerning the properties and characteristics of phosphates. There are not likely to be any revolutionary new types, but how to use the available high-quality products effectively and efficiently will be more important. Application practices that are compatible with the growing practice of conservation tillage, or reduced tillage, must be developed. The remainder of this chapter deals with chemical and physical characteristics of phosphate fertilizers, fertilization practices to best utilize them, chemistry of P in the soil, and physical factors that affect P movement and utilization in the soil.

V. USE OF PHOSPHORUS FERTILIZERS

A. Absorption and Utilization of Phosphorus by Plants

The median concentration of P in the soil solution is approximately 0.05 mg/kg and seldom reaches levels > 0.03 mg/kg in soils that have not had a history of fertilization (Barber et al., 1963; Fried & Shapiro, 1961). Plants absorb most of their P from the soil solution as either the primary orthophosphate ion $H_2PO_4^-$ or the secondary orthophosphate ion HPO_4^{2-}. The $H_2PO_4^-$ ion is more readily absorbed by most plants than HPO_4^{2-}. The former ion is absorbed at a rate that is nearly 10 times faster than the rate of absorption of the latter ion. Other forms of P, both organic and inorganic, may be absorbed by plants under certain conditions but are of little practical significance.

Plants absorb P throughout their entire growth cycle, but young plants absorb it very rapidly when conditions are favorable. Plants often absorb 50% of the seasonal total demand by the time they accumulate 25% of the total seasonal dry matter. The early season response to P fertilization commonly shown by crops is partly explained by this P absorption pattern. Phosphorus is mobile in plants and moves from older tissue into younger tissue if a deficiency develops. As the plant matures, much of the P is translocated from vegetable parts into the seeds and fruits.

Phosphorus nutrition generally has little effect on the quality of grain products except where P is extremely deficient. One common symptom of P deficiency is delayed maturity. Application of P will correct this condition and often results in a lower grain moisture content at harvest. However, the application of P will not hasten maturity where crops are not under a P stress. Olsen (1958) reported a marked decrease in tiller number in rice plants (*Oryza sativa* L.) in P-deficient nutrient solutions. This decrease has also been reported by many researchers on other small grains. The effect of P deficiency on vegetable crops is of particular importance because of its influence on market quality, particularly size and grade. Under P-deficient conditions, the fruit size is often decreased and does not meet acceptable size criteria. The effect of P fertilization on hastening the maturity of vegetables may be of particular importance where timeliness to the market is an economic consideration. Lorenz and Vittum (1980) reported that the maturity of lettuce (*Lactuca sativa* L.) may be hastened by as much as 2 weeks by P fertilization compared with a crop that is grown on a P-deficient soil. Phosphorus fertilization may also eliminate some of the other visual disorders that can affect crop quality. An excellent review of Lorenz (1963) summarizes the effects of P on the quality of vegetables.

It is generally recognized that plant species differ in their response to different levels of soil P. Some crops grow relatively well on soils low in available P, whereas others do not. An excellent example of this is wheat (*Triticum aestivum* L.) vs. corn (*Zea mays* L.). Wheat is much more sensitive to low soil P levels than is corn. Hybrids and cultivars within species may also differ greatly in their ability to absorb P from the soil. Studies in Pennsylvania have shown that certain single-cross corn hybrids contained more than twice as much leaf tissue P at silking compared with other hybrids (Baker et al., 1970). The differences in P accumulation within the plants were found to be genetically controlled. The significance of this relationship to plant breeding and soil fertilization practices needs additional study.

VI. SOIL FACTORS AFFECTING PHOSPHORUS AVAILABILITY

The physical and chemical characteristics of the soil largely control the availability of P to plants and influence the chemical nature of the end products formed when P fertilizers are applied to a soil. These characteristics generally cannot be altered, but an understanding of their influence on P availability is helpful in using P fertilizer efficiently.

A. Chemical Characteristics

1. Forms of Soil Phosphorus

The total P content of soils is relatively low, with many soils containing between 0.02 and 0.10% P. Total P concentration is usually highest in the surface soil and lowest in the lower A or upper B horizon as a result of P cycling by growing plants. Phosphorus is present in inorganic and organic forms. Chlorapatites, fluorapatites, and hydroxyapatites generally are major primary minerals, whereas wavellite and vivianite are major secondary minerals. Some apatites, particularly the sedimentary types, are secondary minerals. Organic P compounds such as inosital phosphates, nucleic acids, and phospholipids account for $< 50\%$ of the organic P in a soil, whereas the remaining organic P compounds occur as other "unidentified P compounds."

2. Soil pH

Soil pH influences the availability of P to plants in two ways. First, the pH of the soil solution largely determines the ion form that is present. Within the pH range 5 to 7.2, $H_2PO_4^-$ is the dominant ionic form, whereas between pH 7.2 and 9, HPO_4^{2-} ion dominates. As indicated previously, the former is absorbed more readily by most plants. Second, the pH of the soil also controls the type and solubility of soil minerals. These minerals can be fertilizer reaction products, secondary or primary minerals. Our major concern is with the reaction products formed when P fertilizers are applied to soils.

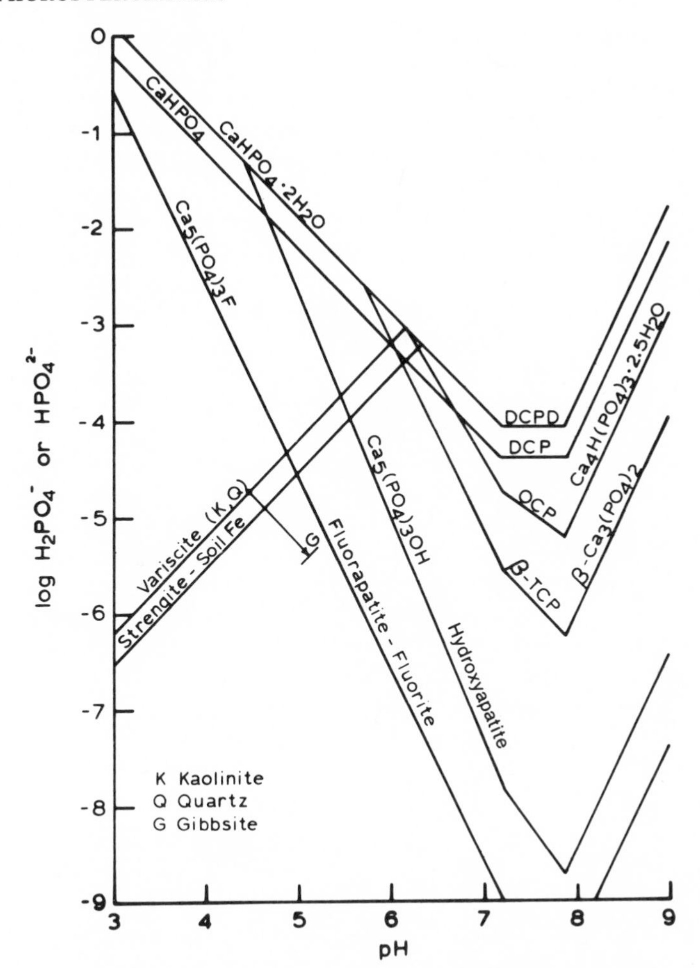

Fig. 9–11. Solubility of calcium phosphates compared with strengite and variscite when Ca^{2+} is $10^{-2.5}$ *M* is fixed by calcite and $CO_2(g)$ at 0.0003 atm (Lindsay, 1979).

When a P fertilizer is added to an acid soil, it reacts with Fe and Al compounds to form complex products that can be quite insoluble and less available to the plant. The compounds formed may be precipitated from solution, adsorbed on the surface of Fe and Al oxides, or adsorbed on clay particles.

In alkaline or calcareous soils, soluble phosphates will also revert to relatively insoluble dicalcium and tricalcium phosphates or even apatite-like compounds. Phosphate ions may also be tied up in unavailable forms on the surface of $CaCO_3$ particles and on Ca-saturated clays. However, these precipitated compounds have large surface areas in contact with the soil solution, and much of the P may be slowly released for plant use.

The relationship between the concentration of solution $H_2PO_4^-$ and HPO_4^{2-} and pH is shown in Fig. 9–11. This diagram shows the solubility of various iron, aluminum, and calcium phosphate minerals with phosphate solubility expressed in terms of predominant ionic species: $H_2PO_4^-$ below pH 7.2 and HPO_4^{2-} above pH 7.2. In acid soils, variscite ($AlPO_4 \cdot 2H_2O$) is the first stable mineral formed after P fertilization, followed by strengite ($FePO_4 \cdot 2H_2O$). If gibbsite is the clay mineral controlling Al solubility, the relative stability of variscite and strengite is reversed. In alkaline soils calcium phosphates are the most stable minerals. Dicalcium phosphate dihydrate (DCPD) is the first stable mineral formed after P fertilization. As time progresses, less soluble minerals of dicalcium phosphate (DCP), octocalcium phosphate (OCP), or β-tricalcium phosphate (β-TCP) are formed. In the pH range of 6 to 6.5, several of the phosphate minerals (variscite, strengite, DCPD, DCP, OCP, and β-TCP) can coexist in a concentration of approximately $10^{-3.25}$ M $H_2PO_4^-$. In this pH range, maximum P availability exists. Figure 9–11 explains why liming acid soils to pH 6.5 increases P availability by reducing Fe and Al activities and why finely divided apatite minerals can supply appreciable quantities of P to plants in acid soils and not in alkaline soils. Above pH 7.8, the solubility of calcium phosphate minerals increases rapidly. This is a result of a decrease in the concentration of Ca^{2+} in soil solution due to precipitation of $CaCO_3$, which in turn will support a higher concentration of HPO_4^{2-} in soil solution.

In alkaline soils the formation of DCPD and DCP from applied P fertilizers is relatively fast (a few weeks' time period). The formation of OCP generally occurs within 3 to 5 months, whereas the formation of β-TCP will occur within 8 to 10 months to 1 or 2 yr. These time periods are relative, however, and may vary considerably from soil to soil. Recent research on calcareous soils has indicated that OCP controls P solubility at $NaHCO_3$–P levels > 32 mg/kg, whereas β-TCP controls P solubility below this level. Soil solution P buffered near the TCP isotherm indicates this mineral, or one of similar solubility, was the reaction product of P fertilizer applied from 3 to 6 yr previously. There is evidence indicating apatite-like minerals can form from fertilizer residues after long periods of time.

B. Physical Characteristics

1. Aeration and Compaction

Soil aeration influences the oxidative state of inorganic compounds, the decomposition of organic matter and release of P, as well as the complex metabolic processes associated with plant growth. As a result, aeration has a definite influence on the availability and absorption of P. An example is encountered in the production of paddy rice. Under anaerobic conditions resulting from flooding, the ferric iron is reduced to the ferrous form. Ferrous phosphates are more soluble than ferric phosphates; consequently, P availability is increased severalfold.

Soil compaction and structure influence P relationships indirectly through effects on aeration. Increased compaction may physically impede root penetration resulting in P being positionally unavailable. In a compacted soil with aggregate condition that promoted the development of a plant with a coarse rooting system, P uptake was decreased substantially

(Wiersum, 1961). This was attributed to a less dense root system that resulted in the P being positionally unavailable.

2. Temperature

Phosphorus deficiencies are more pronounced at low temperatures. For this reason, crops often respond to starter applications of P fertilizer during cool spring weather. This increase in P response as temperature decreases has been attributed to several factors. Among these are decreased translocation of P from the root, decreased root growth, decreased P uptake by the root, decreased mineralization of soil organic P, slower reaction of the fertilizer granule, and decreased diffusion rate of P to the root surface (Powers et al., 1963; Sulton, 1969). One of the main factors affecting P uptake at lower temperatures is reduced root development. Cooper (1973) found that the root/shoot ratio is largest at high and low temperatures, with a minimum ratio occurring in the range of 20 to 25° C, even though root extension rates normally increase with temperatures up to 25 to 30° C. Thick and less-branched roots are developed at low temperatures (Nielsen & Humphries, 1966), but maximum lateral roots occur at different temperatures depending on the species. Maximum lateral root development occurred at 20° C for corn and at 34° C for pine (*Pinus* sp.) trees. Of all nutrients studied, Sulton (1969) concluded that the clearest interaction between temperature and nutrition occurred with P. Sulton attributed lower absorption rates at lower temperatures to the physical and chemical effects on P in the soil and to slower root extension. The combination of the above influences on root development and P movement in the soil results in an inverse relationship between availability and soil temperature. These relationships point to the importance of the use of starter fertilizer under low soil temperatures.

3. Moisture

The major portion of the P moves to the root by diffusion through the water films around the soil particles. As soil moisture decreases, this diffusion path becomes more tortuous, and movement proceeds at a slower rate. Phosphorus uptake by plants has been shown to decrease as moisture tension increases. The relative uptake of P by corn seedlings was 100, 94, 80, 50, and 35 at 1/3, 1/2, 1, 3, and 5 bars of soil moisture, respectively (Olsen et al., 1961). Marais and Wiersma (1975) reported that soybean [*Glycine max* (L.) Merr.] plants with high P tissue levels were influenced less by moisture stress than those with lower P tissue levels. In a long-term (18-yr) experiment in the Midwest, soybeans showed a greater percent yield increase from the application of fertilizer P when a lower amount of rainfall was received for the 12-week period following planting (Barber, 1971). Lange et al. (1963) reported that over a 13-yr period in southern Illinois, the percentage response of corn to fertilizer P increased with decreasing rainfall. Adequate P fertilization has also been reported to increase water use efficiency by plants. For example, Stanberry et al. (1955)

reported that the application of P to an adequate level was associated with a yield increase of 6.5 metric tons/ha and a decrease in water use of 13.6 cm/metric ton.

4. Movement and Losses

Phosphorus is relatively immobile in most soils and does not move appreciably from the point of application. Soluble P rarely moves more than 2 or 3 cm from a fertilizer granule before reaction with soil components essentially stops further movement. Repeated P application to the undisturbed surface of medium-textured soils will result in slow downward movement to 10 to 12 cm. Applications mechanically incorporated in the surface soil normally have little effect on the available P level of the underlying soil. This limited movement of P indicates the need for initially placing fertilizer P in the proper position for maximum effectiveness.

In Canada, Reed (1982) recently documented a process called *biocycling*. He measured deposition of surface applied P to soil depths of 120 cm by this process. This cycling is the result of plant roots releasing P into the soil or dying at depths where they are growing. This results in an increased P content down the entire soil profile. The significance of biocycling to plant growth has not been delineated under different soil and climate conditions.

Because of its immobility in the soil, loss of P from the profile is small except for crop removal and soil erosion. The low concentration in the soil solution means that loss by leaching is small even if considerable drainage occurs. The only soils from which loss of fertilizer P by leaching is potentially significant on a short-term basis are sands and some organic soils that have little tendency to react with P. However, since P tends to be concentrated in the surface soil, it is susceptible to loss by erosion. Phosphorus lost by this means is presently receiving considerable attention because of its influence on surface water quality.

VII. PHOSPHORUS FERTILIZER CHARACTERISTICS

The characteristics of the fertilizer compound added and the nature of the soil environment adjacent to the fertilizer particles largely determine the nature of the reaction products that form in the soil. These fertilizer–soil reaction products serve as sources of P for plants. A knowledge of the nature and persistence of these products should be useful in explaining why fertilizers differ in effectiveness and how they can be used most efficiently on different soils.

Workers at the TVA studied these complex reactions and identified and characterized a number of reaction products formed under varying conditions (Lindsay et al., 1962). These and other studies dealing with the subject are discussed in other sections and are only briefly considered here to illustrate how different fertilizer properties influence their effectiveness and usage.

Table 9–1. Approximate water solubility of selected P-containing fertilizer materials.

Fertilizer	N–P–K	N–P_2O_5–K_2O	% of available water-soluble P
Triple superphosphate	0–20–0	0–46–0	88
Monoammonium phosphate	11–22.7–0	11–52–0	91
Diammonium phosphate	18–20–0	18–46–0	91
	12–5.2–10	12–12–12	92
	6–10.5–20	6–24–24	93
	8N–14P–11.7K–3Zn	8N–32P–14K–3Zn	96
Phosphate rock			0

A. Fertilizer Solubility

In the early history of the phosphate fertilizer industry, there was a wide range in water solubility of P compounds. Therefore, this characteristic received much more attention than any other in studying the effect of different P fertilizers on plant growth. However, in recent years the water solubility of P materials has become much less a significant factor. The reason for this is that virtually all significant commercial phosphate fertilizers now produced and marketed in the USA are sufficiently water soluble to ensure optimum agronomic performance. Typical water solubilities of important P sources of today are shown in Table 9–1.

Phosphorus compounds can be classified into three groups on the basis of solubility: (i) P in the water-soluble form, (ii) P not readily soluble in water but soluble in neutral ammonium citrate solutions, and (iii) P that is insoluble in neutral ammonium citrate.

In the USA, available P is defined as that P obtained by subtracting the citrate-insoluble P from the total P when both are determined by official Association of Official Analytical Chemists (AOAC) methods. The water-soluble fraction of dry fertilizers of equal available P can differ significantly, depending on the source. Consequently, the comparison of fertilizers on the basis of water solubility is of little agronomic significance.

B. Dissolution pH

Studies at TVA have shown that water-soluble phosphates placed in moist soil are quickly wetted by water vapor, and the resulting fertilizer solution moves out and reacts with the soil. The P solution thus formed from the granule or fertilizer droplet creates a distinctive acidic or basic pH environment. In the case of monocalcium phosphate fertilizer (triple superphosphate), the concentrated solution moving from the granule is very acidic. The pH of this solution will be about 1.5, and it will dissolve Fe, Al, Mg, and Ca from the soil that reacts with the released P to form phosphate precipitates of varying solubility. A residue of dicalcium phosphate, both anhydrous and dihydrate, remains at the particle site. In the case of ammonium phosphates, the P component moves out of the granule in a similar front; however, the pH of the saturated solution of monoammonium phosphates is about 3.5, the pH of diammonium phosphate is about 8.0, and that of fluid ammonium polyphosphate (10–14.8–0) (10–34–0) is about 6.2. Little, if any, residue remains at the granule or droplet site from these materials (Sample et al., 1980).

There are significant effects from these pH differences of the various phosphate fertilizers. However, these effects involve only a small portion of the total soil volume, are temporary, and do not have any great influence on overall soil pH. But, they can influence plant nutrition during any given crop season.

In a comparison of MAP and DAP, the acid pH near a MAP granule should produce more soluble P. It is also thought that MAP is initially more mobile than DAP so that it influences a slightly larger volume of soil.

Another possible effect is that MAP's lower pH may dissolve certain soil micronutrients that are otherwise insoluble. These effects have greater importance in alkaline than in acid soils and make MAP in some instances the preferred P fertilizer for high pH soils.

In a comparison of MAP and DAP in calcareous soils, the acidic pH of MAP tends to slow down the rapid reversion to water-insoluble P compounds, keeping it available to plants a greater length of time. Another aspect of the MAP vs. DAP comparison is the possibility of free ammonia from DAP, causing germination or seedling injury when the fertilizer is placed in direct contact with the seed. This is particularly true in calcareous soils because the high pH increases the possibility of free ammonia. Monoammonium phosphate, on the other hand, does not present such a problem and is the preferred source for direct placement of fertilizer with the seed in a band application.

The ammonium polyphosphate fertilizers react extensively with Fe, Al, Mn, and other elements in soils. They also react rapidly with Ca and Mg to form complex compounds. The agronomic significance of this difference between the polyphosphates and orthophosphates is generally regarded to be slight if there is any effect at all.

Fluid P fertilizers applied directly should react with the soil in much the same manner as similar phosphatic compounds in water-soluble solids. Reaction of the concentrated solution or droplets with the soil should be immediate, and zones of increased concentrations of P would develop in the vicinity of the droplets. The latter zones are influenced by rate of application, pattern of distribution, and the extent of mechanical mixing with the soil just as in the case of dry, granular phosphate fertilizers.

C. Granule or Droplet Size Effects

Granule size of commonly sold phosphate fertilizers in the USA may range from approximately 3.36 to 0.841 mm in diam (−6- to +20-mesh size). The size of droplets for application of fluid fertilizers is likely to cover a wider size range than with dry materials. The droplet size can be quite large in certain low-pressure type of applications or can be quite small in certain high-pressure spray applications.

In the case of dry materials, there is a strong interaction between granule size, water solubility, and placement patterns (Engelstad & Terman, 1980). This interaction is related to the fact that (i) water-soluble P fertilizers are converted to less available forms, (therefore, it is better to decrease the contact between P source and soil), and that (ii) water-insoluble sources are more available if they provide more surface area and are widely distributed in the soil. The combined pattern of these opposing factors gives rise to the interaction.

In the early history of granular phosphate fertilizers, there was considerable range in water solubility. However, with today's highly water-soluble phosphates (generally 80% or higher), the rate of dissolution became of less concern than is reversion or reaction of the P with the soil to form less-soluble compounds. During the early 1980s, higher water solubility became a factor again for phosphate fertilizers exported from the USA to Europe. European countries generally specify higher water solubility standards than in the USA, and this has tended to offer a threat to the important export trade. The recent trend toward lower water solubility results from the higher impurities content (Fe, Al, and Mg) of the phosphate rock being mined in Florida and is expected to continue.

The important factor of rate of reaction of P fertilizers with the soil tends to be less with large granules than with small granules. With larger granules, there are large diffusion zones around the granules, and the concentration of P in these diffusion zones tends to delay P fixation and thus provide localized zones of available P. In general, research data show that coarser granules are advantageous for highly water-soluble sources of P.

D. Ammonium Effects

Numerous investigators have found better uptake and utilization of P by crops when ammonium N (NH_4^+-N) compounds were present in the fertilizer or applied with it. The same effect has generally not been observed when the N was applied as the nitrate form. Various researchers have attributed the stimulative effects of NH_4^+-N on P availability to low soil pH in the rhizosphere, increased plant root proliferation, increased metabolic activity of the plant, possible increased absorption per unit of root surface, and to other causes. The net effect of this increased P availability to the plant, especially on P-deficient soils, is for the ammonium phosphates to result in increased early growth response and, in some instances, higher final crop yields when compared with P compounds supplied without the accompanying ammonium salts (Terman, 1971).

A potential disadvantage to ammonium-containing phosphate materials occurs with DAP, particularly in alkaline soils. When DAP dissolves in soil, it releases considerable amounts of free ammonia, which may result in toxicity due to the ammonia itself or to an accumulation of nitrate N (NO_2^--N), which is also toxic to plant growth. The same result can occur with a urea-based fertilizer such as urea–ammonium phosphate. The net result is that either of these two materials, especially in alkaline soils, can result in seedling injury if placed in close proximity to the germinating seed. This same problem does not occur with MAP, because of its acid pH resulting in no accumulation of free ammonia.

E. Phosphorus Source Comparisons

1. Water Solubility

In the early history of comparing P fertilizer materials, a key discussion centered on differences in water solubility as well as differences in particle size and associated salts. With modern-day fertilizers, these same discussions are much less pertinent. In the case of water solubility, nearly all commercially available processed phosphates are of a high degree of water solubility, meaning 80 to 90% or more. The one possible exception to this level of water solubility is the nitric phosphates produced in relatively small amounts in comparison with the superphosphates and ammonium phosphates. Main production and use are in Europe and parts of Asia. Numerous tests have shown the agronomic effectiveness of nitric phosphates is closely related to the degree of water solubility. Modern manufactured nitric phosphates, however, usually are at least 65 to 80% water soluble and, therefore, are agronomically just as effective for most applications as the other dry or fluid phosphate materials. The granule size and associated salts have previously been discussed.

2. Superphosphates and Ammonium Phosphates

The superphosphates and ammonium phosphates are most widely used sources of highly water-soluble phosphates. These materials are used for direct application to the soil and as constituents of granular mixed fertilizers and dry bulk blends. Ammonium phosphates are also used in the formulation of fluid fertilizers, both as true solutions and suspension formulations. The ammonium polyphosphates find their greatest use in the fluid fertilizer industry.

Normal (8.7% P, 20% P_2O_5) and triple (20% P, 46% P_2O_5) superphosphates contain monocalcium phosphate as their primary P constituent and at equal rates of P are generally considered to be of equal agronomic value. Normal superphosphate contains about 10% S as calcium sulfate and may be preferred on soils deficient in S.

The ammonium phosphate fertilizers may contain the MAP or DAP salts alone, in mixtures with the other, or in mixtures with ammonium nitrate and ammonium sulfate. All of the resulting high-analysis fertilizers are considered excellent sources of N and P for crops.

The most popular grade of DAP is 18–20–0 (18–46–0), and this type of fertilizer leads in production and use in the USA and in the world. The use of MAP fertilizers is increasing, with grades ranging from 10–23–0 (10–53–0) to 11–24–0 (11–55–0). The 11–24–0 (11–55–0) material has been demonstrated by TVA to be very effective in dry blends and suspension fertilizers; about 20% of the P_2O_5 is polyphosphate.

3. Polyphosphates

The ammonium polyphosphates are produced primarily in the fluid form, although, as indicated above, they can be produced in solid form. The most typical fluid ammonium polyphosphate fertilizers are 10–14.8–0 (10–34–0) and 11–16.2–0 (11–37–0), with the differences coming due to polyphosphate content and method of manufacture. They typically contain at least 50% of their P in the polyphosphate form, with that in 11–16.2–0 (11–37–0) being higher. These high-analysis fluid fertilizers decrease transportation and application costs and are excellent for making mixtures in either solution or suspension form. In addition, the ability of polyphosphate solutions to sequester certain metallic elements makes them useful in incorporating micronutrients in fluid fertilizers and allows them to carry certain levels of impurities in solution that would otherwise not be possible. Polyphosphate fertilizers have been found to be superior to orthophosphates as carriers of Zn (Mortvedt & Giordano, 1967) and Fe (Mortvedt & Giordano, 1970, 1971) but were poor as carriers of Mn (Giordano & Mortvedt, 1969).

Polyphosphate must be hydrolyzed to orthophosphate before the plants can effectively absorb them. This is generally not a problem with these materials, since hydrolysis occurs once they are placed in the soil. They have, however, been shown to be somewhat slow in hydrolysis at low temperatures, potentially being less effective than orthophosphates (Engelstad & Allen, 1971). Numerous field evaluations of the agronomic effectiveness of ammonium polyphosphates compared with the ammonium orthophosphates have shown them to be equally effective as sources of N and P for plants. Slight differences in effectiveness in some individual research tests are thought to have been associated with their influence on micronutrient solubility. In particular this could have occurred on alkaline soils where there may have been appreciable amounts of Zn or Fe in the polyphosphate materials. Some commercial fluid fertilizers have been found to contain significant amounts of Cu, Fe, Mn, and Zn as impurities. These may contribute to plant nutrition in certain high pH soils.

Ammoniated superphosphates, made by reacting ammonia with superphosphates, are the base of some mixed grade fertilizers. During manufacture, the monocalcium phosphate in superphosphate is largely converted to ammonium phosphate, dicalcium phosphate, calcium ammonium phosphate, and other more basic calcium phosphates. The result is a reduction in water solubility of the P to about 50% in the case of concentrated superphosphate and to < 20% with normal superphosphate. Any variation in the agronomic value of these materials is usually associated with the degree of ammoniation and the related level of water solubility.

4. Fluid Mixed Fertilizers

Fluid mixed fertilizers have become increasingly important in many locations in the USA in recent years. Fluids are easy to handle and apply, are convenient to use in conjunction with irrigation, and lend themselves readily to custom mixing and application. Studies have shown that most fluids are not different in effectiveness from water-soluble solids. Use of suspensions has been growing rapidly in many areas to obtain higher analysis and greater economy than with liquid (solution) fertilizers. The performance of suspensions is equal to that of clear solution materials or dry water-soluble fertilizer materials. Any differences that have been observed can be traced back to either differences in chemical composition or to differences in placement in the soil.

5. *Miscellaneous Phosphates*

TVA has investigated phosphate materials of ultra-high analysis in recent years, but none of these is available commercially. In addition, they have introduced the urea–ammonium orthophosphates and urea–ammonium polyphosphates currently being evaluated in agronomic applications. These materials comprise homogeneous formulations with higher N content, and typical grades would be 28–12–0 (28–28–0) or 35–7.4–0 (35–17–0). Another class of experimental materials are the urea phosphates, 17–19–0 (17–44–0) and materials formulated from urea and urea phosphates. The urea phosphate materials hold promise in dealing with the volatilization loss of ammonia from urea in surface nonincorporated applications, especially in reduced tillage or no-tillage systems. These products could become increasingly important fertilizer sources in the future, particularly in light of the need to improve the efficiency of N utilization.

Among other P sources that have been used as fertilizer are dicalcium phosphate, metaphosphates, basic slag, various calcined phosphates, and unreacted phosphate rock. All of these materials are relatively insoluble in water and range in citrate solubility from slightly to completely soluble. In general they are less profitable than the more soluble phosphates for use in modern intensive agriculture and are used primarily in areas where an abundance of one of the products favors it economically. They are sometimes useful as basic broadcast applications for building up a reserve supply of P in the soil and should be supplemented with small amounts of soluble P for best results in many crops, especially on soils that are not high testing in P.

VIII. PHOSPHORUS FERTILIZATION PRACTICES

Fertilization practices should consist of providing adequate plant available P in the soil throughout the crop growing season. In actual practice, P fertilization must necessarily be adjusted considering the source of P fertilizer, source and quantity of other fertilizer nutrients, the crop and its needs, the yield goal, tillage method and time, fertilizer application method, and other factors that may affect the economics of fertilizer choice and rate.

A. Phosphorus Needs

Determining the need for P is the first step in the fertilization program. The most common method of determining soil P levels is through soil testing by means of chemical extracting agents. The most common extractants include the Bray no. 1 (0.025 *M* HCl + 0.03 *M* NH_4F), the Olsen (0.5 *M* $NaHCO_3$), and the Mehlich or double acid (0.05 *M* HCl + 0.025 *M* H_2SO_4). Such extractants are widely used, and many of them provide an excellent index of P levels in soils, particularly when used in areas of somewhat similar soils for which they are calibrated, as discussed in an earlier chapter.

Interpretation of soil test results to give recommendations on quantities of fertilizer P needed presents a more difficult problem than the actual soil test. Satisfactory interpretation is dependent on correlating soil test levels adequately with actual crop responses in the field to applied fertilizer. Historically, much soil test correlation research has been done in many locations, but more is needed to update fertilizer recommendations based on changes in tillage and techniques of fertilizer application.

The concentration of P determined by the soil test is often interpreted into categories such as low, medium, and high. For most crop and soil conditions, response to fertilizer P will nearly always be obtained on low-testing soils. For soils testing medium, crop response to fertilizer is likely but less frequent; and for soils testing high, crop response is generally infrequent. Therefore, for a soil testing low, recommendation for P normally consists of an amount necessary for optimum economic return from the current year's crop, as determined by field correlation data. Such a rate of P usually results in sustained increase in the soil test level with time. For soils testing in the high range, the most economical rate of P may be that which is sufficient to maintain the present soil test level for most crop and soil conditions. Examples of typical interpretations of soil test P level for three extracting agents are shown in Table 9–2 (Thomas & Peaslee, 1973).

Chemical analysis of plants and plant parts for tissue P concentration is another means of assessing P availability. This method has not been widely used with field crops because of the cost involved and difficulty of interpreting results and the fact that the results are usually obtained too late to permit corrective fertilization of the crop being grown. In recent years, a modified approach to interpreting plant analysis results has been

Table 9–2. Relative categorization of P concentrations into low, medium, and high for three common soil extractants (Thomas & Peaslee, 1973).

Relative soil level	Extractant		
	0.025 *M* H_2SO_4 0.05 *M* HCl	0.03 *M* NH_4F 0.025 *M* HCl	0.5 *M* $NaHCO_3$
	Extractable P in soil, kg/ha		
Low	0–36	0–34	0–11
Medium	37–83	35–67	12–22
High	84+	68+	23+

developed and is being used in some laboratories and by some researchers. It is known as the DRIS, an acronym that stands for diagnosis and recommendation integrated system (Sumner, 1979). This technique takes into account the interaction among nutrients, which is important to a meaningful interpretation of most plant analysis information. The interpretation requires considerable experience and a wide knowledge of the factors that can affect nutrient levels in plant tissues. Analyses can vary greatly according to the part of the plant sampled, plant age, content of other nutrients, weather conditions, crop variety, and other factors. Researchers are continuing to acquire correlation and other interpretive data that should make this technique more useful in the future.

Field tissue tests, sometimes referred to as *quick tests,* have been used as a field technique for measuring P in the crushed tissue or sap extracted from a part of a growing plant. The use of this technique has been largely in the area of troubleshooting or tentative field diagnosis in problem situations and requires a skilled interpreter for reasonable accuracy and success. It is best used as a preliminary indicator of otherwise obvious plant problems and should be confirmed with a follow-up plant analysis and soil test.

In addition to soil tests and plant analysis, another criterion that can be used to estimate P fertilizer needs is the consideration of the level of P in the subsoil. It may vary widely among soils in any given area and can have a definite influence on whether or not crops respond to P fertilizer. It is thought that particularly during dry periods when the surface soil may be too dry for active root absorption of available P, soils with a high subsoil P level may be better able to supply P to the crop. Numerous states adjust their fertilizer recommendations based on subsoil P levels for given geographic areas and may also adjust those recommendations for variations in soil texture, drainage, pH, and other characteristics.

B. Rates of Phosphorus Fertilization

Phosphorus fertilizer rates vary considerably depending on crop and soil factors previously discussed as well as differences in fertilization recommendation philosophies. In general, the fertilizer recommended at high soil test levels is regarded as an insurance measure with the intention that the recommender wishes to be sure that fertilizer does not become a limiting factor and that crop production is at an optimum level when the soil is testing high.

For low-testing soils there is disagreement regarding buildup of P soil test levels for maximum production. The amount that it takes to build different soils into higher test level ranges varies. Peck et al. (1971) reported field application of 4 mg/kg of P was required to change the Bray no. 1 extractable P by 1 mg/kg in Illinois soils. In Kentucky soils, Thomas and Peaslee (1973) reported it took 3 to 6 mg/kg of P to change the soil test level 1 mg/kg according to the Bray extractant. Griffin and Hanna (1967) reported for New Jersey soils it took 2 to 4 mg/kg of P to change extractable P soil test by 1 mg/kg. In a Virginia soil, 6 to 10 mg/kg of P were required to in-

Table 9–3. Approximate nutrient removal of P per harvested crop unit (Colliver, 1981).

Crop	P removal, g/kg
Corn, grain	6.79
Sorghum, grain [*Sorghum bicolor* (L.) Moench.]	7.14
Soybeans, grain	13.3
Wheat, grain	10.0
Wheat, grain + straw [*Agropyron cristatum* (L.) Gaertn.]	13.3
Oats, grain (*Avena sativa* L.)	8.0
Oats, grain + straw [*Arrhenatherum elatius* (L.) Beauv. ex J.&C. Presl.]	12.8
Barley, grain (*Hordeum vulgare* L.)	6.83
Alfalfa and alfalfa-grass, hay (*Medicago sativa* L.)	6.01
Red clover, hay (*Trifolium pratense* L.)	5.01
Bromegrass or tall fescue, hay (*Bromus inermis* Leyss.)	7.01
Bermudagrass, hay [*Cynodon dactylon* (L.) Pers.]	7.01
Corn, silage	1.35
Sorghum, silage [*S. bicolor* (L.) Moench + *S. sudanese* (Piper) Stapf]	1.60
Cotton (*Gossypium hirsutum* L.)	25.4
Peanuts (*Arachis hypogea* L.)	5.51
Sugarbeets (*Beta vulgaris* L.)	0.25
Potatoes (*Solanum tuberosum* L.)	1.60

crease extractable P by 1 mg/kg with the H_2SO_4–HCl extractant (Thomas & Peaslee, 1973), and Rouse (1968) reported an average ratio of about 7:1 for the same extractant on Alabama soils. It is generally regarded that these types of ratios should decrease as the initial level of P increases. Soils having low extractable P require much more added P to increase them to a high level than it does to increase a high-testing soil to a very high-testing one. The economics of soil test buildup have not been shown.

Researchers recommend basing P fertilizer rate on the soil test and other soil factors and maintaining fertilization according to crop removal. See Table 9–3 for estimated amounts of P removal by selected field crops.

C. Time and Method of Application

In theory, P fertilizer should not be applied very far in advance of seeding the crop, because soluble phosphates tend to revert to less available forms in the soil with time. In actual practice, however, the timing of P fertilizer application is dependent on the availability of labor, equipment, and time to do the job within the overall cultural practices for the particular crop to be grown.

Placement of fertilizer P in soil in a position and form that will be readily available to the growing plant is likely much more important than the actual timing of fertilizer application. Because of the limited movement of P in the soil, fertilizer P should initially be placed in the proper position with respect to the plant root system and soil moisture supply.

Uniform broadcast application of fertilizer P is feasible in either dry or fluid form. Subsequent incorporation of the broadcast fertilizer by tillage mixes it into the soil and places a part of the P deep enough in the soil for it to be in a moist zone during at least some portion of the growing season.

With soils containing low levels of available P, the time and method of application may be quite important. Placement in the effective rooting zone (and relatively close to the time of actual crop need) generally results in greater efficiency of P utilization. However, on soils containing medium to high levels of available P, the time and method of application assume less importance. In a rotation experiment in Indiana (Barber, 1969), broadcast P fertilizer was applied once every 4 yr at rates of 98 and 196 kg/ha of P over a 16-yr period. Both rates were effective in maintaining yields through the fourth year following each application. Barber concluded that more flexibility in P application is possible without seriously affecting yield provided that the P is plowed under or mixed deeply into the soil.

It is generally accepted that the efficiency of banded fertilizer application is at least equal to and often greater than that for broadcast application. However, this potential agronomic-economic advantage for banding has been ignored by many growers because most banding is considered an "at planting" operation. The inconvenience of banding fertilizer during

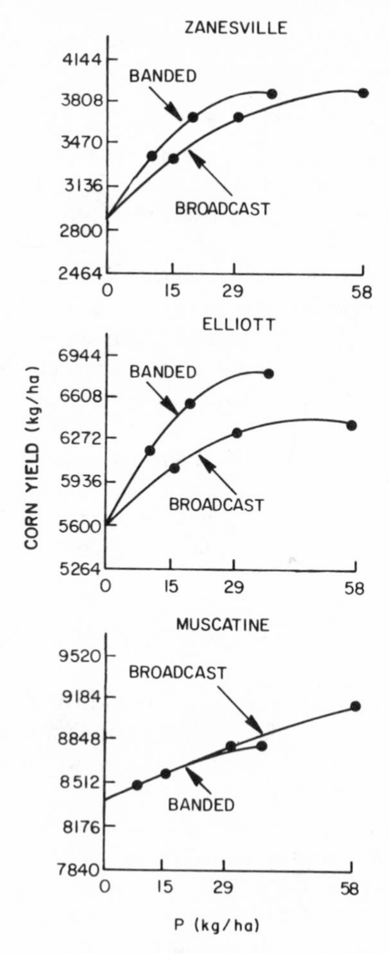

Fig. 9–12. Corn yields on three Illinois soils vs. banded and broadcast rates of P (Welch et al., 1966).

the planting operation has resulted in the preference for broadcasting, especially due to the ease and cost efficiency of custom application. However, with continued increases in crop production cost, many growers now are more receptive to banding to improve fertilizer use efficiency.

Generally, the closer to the time of plant utilization that the nutrient is applied, the greater the efficiency of uptake. Therefore, band application may give greater efficiency because there is less immobilization of P by either microbial, chemical, or physical means.

Work by Welch et al. (1966) in Illinois is shown in Fig. 9–12 in which banded and broadcast P are compared on three soils with corn. The Zanesville (fine-silty, mixed, mesic Typic Fragiudalfs) and Elliott (fine, illitic, mesic Aquic Argiudolls) soils have greater efficiency of P use when banded. In the Muscatine (fine-silty, mixed, mesic Aquic Hapludolls) soil the methods gave equal results. Table 9–4 shows the properties of these three soils, and even though the subsoil P levels were not clearly defined, it is known that Muscatine has higher subsoil P levels than the other two soils. The percentage yield increase due to P was greatest with the Zanesville (very low surface and subsoil P), followed by the Elliott (somewhat better P status), and then by the Muscatine, which would have the best P status of the three considering both top and subsoil. Therefore, the relative efficiency of the two application methods seems to be related to the P fertility status of the soil.

In Table 9–5, Welch et al. (1966) show the amounts of broadcast P required to produce the same corn yield as given rates of banded P. For example, with 20 kg/ha of P banded, the broadcast rate required to give equal yield was 30 kg/ha of P. The calculations show a range of efficiency of 0.49 to 0.88 for the Zanesville and Elliott soils, but the relative efficiency for the Muscatine was generally 1.0 or slightly greater.

As indicated above, band applications of P are frequently applied for

Table 9–4. Properties of three soils differing in response to P fertilization (Welch et al., 1966).

	Zanesville	Elliott	Muscatine
pH	5.8	6.0	6.5
Bray P-1, kg/ha	6	20	18
Subsoil P	--	--	Higher
Percent yield increase of P	51	30	12

Table 9–5. Band vs. broadcast P for equal corn yield and relative efficiency on three soils (Welch et al., 1966).

	Broadcast P					
Banded P	Zanesville		Elliott		Muscatine	
kg/ha	kg/ha	Eff.	kg/ha	Eff.	kg/ha	Eff.
10	15	0.64	20	0.49	10	0.99
20	30	0.65	--	--	19	1.05
29	43	0.70	--	--	26	1.12
39	45	0.88	--	--	32	1.23

Table 9-6. Corn yields vs. P and starter in northwest Iowa on Primghar sl (fine-silty, mixed, mesic Aquic Hapludolls) with a corn-oats-meadow rotation (Webb, 1977).

Plowed down every 3 yr	Starter†	Corn yield 1977	Corn yield 1957-1977
kg/ha		metric tons ha^{-1}	
0	−	5.77	5.39
	+	6.59 (+0.82)	6.21 (+0.82)
22	−	7.28	6.52
	+	7.53 (+0.25)	6.77 (+0.25)
45	−	7.28	6.84
	+	8.03 (+0.75)	7.02 (+0.18)
67	−	7.90	6.84
	+	8.03 (+0.13)	6.90 (+0.06)

† Starter = 129 kg/ha as 5-8.7-8.3 (5-20-10).

increased efficiency of utilization compared with broadcast methods. Row applications of P are frequently applied primarily for early season or starter effects and may be applied over and above those amounts in broadcast applications. Band applications may be a satisfactory means of applying all of the needed P (on soils with relatively high levels of available P). Drilling of the fertilizer with the seed is a very effective means of fertilizing small grains and other crops seeded in close row spacings.

A long-term study from northwest Iowa shows effects of banded starter fertilizer on corn yields (Webb, 1977; Table 9–6). The study is a rotation of corn-oats-meadow and the rates of P are plowed down every 3 yr. Superimposed on the P treatments are either no starter, or a banded starter of 129 kg/ha of 5–8.7–8.3 (5–20–10). The numbers in parentheses are the corn yield increases due to the starter fertilizer. The response to starter over 21 yr is positive, but it is less where the higher P rates have built up soil test P levels.

A second long-term study reported on by Webb and Angstrom (1978) in Table 9–7 is from north central Iowa. The results are very similar to

Table 9-7. Corn yields vs. P and starter in north central Iowa on Webster sicl (fine-loamy, mixed, mesic Typic Haplaquolls) with a corn-oat-meadow rotation (Webb & Angstrom, 1978).

Plowed down every 3 yr	Starter†	Corn yield 1978	Corn yield 1957-1978
kg/ha		metric tons/ha	
0	−	6.59	5.39
	+	7.78 (+1.19)	6.65 (+1.26)
22	−	8.03	7.34
	+	9.72 (+1.69)	7.97 (+0.63)
45	−	8.78	7.65
	+	9.78 (+1.00)	8.03 (+0.38)
67	−	9.03	7.71
	+	8.91 (−0.12)	8.15 (+0.44)

† Starter = 129 kg/ha as 5-8.7-16.7 (5-20-20).

those just discussed for Table 9–6. The main difference is that the average response to banded starter of 129 kg/ha of 5–8.7–16.7 (5–20–20) is even greater over the 2 yr of the study than in the previously discussed example. Even at the highest rate of plowed down P, there is still an average of 0.44 metric tons/ha increase due to starter. Those two studies do not directly compare band to broadcast fertilizer, but they do show that even with adequate broadcast P, there is an additional advantage to adding some banded starter fertilizer under conditions of those studies. Similar results have been reported in other more northern corn-growing areas and are sometimes in contrast to the more southern, warmer locations where starter response is less likely to occur.

For wheat in Nebraska, low soil P levels may require three times as much broadcast P to achieve the same yield as with an appropriate amount of banded (Peterson et al., 1981). At medium to high soil levels, P requirement is the same for both placement methods.

Barber (1974) proposed a method of fertilizer placement that is somewhat of a compromise between banding and broadcasting. He called it *strip placement,* accomplished by banding 5-cm wide strips of fertilizer on the soil surface in 71-cm spacings and then moldboard plowing. During plowing there is limited mixing of fertilizer with the soil, leading to wider distribution than in normal banding but less than in broadcast. This concept is diagrammed in Fig. 9–13. The theory is that strip placement would allow better root exploration of the fertilizer than in ordinary banding, and at the same time minimize soil reaction with fertilizer that may be prevalent in broadcasting where soil-fertilizer contact is greater. Table 9–8 shows 5-yr average corn yields comparing the three placement methods and shows the strip placement having an advantage. This response to strip placement occurred on a soil testing low in available P. Subsequent experiments on higher testing P soils did not show the same advantage. The conclusion is that the potential advantage of strip application is greatest on soils testing low in P.

Another approach to banding fertilizer for improved efficiency is the dual application of N and P materials by injection into soil as described by Murphy et al. (1978). In P-responsive soils, they found agronomic superiority in winter wheat for P dual knifed with N compared with either

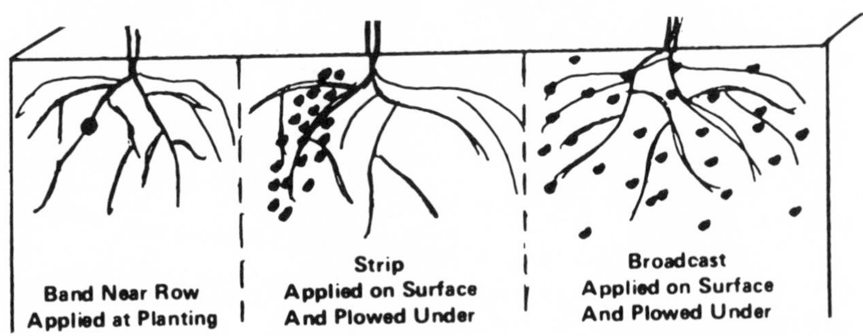

Fig. 9–13. Schematic presentation of distribution of P in soil when applied by three different methods (Barber, 1974).

Table 9–8. Effect of P fertilizer placement on yield of corn and P concentration of the ear leaf at tasseling (Barber, 1974).

Treatment	Yield	Leaf P conc	P soil test†
	metric tons/ha	%	kg/ha
Band by row	7.21	0.26	14
Broadcast and plowed under	7.59	0.27	21
Strip and plowed under	8.28	0.29	37
LSD (0.05)	0.36	0.02	9

† Soil sampled between rows at random during fifth year and tested by Bray P-1 procedure.

broadcast or drill applications. Materials used were either anhydrous ammonia or urea–ammonium nitrate solution and liquid ammonium polyphosphate as fluid materials and solid N-P materials. The positive effect is attributed to deep placement of the N + P nutrients regardless of their form.

Data reported by Leikam et al. (1978) in Table 9–9 show that when N either as anhydrous ammonia or urea-ammonium nitrate solution is dual knifed with ammonium polyphosphate solution, higher grain yields of wheat are achieved as well as higher leaf concentrations of P when compared with methods of N and P application where most of the two nutrients were applied separately. Deep placement of N + P again is the primary reason for the better results.

The interpretation of the apparent advantage of dual placement centers on two most likely effects. Deep placement of N and P for dryland crops in the Great Plains can involve greater nutrient availability due to the fertilizer being placed in soil moisture at lower soil depths. Also, the enhancement of P uptake by the presence of ammonium ions in the soil has been identified as a factor in P absorption for many years. Olson and Drier (1956) reported that NH_4^+-N enhanced P absorption in Nebraska. Riley and Barber (1971) reported that ammonium N influences P absorption in a positive way through a decrease in soil pH in the vicinity of plant roots as the plant absorbs ammonium ions.

Tillage implements and ammonia applicators can be equipped for dual applications of ammonia and fluid fertilizers (Follett et al., 1981; Fig. 9–14). The ammonia and liquid lines must be kept separate to prevent vaporizing ammonia from freezing the fluid line. Separation by 2 or 3 cm at the delivery points of both lines prevents freezing. Vertical separation of the release points (2–3 cm) on shank-type applicators prevents accumulation of solid ammonium polyphosphate compounds around the holes in the fluid line. These result from reactions of the ammonia and liquid ammonium polyphosphate.

In addition to the dual application of ammonia and fluid fertilizer, there is also equipment available for the delivery of dry fertilizer in dual application. Pneumatic systems for the delivery of dry fertilizer through tubes allows the simultaneous application of dry products down the back of the same shank in which ammonia is applied. Dual application of N and P materials, particularly when combined with a tillage operation, provides

Table 9–9. Comparative effectiveness of methods of N–P application on grain yield and leaf concentrations for wheat in Kansas (Leikam et al., 1978).

Method†	Reno County			Ellsworth County			Labette County	Dickinson County	
	metric tons/ha	% N	% P	metric tons/ha	% N	% P	metric tons/ha	% N	% P
No N	1.4	3.80	0.17	2.6	3.91	0.20	1.7	3.58	0.22
Knifed ammonia	1.5	3.44	0.16	2.9	3.90	0.16	2.6	4.14	0.23
Knifed ammonia, knifed APP	2.6	3.74	0.25	3.9	4.74	0.30	3.0	4.50	0.28
Knifed ammonia, broadcast APP	2.6	3.78	0.20	3.6	3.92	0.19	2.9	4.06	0.22
Knifed ammonia, band APP	2.4	3.80	0.19	3.4	4.18	0.21	2.8	4.12	0.24
Knifed UAN	1.7	3.84	0.16	2.7	4.02	0.19	2.5	4.07	0.23
Knifed UAN, knifed APP	2.9	3.87	0.23	3.7	4.62	0.29	3.0	4.02	0.28
Knifed UAN, broadcast APP	2.6	3.82	0.20	3.5	4.02	0.20	3.0	3.92	0.21
Knifed UAN, band APP	2.4	3.84	0.20	3.0	4.06	0.21	3.0	4.07	0.23
Knifed ammonia, knifed APP	2.6	3.74	0.25	3.9	4.74	0.30	3.0	4.50	0.28
Knifed ammonia, knifed APP plus N-SERVE	3.2	4.02	0.23	3.0	4.49	0.30	3.3	4.36	0.33
Soil pH		7.3			6.0		5.5	6.1	
Soil test P, kg/ha		9			11		9	26	

† APP is ammonium polyphosphate supplied as 11–16.2–0 (11–37–0) from TVA at rate of 19 kg/ha P (N constant at 84 kg/ha N). UAN is nonpressure, 28% urea–ammonium nitrate solution. N-SERVE is a nitrification inhibitor product of Dow Chemical Company (Midland, MI). Band means placement of P with seed.

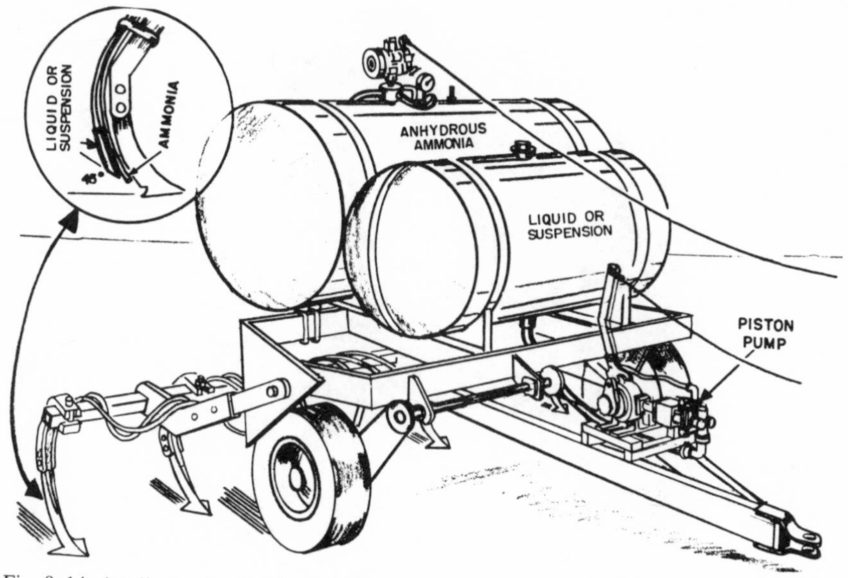

Fig. 9–14. Applicator for the dual-knife application of anhydrous ammonia and fluid fertilizers (courtesy of TVA, [Anon., 1982, p. 72]).

both agronomic and economic advantages to the grower. The trend towards fewer trips across the field by growers will likely result in increasing use of this method of fertilizer application.

In some areas, P application is accomplished by injecting either orthophosphoric acid or ammonium polyphosphate solution into irrigation water. As long as precipitation of P compounds is avoided, distribution of P should be satisfactory. However, P still should be applied early in the growing cycle before seeding and, for most cropping situations, it should be incorporated in the plant root zone. Follett et al. (1981) point out that problems in application of P in irrigation water are at least threefold:

1. Precipitation may be encountered when ammonium polyphosphate-containing liquids are injected into high Ca or Mg water.
2. Phosphorus must be applied very early in the growing cycle for most crops, particularly if there is a definite P need to prevent any early season suppression of yield potentials.
3. Phosphorus applied for row crops through irrigation water may remain on or near the soil surface if not incorporated and be less effective than P applied preplant and incorporated by a tillage operation. The latter problem, however, may be insignificant in the irrigation of crops such as alfalfa and grasses.

D. Cropping and Tillage Systems

In recent years there has been considerable attention given to the P fertilization of crops grown in systems of reduced tillage or no-tillage. Reduced tillage systems are used to reduce soil erosion, conserve water, and to decrease production costs. They may vary from elimination of a single conventional tillage operation to almost complete elimination of tillage except for planting. Such systems offer less opportunity for incorporation of P fertilizers to any appreciable depth in soil. Suggested methods of fertilizer application have included row placement supplemented by the occasional plowing under of a heavy rate of P, band placement with knives or other devices, and surface placement with whatever mixing is provided by tillage operations. Phillips and Young (1973) stated that no-tillage results in keeping the upper part of the soil profile sufficiently moist to ensure P availability for plants. They suggest that incorporation of fertilizer below the soil surface is not necessary. They also suggest that in no-tillage there is proliferation of active roots in the very shallow soil surface that results in sufficient uptake of P so that incorporating to deeper depth is not necessary. However, climate variability and droughty spells may limit surface moisture during critical nutrient uptake periods even if there are large amounts of crop residue on the surface. Subsurface placement of P fertilizer provides a hedge against such eventualities.

From an environmental standpoint, runoff and erosion of surface soils containing large amounts of unincorporated P fertilizer can be a detrimental source of surface water pollution. Randall (1980) recognized the problem of positional unavailability of P that can occur especially in dry years when root activity in the soil surface layer is decreased in the area of major concentration of available P. His results in Minnesota are shown in Table 9–10 with the reduced tillage treatments. Much of the added P along with the P brought up by the roots and recycled into the crop residue was concentrated in the top 0.102 m of the soil profile. With moldboard plowing, P was distributed evenly throughout the top 0.305 m of soil. Continuous chisel plowing resulted in P being incorporated to a depth of 0.152 m after 8 yr but to a depth of only 0.102 m after 4 yr. Randall recommended a preventive measure to minimize low P fertility: Soil P should be built up to high levels throughout the soil plow layer before switching to reduced or conservation tillage systems. He further suggested that a farmer's tillage

Table 9–10. Influence of continuous tillage methods on the distribution of soil P within the 0- to 0.305-m profile at Waseca, MN after 4 yr (1973) and 8 yr (1977) (Randall, 1980).

	Primary tillage			
Depth	Moldboard plow	Chisel plow	Disk	No tillage
m	P_1 soil test, kg/ha			
		1973		
0–0.051	76	110	121	121
0.051–0.102	76	81	63	54
0.102–0.152	74	43	34	36
0.152–0.229	54	20	18	27
0.229–0.305	27	9	9	9
		1977		
0–0.051	63	128	152	155
0.051–0.102	65	116	119	96
0.102–0.152	72	85	56	49
0.152–0.229	78	40	31	38
0.229–0.305	49	20	18	27

system should include moldboard plowing occasionally in a reduced tillage system to redistribute soil P throughout the profile. This could be done perhaps once every 4 or 5 yr for adequate redistribution of P.

REFERENCES

Achorn, F. P., H. L. Kimbrough, and F. J. Myers. 1974. Latest developments in commercial use of the pipe reactor process. Fert. Solutions 18(4):8–9, 12, 14, 16, 20–21.

Achorn, F. P., and D. G. Salladay. 1976. Pipe-cross reactor eliminates the dryer. Farm Chem. 139(7):34, 36, 38.

Achorn, F. P., and D. G. Salladay. 1978. Fluid fertilizers 1978. Proc. Annu. Meet. Fert. Ind. Round Table 28:114–125.

Allgood, H. Y., F. E. Lancaster, Jr., J. A. McCollum, and J. P. Simpson. 1967. A high temperature superphosphoric acid plant. Ind. Eng. Chem. 59(6):18–28.

Anonymous. 1981. News in Brief 4(4):6–7.

Anonymous. 1982. Situation 82. TVA Natl Fertilizer Development Center Bull. Y-74. Tennessee Valley Authority, Muscle Shoals, AL.

Anonymous. 1983a. Chem. Marketing Rep. 31 October, p. 5.

Anonymous. 1983b. World phosphate fertilizer statistics, 1981–82. Phosphorus Potassium. 123(Jan.–Feb.):47–50.

Baker, D. E., A. E. Jarrel, L. E. Marshall, and W. I. Thomas. 1970. Phosphorus uptake from soils by corn hybrids selected for high and low phosphorus accumulation. Agron. J. 62:103–106.

Barber, S. A. 1969. Flexibility in applying phosphorus and potassium. Crops Soils Mag. 21(9):16–17.

Barber, S. A. 1971. Soybeans do respond to fertilizer in a rotation. Better Crops Plant Food 55(2):9–11.

Barber, S. A. 1974. A program for increasing the efficiency of fertilizers. Fert. Solutions 18(2):24–25.

Barber, S. A., J. M. Walker, and E. H. Vasey. 1963. Mechanism for the movement of plant nutrients from the soil and fertilizer to the plant root. J. Agric. Food Chem. 11:204–207.

Colliver, G. W. 1981. Plant nutrient removal guide. Technology Guide Fertilizer/Ag Chemicals Tech. Bull. no. 2. Farmland Industries, Inc., Kansas City, MO, p. 2–6.

Cooper, A. J. 1973. Root temperature and plant growth. Res. Rev. no. 4. Commonwealth Bureau of Horticulture and Plantation Crops, Farnham Royal, U.K.

Engelstad, O. P., and S. E. Allen. 1971. Ammonium pyrophosphate and ammonium orthophosphate as phosphorus sources: Effects of soil temperature, placement, and incubation. Soil Sci. Soc. Am. Proc. 35:1002–1004.

Engelstad, O. P., and G. L. Terman. 1980. Agronomic effectiveness of phosphate fertilizers. p. 311–329. *In* F. E. Khasawneh et al. (ed.) The role of phosphorus in agriculture. American Society of Agronomy, Crop Science Society of America, and Soil Science Society of America. Madison, WI.

Fleming, J. D. 1969. Polyphosphates are revolutionizing fertilizers, Part I. What polyphosphates are. Farm Chem. 132(8);30–36.

Follett, R. H., L. S. Murphy, and R. L. Donahue. 1981. Applying fertilizers and soil amendments. p. 302–392. *In* Fertilizers and soil amendments. Prentice-Hall, Inc., Englewood Cliffs, NJ.

Fried, M. and R. E. Shapiro. 1961. Soil plant relationships in ion uptake. Annu. Rev. Plant Physiol. 12:91–112.

Giordano, P. M., and J. J.Mortvedt. 1969. Response of several corn hybrids to level of water soluble zinc in fertilizers. Soil Sci. Soc. Am. Proc. 33:145–148.

Griffin, G. F., and W. J. Hanna. 1967. Phosphorus fixation and profitable fertilization: I. Fixation in New Jersey soils. Soil Sci. 103:202–208.

International Fertilizer Development Center and United Nations Industrial Development Organization. 1979a. Fertilizers derived from phosphoric acid. p. 187–201. *In* Fertilizer manual. International Fertilizer Development Center and United Nations Industrial Development Organization, Muscle Shoals, AL.

International Fertilizer Development Center and United Nations Industrial Development Organization. 1979b. Nitrophosphates. p. 203–210. *In* Fertilizer manual. International Fertilizer Development Center and United Nations Industrial Development Organization, Muscle Shoals, AL.

Kelso, T. M., J. J. Stumpe and P. C. Williamson. 1968. Production of ammonium polyphosphate from superphosphoric acid. Commer. Fert. 116(3):10–16.

Lange, A. L., L. B. Miller, and P. E. Johnson. 1963. Fertility pays in drought or flood. Better Crops Plant Food 47(1):16–22.

Lee, R. G., M. M. Norton, and H. G. Graham, Jr. 1974. Urea-ammonium phosphate production using the TVA melt-type granulation process. Proc. Annu. Meet. Fert. Ind. Round Table 24:79–86.

Lehr, J. R., and G. H. McClellan. 1974. Phosphate rocks: Important factors in their economic and technical evaluation. CENTO Symp. Min. Benefic. Fert. Miner. Proc. 1973:194–242.

Leikam, D. R., R. E. Lamond, P. J. Gallagher, and L. S. Murphy. 1978. Improving N-P Application. Agrichem. Age 22(3):6.

Lindsay, W. L. 1979. Phosphates. p. 162–205. *In* Chemical equilibria in soils. Wiley-Interscience, New York.

Lindsay, W. L., A. W. Frazier, and H. F. Stephenson. 1962. Identification of reaction products from fertilizers in soils. Soil Sci. Soc. Am. Proc. 26:446–452.

Lorenz, O. A. 1963. Effect of mineral nutrition on quality of vegetables. p. 163–174. *In* Proc. of the 39th Annual Meeting of the Council on Fertilizer Application, Amherst, MA. 6–10 August. National Plant Food Institute, Washington, DC.

Lorenz, O. A., and M. T. Vittum. 1980. Phosphorus nutrition of vegetable crops and sugarbeets. p. 737–762. *In* F. E. Khasawneh et al. (ed.) The role of phosphorus in agriculture. American Society of Agronomy, Crop Science Society of America, and Soil Science Society of America, Madison, WI.

Mann, H. C., K. E. McGill, and T. M. Jones. 1981. Production of ammonium polyphosphate suspension fertilizer—phase I. Paper presented at the 182nd National Meeting of the American Chemical Society, New York. 23–28 August. TVA Library Reference X-559. Tennessee Valley Authority, Muscle Shoals, AL.

Marais, J. N., and D. Wiersma. 1975. Phosphorus uptake by soybeans as influenced by moisture stress in the fertilizer zone. Agron. J. 67:777–781.

McCamy, I. W., M. M. Norton, and B. R. Parker. 1979. TVA's experience with the production of granular NP and NPK fertilizers containing urea. Proc. Annu. Meet. Fert. Ind. Round Table 29:101–107.

Meline, R. S., H. L. Faucett, C. H. Davis, and A. R. Shirley, Jr. 1971. Pilot-plant development of the sulfate recycle nitric phosphate process. Ind. Eng. Chem. Proc. Des. Dev. 10(2):257–264.

Meline, R. S., R. G. Lee, and W. C. Scott. 1972. Use of a pipe reactor in production of liquid fertilizers with very high polyphosphate content. Fert. Solutions 16(2):32–34, 36, 38, 40, 42, 44–45.

Mortvedt, J. J., and P. M. Giordano. 1967. Crop response to zinc oxide applied in liquid and granular fertilizers. J. Agric. Food Chem. 15:118–122.

Mortvedt, J. J., and P. M. Giordano. 1970. Crop response to iron sulfate applied with fluid polyphosphate fertilizers. Fert. Solutions 14(4):22–27.

Mortvedt, J. J., and P. M. Giordano. 1971. Response of grain sorghum to iron sources applied alone or with fertilizers. Agron. J. 63:758–761.

Murphy, L. S., D. R. Leikam, R. E. Lamond, and P. J. Gallagher. 1978. Dual application of N and P—better agronomics and economics? Fert. Solutions 22(4):8.

Nielsen, K. F., and E. C. Humphries. 1966. Effects of root temperature on plant growth. Soil Fert. 29:1–7.

Olsen, K. L. 1958. Mineral deficiency symptoms in rice. Arkansas Exp. Stn. Bull. 605.

Olsen, S. R., F. S. Watanabe, and R. E. Danielson. 1961. Phosphorus adsorption by corn roots as affected by moisture and phosphorus concentration. Soil Sci. Soc. Am. Proc. 25:289–294.

Olson, R. A., and A. F. Drier. 1956. Nitrogen—a key factor in fertilizer phosphorus efficiency. Soil Sci. Soc. Am. Proc. 20:509.

Peck, T. R., L. T. Kurtz, and H. L. S. Tandon. 1971. Changes in Bray P-1 soil phosphorus test values resulting from applications of phosphorus fertilizer. Soil Sci. Soc. Am. Proc. 35:595–598.

Peterson, G. A., D. H. Sanders, P. Grabouski, and U. L. Hooker. 1981. A new look at row and broadcast phosphate recommendations for winter wheat. Agron. J. 74:13–17.

Phillips, S. H., and H. M. Young, Jr. 1973. No-tillage farming. Reiman Associates, Milwaukee, WI, p. 151.

Powers, J. F., D. L. Grunes, W. O. Willis, and G. A. Reichman. 1963. Soil temperature and phosphorus effects upon barley growth. Agron. J. 55:389–392.

Randall, G. W. 1980. Fertilization practices for conservation tillage. p. 78–83. *In* the 32nd Annual Fertilizer and Ag Chemical Dealers Conf., Des Moines, IA. 8–9 January. Iowa State Univ. Press, Ames.

Reed, D. W. L. 1982. Bio-cycling of phosphorus in the soil. Better Crops Plant Food 66(Summer):24–25.

Riley, D., and S. A. Barber. 1971. Effect of ammonium and nitrate fertilization on phosphorus uptake as related to root-induced pH changes at the root-soil interface. Soil Sci. Soc. Am. Proc. 35:301.

Rouse, R. D. 1968. Soil test theory and calibration for cotton, corn, soybeans, and coastal bermudagrass. Agric. Exp. Stn. Bull. 375. Auburn University, Auburn, AL.

Sample, E. C., R. J. Soper, and G. J. Racz. 1980. Reaction of phosphate fertilizers in the soil. p. 263–309. *In* F. E. Khasawneh et al. (ed.) The role of phosphorus in agriculture. American Society of Agronomy, Crop Science Society of Agronomy, and Soil Science Society of America, Madison, WI.

Siegel, M. R., and R. D. Young, 1969. Polyphosphates are revolutionizing fertilizers, Part II. Base materials. Farm Chem. 132(9):41–47.

Slack, A. V. 1968. Phosphoric acid, Vol. I, Part I and Part II. Marcel Dekker, New York.

Stanberry, C. O., C. P. Converse, H. R. Haise, and O. J. Kelley. 1955. Effect of moisture and P variables on alfalfa hay production on the Yuma Mesa. Soil Sci. Soc. Am. Proc. 19:303–310.

Striplin, M. M., Jr. 1969. Phosphorus. p. 47–66. *In* Atomic Energy Commission. Abundant nuclear energy. Atomic Energy Commission Ser. Symp. 14. Clearinghouse for Federal Scientific and Technical Information, Springfield, VA.

Sulton, P. D. 1969. The effect of low temperature on P nutrition of plants—a review. J. Sci. Food Agric. 20:1–3.

Sumner, M. E. 1979. Interpretation of foliar analyses for diagnostic purposes. Agron. J. 71:343–348.

Tennessee Valley Authority. 1978. New developments in fertilizer technology. Twelfth Demonstration. Bull. Y-136. Tennessee Valley Authority, Muscle Shoals, AL, p. 47–51.

Tennessee Valley Authority. 1980. New developments in fertilizer technology. Thirteenth Demonstration. bull. Y-158. Tennessee Valley Authority, Muscle Shoals, AL, p. 50–54.

Terman, G. L. 1971. Phosphate fertilizer sources: agronomic effectiveness in relation to chemical and physical properties. Proc. Fert. Soc. 123:1–39.

Thomas, W. G., and D. E. Peaslee. 1973. Testing soils for phosphorus. *In* L. M. Walsh and J. D. Beaton (ed.) Soil testing and plant analysis. Rev. ed. Soil science Society of America, Madison, WI.

Webb, J. R. 1977. Annual progress report, 1977. Northwest Res. Center, Sutherland and Doon, IA. Iowa State University, Ames.

Webb, J. R., and S. J. Angstrom. 1978. Annual Progress Report. 1978. Northern Res. Center, Clarion-Webster Res. Center, Kanawha, IA. Iowa State University, Ames.

Welch, L. F., D. L. Mulvaney, L. V. Boone, G. E. McKibben, and J. W. Pendleton. 1966. Relative efficiency of broadcast vs. banded phosphorus for corn. Agron. J. 58:283–287.

Whitehurst, B. M. 1974. The production and marketing of superphosphoric acid from North Carolina ore. Proc. Annu. Meet. Fert. Ind. Round Table 24:86–95.

Wiersum, L. K. 1961. Uptake of nitrogen and phosphorus in relation to soil structure and nutrient mobility. Plant Soil 14:62–70.

Young, R. D. 1965. Nitric phosphate processes utilizing supplemental acid. Proc. Annu. Meet. Fert. Ind. Round Table 15:67–72.

Young, R. D., and C. H. Davis. 1980. Phosphate fertilizers and process technology. p. 195–226. *In* F. E. Khasawneh et al. (ed.) The role of phosphorus in agriculture. American Society of Agronomy, Crop Science Society of America, and Soil Science Society of America, Madison, WI.

Young, R. D., and G. C. Hicks. 1967. Production of monoammonium phosphate in a TVA-type ammonium phosphate granulation system. Commer. Fert. 114(2):26–27.

Young, R. D., G. C. Hicks, and C. H. Davis. 1962. TVA process for production of granular diammonium phosphate. J. Agric. Food Chem. 10:442–447.

Young, R. D., W. C. Scott, and R. S. Meline. 1968. Alternatives in production of ammonium polyphosphate materials for use in fluid fertilizer formulations. Proc. Annu. Meet. Fert. Ind. Round Table 18:128–135.

10

Production, Marketing, and Use of Potassium Fertilizers

Stanley A. Barber
Purdue University
West Lafayette, Indiana

Robert D. Munson
Potash & Phosphate Institute
St. Paul, Minnesota

W. B. Dancy[1]
International Minerals and Chemical Corporation
Carlsbad, New Mexico

The production, marketing, and use of potassium fertilizers is very important for crop production. This is especially true in the more humid regions where soils are more often deficient in K. Significant changes have occurred in recent years in production methods, materials produced, marketing procedures, and methods of application of K. This chapter discusses these changes.

I. POTASSIUM FERTILIZER PRODUCTION

Commercial scale production of K fertilizers began in Germany about 1861, approximately 15 yr after Liebig found that K is an essential element for plant growth. German salt deposits were being mined at the time for the production of common salt. Potassium salts, present in some of the run of mine ore, were regarded as contaminants and rejected from the salt production operations as wastes. Thus, when needed for fertilizer K production, the mines were in place. What remained to be accomplished was the construction of refining facilities to upgrade the crude K-containing salts into agriculturally useful products. This challenge was met by German chemists, resulting in the birth of the modern K fertilizer industry. Due to a favorable reserve position and technical expertise, Germany was essentially the only source of agricultural-grade K salts until World War I. This situation began to change at that time when France acquired the Alsatian

[1]Deceased 10 Aug. 1985.

reserves and mines (British Sulphur Corp., 1979). Ore reserves and K brine sources were found in other nations, including the USSR, Canada, the USA, and the Dead Sea of Asia. Significant production shifts began to occur after World War I. Major production sources are now located in Canada and the USSR. The Federal Republic of Germany (FRG), the German Democratic Republic (DGR), the USA, and Israel continue to contribute major quantities of K to the mineral fertilizer industry (Barry, 1980; British Sulphur Corp., 1981). Reserves of soluble K ores and natural K-containing brines are sufficient to meet world K fertilizer requirements for several centuries (Weisner et al., 1980).

Underground K salt deposits are the major raw material source for K fertilizer. Naturally occurring brines (i.e., Asian Dead Sea and Great Salt Lake) that contain dissolved K salts are important secondary sources for commercial K salt production. These brines are located in regions where solar evaporation, the most economical source of energy, can be used to cause the dissolved K salts to recrystallize. Sodium chloride (NaCl) is always present as a major contaminant in natural K salt ores or salts recrystallized by the solar evaporation of naturally occurring brines. All of the underground deposits being worked were formed by the evaporation of seawater.

Seawater also contains dissolved K salts that crystallize concurrently with Mg salts. Thus, K salt deposits generally contain Mg combined as sulfates, chlorides, or both. The quantity of Mg salts present in a given deposit is dependent on the conditions existing when the K and Mg salts were initially crystallized or on conditions prevailing if the deposits were reworked geologically. Some sections of the vast K salt deposit in Saskatchewan Province, for example, contain as much as 20% magnesium chloride (MgCl) as the double salt carnallite ($KCl \cdot MgCl_2 \cdot 6H_2O$), whereas this salt is nearly absent in other sections of the same deposit. Obviously, conditions varied greatly when the salts were being crystallized or when the salts were being reworked geologically. Origin of salt deposits has been extensively studied by numerous earth scientists (Braitsch, 1971).

Potassium salt deposits, with a few minor exceptions, are in locations where markets for $MgCl_2$ and NaCl are virtually nonexistent. Thus, these salts are considered to be unwanted contaminants.

Magnesium sulfate ($MgSO_4$) present in some K deposits as a monohydrated salt called *kieserite* ($MgSO_4 \cdot H_2O$) or as a double salt called *langbeinite* ($K_2SO_4 \cdot 2MgSO_4$) ae recovered as valuable products. Both of these salts are marketed in agricultural regions known to be deficient in Mg, S, or both. These salts are also used as a raw material source for the production of potassium sulfate (K_2SO_4) (MacDonald, 1960).

Potassium chloride (KCl) is the predominant K salt in most world K deposits. With few exceptions, notably in New Mexico, Utah, and Sicily, K mines and refineries are operated primarily to products KCl. Langbeinite is produced as a primary K product by two companies in New Mexico. *Kainite* ($KCl \cdot MgSO \cdot 2.75H_2O$) is produced from natural ore in Sicily and by solar evaporation of Great Salt Lake brine in commercial operations in Utah. Essentially all of the kainite is converted to K_2SO_4 in both locations.

A. Location of Mines and Refineries

The following table shows the principal world sources of K salts (British Sulphur Corp., 1979).

Country	Mine Source
Canada	Saskatchewan Province
USA	New Mexico, Utah, and California
Chile	Caliche Deposits
Israel	Dead Sea
PRC	Lake Qarban
Spain	Catalonia, Navarre
Italy	Sicily
France	Alsace
FRG	Zechstein Basin
GDR	Zechstein Basin
United Kingdom	Yorkshire
USSR	Ukraine, Byelorussia, and Urals

Production is scheduled in the near future from a mine in New Brunswick, Canada, and a solar evaporation operation in Jordan.

B. Mining Systems

Potassium salt deposits that can be mined conventionally range from 233 to about 1050 in depth. Evaporite beds consisting of regular, tabular bodies are mined by the room and pillar method. Continuous mining machines of the boring or rotating type are commonly used to extract ore composed of the chloride salts (a mechanical mixture of KCl and NaCl crystals called *sylvinite*). Ores that contain appreciable quantities of $MgSO_4$ cannot be extracted economically with continuous miners. These ores are extracted by a cycle of undercutting, drilling, and blasting with an ammonium nitrate–fuel oil mixture. Long wall mining, an adaptation from coal mining, is used in many European countries (Singleton, 1978).

C. Refining Systems for Potassium Chloride

1. Separation from Other Components in the Ore

a. Fractional Crystallization—This method of producing agricultural grade KCl (muriate of potash) from crude ore containing this compound is the historic process developed in Germany. It continues to be used in both the FRG and the DGR as well as in other parts of the world to process K salts that are not amenable to treatment by other processes devised more recently. The fractional crystallization process is based on favorable solubility relationships. Stated simply, the solubility of KCl in solutions saturated with KCl and NaCl varies directly with temperature. Sodium chloride solubility remains relatively constant. Brine at ambient temperature (30° C) saturated with KCl and NaCl is heated to about atmospheric boiling temperature (100° C). Crushed ore containing sufficient KCl to saturate the brine with KCl at 100° C is mixed with the hot brine. Essentially all of the KCl present in the ore dissolves, leaving the NaCl present in the ore as an insoluble residue. The NaCl is separated from the hot saturated brine and rejected from the process. Vacuum crystallization is employed to cool the hot saturated brine to ambient temperature to recrystallize the KCl dissolved from the ore. The recrystallized KCl is separated from the cold brine, dried, and sent to product storage. Brine remaining after the separation is recycled to the process. Potash ores frequently contain Mg salts as soluble impurities. These salts distort the KCl–NaCl solubility relationships. For process control reasons, the concentration of soluble Mg salts in the process brine is controlled at a predetermined level by bleeding an appropriate quantity of brine from the process and replacing it with water (Krull, 1928; Wilson, 1969b).

b. Froth Flotation—Although appearing on the K salt processing scene much later than fractional crystallization, the froth flotation method is used to produce a major portion of agricultural grade KCl in North America and a significant portion elsewhere in the world. Froth flotation is a much less energy intensive process compared with fractional crystallization. Other advantages include flexibility and ease of control. In the flotation process, the KCl containing ore is crushed fine enough so that the KCl, NaCl, and other impurities are liberated to provide relatively pure, single mineral grains. The crushed ore, after mixing with brine saturated with the components of the ore, is hydraulically deslimed to separate clays and other finely divided insoluble material from the grains of salts. Deslimed ore is treated with hydrogenated tallow amine hydrochloride (or acetate). The KCl grains are selectively coated with the tallow amine soap. Air bubbles are distributed through the pulp (salts and brine) in shallow, mechanically agitated tanks called flotation cells. Potassium chloride particles adhere to the air bubbles, forming agglomerates that rise to the surface of the cells. The enriched froth is removed continuously from the surface of the cells by mechanical skimmers. Sodium chloride and other salts present in the ore remain on the bottom of the cells. They are continuously removed by gravity flow and sent to filters for separation from the process brine, which is recycled within the process. Potassium chloride particles are filtered to remove the process brine, dried, and sent to product storage or a compaction plant for conversion to granular KCl (Crocker et al., 1969).

c. Heavy Media Separation—This process is used commercially by a major private operator to produce granular KCl from K salt ore. Due to its coarse crystalline structure, Saskatchewan K salt ore does not need to be crushed as fine as other K salt ores to obtain relatively pure grains of KCl and NaCl. As an example, efficient operation of a flotation process dictates that the raw ore should be crushed through sieve openings of about 2.3 mm. Heavy medic separation operates most efficiently on coarse ore. Consequently, for K ores having a coarse crystalline structure such as those mined in Saskatchewan, a primary crushing standard of about 6.5 mm is highly satisfactory. In practice, the ore is crushed sufficiently to separate the 1.6- to 6.5-mm size by screening. This fraction is separated further by heavy media into a granular product meeting agricultural grade K specifications of 49.8% K and a waste tailings product consisting primarily of NaCl. Ore that passes through the 1.6-mm screen is processed by froth flotation. The heavy media process involves mixing saturated brine with enough pulverized magnetite (Fe_3O_4) (sp. gr. = 5.17) to adjust the specific gravity of the mixture to a value intermediate between the specific gravity of KCl (1.98) and NaCl (2.17). Sized particles (1.6 to 6.5 mm) of ore from the primary ore sizing screens are mixed with the homogeneous suspension of magnetite and brine. The resulting slurry is pumped through hydrocyclones that act as centrifugal separators. Potassium chloride crystals overflow the hydrocyclones, and NaCl is discharged at the underflow spout. The magnetite media is recovered by washing both salts with brine. Since the magnetite recovered in the washing steps is too dilute for recirculation, it is concentrated by magnetic separation. The KCl concentrate is debrined, dried, crushed, and screened to sizes specified by the market. After debrining, NaCl is stacked on a waste pile (Dancy, 1968). Heavy media processing is used by the same firm in an operation in New Mexico to separate a complex K ore into its major components (i.e., langbeinite, KCl, and NaCl). The amount of KCl in the ore is so low that the recovered KCl requires additional processing by froth flotation to produce a KCl product meeting the agricultural specification of 49.8% K (Dancy, 1982).

D. Sizing and Granulation

Due to the heterogeneous nature of the fertilizer industry, the K production segment of the industry must be able to produce K products of uniform chemical quality but of variable grain size. For example, the bulk blending segment of the fertilizer industry requires K products in the particle size range of 0.84 to 3.36 mm. Suspension fertilizer manufacturers require K salts in the size range of 0.42 to 0.11 mm. Potassium products in the size range of 1.2 to 0.2 mm are a satisfactory source of K for fertilizer manufacturers that produce complete NPK fertilizers by drum granulation. To supply these diverse markets, nearly all of the K salt plants contain elaborate screening circuits. As an essential auxiliary operation, most K plants have compactor installations to meet the increasing demand for granular KCl. Compaction involves passing a portion of the KCl produced in the primary refinery operation through high-pressure compactor rolls. The dense KCl flake formed by compression and fusion is crushed and screened to the granular KCl size range. Fines created when the flake is crushed are recycled to the compactor feed (Kurtz & Barduhn, 1960).

E. Production of Other Potassium Fertilizer Materials

1. Potassium Sulfate

Most of the commercial plants that produce agricultural grade K_2SO_4 (minimum 41.5% K and maximum 2.5% chloride) react KCl with an available source of SO_4^{2-} ion. Sulfuric acid provides the SO_4^{2-} ion in many operations. Hydrochloric acid (HCl) is produced as a coproduct (Chisso Corp., 1981). Other major operations have magnesium sulfate as kieserite or langbeinite available as SO_4^{2-} ion sources (Wrege & Dancy, 1949). Operations in Utah and Sicily involve decomposition of the mineral kainite to form *schoenite* ($K_2SO_4 \cdot MgSO_4 \cdot 6H_2O$). The K_2SO_4 component of the latter salt is extracted by chemical manipulation.

2. Langbeinite

The agricultural specification of langbeinite is minimum K of 18.3%, minimum Mg of 10.8%, maximum chloride of 2.5%, and minimum S of 22%. In addition to being a relatively chloride-free source of K, langbeinite is also a water-soluble source of Mg and S. These elements are required in many mineral fertilizers.

Processing langbeinite ore that does not contain recoverable quantities of KCl involves mixing crushed ore with enough fresh water to dissolve the NaCl contained in the ore. Most of the langbeinite remains in the solid state and is separated from the resulting brine, dried, and screened, and the individual products are sent to separate storage bins. Ores that contain recoverable quantities of KCl are processed by heavy media and froth flotation as described previously to recover both langbeinite and KCl as agricultural grade K chemicals (Dancy, 1982).

3. Potassium Nitrate

Potassium nitrate (38.2% K and 14% N) is produced by the reaction of KCl and nitric acid and (HNO_3) in two major industrial operations. Hydrochloric acid is produced as a coproduct (Dancy, 1982).

4. Miscellaneous Potassium Fertilizer Materials

Minor quantities of potassium hydroxide, potassium carbonate, and potassium phosphates are used in specialty fertilizers. The pricing structures of these fertilizers are such that the above K compounds, produced for industrial purposes, can be used to formulate certain specialty fertilizers.

II. MARKETING POTASSIUM FERTILIZERS

A. Market Changes and Trends

Marketing of K includes the total system of getting the refined product from the mine to the point of use on the land for improved crop production. Barber et al. (1971) reviewed marketing trends of K up to 1970. Millions of tons of K are needed and used annually, requiring a total system of production, research and education for market development, sales, communications, transportation, storage, and equipment for application in the marketing process.

Market trends in K consumption have shown dramatic expansion. In 1970, world K use was 6.8 million metric tons. Trends in increased K use by country, region, or continent and the world from 1965 to 1980 and 1985 are shown in Table 10–1. In 1980, consumption was > 20 million metric tons, nearly three times greater than in 1960. Estimated consumption for the year 1990 is 35.1 million metric tons of K annually (Anon., 1981). By 2000, use is expected to increase, and projections range from 40 million to 50 million metric tons of K.

World K consumption increased at an annual compounded rate of > 6%/yr between 1965 and 1975. With the adjustments in the world economy to the stress of increased energy costs, the rate of K use was slowed. Between 1975 and 1980, the annual rate of increase in consumption was slightly > 4%/yr.

von Peter (1980) has reported that the World Bank estimates indicate the greatest annual growth rates for K to be in developing countries, with Latin America and the Near East showing K growth rates of 12.8 and 14.5%, respectively, for the period studied. The high cost of energy and disruption of the world economy by recession caused these growth rates to slow during the early 1980s in both the developed and developing countries. But, if world food demands are to be met, the growth rates of K and its use in developing countries will have to increase and remain high.

Gidney (1981) provided production figures for 1981 and projected future production for 1991 by region (Table 10–2). It appears that if production is economical, adequate K can be produced to meet demands. Searls (1981) estimated that 1981 world production was 23.9 million metric tons of K from the world reserve base of 16.9 billion metric tons. Estimated world K resources total about 116 billion metric tons.

B. United States Consumption

As a country, the USA currently ranks second in K use and is a very important market. In 1981, total K use in the USA for all purposes was 5.6 million metric tons, which was 23.3% of world production. Net imports of K accounted for 68% of the total used in 1981 (Searls, 1981). In 1980, 87% of the total K use was used for fertilizers. About 95% of the U.S. domestic production is used for fertilizer.

Consumption of fertilizer K has grown steadily but at a variable rate depending on crop surpluses, world food demand, changing energy costs, grain embargos, and economic conditions. The growth of the market from 1960 through 1980 is shown in Table 10–3. The overall increase in K use during that period was 2.86 times, or 286%. The average compounded rate of growth has been > 5%/yr. The largest consumption year to date was 4.7 million metric tons in 1979.

In 1980 U.S. K production was estimated to be 2.1 million metric tons, which was 36% of that used in fertilizers (Mahan & Stroike, 1980). However, some of that K produced is exported. Most of the U.S. imported K comes from Saskatchewan Province in Canada, which accounts for about 76% of the total U.S. fertilizer use.

Table 10-1. World K fertilizer consumption and percentage (in parentheses) of the market for regions for 1965, 1975, and 1980 and estimates for 1985 (Harre & Harris, 1979).

Year	North America	Latin America	Western Europe	Eastern Europe	USSR	Africa	Asia	Oceania	World
	1000 metric tons K								
1965	2 236.3(24.4)	222.9(2.4)	3 233.2(35.3)	1 263.6(13.8)	1 179.4(12.9)	129.9(1.4)	767.2(8.4)	126.6(1.4)	9 159.1
1975	3 524.3(21.4)	771.6(4.7)	3 819.9(23.2)	2 819.8(17.1)	3 223.7(19.6)	285.4(1.6)	1 811.7(11.0)	197.4(1.2)	16 453.8
1980†	4 947.8(25.6)	1 315.4(6.8)	4 693.3(24.2)	5 953.8 --	--	186.0(1.0)	1 591.8(8.2)	184.5(1.0)	20 132.7
1985	5 539.2(20.8)	1 574.8(5.9)	4 725.1(17.7)	3 812.5(14.3)	7 826.9(29.3)	463.6(1.7)	2 456.3(9.2)	281.8(1.0)	26 680.2

† Figures provided by the Potash & Phosphate Institute, Atlanta.

Table 10-2. Current and future K production by regions and the world (Gidney, 1981).

Year	North America	Latin America	Western Europe	Eastern Europe	Asia	World
			million metric tons K			
1980-1981	7.9	--	4.7	9.5	0.7	22.8
1990-1991	13.9	0.4	4.2	15.5	1.7	35.7

Table 10-3. United States consumption of K fertilizers for periods from 1960 through 1980, with percentage increases during periods (USDA, 1980).

1960	1965	1970	1975	1980
		1000 metric tons K		
1621.4	2134.3	3038.6	3353.2	4639.2
		% increase over previous period		
	31.6	42.4	10.4	38.4

Table 10-4. The K fertilizer use and percentage of the total U.S. use by regions for 1978-1979 (Andrilenas & Rortvedt, 1979).

U.S. geographic region	1000 metric tons K	% of total
Northeast	224.4	4.8
Lake states	777.8	16.7
Corn Belt	2006.7	43.1
Northern Plains	120.5	2.6
Appalacian	475.9	10.2
Southeast	598.6	12.8
Delta states	196.5	4.2
Southern Plains	118.2	2.5
Mountain	33.0	0.7
Pacific	108.4	2.3
Total	4660.1	

The use and percentage of the total use by geographic regions is shown in Table 10–4. This K is distributed as both fluid and dry fertilizers through > 5000 fertilizer plants in the USA.

The eight Corn Belt and Great Lake States account for nearly 60% of the U.S. K use. When the Appalachian and Southeast states are included, these areas account for about 83% of the consumption. Most of the soils in the semiarid Great Plains and West are inherently higher in K. Lower K losses due to leaching and crop removal due to lower yields under dryland production account for the tendency for lower use in these regions.

The average rates of K use on corn (*Zea mays* L.), soybeans [*Glycine max* (L.) Merr.], wheat (*Triticum aestivum* L.), and cotton (*Gossypium hirsutum* L.) for selected states in the USA are shown in Table 10–5. These data are based on USDA information. Because of increased hectarages without greater total use of K, these values actually decreased slightly during the early 1980s. It is expected that K use will continue to increase in the late 1980s and beyond as world food demands increase and markets for U.S. grain products are developed and expanded.

Table 10-5. Average rate of K use on corn, soybeans, wheat, and cotton for selected states in the USA for 1979 (Andrilenas & Rortvedt, 1979).

State	Use on corn	Use on soybeans	Use on wheat	Use on cotton
	kg/ha			
Alabama	--	47.8	--	53.4
Arizona	--	--	--	1.4
Arkansas	--	17.5	--	39.1
California	--	--	--	1.8
Colorado	9.8	--	--	--
Georgia	80.8	61.6	--	72.5
Idaho	--	--	1.1	--
Illinois	91.8	19.8	47.5	--
Indiana	99.3	41.1	53.3	--
Iowa	61.2	5.9	--	--
Kansas	8.8	1.7	2.0	--
Kentucky	72.4	39.3	--	--
Louisiana	--	18.8	--	31.7
Michigan	77.7	--	53.3	--
Minnesota	63.2	10.9	15.8	--
Mississippi	--	26.5	--	19.5
Missouri	58.5	20.1	40.9	47.9
Montana	--	--	0.8	--
Nebraska	9.6	1.8	0.3	--
New Mexico	--	--	--	1.9
North Carolina	90.4	51.2	--	--
North Dakota	--	--	1.6	--
Ohio	71.3	28.2	54.5	--
Oklahoma	--	--	1.8	7.0
Oregon	--	--	0.7	--
Pennsylvania	52.2	--	--	--
South Carolina	--	70.6	--	94.7
South Dakota	--	--	0.4	--
Tennessee	--	39.5	--	72.9
Texas	9.1	--	1.3	2.5
Virginia	81.4	--	--	--
Washington	--	--	0.9	--
Wisconsin	83.6	--	--	--
Avg	63.7	24.4	7.0	11.2

In the central Corn Belt the seasonal marketing of K has shifted. In the late 1960s, peak marketing took place in the spring and fall, but by the late 1970s, there had been a shift to greater fall applications of K. This shift can be readily observed in the 3-yr average monthly sales curves for years 1968–1970 and 1978–1980 for Illinois (Fig. 10–1). In 1981, preliminary USDA estimates K consumption in Illinois was > 0.815 million metric tons.

Crop production and K use in the USA is still on an extensive basis compared with many other parts of the world. When K use on arable land in various countries is compared, it is apparent that annual use per hectare in the USA ranks well below many developed countries where more intensive crop production is practiced (Table 10–6). The U.S. market is still a developing one.

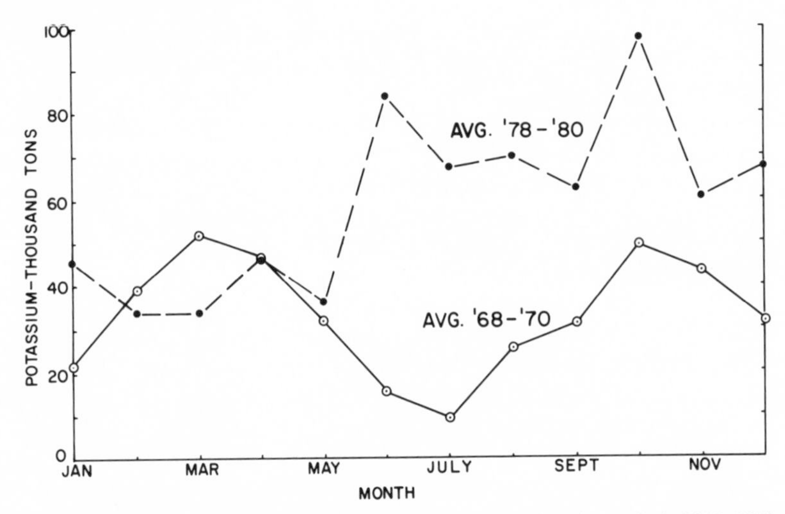

Fig. 10–1. Three-year average monthly sales of K fertilizers for the periods 1968–1970 and 1978–1980 for Illinois.

Table 10–6. Potassium use per hectare of arable agricultural land in some countries and the world in 1979 (FAO, 1981).

	Average K use
	kg/ha
Ireland	160.7
Belgium	154.8
FRG	133.2
Japan	124.5
Netherlands	119.1
DGR	78.2
South Korea	72.0
United Kingdom	55.4
USA	24.6
USSR	15.8
Brazil	14.6
South Africa	7.2
Kenya	2.9
India	2.6
Mexico	2.1
World	13.4

C. Market Development

Market development for K is directly tied to research and educational programs by governments, international centers, university and academic institutions, and privately supported groups. These programs play a key role in market development in every country.

The Food and Agriculture Organization of the United Nations has aided in conducting thousands of trials through their Freedom from Hunger Campaign or Fertilizer Programme in cooperation with developing countries to identify and demonstrate crop response to K. Individual country agencies such as the U.S. Agency for International Development have had action programs of cooperation with developing countries to aid in soil fertility research. More recently, the International Fertilizer Development Center has initiated a program involving K (Munson, 1982). Internationally, private sector leadership in K market development has come from the International Potash Institute and the Potash & Phosphate Institute. These organizations cooperate in sponsoring missions or programs in East and Southeast Asia, the Kali Kenkyu Kai in Japan, the Association for Potash Research in South Korea, and the Instituto da Potassa in Brazil.

Other market development groups in various parts of the world include the Büntehof Agricultural Research Station associated with Kali and Salz A.G. in the FRG, the Societe Commerciale des Potasses et de l'Azote in France, and the Potash Research Institute of India. Additionally, trade associations and individual companies in the countries themselves aid in the research and educational aspects of K development.

These organizations are staffed with well-trained scientists and serve the industry and the agricultural community by assisting in getting the facts on the soil that show crop response to K, the K requirements of high-yielding crops, and the K needs for profitable crop production on various soils around the world.

Educational programs are conducted by extension groups, both academic and private, directly or through publications. Training of dealers, managers, or consultants who deal directly with farmers is a key part of this effort. Educational publications such as *The Potash Review, Better Crops with Plant Food, Agro-Knowledge, News & Views, The Fertilizer Correspondent,* newsletters, folders, and brochures are used to disseminate information. Radio tapes and scripts are available for radio farm directors. News releases on timely topics are prepared for the media. These help inform agricultural leaders on the benefits and profitable use of K fertilizer on crops. Slide sets and scripts, which may be read or broadcast from a tape cassette, and movies are available on different topics for extension groups industry to use in dealer and farmer educational meetings.

Academic societies and market development organizations periodically sponsor symposia, or workshops bringing together research, extension, and industry leaders to evaluate or discuss ideas or topics on K. Published proceedings of such meetings summarize the latest facts on K needs and use.

Research projects are developed and grants are provided to scientists to aid in obtaining basic information on K in soils, fertilizers, and crops. In some cases these are provided by institutes and the industry, whereas in other situations and countries the industry may be government owned. Research laboratories and experimental farms in some cases are owned and operated by the industry. Industry in general has increased its investment in soil fertility research on K. Efforts are being made to determine the maximum yields of crops and the most profitable levels of sustained K use associated with higher yield levels. In the USA, a new organization providing increased support to this effort is the Foundation for Agronomic Research. It has broad support from agricultural industries and is associated with the Potash & Phosphate Institute.

D. Sales

Producers of K fertilizers maintain active sales organizations distributed regionally in the world. In North America each company maintains a system of regional offices and sales personnel to complete contracts, service, and handle technical problems for manufacturer's and dealer customers. Sales people are a key link in the marketing of K. They help determine means and methods of most rapid delivery of the K to dealers in their region. The dealers are the persons completing the final transaction and delivery of K to farmers on their fields.

In Canada, an organization called Canpotax Ltd., has been responsible for the export sales. Based on recent developments, the number of exporters from that country may increase. In the USA, exports are made by the individual companies.

E. Transportation and Storage

In North America, transportation and storage have been undergoing change. Railroads provide the major method of K transportation from the areas of mining and refining. One-hundred car unit trains have been and still are used to transport K (Fig. 10–2). These trains transport K to warehouses or harbors for loading onto ships for world export (Fig. 10–3).

Within North America, unit trains bring K fertilizer to new facilities at Thunder Bay, Ontario, Canada on the Great Lakes. (Fig. 10–4). Cars are dumped and loaded on 27 200–metric ton ore boats for distribution to docks around the Great Lakes (Fig. 10–5). The K fertilizers are then trucked or reloaded into hopper cars for further transport. They also may be loaded into barges for transport through the U.S. river system to points of storage or use.

Because of increases in railroad abandonments in the USA, there has been a large increase in the number of trucks being used to transport K fertilizers both to and from an expanded number of warehouses placed strategically within major K use areas. These truck transports with trailers may haul 36 metric tons or more and can be quickly unloaded at warehouses (Fig. 10–6). Warehouses may store from 18 000 to 50 000 metric tons of K fertilizer. The increased use of trucks and warehouses make K fertilizers more readily accessible and shortens the supply line during peak periods of use and application during the spring and fall season. Larger dealers may have K storage of 5500 metric tons or more at a given site.

Fig. 10–2. Unit trains of 90 metric–ton hopper cars of K fertilizers on their way to world markets (courtesy of International Minerals and Chemical Corporation, Carlsbad, NM).

Fig. 10–3. Potassium products moving into the world export market are transported in large freighters through major sea ports (courtesy of Santa Fe Railway, Spokane, WA).

Fig. 10–4. Potassium fertilizers being loaded onto an ore boat at the Thunder Bay, Ontario, Canada Terminal to be shipped to dock terminals on the Great Lakes for redistribution (courtesy of American, Spokane, WA).

Fig. 10–5. Potassium fertilizer being unloaded from an ore boat at a Great Lakes dock for rail or truck transport to fertilizer plants (courtesy of International Minerals and Chemical Corporation, Carlsbad, NM).

Fig. 10–6. Potassium fertilizer being rapidly unloaded from a truck and trailer transport at a regional warehouse (courtesy of International Minerals and Chemical Corporation, Carlsbad, NM).

III. USE OF POTASSIUM FERTILIZERS

The use of K fertilizer on crops supplements the supply from the soil. The amount needed depends on the forms present in the soil, the amounts of each form, their rate of supply to the plant root, the reaction of the added K with the soil, and environmental and crop factors that affect fertilizer K utilization. Each of these topics is discussed with respect to their influence on fertilizer use.

A. Potassium in the Soil

Soils contain from 0.5 to 2.5% K or 100 to 500 q/ha in the top 15 cm. This is a very large amount relative to the annual K uptake by plants, but all but 1 or 2% is essentially unavailable because it is in rather insoluble minerals. Soil K can be grouped into four categories according to availability. The categories and the approximate amounts in each are 5000 to 25 000 mg/kg mineral, 50 to 750 mg/kg "difficultly available," 40 to 600 mg/kg exchangeable and 1 to 50 mg/kg solution. A few soils may have amounts beyond the ranges given.

There is movement between each form to establish an equilibrium, but the rate is so slow between the mineral and more available forms that equilibrium is usually not maintained. Exchangeable and solution K^+

equilibrate rapidly. Difficultly available K equilibrates very slowly with the exchangeable and solution forms. The movement from the mineral form to any of the other three forms is extremely slow in most soils so that during the growth period of a crop mineral K can be considered essentially unavailable.

Mineral K is held tightly in the structure of feldspars and micas, which are very resistant to weathering. Difficultly available K is present mainly with clay minerals such as illite, vermiculite, and chlorite. Clays release K by separation of clay layers. The small particle size of clays facilitates release of K. Exchangeable K^+, held by electrostatic forces, is very readily exchanged from the solid phase of the soil to solution by other cations.

B. Potassium Supply to the Plant Root

Potassium must be positionally available at the root surface as well as in an available form as K^+ in solution before it will be absorbed into the root by metabolically controlled uptake mechanisms. In most soils, K must move to the root before absorption since the amount intercepted at the root surface by the growing root is extremely small.

The mechanisms of supply to the root are mass flow and diffusion. Mass flow is the transport of K in the convective flow of water to the root resulting from absorption and transpiration of water by the plant. The amount transported by mass flow depends on the amount of water used by the plant and the K content of the water flowing through the soil to the root. Barber et al. (1963) determined the K content of saturation extracts from 142 North Central region U.S. soils and found a mode of 4 mg/kg and a range of 1 to 80 mg/kg. In a survey of the literature, Barber (1962) found K contents of soil saturation extracts varied from 3 to 156 mg/kg. Reisenauer (1964) surveyed values in the literature and for 155 samples found a median value of 41 to 60 mg/kg. The higher values were from arid or saline soils.

If mass flow were to supply all the K to a corn crop, the water moving to the root would need to have a K content in excess of 40 mg/kg. Since many soils of the humid region have only about one-tenth this amount, diffusion is usually considered the most important mechanism supplying K to the root.

Diffusion of K to roots growing in soil occurs over short distances. The diffusivity is slow enough that the concentration gradient extends only about 1 to 4 mm from the root surface during the time of growth of annual plants. The general nature of the depletion of K about corn roots is shown autoradiographically in Fig. 10–7. This autoradiograph was made using ^{86}Rb, an element very similar to K, but having a more convenient radioactive isotope. The behavior of Rb in the soil is very similar to K. This autoradiograph illustrates that only K within a few millimeters of the root can be absorbed. Potassium further away may be in an available form, but it is not available positionally.

Since diffusion is usually the dominant mechanism, factors that affect the rate of supply by diffusion will influence K availability to the plant. The

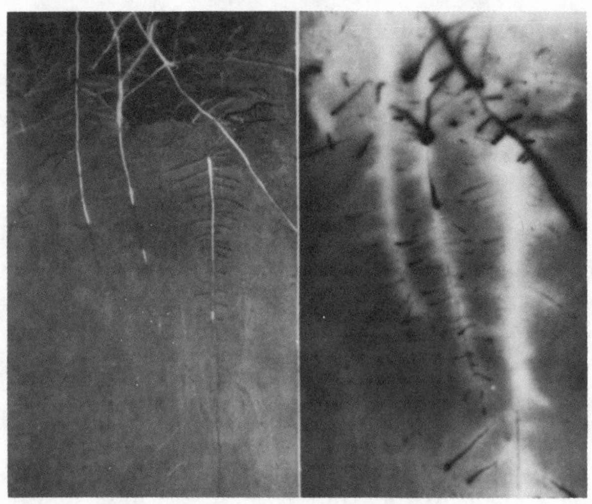

Fig. 10–7. *Left,* photograph of corn roots growing in soil. *Right,* autoradiograph of ^{86}Rb depletion about corn roots. Dark areas are where ^{86}Rb is present; light areas are areas of depletion. Pattern indicates supply to the root by diffusion.

amount of K diffusion will influence K availability to the plant. The amount of K diffusing to the root system depends on (i) the K concentration gradient, (ii) the rate of diffusion, and (iii) the surface area of the root.

The K concentration gradient for diffusion depends on the level of K in the soil and the amount the root reduces the K concentration at its root surface. The latter varies with species and depends on the rate of K absorption by the root compared with the rate of supply by the soil.

The rate of diffusion in the soil, the apparent diffusion coefficient, increases as soil moisture increases. This is due to reduced tortuosity of the diffusion path and a greater volume of water through which the K can diffuse. Increasing salt content as well as increasing the K level in the soil increases the diffusion rate. Potassium diffuses faster through coarse- than fine-textured soils. At equal moisture contents, increasing soil bulk density causes an increase in diffusion up to a bulk density of about 1.5. Above this value, diffusion decreases as bulk density increases because the diffusion path becomes more tortuous.

Crops with a large number of fine roots have the possibility of more K diffusing to the root per gram of roots than crops with coarse roots since they have a much greater root area. The diffusion of K in most soils is apparently rapid enough that the presence of root hairs does not greatly increase K uptake.

Vaidyanathan et al. (1968) found that increasing the moisture level and K saturation increased the rate of diffusion. Place and Barber (1964) found that the increase in diffusion rate of ^{86}Rb with increased moisture was greater the higher the level of exchangeable Rb^+ present in the soil. The measured diffusion coefficients correlated highly with the uptake per unit area of root, indicating that rate of diffusion was controlling uptake.

C. Soil Factors Affecting the Efficiency of Potassium Use

The amount of K fertilizer required by a particular crop will depend on the needs of the crop, the amount present in the soil, and the efficiency with which the crop uses the soil and fertilizer K. Some factors that influence the efficiency of fertilizer K use are (i) the reaction of added fertilizer K with the soil, (ii) the rate of K movement in the soil, (iii) the losses of K from the soil by erosion and leaching, and (iv) soil properties, such as aeration, that influence the rate at which roots will absorb K.

1. Reaction of Fertilizer Potassium with the Soil

Fertilizer K as KCl added to the soil is 100% water soluble. This soluble K^+ equilibrates with the exchangeable cations present in the soil with the result that most (95% or more on many soils) becomes exchangeable K^+ initially. Exchangeable K^+ is readily available for diffusion to the root. The exchange sites minimize K leaching.

Some of the added K slowly moves into a form called *fixed K.* This K is trapped between the clay layers and is very slowly available to plant roots. The amount of fixation depends on the type of clays present in the soil. Illites, vermiculites, and chlorites fix K; montmorillonites and kaolinites usually do not. The fixation is related to the charge density of the clay. Clays with a high charge density on their layers are held together more tightly after K has been trapped between them; hence, clays with a high lattice charge fix the most K.

In some high–charge density clays, the amount of K fixation is reduced because of the presence of hydroxyl Al, Fe interlayer groups that prop the silica layers apart, preventing collapse about the K^+ ions. Soils developed under acidic conditions are more likely to have interlayer groups that prop the layers apart.

When K is added to soils, exchangeable or soil test K may not increase nearly as much as the K added. If crops have not removed K and K is not leached, the reason for the reduced K level is fixation. Because of high K fixation, it is not practical to try to build up available K levels in some soils. In studies on the rate and amount of fixation of K by a number of Indiana soils, Horton found that after 221 days' incubation, high-fixing soils were still fixing K (Horton, 1959). In low-fixing soils only 10 to 25% of that added was fixed when the system had reached equilibrium.

2. Release of Difficultly Available Potassium

Potassium is released from nonexchangeable forms to exchangeable and soluble K^+ when soils are cropped. Measurements of the K potentially available for release have been made by exhaustively cropping the soil. The amount of release that occurs when crops are fertilized with K will likely be much less than that under exhaustive cropping, since the K concentration gradient for movement from nonexchangeable to exchangeable will be less.

Release of K is the opposite reaction to fixation. To get release, one usually must reduce the charge on the clay mineral so that the high charge density will not cause the K to be held tightly. One mechanism for reduction of charge is through oxidation of Fe present in the crystal structure of the clay. Exhaustive cropping experiments conducted by a large number of investigators have shown that soils may release from 0 to 1500 kg of K/ha. Barber and Humbert (1963) summarized the experiments prior to 1962 and found that the soils releasing the largest amounts of K were usually of alluvial or lacustrine origin. The majority of soils released less than 400 kg/ha. This may appear to be an appreciable quantity, but this measurement was made under exhaustive cropping conditions where the plants were suffering from K deficiency. When K fertilizer is added so that the plants are not K deficient, the equilibrium is shifted and less K is likely to be released.

3. Potassium Movement in the Soil

Since K is held as an exchangeable cation, it does not move through the soil readily with water movement in soils that have appreciable cation exchange capacity (CEC) (> 0.05 meq/g). Potassium moves primarily by diffusion. When K is applied to the surface of a loam or finer-textured soil, it will not move > 1 or 2 cm into the soil in the first year. Deeper incorporation would have to come from tillage. Avnimelech et al. (1970) showed that K could be moved into the soil by displacing it with calcium sulfate ($CaSO_4$). This provided a method of moving K into the soil without tillage. For the two soils studied, a sandy loam and a clay loam, the rate of movement of K relative to the rate of movement of saturated $CaSO_4$ was 0.20 and 0.04 for the Sharon sandy loam (coarse-silty, mixed, acid, mesic Typic Udifluvents) and Rehania basaltic clay loam, respectively.

4. Potassium Loss from the Soil

Under most situations, very little K leaches from the soil. If we know the average K content of a soil solution and the amount of water that percolates through a soil in a year, we can estimate annual losses. If we assume a soil solution K content of 4 mg/kg (the mode of measurement on humid region soils made by Barber et al., 1963), for each centimeter of water moving through the soil we would lose 0.4 kg K/ha. Where 25 cm of water percolates in a season, the loss of K would be 10 kg/ha. Soils in the more arid areas may have higher K contents in the soil solution, but they also have little water percolating through the profile.

Losses may be higher on acid soils. On acid soils a higher proportion of the available K is in the soil solution when the companion exchangeable ions are H^+ and Al^{3+} than when they are Ca^{2+} and Mg^{2+}. Higher losses have also been reported on organic soils.

Runoff water usually will contain less K in solution than percolation water; hence, losses of K would be less per unit of water lost from the soil. Since runoff waters remove silt and clay containing exchangeable and mineral K, the total K lost may be much larger than losses of solution K. Barrows and Kilmer (1963) reviewed the research on K losses and reported annual soluble K losses of 0.9 to 8 kg/ha, available K losses of 0.7 to 23 kg/ha, and total K losses of 2.8 to 1110 kg/ha. Also, Munson and Nelson (1963) have prepared an extensive review on the movement of applied K in soils.

D. Soil Properties Affecting Plant Uptake of Potassium

1. Aeration

Lawton (1946) showed that reduced soil aeration reduced K uptake more than it reduced the uptake of other major nutrients. Aeration was reduced by high moisture conditions and soil compaction. Forced aeration under these conditions increased corn growth and its K content. Poor aeration of soils can result from improper tillage practices, particularly tillage of wet soil. Soils that are very dense may be improved by drainage.

2. Soil Moisture

There is a relation between the response of crops to applied K and the average soil moisture present during the growth of the crop. Van Der Paauw (1958) found a relatioin between the number of rainless days and the response of potatoes (*Solanum tuberosum* L.) to applied K. With a greater number of rainless days, there was a larger response to added K. Barber (1960) found a similar relation between the total rainfall for June, July, and August and the response of corn to added K. When the rainfall was 35 to 50 cm, only a small response occurred. Rainfall $>$ 50 cm for the 3-month period gave a large K response that was attributed to a reducion in soil aeration caused by the large amount of water present in the soil.

There is variation from year to year in response to added K at any one level of available soil K measured by a soil test. This variation may be due to the influence of moisture and aeration.

E. Measuring Potassium Needs

Since soils vary in the amount of K they supply for growing crops, methods of predicting this supply, and, hence, calculating the need for K fertilizer are important. Methods include various types of soil tests, tissue analyses for K, and recognition of deficiency symptoms.

The uptake of K by crops growing on soil depends on both the ability of the plant roots to absorb K and the rate at which the soil can supply K to the root surface. Claassen and Barber (1976) developed a mechanistic simulation model that described K uptake by plants growing in soil. They obtained a correlation of $r^2 = 0.87$ between observed and predicted K uptake by corn grown in four soils (Fig. 10–8). Predicted uptake was greater than observed uptake because of competition between roots for K uptake. Cushman (1979) developed a model that also considered competition between roots.

The model describes soil K supply to the root by three parameters: Initial K concentration in the soil solution, buffer power of exchangeable K^+

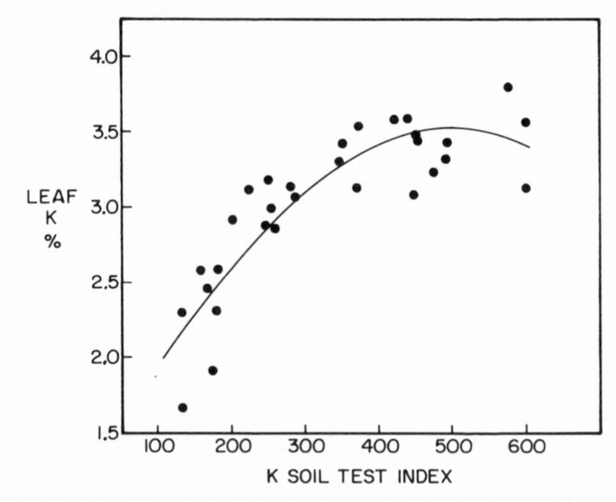

Fig. 10–8. Relation between K in corn leaf and soil test index for K. Leaf sampled was opposite and below ear leaf (James, 1970).

for solution K^+, and apparent diffusion coefficient for K diffusion in the soil. The product of solution K^+ and buffer power represents exchangeable K^+. Since there are three parameters involved, use of only one parameter to describe soil K availability to the plant depends on a correlation among the three parameters. If the soils are derived from similar materials, the three parameters may be closely correlated.

1. Available Potassium Measurement

The K held as exchangeable K^+ plus the smaller amount present in soil solution is the most frequently measured by extracting it from the soil with 1 *M* ammonium acetate (NH_4OAc) adjusted to pH 7.0. Other extractants such as sodium nitrate ($NaNO_3$), HCl, and calcium nitrate [$Ca(NO_3)_2$] have been used. On some soils these extractants remove slightly more K than NH_4OAc; however, for most soils the results are rather similar providing that sufficient extractant is used to replace essentially all of the exchangeable K^+ present.

The methods of evaluating the significance of the amount of available K extracted varies. Many laboratories express the amount as either parts per million (milligrams per kilogram) or as kilograms per hectare, using an arbitrary constant for changing parts per million to kilograms per hectare. Some calculate level of K in terms of soil volume. This is particularly significant where low bulk density organic soils are being evaluated.

Alternatively, exchangeable K^+ can be expressed as a percentage of the CEC of the soil. Where this has been done, it has been necessary to adjust for the size of the exchange capacity to get a suitable measure of availability. A higher percentage saturation is required where exchange capacities are low than where they are high. This method has not generally been as suitable. It infers that K is less available the higher the CEC of the soil.

Solution K^+ has been proposed as a measure of available K. This is used along with its relation to exchangeable K^+ as a measure of available K and is called the *quantity-intensity,* or *Q/I,* approach. The intensity is defined as the immediate availability to the root, whereas the capacity is the ability of the soil to maintain a level of intensity as K is withdrawn from solution by the root. The *intensity parameter* is defined as the relation K activity/Ca activity. Since the ion content of the solution is relatively small compared with the amount of exchangeable ions, the intensity relationship remains approximately constant with dilution of the soil with water (Schofield, 1947). Woodruff (1955) suggested that the relation K/Ca be used as a measure of K availability. Later the relation was expanded to K/(Ca + Mg). The use of the intensity term indicates that uptake by the plant is related to K/Ca. This is not the case in nutrient culture; at a constant K concentration, changing the Ca concentration above a minimum level has little effect on K uptake by the plant. Clark (1978) found that changing solution Ca^{2+} levels in seven steps from 0 to 25.4 mmol/L had very little effect on the K concentration of the shoots of 21-day-old corn even though the Ca concentration increased from 0.4 to 1.8. The uptake is much more closely related to the K concentration or activity in solution (Asher & Ozanne, 1967). Hence, if we could measure the K concentration in the soil solution this would be the appropriate value for immediate availability.

Koch et al. (1970) made *Q/I* measurements on the soil and related them to K uptake by the plant by measuring leaf K content and yield response of corn to applied K. He found the highest correlation was with the quantity factor, or the labile form, of K in the soil. In many soils the exchangeable K^+ would be similar to that measured as labile in these experiments.

In many experiments, exchangeable K^+ has given a high correlation with K uptake by the plant. In the uptake of K, the initial availability is governed by the level in soil solution. However, the K concentration in solution at the root surface is soon reduced to a fraction of the initial level, and the rate of this reduction is related to buffer power and diffusion rate. Thus, if very little K were absorbed, solution K^+ level would probably be a good measure of available K.

When K uptake continues, a K concentration gradient extends out from the root. The distance it extends from the root depends on the diffusion coefficient. Where soils have a low CEC, K is not usually held as tightly by the soil, and diffusion distance is greater. Therefore, using the CEC in the calibration value may be useful where plants absorb intermediate amounts of the exchangeable K^+ present in the soil.

With many crops, root growth is intensive in the topsoil and adjacent roots are close to one another. When this happens, adjacent roots compete for K. Soil K can only move to the root from the half-distance between roots. When this occurs, the total exchangeable K^+ present begins to determine the availability of K to the plant. The data in Table 10–7 were used to calculate K uptake for two different root densities in the soil. In the calculation, total root length per plant remained constant; therefore where r_1, the half distance between roots, was reduced, the total soil per plant was

Table 10–7. Predicted K uptake as affected by interaction of root density and K absorption properties of the soil (Barber, 1981).

Soil property	Soil		
	I	II	III
C, mmol/kg	2	2	2
C_{li}, mM	1	0.2	0.1
b	2	10	20
De, $cm^2/s \times 10^7$	5	1	0.5
Uptake μmol/plant per week			
$r_1 = 0.09$ cm	23	21	20
$r_1 = 0.45$ cm	269	121	71

reduced. The data in Table 10–7 show how strength of bonding of K by the soil can affect K uptake where roots do not compete; however, when they do compete, these differences become minimal. Usually as CEC is increased, relative strength of bonding increases.

The correlation of exchangeable K^+ with K uptake by the plant can vary according to the intensity of cropping. However, for crops that absorb appreciable quantities of K (100 kg/ha or more), the level of exchangeable K^+ will probably become the most reliable indicator of availability.

2. Difficultly Available Potassium

Release of K from difficultly available forms may supply K to the crop, and variation among soils in the amounts released may be important when the K fertility status of the soil is assessed. Because the role of difficultly available K was discussed under section IIIC2, only a brief reference is made here. On many soils, K release does not contribute greatly to the K nutrition of the crop because the soils contain <200 mg/kg of difficultly available K. It is only on high-release soils where measurements in addition to exchangeable K^+ may be needed to satisfactorily measure available K. Experiments conducted to date indicate only a small portion of the soils are of this type.

Tabatabai and Hanway (1969) found that soils released considerable difficultly available K when cropped at their minimal levels of exchangeable K^+. The *minimal level* was defined as the level to which the exchangeable K^+ would fall with intensive cropping. The amounts released during a 758-d cropping period averaged 24 times as much as that present as exchangeable K^+ at the minimal level of exchangeable K^+. The amount released was highly correlated with the minimal level. Most of the subsoils used were at or near the minimal level naturally. Hence, where soils were at the minimal level, exchangeable K^+ was a measure of K supplying power.

Attempts have been made to chemically determine difficultly available K. The most common procedure is boiling the soil for 10 min in 1 M nitric acid (HNO_3), a method proposed by Wood and DeTurk (1941). This method removes amounts of K that are similar to those removed in 8 to 15 successive croppings of the soil.

Jencks (1968) investigated the K released by boiling HNO_3 on 75 West Virginia soils. He found, in general, a good correlation between exchangeable K and K released by boiling HNO_3; however, there was a wide variation between topographic groups. The bottomland soils had a significant negative correlation, whereas the upland, colluvial, and terrace soils had significant positive correlations between these two properties.

Many soils are air-dried between the time of collection in the field and the determination of available K. Air-drying increases the exchangeable K^+ on many soils. On a few soils, particularly those high in exchangeable K^+, it reduces the level of exchangeable K^+. The exchangeable K^+ level of subsoils may be increased several-fold by drying (Hanway et al., 1962). Cropping experiments have shown that the exchangeable K^+ level determined on field-moist soil most nearly predicts the availability of K in the soil (Barber et al., 1961) as it exists in the field.

The change in level of exchangeable K^+ that occurs upon air-drying may be so large on some soils as to make the exchangeable K^+ measured meaningless in terms of availability to crops. For example, on a Brookston silt loam (fine-loamy, mixed, mesic Typic Argiaquoll) subsoil sample from Indiana, the field-moist K was 38 mg/kg, and the air-dry K was 96 mg/kg. A Clyde loam (fine-loamy, mixed, mesic Typic Haplaquoll) surface soil from Iowa had a field-moist K of 83 mg/kg and an air-dry K of 188 mg/kg (Barber et al., 1961). The soils that most frequently show an increase in exchangeable K^+ upon drying are either subsoils or surface soils from the less humid soil regions.

Scott and Smith (1968) have suggested that the release of K upon drying is due to the cracking of edge-weathered micas on drying. These cracks expose interlayer K that can then be released to the exchange sites.

In areas where K release on drying is a serious problem, it is necessary to collect soils in moistureproof containers. The K can then be determined on soil in the field-moist condition, since this is the measurement that reflects the availability of K to the crop.

3. *Subsoil Potassium*

Soil sampling for K availability of soils in the field is usually confined to the top 20 cm, or the plow layer. Although a large percentage of the roots are usually found in the tilled soil layer, the relative level of available K in the subsurface layers will also influence the total supply to the crop since roots will penetrate 100 cm or more into many soils. Most subsoils are lower in available K than the surface layers when available K is measured on field-moist samples (Hanway et al., 1962).

A regional field experiment with K in the North Central region of the USA investigated the contribution of subsoil K to the uptake of K by the plant and the response to added K using a multiple regression approach. Including values for exchangeable K^+ of the subsoil layers generally improved the degree of correlation, but the degree of improvement was small.

Where exchangeable K^+ (field-moist K) is high in the subsoil, it may be expected to contribute to crop uptake of K. However, in the usual case in humid regions, the subsoil is much lower in available K, and the inclusion of the subsoil measurement along with that of the surface soil does not appreciably increase the predictability for response to added K. It is important to realize that the subsoil may have a different level of available K when soils are being sampled. If the surface samples are taken to a depth that includes some subsoil with the surface soil, the resultant value for available K will be lower than that for the surface soil alone.

Schenk and Barber (1980) used a mathematical model to calculate the relative amounts of K absorbed from the tillage layer and the subsoil. Even though 55% of the roots were in the subsoil, the roots in the subsoil absorbed only 10% of the total K absorbed because of the lower K availability in the subsoil.

4. Plant Analysis

The level of available soil K is often reflected in the K content of the plants growing on the soil. Hence, plant analysis can be used as a predictor of the sufficiency of the supply of available soil K. This measurement may be too late to fertilize the present crop, but it can be used as a guide to fertilization of the subsequent crop. A number of factors need to be considered in using plant analysis for measuring K availability. These are:

1. Factors other than K should not be seriously limiting crop yield.
2. The part of the plant sampled and the stage of growth of the crop are important considerations, since K levels vary with plant part and in general K content declines as plant age increases.
3. Potassium concentration in the plant part at a particular sampling date must be calibrated with response to applied K so that the data can be interpreted in terms of the sufficiency of K present in the soil.

The relation that may be obtained because K content of the plant and available soil K is shown in Fig. 10–8. This indicates the relation between K content of the ear leaf of irrigated corn at tasseling and the exchangeable K^+ in a Shano silt loam (coarse-silty, mixed, mesic Xerollic Camborthid) soil in Washington (James, 1970).

Plant analysis is particularly useful on tree crops since, due to the extensive nature of the tree root system, the K level in the top 15 cm of soil frequently does not provide a satisfactory measure of K availability, and a relation with available soil K such as that shown in Fig. 10–8 is not obtained.

6. Deficiency Symptoms

When K deficiency is severe, the growth of the plant is stunted, and deficiency symptoms usually appear on the leaves. With K deficiency, the leaves on most plants become necrotic along the edges and at the tips, whereas the central part of the leaves may remain green. The deficiency is frequently most severe on the older leaves. Potassium is very mobile within the plant so that K is translocated from the older leaves to supply the new growth.

Detailed information and color plates that are helpful in recognizing K deficiency may be found in publications such as that edited by Sprague and Plucknett (1985).

F. Buildup of Available Soil Potassium

There is a level of available soil K above which K application does not increase crop yield. If we use a fertilization practice of building up the available soil K, we should attempt to maintain the soil at this level. The amount of K that must be added to maintain a particular level will depend on the initial level and the degree of K fixation by the soil. As discussed previously, the degree of K fixation depends on the clay minerals present in the soil. The amount of K needed also is affected by crop removal. When the total aboveground portion of the crop is harvested, a large amount of K may be removed. In grain crops with only the grain removed, K removal is much less, because ordinarily the grain has a much lower K content than the remainder of the plant.

An example of the effect of K application on the level of exchangeable K^+ is given in Table 10–8. Lucas (1968) reports only 8% recovery of applied K as exchangeable K^+ on a Genesee (fine-loamy, mixed, nonacid, Mesic Typic Udifluvent) soil, whereas on a Granby sandy loam (sandy, mixed mesic Typic Haplaquoll) recovery was as high as 78%. Hence, there is a wide variability between soils in the degree and rate at which they fix K in an unavailable form. On high-fixing soils, it may not be practical to attempt to build up the level of exchangeable K^+.

G. Potassium Fertilizer Application

1. Rate

The rate of K fertilizer required to give either the maximum or more profitable yield depends on the crop grown, the yield level of the crop, the level of available K in the soil, and the number of successive crops fertilized with the current application. Small grains are usually less responsive to K

Table 10–8. Percent recovery of added K measured as exchangeable K^+ after 221 d (Horton, 1959).

Soil	% recovery
Odell (fine-loamy, mixed, mesic Aquic Argiudolls)	81
Chalmers (fine-loamy, mixed, mesic Typic Argiaquolls)	74
Parr (fine-loamy, mixed, mesic Typic Arguidolls)	84
Crosby (fine, mixed, mesic Aeric Ochraqualfs)	78
Brookston	37
Zanesville (fine-silty, mixed, mesic Typic Fragiudalfs)	77
Vigo (fine-silty, mixed, mesic Typic Glossaqualfs)	85
Bedford (fine-silty, mixed, mesic Typic Fragiudults)	84

than crops such as alfalfa (*Medicago sativa* L.) or corn. Barber (1970) found the average percent response to applied K for four crops grown in rotation over an 18-yr time period to be corn, 29%; soybeans, 24%; wheat, 6%; and alfalfa, red clover (*Trifolium pratense* L.), and bromegrass (*Bromus inermis* Leyss.) hay, 26%.

Crops that remove large amounts of K reduce the level of available soil K and increase the need for K. Whenever the whole crop is removed, large amounts of K are removed because most plants contain 1 to 3% K on a dry weight basis.

The relationship between soil test level and the need for K on corn is illustrated in Fig. 10–9. At the lower soil test values, much larger amounts of K are required to get the maximum yield. Figure 10–9 was made from the regression equation obtained with data from 41 field experiments conducted in the North Central region of the USA on a regional project.

Rate of application required will vary with the yield level of the crop grown. Where other production factors allow for a larger than average yield, the amount of K fertilizer needed will be larger; however, the increase will be much less than the proportional increase in yield. Crops producing more top growth usually produce more roots, which in turn can forage for more K in the soil. Eventually larger applications will be needed, however, to replace the K removed by the larger yields.

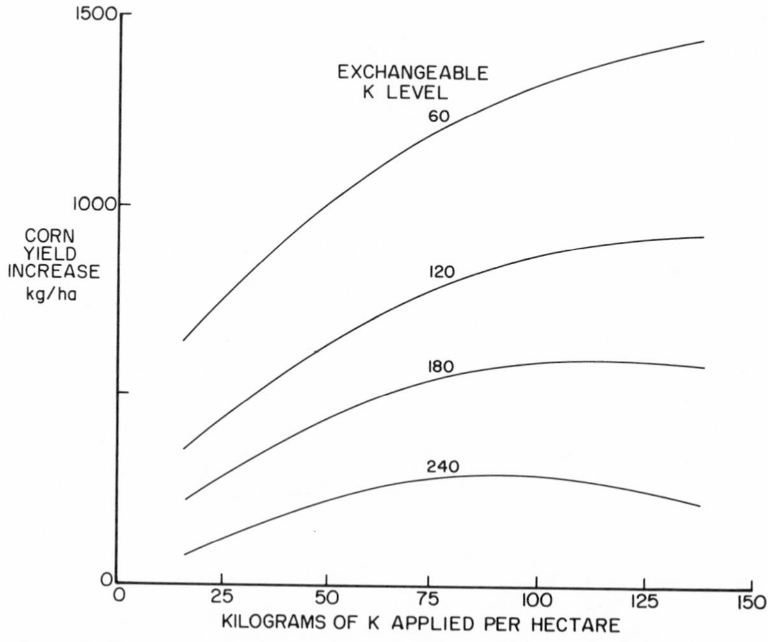

Fig. 10–9. Predicted increase in corn grain yield from different rates of applied K as influenced by level of exchangeable K^+ in field-moist 0- to 15-cm soil layer (North Central regional experiment). Values on the curves are exchangeable K^+ in kilograms per hectare (Hanway et al., 1962).

2. Frequency

The frequency of application of K for optimum yield depends on the crop grown, the relative fixing power of the soil, and the ability of the soil to hold K against loss by leaching. Where the total crop is removed, it is usually necessary to fertilize each crop. Where hay is grown and several cuttings a year are removed, it may be possible to take two or three cuttings from one K application. More efficient use of K would occur with an application after each harvest; however, the cost of application may make it more economical to apply the K less frequently.

Potassium is frequently required in corn-growing areas. Where corn is harvested for grain only, K removals are not large, about 4 kg of K/1000 kg of grain. Roughly one-fifth of the K absorbed is in the grain at harvest. When only the corn grain is removed from the field, it may be possible to fertilize less often than once per crop. Barber (1980) investigated the residual effect of K applied for the first corn crop on the yield and leaf K content of the second and third crop grown following fertilization. On a Chalmers silt loam (fine-loamy, mixed, mesic Typic Argiaquoll), using rates up to 168 kg/ha per year, he found that the effect of the residual K from the first crop on the second crop was about half that of the effect on the first crop. The effect on the third crop was negligible. When rates of 660 kg/ha were used, the K application was effective for the fourth crop. This soil fixed much of the applied K in an unavailable form. On soils that fix little K, the K applications should be effective for several years.

3. Placement

The various ways of applying K for a crop are (i) banding with or near the seed at planting, (ii) broadcasting on the surface, (iii) broadcasting and plowing under, (iv) broadcasting and working in with surface tillage, (v) banding into the surface soil, and (vi) banding into the subsoil. When rates are low, row application has frequently been more efficient than broadcast application. As the rate is increased, the difference between application methods usually disappears. Welch et al. (1966) in Illinois found surface-applied K was 0.33 to 0.47 as efficient as row application for corn. Where K was plowed down, the plowed down application was 0.69 to 0.88 as efficient as row application. The rates applied varied from 0 to 74 kg/ha.

Walker and Parks (1969) in Tennessee compared row and broadcast K applications for corn for 3 yr at rates of 0 to 84 kg/ha. Surface broadcast applications were approximately 73% as effective as row placement. As the rate increased, the difference between broadcast and row applications decreased.

Barber (1976) compared corn yield and corn leaf K composition resulting from three levels of K (28, 56, and 112 kg/ha) applied by three methods. The application methods were broadcast in the fall and plowed under, applied as a 7-cm wide strip every 70 cm on the surface in the fall and plowed under, and applied as a row application. The results summarized over rate are given in Table10–9 (Barber, 1976). The values are the aver-

Table 10–9. Effect of K placement on corn grain yield and K concentration in the ear leaf (Barber, 1976).

Placement†	Yield†	K in ear leaf†
	kg/ha	%
Row	7210	1.33
Broadcast and fall plow	7540	1.37
Strip and fall plow	8140	1.69

† Values are the average of 5 yr results of three rates of application: 28, 56, and 112 kg K/ha.

age for 5 yr where the treatments were applied annually to the same plots. Applying K in the strip gave the greatest K uptake as well as the highest grain yield. When K was applied in the row at planting, the added K contacted too few roots; therefore, K uptake was not as high. When K was broadcast, the K made contact with the whole plowed layer of soil, and part of it was tied up in an unavailable form. The strip treatment was a compromise between the two extremes, and it resulted in greater K availability. The effect of strip application will be dependent on the degree to which soil ties up K in an unavailable form.

Fertilizer K should be distributed throughout enough soil so that a significant (10 or 20% or more) portion of the roots will be in the K fertilized soil. Restricting K application to a fraction of the soil will usually reduce the proportion of added K that moves into an unavailable form.

4. Form of Material

Most of the K fertilizer is used as KCl. Smaller amounts of KNO_3 and K_2SO_4 are used, particularly where chloride may be deleterious to the quality of the crop. Since these forms of K are all water soluble and the K will react with the soil after application, any differences between sources would be due to effects of the different anions that accompany K.

5. Salt Effects of Potassium Fertilizer

When K fertilizer in the form of KCl is applied in the row with the seed at planting time, it may cause salt injury to the germinating seedling. Because of this possibility of injury, methods have been devised to separate the fertilizer band from the seed. The sensitivity to salt injury varies with species. The amount of salt that is harmful will also vary with soil texture and soil moisture conditions. Where rainfall is heavy, the deleterious chlorides may be leached from the vicinity of the seed. Where mixed fertilizers containing N, P, and K are used, the salt injury usually increases because of the soluble salts of N and K. As a general guide, when the amount of N plus K exceeds 50 kg/ha, salt injury may occur on loam and heavier-textured soils where these fertilizers are applied near the seed of corn. Where fertilizer is applied with the seed, amounts in excess of 20 kg/ha may be injurious. Lesser amounts may be harmful on sandy soils. On small grains, the row application of N plus K should not exceed 100 kg/ha in a humid region and no more than 20 kg/ha in a semiarid region.

Table 10-10. The effect of minimum vs. conventional tillage on the K distribution in the soil where K has been applied broadcast (Triplett et al., 1969).

	Minimum tillage		Plowed
Depth	In row	Between row	Sampled in row
cm	mg/kg K		
0-2.5	278a*	197a	190a
2.5-5.0	209b	172a	172a
5.0-7.5	178c	144b	189a
7.5-13	149d	130bc	176a
13-18	121e	114c	137b

* Means not followed by the same letter are significantly different at the 0.05 level.

6. *Effect of Tillage Practice*

Much of the research on K fertilization of crops has been conducted where the soils have been tilled with a moldboard plow to depths of 15 cm or more. Minimum tillage concepts have shown that this plowing may be unnecessary on many soils, and, in fact, little tillage other than enough to get the crop planted may be needed. Under these conditions, methods of placement of K fertilizers are restricted. They may be banded near the row while planting, or they may be broadcast on the surface without tillage incorporation.

Where the K is applied broadcast and minimum tillage is practiced, most of the K will remain in the top few centimeters of soil. The data in Table 10–10 from Triplett et al. (1969) in Ohio are representative of the K distribution in the soil resulting from broadcast application and no tillage for corn production. With conventional tillage where the soil is plowed with a moldboard plow, the K is uniformly distributed throughout the soil to the depth of plowing. Even though most of the available K is near the surface with no tillage, corn is able to get sufficient K, since with no tillage the corn root density is greater near the surface.

Leaf K content of corn grown with no tillage has been lower than for corn grown under conventional tillage. This may be due to either the position of the applied K or poorer aeration of the soil. Sufficient K can be supplied by using a higher rate with no tillage.

REFERENCES

Andrilenas, P., and R. Rortvedt. 1979. 1980 Fertilizer situation. Economics, Statistics, and Cooperatives Service, USDA, Washington, DC.

Anonymous. 1981. Bright days ahead for potash: FC Report from Saskatchewan. Farm Chem. 144(2):13–32.

Asher, C. J., and P. G. Ozanne. 1967. Growth and potassium content of plants in solution culture maintained at constant potassium concentrations. Soil Sci. 103:155–161.

Avnimelech, Y., V. Shaham, and A. Feder. 1970. Elution of potassium in soil. Soil Sci. Soc. Am. Proc. 34:407–411.

Barber, S. A. 1960. The influence of moisture and temperature on phosphorus and potassium availability. Trans. Int. Congr. Soil Sci., 7th 4:435–442.

Barber, S. A. 1962. A diffusion and mass-flow concept of soil nutrient availability. Soil Sci. 93:39–49.

Barber, S. A. 1970. Residual effect of potassium fertilization of continuous corn on Chalmers silt loam. Purdue Univ. Res. Progress Rep. 377.

Barber, S. A. 1976. Efficient fertilizer use. p. 13–30. *In* F. L. Patterson (ed.) Agronomic research for food. Spec. Pub. 26. American Society of Agronomy, Madison, WI.

Barber, S. A. 1980. Twenty-five years of phosphate and potassium fertilization of a crop rotation. Fert. Res. 1:29–36.

Barber, S. A. 1981. Soil chemistry and the availability of plant nutrients. p. 1–12. *In* R. H. Dowdy et al. (ed.) Chemistry in the soil environment. Spec. Pub. 40. American Society of Agronomy, Madison, WI.

Barber, S. A., R. H. Bray, A. C. Caldwell, R. L. Fox, M. Fried, J. J. Hanway, D. Hovland, J. W. Ketcheson, W. M. Laughton, K. Lawton, R. C. Lipps, R. A. Olson, J. T. Pesek, K. Pretty, M. Reed, F. W. Smith, and E. M. Stuckney. 1961. North Central Regional potassium studies: II. Greenhouse experiments with millet. North Central Regional Pub. no. 123. Purdue Univ. Bull. 717.

Barber, S. A., and R. P. Humbert. 1963. Advances in knowledge of potassium relationships in the soil and plant. p. 231–268. *In* McVickar et al. (ed.) Fertilizer technology and usage. Soil Science Society of America, Madison, WI.

Barber, S. A., R. D. Munson, and W. B. Dancy. 1971. Production, marketing and use of potassium fertilizers. p. 303–334. *In* R. A. Olson et al. (ed.) Fertilizer technology and use, 2nd ed. Soil Science Soc. of America, Madison, WI.

Barber, S. A., J. M. Walker, and E. H. Vasey. 1963. Mechanisms for the movement of plant nutrients from the soil and fertilizer to the plant root. J. Agric. Food Chem. 11:204–207.

Barrows, H. L., and V. J. Kilmer. 1963. Plant nutrient losses from soils by water erosion. Adv. Agron. 15:303–316.

Barry, G. S. 1980. Canadian mineral survey. Mineral Policy Sector, Energy, Mines, and Resources Canada, p. 66–68.

Braitsch, O. 1971. Salt deposits: Their origin and composition. Springer-Verlag New York, Inc., New York.

British Sulphur Corporation. 1979. World survey of potash resources, 3rd ed.

British Corporation. 1981. Statistical supplement: Raw materials supply and demand 1979, Vol. 22. British Sulphur Corporation, Ltd., London.

Chisso Corporation. 1981. Technical information of potassium sulfate plant. Chisso Corporation, Tokyo.

Claassen, N., and S. A. Barber. 1976. Simulation model for nutrient uptake from soil by a growing plant root system. Agron. J. 68:961–964.

Clark, R. B. 1978. Differential response of corn inbreds to calcium. Commun. Soil Sci. Plant Anal. 9:729–744.

Conyers, E. S., and E. O. McLean. 1969. Plant uptake and chemical extractions for evaluating potassium release characteristics of soils. Soil Sci. Soc. Am. Proc. 33:226–230.

Crocker, B. S., J. T. Dew, and R. J. Roach. 1969. Contemporary potash plant engineering. Canadian Inst. of Mining Metallurgy Bull.

Cushman, J. H. 1979. An analytical solution to solute transport near root surfaces for low initial concentrations: I. Equation development. Soil Sci. Soc. Am. J. 43:1087–1092.

Dancy, W. B. 1968. Beneficiation of potash ores. Canadian Patent 792 819. Date issued: 20 August.

Dancy, W. B. 1982. Potassium compounds. p. 920–950. *In* Kirk-Othmer: encyclopedia of chemical technology, Vol. 18. 3rd ed. John Wiley & Sons, Inc., New York.

Food and Agricultural Organization. 1981. FAO fertilizer yearbook (1980), Vol. 30. FAO Statistics Series no. 36. Statistics Analysis Serv., Statistics Division, Food and Agricultural Organization of the United Nations, Rome.

Gidney, D. R. 1981. Potash—where in the world will it come from? Fert. Progr. 12(5):20–21, 24–26.

Hanway, J. J., S. A. Barber, R. H. Bray, A. C. Caldwell, M. Fried, L. T. Kurtz, K. Lawton, J. T. Pesek, K. Pretty, M. Reed, and F. W. Smith. 1962. North Central Regional potassium studies: III. Field studies with corn. North Central Regional Pub. no. 135. Iowa Univ. Res. Bull. 503.

Harre, E. A., and G. T. Harris. 1979. World fertilization situation and outlook, 1978–85. International Fertilizer Development Center and National Fertilizer Development Center, Muscle Shoals, AL.

Horton, M. L. 1959. Influence of soil type on potassium fixation. M.S. thesis. Purdue University, Lafayette, IN.

James, D. W. 1970. Potassium in acid region soils. Better Crops Plant Food 54(1):22–25.

Jencks, E. M. 1968. Relationship between exchangeable and boiling nitric acid–extractable potassium in seventy-five West Virginia soil series. Agron. J. 60:636–639.

Koch, J. T., E. R. Orchard, and M. E. Sumner. 1970. Leaf composition and yield response of corn in relation to quantity intensity parameters for potassium. Soil Sci. Soc. Am. Proc. 34:94–97.

Krull, O. 1928. Die Verarbeitung der Kalirohsalze, Enke's Bibliothek fur Chemie Und Technik, Vol. 8. Das Kali, Die Kalisalze, Stuttgart, FRG.

Kurtz, B. E., and A. J. Barduhn. 1960. Compacting granular solids. Chem. Eng. Prog. 56:67–72.

Lawton, K. 1946. The influence of soil aeration on the growth and absorption of nutrients by corn plants. Soil Sci. Soc. Am. Proc. 10:263–268.

Lucas, R. E. 1968. Potassium nutrition of vegetable crops. p. 489–498. *In* V. J. Kilmer et al. (ed.) The role of potassium in agriculture. American Society of Agronomy, Crop Science Society of America, and Soil Science Society of America, Madison, WI.

MacDonald, R. A. 1960. Potash occurrences, processes, and production. Am. Chemical Soc. Monogr. no. 148. Van Nostrand Reinhold Company, Inc., New York.

Mahan, J. N., and H. L. Stroike. 1980. The fertilizer supply: Nitrogen, phosphate, potash 1979–80. Agricultural Stabilization and Conservation Service, U.S. Department of Agriculture, U.S. Government Printing Office, Washington, DC.

Munson, R. D. 1982. Potassium, calcium, and magnesium in the tropics and subtropics. Tech. Bull. IFDC T-23. International Fertilizer Development Center, Muscle Shoals, AL.

Munson, R. D., and W. L. Nelson. 1963. Movement of applied potassium in soils. J. Agric. Food Chem. 11:193–201.

Place, G. A., and S. A. Barber. 1964. The effect of soil moisture and rubidium concentration on diffusion and uptake of rubidium 86. Soil Sci. Soc. Am. Proc. 28:239–243.

Reisenauer, H. 1964. Mineral nutrients in soil solution. p. 507–508. *In* P. L. Altman and D. S. Dittmer (ed.) Environmental biology. Federation of American Societies for Experimental Biology, Bethesda, MD.

Schenk, M. K., and S. A. Barber. 1980. Potassium and phosphorus uptake by corn genotypes grown in the field as influenced by root characteristics. Plant Soil 54:65–76.

Schofield, R. K. 1947. A ratio law governing the equilibrium of cations in the soil solution. Int. Congr. Pure Appl. Chem. Proc., 11th 3:257–261.

Scott, A. D., and S. J. Smith. 1968. Mechanism for potassium release by drying. Soil Sci. Soc. Am. Proc. 32:443–444.

Searls, J. P. 1981. Potash in crop year 1981. Mineral Industrial Surveys, Bureau of Mines, U.S. Department of the Interior, U.S. Government Printing Office, Washington, DC.

Singleton, R. H. 1978. Potash. Mineral commodity profiles MCP-11. Bureau of Mines, U.S. Department of the Interior, U.S. Government Printing Office, Washington, DC.

Sprague, H. B., and D. L. Plucknett (ed.). 1985. Detecting mineral nutrient deficiencies in temperate and tropical soils. Westview Press, Boulder, CO.

Tabatabai, M. A., and J. J. Hanway. 1969. Potassium supplying power of Iowa soils at their "minimal" levels of exchangeable potssium. Soil Sci. Soc. Am. Proc. 33:105–109.

Triplett, G. B., Jr., and D. M. Van Doren, Jr. 1969. Nitrogen, phosphorus and potassium fertilization of non-tilled maize. Agron. J. 61:637–638.

U.S. Department of Agriculture. 1980. Commercial fertilizers. Statistical Reporting Service, Washington, DC.

Vaidyanathan, L. V., M. C. Drew, and P. H. Nye. 1968. The measurement and mechanism of ion diffusion in soils: IV. The concentration dependence of diffusion coefficients of potassium in soils at a range of moisture levels and a method for estimation of the differential diffusion coefficient at any concentration. J. Soil Sci. 19:94–107.

Van Der Paauw, F. 1958. Relations between potash requirements of crops and meteorological conditions. Plant Soil 9:254–268.

von Peter, A. 1980. Fertilizer requirements in developing countries. Proc. Fert. Soc. no. 188.

Walker, W. M., and W. L. Parks. 1969. Effect of soil potassium, potassium fertilizer, and method of placement upon lodging in corn (*Zea mays* L.) Soil Sci. Soc. Am. Proc. 33:909–911.

Weisner, R. C., J. F. Lemons, and L. V. Coppa. 1980. Valuation of potash occurrences within the nuclear waste isolation pilot plant in southwestern New Mexico. Bureau of Mines, U.S. Department of the Interior, U.S. Government Printing Office, Washington, DC.

Welch, L. F., P. E. Johnson, G. E. McKibben, L. V. Boone, and J. W. Pendleton. 1966. Relative efficiency of broadcast versus banded potassium for corn. Agron. J. 58:618–621.

Wilson, W. P. 1969a. A to Z in potash ore processing. Eng. Min. J. 167:86–91.

Wilson, W. P. 1969b. Potassium chloride crystallization. Western Knapp Engineering, San Francisco.

Wood, L. K., and E. E. DeTurk. 1941. The absorption of potassium in soils in non-replaceable forms. Soil Sci. Soc. Am. Proc. 5:152–161.

Woodruff, C. M. 1955. The energies of replacement of calcium by potassium in soils. Soil Sci. Soc. Am. Proc. 19:167–171.

Wrege, E. E., and Dancy, W. B. 1949. Quality by the ton. Chem. Ind. Week 65(1):46–49.

11

Production, Marketing, and Use of Sulfur Products

James D. Beaton
Potash & Phosphate Institute
Cochrane, Alberta, Canada

Robert L. Fox
University of Hawaii
Honolulu, Hawaii

Milton B. Jones
University of California
Hopland, California

The essentiality of S for plant growth has been recognized for approximately 140 yr. Sulfur is a macronutrient, although plants require much less of it than of N or K. The quantities of S required by most crops are comparable with requirements for P or Mg. Deficiencies of S were identified as early as 1900 in the northwestern USA and in 1927 on some soils in Alberta, Canada (Beaton, 1969).

Long before the essentiality of S for plant growth was confirmed, the practical value of applying gypsum was recognized. From 1790 to 1815, much gypsum was imported into the eastern states from mines near Paris (from which the phrase plaster of paris originated). The gypsum was used in conjunction with legumes to enhance the N status of soils (Craven, 1925).

Interest in S as a plant nutrient has increased in recent years. Such interest extends beyond increased crop yields. Food quality, especially low contents of essential S-containing amino acids, is a matter of serious concern in many technologically less-developed areas of the world (Evans et al., 1977). Widespread S deficiency has already been demonstrated in some of these areas (Bromfield, 1972, 1975).

Studies in West Africa suggest that S yields of crops (and thus yields of protein) are restricted by the quantity of S supplied in the rainwater (Bromfield, 1974; Fox et al., 1977). If pollution control measures are effective, atmospheric contributions of S will decrease with time. Thus, there are expectations that decreased atmospheric pollution will lead to even less sulfate (SO_4^{2-}) in rainwater, and this in turn will lead to more intense and expanded S deficiencies than have heretofore been recognized.

Sulfur deficiencies have been demonstrated by field experiments in many nations of the world, including numerous sections of the USA and Canada. These deficiencies seem to be appearing with greater frequency for the following reasons:

1. Increased use of high-analysis fertilizers, which are practically free of S.
2. Greater crop yields with disproportionately greater S utilization.
3. Better control of air pollution, thus decreasing this important source of S.
4. Decreased use of S as a fungicide and insecticide.
5. Improved ability to identify S-deficient soils.
6. Decreased levels of soil organic matter and less dependence on soil organic matter as a source of nutrients.
7. More diligent search for instances of S deficiency by agricultural research workers, representatives of industry, and farmers.

Factors that tend to mitigate against S deficiency are at work also. Among these are decentralization of industry, population shifts from areas of high population density to low density, and the general buildup of population pressure with increased use of manures and fertilizers that do contain S.

Sulfur deficiencies become more numerous and serious when heavy rates of S-free fertilizer are applied. In particular, the need for fertilizing agricultural crops with S is closely related to the amounts of N being applied (Fox, 1976). Figure 11–1 demonstrates the interrelationships between these two nutrients for production of wheat forage and grain in Australia. There was no response to either N or S alone, but growth increased

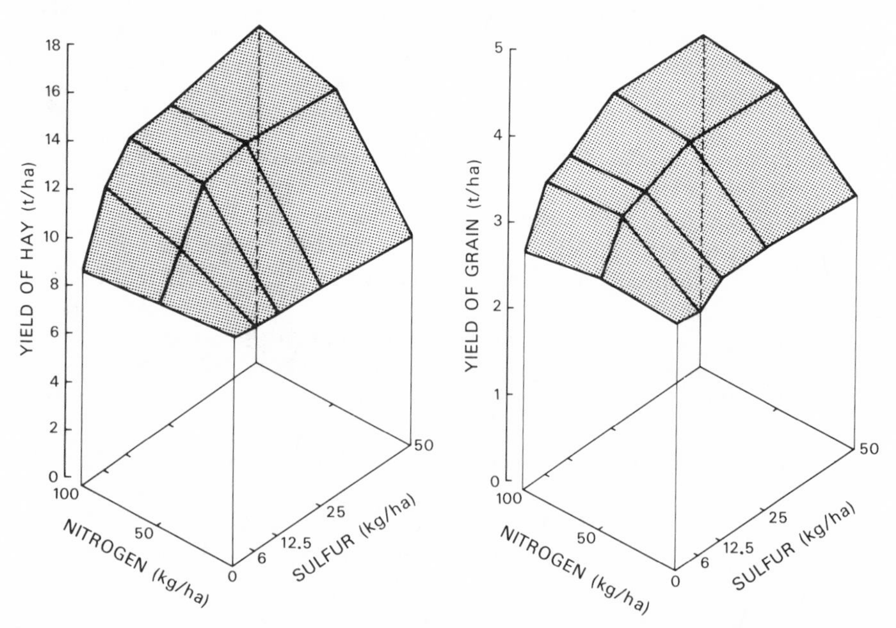

Fig. 11–1. Effect of S and N supply on wheat hay and grain yields (Spencer & Freney, 1980).

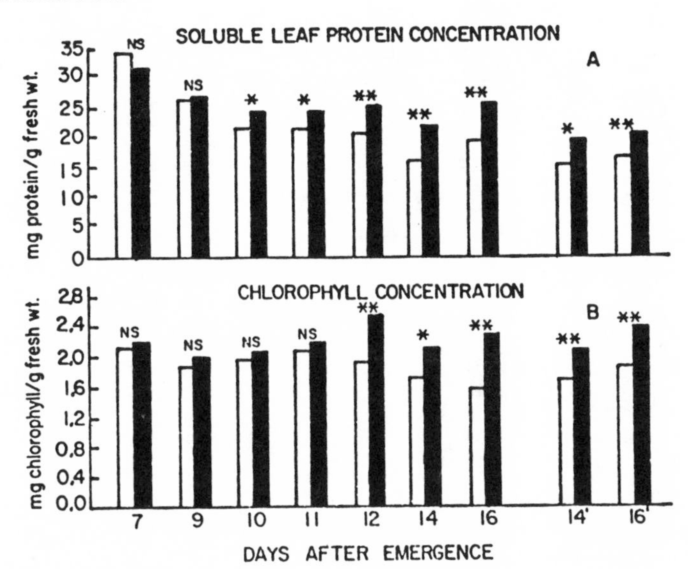

Fig. 11–2. Effect of S deprivation on the concentration of soluble protein and total chlorophyll in leaf blades of S-deprived (□) and normal (■) corn seedlings at various times after emergence. *, ** Significantly different at 0.05 and 0.01 levels of probability, respectively (Friedrich & Schrader, 1978).

substantially when both were applied. The beneficial effect of S increased with the rate of N added.

Inadequate S retards the growth of plants because it is needed for the following essential roles:

1. Component of the amino acids, cystine, cysteine, and methionine, which are required components of proteins.
2. Synthesis of chlorophyll, even though there is no S in this vital plant component. The effect of S deprivation on lowering chlorophyll concentration in leaf blades of corn seedlings is evident in Fig. 11–2.
3. Activation of certain proteolytic enzymes, such as the papainases.
4. Synthesis of certain vitamins (biotin and thiamin or vitamin B_1), glutathione, and coenzyme A.
5. Formation of the glucoside oils of plants such as onion and those in the Cruciferae family.
6. Formation of certain disulfide linkages that are associated with the structural characteristics of protoplasm. (The concentration of sulfhydryl [–HS] groups in tissues of some plants is related to cold and drought resistance.)
7. Formation of a ferredoxin that has an important role in photosynthetic processes (Arnon, 1965; Zaneti & Forti, 1969).
8. Formation of a ferredoxin-like compound that is involved in N_2 fixation by root nodule bacteria (Koch et al., 1970) and free-living N_2-fixing soil bacteria (Bullen, 1968).
9. Activity of ATP sulfurylase, an enzyme concerned with the metabolism of S (Adams & Rinne, 1969; Ellis, 1969).

Table 11-1. Effect of S deprivation on concentration of NO_3^--N and on fresh weight of leaf blades and stems of corn seedlings at various times after emergence (Friedrich & Schrader, 1978).

Days after emergence	NO_3^--N		Fresh wt	
	No S	Plus S	No S	Plus S
	— mg/kg fresh wt —		——— g ———	
		Leaf blades		
7	39	27	2.68	2.76
9	38	24	3.36	3.50
10	45	37	4.16	4.02
11	52	34	4.22	4.01
12	32	19	4.44	4.27
14	46	14	4.57	4.66
16	10	23	4.95	5.24
14‡	29	24	4.75	5.00
16‡	23	12	4.78	5.12
Mean§	37**	25**	4.05NS¶	4.07 NS¶
		Stems†		
7	319	224	1.78	1.85
9	378	248	2.34	2.37
10	346	238	2.91	2.80
11	363	356	3.41	3.35
12	360	267	3.84	4.04
14	369	283	4.34	4.97
16	341	350	5.23	5.96
14‡	416	255	4.50	5.26
16‡	418	312	4.91	5.96
Mean§	354**	280**	3.41**	3.62**

** Means significantly different at 0.01 level of probability. Comparison at individual harvest dates are not statistically significant.
† Includes leaf sheaths, unfurled leaves, and culms.
‡ Includes treatments where SO_4^{2-} was added at day 12.
§ Does not include treatments where SO_4^{2-} was added at day 12.
¶ NS = not significant.

10. Activity of nitrate reductase, the enzyme responsible for conversion of nitrate-N (NO_3^--N) taken up by plants into amino acids and then into protein. Because of lower activity of this enzyme due to S deficiency, there are reductions in soluble proteins and increases in nitrate (NO_3^-) concentration in plant tissues (Fig. 11–2 and Table 11–1).

Crops differ in their internal requirements for S, as is evident from the S contents of several crops listed in Table 11–2. High-yielding crops such as sugar crops, Coastal bermudagrass, orchardgrass, corn, and sorghum contain more S than most other crops. A ready supply of S probably is just as important for vegetable crops, rape, and cotton, although the quantity of S taken up by such crops is generally in the intermediate range. Alfalfa, which has been one of the most S-responsive crops, does not accumulate unusually large amounts of S. Most small grains and grasses require less S. Table 11–2 also shows that the S contents of many crops are similar to the contents of P and Mg.

Table 11-2. Approximate quantity of nutrients contained in various crops.

Crop	Yield	N	P†	K‡	S	Mg
	kg/ha§					
	Grains¶					
Corn (*Zea mays* L.)	12 000	343	50	220	47	71
Grain sorghum [*Sorghum bicolor* (L.) Moench]	8 000	288	54	205	43	40
Wheat (*Triticum aestivum* L.)	5 000	156	32	104	23	25
Barley (*Hordeum vulgare* L.)	5 000	156	26	130	26	--
Oats (*Avena sativa* L.)	3 500	109	21	109	21	21
Rice (*Oryza sativa* L.)	7 000	145	24	144	19	16
	Forage crops¶					
Alfalfa (*Medicago sativa* L.)	12 000	359	34	240	33	33
Clovers (*Trifolium* spp.)	8 000	178	20	147	20	26
Bermudagrass 'Coastal' [*Cynodon dactylon* (L.) Pers.]	22 000	629	71	365	49	55
Orchardgrass (*Dactylis glomerata* L.)	16 000	318	49	293	57	--
	Oil crops¶					
Soybeans [*Glycine max* (L.) Merr.]#	3 500	216	26	117	11	14
Peanuts (*Arachis hypogaea* L.)	3 500	256	23	117	29	33
Rapeseed (*Brassica napus* L.)	2 000	90	16	110	35	--
	Fiber crops¶					
Cotton (*Gossypium hirsutum* L.)	1 500	150	40	90	28	19
	Stimulant crops¶					
Tobacco (*Nicotiana tabacum* L.)	3 500	110	13	187	25	28
	Sugar crops					
Sugarcane (*Saccharum officinarum* L.)¶	70 000	141	20	245	94	115
Sugar beets (*Beta vulgaris* L.)††	70 000	187	29	245	57	--
	Vegetables					
Potatoes (*Solanum tuberosum* L.)‡‡	25 000	208	25	266	28	25
Cabbage (*Brassica oleracea* var. *capitata*)§§	45 000	147	16	119	41	11
Turnip (*Brassica rapa*)††	70 000	133	26	222	45	26
Onions (*Allium cepa* L.)¶¶	45 000	135	27	99	28	13

† P × 2.3 = P_2O_5.
‡ K × 1.2 = K_2O.
§ kg/ha × 0.9 = lb/acre.
¶ All aboveground portions.
# Beans only.
†† Tops and roots.
‡‡ Tops and tubers.
§§ Heads.
¶¶ Tops and bulbs.

The data in Table 11–2 implies that internal S requirements of crops are approximately equal to the S contents given but may not be a useful guide for S fertilization. Cotton, a crop that is very sensitive to S deficiency, is among the lowest of the various crops ranked according to S contents. Nevertheless, it is clear that crops cannot produce beyond certain limits imposed by the S supply. Table 11–2 should provide a reasonable estimate of what the minimum external S supply must be.

One other aspect of the S requirement of crops should be introduced because it bears directly on the use of S fertilizers: the concentration of S in the root environment. External S requirements for near-maximum yields have been estimated for several crops. From the results of these investigations, it seems reasonable to generalize that for many crops this requirement should be approximately 5 mg/L S in solution, but the requirement may be as low as 2 mg/L for some crops (Fox, 1980; Fox et al., 1979). Sulfate in the soil solution is transported to root surfaces in water taken up by plants, a process commonly referred to as *mass flow*.

Sulfur may have beneficial side effects on plant growth in addition to its role as a plant nutrient. The acidifying effects of ammonium thiosulfate, ammonium sulfate, elemental S, and so forth in fertilizer bands in basic or alkaline soils may increase the availability of other essential nutrients, such as P, Mn, and Zn. Large quantities of elemental S or sulfuric acid (H_2SO_4) are usually required for more general pH adjustments. Sulfur and a number of its compounds are useful for reclaiming sodic soils, and many of these same materials are used for treatment of irrigation water to improve water infiltration and percolation.

This topic of production, marketing, and use of S products was reviewed thoroughly by Beaton and Fox (1971). In this chapter we have updated and added new material on the subject.

I. SULFUR SOURCES

A. United States and World Reserves

Sulfur has been known for > 6000 yr, but its widespread use began only about 200 yr ago (Hatch, 1972). As early as 2000 B.C., Egyptians used S to bleach cotton and linen and to formulate various paint pigments and dyes. It was used by the Greeks and Romans as a medicine, a disinfectant, a fumigant, and for religious rites. Sulfur is the "brimstone" of the Bible, and before the time of Christ, the Chinese used S for making gunpowder. The Romans also used it in warfare as an incendiary weapon. Sulfur was popular with the alchemists, and they found it ideal for many of their purposes.

Sulfur constitutes approximately 0.1% of the earth's crust and most commonly occurs as elemental S (brimstone) in deposits associated with calcite, gypsum, and anhydrite; combined S in metal sulfide ores; combined S in mineral sulfates; hydrogen sulfide (H_2S) contaminant in natural gas; organic contaminants in crude oils; pyritic and organic compounds in coal; and organic compounds in tar sands (Comiskey et al., 1969).

World S resources are vast and, for all intents and purposes, unlimited. However, most of them are uneconomical to develop at current prices (Bixby, 1980a). Reserves of conventional S in the USA and elsewhere in the world are compiled in Table 11–3. Values of > 300 million metric tons in the USA and nearly 2800 million metric tons worldwide indicate that conventional reserves will be adequate in the foreseeable future.

In addition to these conventional reserves, about 1100 trillion metric tons of SO_4^{2-} are dissolved in the oceans and deposited on the earth's surface, primarily in the form of gypsum (Hatch, 1972). Sulfur present in coal, oil shale, anhydrite, and gypsum in the USA alone is estimated to be > 33

Table 11-3. Apparent reserves of S from conventional sources in the world (Pearse, 1974).

Area	Petro-leum	Natural gas	Native ore	Dome	Sulfide ores	Pyrite	Total
	million metric tons						
USA	30.5	25.4	101.6	101.6	25.4	25.4	309.9
Canada	5.1	172.7	NS†	--	25.4	25.4	228.6
Mexico	5.1	NS	NS	50.8	10.2	5.1	71.1
Central and South America	30.5	NS	101.6	--	50.8	50.8	233.7
Western Europe	NS	25.4	5.1	--	5.1	25.4	61.0
Eastern Europe and USSR	50.8	50.8	152.4	--	152.4	203.2	609.6
Africa	5.1	NS	NS	--	20.3	20.3	45.7
Near East and South Asia	345.5	508.0	203.2	--	5.1	50.8	1112.6
Far East and Southeast Asia	10.2	10.2	50.8	--	20.3	20.3	111.8
Oceania	--	--	NS	--	5.1	5.1	10.2
Total	482.6	792.5	614.7	152.4	320.0	431.8	2794.1

† NS = not significant.

billion metric tons, more than 10 times world conventional reserves (Pearse, 1974). Canada's Athabasca tar sands in northern Alberta are another potentially very large source of S (Vroom, 1972). They are believed to contain at least 4.5 billion metric tons of S as a 4.5 to 5.0% contaminant of the petroleum. United States oil shales contain about 0.75 to 1.0% S by weight (Rieber et al., 1981).

Sensitivity of utilization of the various S reserves to cost/price relationships is demonstrated in Fig. 11–3. This inverted pyramid shows that the size of reserves and cost of production increase from brimstone at the bottom tip to anhydrite and gypsum, coal, and seawater in the broad upper levels. Normally, if lower cost sources are unable to satisfy demand, the

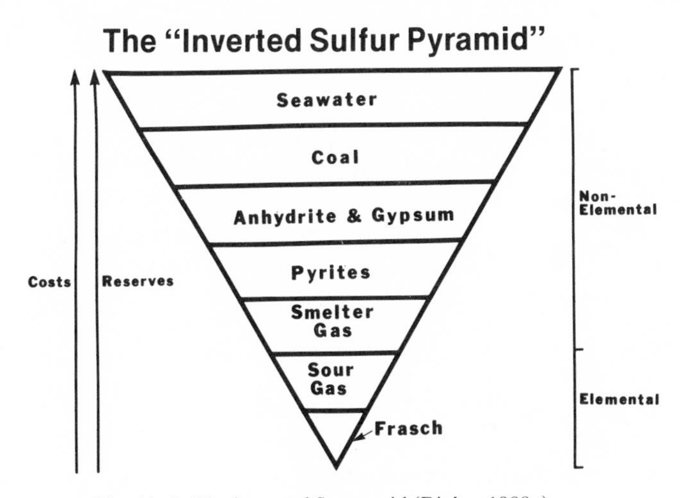

Fig. 11–3. The inverted S pyramid (Bixby, 1980a).

resulting rise in prices makes it economically attractive to develop more expensive sources higher up in the pyramid. A possible exception to this natural relationship might be S from sources where the production costs are subsidized in pollution abatement facilities (Bixby, 1980a).

Sulfur is produced by more diverse and ingenious methods than any other common element (Hatch, 1972). It is dug from pits like coal, recovered from waste gases of smelters (as sulfur dioxide [SO_2]), extracted from natural and refinery gases (as H_2S), and melted underground and brought to the surface as liquid elemental S (the Frasch process). Frasch S is the most important single source of S in the USA (Rieber et al., 1981), but its importance is expected to begin a gradual decline over the next several years until the deposits are depleted or uneconomical to mine (Bixby, 1980a). Forecasts of the amounts of S to be provided from the various U.S. sources in 1985 and 1990 are shown in Table 11–4.

Table 11–4. Estimated U.S. S supply outlook by region and source, 1985 and 1990 (Rieber et al., 1981).

Source	1985	1990
	million metric tons	
	Northeast	
Refinery S	0.71	0.91
Recovered acid	0.10	0.20
Total	0.81	1.11
	Southeast	
Recovered acid	0.30	0.41
	Gulf Coast	
Frasch	4.10	3.56
Refinery S	1.73	2.34
Sour gas S	1.62	2.03
Total	7.45	7.93
	Midwest	
Refinery S	0.51	0.71
Recovered acid	0.20	0.30
Coal gas–flue gas	0.20	0.20
Total	0.91	1.21
	Midcontinent†	
Frasch	3.05	3.05
Refinery S	0.20	0.30
Sour gas S	1.93	4.57
Recovered acid	1.22	1.63
Total	6.40	9.55
	West Coast	
Refinery S	0.61	0.81
Recovered acid	nil	nil
Total	0.61	0.81
U.S. total	16.46	21.03

† Including southwest and intermountain USA.

B. Energy Requirements

Because of the present dominant position of Frasch S in the USA and since more information is readily available about energy consumed in its production, this discussion focuses on this method of S mining. Production costs and energy usage are closely related to the amounts of superheated water at a temperature of 160 to 163° C that must be forced under a pressure of 8.78 to 14.06 kg/cm^2 down a borehole into the underlying S-bearing formation (Ellison, 1971; Pearse, 1974). The quantities of water needed range from about 5600 to 26 000 L/metric ton of S depending on the characteristics of the deposit and the age of the mine (Ellison, 1971).

Energy requirements range from about 4.8 to 9.5 × 10^9 J/metric ton of elemental S (Rieber et al., 1981). It averages about 9.3 × 10^9 J/metric ton of S (Blouin & Davis, 1975). Most of this is heat energy, generally from natural gas.

According to Blouin and Davis (1975), S recovery from sour gas requires only about 0.35 × 10^9 J/metric ton of S. Desulfurization of oil is energy intensive, consuming approximately 31.4 × 10^9 J/metric ton of S. Production of S from the roasting of pyrites utilizes almost 0.5 × 10^9 J/metric ton, just slightly more than that used in sour gas recovery.

II. SULFUR FERTILIZERS

Many fertilizer materials contain significant quantities of S. Brief descriptions of the most popular S sources are listed in Table 11–5.

A. Fertilizers Containing Elemental Sulfur

1. Agricultural, Flake, and Porous Sulfur Granules

The use of elemental S to reduce soil pH and to reclaim sodic soils is well known. However, problems of dustiness, unpleasantness, and hazards of fire and explosion have limited the utility of agricultural grade S and to a lesser extent the flake form for fertilizer purposes. Modifications are possible to overcome these serious drawbacks and at the same time take advantage of the obvious economics in using practically pure S. Unfortunately, these changes, which mostly entail larger or more durable particles, are usually made at the expense of agronomic effectiveness.

Pure S is easily prilled, but the product is unsuitable for normal fertilizer uses because it is virtually insoluble and has low surface area. These properties seriously delay the rate at which it will be converted to plant available SO_4^{2-}. One way of improving the effectiveness of prilled elemental S is to form porous irregular granules with much greater surface areas. This is the approach taken with products such as Popcorn Sulfur (Unocal Company of California, Los Angeles, CA) and Poro-Sul (Smith and Ardussi, Inc., Seattle, WA) The performance of flake S can, of course, be controlled to some degree by changing flake size and thickness.

Table 11-5. Sulfur-containing fertilizer materials (Bixby & Kilmer, 1975).

Sulfur as S	Sulfur as SO_4^{2-}
Agricultural S	Ammonium sulfate
Flake S	Potassium sulfate
Granular S-bentonite	Calcium sulfate
Phosphate rock–S	Magnesium sulfate
Normal superphosphate + S	Potassium–magnesium sulfate
Triple superphosphate + S	Normal and enriched superphosphate
DAP† + MAP‡ + S	Ammonium nitrate–sulfate
Ammonium polyphosphate + S	Ammonium sulfate–nitrate
AP-UP§ + S	Ammonium phosphate–sulfate
NPK grades + S	Ammonium phosphate–sulfate gypsum
Urea-S	Ammonium phosphate–sulfate urea
Potash-urea-S	Several micronutrient salts
S-coated potash	Urea-sulfate
S-coated urea	Urea-ammonium sulfate
Ammonia–S	Sulfuric acid
S suspensions	NPKS
Micronutrients in S	
Sulfur in other forms	
Ammonium thiosulfate	
Ammonium bisulfite	
Ammonium polysulfide	
Sulfur dioxide	
Certain ultra–high analysis materials	
Lime-S	

† DAP = diammonium phosphate.
‡ MAP = monoammonium phosphate.
§ AP–UP = ammonium phosphate–urea phosphate.

Flake S and the porous S products are used for direct application and bulk blending. These sources are probably incapable of correcting S deficiencies during the first growing season after application unless approximately 25% of the particles are < 60-mesh size. On severely S-deficient soils and ones of low or unknown S oxidizing capacity, these products should be used in conjunction with some soluble SO_4^{2-}. Also, under these kinds of conditions they should be applied and incorporated into the soil as far in advance of seeding as possible.

2. *Water-degradeable Granular Sulfur-Bentonite*

Another means of improving the effectiveness of granular elemental S products is to incorporate 5 to 10% by weight of a swelling clay such as bentonite. This fertilizer imbibes soil moisture and disintegrates into finely divided S, which is more readily converted to SO_4^{2-}. One manufacturing procedure involves cooling the molten S-bentonite mixture on stainless steel slating belts, followed by crushing and screening. A product with similar constituents but formed by prilling is currently being evaluated.

Granular water-degradeable S-bentonite fertilizers are manufactured at several locations in the USA and Canada. Agri-Sul (Agri-Sul Inc., Mineola, TX), Degra-Sul (Degra-Sul Fertilizer Production Ltd., Calgary, Alberta), Disintegrating Flake (Montana Sulphur and Chemical Co., Billings, MT), Disper-Sul (Chemical Enterprises, Inc., Houston, TX), Terasul (Canadyne Chemical, Burlington, Ontario), and Tiger. 90 (Tiger Chemical Ltd., Calgary, Alberta) are commercial examples. They are used for both direct application and blending with other compatible dry sources of N, P, and K. Some of these products are dusty, especially when they are subjected to rough and excessive handling with augers. A surfactant may be sprayed on the granules to reduce dusting and provide wettability.

Because of uncertain effectiveness of these S sources during the first growing season after fertilization, the same provisions should be made that were outlined for agricultural, flake, and porous S materials. Repeated use of elemental S-containing fertilizers tends to gradually enlarge the population of S-oxidizing microorganisms, resulting in a corresponding increase in the rate of SO_4^{2-} formation.

3. Sulfur Suspensions

Elemental S can be readily used in suspensions. The Tennessee Valley Authority (TVA) has used −20-mesh, crushed lump S to make suspensions containing 60% S and 1% clay. Complete formulas, such as 9–8–15–10S (9–18–18–10S) and 12–5–10–20S (12–12–12–20S), have been prepared from 12–18–0 (12–40–0), urea–ammonium nitrate (UAN) solutions, potassium chloride (KCl), and S.

One commercial firm has successfully made and applied a 60% S slurry. A portable mixer suitable for field use and equipped with a motor-driven impeller was used to make the slurry, which was then pumped to field applicators and sprayed onto the soil using existing liquid fertilizer equipment. The S slurry can also serve as a carrier for micronutrients.

Another company provides instructions on how to formulate a 52% S suspension from −100-mesh S and either 1% predispersed liquid clay or 2% regular clay. Surfactant at a rate of 1% and a mixing time of 3 min are also called for in the formulation.

Relatively inexpensive, crude elemental lump S reduced to the desired size by suspension equipment has not proved entirely satisfactory. Dry ground S is more efficient but has obvious drawbacks. Evaluations are being made of an alternate source of S provided from ground S-bentonite products or fines from their manufacturing processes. This material is less dusty than flowers of S, yet it has a sufficiently fine particle size to give a good short-term suspension.

Although more costly, S suspensions intended for fungicidal purposes can be diverted to fertilizer uses. One product of this type is a cream-yellow liquid containing 40 to 54% S, with a particle size of only 1 to 2 μm.

4. Phosphate-Sulfur

Monoammonium and diammonium phosphates containing from about 5 to 20% S can be made by applying a hydraulic spray of elemental S at 1.4 kg/cm^2 during drum or pan granulation. At TVA, a product containing 12% N, 23% P, and 15% S (12–52–0–15S) was prepared by spraying liquid S onto a rolling bed of granular ammonium polyphosphate. On the basis of limited experience with these materials, it appears that they are excellent sources of both P and S (Beaton et al., 1968–1969). They should be satisfactory for bulk blending with other granular fertilizers or for direct application, particularly for topdressing legumes when both P and S are required.

A granular concentrated superphosphate-S fertilizer with an analysis of 0–18–0–20S (0–40–0–20S) was produced experimentally at TVA. The process involved granulating elemental S and concentrated superphosphate with the aid of steam and water in a drum granulator. A similar material analyzing 0–15–0–28S (0–35–0–28S) was made for a short time in the western USA. In spite of enthusiastic acceptance, production was terminated because of fire and explosion hazards connected with the grinding of oversized granules. If there were sufficient economic incentive, this product would likely be manufactured again, but in an inert atmosphere.

Sulfur fortification of normal superphosphate is popular in Australia and New Zealand. In Australia these specially prepared high S fertilizers are supplied in two grades—S.F. 25 and S.F. 45, with the numerals indicating total S concentration (Bell, 1975). They are manufactured by metering liquid S into the acidulation mixer. These materials are homogeneous, containing both SO_4^{2-} for fast release and from 18 to 37% elemental S for longer-term release in soils where leaching losses of SO_4^{2-} are serious. The two S-fortified superphosphates commonly produced in New Zealand contain 10 and 20% elemental S and provide one-half and one-third, respectively, of the total S as SO_4^{2-} (Muller et al., 1975).

There has been periodic activity and interest, dating back to at least 1916, in the preparation of phosphate rock–S mixtures. Evaluation of their agronomic effectiveness is incomplete. Biosuper (Australian Mineral Development Lab., Frewville, S. Australia), a granular phosphate rock–S product containing thiobacilli has been under investigation in Australia since the mid- to late 1960s. This material contains about 16% S, and the preferred ratio of phosphate rock to S is 5:1 (Swaby, 1975).

As a result of the activity of the thiobacilli, sulfuric acid is formed, which partially reacts with the phosphate rock to form superphosphate. Thiobacilli do not form spores, and death of the inoculum by dehydration during storage of Biosuper is a problem. The surviving thiobacilli will, however, accelerate S oxidation and formation of soluble P, particularly in soils deficient in native S-oxidizing organisms. Biosuper seems to be well suited for topdressing pastures in tropical areas of Australia and other countries where the annual precipitation exceeds 635 mm.

5. *Urea-Sulfur*

The complete miscibility of urea and liquid S facilitates production of a homogeneous granular product with excellent storage and handling properties. A 40–0–0–10S prilled material was manufactured and marketed on a limited scale in the mid-1960s in the western USA. Although all other properties were favorable, its S component had limited availability for early growing season response.

TVA prepared a granular urea-S analyzing about 44–0–0–4S by spraying molten S and urea melt as separate streams into a pilot plant pan granulator. The desired fineness of S particles was achieved by atomization.

In the early 1980s, a drum granulated urea-S became commercially available in western Canada. This homogeneous fertilizer material has a typical analysis of 36–0–0–20S. It has superb storage and handling characteristics. As might be expected due to differences in population and activity of S-oxidizing organisms at the experimental sites, field results have been variable, but in general they show that the S component becomes available in a reasonable period of time.

6. Sulfur Coatings

One way of achieving controlled release of soluble forms of plant nutrients is to coat them with relatively insoluble and affordable materials such as elemental S. TVA first made S-coated urea (SCU) in laboratory and bench-scale tests conducted about 1961. This controlled-release N fertilizer as currently produced is 77 to 82% urea and 14 to 18% S coating. At least three firms manufacture SCU products—one each in the USA, Canada, and the U.K. Sulfur used in these protective coatings will become available for plant uptake.

At TVA, S coatings for controlled release purposes have also been successfully applied to urea–ammonium phosphate granules and granular KCl.

Sulfur coatings may also be used to intentionally supply plant nutrient S. For example, a product analyzing 40–0–0–10S made by coating urea with finely divided S using oil and calcium lignosulfonate as a binder is being used on rice and Coastal bermudagrass in the west south central states, in Mississippi, and in Mexico.

7. Other Fertilizers Combined with Elemental Sulfur

A few other fertilizer products of only minor importance have been made by adding elemental S in one way or another (e.g., anhydrous ammonia–S and micronutrient-S fusions).

B. Fertilizers Containing Sulfate

Sulfate-containing fertilizers provide most of the fertilizer S applied to soils. These materials have the advantage of supplying S in a form that is immediately available for plant uptake. In areas with high leaching losses of plant nutrients, fast-acting sulfates may not be as effective as sources containing elemental S, which slowly convert to SO_4^{2-}.

1. Ammonium Sulfate

Ammonium sulfate is one of the oldest of the N- and S-containing fertilizers, and it still remains popular. In 1975 to 1976, it accounted for approximately 5.2 million metric tons of plant nutrient S worldwide (Bixby & Kilmer, 1975). Approximately 58% of it is synthetic, 16% is a by-product from metallurgical operations, and 26% is a coproduct from manufacture of caprolactam, the raw material for nylon 6, a synthetic fiber.

The popularity of ammonium sulfate (21–0–0–24S) has been due largely to its competitive price as a source of N. Part of its production cost is frequently charged against the primary product from which it results. At times there is virtually no charge for the S values in ammonium sulfate, but this is changing because of the increasing attention being given to plant nutrient S.

The uses of ammonium sulfate as a solid fertilizer are well known. It is applied alone or blended with other fertilizer materials. Segregation problems can occur in bulk blends when its product size is not well matched with those of N, P, and K materials. This difficulty can be minimized by carefully controlling uniformity and sizing of ammonium sulfate used in blends.

When ammonium sulfate is used for direct application as a N source, much more S is applied incidentally than is required by most crops. In addition to this N/S imbalance, excessive soil acidity can develop when frequent high rates are applied to poorly buffered soils.

Ammonium sulfate can supply up to 10% S in suspensions. When it is used in suspension blends high in K, S concentrations of only about 1 to 2% are possible.

It can also be used in clear liquids. For certain local situations involving short hauling distances and low costs for product, ammonium sulfate is used to make solutions analyzing about 8% N and 9% S. Sulfur concentrations in solutions based on ammonium sulfate and UAN solution can vary from 1 to 9%. An example of this combination is a 25–0–0–3.5S solution sold in the eastern USA. In NPS blends formulated with ammonium sulfate, the usual S concentrations range from 1 to 3%.

2. *Ammonium Nitrate–Sulfate*

Ammonium nitrate–sulfate with a grade of 30–0–0–06.5S is manufactured in Washington State. It is made by granulating a slurry of the double salt. At TVA, similar products analyzing 30–0–0–5S and 27–0–0–11S were made by neutralizing nitric acid (HNO_3) and sulfuric acid (H_2SO_4) with ammonia, and concentrating and feeding the product onto a tumbling bed of recycled product on a pan granulator.

A double salt of ammonium nitrate and ammonium sulfate present in equal proportions was formerly produced as a synthetic product in the Federal Republic of Germany. It was known as Leunasalpeter and had a grade of 26–0–0–13S. This S source has several advantages, including less hygroscopicity, a satisfactory N/S ratio for direct application purposes, and a combination of ammoniacal and nitrate forms of N, making it suitable for both spring and fall fertilization. In the U.S. Pacific Northwest, ammonium nitrate–sulfate is widely used on forage and grass seed crops and on fall-seeded small grains.

3. Urea–Ammonium Sulfate

For a few years beginning in 1976 a granular urea–ammonium sulfate fertilizer with a grade of 40–0–0–6S was commercially produced in western Canada. It was made by coating ammonium sulfate fines with urea in a spherodizing drum. Although this S source was replaced by a new urea–elemental S material, which was mentioned previously, a second manufacturer began production in 1984 of a similar material analyzing 38–0–0–7S.

Granular urea–ammonium sulfate has been made by a variety of ways at TVA. Grades ranging from 40–0–0–14S to 30–0–0–13S have been produced by oil prilling. It was also made by coating ammonium sulfate fines with urea in a granulator and by air prilling.

Urea–ammonium sulfate granules tend to be more resistant to physical breakdown and less hygroscopic than urea prills. Its physical properties can be further improved by the addition of gypsum, which forms a complex with urea. The N/S ratio may be varied from 3:1 to 7:1, resulting in considerable flexibility in the correction of N and S deficiencies in most soils. Uses in the western USA were similar to those described earlier for ammonium nitrate–sulfate.

Mechanical mixtures of urea and ammonium sulfate with a grade of 34–0–0–11S have been sold in western Canada since 1967, and they are still well accepted. Because of segregation problems caused by unsatisfactory product sizing of ammonium sulfate, which were mentioned earlier, these blends should be used soon after they are prepared and not stored in fertilizer dealers' bins or on farms. Also, the amount of handling should be minimized.

TVA has demonstrated the feasibility of producing a 29–0–0–5S urea–ammonium sulfate suspension. It is made by reacting anhydrous ammonia, H_2SO_4, 75% urea solution, and water in a tank-type boiling reactor (National Fertilizer Development Center, 1980). Two cooling stages are used, and 2% attapulgite gelling clay is added during the first cooling stage. Handling and storage properties are good, and the N/S ratio of nearly 6:1 is favorable. This product may be used either for direct application or as a base suspension in the manufacture of suspension mixtures by a simple cold-mixing procedure.

4. Ammonium Phosphate–Sulfate

The most common grade of ammonium phosphate–sulfate analyzes 16–9–0–14S (16–20–0–14S). It is composed of about 40% monoammonium phosphate and 60% ammonium sulfate. Other products of this type include 13–16–0–20S (13–39–0–20S), 19–4–0–20S (19–9–0–20S), and 23–9–0–7S (23–20–0–7S). The latter contains some urea.

They are made by several processes, including reacting a mixture of phosphoric acid (H_3PO_4) and sulfuric acid with ammonia, and introducing ammonium sulfate solutions and H_2SO_4 into a H_3PO_4 plant reaction circuit.

Direct application of 16–9–20–14S (16–20–0–14S) to forage crops, particularly legumes, is practiced in many areas. It is also popular for in-row applications on small grains and rapeseed/canola seeded on previously fallowed land. This product is frequently used for formulating bulk blends.

A granular product with a grade of 15–13–0–8S (15–30–0–8S) is currently manufactured in the southeastern USA. Ammonium phosphate, ammonium sulfate, and gypsum are used to make this product, which is suitable for both bulk blending and direct application.

5. Gypsum

Gypsum ($CaSO_4 \cdot 2H_2O$) is used as a source of S in the localities where it is mined or produced as a by-product. Its low nutrient analysis, 13 to 19% S, limits the extent to which it can be profitably used. Although finely divided gypsum is difficult to apply, two U.S. manufacturers now make an easily spreadable granular form. Gypsum has the added benefit of supplying Ca for either plant nutrient or soil amendment purposes.

There is a vast supply of this mineral from mines situated in several parts of the USA and Canada. Appreciable quantities are also produced as a by-product in the manufacture of wet-process H_3PO_4.

6. Normal Superphosphate

Normal superphosphate was at one time a major source of P and still is prominent in Australasia due to its 10 to 14% S content. Normal superphosphate is composed of approximately 50% by weight each of monocalcium phosphate and gypsum or its lower hydrate. The occurrence of S deficiencies has been delayed in many areas of the world because of the involuntary addition of S when normal superphosphate was used to supply P. Its Ca content, ranging from 18 to 21%, can be important in soils low in this nutrient.

7. Potassium Sulfate and Potassium–Magnesium Sulfate

Potassium sulfate analyzing 0–0–42–17S (0–0–50–17S) and potassium–magnesium sulfate with a grade of 0–0–18–22S (0–0–22–22S) plus 11% Mg are important sources of S. They are particularly useful when low levels of chloride are desired, as is often the case for crops such as tobacco, potatoes, avocado (*Persea americana* Mill.), peach [*Prunus persica* (L.) Batsch], some legumes and turf grass. These two products also supply K, S, or Mg individually or remedy multiple deficiencies of these elements.

Both materials are suitable for direct application, bulk blending, and inclusion in suspensions. Potassium sulfate is least abrasive of the two products. The abrasion problem with potassium–magnesium sulfate has been largely overcome in a modified product that is ground to −20-mesh size and then regranulated.

8. Magnesium Sulfate and Sulfates of Micronutrients

Magnesium sulfate containing 13% S and 9.8% Mg has limited use as a source of Mg in clear liquid fertilizers and foliar sprays (Beaton & Bixby,

1973). Significant amounts of S will also be provided when it is used to supply Mg.

Micronutrient sulfate salts are also incidental carriers of S. For example, in the group consisting of Cu, Fe, Mn, and Zn, concentration of S varies between 13 and 21%.

9. Sulfuric Acid and Urea–Sulfuric Acid

Use of sulfuric acid (H_2SO_4) as a soil amendment is discussed later. The usual concentration of acid used for this purpose is 93%, and it contains about 30% S. Thus, there will be substantial additions of S when arid lands are treated with H_2SO_4.

Personal hazards have restricted the use of H_2SO_4. A simple, rapid, and economic process has been developed in California to combine urea and H_2SO_4 into a product with greatly reduced destructivity of human tissue without neutralizing any of the acidity of the sulfuric acid (Gregory, 1983). A mild skin irritation may occur after prolonged exposure to this product, and it will sting on contact with a cut or sore on the skin.

The reaction between urea and concentrated H_2SO_4 is strongly exothermic, and measures must be taken to dissipate the heat evolved in the process (Verdegaal & Verdegaal, 1982). The resulting end product is a liquid that remains in the fluid state to temperatures of −2.2 to 3.3° C, depending on the formulation. Urea–H_2SO_4 fertilizers have a pH of < 1, and stainless steel and thermoplastics such as polyethylene, polyvinyl chloride, polyvinylidene fluoride, and polypropylene must be used for manufacturing, storing, and handling them. Workers must use rubber protection equipment, especially footgear.

Typical urea–H_2SO_4 formulations, amendments, and fertilizers have grades of 10–0–0–18S, 15–0–0–16S, and 28–0–0–9S. Urea–Sulfuric acid fertilizers can be applied in any manner that conventional fertilizers are applied, provided that suitable corrosion resistant equipment is used. They can be broadcast before planting or shanked in as a sidedressing. In addition, they can be applied in surface irrigation water or injected into circle pivots, linear moving systems, and drip systems. Other fertilizer materials such as 10–15–0 (10–34–0) and H_3PO_4 are compatible with urea–H_2SO_4 fertilizers.

C. Fertilizers Containing Other Forms of Sulfur

Two materials containing reduced or partially oxidized S compounds are commonly used in the rapidly expanding market for fluid fertilizers. These are ammonium thiosulfate and ammonium polysulfide. Use of ammonium bisulfite, a once popular S source in certain localized areas of the Pacific Northwest, has diminished in recent years and became almost nil in 1982. Sulfur dioxide is a high-analysis liquid that is not used for soil applications because of technical difficulties, but it is commonly used for water treatment in California and Arizona.

1. Ammonium Thiosulfate

Ammonium thiosulfate solution, 12–0–0–26S, is the most widely used S product in clear liquid fertilizers. It is prepared by reaction of SO_2 and aqueous ammonia, followed by further reaction with elemental S. Several variations of this basic process exist.

This material is compatible in any proportion with neutral to slightly acidic (not $< pH = 5.8$) orthophosphate and polyphosphates. It can also be readily blended with aqueous ammonia and with UAN solutions. It is not suitable for mixing with anhydrous ammonia.

The versatility of ammonium thiosulfate is such that it can be used to provide up to 10 to 12% S in a wide variety of N–S, N–P–S, and N–P–K–S formulations. Trial blends should be prepared, especially when K is a component, before going to large-scale mixing.

It is essentially noncorrosive and may be stored in mild steel and aluminum containers. However, contact with tin, copper, brass, or other copper alloys should be avoided. Storage temperatures of between -1 and 38° C are permissible.

Ammonium thiosulfate is well adapted to the many methods of applying fertilizer solutions, and it can be applied through open ditch and sprinkler irrigation systems. It has favorable soil amendment properties also, but there are less expensive alternatives.

2. Ammonium Polysulfide and Calcium Polysulfide

Ammonium polysulfide is a highly alkaline (pH 10) reddish brown solution with a strong odor of hydrogen sulfide (H_2S). It analyzes 20–0–0–45S and is made by reacting aqua ammonia with H_2S and elemental S in the presence of heat and agitation.

Ammonium polysulfide is compatible with anhydrous ammonia, and at volumes $> 10\%$ it is stable in both aqua ammonia and urea–ammonium nitrate solutions. Blending with acidic solutions should be avoided because it will decompose in them, releasing H_2S and colloidal S. Generally, it is considered unsuitable for mixing with phosphate solutions, although there are exceptions. Complete N–P–K–S solutions containing up to almost 6% K are possible with this product as a component.

Storage and handling equipment should be made of black iron, stainless steel, aluminum, or certain poly-type plastics. Brass and bronze metal parts are subject to severe corrosion. It can be stored at temperatures ranging from -1 to 38° C. This product should be stored at a pressure of 0.035 kg/cm^2 to prevent loss of ammonia and subsequent precipitation of S.

It may be applied directly in open ditch irrigation systems, but addition through sprinkler systems is not recommended. This material in mixtures of aqua ammonia or anhydrous ammonia is injected into soil using anhydrous ammonia applicators. It can be applied directly to the soil surface if diluted with water to lower the N concentration to $< 5\%$. Urea–ammonium polysulfide solution, 25–0–0–35S, is applied in irrigation runs to many crops in the southwestern USA. At one time in this region of the

USA, a blend of ammonium polysulfide and anhydrous ammonia analyzing 55–0–0–20S was specially formulated for soil injection.

Although ammonium polysulfide is an excellent source of plant nutrient S, its primary uses are as a soil amendment and for water treatment. It is not as convenient or as pleasant to handle as is ammonium thiosulfate. For example, after equipment is used for handling and distributing ammonium polysulfide, it must be thoroughly cleansed with water or anhydrous ammonia.

A 29% solution of calcium polysulfide is used similarly to ammonium polysulfide as a soil amendment and for water treatment. It contains 22% S and 7% Ca, has a pH of 11 to 12, and has a strong odor of H_2S. The amendment action of 100 kg of calcium polysulfide is equivalent to 147 kg of 100% gypsum or 27.6 kg of elemental S.

Calcium polysulfide may be applied at rates of 120 to 600 kg/ha through flood or furrow irrigation systems. Higher amounts may be used under certain conditions. When calcium polysulfide is applied to growing crops, rates in a single application should not exceed 120 to 180 kg/ha.

3. *Ammonium Bisulfite*

Ammonium bisulfite, a low analysis (8.5–0–0–17S) solution with a strong odor of sulfur dioxide, was once an important source of S in a localized region of the Pacific Northwest. This material is well suited to mixing with aqueous ammonia and urea–ammonium nitrate solutions. Similar to ammonium polysulfide, it should not be mixed with acidic solutions. In most other respects it is used in the same ways as ammonium thiosulfate.

Mild steel, aluminum, and plastic tanks are suitable for storing ammonium bisulfite.

4. *Sulfur Dioxide*

Liquid SO_2 contains 50% S and for a time was considered a potentially useful high analysis S fertilizer. It is kept under pressure and for this reason has to be applied with anhydrous ammonia injection equipment. One of its major disadvantages is incompatibility with anhydrous ammonia, necessitating a separate tank and pump on application rigs.

Despite its lack of acceptance for fertilizer purposes, SO_2 is used as a soil amendment and for water treatment. It is produced from elemental S in specially designed field burners for addition to irrigation waters in Arizona and California. More is said about this aspect of its use in section VI regarding S soil amendments.

For more technological and agronomic information on the use of the principal S fertilizers, the recommendation tables drawn up by Bixby and Beaton (1970) and Beaton and Fox (1971) are useful. Considerable detail on mixing techniques for supplying S in liquids and suspensions is given in reviews by Beaton and Bixby (1973) and Bixby (1978). A comprehensive list of S fertilizer materials along with producers and suppliers was compiled by Bixby (1980b).

III. MARKETING SULFUR FERTILIZERS

Need for fertilizer S in the USA and Canada has grown considerably in recent years. In 1960, S deficiencies had been confirmed in field experiments in 13 states, and by the early 1980s, this number had increased to 35. Five provinces in Canada now have recognized shortages of this nutrient.

A. Consumption in the United States and Canada

Despite the widespread occurrences of S deficiencies in the USA and Canada, use of fertilizer S is far below crop requirements. Approximately 1 million metric tons of S were applied in the USA in 1972–1973 compared with calculated crop requirements ranging between 2.3 and 3.6 million metric tons of S (Beaton et al., 1974). Although both usage and crop needs of S are expected to be somewhat greater now, the gap between them has probably not narrowed. It should be recognized that not all of this fertilizer S consumption was planned; much of it was applied incidentally as a component of N–P–K fertilizer materials.

Crop responses to fertilizer S are common in the four western Canadian provinces of British Columbia, Alberta, Saskatchewan, and Manitoba. Annual additions of fertilizer S in these four provinces during the 11 yr from 1969 to 1979 ranged between 18 800 and 44 000 metric tons (Beaton, 1980). The 39 600 metric tons applied in 1979 are considerably lower than the estimated 0.15 to 0.2 million metric tons needed for the seven main crops grown in the three Prairie Provinces. Very little of this S use was deliberate, but rather it was incidental mainly in the form of popular materials such as ammonium sulfate, ammonium phosphate–sulfate, ammonium phosphates, and the various 1:1 and 2:1 nitrogen phosphates based on ammonium phosphate.

These deficits between crop requirements for S and the amounts applied in fertilizer are probably not as large as indicated. In some regions of the USA and Canada, there are undoubtedly substantial S contributions in wet and dry deposition from the atmosphere and in irrigation waters. Sulfates in irrigation water are often more than adequate for crop requirements.

B. Market Development

Market development for S is carried out by many fertilizer manufacturers in individual programs and by their support of research and education organizations such as the Sulphur Institute and the Sulphur Development Institute of Canada. Other organizations aiding in collective development programs are the Potash & Phosphate Institute and its affiliated Foundation for Agronomic Research. These industry-financed institutes are staffed with highly qualified scientists that serve the fertilizer industry, universities, government, and agriculture in the USA, Canada, and elsewhere in the world.

The industry-sponsored organizations document the need for S and its sound use in crop and livestock production. They fund research projects at universities and other research establishments to develop this kind of information and to obtain basic knowledge about S in soils, plants, and animals. Assistance is also given to demonstrations of the need for and benefits from using S in agriculture.

Strong educational programs are a major part of the market development activities of the industry-supported organizations. There is extensive use of publications such as the *Sulphur Institute Journal* (now restructured and renamed *Sulphur in Agriculture*), technical bulletins, newsletters, and leaflets. Visual aids, including 35-mm color slides and color films are also important educational tools. All of these items are designed to inform agricultural leaders among growers; in research, extension and instruction; bankers; and agribusiness of significant findings on the role of S in profitable crop and livestock production. Timely press releases on this same topic are sent to various farm publications.

Staff of the Sulphur Institute and to a lesser extent the Potash & Phosphate Institute frequently make presentations on S-related topics at farm meetings, crop production seminars, fertilizer dealer meetings, fertilizer association conferences, and at meetings of professional associations. These organizations periodically hold symposia, workshops, and other educational meetings to assemble researchers, educators, administrators, and members of industry to report on and evaluate problems and progress made in S research. Specific S-related topics may also be reviewed and discussed at these meetings.

In addition to research grants offered to universities and other domestic research agencies, some financial and technical assistance is given to international agencies such as the United Nation's Food and Agriculture Organization's fertilizer program.

Company-identified market development activities may take a number of directions. The principal ones are in-house research and development projects; exhibits at trade shows and other meetings of the various state, regional, and national fertilizer associations; and provision of funding and materials for research projects at universities and government research agencies. Other approaches used are direct mailers on S-related topics to farmers and agriculturists and advertisements in farm newspapers and magazines, on radio, and perhaps on television.

In some cases, qualified fertilizer company personnel prepare technical literature, write educational articles for farm publications, and give talks on subjects on various aspects of S in crop and livestock production.

C. Marketing Programs

Sulfur fertilizer manufacturers market their products in different ways depending on how the individual firm is structured. Producers who have company-owned or franchised retail dealerships (or both) will sell S-containing materials through these outlets. Manufacturers with and without such retailing capability may also sell to wholesale distributors, who in turn will service retailers. Several manufacturers of special S-containing materials do not have any retailing facilities. They market directly to dealers, and substantial quantities may also be handled through wholesalers. Rarely are S fertilizer products sold directly from the manufacturer to the grower.

By purchasing S fertilizers through retail outlets, farmers have access to both local and broader expertise on the use of and benefits from applying S. They will also be able to have fertilizer S, where required, blended with common N–P–K materials.

D. Marketing Problems

One serious complication in the marketing of fertilizer S is the existence of large supplies of by-product ammonium sulfate. The low monetary value often credited to the S component of ammonium sulfate makes it difficult for other more costly S-containing products to be price competitive.

There have been marketing problems with several of the elemental S-containing sources. Brief accounts of these difficulties are given in the discussions on concentrated superphosphate-S and water-degradeable granular S-bentonite materials in section IIA2. More serious consequences of the dust problem resulting in destruction of a manufacturing plant in Alberta in early 1980 were reported by Beaton (1980).

IV. FORMS OF SULFUR IN SOILS AND THEIR UTILIZATION BY PLANTS

Soils contain both inorganic and organic S. As far as we know, plants utilize the inorganic SO_4^{2-} form of S almost exclusively. This is not to imply that organic forms of S are unimportant; many soils of the humid and subhumid temperate zone contain very small quantities of inorganic S. In such soils plants quickly utilize (or percolating waters readily leach from the root zone) S mineralized through the slow and continuous oxidation of soil organic matter, together with S that accrues as dissolved SO_2 in rainwater and SO_2 absorbed directly by soils from the atmosphere.

Even though the immediate source of S for plants is inorganic SO_4^{2-}, organic S is the original source for most of the SO_4^{2-} used by plants. Because of this, organic S has been referred to as *reserve S* (Bardsley & Lancaster, 1960), a term generally appropriate for soils that contain relatively little inorganic S in relation to organic S.

Some soils contain so much SO_4^{2-} in relation to organic S that the term reserve S seems inappropriate to describe the organic fraction. Two cases serve as examples: gypsum-rich soils of arid and semiarid regions and soils developed in highly weathered volcanic ash materials with very high capacities to sorb SO_4^{2-}. Such are the Andepts of Hawaii, which may contain as much as 0.7% SO_4^{2-}-S in the soil material below 30 cm (Hasan et al., 1970).

Sulfate-sulfur is present in soils in three forms: in solution, as a precipitate, and adsorbed onto colloid surfaces. Gypsum is the most frequently encountered precipitate of SO_4^{2-} in soils. It is a sparingly soluble salt (ca. 450 μg S/mL in water). The concentration of SO_4^{2-} in equilibrium with solid phase gypsum dihydrate is several times greater than the level plants require for good nutrition.

Most highly weathered soils contain sorbed SO_4^{2-}, some of which contain great quantities, whereas soils with properties dominated by the layer-silicate clay minerals (they contain little of the hydrated oxides of Fe or Al) sorb little SO_4^{2-}. For example, 58% of readily extractable S in a group of soils from Brazil was sorbed SO_4^{2-}, but only 2% in soils from Iowa was sorbed SO_4^{2-} (Neptune et al., 1975). Some uncertainty exists about the nature and availability of this SO_4^{2-}. It is usually thought to be retained on sesquioxide surfaces by some kind of ligand exchange mechanism. Adams and Rawajfih (1977), however, have demonstrated that basaluminite and alunite could be the cause of SO_4^{2-} retention by acid soils.

The concentration of SO_4^{2-} in soil solutions is highly variable, especially in soils that have little capacity to sorb SO_4^{2-}. Such soils are relatively unbuffered against depletion of SO_4^{2-} by plant uptake and leaching or against accumulations resulting from organic S mineralization, fertilizer applications, or moisture depletion. A determination of the apparent solubility of SO_4^{2-} in profile samples of 11 soils from Puerto Rico, all with an appreciable capacity to sorb SO_4^{2-}, gave a range of 0.4 to 24 μg/mL (Fox, 1982).

If soils have an appreciable capacity to sorb SO_4^{2-}, the concentration of SO_4^{2-} in solution will depend to a great degree on the percentage saturation of the sorption complex with SO_4^{2-}.

Data collected on soil profile samples primarily from highly weathered soils of Hawaii (Hasan et al., 1970) and Puerto Rico (Fox, 1982) suggest that if soils are to have 5 μg/mL of SO_4^{2-} in solution, the SO_4^{2-} sorption capacity should be, on average, 60 to 80% saturated. Not much SO_4^{2-} is required to attain this for most surface soil materials, because the SO_4^{2-} sorption capacity is usually relatively low for the surface horizon. This is demonstrated for an Ultisol of Puerto Rico in Table 11–6. Note that 0.6 μg S/mL in solution, probably a deficient condition, was associated with 10 mg S/kg of soil and a SO_4^{2-} saturation percentage of only 20.

Table 11–6. Status of SO_4^{2-}-S in selected depth increments of Daguey soil† from Puerto Rico (Fox, 1980).

Depth increment	Sorbed SO_4^{2-}	SO_4^{2-} in solution	SO_4^{2-} sorption maximum	SO_4^{2-} saturation
cm	mg S/kg	μg S/mL	mg S/kg	%
0–10	10	0.6	51	20
10–30	170	3.0	435	39
30–60	250	1.5	548	46
60–95	373	1.5	671	56

† Clayey, oxidic, isohyperthermic Orthoxic Tropohumult.

V. REACTIONS OF SULFUR FERTILIZERS IN SOILS

Many factors influence the efficiency of S fertilizers. In very general terms, the effects these factors have on fertilizer efficiency will depend on the kind of fertilizer materials involved, the nature of the soil, crops used, weather conditions, and time.

A. Elemental Sulfur

Because SO_4^{2-} is the form of S taken up by plants, S in lower oxidation states must be oxidized before it is used. Elemental S is converted to SO_4^{2-} by biological oxidation. Although autotrophic and heterotropic bacteria, fungi, and actinomycetes are capable of oxidizing elemental S, the autotrophic thiobacilli are the most important. Some soils are underpopulated with S-oxidizing organisms. When elemental S is added to such soils, some time will elapse while the microbial population builds up. Elemental S oxidation is affected by particle size. The relationship between surface area of the S particles and S uptake by corn growing in a S-deficient soil was almost perfectly linear (Fig. 11–4).

Temperatures between 30 and 40° C and soil moisture contents higher than field capacity favor elemental S oxidation. Oxidation proceeds most rapidly when the elemental S is mixed with the soil. If elemental S is banded in acid sandy soils, adding lime in the band will speed oxidation (Table 11–7).

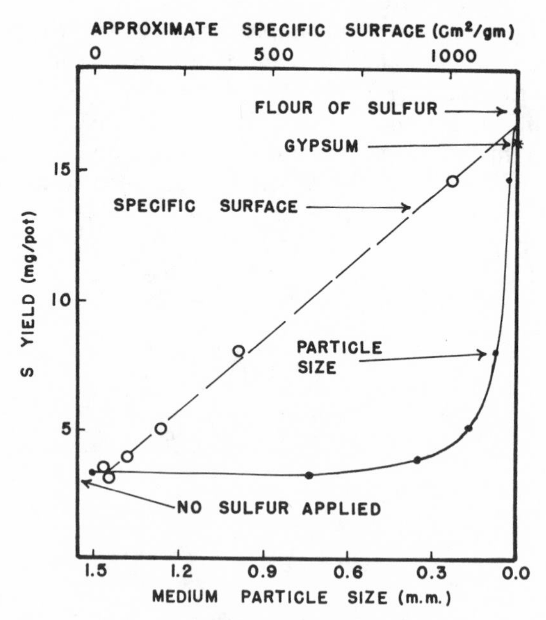

Fig. 11–4. Sulfur yield in corn plants 32 d after planting in relation to the specific surface of S particles. The rate of S addition was 5 mg/kg in all treatments (Fox et al., 1964).

Table 11-7. Sulfur yield in corn plants 46 d after planting as influenced by placement of different S-containing materials and liming (Fox et al., 1964).

S treatment	S yield Unlimed	S yield Limed
	mg/pot	
	Fertilizer mixed	
None	4.3	3.3
Gypsum	14.1	--
S flour	15.5	14.7
	Fertilizer banded	
Gypsum	15.4	--
S flour	8.4	14.3

B. Other Forms of Sulfur

Thiosulfate, sulfites, and sulfur dioxide are converted chemically and biologically to SO_4^{2-}. Because these materials are usually applied to the soil in fluid form, a high degree of contact with the soil is expected. Several of these materials undergo transformations, resulting in the formation of colloidal S when they are added to soil. Thiosulfates are chemically unstable in acid solution, yielding SO_2 and elemental S. The polysulfides are chemically unstable and form colloidal S and sulfides when added to soil. These soluble sulfides rapidly oxidize chemically to elemental S. Oxidation of elemental S formed from these fluid products probably proceeds biologically and is influenced by temperature and moisture in the same way as elemental S applied directly to soils. Particle size of the elemental S formed during these transformations is believed to be ideal for subsequent oxidation.

Sulfur dioxide reacts with water in the soil to form sulfurous acid, which in turn reacts with soil constituents to form sulfite salts. These oxidize to SO_4^{2-} either chemically or biologically.

C. Sulfates

Regardless of the form of S fertilizer applied to soils, the end product in well-aerated soils is SO_4^{2-}. The SO_4^{2-} will form sparingly soluble salts with cations such as Ca^{2+} in the soil. Sulfates coprecipitated with $CaCO_3$ in soils probably are relatively unavailable to plants. In humid regions, SO_4^{2-} usually leaches into the subsoil where it may accumulate; however, soluble SO_4^{2-} salts frequently leach through the profile below the root zone. Sulfate is less mobile in soils than NO_3^- and Cl^- and is less subject to leaching losses. This is one reason why the fertilizer requirements for S tend to be less than would be expected relative to N.

Many highly weathered soils have a considerable capacity to adsorb SO_4^{2-}. Sulfate sorption by soils is concentration dependent. Fertilizers that greatly increase SO_4^{2-} concentration in the soil enhance SO_4^{2-} sorption. The reaction is reversible. Plants can utilize adsorbed SO_4^{2-}. Although the relationships are far from perfect, evidence suggests that the availability of adsorbed SO_4^{2-} is only slightly less than soluble SO_4^{2-} or at least is in keeping with its solubility in the soil system (Barrow, 1969; Fox, 1980; Hasan et al., 1970; Hogg & Toxopeus, 1966). Adsorption of SO_4^{2-} should help to conserve S fertilizers and native soil S by reducing leaching losses and regulating the movement of SO_4^{2-} to the roots of plants. Liming and phosphate fertilization reduce the capacity of soils to sorb SO_4^{2-}. The short-term effect of these practices may be increased availability of fertilizer or soil sulfates because of increased SO_4^{2-} solubility. The long-term effects are likely to be increased depletion of S in surface horizons because of greater leaching.

Sulfate in strongly anaerobic soils may be reduced by bacteria to form H_2S, which in turn reacts with heavy metals to produce highly insoluble sulfides. If the system later becomes oxidized, these sulfides oxidize to elemental S, which is converted by biological oxidation to sulfuric acid, causing very acidic conditions.

VI. PREDICTING AND DIAGNOSING THE NEED FOR SULFUR FERTILIZERS

A. Available Soil Test Levels

Efforts to develop S soil tests have been largely concentrated on measuring SO_4^{2-} as an estimate of available S (Ensminger & Freney, 1966). Some recommended extraction solutions have been (i) monocalcium orthophosphate [$Ca(H_2PO_4)_2$] or monopotassium orthophosphate (KH_2PO_4), (ii) neutral salt solutions, (iii) hot water, (iv) sodium acetate (NaOAc, pH 4.8), and (v) sodium bicarbonate ($NaHCO_3$). The first two measure soluble plus some adsorbed SO_4^{2-}, and the remaining procedures also include some organic S. Solutions containing phosphate have given the highest correlations of extractable S related to crop yields or S uptake by plants. In pot tests, correlation coefficients *(r)* of 0.8 and 0.9 have been reported (Barrow, 1967; Scott, 1981), but in the field an *r* of 0.4 is more usual (Hoeft et al., 1973; Probert & Jones, 1977).

Some workers have found no significant relationships of soil test values to crop yields or S uptake in the field (Spencer & Glendinning, 1980). Soil tests appear to be of most value on sandy soils. For example, Reneau and Hawkins (1980) observed a response to S in Virginia on corn and soybeans grown on Coastal Plain soils. These are moderately well to well-drained soils, low in organic matter, belong to the fine loamy or coarse-textured families, and have $Ca(H_2PO_4)_2$–extractable soil S concentrations of 6 to 7 kg S/ha or lower. Soils with the same soil characteristics but with soil S concentrations between 7 and 15 kg/ha would be expected to respond to applications in dry years but not in wet years. This response appears to be related to soil moisture conditions and the depth and concentrations of extractable S in subsurface horizons.

Brogan and Murphy (1980) in Ireland reported that soil testing for S was not promising. They observed, however, that soils with $> 50\%$ sand and $< 3\%$ organic C were likely to respond to S.

B. Plant Analysis

Four criteria of assaying the S status of plants have been most widely used: (i) total S, (ii) SO_4^{2-}-S, (iii) N/S ratios, and (iv) SO_4^{2-}-S/total S ratio.

Total S is an obvious choice since concentrations are directly related to S supply. However, critical values are affected by plant part sampled, stage of growth, and amount of defoliation previous to sampling, as in the case of a grazed or mowed pasture crop (Jones et al., 1981; Ensminger & Freney, 1966).

Sulfate-sulfur critical values do not change as much as total S with the above variables and critical N/S, and SO_4^{2-}-S/total S ratios change even less. However, the latter two methods require two analyses that increase the cost and the variability. The disadvantages of these four criteria are that time of sampling is restricted to a relatively short period, they indicate only the S status of the plants at the time of sampling, and then only if other nutrient or environmental factors are not more limiting than S. Plant analysis can be useful if time of sampling, plant part, and chemical analysis are properly chosen.

To put the S status in perspective with respect to other nutrients, the diagnosis and recommendation integrated system (DRIS) has been suggested (Sumner, 1981). This appears to have merit where a large data base for yields and chemical composition can be obtained on such crops as corn, soybeans, and alfalfa. However, on pasture crops, yield data from thousands of sites are more difficult and expensive to obtain and are not yet available.

C. Contributions from Atmospheric Sources and Irrigation Waters

Global emissions of S resulting from industrial activities are estimated at 100 to 170 million metric tons annually (Husar et al., 1977). More than 26 million metric tons of SO_2 (13 million metric tons of S) were estimated to have been emitted into the atmosphere of the USA in 1980. Total annual emissions of sulfur oxides in Canada are approximately 7 million metric tons or 3.5 million metric tons of S (Canada Dep. of the Environment, 1978; Summers & Whelpdale, 1975).

Sulfur enters the atmosphere as SO_2, H_2S, H_2SO_4, various sulfate compounds, and as volatile organic compounds such as dimethyl sulfide, dimethyl disulfide, and methyl mercaptan. In unpolluted air, S occurs primarily in three compounds: SO_4^{2-} in aerosols and SO_2 and H_2S gases (Tabatabai & La Flen, 1976). Hydrogen sulfide in the air is normally oxidized to SO_2, which in turn is oxidized to SO_3. Sulfur trioxide dissolves in water droplets to form H_2SO_4, which may react further to form SO_4^{2-} salts such as $(NH_4)_2SO_4$ or $CaSO_4$.

Sulfuric acid and the SO_4^{2-} salts present in aerosols are removed by precipitation and, to a lesser extent, gravitational settling. Sulfur dioxide that enters the air has a residence time of only about 5 days to 2 weeks before it is removed by these mechanisms (Tabatabai & La Flen, 1976).

The most important processes that remove pollutants from the atmosphere (Overrein, 1977) are:

1. Removal by precipitation of gases and aerosols.
2. Absorption of gases by vegetation, soil, and water.
3. Impaction of particles on vegetation and ground.

The first of these processes is called *wet deposition* and the other two *dry deposition*.

1. Wet Deposition

Atmospheric deposition of S in North America generally increases from west to east as it is moved by prevailing winds across the North American continent. In 1980 some areas of Idaho, Nevada, and Utah received < 0.5 kg S/ha whereas the southern Great Lakes and Ohio Valley received 12.5 kg S/ha. (Fig. 11–5). There are smaller areas on the downwind side of industrial complexes where the deposition of S is much higher than the maximum shown.

Amounts of S deposited annually in precipitation in Alberta and Minnesota ranged from 2.2 to 220 kg/ha (Hoeft et al., 1972). They found at 20 locations in Wisconsin from 1969 to 1971 that average annual S in precipitation measured 2 km from an industrial site was 168 kg/ha, whereas at 9 urban and 13 rural sites it was 42 and 16 kg/ha, respectively. The annual addition of S by precipitation at six sites in Iowa varied from 13 kg/ha in northeastern parts of the state to 17 kg/ha in north central Iowa (Tabatabai & La Flen, 1976). These values for Iowa are comparable with those reported for rural Wisconsin.

The wet deposition figures above for Iowa and Wisconsin plus values reported in a number of other states are listed in Table 11–8, which was summarized by Terman (1978). In states bordering the Gulf and southern Atlantic coasts at points not affected appreciably by local industry, the quantities of S in precipitation were usually < 6 kg of S/ha per year. Greater amounts ranging from 10 to 30 kg/ha per year were recorded near local industries and generally over northern Alabama, Kentucky, Tennessee, and Virginia.

An average yearly rainwater return of 28 kg of S/ha has been reported in Ontario along the northeastern shore of Lake Ontario (Rutherford, 1967). In adjoining U.S. states, the amounts of S in rainfall ranged from 9 kg of S/ha in northern Michigan (Cressman & Davis, 1962) to 55 kg of S/ha in New York (Leland, 1952).

In spite of the estimated 0.275 million metric tons of S, primarily in the form of SO_2, emitted yearly to the atmosphere in Alberta, there is surprisingly low washout of only about 1.4 kg of S/ha per year (Table 11–9). Rain falling in remote unpopulated areas of Alberta is characterized by very low

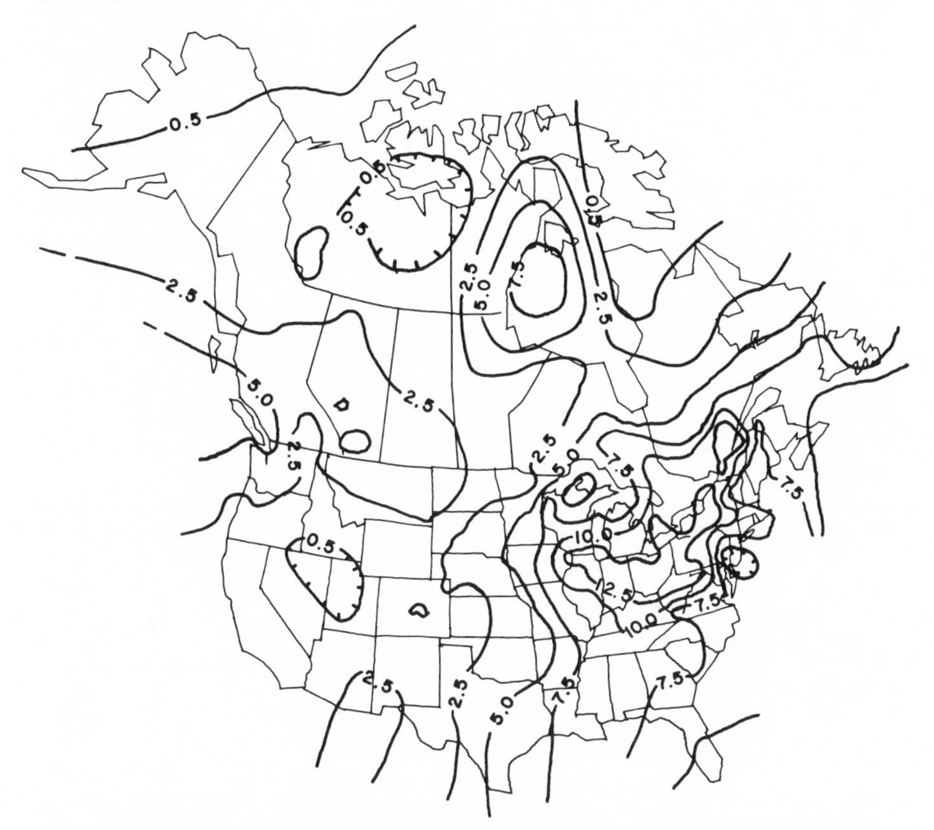

Fig. 11–5. Sulfur deposition isopleth map in kilograms per hectare for January–December 1980 (J. H. Gibson and C. V. Baker, unpublished data, Natural Resource Ecology Laboratory, Colorado State Univ., Fort Collins).

SO_4^{2-}-S content, typically 0.2 mg/L. This level approximates that found in fresh snow and is representative of the global background level of SO_4^{2-}-S resulting from natural sources. Median values for 1973 and 1974 precipitation in central Alberta were 0.7 and 0.5 mg/L, respectively. The annual average SO_4^{2-}-S concentrations in Iowa precipitation, supposedly uninfluenced by industrial emissions, were much higher ranging from 1.08 to 2.31 mg/L.

Sulfur content of precipitation in the tropics is frequently low, as exemplified by a mean value of 1.14 kg/ha per year detected during the rainy season in northern Nigeria (Bromfield, 1974).

From this brief review and many other reports in the literature, it seems that rain and snow usually contain small amounts of S in regions remote from industry, as little as 1 or 2 kg of S/ha per year. Near urban and industrial areas precipitation may contain from 20 to > 100 kg of S/ha per year. Ten kilograms of S/ha per year in precipitation have been used as a rough guide for delineating areas of potential S deficiency.

Table 11-8. Amounts of S found in precipitation in various U.S. states (Terman, 1978).

State	Location in state	No. of sites	Years	Major source	Amounts/yr Range	Avg
			Southern states		kg S/ha	
Alabama	Prattville	1	1954-1955	General	3.4- 4.0	3.7
	Muscle Shoals	19	1954	General	3.1- 6.8	5.4
	Muscle Shoals	20	1955	Steam Plant	6.4-16.2	11.9
	Muscle Shoals	23	1956	Steam Plant	6.5-16.6	11.0
Arkansas	NW and SE	2	1954-1955	General	2.6- 5.3	3.7
Florida	Cantonment	1	1953-1955	Industry	14.8-52.6	33.5
	Gainesville	1	1953-1955	Urban	8.0- 9.6	8.8
	Others	5	1953-1955	General	1.8- 4.4	3.2
Georgia	Various	6	1954-1955	General	2.8-13.9	7.7
Kentucky	Various	6	1954-1955	General	4.1-22.3	13.1
Louisiana	Various	5	1954-1955	General	2.1-13.9	9.0
Mississippi	Various	7	1953-1955	General	0.8-10.1	5.0
North Carolina	Statesville	1	1953-1955	Industry	12.7-19.4	15.5
	Others	15	1953-1955	General	3.1-14.6	6.0
	Clyde	1	1969-1973	Industry	10.6-43.3	22.1
Oklahoma	Stillwater	1	1927-1942	General	6.9-14.2	9.7
South Carolina	Various	4	1953-1955	General	3.1-17.7	7.5
Tennessee	Various	7	1955	General	10.5-21.4	14.2
	Various	5	1971-1972	General	13.9-20.0	17.1
Texas	Beaumont	1	1954-1955	Industry	10.0-14.1	12.1
	Others	4	1954-1955	General	3.1- 7.5	5.7
Virginia	Norfolk	1	1954-1955	Industry	33.4-37.0	35.2
	Others	14	1954	General	8.4-26.9	15.0
	Others	17	1955	General	13.3-34.5	21.2
	Various	16	1953-1956	General	14.2-37.5	21.4
			Northern states			
Indiana	Gary	1	1946-1947	Industry	--	142.2
	Others	10	1946-1947	General	22.4-37.0	30.0
Michigan	Various	5	1959-1960	Industry	9.0-14.0	11.3
Minnesota	St. Paul	1	1962-1967	Urban and Industry	--	16.5
	Others	3	1962-1967	Urban and Industry	4.1-10.1	7.7
Nebraska	Various	7	1953-1954	General	2.6-15.9	7.2
New York	Ithaca	1	1931-1949	Urban and Industry	34.7-76.2	54.9
Wisconsin	Industrial site	1	1969-1971	Industry	--	168.0
	Urban	9	1969-1971	Urban	--	42.0
	Rural	13	1969-1971	General	--	16.0

Table 11-9. Total yearly average SO_4^{2-}-S deposition in central Alberta, Canada (Klemm, 1977).

Source of deposition	kg SO_4^{2-}-S/ha
Snow (1 November-13 February)	0.45
Rain (June-August)	0.90
Dry fallout (May-November)	0.78
Total	2.13

2. Dry Deposition

a. Lead Peroxide Absorption—Lead peroxide (PbO_2) cylinders protected from rainfall have been commonly used to measure the quantity of SO_2 that could be absorbed from the atmosphere. Terman (1978) reviewed this subject, and most of his information is summarized in Table 11–10. Absorption of about 10 kg of S/ha per year or less appeared to be characteristic of rural areas unaffected by urban and industrial activities. Quantities in the 40 to 60 kg/ha per year range were associated with urban centers, whereas amounts of > 300 kg/ha per year were indicative of emissions from industrial point sources.

b. Soil Absorption—Emissions of SO_2 can be directly absorbed by soils, and in some areas such as Alberta it may be the most important mechanism of SO_2 deposition (Nyborg & Walker, 1977). Absorption of SO_2 by bare and cropped soils at two locations in Alberta varied between 20 and 38 kg of S/ha per year. The data in Table 11–9 are supportive since it can be seen that dry deposition was equal in importance to rainout. In a 6-yr study in Sweden exposing potted bare soil, under protective canopies, positioned at distances of 1 to 7 km downwind from a SO_2 source, the ratios of S absorbed directly from the air by soils exceeded that in precipitation by factors of 7 to 2 (Johansson, 1959).

Direct absorption of SO_2 from the air by soils and measurement of subsequent acidification is an indirect method for monitoring SO_2 emissions. This approach is useful but can be misleading because (i) there can be natural fluctuations and variability in soil pH under field conditions, (ii) there can be different buffering capacities in soil, (iii) precipitation affected by S emissions is not always acidic due to formation of neutral SO_4^{2-} salts, and (iv) environmental factors such as soil temperature and moisture will affect the processes of ammonification and nitrification, which tend to dominate soil pH in the short term.

Table 11–10. Amounts of atmospheric SO_2 absorbed by PbO_2 cylinders (Terman, 1978).

State	Location in state (no. of locations)	Years	Major source	Amounts of S, kg/ha per year	
				Range	Avg
Alabama	Muscle Shoals (7)	1956	Power Plant	6–30	14
Minnesota	Minneapolis	1937	Urban and industry		460
	Rural site		General		3
	Duluth	1962–1967	General		11
	St. Paul		Urban and industry		41
	Rural sites (2)				7
Virginia	Norfolk	1955	Urban and industry		61
	Others (15)		General	14–39	23
Wisconsin	Alma	1969–1971	Industrial point source		380
	Urban (9)		General		28
	Rural (13)		General		14

c. Plant Absorption—Atmospheric SO_2 can reach plants directly by entering leaf stomata or by being absorbed in moisture on leaves. In either case, SO_3^{2-}-S and then SO_4^{2-}-S are soon formed. Low doses of SO_2 may be harmless or even beneficial. Excessive uptake of SO_2 by vegetation will, however, cause increasingly severe damage, varying from reversible to irreversible. Different degrees of injury may occur on sensitive species upon exposure to SO_2 concentrations above about 0.2 to 0.5 μL/L for 1 h.

Cotton plants well nourished with S in solution cultures still absorbed about 30% of their S from the atmosphere (Olsen, 1957). When S deficient, these plants absorbed up to 90% of their S from air containing 0.01 to 0.05 μL/L of SO_2. However, absorption of SO_2 from the air did not completely satisfy their S requirements. Oats and rapeseed grown in soil treated with ^{35}S-labeled fertilizer absorbed about 50% of their S directly from the air at low S supply but much less at high S supply (Siman & Jansson, 1976).

Faller (1970–1971) demonstrated that tobacco, sunflower (*Helianthus annuus* L.), and corn grown in nutrient solution with S in very low concentration responded to increasing SO_2 concentrations in controlled atmospheres up to 0.525, 0.35, and 0.175 μL/L, respectively. In later trials, Faller (1971) showed that tobacco plants were able to utilize atmospheric SO_2 as the only source of plant nutrient S and that this gaseous form was fully equivalent to SO_4^{2-} supplied through roots.

The large contributions that atmospheric S can make to the total S content of cotton and tall fescue (*Festuca arundinacea* Schreb.) are apparent in Table 11–11. Cotton at the Colbert site and in sand, both very deficient in S, obtained from 84 to 98% of its S requirement by absorption of SO_2 from the atmosphere. Fescue and cotton acted as sinks for atmospheric S when the S supply in the growth medium was low in relation to

Table 11–11. Vegetative production, S content, and source of plant S for cotton and fescue grown at four sites located at varying distances from coal-fired plants in Alabama (Noggle & Jones, 1979).

Growth site	Oven-dry wt	S conc	Total S	Source of plant S			Proportion of plant S from atmosphere
				Soil	Atmosphere		
	g	%	mg	mg	mg	mg/100 g, dry wt	%
			Cotton				
Crossville	117	0.12	152	152	0	0	0
Widows Creek	154	0.19	332	139	193	125	58
Colbert (soil)	356	0.28	1018	139	854	240	84
Colbert (sand)	186	0.25	494	9	485	260	98
Greenhouse	160	0.18	294	129	164	102	56
			Fescue				
Crossville	42	0.13	57	57	0	0	0
Widows Creek	60	0.19	115	76	39	65	34
Colbert (soil)	50	0.17	86	57	29	58	34
Colbert (sand)	44	0.18	81	8	73	165	90
Greenhouse	14	0.14	21	13	8	57	38

their S needs. From 75 to 90% of the total S accumulation from the atmosphere was in the form of SO_2.

Terman (1978) suggested that up to half of global plant needs for S could be absorbed as SO_2. Dry deposition of atmospheric SO_2 can be an important source of S, particularly when plant roots are inadequately supplied with SO_4^{2-} but apparently not unless there is an active source close at hand. Bromfield (1974) measured only minor quantities of S as dry deposition in northern Nigeria.

3. Irrigation Waters

Sulfur in waters used for irrigation varies widely, depending on geologic formations and other input sources contributing to the waters. For example, the upper Yakima River in Washington State contributes only 0.7 mg/L and crops respond to S fertilization, but below the Yakima River the water contributes 1.4 mg/L and irrigated crops do not respond to S. At the mouth of the Yakima River the water contains 5.1 mg/L (Dow, 1976). Water from the Colorado River and many wells is also high in SO_4^{2-}.

In California, streams flowing out of the granite formations of the Sierra Nevada Mountains are generally low, ranging from 0.4 to 4.4 mg/L. Streams from the Coastal Range are higher, ranging from 1.6 in the north to 138 mg/L in the south (Martin, 1958).

Generally, irrigation waters carrying more than 4 to 6 mg/L of S will supply enough S for most crops. The applicability of this guideline will of course depend on the interaction of several factors, including (i) the amount of water applied per growing season, (ii) crop requirements, and (iii) yield levels.

D. Leaching Losses

Sulfates in the soil solution are subject to leaching losses, particularly under high rainfall conditions and when soils have low adsorption capacities for this anionic form of S. The actual amounts of SO_4^{2-} leached will vary depending on soil properties, precipitation distribution patterns, and ground cover, etc. In the northeastern USA, leaching losses three to six times greater than those from crop removal have been observed. Sulfur losses by this mechanism in the Ohio River basin have been estimated to be 39 kg of S/ha per year.

E. Crop Production Systems

1. Yield Level

Whether a crop will respond to S fertilization depends on the available soil S plus inputs from the atmosphere and irrigation less the losses, usually by leaching. For example, the average alfalfa hay yield for California in 1980 was 14.3 metric tons/ha (McGregor & Heitlinger, 1980). This tonnage of hay, with an average S concentration of 0.25%, would remove 26 kg S/

ha, which is a conservative estimate. Much higher yields of 27 metric tons would remove at least 80 kg S/ha. Other high-yielding crops remove amounts of similar magnitude.

2. Sulfur Cycling

The S pools and flow rates from one pool to another vary from one site to another and between crops. Figure 11–6 shows an example of pasture stocked at 10 sheep/ha in New South Wales, Australia (Till & Blair, 1974). These authors and Devaud and McFarlane (1980) indicate the S cycle is practically a closed system under their environmental conditions. However, under California conditions, where most of the rain falls during the winter when plant growth is very slow, considerable available S is lost by leaching. Also, supplemental feeding is a common practice in the USA and in some Mediterranean areas (Katznelson, 1977). Allowances for these changes plus atmospheric deposition have been included in Fig. 11–6.

Another modification that should be considered is the transfer of nutrients from the pasture to resting areas. Studies with sheep in Australia indicate that 22% of the feces were deposited on 3% of the pasture area and 34% were deposited on 10% of the area (Hilder, 1964). This means that some areas of the pasture receive only a very small return of the S removed by grazing animals.

These examples demonstrate the value of S cycling studies, which can be helpful in making efficient use of fertilizers while achieving high crop production.

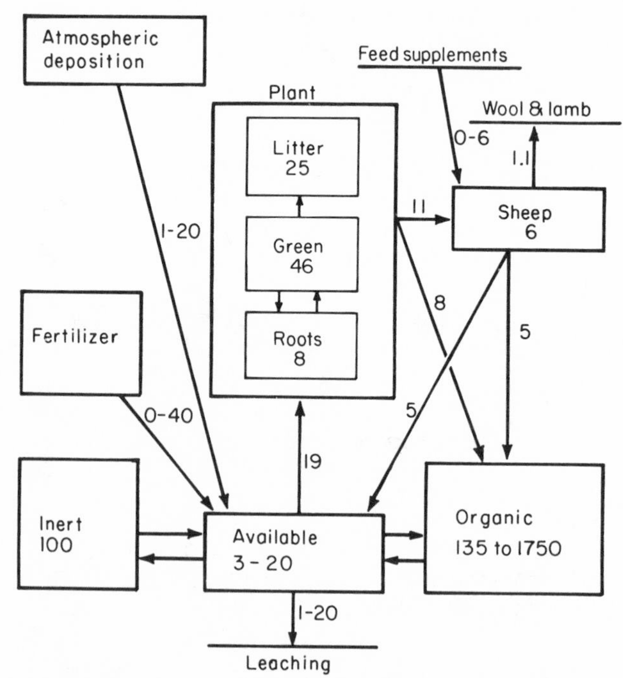

Fig. 11–6. Sulfur cycling in a pasture grazed at 10 sheep/ha. Arrows indicate flows in kilograms of S/ha per year (Till & Blair, 1974).

VII. USE OF SULFUR PRODUCTS AS SOIL AMENDMENTS

It has been known for many years that S and a number of its compounds can be used to improve calcareous and sodic soils and waters. These S-containing chemical materials include gypsum, elemental S, H_2SO_4, ammonium and calcium polysulfides, SO_2, ammonium thiosulfate, ammonium sulfate, ammonium bisulfite, and iron and aluminum sulfates. Most of these substances were described previously in this chapter.

Use of these various S materials in irrigated arid land agriculture has been reviewed by Tisdale (1970), Stroehlein et al. (1978), and Stromberg and Tisdale (1979). Brief comments will be made about the beneficial effects that these S compounds have on soils and waters.

A. Improvement of Nutrient Availability in Calcareous Soils

There are often excessive amounts of calcium carbonate in surface and subsoils in arid and semiarid regions. Land leveling for irrigation purposes may expose calcareous subsoils that are unfavorable media for optimum plant growth. Availability of P, K, B, Cu, Fe, Mn, and Zn is usually impaired by the presence of $CaCO_3$ and the associated high pH of about 8.0 to 8.2.

Table 11–12 illustrates how the productivity of a calcareous soil for growth of grain sorghum [*Sorghum bicolor* (L.) Moench] was greatly improved by the addition of either 560 kg/ha of H_2SO_4 or a similar rate of Fe in the form of $FeSO_4$.

A comparison is made in Table 11–13 of the effects of three S sources and the application of a water-soluble P fertilizer on P uptake by lettuce (*Lactuca sativa* L.) grown on four calcareous soils. It is apparent that the S treatments, especially H_2SO_4, greatly increased P uptake. Although the data are not presented here, on the average, for every 0.1 unit decrease in soil pH, available soil test P increased 3.2 mg/kg.

Another interesting example of the effect of acidification resulting from S additions on nutrient availability was reported by Hassan and Olson (1966). They applied finely divided elemental S at rates of 0, 50, 500, and

Table 11-12. Effects of $FeSO_4$ and H_2SO_4 on grain sorghum yields on a calcareous Texas soil (Mathers, 1970).†

	Fe, kg/ha		
H_2SO_4	0	112	560
	kg/ha		
0	434d*	1460bc	2275ab
112	605cd	1538ab	2474a
560	2169ab	2429ab	2230ab
5600	1885ab	1971ab	1810ab

* Yield values followed by the same letter are not significantly different at the 0.05 level.

† Standard error of the mean was 304.

Table 11-13. Average P uptake by four lettuce crops grown on four calcareous California soils (Clement, 1978).

Soil treatments	P uptake, mg/pot†				
	Salinas clay‡	Meloland fine sandy loam§	Ripley silty clay¶	Panoche clay loam#	Treatment means
Sulfuric acid	99.5c*	103.8c	63.8cd	107.3cd	93.9cd
Soil S	83.2b	98.7c	49.0bc	85.2bc	78.9bc
Popcorn S	80.5b	77.8b	48.0b	66.2ab	68.9b
Monocalcium phosphate††	109.8c	130.7d	68.0d	129.8d	109.2d
Control	38.8a	48.8a	27.6a	39.1a	38.3a
LSD (%)	15.3	16.1	15.7	30.6	21.4

* Means in the same column that do not have a common letter are significantly different at the 0.05 level (Duncan's multiple range test).
† Uptake from S treatments averaged for two rates of 168 and 336 kg of S/ha.
‡ Fine-loamy, mixed, thermic Pachic Haploxerolls.
§ Coarse-loamy, over clayey, mixed (calcareous), hyperthermic Typic Torrifluvents.
¶ Coarse-silty, over sandy or sandy-skeletal, mixed (calcareous), hyperthermic Typic Torrifluvents.
# Fine-loamy, mixed (calcareous), thermic Typic Torriorthents.
†† Applied at a rate of 56 kg of P/ha.

5000 mg/kg to three soils ranging from acid to calcareous. These S dressings increased nutrient uptake by corn in the greenhouse as follows: P on neutral and calcareous soils, S and Mn (up to modest rates of S) on all soils, Zn in early growth stages on all soils with slight benefit extending over the late harvest on the calcareous soil, Fe on the calcareous soil, and Cu until early harvest with all soils and through late harvest with the neutral and calcareous soils.

B. Reclamation of Sodic Soils

High content of exchangeable Na^+, as in sodic soils, tends to deflocculate colloidal materials, causing soils to lose their granular structure. Such soils become almost impervious to water and air because of the loss of large pores resulting from the breakdown of soil aggregates. Root penetration is impeded, hard clods form when the soil dries, and preparation of a good seedbed becomes difficult. Surface crusting frequently results in poor germination and uneven stands.

Crop growth is reduced because of these poor physical properties. Also, the nutrition of sensitive crops such as tree fruits, vines, and some annuals will be adversely affected by exchangeable Na^+ before significant deterioration of soil physical properties has occurred. High pH values of > 8.0, which are typical of Na-affected soils, lower the availability of most plant nutrients except Mo. Typical yield reductions in sodic soils are given in Table 11–14.

Reclamation of sodic soils involves replacement of the exchangeable Na^+ with a desirable cation such as Ca^{2+}, followed by removal of the displaced Na^+ by leaching. The addition of chemicals where needed to supply

Table 11-14. Effect of exchangeable Na^+ on crop yield (Velasco, 1981).

Type of soil	Na^+	Avg decrease in crop yield
	%	
Slightly sodic	7-15	20-40
Moderately sodic	15-20	40-60
Very sodic	20-30	60-80
Extremely sodic	>30	>80

Ca^{2+} or to solubilize natural supplies of calcium carbonate and other Ca-containing minerals in soils is a common remedial treatment for reclaiming sodic soils. Sulfur and a number of its compounds are successfully used in the reclamation process.

Gypsum is an obvious choice when soils do not contain free $CaCO_3$, and it has a long history of successful use. Calcium polysulfide is another Ca-containing compound used in treatment of sodic soils.

On soils containing free $CaCO_3$, any of the acid-forming S compounds can be used to make available this natural source of Ca. Sulfuric acid added to soils will react quickly with native $CaCO_3$. When elemental S, calcium polysulfide, ammonium polysulfide, SO_2, or ammonium thiosulfate are added to soil, the S components must first be converted to H_2SO_4 by S-oxidizing microorganisms before the solubilization of soil $CaCO_3$ occurs.

A substantial reduction in exchangeable Na^+ in surface soil following the application of H_2SO_4 is shown in Fig. 11-7. Increasing amounts of leaching water also enhance the removal of exchangeable Na^+. The extent of Na^+ removal also depends on acid rates and properties of soils and waters (Miyamoto et al., 1975).

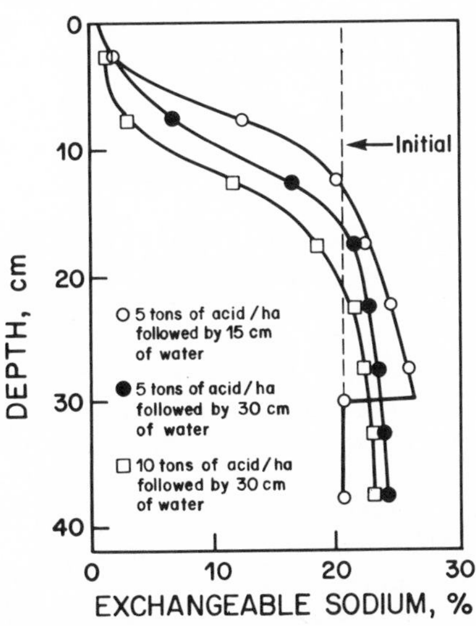

Fig. 11-7. Distribution of exchangeable Na^+ after application of H_2SO_4 and leaching water to Pima clay loam. (Stroehlein et al., 1978).

Table 11–15. Effect of surface-applied acid on the time required to pass 30 cm of water through Na-affected soils in laboratory tests (Stroehlein et al., 1978).

Soils	Na^+	CEC†	Sulfuric acid, metric tons/ha‡				
			0	1	3	5	10
	%	meq/100 g	d				
Stewart clay loam (loamy, mixed, thermic, shallow Typic Durorthids)	100	45	∞	∞	>40	39	28
Pima clay loam (fine-silty, mixed [calcareous], thermic Typic Torrifluvents)	21	40	39	34	8	3	3
Gothard clay loam (fine-loamy, mixed, thermic Typic Natrargids)	17	42	8	2	1	0.6	--
McAlister loam (fine-loamy, mixed, thermic Ustollic Haplargids)	12	28	3	2	1	1	--

† CEC = cation exchange capacity.
‡ Metric tons/ha × 0.45 = U.S. tons/acre.

Soil structure improvement is important in the reclamation of Na-affected soils since structure influences the rate of water infiltration into the soil and also movement of water in the soil profile. The beneficial effect of surface-applied sulfuric acid on water infiltration rates is illustrated in Table 11–15.

There are many other dramatic examples of the ameliorating effect of S treatments on sodic soils in the USA, Spain (Velasco, 1981), and other areas of the world. For example, in some trials application of elemental S, H_2SO_4, and ammonium polysulfide to sodic soils in California, Montana, Nevada, North Dakota, and Wyoming raised crop yields between 8.5 and 400%. These and other S-containing amendments are expected to have favorable effects on crop growth on most problem sodic soils in the mountain states and in other parts of the USA and Canada (Cairns & Beaton, 1976).

C. Water Treatment

For convenience of application, gypsum, H_2SO_4, urea–H_2SO_4, SO_2, ammonium polysulfide, calcium polysulfide, ammonium thiosulfate, and ammonium bisulfite can be successfully added to arid land soils through surface-applied irrigation waters. Some of these materials are also suitable for use in sprinkler and drip irrigation systems. It has been observed that these soil-amending substances when added to water will greatly aid in maintaining water transmission properties of soils (McGeorge et al., 1956, p. 29; Howard, 1968; Robinson et al., 1968). Figure 11–8 shows how additions of gypsum and calcium polysulfide to low salt irrigation water influenced flow rates. Although each increment of gypsum improved flow rates, it took 60 mg/L or more to give a statistically significant increase in flow rate compared with no treatment of low salt water. Calcium polysulfide was also effective in improving flow rates, especially at low treat-

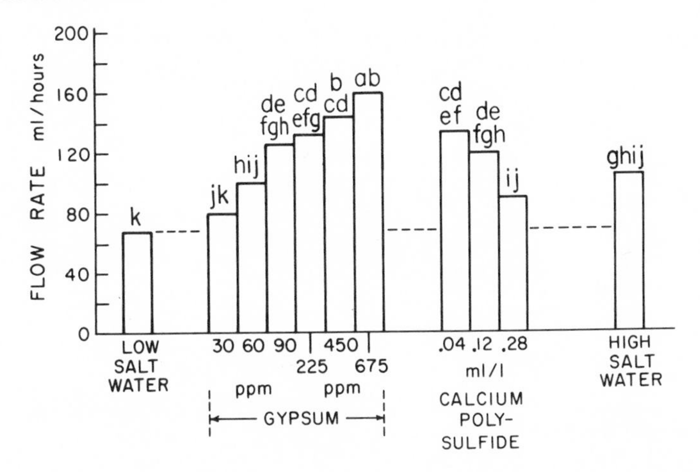

Fig. 11–8. Flow rate through Hanford (coarse-loamy, mixed, nonacid, thermic, Typic Xerothents) sandy loam with gypsum and calcium polysulfide added to low salt irrigation water (Mohammed et al., 1979).

ment rates. The decline in effectiveness of calcium polysulfide at the higher rates is believed to be related to the sealing of soil pores by the colloidal S that forms when polysulfide materials are placed in water.

There is still some uncertainty about the soil and water conditions under which the various amendments will have a favorable effect on water transmission properties. The most pronounced benefits are obtained with high pH and low-salt water (i.e., conductivity about 0.2 mmho and a pH > 8.4. Such water will likely contain < 20 mg/L of Ca^{2+} and be dominated by Na^+, CO_3^{2-}, and HCO_3^-. The mechanism responsible for improvement in water movement is still uncertain, but it is probably related to creating and maintaining flocculation of soil particles at the soil surface.

D. Acidification in Fertilizer Bands

In many calcareous and high pH soils, it is uneconomical to use the amounts of acidifying materials required to fully neutralize soil in the entire rooting volume. McGeorge (1945) suggested that it is not necessary to neutralize all of the soil alkalinity. More practically, soil zones favorable for root growth and nutrient uptake can be created by confining the S treatments to bands, furrows, auger holes, etc. A decline in soil pH in the vicinity of a banded mixture of ammonium thiosulfate and ammonium polyphosphate is evident in Fig. 11–9. Accompanying this soil acidification was an increase in available soil levels of Fe and Mn.

The observed benefits from banding acid-forming materials for various vegetable crops in Arizona and Colorado and for field crops in Kansas are probably the result of more favorable conditions in the zone of fertilizer placement rather than a direct nutritional benefit from S (Beaton & Fox, 1971).

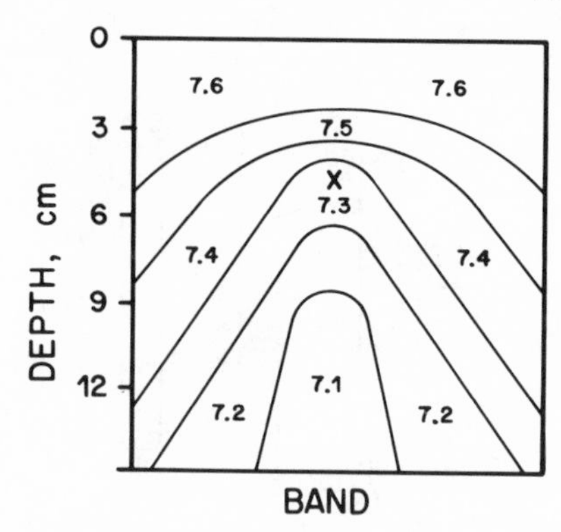

Fig. 11–9. Application of P-S solution lowers soil pH in vicinity of the fertilizer band. Point of application denoted by *X* (Leiker, 1979).

REFERENCES

Adams, C. A., and R. W. Rinne. 1969. Influence of age and sulfur metabolism on ATP sulfurylase activity in the soybean and survey of selected species. Plant Physiol. 44:1241–1246.

Adams, F., and Z. Rawajfih. 1977. Basaluminite and alunite: A possible cause of sulfate retention by acid soils. Soil Sci. Soc. Am. J. 41:686–692.

Arnon, D. I. 1965. Ferredoxin and photosynthesis. Science (Washington, D.C.) 149:1460–1470.

Bardsley, C. E., and J. D. Lancaster. 1960. Determination of reserve sulfur and soluble sulfates in soils. Soil Sci. Soc. Am. Proc. 24:265–268.

Barrow, N. J. 1967. Studies on extraction and on availability to plants of adsorbed plus soluble sulfate. Soil Sci. 104:242–249.

Barrow, N. J. 1969. Effects of adsorption of sulfate by soils on the amount of sulfate present and its availability to plants. Soil Sci. 108:193–201.

Beaton, J. D. 1969. The importance of sulphur in plant nutrition. Agrichem. West 12(1):4–6, 27.

Beaton, J. D. 1980. The role of NPKS. p. 325–445. *In* Soils and land resources. Prairie Production Symp. Canadian Wheat Board Advisory Comm., Saskatoon, Saskatchewan, 29–31 October. Canadian Wheat Board, Winnipeg, Manitoba, Canada.

Beaton, J. D., and D. W. Bixby. 1973. Mixing techniques of secondary elements with liquids and suspensions. p. 15–20. *In* 1973 Nat. Fertilizer Solutions Assoc. Roundup Proc., St. Louis, MO.24–25 July. The National Fertilizer Solutions Assoc., Peoria, IL.

Beaton, J. D., D. W. Bixby, S. L. Tisdale, and J. S. Platou. 1974. Fertilizer sulphur—status and potential in the U.S. Tech. Bull. 21. Sulphur Institute, Washington, DC.

Beaton, J. D., and R. L. Fox. 1971. Production, marketing and use of sulfur products. p. 335–379. *In* R. A. Olson et al. (ed.) Fertilizer technology and use. 2nd ed. Soil Science Society of America, Madison, WI.

Beaton, J. D., W. A. Hubbard, R. C. Speer, and R. T. Gardiner. 1968–1969. Ammonium phosphate-sulphur and sulphur gypsum: New granular sulphur sources for alfalfa. Sulphur Inst. J. 4(4):4–8.

Bell, G. K. 1975. Types of sulphur fertilizers made in Australia and their place in agriculture. p. 203–212. *In* K. D. McLachlan (ed.) Sulphur in Australasian agriculture. Sydney University Press, Sydney, New South Wales, Australia.

Bixby, D. W. 1978. Addition of sulfur in all types of fertilizer, including fluids. Proc. Annu. Meet. Fert. Ind. Round Table 28:125–132.

Bixby, D. W. 1980a. Sulfur requirements of the phosphate fertilizer industry. p. 129–150. *In* F. E. Khasawneh et al. (ed.) The role of phosphorus in agriculture. American Society of Agronomy, Crop Science Society of America, and Soil Science Society of America, Madison, WI.

Bixby, D. W. 1980b. Sulphur sources: For fertilizers. Sulphur Institute, Washington, DC.

Bixby, D. W., and J. D. Beaton, 1970. Sulphur containing fertilizers: Properties and application. Tech. Bull. 17. Sulphur Institute, Washington, DC.

Bixby, D. W., and V. J. Kilmer. 1975. Sulphur resources available to the fertilizer industry, the fertilizers made, their role and the problems involved in their manufacture, p. 231–241. *In* K. D. McLachlan (ed.) Sulphur in Australasian agriculture. Sydney University Press, Sydney, New South Wales, Australia.

Blouin, G. M., and C. H. Davis. 1975. Energy requirements for the production and distribution of chemical fertilizers in the United States. p. 51–67. *In* Proc. of the Conf. Workshop Southern Regional Education Board. 1–3 October. Southern Regional Education Board, Atlanta, GA.

Brogan, J. C., and M. D. Murphy. 1980. Sulfur nutrition in Ireland. Sulphur Agric. 4:2–6, 22.

Bromfield, A. R. 1972. Sulfur in Northern Nigerian soils 1. The effects of cultivation and fertilizer on total S and sulfate patterns in soil profiles. J. Agric. Sci. 78:465–470.

Bromfield, A. R. 1974. The deposition of sulfur in the rainwater of northern Nigeria. Tellus 26:408–411.

Bromfield, A. R. 1975. Effects of ground rock phosphate-sulphur mixture on yield and nutrient uptake of groundnuts *(Arachis hypogaea)* in Northern Nigeria. Exp. Agric. 11:265–272.

Bullen, W. A. 1968. Notes on biochemistry of nitrogen fixation. *In* 155th Am. Chemical Soc. Natl. Meeting, San Francisco. Abstr. 39. Agricultural and Food Chemistry Div. 31 March–5 April. American Chemical Society, Washington, DC.

Cairns, R. R., and J. D. Beaton. 1976. Improving a solonetzic soil by nitrogen-sulfur materials. Sulphur Inst. J. 12(3–4):10–12.

Canada Department of the Environment. 1978. A nationwide inventory of emissions air contaminants (1974). Rep. EPS 3-AP-78-2. Canada Department of the Environment, Downsview, Ontario.

Clement, L. 1978. Sulphur increases availability of phosphorus in calcareous soils. Sulphur Agric. 2:9–12.

Comiskey, P. T., L. B. Gittinger, Jr., L. F. Good, F. L. Jackson, L. A. Nelson, Jr., and T. K. Wiwiorowski. 1969. Sulfur. p. 337–364. *In* A. Standen (ed.) Kirk-Othmer encyclopedia of chemical technology, Vol. 19. 2nd ed. Interscience, New York.

Craven, A. O. 1925. Soil exhaustion as a factor in the agricultural history of Virginia and Maryland 1606–1860. Ill. Stud. Soc. Sci. 13(1).

Cressman, H. K., and J. F. Davis. 1962. Sources of sulphur for crop plants in Michigan and effect of sulphur fertilization on plant growth and composition. Agron. J. 54:341–344.

Devaud, E., and J. D. McFarlane. 1980. The fate of radioactive sulfur applied to grazed irrigated lucerne. Aust. J. Agric. Res. 31:887–897.

Dow, A. J. 1976. Sulfur fertilization of irrigated soils in Washington State. Sulfur Inst. J. 12(1):13–15.

Ellis, R. J. 1969. Sulfate activation in higher plants. Planta 88:34–42.

Ellison, S. P., Jr. 1971. Sulfur in Texas. Bureau of Economic Geology Handb. no. 2. University of Texas, Austin.

Ensminger, L. E., and J. R. Freney. 1966. Diagnostic techniques for determining sulfur deficiencies in crops and soils. Soil Sci. 101:283–290.

Evans, I. M., D. Boulter, R. L. Fox, and B. T. Kang. 1977. The effects of sulfur fertilizers on the content of sulpho-amino acids in seeds of cowpea *(Vigna unguiculata)*. J. Sci. Food Agric. 28:161–166.

Faller, N. 1970–1971. Effects of atmospheric SO_2 on plants. Sulphur Inst. J. 6(4):5–7.

Faller, N. 1971. Plant nutrient sulphur—SO_2 vs. SO_4. Sulphur Inst. J. 7(2):5–6.

Fox, R. L. 1976. Sulfur and nitrogen requirements of sugarcane. Agron. J. 68:891–896.

Fox, R. L. 1980. Responses to sulphur by crops growing in highly weathered soils. Sulphur Agric. 4:16–22.

Fox, R. L. 1982. Some highly weathered soils of Puerto Rico: 3. Chemical properties. Geoderma 27:139–176.

Fox, R. L., H. M. Atesalp, D. H. Kampbell, and H. F. Rhoades. 1964. Factors influencing the availability of sulfur fertilizers to alfalfa and corn. Soil Sci. Soc. Am. J. 28:406–408.

Fox, R. L., B. T. Kang, and D. Nagju. 1977. Sulfur requirements of cowpea and implications for production in the tropics. Agron. J. 69:201–205.

Fox, R. L., B. T. Kang, and G. Wilson. 1979. A comparative study of the sulfur nutrition of banana and plantain. Fruits 34:525–534.

Friedrich, J. W., and L. E. Schrader. 1978. Sulfur deprivation and nitrogen metabolism in maize seedlings. Plant Physiol. 61:900–903.

Gregory, J. R. 1983. Urea–sulfuric acid fertilizers. *In* 168th American Chemical Soc. Natl. Meeting, Washington, DC. Abst. 25. 28 August–2 September. American Chemical Society, Washington, DC.

Hasan, S. M., R. L. Fox, and C. C. Boyd. 1970. Solubility and availability of sorbed sulfate in Hawaiian soils. Soil Sci. Soc. Am. Proc. 34:897–901.

Hassan, N., and R. A. Olson. 1966. Influence of applied sulfur on availability of soil nutrients for corn (*Zea mays* L.) nutrition. Soil Sci. Soc. Am. J. 30:284–286.

Hatch, L. F. 1972. What makes sulfur unique? Hydrocarbon Process. 51(7):75–88.

Hilder, E. J. 1964. The distribution of plant nutrients by sheep at pasture. Proc. Aust. Soc. Anim. Prod. 5:241–248.

Hoeft, R. G., D. R. Keeney, and L. M. Walsh. 1972. Nitrogen and sulfur in precipitation and sulfur dioxide in the atmosphere in Wisconsin. J. Environ. Qual. 1:203–208.

Hoeft, R. G., L. M. Walsh, and D. R. Keeney. 1973. Evaluation of various extractants for available soil sulfur. Soil Sci. Soc. Am. Proc. 37:401–404.

Hogg, D. E., and M. R. J. Toxopeus. 1966. Studies on the retention of sulphur by yellow-brown pumice soils. N.Z. J. Agric. Res. 9:93–97.

Howard, F. K. 1968. Liquid fertilizer in the Imperial Valley. Fert. Solution 12(3):32–33, 35.

Husar, R. B., J. P. Lodge, and D. J. Moore. 1977. Sulfur in the atmosphere. Proc. Int. Symp. Atmos. Environ. 12:1–796.

Johansson, O. 1959. On sulphur problems in Swedish agriculture. Lantbrukshoegsk. Ann. 25:257–61.

Jones, M. B., J. E. Ruckman, W. A. Williams, and R. L. Koenigs. 1981. Sulfur diagnostic criteria as affected by age and defoliation of subclover. Agron. J. 72:1043–1046.

Katznelson, J. 1977. Phosphorus in the soil-plant-animal ecosystem: an introduction to a model. Oecologia 26:325–334.

Klemm, R. F. 1977. Sulfur in Alberta precipitation—consequences of three years of survey. p. 293–305. *In* H. S. Sandhu and M. Nyborg (ed.) Proc. of the Alberta Sulphur Gas Workshop III. Research Secretariat Alberta Environment, Edmonton, Alberta, Canada.

Koch, B., P. Wong, S. A. Russell, R. Howard, and H. J. Evans. 1970. Purification and some properties of a non-haem iron protein from the bacteroids of soya-bean (*Glycine max* Merr.) nodules. Biochem J. 118:773–781.

Leiker, W. J. 1970. Plant responses to sulfur applications. M.S. thesis. Kansas State University, Manhattan.

Leland, E. W. 1952. Nitrogen and sulphur in the precipitation at Ithaca, N.Y. Agron. J. 44:172–175.

Martin, W. E. 1958. Sulfur deficiency widespread. Calif. Agric. 12(11):10–12.

Mathers, A. C. 1970. Effect of ferrous sulfate and sulfuric acid on grain sorghum yields. Agron. J. 62:555–556.

McGeorge, W. T. 1945. Sulphur: A soil corrective and soil builder. Arizona Agric. Exp. Stn. Tech. Bull. 201.

McGeorge, W. T., E. L. Breazeale, and J. L. Abbott. 1956. Polysulfides as soil conditioners. Arizona Agric. Exp. Stn. Tech. Bull. 131.

McGregor, R. A., and R. H. Hettinger. 1980. Field crop statistics—California. California Crop and Livestock Reporting Service, Sacramento.

Miyamoto, S., R. J. Prather, and J. L. Stroehlein. 1975. Sulfuric acid and leaching requirement for reclaiming sodium-affected soils. Plant Soil 43:573–585.

Mohammed, E. T. Y., J. Letey, and R. Branson. 1979. Sulphur compounds in water treatment—effect on infiltration rate. Sulphur Agric. 3:7–11.

Muller, F. B., G. McSweeney, and J. Rogers. 1975. Sulphur in New Zealand fertilizers. p. 221–230. *In* K. D. McLachlan (ed.) Sulphur in Australasian agriculture. Sydney University Press, Sydney, New South Wales, Australia.

National Fertilizer Development Center. 1980. Production of urea-ammonium sulfate suspension fertilizer p. 61–64. *In* New developments in fertilizer technology. 13th Demonstration of the National Fertilizer Development Center, Muscle Shoals, AL. 7–8 October. National Fertilizer Development Center, Muscle Shoals, AL.

Neptune, A. M. L., M. A. Tabatabai, and J. J. Hanway. 1975. Sulfur fractions and carbon-nitrogen-phosphorus-sulfur relationships in some Brazilian and Iowa soils. Soil Sci. Soc. Am. J. 39:51–55.

Noggle, J. C., and H. C. Jones. 1979. Accumulation of atmospheric sulfur by plants and sulfur-supplying capacity of soils. Interagency Energy/Environment R & D Program Rep. EPA-600/7-79-109 and TVA/OnR-79-10.

Nyborg, M., and D. R. Walker. 1977. An overview—the effects of sulphur dioxide emission on the acidity and sulphur content of soils. p. 152–182. *In* H. S. Sandhu and M. Nyborg (ed.) Proc. of the Alberta Sulphur Gas Workshop III. Research Secretariat Alberta Environment, Edmonton, Alberta, Canada.

Olsen, R. A. 1957. Absorption of sulfur dioxide from the atmosphere by cotton plants. Soil Sci. 84:107–111.

Overrein, L. N. 1977. Acid precipitation—impacts on the natural environment. p. 1–26. *In* H. S. Sandhu and M. Nyborg (ed.) Proc. of the Alberta Sulphur Gas Workshop III. Research Secretariat Alberta Environment, Edmonton, Alberta, Canada.

Pearse, G. H. K. 1974. Sulphur—economics and new uses. p. S1–S7. *In* Proc. of the Canadian Sulphur Symp. Calgary, Alberta. 30 May–1 June. Univ. of Calgary, Calgary, Alberta, Canada.

Probert, M. E., and R. K. Jones. 1977. The use of soil analysis for predicting the response to sulphur of pasture legumes in the Australian tropics. Aust. J. Soil Res. 15:137–146.

Reneau, R. B., Jr., and G. W. Hawkins. 1980. Corn and soybeans respond to sulfur in Virginia. Sulfur Agric. 4:7–11.

Rieber, M., R. Fuller, and B. Okech. 1981. Sulfur pollution control: Phase I. The disposal program (sections 1–4). Bur. Mines Open File Rep. (U.S.) 94(1)–81.

Robinson, F. E., D. W. Cudney, and J. P. Jones. 1968. Evaluation of soil amendments in Imperial Valley. California Agric. 22(12):10–11.

Rutherford, G. K. 1967. A preliminary study of the composition of precipitation in S. E. Ontario. Can. J. Earth Sci. 4:1151–1160.

Scott, N. M. 1981. Evaluation of sulphate status of soils by plant and soil tests. J. Sci. Food Agric. 32:193–199.

Siman, G., and S. L. Jansson. 1976. Sulphur exchange between soil and atmosphere, with special attention to sulphur release directly to the atmosphere 2. The role of vegetation in sulphur exchange between soil and atmosphere. Swedish J. Agric. Res. 6:135–144.

Spencer, K., and J. R. Freney. 1980. Assessing the sulfur status of field-grown wheat by plant analysis. Agron. J. 72:469–472.

Spencer, K., and J. S. Glendinning. 1980. Critical soil test values for predicting the phosphorus and sulfur status of subhumid temperate pastures. Aust. J. Soil Res. 18:435–445.

Stroehlein, J. L., S. Miyamoto, and J. Ryan. 1978. Sulfuric acid for improving irrigation waters and reclaiming sodic soils. Agric. Engineering and Soil Sci. Bull. 78–5. University of Arizona, Tucson.

Stromberg, L. K., and S. L. Tisdale. 1979. Treating irrigated arid-land soils with acid-forming sulphur compounds. Tech. Bull. 24. Sulphur Institute, Washington, DC.

Summers, P. W., and D. M. Whelpdale. 1975. Acid precipitation in Canada. Proc. Int. Symp. Acid Precip. For. Ecosyst. 1st 1975:30.

Sumner, M. E. 1981. Diagnosing the sulfur requirements of corn and wheat using foliar analysis. Soil Sci. Soc. Am. J. 45:87–90.

Swaby, R. J. 1975. Biosuper-biological superphosphate. p. 213–220. *In* K. D. McLachlan (ed.) Sulphur in Australasian agriculture. Sydney University Press, Sydney, New South Wales, Australia.

Tabatabai, M. A., and J. M. La Flen. 1976. Nutrient content of precipitation over Iowa. Water Air Soil Pollut. 6:361–373.

Terman, G. L. 1978. Atmospheric sulphur—the agronomic aspects. Tech. Bull. 23. Sulphur Institute, Washington, DC.

Till, R. A., and G. J. Blair. 1974. Sulfur and phosphorus cycling. Trans. Int. Congr. Soil Sci., 10th 2:153–157.

Tisdale, S. L. 1970. The use of sulphur compounds in irrigated arid-land agriculture. Sulphur Inst. J. 6(1):2–7.

Velasco, I. 1981. Improving the sodic soils of Spain. Sulphur Agric. 5:2–4.

Verdegaal, W. J., and G. F. Verdegaal. 1982. Process for making liquid fertilizer. U.S. Patent 4 310 343. Date issued: 12 January.

Vroom, A. H. 1972. New uses for sulfur: The Canadian viewpoint. Hydrocarbon Processing 51(7):79–85.

Zaneti, G., and G. Forti. 1969. Reactivity of the sulfhydryl groups of ferrodoxin nicotinamide adenine dinucleotide phosphate reductase. J. Biol. Chem. 244(17):4757–4760: [Chem. Abstr. 71(9):40.]

12

Production, Marketing, and Use of Calcium, Magnesium, and Micronutrient Fertilizers

John J. Mortvedt
National Fertilizer Development Center
Tennessee Valley Authority
Muscle Shoals, Alabama

F. R. Cox
North Carolina State University
Raleigh, North Carolina

Increased knowledge acquired since 1960 about plant requirements as well as sources and effective methods of applying secondary nutrients and micronutrients has resulted in increased use in the USA. Improvements in soil and plant analyses have provided additional knowledge of specific crop needs and also of the wide variations in crop response to secondary nutrients and micronutrients found under field conditions. Higher rates of mixed NPK fertilizers and higher crop yields have increased the need for lime and may have increased the need for micronutrients.

In the first edition of this book, separate chapters covered secondary and micronutrient fertilizer production (Tisdale & Cunningham, 1963) and plant requirements, soil pH effects, and methods of application (Berger & Pratt, 1963). These subjects, except S, which was discussed separately, were consolidated in the second edition (Mortvedt & Cunningham, 1971). Several books on micronutrients (Sauchelli, 1969; Mortvedt et al., 1972; Davis, 1980) have been published; thus, this discussion will not include all of the subject matter covered in depth in the above references.

I. PRODUCTION METHODS

A. Nutrient Sources

Production methods of Ca and Mg fertilizers are not included in this section because dolomitic limestone generally is used to supply these nutrients; supplying Ca and Mg with sources other than limestone is discussed later.

Less than 10% of the industrial production of micronutrient elements in the USA is for agricultural use. For example, about 90% of the Mn imported into the USA is used for steel production. Because of the relatively minor fraction used in agriculture, it is difficult for those in metal industries to understand the seasonal need for fertilizers compared with year-round industrial needs. Therefore, the fertilizer industry is faced with the problem of obtaining sufficient micronutrient sources to meet demands each spring. Storage due to stockpiling increases costs.

1. Inorganic Sources

Inorganic sources include naturally occurring ores, manufactured oxides, carbonates, and metallic salts such as sulfates, chlorides, and nitrates. Some oxides, such as cuprous oxide (Cu_2O), may be used as mined, but plant availability of other oxides, such as naturally occurring manganese dioxide (MnO_2), is so low their use is not recommended. Inorganic sources usually are the least expensive per unit of micronutrient, but they are not always the most effective for soil application.

Sulfates are by far the most common of the metallic salts, and their physical properties make them suitable for use with mixed fertilizers. Although these sources also add plant-available S to soils, the amount of S applied at most recommended micronutrient rates is low. Ground ores or oxides are treated with sulfuric acid (H_2SO_4) to produce crystalline sulfate materials. Some of these products can be made in granular form by extrusion, compaction, or granulation by several methods. Oxysulfates, especially of Zn and Mn, are produced by partial acidulation of oxides.

Metal-ammonia complexes also are used as micronutrient sources. Ammoniated zinc sulfate ($ZnSO_4$) solutions are marketed widely and are compatible with most fluid fertilizers. The most common product contains 10% N, 10% Zn, and 5% S.

Borax ores found in old geological lake deposits are mined, milled, and refined to remove impurities. The most common source of ores was $Na_2B_4O_7 \cdot 10H_2O$, but products containing fewer waters of hydration now are used more widely. Granular borates are produced for use with bulk-blended fertilizers. Finely ground borates are produced for foliar sprays and for incorporation with fluid fertilizers.

Molybdenum sources are produced from molybdenite (MoS_2), which is mined in Colorado, and from by-products of Cu mining in Utah (Hurlbert, 1952). Both sodium molybdate (Na_2MoO_4) and ammonium molybdate [$(NH_4)_2MoO_4$], the main fertilizer sources of Mo, are produced by ore purification, roasting to form oxides, and through subsequent solubilization reactions.

Zinc sources are produced by several processes. Metallic Zn (dross) is milled and roasted to form zinc oxide (ZnO). The French process produces fine particles of pure ZnO that are used as a paint pigment, but the material is too expensive for fertilizer use. Sphalerite (ZnS) also is roasted to form ZnO. Zinc sulfate ($ZnSO_4 \cdot H_2O$) is produced by reacting ZnO with H_2SO_4. Partially acidified products also are sold; amounts of water-soluble Zn are directly related to the proportion of $ZnSO_4$ present in those products.

2. Organic Sources

Organic sources include synthetic or natural chelates, natural organic complexes, and combinations of the above. Effectiveness of these sources for plant utilization varies widely. Synthetic chelates generally are more expensive than other sources but may be more effective under certain soil conditions.

Chelates are formed by combining a chelating agent with a metal through coordinate bonds. Chelates may be synthetic (manufactured) or natural (from sugars or from natural products). A chelating agent is a compound containing donor atoms that can combine with a single metal ion to form a cyclic structure called a *chelation complex* or, more simply, a *chelate.* The descriptive word "chelate," derived from the Greek "chela" or "claw," was proposed for this kind of molecular structure in 1920 (McCrary & Howard, 1979). The stability of the metal-chelate bond generally determines plant availability of the applied nutrient. An effective chelate is one in which the rate of substitution of the chelated micronutrient for cations already in the soil is low, thus maintaining the applied nutrient in chelate form for sufficient time to be absorbed by plant roots.

Some chelating agents used for production of micronutrient chelates are:

1. Ethylenediaminetetraacetic acid (EDTA)
2. *N*-hydroxyethylethylenediaminetriacetic acid (HEDTA)
3. Diethylenetriaminepentaacetic acid (DTPA)
4. Ethylenediamine di(*o*-hydroxyphenylacetic acid) (EDDHA)
5. Nitrilotriacetic acid (NTA)
6. Glucoheptonic acid
7. Citric acid

The most common chelating agent used as a micronutrient source is EDTA. There are seven commonly known synthesis routes to EDTA, but four processes are commercially used (Strauss et al., 1983). Two methods are two-step processes involving production of a cyanomethyl derivative of formaldehyde and subsequent saponification by sodium hydroxide (NaOH). The other methods go directly from ethylenediamine to the final product in a single step, using sodium cyanide (NaCN) or hydrogen cyanide (HCN). Metal chelates generally are produced by reaction of the Na chelate salt with a metal; metal chlorides are preferred because metal sulfates result in precipitation of sodium sulfate (Na_2SO_4). Another process increases the pH of the chelate to between 4 and 5 with sodium hydroxide (NaOH), potassium hydroxide (KOH), or ammonium hydroxide (NH_4OH). Metal oxides, which are less costly than the chlorides, may be dissolved in this chelate. After complete solubilization, more caustic is added to bring the pH to between 7 and 8. The caustic source determines the amount of metal that can be carried in the final solution. For example, the highest concentration of Zn in EDTA solutions is 6.5 to 7.0% with NaOH, 8.0 to 9.0% with KOH, or 9.0 to 9.5% with NH_4OH.

Many chelates are sold in liquid form in the USA because production costs per unit of micronutrient are lower than those of the powder form, which requires drying. Liquid chelates are sold mainly for mixing with fluid fertilizers. Dry chelates are incorporated in some granular fertilizers, but their use is generally restricted to specialty products for lawns and gardens.

Natural organic complexes are made by reacting metallic salts with organic by-products of the wood pulp industry. Several classes of these complexes are the lignosulfonates, phenols, and polyflavonoids. The type of chemical bonding of the metals to the organic components in the above products is not well understood. Some of the bonds may be similar to those in chelates, but the remainder are not well defined, hence, the term *complexes.*

Digest liquors from the wood pulp industry contain organic complexes in the Ca or Na form. These liquors are concentrated to about 40% solids, and a micronutrient source is added to form a metal complex. Oxides may be added in this process, but sulfates are preferred because they are soluble. Digest liquors from the Sulfite process contain lignosulfonates that complex metals. Liquors from the Kraft process of pulp production must be sulfonated to improve their complexing capacity, which increases their cost. The most popular complexes are of Zn and Fe, although limited amounts of Cu and Mu complexes are produced. Products containing more than one micronutrient are made for foliar sprays or for application with irrigation water.

3. Fritted Glasses

Fritted micronutrients are glassy products whose solubility is controlled by particle size and change in matrix composition. The powdered raw materials are dried, mixed with silicates or phosphates, and melted in a furnace. This molten matrix then is quenched, dried, and milled. Products may be compacted or granulated prior to bagging and storage. More than one micronutrient may be included to provide custom mixes for various crops.

4. Industrial By-products

Many by-products are being marketed as fertilizers since regulations restrict indiscriminate disposal of industrial wastes. The most common micronutrient industrial by-products contain Mn or Zn. Use of industrial by-products as Zn sources has increased dramatically; flue dusts and baghouse dusts from galvanizing, pigment, rubber, battery, and other industries are used widely. Leaching processes are used to remove many impurities from these products, but some ZnO products are used without purification. Zinc chlorides, nitrates, sulfates and oxysulfates, and manganese sulfate ($MnSO_4$) also are available as industrial byproducts.

Some by-products may also contain appreciable amounts of heavy metals such as Cd, Cr, Ni, and Pb. Plant availability of these heavy metals is not well known, but since their soil application rate is very low, their effects should be minimal. For example, applying ZnO (70% Zn) containing 1000 mg/kg of Pb to soil at a rate of 5 kg of Zn/ha would result in a rate of 0.007 kg of Pb/ha. Because these heavy metals are relatively immobile in soil, they will accumulate in the surface layer and may eventually build up to significant levels if such by-products are applied to a field for a number of years.

B. Incorporation With Mixed Fertilizers

Recommended rates of micronutrients usually are < 10 kg/ha. This makes it difficult to apply micronutrient sources separately and uniformly in the field. Therefore, both granular and fluid NPK fertilizers are used widely as carriers of micronutrients. Including micronutrients in mixed fertilizers is convenient and allows more uniform distribution with conventional application equipment. Costs also are reduced by eliminating the separate application of micronutrients.

Chemical reactions between the micronutrient source and one or more components of the macronutrient fertilizer may occur during manufacture, in storage, or after soil application. Reaction products may differ markedly according to sources, pH of the mixture, and temperature (Lehr et al., 1967). Plant availability of the reaction products may differ from that of the original sources. Therefore, care must be taken to select those micronutrient sources and mixed fertilizers that, when combined, result in products that supply micronutrients in a plant-available form.

Micronutrient sources can be added to mixed fertilizers by incorporation during the manufacture of granular fertilizers, coating granular fertilizers, bulk blending with granular fertilizers, or mixing with fluid fertilizers. Advantages and disadvantages of each method are discussed in the following sections.

1. Incorporation During Manufacture of Granular Fertilizers

Incorporation uniformly distributes micronutrients throughout NPK fertilizers. During manufacture, the micronutrient source is in intimate contact with the mixed fertilizer components under conditions of high temperature and moisture. This enhances the rate of chemical reactions that may decrease the plant availability of some incorporated micronutrient sources. For example, when ZnEDTA (or any synthetic chelate) is mixed with superphosphate before ammoniation, acid decomposition of the chelate molecule results in decreased availability of the applied Zn (Ellis et al., 1965). When the process is changed by adding the ZnEDTA with the ammoniating solution, plant availability of Zn is increased (Brinkerhoff et al., 1966).

Granulation of either $ZnSO_4$ or ZnO with orthophosphate fertilizers having saturated salt solutions of relatively high pH results in Zn products low in initial availability to plants. Yet ZnEDTA remains available to plants when granulated with the same fertilizers (Mortvedt & Giordano,

1969). The reaction products of Zn salts and orthophosphate fertilizers, zinc ammonium phosphate ($ZnNH_4PO_4$), and zinc phosphate tetrahydrate [$Zn_3(PO_4)_2 \cdot 4H_2O$] were available for corn (*Zea mays* L.) when finely divided and mixed with soil but not when applied as granules (Allen & Terman, 1966). Also, ZnO was as effective as $ZnSO_4$ for corn when both sources were applied as fine powders, but only $ZnSO_4$ was effective as a granular material. Distribution of applied Zn in the soil thus may be more important than the water solubility of the Zn sources. Incorporating $ZnSO_4$ with superphosphates before ammoniation results in decreased water solubility of Zn (Jackson et al., 1962). Initial plant availability of applied Zn decreases with level of water-soluble Zn in ammoniated phosphates (Mortvedt, 1968). Other problems encountered with incorporation of micronutrients during manufacture of mixed fertilizers are discussed elsewhere (Achorn & Mortvedt, 1977).

The greatest disadvantage of incorporation during manufacture is that special handling and storage of numerous tons of special fertilizers increase distribution costs. Furthermore, micronutrient concentrations in the fertilizers cannot be changed to meet customer needs. This production method is thus usually limited to fertilizers that can be used on large land areas.

2. Coating Granular Fertilizers

As with incorporation during manufacture, coating micronutrients onto granular fertilizers can result in uniform application. The coating procedure consists of dry mixing the granular fertilizers with a finely ground (−100-mesh size) micronutrient source, then spraying a liquid binder and continuing the mixing process. The total cycle takes about 5 min per batch in small rotary mixers and longer periods for larger mixers. Mechanisms of binding action by the liquid are mechanical or by promoting formation of reaction products on the surface of the fertilizer granules (Silverberg et al., 1972).

The coating agent must be inexpensive, must adhere to the fertilizer during handling, and must not cause undesirable physical properties. Water, oils, waxes, ammonium polyphosphate, and urea–ammonium nitrate solutions have been used as binding agents. Water may be used where an increase in granule surface moisture does not promote caking. Oil should not be added to mixtures containing ammonium nitrate (NH_4NO_3) because of the explosion hazard, and not $> 1\%$ by weight of oil should be used in other mixtures to prevent oil from seeping through the fertilizer bags. Fertilizer solutions as binding agents are preferred because the fertilizer grade is not decreased appreciably. Some binding materials are unsatisfactory because they do not maintain the micronutrient coatings during bagging, storage, and handling operations, resulting in segregation of finely ground micronutrient sources from fertilizer granules.

The required amount of coating agent ranges from 1 to 3% and varies with the amount and physical state of the micronutrient source and the granular fertilizer. Fertilizer salt solutions and water provide chemical bonds; oils provide mechanical adherence. Fertilizer solution binders may promote caking, whereas some lighter oils penetrate and thereby weaken bags. In the former case, conditioning agents may be added to prevent caking.

Agronomic effectiveness of micronutrients coated onto soluble granular fertilizers should be similar to that with incorporation during manufacture. Ellis et al. (1965) reported that navy bean (*Phaseolus vulgaris* L.) yields were similar with either $ZnSO_4$ or ZnO incorporated during manufacture or coated onto a granular mixed fertilizer, but Zn uptake was higher with $ZnSO_4$ (Table 12–1). They also reported that ZnEDTA remained completely water soluble when coated onto a NPK fertilizer in conjunction with $MnSO_4$, whereas it was only 40% water soluble when coated with MnO onto the same fertilizer. Yields of navy beans also were lower with the latter product.

The nature of the binder does not appear to affect Zn availability to plants. Dissolution of Zn coated onto ammonium polyphosphate or monoammonium phosphate ($NH_4H_2PO_4$) granules as $ZnSO_4$ or ZnEDTA was similar in moist soil whether the coating agent was a fertilizer salt solution or a mixture of diesel oil and soft wax (J. J. Mortvedt and P. M. Giordano, 1966, unpublished data). Use of oil to coat ZnO and MnO onto 8–14–13 (8–32–16) fertilizer granules did not have a significant effect on Zn and Mn uptake by corn and soybeans [*Glycine max* (L.) Merr.] in field experiments (Jones, 1969).

One disadvantage of coating micronutrient sources onto granular fertilizers is higher production costs due to the batch process, which requires more handling. Coating provides more flexibility than incorporation in obtaining the recommended grades for separate fields, but it is not widely used in the USA.

3. *Bulk Blending with Granular Fertilizers*

The main advantage of bulk blending is its flexibility to make small batches of many grades. Thus, micronutrient sources can be blended with mixed fertilizers to produce grades that will provide the recommended micronutrient rate for a given situation. Blending should be done just prior to application, so that there is less chance for chemical reactions to occur that

Table 12–1. Yield and Zn concentration of navy beans as affected by Zn source and method of inclusion with a granular NPK fertilizer (Ellis et al., 1965).

Zn source†	Method of inclusion	Yield	Zn in plants
		kg/ha	mg/kg
--	--	12 320	20
$ZnSO_4$	Blended	16 580	40
$ZnSO_4$	Incorporated	16 350	31
$ZnSO_4$	Coated	16 690	34
ZnO	Incorporated	16 240	23
ZnO	Coated	16 690	26
LSD (0.05)	--	1 680	3

† Fertilizers banded at planting to supply 3 kg of Zn/ha.

would decrease availability to plants. A procedure to put the proper amount of micronutrient in each batch is required, and a longer mixing time to obtain a uniform mix is desirable. Extra bin space is eliminated with the special batches are prepared just before delivery. The popularity of this method of applying micronutrients has increased in recent years but now seems to have reached a plateau. Results of a 1984 survey conducted by TVA and the Association of American Plant Food Control Officials (AAPFCO) showed that there were 5200 bulk-blending plants in the USA with a total annual production of 20 million metric tons of fertilizer (Hargett & Berry, 1985). Micronutrients were being added to bulk blends at 73% of these plants in 1984 compared with 45% in 1974.

The main disadvantage in applying micronutrients with bulk-blended fertilizers is that segregation of nutrients can occur during the blending operation and subsequent handling. Segregation results in nonuniform application of nutrient, which is critical with micronutrients since their application rates are quite low. Studies have shown that the chief cause of segregation is differences in particle size, although particle shape and density also have an effect (Silverberg et al., 1972). Mechanical devices to minimize coning and segregation of materials during handling and storage are available (Achorn & Mortvedt, 1977). A comparison of micronutrient distribution in bags with incorporated ZnO and bulk-blended MnO is shown in Table 12–2. Concentrations of Zn were uniform in all bags, but those of Mn were unacceptably variable. Obviously, fine MnO had segregated from the granular fertilizer during blending and bagging operations.

Incorporation of a micronutrient source with one component of a bulk blend also has been used. This requires a higher micronutrient concentration in the particular component and thus reduces the number of fertilizer granules that contain the micronutrient. The latter factor can be important for elements of limited mobility in soil. Corn forage yields in a greenhouse experiment decreased as the Zn concentration in superphosphate increased from 0.5 to 8% Zn, with the Zn application rate held constant (Giordano & Mortvedt, 1966). The number of Zn-containing granules required to supply 3 mg of Zn was 32, 8, and 2 when the Zn concentration was 0.5, 2.0, and 8.0%, respectively. After movement of Zn from these granules in soil was measured, the volume of soil affected by applied Zn was calculated to be 5.0, 1.2, and 0.3% of the 3 kg of soil in each pot.

Table 12–2. Range in micronutrient concentrations in four bags of granular fertilizers after incorporating ZnO in granular MAP and blending this product with granular KCl and powdered MnO (Hignett, 1964).

	Test 1		Test 2	
Bag no.	% Mn	% Zn	% Mn	% Zn
1	3.0	0.7	2.9	1.4
2	0.7	0.6	0.9	1.6
3	2.9	0.6	9.8	1.1
4	0.6	0.6	4.9	1.4
Intended conc	3.0	0.5	3.0	1.2

Applying granular micronutrients in bulk blends of mixed fertilizers is perhaps the most popular application method. Using granular sources helps prevent segregation from the mixed fertilizer because of more equal particle sizes, but the granule size of micronutrient sources reduces the number of application sites in the soil. For example, the number of granule sites may be < 20/m^2 when granular $ZnSO_4$ is applied at a rate of 1 kg of Zn/ha. In contrast, if $ZnSO_4$ were incorporated with a mixed fertilizer to contain 2% Zn, the number of granule sites would be about 350/m^2 at the same Zn rate. Application of granular sodium tetraborate ($Na_2B_4O_7$) also results in high B concentrations in soil near the granule site, which could be toxic to nearby plant roots (Mortvedt & Osborn, 1965). Granular oxides of micronutrients may not be available to plants because they are relatively water insoluble, and their specific surface is greatly reduced in granule form. Granular ZnO was ineffective in providing Zn for corn (Allen & Terman, 1966), and granular MnO did not provide available Mn for oats (*Avena sativa* L.) (Mortvedt, 1977) in greenhouse tests. Therefore, granular oxides may not be recommended for use in bulk blends, especially to correct micronutrient deficiencies in young plants.

4. Mixing with Fluid Fertilizers

Application of micronutrients with fluid fertilizers generally assures uniform analysis and application. Because it is relatively easy to mix desired amounts of micronutrients with fluid fertilizers just before application, this method of applying micronutrients has become popular in the USA. Clear liquids are used widely as starter fertilizers, and some micronutrients, especially certain Zn sources, are easily applied with these products.

Solubility of metallic micronutrient sources is greater in polyphosphate than in orthophosphate clear liquids. Solubility is higher in 11–16–0 (11–37–0) made from electric furnace superphosphoric acid than in 10–15–0 (10–34–0) made from wet-process superphosphoric acid and is still lower in 8–11–0 (8–24–0) made from an orthophosphate base solution (Table 12–3). Solubility of these micronutrients in polyphosphate solutions is related to pH and polyphosphate content; thus, it may not be possible to achieve these levels in fluids with low polyphosphate contents.

Because of their high solubility, enough B and Mo may be incorporated with fluid fertilizers to correct severe deficiencies at usual fertilizer application rates. However, amounts of Cu, Fe, Mn, or Zn supplied as inorganic sources with fluids may not be sufficient to correct deficiencies. Organic sources may be required to provide adequate amounts with solution fertilizers, but some of these products are not compatible with all fluid fertilizers. A jar test should be made to determine compatibility before any micronutrient source is mixed with a fluid fertilizer.

The amounts of some micronutrients that can be applied with clear liquids at the desired N and P rates are limited. Suspensions may be used if greater amounts of micronutrients are desired. Suspensions have an advantage over clear liquids since complete solution of the micronutrient is

Table 12-3. Solubility of some micronutrient sources in ammonium orthophosphate or polyphosphate solutions (Achorn, 1969; Silverberg et al., 1972).

Micronutrient source	Solubility of micronutrient in		
	Orthophosphate	Polyphosphate	
		Wet-process	Electric furnace
	% element by weight		
$Na_2B_4O_7 \cdot 10H_2O$	0.90	0.90	0.9
CuO	0.03	0.53	0.7
$CuSO_4 \cdot 5H_2O$	0.13	1.13	1.5
$Fe_2(SO_4)_3 \cdot 9H_2O$	0.08	0.80	1.0
MnO	<0.02	0.15†	0.2†
$MnSO_4 \cdot H_2O$	<0.02	0.15†	0.2†
$Na_2MoO_4 \cdot 2H_2O$	0.05‡	0.38‡	0.5‡
ZnO	0.05	2.25	3.0
$ZnSO_4 \cdot H_2O$	0.05	1.50	3.0§

† Precipitates formed after several days.
‡ Largest amount tested.
§ Solution pH of 6.0.

not required. Micronutrient sources of low water solubility, such as oxides, also may be used in suspensions.

Incorporation is done while suspensions are being made or just before application in the field. Particles of micronutrient sources should be fine enough not to settle in the suspension or clog spray nozzles. Particle sizes finer than 60-mesh size are suggested for incorporation with suspensions (Mortvedt, 1979). Incorporating micronutrients with suspensions must be done in a manner to prevent the formation of aggregates, which are very difficult to disperse. One of the best ways to incorporate dry, fine powders directly into suspensions is through use of an injector or an eductor (Silverberg et al., 1972).

II. MARKETING

A. Philosophies of Use

Liming materials are applied to decrease soil acidity, but they also usually supply sufficient Ca and Mg to meet crop needs at optimum soil pH levels. One philosophy of liming is that the soil pH should be increased to about 5.5 to neutralize exchangeable Al^{3+}, whereas others recommend increasing soil pH levels to between 6.0 and 7.0 for optimum crop production. Dolomitic lime should be used, if possible, for liming soils that are low in exchangeable Mg^{2+}; other Mg fertilizers then may not be required to meet Mg needs of crops. Additional Ca is needed for a few crops even though the pH is in the proper range.

The "shotgun" approach has been used to add low amounts of more than one, and sometimes all, micronutrients to the soil. This method, designed to supply all micronutrients removed by a crop, could be considered as an insurance or maintenance program. "Premium" fertilizers containing all micronutrients were promoted in this manner; usually this program did not consider specific crop needs or levels of available micronutrients in soil. Now that soil and plant analyses can more accurately assess micronutrient levels, it is preferable to apply only those nutrients required to reach a specified yield goal for a crop in a given field. This practice helps prevent excessive application of needed micronutrients and eliminates use of those nutrients already present in adequate amounts. More precise recommendations also protect against antagonisms encountered among nutrients in plant nutrition due to unbalanced ratios in soil.

With the development and increased use of bulk-blended granular fertilizers and fluid fertilizers, there is a trend to formulate grades for specific fields at the fertilizer dealer level. These methods increase flexibility in meeting needs and have changed micronutrient marketing methods in the USA.

Manufacturers sell most micronutrients to wholesale distributors or directly to fertilizer dealers, who prepare the requested formulations for farmers at their local plants. A number of granular fertilizers containing micronutrients also are made at NPK manufacturing plants. The premium fertilizers are made for a specific crop over a wide geographic area. For example, a premium fertilizer for corn in the Corn Belt may contain 1 to 2% Zn, 0.5% Mn, and possibly very low amounts of the other micronutrients. Premium fertilizers for soybeans contain higher levels of Mn than Zn, whereas those for cotton (*Gossypium hirsutum* L.) contain mainly B and Mn. Because numerous tons of a specific grade are needed for profitable production, the demand for such a grade must be estimated beforehand.

Foliar sprays containing one or more micronutrients also are sold by many fertilizer dealers. Advantages of foliar sprays are (i) application rates are much lower than for soil application, (ii) uniform distribution is easily obtained, (iii) response to the applied nutrient is almost immediate so that deficiencies can be corrected during the growing season, and (iv) suspected deficiencies can be more easily diagnosed with spray trials. Disadvantages are: nutrient demand is often high when plants are small and leaf surface is insufficient for foliar absorption, leaf burn may result if salt concentrations are high; it may be too late to correct the deficiency and still obtain maximum yields; there is little residual effect from foliar sprays, and extra application costs may be required. It is often convenient to combine a micronutrient with an insecticide or fungicide. Examples include B with an insecticide for cotton and Zn with an insecticide for rice (*Oryza sativa* L.).

Soil and plant analyses are widely used to diagnose nutrient deficiencies. Many fertilizer dealers take soil and plant tissue samples for their customers and have the samples analyzed by university or commercial laboratories. The dealer then discusses the resulting recommendations with the farmer, and they develop a fertilizer plan. Results of several surveys have shown that farmers depend heavily on recommendations made by their local fertilizer dealer. Therefore, much effort is made to educate dealers on the most effective means of providing fertilizers for the farmer.

B. Consumption Data

1. Liming Materials

Use of agricultural limestone in the USA was very low until 1935 when new federal soil conservation programs provided financial assistance to farmers for applying lime, and annual lime use increased > 10-fold in 15 yr (Barber, 1967). Since then, it has continued to increase steadily from about 21 million metric tons in 1960 to more than 32 million metric tons in 1980 (Table 12–4). This may be related to more educational programs, better application equipment, and a greater number of lime vendors. A large amount of liming materials now is applied by custom applicators with bulk handling equipment. Increased N use results in an increase in soil acidity, which requires more frequent lime applications. Effects of lime applications are discussed in chapter 4 of this book. Use of gypsum and Mg-containing materials for direct soil application has not increased appreciably since 1970 (Table 12–4). Use of dolomitic lime for soils requiring Mg applications may account for variable consumption of Mg materials since 1960.

2. Micronutrients

The Crop Reporting Board of the Statistical Reporting Service of the U.S. Department of Agriculture began publishing summaries of micronutrient use in 1968. Data in Table 12–5 were obtained from the known pro-

Table 12-4. Amounts of agricultural lime, gypsum, and Mg materials used for direct application to soils in the USA.

Year	Lime†	Gypsum‡	Mg materials‡
	1000 metric tons		
1960	21 076	1 300	--
1965	26 439	1 410	4.6
1970	34 423	1 150	1.8
1975	32 262	1 630	3.6
1980	32 648	1 860	5.0

† Levenson et al. (1981).
‡ Statistical Reporting Service (1961, 1966, 1971, 1976, 1981).

Table 12-5. Amounts of micronutrients sold in the USA (Statistical Reporting Service, 1968, 1972, 1975, 1978, 1981).

Year	Cu	Fe	Mn	Mo	Zn
	1000 metric tons†				
1967-1968	2.2	3.1	10.5	0.07	13.1
1971-1972	0.6	1.2	11.2	0.09	14.4
1974-1975	0.5	1.9	10.3	0.13	12.7
1977-1978	1.2	3.9	16.8	0.10	33.1
1980-1981	1.4	6.4	12.7	0.01	39.9

† Amounts expressed on elemental basis.

ducers of Cu, Fe, Mn, Mo, and Zn sources for use in both mixed and direct application in each region of the USA since 1967. Boron use data were not included since there were so few producers of fertilizer B. However, recent reports have estimated 1982 use at 3037 metric tons of B (Lyday, 1982).

Amounts of Cu and Fe used in the USA decreased from 1968 to 1972 and increased thereafter (Table 12–5), but the amount of Cu sold in 1980 still was lower than in 1967. Decreased use of Cu was related to lower consumption in Florida, where recommended rates were decreased. Toxicities to plants can result if Cu levels are too high, especially on acid, sandy soils. Amounts of Zn and Mn are the highest because these micronutrients are recommended for such large-area crops as corn, wheat (*Triticum aestivum* L.), soybeans, and rice. Rapid increases in Zn usage since 1975 are related to actual increases in consumption as well as to the inclusion of more producers reporting Zn sources, including some industrial by-products.

The figures in Table 12–5 are reported on an elemental basis; thus, the actual amount of materials used is much higher. This method of reporting was necessary since the nutrient content of the micronutrient sources varies widely. The tonnage of micronutrient materials sold in this country is only about 1% of the tonnage of NPK materials.

C. Regulations

Fertilizer regulations are governed by each state in the USA. In 1963, a general proposal was made that minimum percentages for the allowable levels of micronutrients be established and guaranteed on fertilizer labels. A uniform state fertilizer bill for general standardization was proposed by the AAPFCO for legislation by each state to simplify interstate shipping of fertilizers. The suggested minimum percentages acceptable for registration of each nutrient are shown in Table 12–6. Control officials emphasized that these were not to be considered as recommended levels; they are simply levels that can be determined accurately in control laboratories. One

Table 12–6. Suggested minimum percentages of secondary and micronutrients that can be guaranteed on a fertilizer label and also allowable deficiencies in fertilizer grades (Assoc. of Am. Plant Food Control Officials, 1968).

Nutrient	Minimum guarantee, %	Allowable deficiency†
Ca	1.00	0.2% + 5% of amount guaranteed
Mg	0.50	0.2% + 5% of amount guaranteed
S	1.00	0.2% + 5% of amount guaranteed
B	0.02	0.003% + 15% of amount guaranteed
Cu	0.05	0.005% + 10% of amount guaranteed
Fe	0.10	0.005% + 10% of amount guaranteed
Mn	0.05	0.005% + 10% of amount guaranteed
Mo	0.0005	0.0001% + 30% of amount guaranteed
Zn	0.05	0.005% + 10% of amount guaranteed

† Percentage of element below the guaranteed percentage on the fertilizer label that is allowed before the element is considered deficient in the fertilizer grade.

disadvantage of allowing such low levels of micronutrients to be guaranteed on fertilizer labels is that these amounts applied to soils at the usual NPK rates are too low to be of much benefit to crops if there are micronutrient deficiencies. Some states have set minimum guarantee levels more in line with their recommended levels of these elements.

This bill also states that such elements shall be deemed deficient in the fertilizer formulation if they are lower than the guarantee on the label by an amount exceeding the values shown in Table 12–6. For example, the allowable deficiency would be 0.205% Zn in a fertilizer grade containing 2.0% Zn. Guarantees and allowances vary by states. Increasing numbers of fertilizers are being offered with a guarantee for micronutrient content. Frequency of deficiencies appears to be in line with those found for the macronutrients.

Another suggestion was that the fertilizer laws require a warning or caution statement on the label for any product containing $>0.03\%$ B or 0.001% Mo in water-soluble form. This is to prevent toxicity of B to the plant through overuse and as a safeguard against Mo toxicity to animals resulting from consumption of forage with a high Mo concentration. To date, 38 states have adopted portions of this bill in their fertilizer laws.

III. METHODS AND RATES OF APPLICATION

The use of secondary and micronutrient fertilizers is governed by principles affecting both the need for and application of these nutrients. Needs are determined by crop requirements and soil conditions, which include amounts of a given nutrient present and reactions that affect their availability to plants. Efficient and effective use of any fertilizer depend on the source, rate, and method of application. Also, the philosophy of whether the crop or the soil is being fertilized must be considered. Soil fertilization with certain nutrients also may result in substantial residual effects, and these need to be evaluated in long-term studies. The focus in this section is on rate and method of application, but sources and time of application also are included as applicable. Major emphasis is on research results published since 1971, and residual effects are discussed briefly.

A. Calcium

The rather large amount of exchangeable Ca^{2+} in the soil that normally dominates the exchange complex satisfies the Ca requirements of most crops. Under acid conditions the amount of exchangeable Ca^{2+} in soil decreases, whereas exchangeable Al^{3+} increases, and Al toxicity may eventually cause reduction in plant growth. Correction of this acid condition is accomplished by liming with either calcitic or dolomitic lime, which maintains an abundant supply of Ca in the soil.

An international symposium on Ca nutrition of economic crops was held in 1977 (Shear, 1979). The proceedings provide excellent information on the nature of the problem and approaches as to correction, especially on horticultural crops.

Calcium deficiencies have been reported in some crops even though soil acidity may not have been extreme. Such deficiencies are usually localized in fruits, storage organs, or shoot tips—plant parts that are naturally low in Ca. Some examples are poor kernel formation in peanuts (*Arachis hypogaea* L.), bitter pit in apples (*Malus sylvestris* Mill.), blossom end rot in tomatoes (*Lycopersicon esculentum* Mill.), tipburn in lettuce (*Lactuca sativa* L.), and black heart of celery *(Apium graveolens)*. Many other examples existing in fruits and vegetables have been listed by Shear (1975).

The field crop most likely to show Ca deficiency due to the need for a liberal supply of Ca for proper development of its subterranean fruit is peanuts. Walker (1975) reviewed the Ca requirement of peanuts and reported lime to be somewhat effective, but the more soluble $CaSO_4$ (gypsum, or "landplaster") was superior. Gypsum is usually applied at 600 to 800 kg/ha just before fruiting in a 45-cm band over the row. Only a few definitive studies have been conducted to establish the optimum rate of gypsum for peanuts. Yield responses likely are mediated by environmental conditions and are not guaranteed, even under low soil Ca levels. Furthermore, it has not been shown that the Ca rate should be increased with decreasing levels of soil Ca. Essentially full yield response has been obtained with gypsum rates as low as 200 kg/ha (Cox, 1972), but the recommended rates usually are higher for several reasons. The yield and grade may continue to increase with increasing rate of applied Ca; Walker et al. (1979) found this to be true even above 600 kg of gypsum/ha. Thus, it is economical to apply the higher rates because of the increased value of the crop. Adequate Ca also ensures good quality seed that germinate well (Cox et al., 1976), and high Ca rates may help reduce certain forms of pod rot disease (Hallock & Garren, 1968; Walker et al., 1979).

The term *gypsum* is used in a general sense to mean a material with a calcium sulfate base. Little pure $CaSO_4 \cdot 2H_2O$ is sold; most commercial sources range from 50 to 90% $CaSO_4$. Textures range from very fine to granular, and some by-product precipitated materials vary in particle size due to lumps forming in the damp material. Because fine materials react more quickly than coarser materials (Daughtry & Cox, 1974; Keisling & Walker, 1978), fine materials may have an advantage for peanuts if applied at the typical early flowering stage. Coarser materials are often applied by bulk spreading very early in the growing season, which is the only period when such equipment may be used without damaging plants. All materials seem to be effective (Hallock & Allison, 1980), but the broadcast rate must be about twice the banded rate due to the extra material that covers the area between the rows.

Although generally considered less effective than gypsum, limestone was a satisfactory source of Ca for 'Florunner' peanuts (Adams & Hartzog, 1980). The rate needs to be higher than that for gypsum (Adams & Hartzog, 1979), and lime must be incorporated at a very shallow depth to affect primarily the fruiting zone in the soil. Sullivan et al. (1974), however, found this treatment completely ineffective with cv. NC5.

Foliar applications of Ca are not effective with peanuts since there is little xylem flow to carry the nutrient to the developing fruit. A similar problem exists with head lettuce. Misaghi et al. (1981) found foliar sprays of either calcium chloride ($CaCl_2$) or calcium nitrate [$Ca(NO_3)_2$] ineffective in reducing tipburn incidence and severity. The deficiency occurs on the inner and middle leaves, which are not exposed after head formation. Soil applications of $Ca(NO_3)_2$ were not effective, but $CaCl_2$ at 224 kg/ha or greater increased tissue Ca and reduced tipburn in one of six trials.

Among the horticultural crops showing Ca deficiency, apples have been studied most extensively. Recently, Lidster et al. (1978a, 1978b) studied effects of prebloom and summer foliar applications of Ca as well as a postharvest dip of the fruit in reducing breakdown development in stored 'Spartan' apples. They found prebloom $CaCl_2$ sprays to be ineffective, but summer sprays (0.6% $CaCl_2$) applied four times at 2-week intervals beginning in mid-July increased fruit Ca levels and decreased breakdown development significantly. A postharvest dip in 4% $CaCl_2$ was even more effective, but the most effective method was to combine summer sprays with the postharvest dip.

Application of lime and minimizing the K supply have been used to reduce Ca deficiency, termed *metsubure,* in taro [*Colocasia esculenta* (L.) Schott] corn (Tanabe & Ikeda, 1980, 1981). Both treatments affected the K/Ca ratio in the corms. There was no effect of N source as sometimes occurs with other crops, but varietal differences were noted. There appears to be great potential to develop varieties for tolerance to low Ca conditions for this crop as well as for many others.

B. Magnesium

Availability and plant uptake of Mg, like Ca, decreases under acid soil conditions due to a pH effect and also due to the direct loss of Mg from the root zone. Increasing the pH of a soil, even with a calcitic source of lime, often increases the uptake of Mg (Christenson et al., 1973). However, the primary method to ensure an adequate level of Mg in acid soils is by applying dolomitic lime. This has been the dominant source of Mg, even in the South Atlantic states, where Mg deficiencies should be most severe. The most common fertilizer sources of Mg are MgO and the sulfate form, usually applied as potassium sulfate–magnesium sulfate ($K_2SO_4 \cdot MgSO_4$).

The status of Mg research in the southeastern USA as well as factors affecting the availability of Mg to plants and animals were discussed in a symposium in 1972 (Jones et al., 1972). Grass tetany, a Mg deficiency occurring during lactation in ruminant animals, is the most important problem associated with Mg nutrition. Some low Mg levels in plants also were noted, but most of these were predictable by soil test data. Research results indicated only a small number of yield responses to Mg fertilization.

Soil test interpretations may have changed somewhat since that symposium was held in 1972. At that time, only one state based Mg availability on the cation saturation ratio in soil. This method is becoming more popular, but little substantiating field research has been published. In greenhouse work, McLean and Carbonell (1972) found that 6 to 10% Mg saturation was an adequate level for most Ohio soils, whereas $<5\%$ Mg saturation is considered deficient and $< 10\%$ is low in North Carolina.

General recommendations for Mg based on soil tests are to use dolomite if lime is needed or Mg at 20 to 40 kg/ha if it must be applied in NPK fertilizers. There have been no recent studies, however, to verify these recommendations. Gallaher et al. (1975) in a 3-yr Mg rate study with grain sorghum [*Sorghum bicolor* (L.) Moench], found a positive yield response the first year, a trend in the second, and no response in the 3rd yr. The maximum Mg rate (68 kg/ha as $MgSO_4$) was required the 1st yr and only 34 kg/ha the second. Klein et al. (1981) reported that 44 kg/ha of $MgSO_4$ was the optimum rate for potatoes (*Solanum tuberosum* L.). This rate had little effect on yield but improved quality by reducing discoloration, which was associated with lower amounts of phenols and more lipids and phospholipids in the tubers. Jones and Jones (1978) compared $MgSO_4$ with dolomite at 20 kg of Mg/ha for tomato production. Yields were increased by both sources in one of three crops, but the sulfate form was more effective than dolomite in increasing leaf Mg in all crops.

Application of $MgSO_4$ does not always increase tissue Mg concentration appreciably. In three Mg rate studies in Kansas, corn leaf Mg levels at silking ranged from 0.10 to 0.12% and showed little response up to 270 kg/ha of Mg as $MgSO_4$ (Whitney, 1973). Lack of a response seemed to be associated with very high Ca activities in these soils.

C. Boron

Sodium tetraborate is the main source of fertilizer B. Levels of hydration among the materials available result in B concentrations ranging from 11 to 20%. The most concentrated form is especially adapted for foliar sprays. Boron may be applied to either the soil or the foliage to correct a deficiency; Gupta and Cutcliffe (1978) found both methods to be effective in controlling brown heart in rutabaga (*Brassica napobrassica* L.). Several foliar applications at a low rate were more effective than a single application at a higher rate; this has been shown with other crops and is apparently due to the immobility of B in leaf tissue (Anderson & Ohki, 1972).

The rate of B required to correct a deficiency in annual crops is low. Murphy and Lancaster (1971) obtained maximum cotton yields with either 0.5 kg of B/ha applied to the foliage (fives times at 0.1 kg/ha each) or > 0.3 kg/ha applied to soil. Hill and Morrill (1974) found a similar rate sufficient to maintain the quality of peanuts; B could be applied any time up to 60 d after planting and still be effective for that crop. Rates > 1.0 kg/ha may create B toxicity. Blamey and Chapman (1979) noted yield reductions even before observing B toxicity symptoms in peanuts when high rates were applied to a sandy loam. Rates of 2 to 3 kg/ha are typically applied to alfalfa (*Medicago sativa* L.) when grown on finer-textured soils.

D. Chlorine

Although Cl is readily leached from surface soils in regions of high rainfall, its supply is usually replenished by applications of potassium chloride (KCl), the predominant K source. In regions of low rainfall, Cl may accumulate. Deficiencies of this element, therefore, are very rare, and it is much more common to find effects of excess Cl. Toxicities of Cl may reduce crop yields, especially under saline conditions. Very high levels of Cl are more likely to affect crop quality than yields under nonsaline conditions. An example of an affected crop is flue-cured tobacco (*Nicotiana tabacum* L.). A high tissue level of Cl is undesirable, because it is associated with a reduction in the burn rate and high moisture equilibrium values. Ishizaki and Akiya (1978) reconfirmed this relationship but also noted that the Cl limit for acceptable burning quality decreased with increasing K concentration.

E. Copper

Factors affecting use of Cu have been reviewed recently in a symposium on Cu in soils and plants (Loneragan et al., 1981). Rates, methods, and sources of Cu for application were discussed by Gartrell (1981) and Gilkes (1981). Even though soil and foliar applications are both effective, soil applications are more common, with recommended rates ranging from < 1 to > 20 kg of Cu/ha. The placement geometry of applied Cu is important; if the Cu source is concentrated and the granules large, few roots have access to the applied Cu. Unless applied Cu is mixed with soil, it may not be available to plant roots, especially if the surface soil remains dry. Efficiency is increased by applying Cu either in solution or as a fine material and either banding or thoroughly incorporating it.

Broadcast applications were more effective than band applications of Cu in increasing the yield and leaf Cu concentration of cucumbers (*Cucumis sativus* L.) (Navarro & Locascio, 1980). A rate of only 2.2 kg/ha increased yields from 11 to 21 metric tons/ha. Barnes and Cox (1973) applied Cu in solution over dormant wheat during the winter and found 3 to 6 kg of Cu/ha to be adequate, even on soils of high organic matter content. These are typical rates and have substantial residual effects.

Foliar application of Cu in solution will cause leaf burn on growing tissue, such as wheat that has broken dormancy. However, damage is minimal, and the treatment is effective if the Cu rate is kept low (0.25–0.5 kg/ha). For higher rates, hydrated lime is added to form a suspension. Basic suspensions or dusts also have been applied to seed with some success, but neither the foliar nor the seed treatment has an appreciable residual effect.

The most common Cu source is copper sulfate ($CuSO_4$). Several studies have compared the effectiveness of several Cu sources, but no differences have been found (Gilkes, 1981; Barnes & Cox, 1973).

F. Iron

Iron deficiencies occur on high pH soils. This condition also may be aggravated by application of alkaline irrigation water. High bicarbonate levels intensify Fe deficiencies, and soil applications of Fe as sulfate sources at rates similar to those for other micronutrients have not been effective. Westfall et al. (1971) found ferrous sulfate ($FeSO_4$) and ferric sulfate [$Fe_2(SO_4)_3$] equally effective for rice production, with rates of 50 to 100 kg of Fe/ha required for maximum yields. Patel et al. (1977) obtained no increase in rice yields from 40 kg of Fe/ha as $FeSO_4$, whereas other treatments were effective. Thomas and Mathers (1979) reported that farmyard manure was more effective than $FeSO_4$ in correcting Fe chlorosis of grain sorghum on a calcareous, Fe-deficient soil.

Other approaches used to increase Fe uptake and utilization are acidulation of the soil, complexing of the Fe by an organic ligand or polyphosphate, and application of gypsum. Acidulation corrected Fe deficiency in paddy rice (Patel et al., 1977), but this approach may not be practical in highly calcareous soils due to the enormous quantities of S required. A modification of this approach was tried by Wallace and Mueller (1978), who found that spot acidulation of as little as 0.4% of the calcium carbonate ($CaCO_3$) under soybean seedlings prevented Fe chlorosis.

Some of the response with soil acidulation may be due to increased S availability rather than increased Fe availability. Dungarwol et al. (1974) found that both acidulation of the soil with S and spraying the plants with a 0.1% H_2SO_4 were effective in correcting chlorosis and increasing yields of peanuts. These treatments increased leaf S and decreased leaf Fe. Leaf Fe was very high, however, and the authors postulated that the treatments increased the activity of Fe in the plant.

Organic Fe chelates or natural organic complexes are used to maintain Fe in an available state for a longer period. The most stable, and thus most successful, of these has been FeEDDHA. Low rates of this chelate have been effective in peanut production (Hartzook et al., 1971) but only if banded 40 to 50 d after emergence. Moraghan (1979) also was able to correct a Mn/Fe imbalance in flax (*Linum usitatissimum* L.) with an application of FeEDDHA. However, Westfall et al. (1971) found low rates of FeEDDHA less effective than high rates of $Fe_2(SO_4)_3$ for rice.

Complexing with polyphosphate fertilizers also increases the availability of fertilizer Fe. Mortvedt and Giordano (1971) showed this effect with both sulfate and chelated Fe sources, but FeEDDHA was more effective than $FeSO_4$ or $Fe_2(SO_4)_3$ at the same Fe rates.

The use of gypsum to increase Fe uptake under greenhouse conditions has been reported by Olsen and Watanabe (1979). Low plant Fe has been associated with high Mo. Gypsum applications decreased plant Mo considerably and increased plant Fe slightly in grain sorghum. This effect may account for part of the beneficial effect of gypsum noted on alkaline soils.

Because of difficulties in correcting Fe deficiencies by soil treatments, other methods generally are used. Foliar applications of Fe with about 3% $FeSO_4$ solution have been effective on some crops. Seeliger and Moss (1976) found a good response with peas (*Pisum sativum* L.) sprayed twice at 3-week intervals. In contrast, Patel et al. (1977) found that foliar-applied $FeSO_4$ was ineffective for rice.

Genetic variation also is used to overcome the problem of Fe deficiency. Hartzook et al. (1974) found that yields of Fe-efficient cultivars of peanuts were just as high as those of inefficient cultivars that were fertilized with FeEDDHA. This approach should be exploited for many other crops.

G. Manganese

Manganese deficiencies occur on high-pH soils and also on soils which are inherently low in Mn. The degree of Mn deficiency primarily affected by the above factors are seldom considered when the proper rate and method of Mn application are selected. For example, a case of extreme deficiency was found by Reuter et al. (1973), who studied means of correcting Mn deficiency on a soil containing > 8% $CaCO_3$. Manganese applied to the soil increased yields, with 6 kg/ha being as effective as larger rates. Yet deficiency symptoms reappeared late in the season, and a foliar application resulted in a further increase in yields. Apparently, the soil treatment was not effective for the entire period because of oxidation to unavailable forms, and a combination of methods was needed to achieve maximum yields. However, most Mn deficiencies are not as severe as the above and can be corrected by either soil or foliar applications. Even so, because the deficiency still affects the rate of Mn required for correction, recommended rates will not always be the same.

The rate of Mn required to correct a deficiency depends on the method of application, with increasing rates required for foliar, banded, and broadcast applications, respectively. Alley et al. (1978) obtained maximum soybean yields with 2.2 kg/ha foliar, 5.6 kg/ha banded, and 11 to 22 kg/ha broadcast applications on an Atlantic Coastal Plain soil, with the optimum broadcast rate depending on soil pH. Except for lower foliar rates, similar results were obtained by Randall et al. (1975), with maximum yields at 0.56 kg/ha banded and 17 kg/ha broadcast applications. In both of these studies the lowest foliar rate examined was effective; therefore, the actual minimum rate is unknown. Alley et al. (1978) split their foliar rate and found two applications to be more effective than one, but Boswell et al. (1981) found no differences when Mn was applied as broadcast preplant, banded at planting, sidedress, or broadcast preplant plus foliar spray. Soybean yield responses occurred in only 1 of 3 yr and maximized at 11 kg of Mn/ha. In an adjacent study on the same soil, Shuman et al. (1979) reported responses to broadcast preplant Mn in 2 of 3 yr. Soybean yields were highest at the 5.6-kg rate one year and at 11.2 kg the other. Differences among years appeared related to soil test Mn level.

Shuman et al. (1979) also reported that effectiveness of soil-applied Mn sources decreased in the order $MnSO_4$, manganese oxide (MnO), a Mn frit, and then a chelate that was ineffective. Voth and Christenson (1980) obtained similar results with sugar beets (*Beta vulgaris* L.), soybeans, and navy beans. Likewise, Knezek and Davis (1971) obtained good responses to $MnSO_4$ and MnO but emphasized that the particle size was important, especially for MnO, which must be finely ground to be effective.

Mixing or blending Mn sources with certain fertilizers also may increase availability of Mn. Mortvedt and Giordano (1975) found increased Mn uptake by oats when either $MnSO_4$ or MnO was mixed with phosphate fertilizers. In general, granular phosphates were more effective Mn carriers than fluid polyphosphates. Voth and Christenson (1980) noted that acid-forming fertilizers not only increased availability of fertilizer Mn but also increased that of native soil Mn.

H. Molybdenum

Plant availability of Mo, unlike that of other micronutrients, decreases with increasing soil acidity. Thus, liming to correct acidity improves Mo nutrition of plants. Another condition affecting Mo availability is the Fe status of soil. Soils that are high in Fe, especially noncrystalline Fe on clay surfaces, tend to have lower levels of available Mo.

The optimum Mo rate depends on the application method. Molybdenum may be applied to soil, sprayed on foliage, or put on seed prior to planting. The application rate is very low, especially for the latter two methods. Parker (1978) found that about 1 kg of Mo/ha was needed for soybeans if it was band applied before planting, but only 25 g/ha was required for either foliar or seed applications. Their foliar spray was split into two applications. Other work (Boswell & Anderson, 1969) has shown that foliar sprays of Mo should be applied prior to the bloom stage.

Legumes are especially sensitive to Mo deficiency, but the deficiency symptoms exhibited are those of N deficiency. Parker and Harris (1977) found that the primary function of Mo for soybeans was to correct N deficiency; there was enough soil Mo for nitrate reduction, but not enough for the rhizobial symbiont. They used both seed and foliar treatments at 17 to 34 g of Mo/ha.

Giddens and Perkins (1972) studied the response of alfalfa to Mo applications on two highly oxidized Piedmont soils, a Typic Hapludult and a Rhodic Paleudult. They obtained responses to 100 g of Mo/ha even though the soils were limed before planting each year during the study.

Obtaining uniform distribution of small rates applied to soil is difficult. Giddens and Perkins (1972) applied to Na_2MoO_4 as a solution to broadcast it uniformly. In Australia, molybdenum trioxide (MoO_3) has been incorporated into phosphate rock pellets to facilitate distribution. Kerridge et al. (1973) found this technique as effective as a direct application of MoO_3 at 100 g of Mo/ha for pasture legumes. Another method is to incorporate MoO_3 in a seed pellet. Johansen et al. (1977) found no difference among methods when Mo was applied in a seed pellet or to the soil surface as MoO_3 or Na_2MoO_4 for pasture legumes. They also found the Mo requirement among several sites varied from nil to 200 g/ha, and one application was effective for 3 yr. This study also confirms earlier observations that there are no differences among commonly used Mo sources.

Several interesting interactions concerning the Mo nutrition of burley tobacco have been studied. This crop usually is grown under acid soil conditions and may receive rather heavy applications of sulfate in the fertilizer. Sims et al. (1979) found that application of sulfates tended to reduce yields at low soil Mo levels but increased yields at high soil Mo levels. Sims and Atkinson (1976) found that applied Mo may be rendered partially unavailable in a soil of pH < 5.4, because there was a positive response to Mo fertilization when the soil was limed. These studies also indicated that soil application of about 0.5 kg of Mo/ha applied to the soil was adequate.

I. Zinc

Zinc has been applied to both soils and plants by several methods. Soil application methods include broadcast incorporated, surface broadcast, and banded. Plant methods include seed coatings, root dips, and foliar sprays. Several sources of Zn are sold—sulfate, oxide, chelates, natural organic complexes, and frits—and a number of options are available to correct Zn deficiencies.

For corn production, soil application of Zn has proved more effective than plant methods. Tanner and Grant (1973) found soil fertilization more effective than seed coatings, and Grewel and Singh (1975) found a similar trend when comparing soil and foliar applications. Hawkins et al. (1973) found that disked-in and plowed-down $ZnSO_4$ were equally effective, and the optimum Zn rate was 6.7 kg/ha. Maximum yields with broadcast $ZnSO_4$ in a regional study were obtained with 3 to 9 kg of Zn/ha (Cox & Wear, 1977). Schnappinger et al. (1972) broadcast both $ZnSO_4$ and ZnEDTA in solution prior to incorporation and found 14 kg of Zn/ha as the optimum rate with $ZnSO_4$; the highest rate of ZnEDTA (4.4 kg of Zn/ha) was not sufficient for maximum corn yields.

Considerable research also has been conducted on Zn deficiency of rice. Randhawa et al. (1978) estimated that about 2 million ha of rice soils in Asia are believed to be Zn deficient. For paddy rice, Yoshida et al. (1973) found that dipping transplant roots in a 1 to 2% ZnO suspension was effective in controlling Zn deficiency. Mikkelson and Brandon (1975) found ZnO somewhat less effective than $ZnSO_4$, but ZnO was still economical due to its lower cost. An effective rate for both sources was about 8 kg/ha. In contrast, Amer et al. (1980) reported that ZnO was superior to $ZnSO_4$ and that 5 kg of Zn/ha was sufficient for rice in Egypt. In the two reports just cited, applying Zn to the soil surface before flooding or to the surface water after transplanting were equally effective.

Zinc chelates are effective for rice production (Kang & Okoro, 1976) as well as navy bean and corn production. Boawn (1973) concluded that only about half as much Zn was required for corn and beans if ZnEDTA was the source rather than $ZnSO_4$. Results indicated that ZnEDTA might be applied effectively through sprinkler irrigation, but little response would be expected from $ZnSO_4$ applied in this manner. This hypothesis is supported by the work of Halvorson and Lindsay (1977), who suggested that chelation aids in the transport and movement of metal ions to plant roots.

The optimum method, source, and rate of Zn application varies for other crops as well. For wheat, Trehan and Sekhon (1976) obtained maximum response with about 3 kg of Zn/ha banded or 9 kg/ha broadcast as $ZnSO_4$, but Zn frits were less effective. Tehran and Grewel (1981) reported that dipping potato tuber pieces in a 2% ZnO suspension gave greater increases in yield and tuber size than several other methods. Foliar sprays have been used effectively with grapes (*Vitis vinifera* L.) (Christenson, 1980) and pecans [*Carya illinoensis* (Wangenh.) K. Koch] (Smith et al., 1979) but only when applied at certain times or in a particular manner. For pecans an aerial application was not effective even though the foliar rate was 6 kg of Zn/ha.

REFERENCES

Achorn, F. P. 1969. Use of micronutrients in fluid fertilizers. p. 5–8. *In* Proc. Natl. Fertilizer Solutions Assoc., Kansas City, MO. 22–23 July. National Fertilizer Solutions Association, Peoria, IL.

Achorn, F. P., and J. J. Mortvedt. 1977. Addition of secondary and micronutrients to granular fertilizers. p. 304–332. *In* Proc. Int. Conf. on Granular Fertilizers and Their Production, London. 13–15 November. British Sulfur Corporation, London.

Adams, F., and D. Hartzog. 1979. Effects of a lime slurry on soil pH, exchangeable calcium, and peanut yields. Peanut Sci. 6:73–76.

Adams, F., and D. Hartzog. 1980. The nature of yield responses of Florunner peanuts to lime. Peanut Sci. 7:120–123.

Allen, S. E., and G. L. Terman. 1966. Response of maize and sudangrass to zinc in granular micronutrients. p. 255–266. *In* G. V. Jacks (ed.) Soil chemistry and fertility. Trans J. Meet. Comm. 2,4. Int. Soc. Soil Sci. University Press, Aberdeen, Scotland.

Alley, M. M., C. I. Rich, G. W. Hawkins, and D. C. Martens. 1978. Correction of Mn deficiency in soybeans. Agron. J. 70:35–38.

Amer, F., A. I. Rezk, and H. M. Khalid. 1980. Fertilizer zinc efficiency in flooded calcareous soils. Soil Sci. Soc. Am. J. 44:1025–1030.

Anderson, O. E., and K. Ohki. 1972. Cotton growth response and B distribution from foliar application of B. Agron. J. 64:665–667.

Association of American Plant Food Control Officials. 1968. Uniform state fertilizer bill. AAPFCO Pub. no. 31. Available from H. P. Moore (Treas.), 8995 E. Main St., Reynoldsburg, OH 43068.

Barber, S. A. 1967. Liming materials and practices. *In* R. W. Pearson and F. Adams (ed.) Soil acidity and liming. Agronomy 12: 125–160.

Barnes, J. S., and F. R. Cox. 1973. Effects of copper sources on wheat and soybeans grown on organic soils. Agron. J. 65:705–708.

Berger, K. C., and P. F. Pratt. 1963. Advances in secondary and micronutrient fertilization p. 287–340. *In* McVickar et al. (ed.) Fertilizer technology and usage. Soil Science Society of America, Madison, WI.

Blamey, F. P. C., and J. Chapman. 1979. Boron toxicity in Spanish groundnuts. Agrochemophysica 11:57–59.

Boawn, L. C. 1973. Comparison of zinc sulfate and zinc EDTA as zinc fertilizer sources. Soil Sci. Soc. Am. Proc. 37:111–115.

Boswell, F. C., and O. E. Anderson. 1969. Effect of time of molybdenum application on soybean yield and on nitrogen, oil and molybdenum contents. Agron. J. 61:58–60.

Boswell, F. C., K. Ohki, M. B. Parker, L. M. Shuman, and D. O. Wilson. 1981. Methods and rates of applied manganese for soybeans. Agron. J. 73:909–912.

Brinkerhoff, F., B. G. Ellis, J. F. Davis, and J. Melton. 1966. Field and laboratory studies with zinc fertilization of pea beans and corn in 1965. Mich. Agric. Exp. Stn. Q. Bull. 48:344–356.

Christenson, D. R., R. P. White, and E. C. Doll. 1973. Yields and magnesium uptake by plants as affected by soil pH and calcium levels. Agron. J. 65:205–206.

Christensen, P. 1980. Timing of zinc foliar spray. I. Effects of application intervals preceding and during the bloom and fruit-set stages. II. Effects of day vs. night application. Am. J. Enol. Vitic. 31:53–59.

Cox, F. R. 1972. Effect of calcium sources and a fungicide on peanut production. J. Am. Peanut Res. Educ. Assoc. 4:122–129.

Cox, F. R., G. A. Sullivan, and C. K. Martin. 1976. Effect of calcium and irrigation treatments on peanut yield, grade, and seed quality. Peanut Sci. 3:81–85.

Cox, F. R., and J. I. Wear. 1977. Diagnosis and correction of zinc problems in corn and rice production. Southern Cooperative Ser. Bull. 222. North Carolina Agric. Exp. Stn., Raleigh.

Daughtry, J. A., and F. R. Cox. 1974. Effect of calcium source, rate, and time of application on soil calcium level and yield of peanuts (*Arachis hypogaea* L.). Peanut Sci. 1:68–72.

Davis, B. E. (ed.) 1980. Applied soil trace elements. John Wiley & Sons, Inc., New York.

Dungarwol, H. S., P. N. Mathur, and H. G. Singh. 1974. Effect of foliar sprays of sulfuric acid with and without elemental sulphur in the prevention of chlorosis in peanut (*Arachis hypogaea* L.). Commun. Soil Sci. Plant Anal. 5:331–339.

Ellis, B. G., J. F. Davis, and W. H. Judy. 1965. Effect of method of incorporation of zinc in fertilizer on zinc uptake and yield of pea beans. Soil Sci. Soc. Am. Proc. 29:635–636.

Gallaher, R. N., H. B. Harris, O. E. Anderson, and J. W. Dobson, Jr. 1975. Hybrid grain sorghum response to magnesium fertilization. Agron. J. 67:297–300.

Gartrell, J. W. 1981. Distribution and correction of copper deficiency in crops and pastures. p. 313–349. *In* J. F. Loneragan et al. (ed.) Copper in soils and plants. Academic Press, Inc., New York.

Giddens, J., and H. F. Perkins. 1972. Essentiality of molybdenum for alfalfa on highly oxidized Piedmont soils. Agron. J. 64:819–820.

Gilkes, R. J. 1981. Behaviour of Cu additives-fertilizers. p. 97–117. *In* J. F. Loneragan et al. (ed.) Copper in soils and plants. Academic Press, Inc., New York.

Giordano, P. M., and J. J. Mortvedt. 1966. Zinc availability for corn as related to source and concentration in macronutrient carriers. Soil Sci. Soc. Am. Proc. 30:649–653.

Grewel, J. S., and C. Singh. 1975. Evaluation of methods and sources of micronutrient application to wheat and maize. Indian J. Agric. Sci. 45:63–67.

Gupta, U. C., and J. A. Cutcliffe. 1978. Effects of methods of boron application on leaf tissue concentration of boron and control of brownheart in rutabaga. Can. J. Plant Sci. 58:63–68.

Hallock, D. L., and H. H. Allison. 1980. Effects of three Ca sources on peanuts. I. Productivity and seed quality. Peanut Sci. 7:19–25.

Hallock, D. L., and K. H. Garren. 1968. Pod breakdown, yield, and grade of Virginia type peanuts as effected by Ca, Mg, and K sulfates. Agron. J. 60:253–257.

Halvorson, A. D., and W. L. Lindsay. 1977. The critical Zn^{2+} concentration for corn and the nonabsorption of chelated zinc. Soil Sci. Soc. Am. J. 41:531–534.

Hargett, N. L. and J. T. Berry. 1985. Trends in fertilizer distribution in the U.S. p. 1. *In* Proc. Annu. Meeting Southern Feed, Fertilizer and Pesticide Control Officials. 16–19 June. Mobile, AL.

Hartzook, A., M. Fichman, and D. Karstadt. 1971. The treatment of iron deficiency in peanuts cultivated in basic and calcareous soils. Oleagineux 26:391–395.

Hartzook, A., D. Karstadt, M. Naveh, and S. Feldman. 1974. Differential iron absorption efficiency of peanut (*Arachis hypogaea* L.) cultivars grown on calcareous soils. Agron. J. 66:114–115.

Hawkins, G. W., D. C. Martens, and G. D. McCart. 1973. Response of corn to plowed-down and disked-zinc as zinc sulfate. Commun. Soil Sci. Plant Anal. 4:407–412.

Hignett, T. P. 1964. Supplying micronutrients in solid bulk-blended fertilizers. Commer. Fert. Plant Food Ind. 108(1):23–25.

Hill, W. E., and L. G. Morrill. 1974. Assessing boron needs for improving peanut yield and quality. Soil Sci. Soc. Am. Proc. 38:791–794.

Hurlbert, C. S., Jr. 1952. Dana's manual of mineralogy. 16th ed. John Wiley & Sons, Inc., New York.

Ishizaki, H., and T. Akiya. 1978. Effects of chlorine containing fertilizers on the growth and quality of tobacco. Damages caused by excessive application. JARQ 12:1–6.

Jackson, W. A., N. A. Heinly, and J. H. Caro. 1962. Solubility status of zinc carriers intermixed with N-P-K fertilizers. J. Agric. Food Chem. 10:361–364.

Johansen, C., P. C. Kerridge, P. E. Luck, B. G. Cook, K. F. Lowe, and H. Ostrowski. 1977. The residual effect of molybdenum fertilizer on growth of tropical pasture legumes in a subtropical environment. Aust. J. Exp. Agric. Anim. Husb. 17:961–968.

Jones, J. B., Jr. 1969. Effect of oil coating a fertilizer on yield and the uptake of manganese and zinc by corn and soybeans. Agron. J. 61:476–477.

Jones, J. B., Jr., M. C. Blount, and S. R. Wilkinson. 1972. Magnesium in the environment. Taylor County Printing Company, Reynolds, GA.

Jones, U. S., and T. L. Jones. 1978. Influence of polyethylene mulch and magnesium salts on tomatoes growing on loamy sand. Soil Sci. Soc. Am. J. 42:918–922.

Kang, B. T., and E. G. Okoro. 1976. Response of flooded rice grown on a vertisol from northern Nigeria to zinc sources and methods of application. Plant Soil. 44:15–25.

Keisling, T. C., and M. E. Walker. 1978. Calcium-supplying characteristics of two gypsum materials on southeastern Coastal Plain soils. Soil Sci. Soc. Am. J. 42:513–517.

Kerridge, P. C., B. G. Cook, and M. I. Everett. 1973. Application of molybdenum trioxide in the seed pellet for subtropical pasture legumes. Trop. Grassl. 7:229–232.

Klein, L. B., S. Chandra, and N. I. Mordy. 1981. Effect of magnesium fertilization on the quality of potatoes: yield, discoloration, phenols, and lipids. J. Agric. Food Chem. 29:384–387.

Knezek, B. D., and J. F. Davis. 1971. Relative effectiveness of manganese sulfate and manganous oxide applied on organic soil. Commun. Soil Sci. Plant Anal. 2:17–21.

Lehr, J. R., E. H. Brown, A. W. Frazier, J. P. Smith, and R. D. Thrasher. 1967. Crystallographic properties of fertilizer compounds. TVA Bull. no. 6. Tennessee Valley Authority, Muscle Shoals, AL.

Levenson, N., O. Kamaturi, and A. Leder. 1981. Limestone and Lime. p. 746.1001 I. Chemical Economics Handbook. Stanford Research Institute. Menlo Park, CA.

Lidster, P. D., S. W. Porritt, and G. W. Eaton. 1978a. Effects of prebloom or summer applications of boron, strontium, and calcium on breakdown development in stored Spartan apples. Can J. Plant Sci. 58:283–285.

Lidster, P. D., S. W. Porritt, and G. W. Eaton. 1978b. The effect of fruit size and method of calcium chloride application on fruit calcium content and storage breakdown in Spartan apple. Can. J. Plant Sci. 58:357–362.

Loneragan, J. F., A. D. Robson, and R. D. Graham. 1981. Copper in soils and plants. Academic Press, Inc., New York.

Lyday, P. A. 1982. Boron. 145–150 p. *In* Minerals yearbook. Bureau of Mines, U.S. Department of Interior, U.S. Government Printing Office, Washington, DC.

McCrary, A. L., and W. L. Howard. 1979. Chelating agents. p. 339–368. *In* Kirk-Othmer Encyclopedia of chemical technology, Vol. 5. 3rd. ed. John Wiley & Sons, Inc., New York.

McLean, E. O., and M. D. Carbonell. 1972. Calcium, magnesium, and potassium saturation ratios in two soils and their effects upon yields and nutrient contents of German millet and alfalfa. Soc. Sci. Soc. Am. Proc. 36:927–930.

Mikkelson, D. S., and D. M. Brandon. 1975. Zinc deficiency in California rice. California Agric. 29:8–9.

Misaghi, I. J., C. A. Matyac, and R. G. Grogran. 1981. Soil and foliar applications of calcium chloride and calcium nitrate to control tipburn of head lettuce. Plant Dis. 65:821–822.

Moraghan, J. T. 1979. Manganese toxicity in flax growing on certain calcareous soils low in available iron. Soil Sci. Soc. Am. J. 43:1177–1180.

Mortvedt, J. J. 1968. Crop response to applied zinc in ammoniated fertilizers. J. Agric. Food Chem. 16:241–45.

Mortvedt, J. J. 1977. Micronutrients with granular fertilizers. Custom Applicator 7(3):20–5.

Mortvedt, J. J. 1979. Suspensions as a micronutrient carrier. Custom Applicator 9(3):44–52.

Mortvedt, J. J., and H. G. Cunningham. 1971. Production marketing and use of other secondary and micronutrient fertilizers p. 413–454. *In* R. A. Olson et al. (ed.) Fertilizer technology and use. 2nd ed. Soil Science Society of America, Madison, WI.

Mortvedt, J. J., and P. M. Giordano. 1969. Availability to corn of zinc applied with various macronutrient fertilizers. Soil Sci. 108:180–187.

Mortvedt, J. J., and P. M. Giordano. 1971. Response of grain sorghum to iron sources applied alone or with fertilizers. Agron. J. 63:758–761.

Mortvedt, J. J., and P. M. Giordano. 1975. Crop response to manganese sources applied with ortho- and polyphosphate fertilizers. Soil Sci. Soc. Am. Proc. 39:782–787.

Mortvedt, J. J., P. M. Giordano, and W. L. Lindsay (ed.) 1972. Micronutrients in agriculture. Soil Science Society of America, Madison, WI.

Mortvedt, J. J., and G. Osborn. 1965. Boron concentration adjacent to fertilizer granules in soil and its effect on root growth. Soil Sci. Soc. Am. Proc. 29:187–191.

Murphy, B. C., and J. D. Lancaster. 1971. Response of cotton to boron. Agron. J. 63:539–540.

Navaroo, A. A., and S. J. Locascio. 1980. Copper nutrition of cucumber (*Cucumis sativus* L.) as influenced by fertilizer placement, phosphorus rate, and phosphorus source. Proc. Soil Crop Sci. Soc. Fla. 39:16–19.

Olsen, S. R., and F. S. Watanabe. 1979. Interaction of added gypsum in alkaline soils with uptake of iron, molybdenum, manganese, and zinc by sorghum. Soil Sci. Am. J. 43:125–130.

Parker, M. B. 1978. Molybdenum studies on soybeans. Georgia Agric. Exp. Stn. Res. Bull. 215.

Parker, M. B., and H. B. Harris. 1977. Yield and leaf nitrogen of nodulating and nonnodulating soybeans as affected by nitrogen and molybdenum. Agron. J. 69:551–554.

Patel, G. J., B. V. Ramakrishnayya, and B. K. Patel. 1977. Effect of soil and foliar application of ferrous sulfate and of acidulation of soil on iron chlorosis of paddy seedlings in Goradu soil nurseries in India. Plant Soil. 46:209–219.

Randall, G. W., E. E. Schulte, and R. B. Corey. 1975. Effect of soil and foliar-applied manganese on the micronutrient content and yield of soybeans. Agron. J. 67:502–507.

Randhawa, N. S., M. K. Sinha, and P. N. Takkar. 1978. Micronutrients. p. 581–603. *In* Soils and rice. International Rice Research Institute, Los Baños, Philippines.

Reuter, D. J., T. G. Heard, and A. M. Alston. 1973. Correction of manganese deficiency in barley crops on calcareous soils, Part 1. Manganous sulfate applied at sowing and as foliar sprays. Aust. J. Exp. Anim. Husb. 13:434–439.

Sanchelli, V. 1969. Trace elements in agriculture. Van Nostrand Reinhold Company, Inc., New York.

Schnappinger, M. G., Jr., D. C. Martens, G. E. Hawkins, D. F. Amos, and G. D. McCart. 1972. Response of corn to residual and applied zinc as $ZnSO_4$ and ZnEDTA in field investigations. Agron. J. 64:64–66.

Seeliger, M. T., and D. E. Moss. 1976. Correction of iron deficiency in peas by foliar sprays. Aust. J. Exp. Agric. Anim. Husb. 16:758–760.

Shear, C. B. 1975. Calcium-related disorders of fruits and vegetables. Hortic. Sci. 10:361–365.

Shear, C. B. 1979. (ed.) International symposium on calcium nutrition of economic crops. Commun. Soil Sci. Plant Anal. 10:1–501.

Shuman, L. M., F. C. Boswell, K. Ohki, M. B. Parker, and D. O. Wilson. 1979. Soybean yield, leaf manganese, and soil manganese as affected by sources and rates of manganese and soil pH. Agron. J. 71:989–991.

Silverberg, J., R. D. Young, and G. Hoffmeister, Jr. 1972. Preparation of fertilizers containing micronutrients. p. 431–458. *In* J. J. Mortvedt et al. (ed.) Micronutrients in agriculture. Soil Sci. Soc. Am., Madison, WI.

Sims, J. L., and W. O. Atkinson. 1976. Lime, molybdenum, and nitrogen source effects on yield and selected chemical components of burley tobacco. Tob. Sci. 20:174–177.

Sims, J. L., J. E. Leggett, and U. R. Pal. 1979. Molybdenum and sulfur interaction effects on growth, yield, and selected chemical constituents of burley tobacco. Agron. J. 71:75–78.

Smith, M. W., J. B. Storey, and P. N. Westfall. 1979. The influence of two methods of foliar application of zinc and adjuvant solutions on leaf zinc concentration in pecan. Hortic. Sci. 14:718–719.

Statistical Reporting Service. 1961. Consumption of commercial fertilizers in the U.S. Annual rep. Statistical Reporting Service, U.S. Department of Agriculture, U.S. Government Printing Office, Washington, DC.

Statistical Reporting Service. 1966. Consumption of commercial fertilizers in the U.S. Annual rep. Statistical Reporting Service, U.S. Department of Agriculture, U.S. Government Printing Office, Washington, DC.

Statistical Reporting Service. 1968. Consumption of commercial fertilizers in the U.S. Annual rep. Statistical Reporting Service, U.S. Department of Agriculture, U.S. Government Printing Office, Washington, DC.

Statistical Reporting Service. 1971. Consumption of commercial fertilizers in the U.S. Annual rep. Statistical Reporting Service, U.S. Department of Agriculture, U.S. Government Printing Office, Washington, DC.

Statistical Reporting Service. 1972. Consumption of commercial fertilizers in the U.S. Annual rep. Statistical Reporting Service, U.S. Department of Agriculture, U.S. Government Printing Office, Washington, DC.

Statistical Reporting Service. 1975. Consumption of commercial fertilizers in the U.S. Annual rep. Statistical Reporting Service, U.S. Department of Agriculture, U.S. Government Printing Office, Washington, DC.

Statistical Reporting Service. 1976. Consumption of commercial fertilizers in the U.S. Annual rep. Statistical Reporting Service, U.S. Department of Agriculture, U.S. Government Printing Office, Washington, DC.

Statistical Reporting Service. 1978. Consumption of commercial fertilizers in the U.S. Annual rep. Statistical Reporting Service, U.S. Department of Agriculture, U.S. Government Printing Office, Washington, DC.

Statistical Reporting Service. 1981. Consumption of commercial fertilizers in the U.S. Annual rep. Statistical Reporting Service, U.S. Department of Agriculture, U.S. Government Printing Office, Washington, DC.

Strauss, E., A. L. Waddams, and O. Kamaturi. 1983. Chelating agents. p. 512.5020L-Q. Chemical economics handbook. Stanford Research Institute, Menlo Park, CA.

Sullivan, G. A., G. L. Jones, and R. P. Moore. 1974. Effects of dolomitic limestone, gypsum, and potassium on yield and seed quality of peanuts. Peanut Sci. 1:73–77.

Tanabe, I., and K. Ikeda. 1980. On the "Metsubure" symptoms of taro corms: II. The effects of potassium application on the "Metsubure" corm formation of taro. Soil Sci. Plant Nutr. 26:461–468.

Tanabe, I., and K. Ikeda. 1981. The "Metsubure" symptoms of taro corms: IV. Suggestion for a fertilization technique to minimize the occurrence of "Metsubure" corms in taro. Soil Sci. Plant Nutr. 27:9–17.

Tanner, P. D., and P. M. Grant. 1973. Effectiveness of zincated fertilizers for young maize as influenced by fertilizer pH and method of applying zinc. Rhod. J. Agric. Res. 11:69–75.

Tehran, S. P., and J. S. Grewal. 1981. Comparative efficiency of methods of application of zinc for potato. Indian J. Agric. Sci. 51:240–243.

Thomas, J. D., and A. C. Mathers. 1979. Manure and iron effects on sorghum growth on iron-deficient soil. Agron. J. 71:792–794.

Tisdale, S. L., and H. G. Cunningham. 1963. Advances in manufacturing of secondary and micronutrient fertilizers. p. 269–286. *In* MacVickar et al. (ed.) Fertilizer technology and usage. Soil Science Society of America, Madison, WI.

Trehan, S. P., and G. S. Sekhon. 1976. An efficient method, rate, and source of zinc application to maize and wheat. J. Res. Punjab Agric. Univ. 14:411–18.

Voth, R. D., and D. R. Christenson. 1980. Effect of fertilizer reaction and placement on availability of manganese. Agron. J. 72:769–773.

Walker, M. E. 1975. Calcium requirements of peanuts. Commun. Soil Sci. Plant Anal. 6:299–313.

Walker, M. E., R. A. Flowers, R. J. Henning, T. C. Keisling, and B. G. Mullinix. 1979. Response of Early Bunch peanuts to calcium and potassium fertilization. Peanut Sci. 6:119–123.

Wallace, A., and R. T. Mueller. 1978. Complete neutralization of a portion of calcareous soil as a means of preventing iron chlorosis. Agron. J. 70:888–890.

Westfall, D. G., W. B. Anderson, and R. J. Hodges. 1971. Iron and zinc response of chlorotic rice grown on calcareous soils. Agron. J. 63:702–705.

Whitney, D. A. 1973. Magnesium application to corn on irrigated sandy soils in Kansas. Agron. Abstr. American Society of Agronomy, Madison, WI, p. 105.

Yoshida, S., J. S. Ahn, and D. A. Forro. 1973. Occurrence, diagnosis, and correction of zinc deficiency of lowland rice. Soil Sci. Plant Nutr. (Tokyo) 19:83–93.

13

Production, Marketing, and Use of Solid, Solution, and Suspension Fertilizers

Frank P. Achorn and Hubert L. Balay

Tennessee Valley Authority Muscle Shoals, Alabama

In 1980, 21 million metric tons of fertilizer mixtures was produced in the USA (Hargett & Berry, 1980). This represents about 45% of the total fertilizer sold in this country. These mixtures are produced and marketed as bulk blends (dry mixtures of granular fertilizers), granular homogeneous mixtures, and fluid mixtures. Statistical data show that about 42% of all the mixtures are bulk blends, 37% are granular homogeneous mixtures, and 21% are fluid mixtures.

Since 1970, bulk blends have become popular but the rate of growth in their use has leveled off since 1980. Although there are now fewer regional granulation plants than in the past for the production of granular homogeneous NPK mixtures, the tonnage produced is the same as that produced in 1970.

The boom in use of fluid fertilizers appears to be primarily in the use of suspension mixtures. Suspensions are popular with both farmer and dealer because they provide an excellent way to uniformly apply fertilizer-herbicide mixtures and small amounts of micronutrients. Also, it is possible to accurately place fluid fertilizer in the row. Production of high-analysis grades of suspensions is also possible.

Delivered cost to the farmer on a nutrient basis appears to be about the same for all fertilizer except for solutions; these have special properties, are easier to handle and apply then other kinds of fertilizer, and usually bring a premium price.

The general tendency in marketing fertilizers is to supply the farmer a wide variety of ratios of primary nutrients and a selection of micronutrients. In many instances fertilizer is also combined with a pesticide and applied as a fertilizer-pesticide mixture.

Solid and fluid mixtures are sold in bulk, and most dealers are equipped to produce, store, transport, and rapidly apply bulk materials. This has streamlined fertilizer marketing and distribution and has decreased the amount of labor required to deliver and apply fertilizer mixtures. It also has allowed production of tailor-made mixtures to satisfy specific requirements of the farmer. All of these improvements have contributed to increased production of food and fiber in the USA.

I. BULK-BLENDING PLANTS

A survey by TVA and the Association of American Plant Food Control Officials conducted in 1980 shows that there are about 5200 bulk-blending plants in the USA (Hargett & Pay, 1980). In most bulk-blending plants, raw materials are removed from railway cars and conveyed to storage. A blend is prepared by removing these raw materials from storage, mixing them, and conveying the mixture to a bulk applicator. Sometimes the plant is equipped with a bagging machine to bag some of the product.

The mixer most frequently used in bulk blending plants is the concrete-type rotary mixer like that shown in Fig 13–1. It is similar to a conventional concrete mixer, and in many instances a concrete mixer of the type used to transport concrete from the mixing plant to the use area is used. This type of mixer is popular because of its low cost and the resulting low total investment cost of the plant. The mixer is usually loaded and discharged with belt conveyors similar to those shown in Fig. 13–2, although augers are often used. Storage buildings are sometimes made of prestressed concrete slabs. Holes through which the bins are filled are cut in the roof above each bin. The bins have concrete floors and walls. Buildings often have wooden walls, but the floors are always concrete. Materials are usually removed from storage by front-end loaders and weighed in hopper scales mounted above either the boot of an elevator or a conveyor as shown in Fig. 13–2. With the arrangement shown in Fig. 13–2, only one operator is needed to remove materials from storage, weigh and mix them, and discharge the mixer.

Other plants have elevated holding bins that are filled by use of a front-end loader. A sketch of a plant of this type is shown in Fig. 13–3; this plant is completely automated. Office personnel use an inexpensive computer to calculate least cost formulations using materials available in the plant. Nutrient requirements needed per acre (0.405 ha) along with the

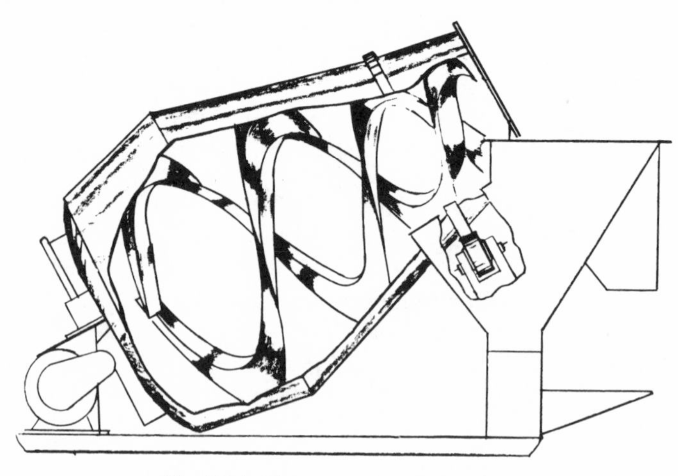

Fig. 13–1. Concrete-type rotary mixer.

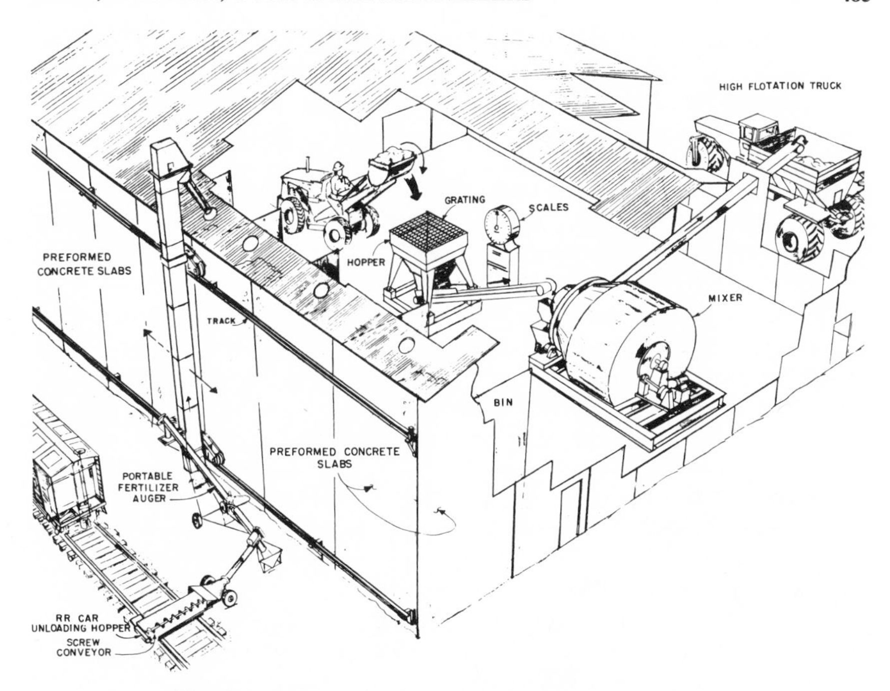

Fig. 13–2. Bulk blend plant with concrete-type rotary mixer.

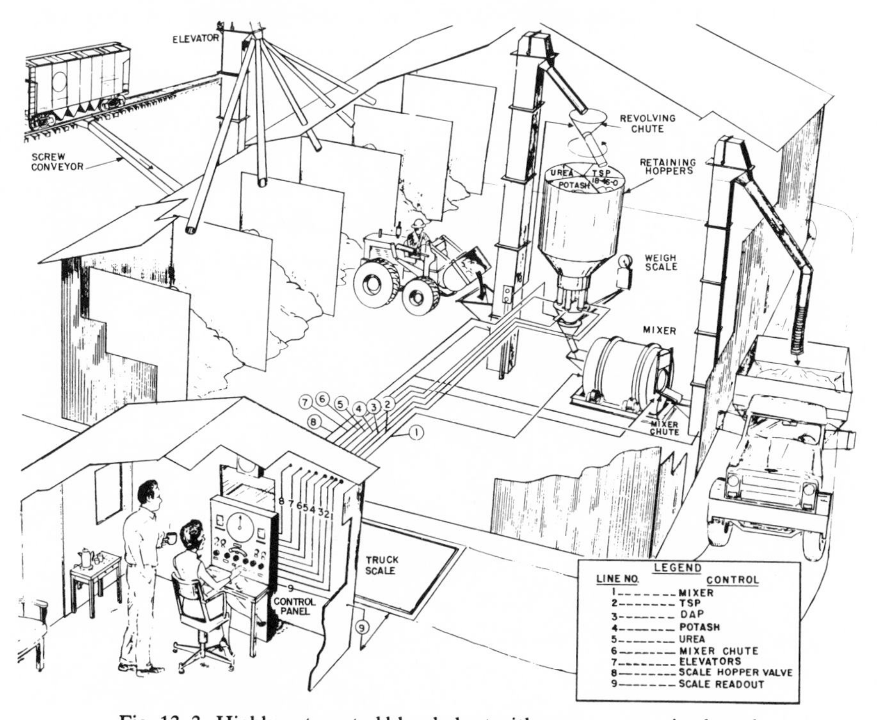

Fig. 13–3. Highly automated blend plant with remote operating board.

price of materials available to the blender are entered into the computer. The computer then calculates the least cost formulation and prints out the required applicator setting and speed to correctly apply the mixture. Cards punched with formulations are used to operate the automatic mix unit. Holding hoppers above the weigh hopper are kept filled using a front-end loader. Level indicators show the amount of material in the hoppers. Usually the scale dial is located in the office. The punch card generates signals to pneumatic valves that control the amount of each material used in the blend. After a batch is assembled in the weigh hopper, it is automatically discharged into the mixer. Only the operator who maintains the level of the material in the retaining hopper must be in the mixing area. Company personnel control the amount and type of materials used in the mixer from the office. Office personnel not only control the mixer but also weigh the applicator before and after filling and print the bill of lading. The scale output is delivered automatically to the computer so that the amount of materials removed from storage can be recorded. Using this data, the computer controls inventories and signals when materials need to be replenished.

The mixer shown in Fig. 13–4 is a horizontal axis type. It is different from the concrete-type mixer in that it is filled on one end and discharged from the other. Also its axis of rotation is horizontal to the floor, whereas the concrete mixer has an inclined axis. Both types of mixers have capacities that vary from 4.5 to 9.1 metric tons. Usually a mixer of 5–metric ton capacity can produce blends at 45 metric tons/h.

When blending first became popular, mixing efficiency tests with the concrete-type mixer were made in which a 15–7–12 (15–15–15) grade was formulated and samples collected as the mixer discharged. Analyses of the samples indicated that the grade varied from 17–6–14 (17–13–17) to 14–10–

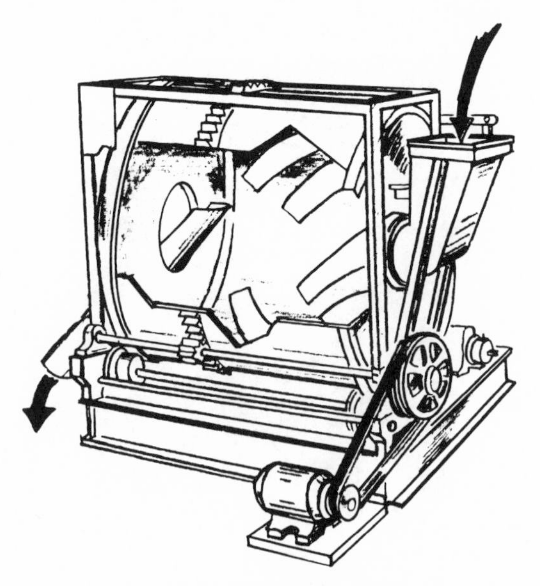

Fig. 13–4. Horizontal axis–type mixer.

10 (14–23–12). Companies that fabricate concrete-type mixers for fertilizer mixing have improved the flight design and modified the shape of the mixer since the 1970s. These changes have improved mixing efficiency. In a test with a modified concrete-type mixer a 15–4–8 (15–10–10) grade was blended. The average grade of all samples from the mixer was 14.3–4.6–8.3 (14.3–10.5–10.1), and the standard deviation ranged from 0.16 for N to 1.35 for K. This deviation would meet the requirements of most state control agencies.

Mixing efficiency tests were also made with a horizontal axis rotary mixer (Fig. 13–4). The tests showed that when a 15–7–12 (15–15–15) grade was mixed, the average grade of several samples was 15.1–6.4–12.5 (15.1–14.7–15.1), indicating satisfactory mixing efficiency.

Many other types of mixers, such as augers, blending towers, and mixing cones, are used for blending. All of these mixers can be made to function satisfactorily. However, rotary types such as those shown in Fig. 13–1 and 13–4 are most popular. Also it is easier to impregnate the granules with pesticides or to coat them with micronutrients in this type of mixer. Figure 13–5 shows a rotary mixer adapted to either coat micronutrients or impregnate pesticides onto the surface of the granules in a bulk blend.

The six raw materials most frequently used by the bulk blending industry are shown in Table 13–1.

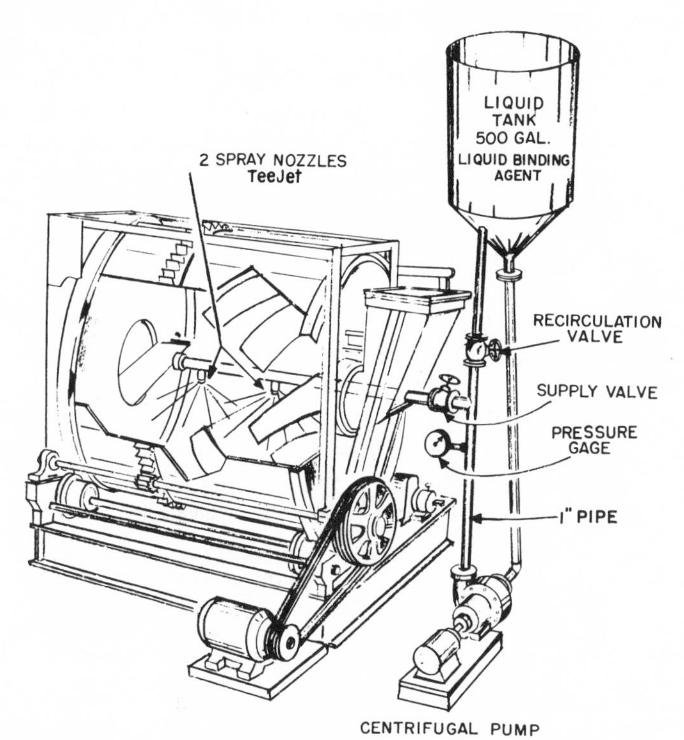

Fig. 13–5. Rotary mixer adapted for coating micronutrients or impregnating pesticides onto granular blends.

Table 13-1. Raw materials most frequently used in bulk blending.

Material	Percent of total material consumed in blending
DAP, 18-20-0 (18-6-0)	25.9
TSP, 0-20-0 (0-46-0)	17.6
Ammonium nitrate	12.0
Urea	6.5
Potash	33.3
Granular ammonium sulfate	4.7

Table 13-2. Particle size analysis of typical commercial DAP.

Screen size (Tyler screens)	Cumulative wt, %
6	<1
8	36
10	86
14	99
20	100

Diammonium phosphate (DAP) typifies the type of product desired for bulk blending. Most commercial DAP products are granular, freeflowing, and of proper size for easy handling and uniform application. A typical screen analysis of commercial DAP is shown in Table 13–2.

Particle size distribution is the main consideration in production of nonsegregating blends (Hoffmeister, 1975). TVA field experience shows that if the particle size fractions of the materials used in blends are within 10% of the average screen analysis of all the materials used in the blend, the material will remain mixed when handled and applied properly. The average screen analyses of the conventional materials used in blending are shown in Table 13–3. These analyses are from several producers, and the size specifications vary among producers. The screen analysis of coarse potassium chloride (KCl) shows the widest variation and the screen analysis of DAP the least. Screen analyses, plotted in Fig. 13–6, show that most materials, except for coarse KCl and granular urea (large), have similar particle sizes. A few of the materials do not meet the 10% variance in particle size specification.

The screen analyses in Fig. 13–6 indicate that there is little likelihood that a blend of DAP, granular triple superphosphate (TSP), and KCl

Table 13-3. Particle size range of conventional blended material.

	Cumulative screen analysis, % retained on Tyler screens				
Material	6	8	10	14	20
DAP, 18-20-0 (18-46-0)	<1	36	86	99	100
Granular potash, 0-0-50 (0-0-60)	5	37	78	95	98
Coarse potash, 0-0-50 (0-0-60)	--	6	31	73	94
Granular TSP, 0-20-0 (0-46-0)	1	21	80	97	100
Prilled ammonium nitrate (33.5-0-0)	<1	10	74	97	100
Urea prilled (46-0-0)	<1	6	68	97	100

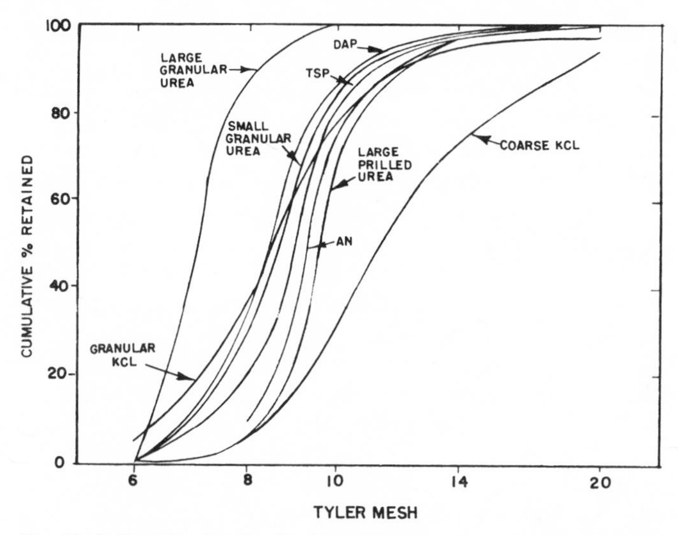

Fig. 13–6. Particle size distribution of fertilizer materials for bulk blends.

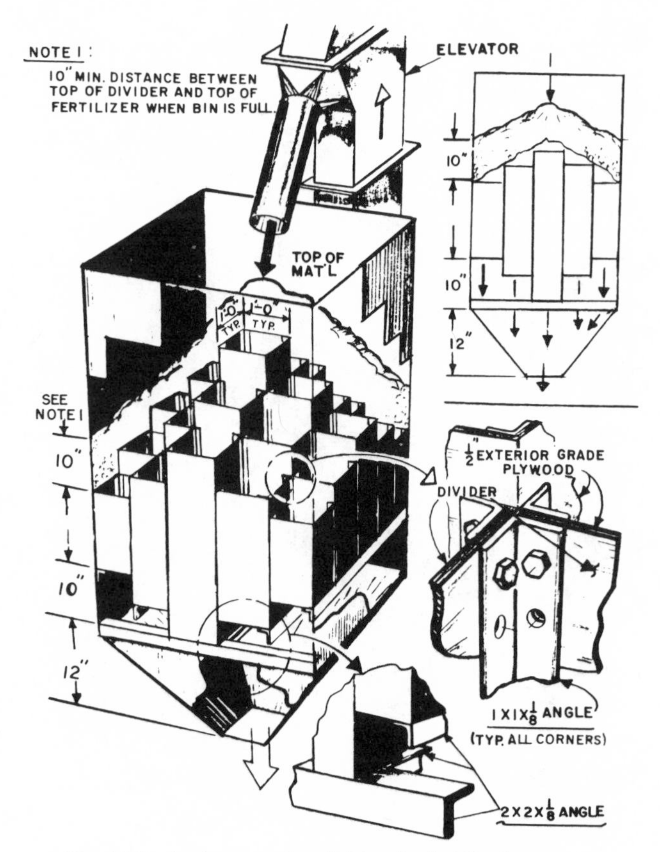

Fig. 13–7. Antisegregation dividers for holding hopper.

would segregate. These data show that the maximum deviation is 10% or less when these materials are used.

Segregation usually occurs by allowing materials to cone when they are discharged into a pile, bin, or applicator. Usually the fine materials collect in the center of the pile and the larger materials at the outer edges of the pile. It is not always practical to select materials that are properly sized; therefore, segregation due to coning should be minimized by installing devices such as holding hopper dividers like those shown in Fig. 13–7. A flexible spout like that in Fig. 13–8 should also be used. With these devices the coning effect is minimized, and less segregation occurs in the blend.

Many materials have recently become available for bulk blending. Probably the two most popular new materials are monoammonium phosphate (MAP) of 11–23–0 (11–52–0 or 11–53–0) grade and granular urea. All of the granular MAP products have excellent particle size distributions that are compatible with most of the other materials used in blending. Some of the granular ureas now available also meet this criteria. However, other commercial granular ureas are too large. The TVA has recently developed a "falling curtain urea granulation process" by which granular urea of a suitable size distribution for blending can be produced. The proc-

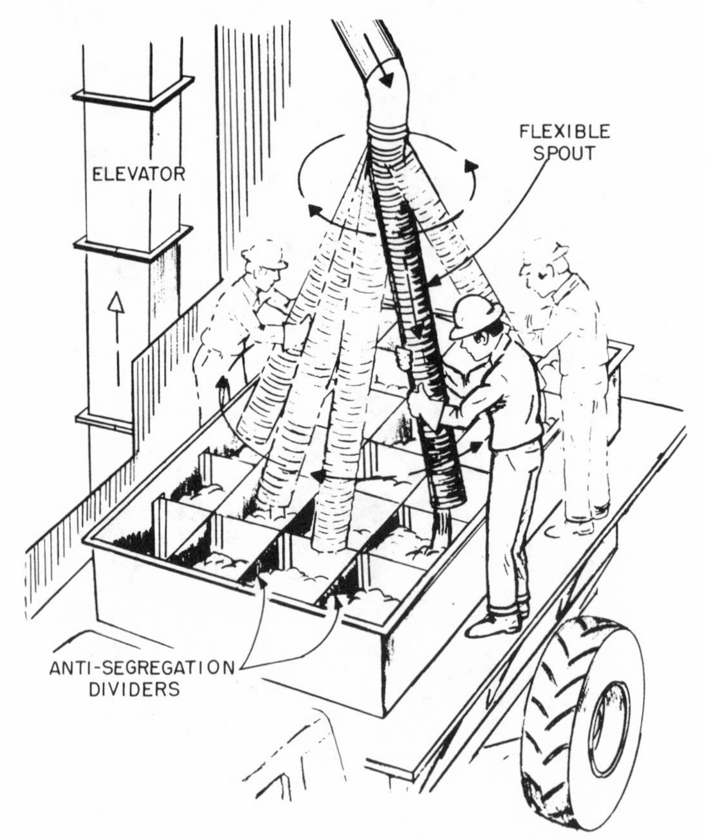

Fig. 13–8. Flexible spout for loading granular fertilizer in applicator.

ess can also be used to produce large-sized urea; this product is often used for forest fertilization by airplane.

It is expected that bulk blending will continue to be one of the most popular ways to mix plant nutrients in the USA. Because of the simplicity of operation and the ready availability of suitable materials, bulk blending will likely become popular in other countries. Bulk blending is particularly well suited for developing countries because of the low plant investment cost.

II. GRANULATION PLANTS

Most granulation plants in the USA use a TVA rotary ammoniator-granulator like that shown in Fig. 13–9.

Granulation plants may be classified as either primary plants or regional ammoniation-granulation plants. Primary plants usually produce granular materials for use in bulk blending and for direct application. Typical products of primary plants are DAP and MAP. These plants are large, with capacities as great as 1800 metric tons/d. Regional ammoniation-granulation plants are smaller plants that usually produce only homogeneous NPK mixtures; typical capacity is 32 metric tons/h, with an annual capacity of about 0.1 million metric tons. Figure 13–10 shows a flow diagram of a conventional ammoniation-granulation plant of this type. Usually fine-sized or powdered materials such as run-of-pile superphosphate (either triple [TSP], normal superphosphate [NSP], or both) fine crystalline by-product ammonium sulfate [$(NH_4)_2SO_4$], and standard KCl are granu-

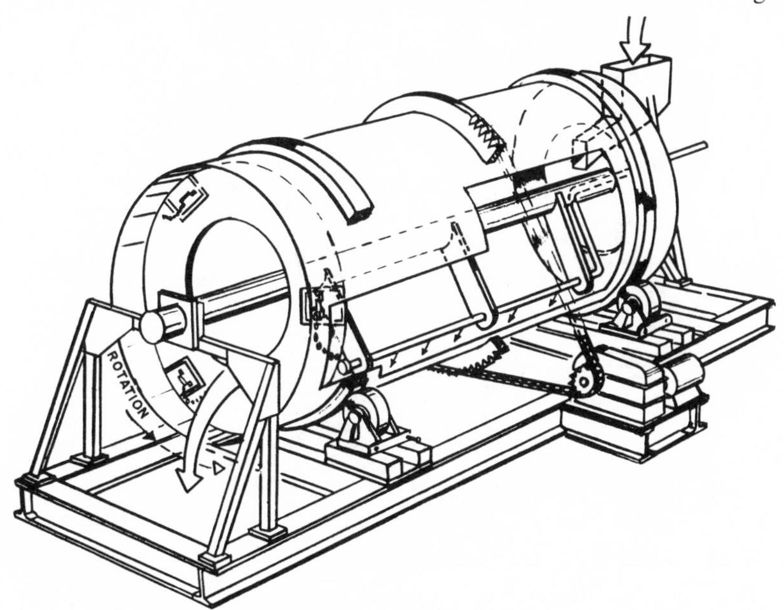

Fig. 13–9. TVA rotary ammoniator-granulator.

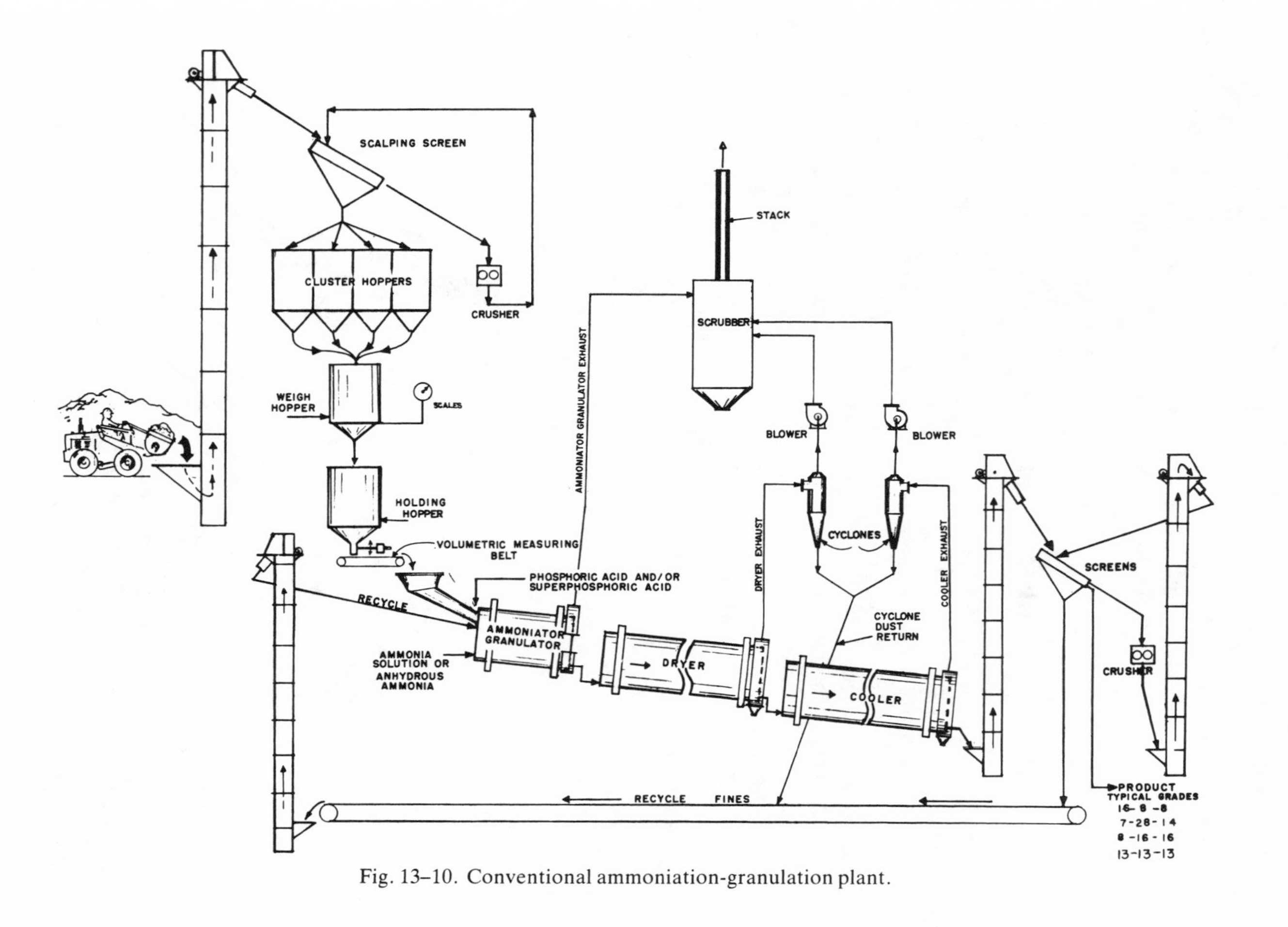

Fig. 13–10. Conventional ammoniation-granulation plant.

lated in a TVA rotary granulator by introducing phosphoric acid (H_3PO_4) and/or sulfuric acid (H_2SO_4) and ammonia and/or ammonia-containing N solution to the granulator. Chemical heat from the ammonia-acid reaction is usually sufficient to cause the salts produced by the chemical reaction and the solids to be granulated into a homogeneous product. Sometimes steam must be added to the granulator to supply heat and liquid phase to promote granulation. Product from the granulator usually contains from 2 to 5% moisture, and the product must be dried and cooled before it will store satisfactorily. Drying is usually accomplished in an oil- or gas-fired rotary cocurrent drier. Cooling is usually accomplished in a rotary cooler, although other types of coolers, such as fluid bed coolers, are sometimes used. Exhaust gases from the drier and cooler pass through cyclone dust collectors where most of the dust is removed from the gas and recirculated to the granulator. Gases from the cyclones are further cleaned in a scrubber that uses water as a scrubbing medium, although in some instances acid is used to scrub these gases. Liquor from the scrubbing operation is usually returned to the ammoniator-granulator. Some plants use bag filters instead of water or acid scrubbers to remove particulate from the exhaust gases.

Product from the cooler is screened and oversize is crushed and recirculated to the granulator. Gases from the cyclones are further cleaned in a scrubber that uses water as a scrubbing medium, although in some instances acid is used to scrub these gases. Liquor from the scrubbing operation is usually returned to the ammoniator-granulator. Some plants use bag filters instead of water or acid scrubbers to remove particulate from the exhaust gases.

Product from the cooler is screened and oversize is crushed and recirculated to the screen for product size removal. Fines from the screens are recirculated to the granulator. Product is conveyed to storage and stored in bulk. Typically about 60% of the product produced in these plants is shipped to fertilizer dealers in bulk. These dealers sell the product in bulk and often supply applicators to the farmer for applying the granules. The remaining 40% of the product is bagged. Typical grades of products produced in regional NPK granulation plants are 16–4–7 (16–8–8), 7–9–12 (7–20–14), 13–6–11 (13–13–13), 6–10–20 (6–24–24), and 8–7–13 (8–16–16).

Regional NPK granulation plants often have difficulty obtaining fuel at a reasonable cost to dry their products. The pipe-cross reactor developed by TVA uses the chemical heat produced when phosphoric and/or sulfuric acids react with ammonia to produce an ammonium phosphate sulfate melt, which granulates and dries the product. This melt also granulates other materials in the formulation. With this type of granulation, it is possible to significantly decrease the amount of energy required to dry the granules. The pipe-cross reactor is installed inside the granulator as shown in Fig. 13–11. This sketch shows how the melt spraying from the reactor mixes with the dry raw material feed plus dry recycle from the screens and cyclones in the ammoniator-granulator. With this process, it is possible to produce granular NPK mixtures without use of fossil fuel.

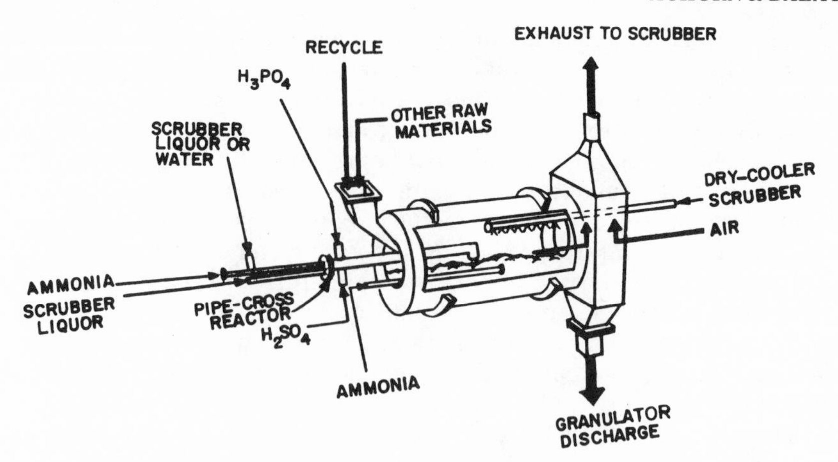

Fig. 13–11. Pipe-cross reactor for granulation plants.

Granular homogeneous NPK fertilizer mixtures continue to be popular in the marketplace because they do not segregate during transport, storage, or application. These mixtures also absorb herbicides uniformly and are excellent materials for the combined application of granular fertilizer-pesticide mixtures.

III. FLUID MIXED FERTILIZERS

The fluid mixed fertilizer business in the USA is composed of two major segments, solutions and suspensions. Solution mixtures usually are free of solids and clear enough to see through. Suspensions are of higher concentration and have small crystals of plant nutrients suspended by a gelling clay, such as attapulgite or sodium bentonite, in saturated fertilizer solutions. Because the clays are usually low in cost, it is practical to use them to produce suspension mixtures.

The objectives of fluid mixed fertilizer production and marketing are much the same as those for solid materials, namely, to produce in a plant with low operating cost that does not emit pollutants, a homogeneous fertilizer that has low raw material cost, high nutrient concentration, product versatility, and can be applied uniformly (Slack & Achorn, 1973).

Many raw materials can be used to produce fluid mixed fertilizers, and the fluid fertilizer industry has proved itself very versatile by accepting new raw materials as they become economical (Achorn & Lewis, 1974; TVA, 1976).

A Solution Fertilizer Mixtures

United States statistical data show that $> 50\%$ of all fluid mixed fertilizers are solution fertilizer mixtures.

The solution marketing system usually consists of regional fertilizer plants that react wet-process superphosphoric acid and ammonia in a TVA pipe reactor to produce an ammonium polyphosphate (APP) base solution, such as the 10–15–0 (10–34–0) grade (Meline et al., 1972). Superphosphoric acid is produced by concentrating merchant-grade orthophosphoric acid containing 24% P (54% P_2O_5) to superphosphoric acid containing 30 to 31% P (68–70% P_2O_5) by heating the merchant-grade in a vacuum or atmospheric concentrator. In superphosphoric acid, 20 to 35% of the P is present as polyphosphate. Physical and chemical characteristics of a typical superphosphoric acid are shown in Table 13–4.

Nearly all of the polyphosphoric acid in wet-process superphosphoric acid is in the form of pyrophosphoric acid ($H_4P_2O_7$). The remaining acid is present as orthophosphoric acid (H_3PO_4). Conversion to superphosphoric acid increases cost of the acid, but the long shipping distance in the USA generates freight savings that reduce this disadvantage.

More than 130 U.S. plants use the TVA pipe reactor process to produce an estimated 1.4 million metric tons of 10–15–0 (10–34–0) or 11–16–0 (11–37–0) APP solution annually. Other countries such as Belgium, France, and the USSR also use this process to produce APP solution.

Two typical APP solution plants are shown in Fig. 13–12. The plant on the left has a combination mix tank–cooler. The upper part of the tank consists of a cooling section filled with plastic ring or saddle packing. Liquid is recirculated downward through this section while air is drawn upward through the section to partially cool the liquid by evaporating water from it. Liquid from the bottom of the tank is passed through a shell and tube heat exchanger, where it is cooled further by expanding anhydrous ammonia before being pumped to storage. Anhydrous ammonia passing through this same heat exchanger is vaporized and passed into the T section of the pipe reactor. Superphosphoric acid is pumped by a positive displacement pump into the same T where the acid and gaseous ammonia react to form a hot melt at 316° C. Usually some ammonia is added to the recirculation line to adjust pH of the liquid to about 6.0. Water content is controlled by adding water to the recirculation line. Product quality control is by pH and density.

The plant at the bottom in Fig. 13–12 uses a separate mix tank and evaporative cooler. The mix tank is used for receiving hot melt from the pipe reactor and mixing it with recirculating liquid. With a separate mix

Table 13-4. Physical and chemical characteristics of a typical superphosphoric acid.

	Percent by wt
Total P	30.6
Polyphosphate, % of total P	30.0
S	1.3
Al	0.6
Fe	0.7
Mg	0.2
F	0.27
Solids (insoluble in methyl alcohol [CH_3OH])	0.15
Solids (insoluble in water)	0.01
sp. gr., 24°C	1.96
Viscosity, kPa/s, 52°C	4.0 (400 cP)

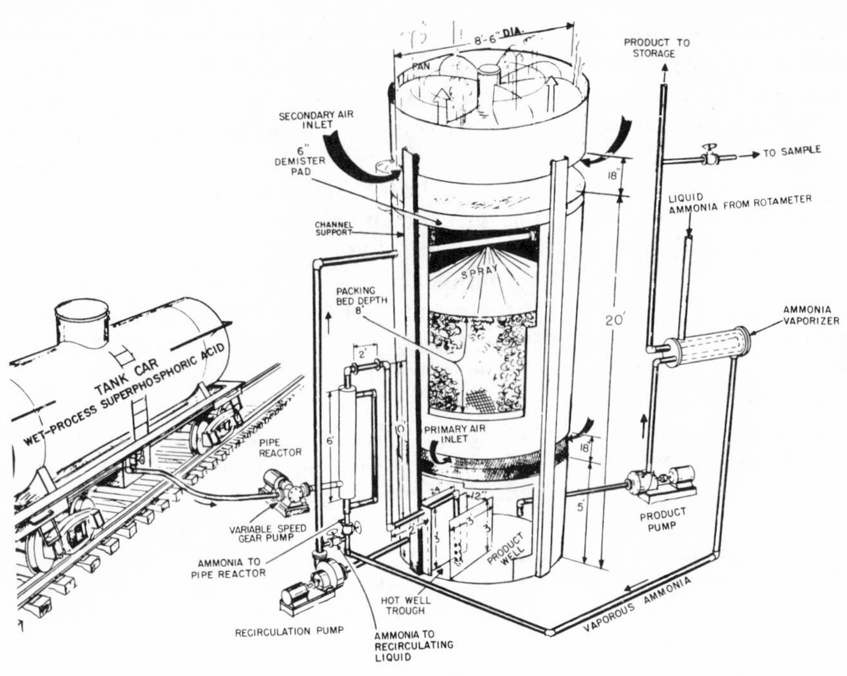

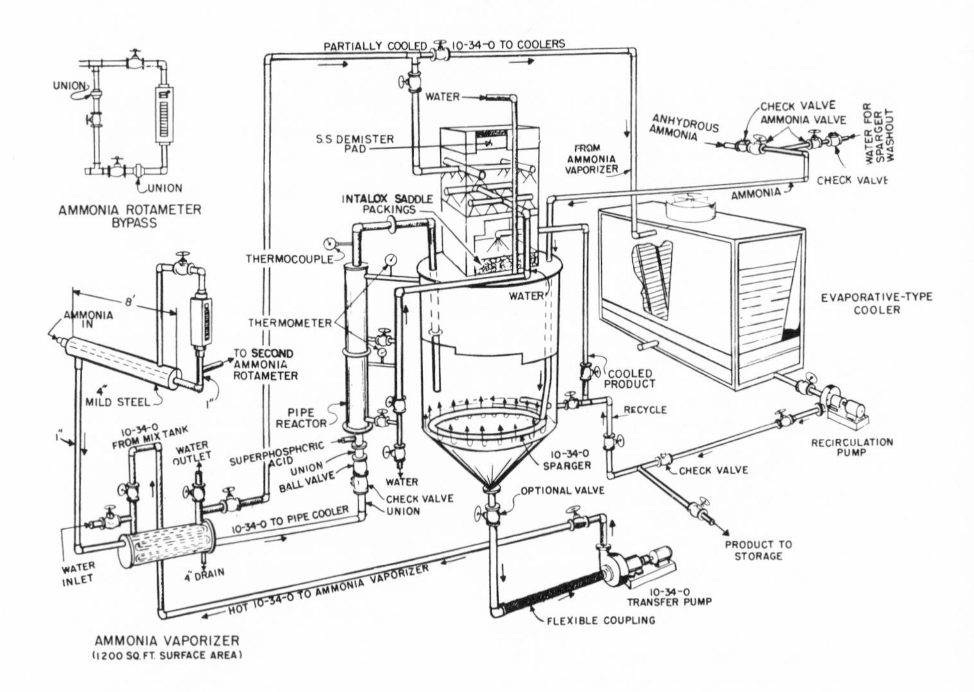

Fig. 13–12. Typical pipe-reactor plants for production of ammonium polyphosphate solution.

tank, it is possible to maintain a high liquid temperature of 82° C, which enhances mixing of the melt in the liquid and also provides extra heat for evaporating and superheating the anhydrous ammonia used in the pipe.

Data indicate that plants with separate mix tanks usually produce an APP solution of slightly higher polyphosphate content than that produced with the combination mix tank–cooler. Probably one reason this occurs is because of the higher temperature of the ammonia used in the reactor. Ammonia not added to the pipe reactor is added as liquid ammonia to the mix tank. Usually about 60% of the ammonia is added to the reactor and about 40% to the mix tank. This tank is equipped with a small scrubber in which a small amount of product from the cooler is used to scrub exit gases from the mix tank.

Both plants have efficient, inexpensive, evaporative product coolers. Few contaminants are discharged from either of the plants, but there is considerable discharge of water as steam.

Physical and chemical characteristics of a typical 10–15–0 (10–34–0) APP solution made by this process are shown in Table 13–5.

A large amount of 10–15–0 (10–34–0) APP solution is used for direct application in the wheat belt and other areas where K is not required, but most 10–15–0 (10–34–0) or 11–16–0 (11–37–0) solutions are used in small mix plants to produce NPK solution mixtures. A typical small mix plant is shown in Fig. 13–13. The APP solution is mixed with N solution (urea–ammonium nitrate solution) and KCl to produce such clear solution grades as 7–9–6 (7–21–7), 8–4–7 (8–8–8), and 21–3–0 (21–7–0).

Sometimes the liquid mix plant produces a KCl base solution, such as a 2–3–10 (2–6–12) or 4–5–9 (4–11–11) grade, and transports it to a satellite station where it is mixed with N solution containing 28 to 32% and 10–15–0 (10–34–0) APP solution to produce NPK solution mixtures. Sometimes such mixtures are made in the applicator tank of the satellite station.

B. Suspension Fertilizer Mixtures

Use of suspensions has grown rapidly in the USA since 1980. Some of the main reasons for the increased popularity of suspensions are the same as those for solutions. However, suspensions have advantages over solution mixtures. Some of these are:

1. The cost is lower because less pure materials can be used to produce suspensions.

Table 13-5. Physical and chemical characteristics of a typical 10-15-0 (10-34-0) APP solution.

	Percent by wt
N	10
P	15
Polyphosphate, % of total P	65-70
Viscosity, kPa/s, 24 °C	0.75 (75 cP)
sp. gr., 27 °C	1.400
Salt-out temperature, °C	< −18

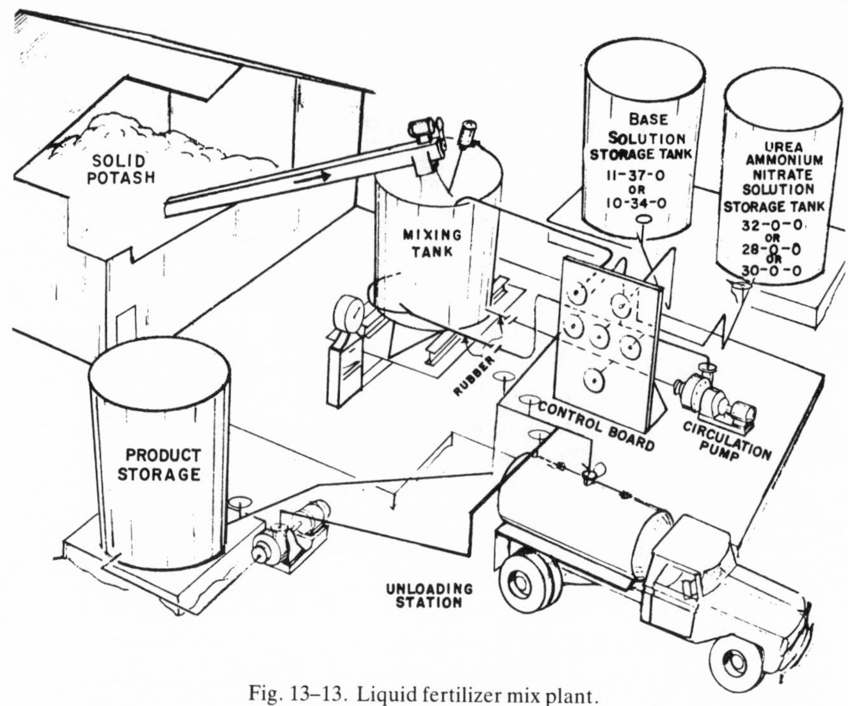

Fig. 13–13. Liquid fertilizer mix plant.

2. Higher analysis grades can be produced. This is especially true when grades containing K are to be made.
3. Larger quantities of micronutrients can usually be suspended than can be dissolved in solution fertilizers.
4. Powdered herbicides and insecticides not normally soluble in solution fertilizers can be suspended and uniformly distributed throughout the mixture.

The main reason for the early production of suspensions was to produce high K grades for use in regions requiring K. The most popular suspension grades are those having high K content, such as 3–4–22 (3–9–27), 3–4–25 (3–10–30), 4–5–20 (4–12–24), 6–8–15 (6–18–18), and 7–9–17 (7–21–21). Recently high N suspensions, such as 14–6–12 (14–14–14), 20–2–8 (20–5–10), 21–3–6 (21–7–7), and 24–4–0 (24–8–0), have become popular.

Currently, the most popular P material for suspensions is APP solution. However, economic pressures have encouraged use of other materials. The next most popular P sources are solid MAP and DAP.

Companies using 10–15–0 (10–34–0) APP solution or orthophosphate base suspension of 13–17–0 (13–38–0), 12–15–0 (12–35–0), or 11–14–0 (11–33–0) grade use a simple plant such as the one shown in Fig. 13–14 to make mixtures. In this plant a 10–15–0 (10–34–0) or 13–17–0 (13–38–0) P base is mixed with nonpressure N solution containing 28 to 32% N and K. All materials in the formulation are weighed and mixed in the scale-mounted mix tank. This plant has a relatively low cost. A number of such plants are located throughout the USA.

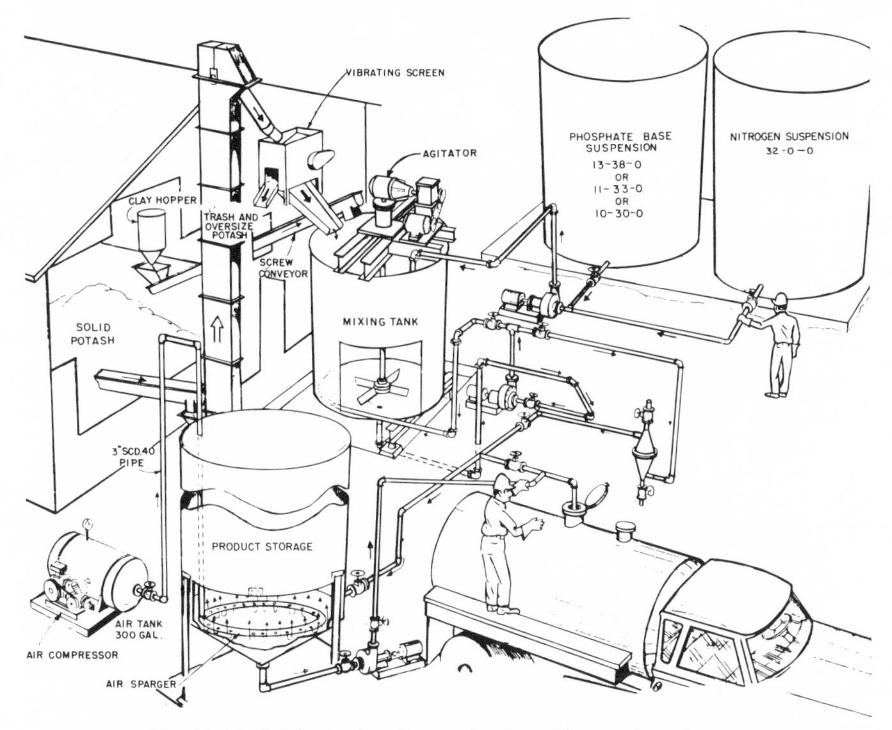

Fig. 13–14. Cold-mix plant for production of suspension mixtures.

Sometimes a batch ammoniation process is used to produce base suspensions or NPK mixtures from merchant-grade H_3PO_4 (23–24% P), anhydrous ammonia, and KCl. A sketch of a typical batch plant using H_3PO_4 and ammonia is shown in Fig. 13–15. This process includes a type 316 stainless steel mix plant, a stainless steel recirculation pump, and a wooden evaporative cooler, which provides rapid cooling. Acid is ammoniated to a N/P weight ratio of 0.76:1. With this procedure, all crystals in the suspension are small DAP crystals.

1. Use of MAP and DAP for Suspensions

Materials such as granular and powdered MAP, granular DAP, and granular and prilled urea have been used to produce suspensions.

Some firms fabricate their own mix tank for these suspensions. A typical mixing tank of this type is shown in Fig. 13–16. Several readily available turn-key plants can be purchased from engineering companies. These plants are prefabricated and shipped to the site on flat-bottom trucks or railway cars. They can be easily installed in a short time.

A dominant feature of these plants is the large pumps used to recirculate liquid within the mix tank. Some of the plants have liquid grinders, and others have high-intensity agitators. All tanks are equipped for adding ammonia and some for adding H_3PO_4. All can convert solid materials such as MAP, DAP, APP, ammonium sulfate, and urea to fluids quickly.

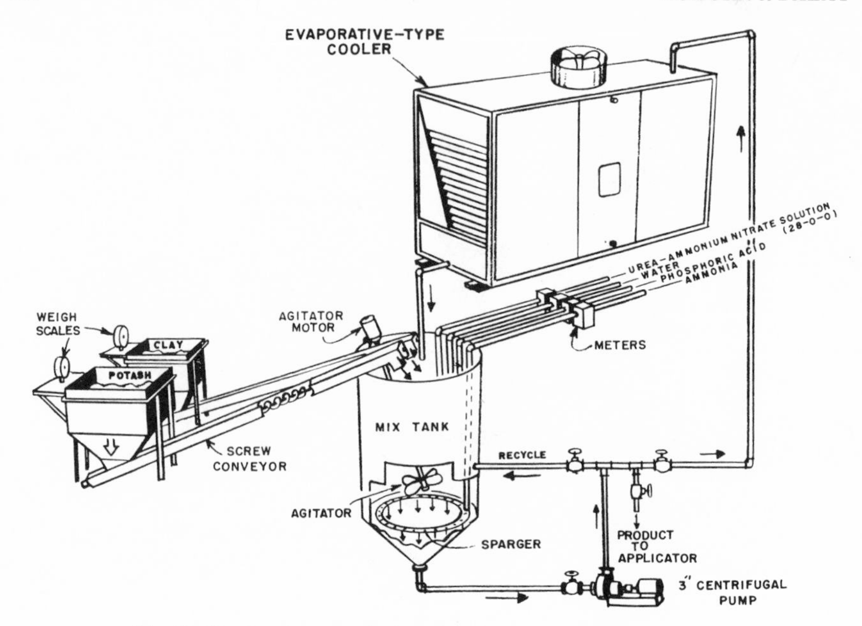

Fig. 13–15. Batch ammoniation plant for NP and PK suspensions.

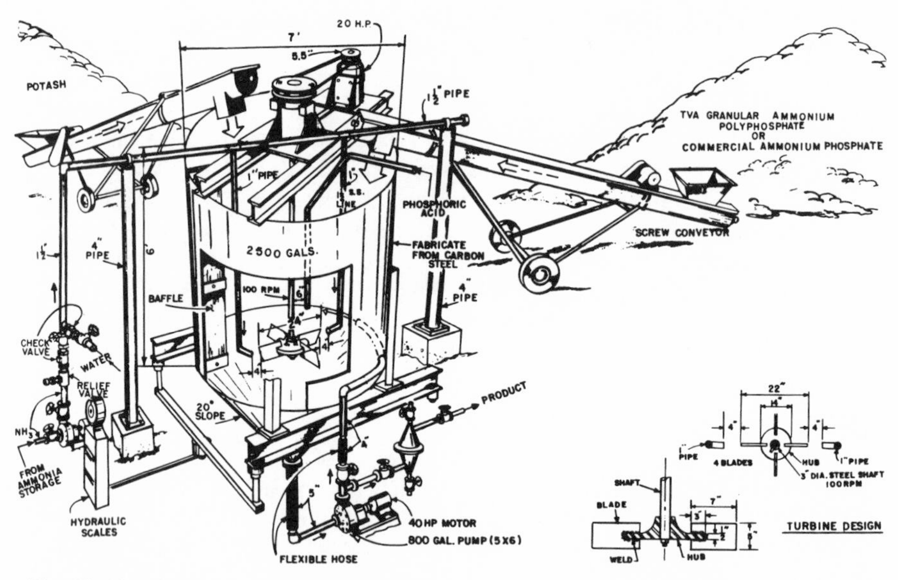

Fig. 13–16. Mix tank for production of 12–metric ton batch of suspensions from granular ammonium polyphosphate or commercial granular ammonium phosphates.

Most producers of suspensions prefer to produce mixtures having the highest possible grade without exceeding practical viscosity limits. Application tests show that when viscosity exceeds 10 kPa/s (1000 cP) at 0° C (Brookfield RVT viscometer, 100 rpm, no. 3 spindle), difficulty in pumping and handling the suspension can occur. Crystals formed in the suspen-

Table 13-6. Typical formulations for production of 11-14-0 (11-33-0) base suspension from MAB, DAP, or APP.

Phospahte source	MAP 11-23-0 (11-52-0)	DAP 18-20-0 (18-46-0)	APP 11-24-0 (11-55-0)
	kg/metric ton product		
MAP	634	--	--
DAP	--	502	--
APP	--	--	600
H_3PO_4 (24% P)	--	184	--
Anhydrous ammonia	49	24	54
Water	302	275	331
Gelling clay	15	15	15

sion should be small and light enough to avoid excessive settling and plugging of nozzles during application. When suspensions are produced, it would seem best to adjust the N/P weight ratio to the point of highest solubility (0.67:1). However, at this ratio the lightest and smallest crystals are not produced. It is important that the suspensions be ammoniated to a N/P weight ratio or 0.76:1 so that all of the crystals in the suspension are DAP. Diammonium phosphate crystals have less tendency to plug nozzles than do MAP crystals. Also, DAP is lighter than MAP (sp. gr. 1.619 vs. 1.803), and settles less during storage and application.

Typical formulations for production of 11–14–0 (11–33–0) base suspension from MAP, DAP, or APP are shown in Table 13–6.

When either MAP or APP is used, only anhydrous ammonia is required to adjust the N/P ratio, but when DAP is used, H_3PO_4 must be added to lower the N/P ratio and improve solubility. Also, extra H_3PO_4 and ammonia must be added to DAP to provide enough heat to disintegrate the granules.

IV. SATELLITE PLANTS

Some companies produce NPK base grades of 3–4–25 (3–10–30), 4–5–20 (4–12–24), and 7–9–17 (7–21–21) and NP base grades of 13–17–0 (13–38–0) or 11–14–0 (11–33–0) suspensions or 10–15–0 (10–34–0) solutions and ship them along with N solutions to satellite stations such as that shown in Fig. 13–17. At these plants N solutions are combined with the base grades in a small mix tank mounted on scales to produce complete grades. With these materials it is possible for a small merchant to install an inexpensive plant for producing NPK suspensions or solution mixtures.

Storage tests show that cone-bottom storage tanks are ideal for storage of suspensions at satellite plants; however, they are more expensive than flat-bottom tanks. Air spargers consisting of open-end pipes are installed to blow high-pressure air onto the bottom of the tanks. Experience has shown that it is preferable to agitate suspensions once daily by blasting 1200 L of air at a gauge pressure of 6.89×10^5 N/m^2 (100 psig) into the suspension. This keeps the suspension well mixed.

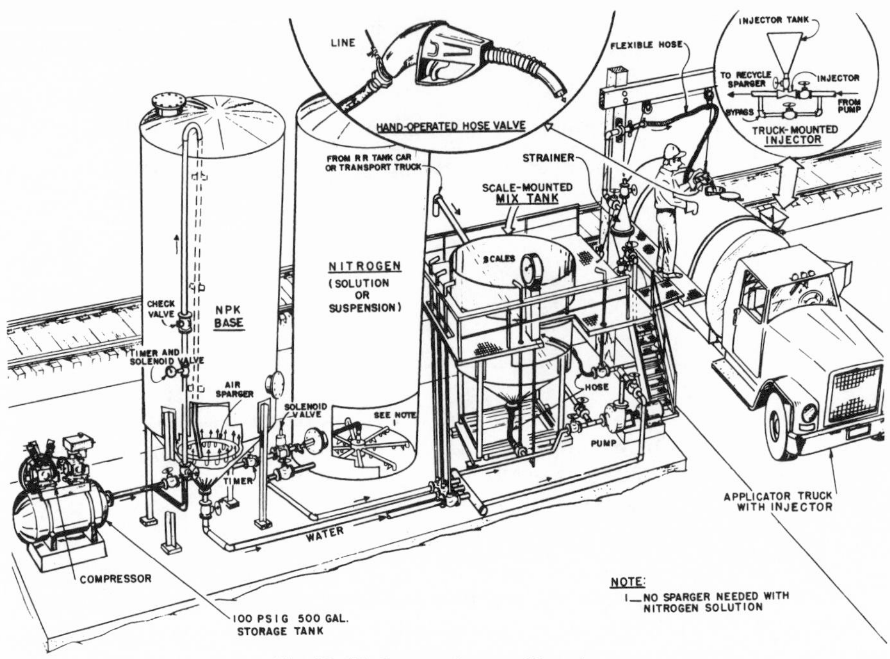

Fig. 13–17. Suspension satellite plant.

REFERENCES

Achorn, F. P., and J. S. Lewis. 1974. Alternate sources of materials for fluid fertilizer industry. p. 8–13. *In* NFSA Round-Up, St. Louis, MO. 24–25 July. Tennessee Valley Authority, Muscle Shoals, AL.

Hargett, L., and T. Berry. 1980 fertilizer summary data. Tennessee Valley Authority, Muscle Shoals, AL.

Hargett, N. L., and R. L. Pay. 1980 TVA-AAPFCO survey. Proc. Annu. Meet.—Fert. Ind. Round Table. 1980: 82–95.

Hoffmeister, George. 1975. Use of urea in bulk blends. Proc. Annu. Meet.—Fert. Ind. Round Table 1975:212–227.

Meline, R. S., R. G. Lee, and W. C. Scott, Jr. 1972. Production of liquid fertilizers with very high polyphosphate contents. 16:(2)32–45.

Slack, A. V., and F. P. Achorn. 1973. New developments in manufacture and use of liquid fertilisers. Fertiliser Society, London.

Tennessee Valley Authority. 1976. Phosphate rock suspensions. p. 70–74. *In* New developments in fertilizer technology. Tennessee Valley Authority, Muscle Shoals, AL.

14

Organic Sources of Nutrients[1]

J. F. Power

Agricultural Research Service, USDA
University of Nebraska
Lincoln, Nebraska

R. I. Papendick

Agricultural Research Service, USDA
Washington State University
Pullman, Washington

Nutrients required for crop production have been provided partially or completely from organic sources since the beginning of agriculture. Even today, with tens of millions of tons of fertilizer used annually in the USA, organic materials continue to be a major source of nutrients for crop production. This is illustrated for N in Table 14–1 by estimates of the N balance sheet for crop production enterprises in the USA in 1977 (Power, 1981). Crop growth took up an estimated 11.9 million metric tons of N that year, but farmers returned N to the soil estimated at 3 million metric tons in crop residues, 1.4 million metric tons in animal manures, and 7.2 million metric tons in biologically fixed N_2, in addition to 9.5 million metric tons as fertilizer N. Thus, organic sources supplied an estimated 55% of the N added to the soil for crop production.

Interest in organic sources of nutrients has intensified considerably since the energy crisis of the 1970s. Much of this increased interest results from large increases in the cost of commercial fertilizers and from projected continued increases in fertilizer prices and uncertainty of supply in future years. Thus, many farmers are looking for alternative methods for maintaining or improving soil fertility.

Prior to World War II, animal manures and legumes provided the basis for management of soil fertility. These sources of nutrients are still of major importance, but changing agricultural practices require us to investigate other ways of utilizing these materials and to consider the utilization of other organic materials. An example of some of these changes in agricul-

[1]Contribution from Agricultural Research Service, U.S. Department of Agriculture, in cooperation with the Nebraska Agric. Exp. Stn., Lincoln, NE. Published with the approval of the Director as Paper no. 6917, Journal Series, Nebraska Agric. Exp. Stn.

Table 14-1. Estimated N used for crop production in the USA in 1977 and N returned to the soil (Power, 1981).

	Million metric tons
N in 1977 crops	
In harvested crops	7.6
In crop aftermath (tops only)	4.3
Total	11.9
N returned to cropland	
In crop residues	3.0
In manure and organic waste	1.4
As symbiotically fixed N_2	7.2
As fertilizers	9.5
Total	21.1

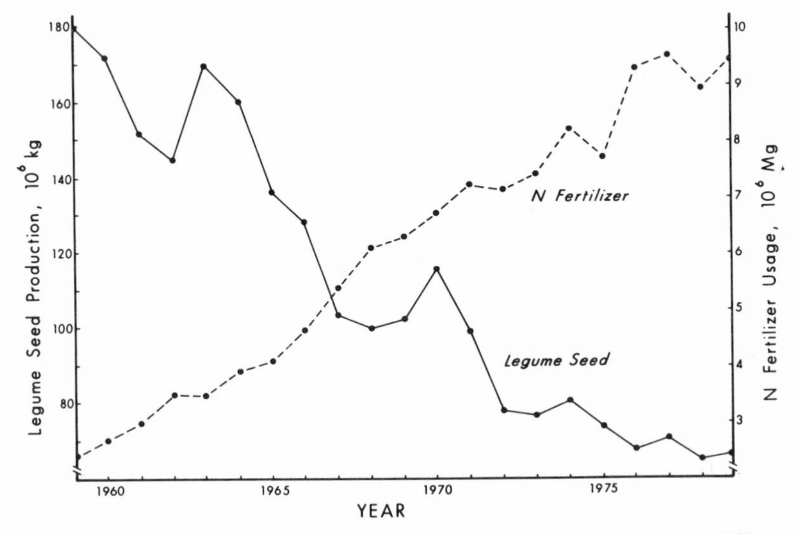

Fig. 14–1. Legume seed production and fertilizer N use in the USA, 1959–1979.

tural practices is illustrated in Fig. 14–1, showing the changes with time of both forage legume seed production and N fertilizer usage in the USA (Power, 1981). Available statistical data on legume seed production are probably more accurate than are those on land area of legume production but seed production and hectarage would probably be highly correlated. Seed production of legumes has steadily decreased since 1950, whereas fertilizer N usage has shown a corresponding increase. This relationship depicts very well the substitution of inorganic N sources for biologically fixed N_2 sources in American agriculture during this time period.

Other changes in agriculture have also occurred. Prior to World War II, most livestock was produced in small herds on family farms. Since the 1950s, livestock feeding operations have specialized to the extent that a major part of the beef cattle *(Bos taurus)* and swine *(Sus domesticus)* feeding operations, and almost all commercial poultry operations, are conducted on an intensified, confined basis. Thus, disposal of the manure produced by these operations is considered as much a liability as an asset, and the manure is often utilized inefficiently. Likewise, in urban areas, disposal of sewage sludge, refuse, and other urban wastes is also often viewed as a major problem. Thus, techniques adapted to American agriculture prior to the 1950s may no longer be useful.

This chapter presents an analysis of the role of organic sources in providing nutrients for crops. The organic sources are divided into (i) crop residues, (ii) animal and green manures, (iii) urban wastes, (iv) other organic sources, and (v) biologically fixed N_2. Information is presented on the quantity, composition, availability, present usage, and problems or opportunities associated with use of nutrients from each of these sources. Information is confined to discussions on N, P, and K.

I. NUTRIENT SOURCES AND QUANTITIES

The quantity of organic materials available for use as soil amendments can only be estimated. One of the most comprehensive estimates was provided in the USDA report *Improving Soils with Organic Wastes* (Miller & McCormack, 1978), which was prepared as a working paper to "evaluate organic wastes and identify some of the ways to improve their use as production resources." Estimates of annual production, N content, and quantity of N from the most important organic wastes applied to the land are given in Table 14–2. Crop residues and animal manures account for about 75% of the organic wastes produced annually, and an estimated 72% of these are presently being returned to the land. Municipal refuse accounts for about 18% of the total, very little of which is returned to the soil. Of the remaining organic wastes listed, sewage sludge is utilized to the greatest extent (23%) for land application.

Estimates of the amount of N, P, and K in these wastes is given in Table 14–3, also taken from the above USDA report (Miller & McCormack, 1978). Animal manures and crop residues account for about 93, 66, and

Table 14-2. Annual production, N content, and N usage of organic wastes in the USA (Miller & McCormack, 1978).

Organic waste	Annual production	Total N content	N used in 1978
	1000 metric tons		
Animal manure	156 275	6 876	4 980
Crop residues	384 961	4 287	3 056
Sewage sludge	3 902	167	39
Food processing	2 858	29	5
Industrial organics	7 337	No data	No data
Logging and milling	31 893	Negligible	Negligible
Municipal refuse†	129 485	725	0
Total	716 711	12 084	8 080
Commercial fertilizer (1978)			9 501

† Does not include municipal wastewater because quantity is unknown (estimated to be 10^{13} to 10^{14} L, with 4×10^5 to 4×10^6 metric tons N, 10 to 15% of which may be applied to the land).

Table 14-3. Total amount produced and amount of N, P, and K returned to the land annually as organic wastes in the USA (Miller & McCormack, 1978).

Organic material	Total produced			Returned to the land		
	N	P	K	N	P	K
	1000 metric tons/yr					
Animal manure	6 876	1 696	3 750	4 980	1 527	3 375
Crop residues	4 287	594	4 705	3 056	396	3 129
Sewage sludge	167	81	16	39	19	4
Food processing	29	5	45	5	1	6
Industrial organic	Data not available					
Logging and wood	Negligible					
Municipal sludge†	725	1 088	488	Negligible		
Total	12 084	3 464	8 964	8 080	1 943	6 514
Commercial fertilizer	9 501	2 190	4 335			

† Does not include municipal wastewater because quantity is unknown (estimated to be 10^{13} to 10^{14} L, with 4×10^5 to 4×10^6 metric tons N, 10 to 15% of which may be applied to the land).

94% of the N, P, and K, respectively, contained in all sources of organic wastes. It is estimated that 67, 56, and 73% of the N, P, and K, respectively, available in organic wastes is returned to the soil annually. This is equivalent to 85, 89, and 150% of the amount of N, P, and K, respectively, applied annually as commercial fertilizers.

A. Crop Residues

Estimated annual U.S. production of crop residues, listed by various crops, is given in Table 14–4, along with estimated total N, P, and K in these residues. Field corn accounts for > 50% of the residue biomass pro-

Table 14-4. Annual production and N, P, and K content of crop residues in the USA (Miller & McCormack, 1978).

Crop	Annual production	Nutrient content		
		N	P	K
	1000 metric tons/yr			
Barley (*Hordeum vulgare* L.)	7 592	57.0	8.3	94.9
Cotton (*Gossypium hirsutum* L.)	3 729	65.3	8.2	54.1
Dry beans (*Phaseolus* sp.)	457	4.1	0.5	5.6
Field corn (*Zea mays* L.)	198 504	2 203.4	357.3	2 640.1
Flax (*Linum usitatissimum* L.)	833	9.6	0.8	13.5
Oats (*Avena sativa* L.)	12 419	78.2	19.8	204.9
Peanut hay (*Arachis hypogaea* L.)	2 292	36.7	2.9	28.7
Rice (*Oryza sativa* L.)	7 755	46.5	7.0	90.0
Rye (*Secale cereale* L.)	925	4.6	1.1	6.5
Seed grass	504	4.7	0.9	9.6
Sorghum [*Sorghum bicolor* (L.) Moench]	16 645	179.7	25.0	219.7
Soybeans [*Glycine max* (L.) Merr.]	42 389	953.8	93.2	445.1
Sugarbeets (*Beta vulgaris* L.)	667	17.1	1.6	3.0
Sugarcane (*Saccharum officinarum* L.)	29	0.3	0.1	0.4
Wheat (*Triticum aestivum* L.)	90 159	604.1	63.1	874.6
Total	384 899	4 265.0	590.0	4 690.5

duced annually in the USA, and corn, wheat, and soybeans combined account for an estimated 86% of the annual quantity of residues produced.

The total quantity of N, P, and K in crop residues is also largely accounted for by corn, wheat, and soybeans. These three crops comprise 88, 87, and 84% of the total N, P, and K, respectively, in all crop residues. Corn alone accounts for > 50% of each of these nutrients contained in crop residues.

Crop residues are frequently utilized for a number of purposes. Historically, wheat and oat straw have been used for livestock bedding. Some residues are also used as livestock feeds. In both instances, however, most of the nutrients contained in residues used for these purposes may be returned to the soil in the form of animal manures if the manure is properly handled. Another use of crop residues is for fuel, in which case little or none of the N, P, and K is returned to the soil. With present fossil energy shortages, the potential for using crop residues as a source of fuel is being considered more seriously. Larson et al. (1978) pointed out that removal of crop residues may result not only in the loss to the soil of the nutrients contained in the residues but also in an equal or greater loss of nutrients through accelerated soil erosion.

Crop residue management practices play a major role in regulating the availability of the nutrients contained in crop residues. The quantity of residues returned to the soil can affect the amount of N and other nutrients immobilized in soil organic matter. For example, Black (1973) with wheat and Barnhardt et al. (1978) with corn have shown that soil organic N levels may be reduced by continued harvesting and removal of crop residue. Often, mineralizable N in the soil follows similar trends. Soil organic N content, on the other hand, is affected only slightly by N fertilization (Barnhardt et al., 1978; Haas et al., 1976). Increased lignin content of the crop residues may enhance immobilization of N in soil organic matter (DeHaan, 1977; Shi et al., 1980). Thus, it is apparent that the quantity and composition of the crop residues returned to the soil are important variables in regulating the immobilization and subsequent mineralization of the nutrients contained in the residues.

Residue management practices, especially those resulting from tillage, also play a very important role in regulating the recycling of nutrients immobilized in crop residues. Because much of the K in residues is water soluble, it is readily leached into the soil and undergoes typical cation exchange reactions. Likewise, much of the P in crop residues is readily hydrolyzed to orthophosphate forms, which are utilized by soil microorganisms and recycled much like N or undergo typical chemical reactions with the mineral components of the soil. Almost all of the N in crop residues, however, is utilized by soil microorganisms and is eventually recycled as the byproduct of microbial activity. Thus, much of the discussion on the effects of tillage or residue placement and nutrient availability is centered on N availability.

Through tillage, crop residues may be left on the soil surface, mixed in the upper few centimeters of soil, or buried beneath the soil surface. Examples of tillage practices that result in each of these types of residue placement are no tillage (weeds controlled with herbicides), reduced tillage (disk, subsurface sweeps, chisels, etc.), and moldboard plowing, respectively. With plowing, not only are crop residues buried, but the surface soil is also inverted and pulverized. Initially, plowing and, frequently, other forms of tillage result in reduced bulk density, increased soil porosity, and reduced volumetric water content (water-filled pore space). All of these factors may hasten decomposition of organic C in crop residues, with an accompanying increase in rate of decomposition of N compounds (Doran & Power, 1983). At the other extreme, no tillage results in lower soil porosity, cooler spring soil temperatures, greater water content, and slower rates of decomposition of organic debris. Consequently, with no tillage, organic C and N accumulate near the surface, rate of mineralization of N is reduced, and available P and K levels near the surface are often increased but are reduced at greater depths (Blevins et al., 1977; Doran, 1980). The net effect is that tillage enhances the rate of nutrient recycling and consequently reduces the total pool of N in the soil. On the other hand, no tillage conserves organic C and N near the soil surface, both as soil organic matter and as microbial biomass. These effects are shown by data from Doran and Power (1983) in Table 14–5. Douglas et al. (1980) show that the effects of residue placement on the recycling of S parallel effects on N recycling.

The availability of N in crop residues that are returned to the soil varies widely, depending on the type of residue, N content, climate, and management practices. In recent years, a number of experiments, utilizing N isotope–tagged residues were conducted to measure mineralization and uptake of the isotope. For example, Krishnappa and Shinde (1980) found 7.2 to 20.6% of the N in rice straw was recovered by the following crop. Yaacob and Blair (1980) found that an average of 15.4% of the isotope in tagged soybean residues was taken up by the next crop. Badzhov and Ikonomova (1971) found that 8 to 19% of the N isotope in oat straw was mineralized in 26 weeks. These values for N recovery are typical, but residue management practices and ambient conditions can result in wide divergence.

Hooker et al. (1982) showed that burning or removal of wheat or sorghum residues for 11 yr compared with incorporation had no effects on the concentration of soluble P, Zn, Na, Ca, or Mg in the surface 15 cm of soil; however, burning or removal decreased soil organic C and exchangeable K^+ while increasing soil pH and the leaching of nitrate-N. The subject of the effects of residue management on the recycling of nutrients was discussed in more detail by Power (1981).

Table 14–5. Percent change in various soil properties and microbial parameters resulting from no tillage compared with plowing (Doran, 1980; Doran & Power, 1983).

Soil depth	Total N	Total C	Mineralizable N	Aerobic organisms	Facultative anaerobes	Microbial biomass	Soil water
cm	% change plow to no tillage						
0–7.5	+34	+42	+35	+32	+59	+29	+47
7.5–15.0	− 2	− 7	− 4	−27	+18	− 5	− 2

Crop residues serve as a recyclable reservoir for the conservation of nutrients. Through selection of management practices in combination with the ambient environment the farmer can exercise considerable control over the size of this reservoir and the rate of turnover of nutrients within the reservoir. As has been mentioned previously, turnover rates can be accelerated by increasing opportunities for decomposition of organic matter. Thus, tillage increases air-filled pore space, and the microbiological decomposition of crop residues and other organic materials is enhanced. Likewise, increased soil temperature and a more aerobic environment also increase microbial decomposition rates of crop residues. Size of the nutrient reservoir created by crop residues is generally reduced as turnover rate is increased. Also, size is reduced by the physical removal of crop residues. In selecting management systems, frequently we are attempting to control mineralization rate of N and other nutrients to equal or exceed the rate of nutrient uptake required for optimum crop growth. This feat is sometimes very difficult to achieve.

B. Animal and Green Manures

1. Animal Manure

Animal manures have been regarded as a primary source of nutrients since the earliest of civilizations. Even today, in many of the less-developed parts of the world, animal manures still form the basis for soil fertility programs. In the USA, however, centralization of many livestock enterprises into confined operations has resulted in problems of disposal of animal wastes. Much research has been conducted since the 1960s to develop technology to deal with the disposal problem.

Production of animal manures in the USA is estimated at 156 million metric tons (dry) annually (Table 14–6). Approximately 58 and 17% of this total is produced by beef cattle and dairy cattle, respectively; however, an estimated 75% of the manure from beef cattle and 35% from dairy cattle are excreted on pasture and are not available for application to cultivated

Table 14–6. Annual production of animal manure in the USA by class of livestock (Miller & McCormack, 1978).

Class of livestock	Annual production	Excreted on pasture	Available for cropland use
	1000 metric tons	%	1000 metric tons
Dairy cattle	27 074	35	17 598
Beef cattle	91 066	75	22 766
Horses (*Equus caballus*)	20 800	71	6 032
Sheep (*Ovis aries*) and goats (*Capra hircus*)	1 486	87	193
Swine	6 740	20	5 392
Broilers (*Gallus domesticus*)	3 595	1	3 559
Other chickens	3 945	3	3 826
Turkeys (*Meleagris gallopavo*)	1 663	32	1 131
Ducks (*Anas platyrhyncha*)	38	22	33
Total	156 406	61	60 530

land. Thus, about 61 million metric tons (dry) of manure are available annually for use on cultivated land, of which about 38 and 29% are derived from beef cattle and dairy cattle, respectively.

An estimated national average of 73% of the manure produced in confinement is applied to the land (Table 14–7). Estimates of the amount of N, P, and K applied to the land annually in the form of manure are also provided in Table 14–7, according to various agricultural regions of the nation. An estimated total of 0.787, 0.572, and 1.093 million metric tons of N, P, and K, respectively, are applied to cropland annually as manure. This is equivalent to about 8, 21, and 20% of the N, P, and K, respectively, applied annually as commercial fertilizers. Thus, with efficient utilization, it is apparent that manures as well as crop residues can provide a significant part of the total quantity of nutrients required for U.S. agriculture that would otherwise need to be supplied with commercial fertilizer.

The data in Tables 14–6 and 14–7 on quantity and geographical distribution of the manure available for use on cultivated land illustrate major regional differences. Intensive dairy farming is found in the Northeast and Great Lake states, where only a small percent of the land is cultivated. Beef cattle and swine operations predominate in the Corn Belt and Northern and Southern Great Plains, where 50 to > 90% of the land is cultivated. Thus, the quantity of N, P, and K from manure applied per unit of cultivated land is several times greater in the dairy regions than in the beef cattle and swine regions. In the Northeast, for instance, N, P, and K applied in manure equals 31, 40, and 36%, respectively, of the N, P, and K applied as commercial fertilizers in that region, whereas comparable figures for the Corn Belt are only 4, 11, and 9%.

The availability of nutrients applied in animal manures varies widely, depending on such factors as source and composition of the manure; rate, method, and time of application; soil type and climate; and cropping system. Also, proper handling of manures to avoid excessive losses of nutri-

Table 14–7. Total N, P, and K from animal manures applied to the land by region in the USA annually (Miller & McCormack, 1978).

Region	Manure produced in confinement	Applied to the land	N applied	P applied	K applied
	1000 metric tons (dry)	%	—— 1000 metric tons ——		
Northeast	6 788	90	135.3	81.9	1 139.2
Great Lake states	10 417	97	142.7	99.9	234.9
Corn Belt	12 069	56	125.8	93.1	180.9
Northern Great Plains	7 759	57	48.2	50.0	116.6
Appalachian states	3 718	79	57.9	45.8	72.1
Southeast	2 910	71	49.0	46.8	50.3
Delta states	1 782	76	32.4	33.4	34.0
Southern Great Plains	4 389	87	60.0	45.0	98.2
Mountain states	4 621	71	66.3	32.6	80.9
Pacific states	6 160	73	78.6	51.2	102.7
Hawaii, Puerto Rico, Alaska	117	19	0.5	0.3	0.4
National	60 640	73	786.8	571.9	1 093.3

ents is very important. This subject is discussed elsewhere (Gilmour et al., 1977; USDA, 1979).

A thorough review of the subject of animal wastes, their use on cropland, and the recycling of nutrients is given by Gilmour et al. (1977). Included in these discussions is information on the availability and recycling from manures of N, P, and K as well as information on concentrations and availability of a number of the microelements. Availability of N from beef cattle (feedlot) manure was calculated by Herron and Erhart (1965) by the equation

$$\% \text{ N available} = 11.15 + 25.7t - 2.9t^2,$$

where t is time in years since manure application. Equations of this nature may be used to estimate adequacy of manure treatments for crop production.

Much of the research information available on utilization of animal manures has been used to develop a manual for evaluating agronomic and environmental effects of animal wastes. This manual (USDA, 1979) was prepared for use by technicians in action agencies and other organizations. From numerous experiments, Table 14–8 was prepared to provide a basis for estimating rates of decay and recycling of N contained in animal wastes. Typical values for total N content of various types of manures are given, along with the fraction of N that would be mobilized and become available for plant growth for up to four growing seasons after application. The rate of decay during the year of application is somewhat related to total N content of the manure with several exceptions. One such exception is for swine manure, with a decay constant of 0.9 for the first year, even though total N content is only 2.8%. This rapid decay would suggest some unique biochemical properties for swine manure. Decay constants the second year are only a fraction of those for the first year and are even smaller in the third and fourth years. Thus, residual availability of N in manure becomes very small within several years. However, when manure is applied at relatively large rates for a period of years, the large mass of N applied may result in rather prolonged residual availability of N from such treatments.

Table 14-8. Decay constants to estimate animal manure N availability to crops (12 months/yr basis) (USDA, 1979).

Source of manure	N content (dry basis)	Years after application			
		1	2	3	4
	%	Fraction decayed/yr			
Poultry (hens)	4.5	0.90	0.10	0.05	0.05
Poultry (broiler)	3.8	0.75	0.05	0.05	0.05
Swine	2.8	0.90	0.04	0.02	0.02
Dairy, fresh	3.5	0.50	0.15	0.05	0.05
Dairy, anaerobic	2.0	0.30	0.08	0.07	0.05
Beef feeder, fresh	3.5	0.75	0.15	0.10	0.05
Beef feeder, corral	2.5	0.40	0.25	0.06	0.03
Beef feeder, corral	1.5	0.35	0.15	0.10	0.05
Beef feeder, corral	1.0	0.20	0.10	0.05	0.05

2. Green Manure Crops

Information on the availability of N in green manure crops is much more limited; however, the information available indicates that N content of the green manure is very important in regulating N availability. Halvorson and Hartman (1975) found that average sugarbeet yield and sucrose content could be maintained equally well over a 19-yr period with either alfalfa (*Medicago sativa* L.), yellow sweet clover (*Melilotus officinalis* Lam.), 22.4 metric tons of manure/ha, or 112 kg of fertilizer N/ha. Shi et al. (1980) found that lignin content of green manures also affected N availability. In comparing azolla *(Azolla imbricate)* with Chinese milk vetch (*Astragalus sinicus* L.) and water hyacinth [*Eichhornia crassipes*(Mart) Solms-Laub.], they found that the availability of N in the latter two species was greater than that of native soil N, but the availability of N from azolla (higher lignin content and lower C/N ratio) was less than that of native soil N. In another study in the People's Republic of China, Huang et al. (1981) found that N in green manures was less available to rice than N in ammonium sulfate initially; however, from tillering until maturity of rice, the two sources were of equal availability. They found that N fertilizer added with the green manure enhanced uptake of N from both sources.

Green manure crops have been used very little in U.S. agriculture since the 1950s. There has been a corresponding lack of detailed research information on production and efficient utilization of N from green manures, especially utilizing modern isotopic and other improved research techniques and under low fertilizer input cropping systems.

With the present structure of agriculture, we can expect little improvement in the utilization of nutrients from animal manures; however, in local regions where such materials are plentiful, some significant advancements may be possible. On the other hand, the potential for utilizing green manures for supplying N, regulating nutrient availability, and improving soil in modern agriculture needs investigation. With increasing costs of fertilizer materials and increased concern for the control of soil erosion and enhancement of soil productivity, development of cropping systems utilizing more green manures is a much needed area of research. In regions of limited rainfall, this research may be complicated by the need to replenish the soil water utilized by the green manure crop.

C. Urban Wastes

1. Municipal Refuse

Urban wastes with potential for application to the soil are comprised primarily of municipal refuse and sewage sludges. Although municipal refuse accounts for about 18% of the total organic wastes in the nation (Table 14–2), direct usage on agricultural lands is almost nil. Several reasons for this include economics of collecting and moving the materials, difficulty and expense of sorting out nonbiodegradable materials, variability in material, and characteristically low N content. Disposition of much of this refuse is accomplished by either burning or burying. Attempts to utilize this refuse for direct application to croplands have generally been unsuccessful (Volk, 1976).

A recommended technology for utilization of municipal refuse is that of mixing municipal refuse with sewage sludge and composting. A number of municipalities have initiated such programs since 1970 with reasonable success. Although equipment and techniques are generally available, the practicality and economics of composting operations vary widely and must be developed on a case-by-case basis. During the composting operation, bulk is often reduced well over 50% primarily because of the oxidation of organic C; however, the final chemical composition of the compost may vary widely, depending on the waste materials used, techniques employed, ambient conditions, etc. Consequently, the availability of nutrients in the compost varies widely.

2. Sewage Sludge and Municipal Wastewater

Although the annual tonnage of sewage sludge is only a fraction of that of municipal refuse, much more sludge is presently used for land application (Table 14–2). Impetus for land application of sewage sludge resulted from passage of federal legislation relative to water quality in both 1972 and 1977. Usually, sewage sludge is stabilized and concentrated before it is applied to the land. Sludge is applied to the land as either a liquid or slurry, as dry or partially dried cake, or as dry compost. The preferred method of application is often dictated by economics, transportation distances, equipment available, and soil type.

Tonnage of N, P, and K in sludges is presented by region of the USA in Table 14–9. As expected, tonnages are greatest in the most populated regions. Tonnages of N, P, and K were calculated from a typical sludge concentration of 4.0, 2.0, and 0.4% for N, P, and K, respectively. The amount of N, P, and K contained in the sludge equals about 0.6, 3.5, and 0.4% of

Table 14-9. Annual production of sewage sludge, N, P, and K content of sludge, and percent applied to the land by region within the USA (Miller & McCormack, 1978).

Region	Annual production	Nutrient content N	P	K	Applied to land
	1000 metric tons/yr				%
Northeast	964.4	38.6	19.3	3.8	5.4
Great Lake states	396.5	15.9	7.9	1.6	33.6
Corn Belt	800.9	32.1	16.0	3.2	39.8
Northern Great Plains	67.9	2.7	1.3	0.3	9.0
Appalachian states	287.1	11.5	5.7	1.2	6.6
Southeast	275.9	11.1	5.5	1.1	16.6
Delta states	111.6	4.5	2.2	0.4	10.1
Southern Great Plains	263.9	10.5	5.3	1.1	26.9
Mountain states	171.9	6.9	3.5	0.7	28.6
Pacific states	479.2	19.2	9.6	2.0	37.7
Hawaii, Puerto Rico, Alaska	19.4	0.8	0.4	0.1	0.2
Total	3838.8	153.7	76.8	15.3	23.1

these elements, respectively, applied as commercial fertilizers; however, only about 23% of the sludge produced is presently used on the land (Table 14–9).

The availability of nutrients in sludge varies with a number of factors. Gilmour et al. (1977) recently reviewed available literature and showed that an average of about 40% of the N in sludge was recovered by a growing crop the first year after application, although individual experiments showed wide variation. Phosphorus and K would generally be equally or more available, depending on properties of the soil and sludge materials and on climatic conditions.

A major problem with sewage sludge in some industrial areas is that of contamination with toxic heavy metals or other toxic organic materials. Numerous studies have been conducted to determine the abundance and fate of heavy metals when applied to soils in sewage sludge (Chaney & Giordano, 1977). Techniques such as waste stream pretreatment are available to alleviate this problem to some extent, but the presence of heavy metals, if not controlled, will continue to limit the use of a significant quantity of the sewage sludges that are produced. Improved quality of sludge and development of improved handling techniques, such as through compost processing, would increase the feasibility for using sewage sludge as a major source of plant nutrients, particularly for truck farms and croplands nearest metropolitan areas.

Municipal wastewaters are a locally important source of plant nutrients near certain cities. Effluents from municipal waste treatment plants have been applied to the land near Melbourne, Australia, and a few other cities around the world since the late 19th century. In the USA, these wastewaters are presently being utilized for irrigation where the growing season is relatively long and periods of low precipitation are common. In the USA, communities in the Southwest meet these requirements best. Probably about 12 to 15% of these wastewaters are presently used for land application.

Accurate information on quantity is lacking because unknown quantities of water from storm sewers frequently are mixed with sewage effluent before release from the treatment plant. Best estimates indicate that between 10^{13} and 10^{14} L of municipal wastewaters are released annually. Total N and P concentrations also vary widely but may average near 40 mg total N and 10 mg total P/L (ca. 60% of the N usually exists as ammonium-N). If 10^{14} L are produced annually, then 4×10^6 and 1×10^6 metric tons of total N and total P, respectively, are contained in these wastewaters. If 10^{13} L are produced, the quantity of N and P is 1 order of magnitude less. The reader is referred to Bouwer and Chaney (1974) for a more complete discussion of this subject.

D. Other Organic Sources

Included in this category are organic materials, such as by-products of the wood industry, by-products of the food processing industry, and various industrial by-products. Annual tonnage of these materials in the USA comprises only a small part of the total quantity of organic materials available (Table 14–2), but in certain local areas, these materials may be very important. Total N, P, and K content is listed (Table 14–10) only for food processing wastes, because for the other materials either adequate data are not available or very little of these materials is applied to cropland. Generally, by-products of the wood industry are produced primarily in regions with very little cultivated land. Most of these materials are extremely low in total N and exhibit few, if any, beneficial nutritional effects when applied to the soil. Generally, wood by-products are better utilized as a source of fuel. A small percentage of these wood by-products are used as soil amendments (Allison, 1965), especially for mulching and ornamental purposes. Because of the extremely high C/N ratio of wood by-products (often > 400:1), care must be exercised that their application to the soil does not create severe N deficiencies for the crop.

Organic wastes are also produced by such industries as paper manufacture, soap and detergent manufacture, distilleries and wineries, leather tanning, petroleum refining, and other processes. Of this listing, the paper industry is a major source of organic wastes (Bauer et al., 1977). The nature of the organic wastes produced by industry varies widely, from concentrated sludges or composts to very dilute liquors. Consequently, even within an industry, it is difficult to arrive at a characteristic value for nutrient composition of wastes. For the paper industry, however, the wastes are almost certain to have a very low N content, thereby severely limiting their usefulness as an organic source of nutrients. Also, the presence of toxic materials (especially B from glues) in organic wastes from the paper industry is always a hazard to land application.

The group of industrial wastes with the most potential for use as a source of nutrients for crops is the waste from the food processing industry (Table 14–10). Again, the physical nature of these wastes varies widely, from solids (meat packing) to very dilute liquors (dairy or canning industries). Of the fruits and vegetables processed annually, about 75% of the organic wastes produced come from processing of citrus (*Citrus* sp.) fruits, potatoes (*Solanum tuberosum* L.), sweet corn, and grapes (*Vitis* sp.); however, N, P, and K in processing wastes of all fruits and vegetables amount to only about 20 600, 4000, and 38 300 metric tons annually, respectively. Thus, although food processing wastes, where available, may have some value as nutrient sources, their use is restricted to localized areas near the processing plants. In addition, a large percentage of the food processing

Table 14–10. Annual production and N, P, and K content of organic wastes originating from fruit and vegetable processing in the USA (Miller & McCormack, 1978).

Source	Annual production	Nutrient content		
		N	P	K
	100 metric tons/yr			
Fruits	1775	18.1	3.6	36.3
Vegetables	1050	2.5	0.4	2.0
Total	2825	20.6	4.0	38.3

wastes is possibly better used as animal feed, wherein the nutrients would then be applied to the land as manure. Presently, only about 11% of these wastes are directly applied to land.

E. Biologically Fixed Dinitrogen

In Table 14–1, an estimate was made that total biological N_2 fixation in U.S. agriculture was approximately 7.2 million metric tons annually. This value was calculated from data on the area producing various legumes in 1977 (USDA, 1978) and from estimates (based on published literature) of annual N_2 fixation by various legume species. Regardless of the accuracy of this method of calculation, the magnitude of this value emphasizes the fact that biologically fixed N_2 is a major input into agricultural systems. Since the bulk of the fixed N_2 is immobilized within the legume plant itself, mineralization of this N is of major significance.

Data in Fig. 14–1 suggest that legume hectarage in the USA in 1977 was only about one-third of that in 1959. Most of this reduction in legume production was at the expense of species of trefoils *(Lotus)*, lespedeza *(Lespedeza)*, and various clovers *(Trifolium)*. The area devoted to alfalfa has remained relatively constant (USDA, 1978). Since alfalfa is generally considered to be one of the most efficient legumes for N_2 fixation, the decline in quantity of biologically fixed N_2 in U.S. agriculture since 1959 is probably somewhat less than is the quantity of legume seed produced annually.

Numerous estimates of the amount of N_2 fixed annually by legumes have appeared in the literature. For alfalfa, these values range from 50 to > 500 kg N/ha; however, under most farming systems, the legume crop is harvested annually, and much of the N_2 fixed is removed in the harvest. Thus, as early as 1917, Swanson (1917) concluded that alfalfa grown for hay in Kansas would halt the decline in soil organic N content that resulted from cultivation of grain crops but generally would do little to increase soil organic matter levels. Haas et al. (1976) found that soil organic N levels could be increased slightly if the foliage were cut and returned to the soil surface but not if the foliage were removed. Evans and Barber (1977) estimated typical N_2 fixation values for a number of legumes (Table 14–11).

The availability of N from legumes grown in rotation with grain crops has been studied frequently. Baldock and Musgrave (1980) concluded that 2 yr of alfalfa in a New York soil increased potentially available soil N by

Table 14-11. Rates of biological N_2 fixation for various legumes (Evans & Barber, 1977).

Legume crop	N_2 fixed
	kg/ha per year
Soybeans	57-94
Cowpeas [*Vigna unguiculata* (L.) Walp. ssp. *unguiculata*]	84
Clover (*Trifolium* sp.)	104-160
Alfalfa	128-600
Lupins (*Lupinus* sp.)	150-169

135 kg N/ha. They found that the effects of N from legumes, manures, and inorganic fertilizers were additive. Likewise, Halvorson and Hartman (1975) found that alfalfa or sweet clover would provide available N to sugarbeets equivalent to 112 kg fertilizer N/ha. However, in water-deficient regions, Haas et al. (1976) found that legumes often reduced grain crop production because legumes often depleted subsoil water and restricted future crop growth.

Since the energy crisis of the 1970s, a few studies have been conducted in which N isotopes have been used to determine the source of N in legume production systems. Heichel et al. (1981) found that alfalfa fixed 148 kg N_2/ha in the seedling year in Minnesota. Bartholomew (1971) summarized data on isotopic N recovery by the crop following a legume taken from a number of experiments. Generally, recovery of N_2 fixed in legumes was in the 50 to 90% range under greenhouse conditions and 10 to 50% during the first year under field conditions.

A major limitation to the use of legumes for N_2 fixation in cropping systems has been that this practice detracts from both space and time for the production of other higher value crops and may reduce farm profits. Deep-rooted crops, such as perennial alfalfa, can also remove much of the subsoil water and cause a yield loss for the following grain crop, particularly in a dry year (Haas et al., 1976). Legume crops are often expensive to establish and may be subject to pest infestations that can thin or destroy the crop. Nevertheless, with improved technology for water conservation, such as improved snow trapping, and runoff and evaporation control, there may be opportunities for using legumes to partially replace fallow in dryland areas as a source of N for the following grain crop.

More extensive use could possibly be made of adaptable grain legumes in some cropping systems to reduce fertilizer N requirements of the rotation. Legumes such as peas (*Pisum sativum* L.), lentils (*Lens culinaris* Medik.), and soybeans, for example, are capable of fixing all or part of the N_2 for the seed crop plus leaving some residual N for a following nonlegume grain crop. The potential for expanded use of grain legumes with improved fixation capacity should be explored for cropping systems in general.

Special mention should be made of N_2 fixation by soybeans. The area of land devoted to soybean production in the USA increased steadily since the 1940s, from nearly zero in 1945 to almost 20 million ha in 1985. Estimates of N_2 fixation by soybeans are frequently in the 100 to 200 kg N/ha range. Thus, soybeans probably add several million tons of N annually to the balance sheet in Table 14–1. However, 100 to 150 kg N/ha are commonly removed in the seed crop; therefore, net additions to the soil may often be < 50 kg N/ha. In fertile soils, more N may be removed in the seed than is added by N_2 fixation, leaving a net deficit in N. However, because of the extremely large number of hectares involved, even at this rate of fixation, soybean production can significantly affect soil N levels on the national basis.

Considerable interest is being expressed in organic farming systems wherein nutrient requirements of crops are provided from organic sources rather than from synthetic or processed sources (USDA, 1980). In organic farming systems, legumes, manures, composts, and crop residues are the major sources of added nutrients. Availability and release of nutrients are controlled through careful soil management. Although the concept of controlled nutrient release through organic farming has been proved in practice, little scientific information is available at this time on the chemistry and biochemistry of the nutrient transformations involved. The information available regarding N transformations was recently summarized by Power and Doran (1984) and is beyond the scope of this publication.

II. CONCLUSIONS

In this chapter, we have attempted to identify the various organic sources of nutrients that can be used for crop production and to provide information on the quantity of material available, typical composition of the material, and the potentials and hazards for use on the land. From this discussion, it is apparent that our primary sources of organic nutrients in U.S. agriculture are crop residues and animal manures, plus N_2 fixed biologically. Research has been conducted since well before the 20th century on the use of animal manures, but research needs have changed as our animal production systems have changed.

The effects of crop residue management practices on soil erosion, soil-water relations, and crop yields have been studied extensively since the 1930s. Only since the early 1970s, however, have we developed much research on the effects of tillage and residue management on nutrient cycling. We are now realizing that we can exercise considerable control over the microenvironment of the soil by our choice of residue management practices and that the microenvironment we create regulates, to a large extent, the nature and intensity of microbial activity that occurs in the soil. The cycling of N, and much of the P and S depends on the nature and intensity of microbial activity in the soil ecosystem. Thus, we must make a better evaluation of how we can exercise control of nutrient cycling through our choice of management practices, especially residue management.

The need for more investigations into the development of cropping systems that better utilize biologically fixed N_2 is evident. With the widespread usage of fertilizer N since the 1950s, research on ways to include legumes in cropping systems has declined. With the development of new legume cultivars, however, it should be possible to develop systems with legumes used not only in rotation with grain crops but also as cover crops and catch crops or for other specialized uses. As an example, if we can develop a rapid-growing legume that will fix 50 kg N_2/ha during the cool weather of the spring in the Corn Belt and then plant corn directly into this legume stand, we have a potential for reducing fertilizer N needs and also providing soil erosion control. Research to develop such a system has hardly begun, however.

Although the use of other organic sources of N (sewage sludge, food processing wastes, etc.) may have little economic significance on a national scale, we have recently conducted considerable research on their utilization in agriculture as an environmentally acceptable means of disposal. In some instances, there is potential for increased usage; however, because of the limited quantities of such materials and because of their concentration in very localized regions, widespread usage of these wastes as sources of nutrients is not possible.

These organic sources of nutrients can be viewed as alternatives to the use of fertilizers. If properly managed, crops can efficiently take up many of the nutrients contained in these organic materials. Changing economic situations suggest that increased research on the use of organic materials in crop nutrition should be intensified in the future.

REFERENCES

Allison, F. E. 1965. Decomposition of wood and bark sawdust in soil, nitrogen requirements, and effects on plants. USDA Tech. Bull. no. 1132. U.S. Government Printing Office, Washington, DC.

Badzhov, K., and E. Ikonomova. 1971. ^{15}N for studying nitrogen transformations in soil, nitrogen nutrition of plants, and in assessing available nitrogen in soil. p. 21–32. *In* Nitrogen-15 in soil-plant studies. International Atomic Energy Agency, Vienna.

Baldock, J. O., and R. B. Musgrave. 1980. Manure and mineral fertilizer effects in continuous and rotational crop sequences in central New York. Agron. J. 72:511–518.

Barnhardt, S. L., W. D. Shrader, and J. R. Webb. 1978. Comparison of soil properties under continuous corn grain and silage cropping systems. Agron. J. 70:835–837.

Bartholomew, W. V. 1971. ^{15}N research on the availability and crop use of nitrogen. p. 1–20. *In* Nitrogen-15 in soil-plant studies. International Atomic Energy Agency, Vienna.

Bauer, D. H., D. E. Ross, and E. T. Conrad. 1977. Land cultivation of industrial wastewaters and sludges. p. 147–156. *In* R. C. Loehr (ed.) Food, fertilizer, and agricultural residues. Ann Arbor Science, Ann Arbor, MI.

Black, A. L. 1973. Soil properties associated with crop residue management in a wheat-fallow rotation. Soil Sci. Soc. Am. Proc. 37:943–946.

Blevins, R. L., G. W. Thomas, and P. L. Cornelius. 1977. Influence of no-tillage and nitrogen fertilization on certain soil properties after 5 years of continuous corn. Agron. J. 69:383–386.

Bouwer, H., and R. L. Chaney. 1974. Land treatment of wastewater. Adv. Agron. 26:133–176.

Chaney, R. L., and P. M. Giordano. 1977. Microelements as related to plant deficiencies and toxicities. p. 235–282. *In* L. F. Elliott and F. J. Stevenson (ed.) Soils for management of organic wastes and waste waters. American Society of Agronomy, Crop Science Society of America, and Soil Science Society of America, Madison, WI.

DeHaan, S. 1977. Humus, its formation, its relation with the mineral part of the soil, and its significance for soil productivity. p. 21–30. *In* Soil organic matter studies, Vol. I. International Atomic Energy Agency, Vienna.

Doran, J. W. 1980. Soil microbial and biochemical changes associated with reduced tillage. Soil Sci. Soc. Am. J. 44:765–771.

Doran, J. W., and J. F. Power. 1983. The effects of tillage on the nitrogen cycle in corn and wheat production. p. 441–455. *In* R. Lowrance et al. (ed.) Nutrient cycling in agricultural ecosystems. Spec. Pub. 23. University of Georgia, Athens.

Douglas, C. L., R. R. Allmaras, P. E. Rasmussen, R. E. Ramig, and N. C. Roager, Jr. 1980. Wheat straw composition and placement effects on decomposition in dryland agriculture in the Pacific Northwest. Soil Sci. Soc. Am. J. 44:833–837.

Evans, H. J., and L. E. Barber. 1977. Biological nitrogen fixation for food and fiber production. Science (Washington, D.C.) 197:332–339.

Gilmour, C. M., F. E. Broadbent, and S. M. Beck. 1977. Recycling of carbon and nitrogen through land disposal of various wastes. p. 173–196. *In* L. F. Elliott and F. J. Stevenson (ed.) Soils for management of organic wastes and waste waters. American Society of Agronomy, Crop Science Society of America, and Soil Science Society of America, Madison, WI.

Haas, J. J., J. F. Power, and G. A. Reichman. 1976. Effect of crops and fertilizer on soil nitrogen, carbon, and water content, and on succeeding wheat yields and quality. ARS-NC-38. ARS-USDA. North Central Region, Peoria, IL.

Halvorson, A. D., and G. P. Hartman. 1975. Long-term nitrogen rates and sources influence sugarbeet yield and quality. Agron. J. 67:389–393.

Heichel, G. H., D. K. Barnes, and C. P. Vance. 1981. Nitrogen fixation by alfalfa in the seeding year. Crop Sci. 21:330–335.

Herron, G. M., and A. B. Erhart. 1965. Value of manure on an irrigated calcareous soil. Soil Sci. Soc. Am. Proc. 29:278–281.

Hooker, M. L., G. M. Herron, and P. Penas. 1982. Effects of residue burning, removal, and incorporation on irrigated cereal crop yields and soil chemical properties. Soil Sci. Soc. Am. J. 46:122–126.

Huang, D. M., J. Goa, and P. Zhu. 1981. The transformation and distribution of organic and inorganic fertilizer nitrogen in rice-soil system. T'u Jang Hsueh Pao 18:107–121.

Krishnappa, A. M., and J. E. Shinde. 1980. The turnover of ^{15}N labelled straw and FYM in a rice-rice rotation under flooded conditions at two locations. J. Indian Soc. Soil Sci. 28:474–479.

Larson, W. E., R. F. Holt, and C. W. Carlson. 1978. Residues for soil conservation. p. 1–16. *In* W. R. Oschwald (ed.) Crop residue management systems. Spec. Pub. 32. American Society of Agronomy, Madison, WI.

Miller, R. H., and D. E. McCormack (coleaders). 1978. Improving soils with organic wastes. USDA Spec. Task Force, U.S. Department of Agriculture, U.S. Government Printing Office, Washington, DC.

Power, J. F. 1981. Nitrogen in the cultivated ecosystem. p. 529–546. *In* F. E. Clark and T. Rosswall (ed.) Terrestrial nitrogen cycles—processes, ecosystem strategies and management impacts. Ecol. Bull. no. 33. Swedish Natural Science Research Council, Stockholm.

Power, J. F., and J. W. Doran. 1984. Nitrogen use in organic farming. p. 585–598. *In* R. D. Hauck (ed.) Nitrogen in crop production. American Society of Agronomy, Crop Science Society of America, and Soil Science Society of America, Madison, WI.

Shi, S., Q. Wen, and H. Liao. 1980. The availability of nitrogen of green manures in relation to their chemical composition. T'u Jang Hsueh Pao 17:240–246.

Swanson, C. O. 1917. The effect of prolonged growing of alfalfa on the nitrogen content of the soil. J. Am. Soc. Agron. 9:305–314.

U.S. Department of Agriculture. 1978. Agricultural statistics. U.S. Department of Agriculture, U.S. Government Printing Office, Washington, DC.

U.S. Department of Agriculture. 1979. Animal waste utilization on cropland and pastureland. USDA Utilization Res. Rep. no. 6. U.S. Department of Agriculture, U.S. Government Printing Office, Washington, DC.

U.S. Department of Agriculture. 1980. Report and recommendations on organic farming. U.S. Department of Agriculture, U.S. Government Printing Office, Washington, DC.

Volk, V. V. 1976. Application of trash and garbage to agricultural lands. p. 154–164. *In* Land application of waste materials. Soil Conservation Society of America, Ankeny, IA.

Yaacob, O., and G. J. Blair. 1980. Mineralization of ^{15}N labelled legume residues in soils with different nitrogen contents and its uptake by Rhodes grass. Plant Soil 57:237–248.

15

Modern Techniques in Fertilizer Application[1]

Gyles W. Randall

University of Minnesota,
Southern Experiment Station
Waseca, Minnesota

K. L. Wells

University of Kentucky
Lexington, Kentucky

John J. Hanway

Iowa State University
Ames, Iowa

Techniques of fertilizer application have changed markedly through the years. Increases in rates of fertilizer application, crop area fertilized, and farm size created the need for larger and faster application methods. In addition, the fertilizer industry has had to meet the needs of the growers as they change their farm operations. Such major changes as reduced tillage, increased irrigation, foliar application, and fertilizer-pesticide combinations have contributed greatly to changes in the fertilizer industry. Improvements and changes in fertilizer materials have also necessitated changes in application techniques. In this chapter, we describe some of the modern techniques being used in fertilizer application, how these techniques may affect fertilizer efficiency, and some changes in application methods needed to meet the agronomic and economic requirements of the future.

I. NEED FOR IMPROVED TECHNIQUES

A. Fertilizer Efficiency

A major goal of many agronomists and crop producers alike is to improve fertilizer efficiency. In other words, the goal is to produce greater crop yield per unit of fertilizer applied. Economists have predicted that fer-

[1]Joint contribution from the Minnesota Agric. Exp. Stn., Univ. of Minnesota; Dep. of Agronomy, Univ. of Kentucky; and the Dep. of Agronomy, Iowa State Univ. Submitted with the approval of the Director, Minnesota Agric. Exp. Stn. as Scientific Journal Series Paper no. 13 767.

tilizer will undoubtedly become more expensive in the future due to escalating costs of natural gas, labor, mining, exploration, processing, and transportation. Some projections indicate that fertilizer prices will increase at a faster rate than crop commodity prices. Since these increased fertilizer costs are passed onto farmers, it becomes important for them to maximize fertilizer efficiency and, thus, increase profit.

New, modern application techniques provide an opportunity to increase fertilizer efficiency. For instance, applicators that will inject or band all fertilizer materials into the soil may play a significant role in conservation tillage systems. More precise metering and application of fertilizer through irrigation systems will surely improve fertilizer efficiency in irrigated, coarse-textured soils. In many cases, though, adoption of these new techniques will require modification of application equipment, new application equipment, or both, along with more intensive management by the farmer and the fertilizer dealer.

B. Environmental Considerations

Concern is being continually registered by the public toward the impact of fertilizers on the environment. Modern application techniques can play a significant role in preserving the quality of the environment while still producing optimum crop yields. For instance, subsurface application of P will significantly reduce the runoff of orthophosphate from no-till corn (*Zea mays* L.). Improved N application techniques by either timing or placement can markedly reduce the potential nitrate (NO_3^-) contamination of waters. Techniques to reduce environmental concern almost always result in improved fertilizer efficiency.

C. Time and Energy Considerations

Farmers are continually striving to reduce the time and energy expenditures necessary for maximum economic yield. Fertilizer suppliers are also trying to increase the pace of their operations while reducing the energy needed. Changes in application technique, such as bulk spreading of urea or urea–ammonium nitrate (UAN) mixtures, have allowed both the farmer and the supplier to gain time and energy efficiencies. However, some of these techniques that reduce the time and energy needed for application may not and often do not result in improved fertilizer efficiency or reduced environmental concern. Thus, the challenge is to maximize efficiency and minimize environmental concern while reducing time and energy needs, which necessitates the development of improved application equipment and techniques.

II. FERTILIZER EFFICIENCY AS INFLUENCED BY APPLICATION METHODS

Broadly, three objectives are involved in fertilizer placement. These are (i) to result in efficient fertilizer use by the plant, (ii) to prevent fertilizer salt injury to plants, and (iii) to provide an economical and convenient operation. The achievement of these objectives is determined to a large extent by certain general principles, summarized as follows:

1. Soil conditions (pH, texture, cation exchange capacity [CEC], organic matter content, bulk density, moisture, temperature, chemical content) will to a large extent govern fertilizer efficiency, that is, once a fertilizer has been added to soil, it reacts in and with soil to varying degrees as influenced primarily by the above soil factors.
2. Fertilizers have inherent chemical and physical properties that influence the degree to which they react in and with soil, such as the molecular form of a nutrient (ammonium [NH_4^+] vs. NO_3^-), chemically prepared polynutrient fertilizer (ammonium phosphate [$(NH_4)_2HPO_4$]), or a physical blend of mononutrient fertilizers (ammonium nitrate + potassium chloride + calcium phosphate [NH_4NO_3 + KCl + $Ca(H_2PO_4)_2$]), and particle size.
3. Plants take up nutrients dissolved in the soil solution largely by diffusion and mass flow and to a lesser extent from direct contact of a root with mineral or organic nutrient–containing materials.
4. The absolute amount of nutrients taken up by the plant at any one time depends on (i) the size of the plant, that is, its root system (the larger the root system, the more surface area there is to absorb nutrients and the more contact there is between soil and plant); (ii) the amount of nutrient in the soil in a form that can be extracted by plant roots; and (iii) the temperature of both soil and air. Plant metabolism is directly related to temperature. At low temperature, metabolism is greatly reduced and nutrient uptake is slow.

The challenge is how to effectively manipulate or manage these principles to do the best job of achieving the objectives of fertilizer placement under a given set of conditions. The given conditions may include such factors as the crop to be grown, variety of the crop, field to be used, cultural practices to be used, and availability of capital for investment in the crop enterprise. Before a decision on fertilization can be made, definitive information on the soil in a particular field is needed: Is it of uniform type? What are the physical and chemical characteristics of the rooting zone? What is the fertility level? Soil testing is certainly an important practice to use to obtain this latter information. A more rational decision can be made concerning fertilizer and lime usage on the basis of soil test results. Such results provide information pertinent to the first principle discussed—soil conditions.

Eventually a method of fertilizer placement must be chosen. Whether the fertilizer is broadcast and plowed down, applied in the row, deep banded, foliar applied, or is applied using some combination of these methods depends on the set of conditions. What were the soil test results for pH, P, and K? What is the soil microclimate? Is fertility being built or maintained? Is the soil acid or alkaline, and to what degree? What is the root system of the crop grown—taproot or lateral? What volume of soil will the roots contact? What is the potential for surface runoff? What fertilizer materials are available? What are their physical and chemical characteristics? What kind of equipment is available for spreading? What is the price

Table 15–1. Location and method of fertilizer placement.

Location of placement	Method of placement
Surface	Broadcast
	Stripped
	Sidedressed
	Irrigation
Subsurface	Broadcast, plowed under
	Broadcast, disked in
	Sidedressed
	Row application with seed
	Banded apart from seed
	Irrigation
Directly onto plant	Foliar sprays

of fertilizer? The final decision often reflects the various degrees of influence of the specific given set of conditions.

Basically, all fertilizers are applied either to the soil surface or under the surface. (Only a few situations exist in which they are applied directly to the plant.) Several methods are used to make surface or subsurface application. Many management systems often combine these methods. Application methods are outlined in Table 15–1. In terms of the nature of placement involved, these methods can be generally considered as either *broadcast* or *banded.*

A. Band Versus Broadcast Placement

To more clearly understand the advantages and disadvantages associated with either banding or broadcasting fertilizers, a clear understanding of how fertilizer nutrients react when placed with soil is helpful. Because of these reactions, the method of placement is often dependent on the nature of the nutrient source and the crop being grown.

1. Nitrogen

Nitrogen is commonly added in the NH_4^+, ammonia (NH_3), NO_3^- or urea chemical forms. It is utilized by plants as NO_3^- or NH_4^+. Commonly used fertilizer forms of these sources are readily soluble in the soil solution. The non-NO_3^- forms are subject to biological oxidation to NO_3^- in well-drained soils. The NO_3^- form, being an anion and thus unreactive with the cation exchange complex of soil, is readily mobile in the soil solution and moves primarily vertically as soil water moves. Thus, placement of such a mobile ion is related primarily to its effect on row spacing, germination, seedling damage, and potential losses. When NH_4^+ is added to soil, it initially reacts with the cation exchange complex to become adsorbed to the surface of clay particles. Thus, mobility is diminished and uptake by the plant is restricted largely to absorption of solution NH_4^+, which is in equilibrium with that of the exchange complex. However, during the portion of the annual season that soil conditions are favorable for aerobic biologic activity, most available soil NH_4^+ will be transformed to the mobile form, NO_3^-.

Ammonia is applied as a pressurized liquid that immediately vaporizes as the liquid emerges from the applicator. For this reason, placement of NH_3 has to be made subsurface to prevent loss of the NH_3 vapor to the atmosphere. The NH_3 readily reacts in the soil with the exchange complex or water to form NH_4^+, which subsequently undergoes biologic oxidation to NO_3^-. A method sometimes used is the *cold flow method,* whereby NH_3 is supercooled to liquid form and metered into soil ahead of an incorporating implement.

Urea, being nonionic, must be transformed to NH_4^+ before it is available for plant use. This reaction readily takes place in warm moist soil in the presence of the naturally occurring enzyme urease to form ammonium carbonate [$(NH_4)_2CO_3$]. Unless a placement method is used to incorporate urea into the soil or unless there is sufficient moisture for it to be readily moved into the soil, there is potential risk for losses of N, since $(NH_4)_2CO_3$ readily breaks down to NH_3, which can be lost to the atmosphere, especially in an alkaline medium. Losses due to this mechanism can be sizeable (Fenn & Miyamoto, 1981; Fenn et al., 1981a, 1981b). Since high concentrations of NH_3 can accumulate in the zone of urea hydrolysis, care must be taken to minimize risk of root toxicity by using low rates of urea, banding it away from the row, or diluting the NH_3 accumulation by broadcasting and incorporating. Creamer and Fox (1980) reported NH_4^+ toxicity resulting from band applications of urea or diammonium phosphate (DAP) and concluded that plant toxic accumulations of NH_3 occur near band applications of urea in all except very acid soils. They also reported that cool temperature (10 vs. 20° C) lowered nitrification of NH_4^+ enough to result in NH_3 persisting around the hydrolysis zone for longer periods of time. The NH_4^+ formed from hydrolysis of urea undergoes biologic oxidation to NO_3^- just as that from any other NH_4^+ source.

2. Phosphorus

Phosphorus is commonly applied in the orthomolecular form (PO_4^{3-}), polymolecular form, or mineral form (apatite). However, the ortho-form is primarily used. Dependent on the associated molecular cation and the degree to which the three negative charges of the orthophosphate anion are saturated by the associated cation, solubility of orthophosphatic fertilizer salts can widely vary. Also, soluble orthophosphate in soil reacts rapidly with cations, hydrous oxide coatings, or cations in solution to form reaction products of varying solubility. Since P is absorbed by plants as $H_2PO_4^-$ or HPO_4^{2-}, it is these ionic forms that need to be present in the soil solution for the plant to utilize it. The degree to which soil P reactions take place and the reaction products formed are related largely to soil pH and to relative amounts of precipitating cations on soil particle surfaces or in soil solution. Even so, orthophosphates react rapidly when added to most agricultural soils to form compounds much less soluble—and therefore less available to the plant—than of the ortho-salt added. This greatly reduces the efficiency of fertilizer P added to the crop being grown in that particular year. Hence, placement of such a material is of considerable importance.

Availability of orthophosphate is related to the solubility of the salt used and to the chemical nature of the soil to which the salts are added. By broadcasting with a water-soluble salt, solubility effects can be partially modified by the particle size of the salt added (Engelstad & Terman, 1966). The larger the particle size, the less contact the salt has with soil; consequently, due to dissolution rate of the salt, a sphere containing a greater concentration of soluble P than that from finely divided particles is maintained. These spheres of soluble P then exist as potential feeding sites for plant roots as they penetrate throughout the soil. Banding soluble phosphatic salts results in a similar effect.

From a placement standpoint with respect to P, the objective is to restrict contact of the added soluble salt with soil to realize greater efficiency of the material by reducing phosphate fixation. Even under best conditions, no more than 20 to 30% of added P is likely to be taken up by a crop during the first growing season after addition of the fertilizer.

Since P is so immobile in soil, it must be placed in such a way that plant roots will come in contact with it. Physiologically, P stimulates early root proliferation of seedlings, which directly affects growth rate (dry matter accumulation) of the plant. This early growth stimulation usually results in higher yields of short-season crops and of full-season crops grown in locations where length of growing season is marginal for such crops. For this reason, it is of special importance that P be placed in such manner that as seedling roots grow, they come in contact with the fertilizer. This implies a band or row application of a soluble P salt. The amount to be applied in this manner depends on the level of plant-available P present in the particular soil. In soils low in P saturation (high P fixation capacity), soluble P added will be utilized more efficiently from a band placement method as long as the band is located in such position that plant roots contact it. If broadcast applications of soluble P are made on such soils, granulated fertilizers will be more efficient than pulverized fertilizers due to the solubility pattern previously explained.

If insoluble forms of P (e.g., phosphate rock [apatite]) are used, it should be placed in such a way as to enhance reaction (maximize contact with soil). Enhancement can be accomplished by increasing the surface area of the material (finely ground) and adding it to the soil in such a way as to maximize contact (broadcast and disk in). For this reason, efficiency of insoluble phosphates can be increased by using finely ground material and thoroughly mixing them in the soil. This effectively increases P solubility of such materials in the soil, that is, the resultant reaction products are more soluble than the material added.

One point that should be made here is the interaction between N and P. Evidence has been accumulated to show that P efficiency is improved by adding it with N. For example, the reason DAP has been shown in some experiments to be more efficient than superphosphate is thought to be due to the presence of NH_4^+ at the same fertilizer site with P. Band application of phosphatic fertilizers has been thoroughly reviewed by Richards (1977), and this publication is recommended for those desiring more information on the topic.

3. Potassium

Potassium is commonly applied to soil as soluble Cl^-, sulfate (SO_4^{2-}), or nitrate salts. Upon dissolution in the soil solution, K exists as the cation K^+, which reacts with the soil cation exchange complex to become relatively immobile. The K^+ ion can also react at interlayer exchange sites of expanded clay particles and become fixed or unavailable to plants. Most added K^+, however, is likely to become adsorbed to cation exchange surfaces and become available for plant use from solution and by contact exchange. The important aspect of K fertilization with respect to placement is that it is relatively immobile and does not react in most cases to become unavailable to plants; therefore, there is no consistent effect of placement on efficient plant utilization of added K^+. As a soluble salt, however, it can contribute to salt injury during germination and seedling growth. For this reason, care must be taken when a seed-salt contact placement is being used. For the same reason, when large amounts are being used, placement should be made to minimize salt damage (band away from seed or broadcast).

4. Secondary Nutrients

The secondary elements Ca and Mg are slightly mobile in the soil. Calcium and Mg attach to the exchange complex of the soil and act much like K. Plant uptake mechanisms are also similar to K. Dolomitic lime, a common source for both Ca and Mg, is usually broadcast-applied because of the higher application rates often needed to reduce soil acidity. Gypsum ($CaSO_4$) also provides Ca and is generally broadcast-applied. Magnesium can be supplied as potassium-magnesium sulfate ($K_2SO_4 \cdot 2\,MgSO_4$) and is also broadcast applied. However, row applications of Mg to shallow-rooted crops such as potatoes (*Solanum tuberosum* L.) are sometimes more efficient.

Sulfur in the SO_4^{2-} form is quite mobile and is usually broadcast-applied as gypsum or elemental S. The elemental S is oxidized to the SO_4^{2-} form, which is the form taken up by plants. Special instances occur where it is desirable to localize or band-place the S fertilizer. Liquid formulations of S, such as ammonium thiosulfate [$(NH_4)_2S_2O_3$], are becoming more frequently added to liquid fertilizers for band placement to row crops. Highly acidic or acid-forming compounds of S (e.g., sulfuric acid [H_2SO_4] and ferrous sulfate [$FeSO_4$]) are sometimes band-applied to highly calcareous soils where acidification of a band or small soil area within the plant rooting zone is necessary.

5. Micronutrients

Micronutrient fertilization is quite complex due to the many different forms of micronutrients available, the crop management system, and the uptake mechanisms and specific needs of the crop. As a result application methods vary widely. Soluble nutrients, such as B, and the sulfate sources of Zn, Cu, and Mn are often broadcast on the soil surface. The more insoluble sources are added as solid compounds or in some suspended form to the macronutrient carriers that are either solid, liquid, or suspension grades. Relatively expensive materials are generally banded to row crops for economical reasons. On the other hand, specific micronutrient needs of intensive, high-value citrus (*Citrus* spp.) and vegetable crops are usually met through broadcast, foliar applications. When the micronutrient is applied with the macronutrient, the application method is usually determined by the nature and relative efficiency of the macronutrient rather than the micronutrient. An example is the common introduction of relatively small amounts of micronutrients into banded starter fertilizers (see section IIC2). Murphy and Walsh (1972) provide a detailed discussion of the fertilizer sources and the application methods used to provide improved efficiency of each of the micronutrients.

B. Dual Placement of Nitrogen and Phosphorus

Considerable attention has been given in recent years to simultaneously injecting anhydrous NH_3 and liquid $(NH_4)_2HPO_4$ solutions in a band 10 to 15 cm below the surface. Subsequent tillage after this dual placement does not disturb these bands; thus, the practice allows for once-over, preplant fertilization with the band advantage for phosphate efficiency and the added advantage for phosphate efficiency of having N and P at the same site. Use of anhydrous NH_3 also gives an economical advantage to this practice. Much of the developmental work for this practice has been conducted by Murphy (1981) and Leikam et al. (1979) in Kansas. Their results show significant advantage from preplant dual application on yield and P uptake on low P soils. Their work has shown the increased P uptake to be related to use of NH_4^+ and that it was necessary for NH_4^+ and P to be in the same soil retention zone to obtain best results.

Lamond and Moyer (1983) found in Kansas trials on soils testing low to medium in P that knifed (subsurface band) fertilization of UAN potassium tripolyphosphate (0–11–21 [0–25–25]) or ammonium polyphosphate (10–15–0 [10–34–0]), and KCl (0–0–50 [0–0–60]) significantly increased tall fescue (*Festuca arundinacea* Schreb.) yields, N content, and N, P, and K uptake compared with surface broadcast fertilization. Higher N uptake indicated higher availability and recovery of N, possibly due to less N volatilization, reduced N tie-up by soil microorganisms, or both. When forage yields were averaged across all fertilizer treatments, the knifed method of application increased yields by 9% in one experiment and 24% in the other.

Wisconsin data (Fixen & Wolkowski, 1981) compared dual N-P placement of anhydrous NH_3 and ammonium polyphosphate with a 5 by 5 cm band placement of N-P for corn grown on a Plano (fine-silty, mixed, mesic Typic Argiudolls) silt loam in 1980 and 1981. Little advantage was shown for the dual N-P application over the conventionally used N-P starter in a 5 by 5 cm band.

The dual N-P application concept was summarized in Nebraska by Penas (1981). He indicated the practice has been found to work best with wheat (*Triticum aestivum* L.) and that the preplant application should be made in a 25- to 50-cm band spacing, with a 30- to 38-cm spacing being ideal.

C. Time of Application

The time at which fertilizer is applied does not always show a direct cause and effect relationship between fertilizer efficiency and method of application. However, the trend toward large farming operations of 300 to 1000 ha that may require seedbed preparation and planting in a relatively short period of time has placed increased emphasis on applying fertilizer prior to the annual spring planting rush. The fertilizer industry has also shown much interest in this as an attempt to spread out their distribution season in such a manner as to minimize shortage of local fertilizer supplies due to transportation problems usually encountered during the rush spring season.

Apart from the convenience and economics usually associated with the purchase and application of fertilizers 8 to 30 weeks ahead of an intended crop, consideration should largely be given to climatic factors involved, mobility of the fertilizer nutrients to be applied, and soil test levels of nutrients.

1. Climatic Factors

Soil temperature, soil moisture content, and air temperature should also be considered in determining when to apply fertilizers. Perhaps the climatic effect of greatest concern is the effect of temperature on nitrification. As previously mentioned, NH_4^+ will react with soil strongly enough to withstand leaching. However, nitrifying bacteria in soil oxidize NH_4^+ to NO_3^-, in which form it is subject to potential loss either by leaching or denitrification. The rate of nitrification is temperature dependent, being almost static at the freezing point, but increasing to the point that at around 10° C the conversion rate is considered to be rapid. To lower the risk of leaching or denitrification losses, one must not apply N so far ahead of need by the crop that losses would likely be encountered. For this reason, fall application of N for corn is a risky practice except for climatic areas where air and soil temperatures are low enough by late fall or early winter that nitrification is negligible and where they stay that low until late winter or early spring. The middle and northern Corn Belt is the largest N-using area where this practice can feasibly be used. Fall application to coarse-textured soils even under these cold conditions is still not recommended due to the high leaching potential. Injection of anhydrous NH_3 to normal depths of 15 to 25 cm also represents a mechanism to delay nitrification since soil temperature at such depths is cooler than at the surface. The highly concentrated NH_3 band is also sufficient to delay rapid conversion to NO_3^- by temporarily inhibiting the nitrifying organisms.

In recent years, nitrification inhibitors have been developed and are commercially available for use as additives to N fertilizers as a means of suppressing nitrification by decreasing activity of the nitrifying bacteria *Nitrosomonas*. Reports from their use indicate greatest advantage when used on soils that have a substantial potential for denitrification during spring and early summer.

2. Mobility of Fertilizer Nutrients

Since the immobile fertilizer nutrients move such short distances from the soil-fertilizer reaction zone, plant uptake can be insufficient in soils cool enough in the rooting zone to delay root growth and thus reduce the area of absorbing root surfaces. Band or seed placement is often used in such soils to help overcome low mobility. Phosphorus and, to a lesser extent, K are the nutrients most often used in this manner. A small amount is applied in the row or in a band at the time of planting to improve early start of seedlings and is therefore called *starter fertilizer*.

3. Soil Test Levels

At high soil test levels, it would generally be immaterial as to when fertilizer is applied. In fact, at high soil test levels, fertilizers may not even be justified on an annual cost-and-return basis except for those who place value on maintaining soil test levels in the high range. However, at low soil test levels of the immobile nutrients, rates normally recommended for broadcasting can be reduced 50% if banded. Banding, of course, usually implies application near the row at planting, meaning that if the lower rates made possible by banding are to be used application would not be done until planting.

D. Residual Nutrient Level

When residual nutrient levels in soil have been built to or naturally occur at high levels, little agronomic growth or yield response can be expected from further application of the fertilizer nutrient being considered. Thus, under these soil test conditions, the method of fertilizer application is generally of little significance.

High residual nutrient levels have received considerable attention since the mid-1970s and was the subject of a special symposium held during the 1981 annual meetings of the American Society of Agronomy. During this symposium, Cope and Khasawneh (1981) reported that there has been a general and often substantial increase in residual soil test levels of P and K in the USA during recent years. Crop response data from field studies in the southeastern, midwestern, and western U.S. states consistently showed little or no yield increase to application of P and K to soils testing high or above (Parks et al., 1981; Randall, 1981b; Westfall, 1981). Parks and Walker (1969) also reported from field studies with corn that as residual soil levels of K increase, the effect of band placement of fertilizer K decreases. Cope (1981) reported on extensive, long-term residual soil test studies conducted in Alabama with several crops and showed that P-K rates that gave the highest economic returns also resulted in increased residual soil test values. Further, such increased residual levels occurred despite annual crop removal values often exceeding economically optimum P-K rates. Olson et al. (1982) reported on a comprehensive 8-yr field study conducted on four major soils in Nebraska with corn. Treatments tested at each site were fertilizer recommendations made by five soil testing laboratories on the basis of either cation balance, nutrient replacement, or nutrient sufficiency philosophies. Despite large differences in amount and kinds of fertilizers recommended on the basis of these three philosophies, no differences in yield were obtained. These results substantiate use of the nutrient sufficiency philosophy both agronomically and economically when the rate of fertilizer to use at a given residual nutrient level is determined.

Peaslee (1978) reported on field studies conducted in Kentucky to determine P fertilizer rates necessary to get near maximum yields at various residual levels of available soil P. His results, shown in Table 15–2 and Fig. 15–1 and 15–2 show that at Bray and Kurtz no. 1 P extraction levels > 44 kg/ha, no further growth response from corn, soybeans [*Glycine max* (L.) Merr.], alfalfa (*Medicago sativa* L.), or clover (*Trifolium* spp.)–grass was obtained.

A commonly followed rule regarding band and broadcast rates of P is that where a P response is likely (low residual soil P, cool soil tempera-

Table 15–2. Effect of residual soil test P level on fertilizer P rates for near-maximum yields (Peaslee, 1978).

Bray and Kurtz no. 1 P level	Fertilizer P required for 0.95 maximum yield			
	Corn	Soybeans	Alfalfa	Clover-grass
kg/ha	kg/ha per year			
0	55	35	62	47
6	45	27	55	40
11	37	17	42	32
22	25	5	27	12
33	12	--	12	--
44	--	--	--	--

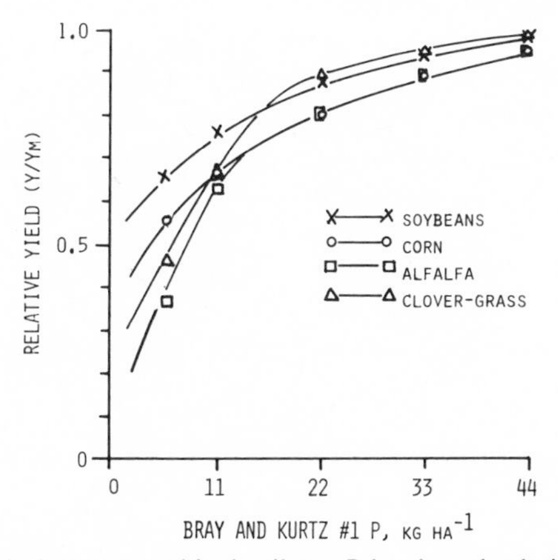

Fig. 15–1. Relationship between residual soil test P levels and relative yield of corn, soybeans, alfalfa, and grass-clover (Peaslee, 1978).

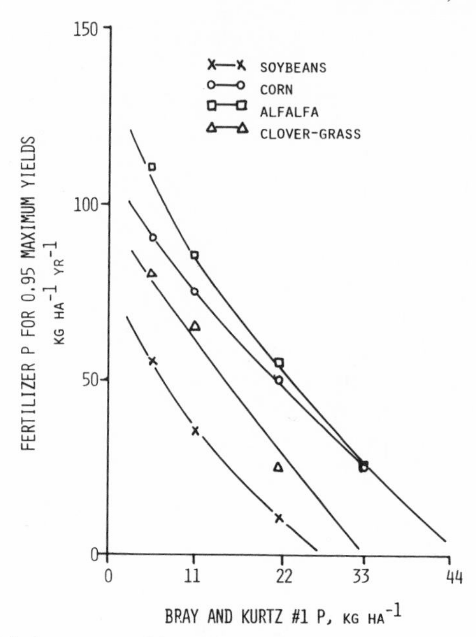

Fig. 15–2. Relationship between residual soil test P levels and fertilizer P required for near-maximum crop yields (Peaslee, 1978).

tures, or both), an economic efficiency can be obtained by banding fertilizer P; banded P is generally considered to be 1.5 to 2 times more effective than broadcast P (i.e., P rates can be cut by one-third to one-half if banded). A Nebraska study on winter wheat was recently reported by Peterson et al. (1981) to test this practice, which they reported to be popular in the High Plains area of Nebraska. After testing several rates of P applied either broadcast or drilled in the row with the seed, they concluded that the relative efficiency of band vs. broadcast P varied with residual soil P levels (Table 15–3). For low-testing soils the relative efficiency of banded P was about 3:1 compared with broadcast P, whereas for medium-testing soils the relative efficiency ratio was 1:1. Figure 15–3 shows the relationship they found between amounts of row and broadcast rates of fertilizer P required for equal yields of winter wheat grown at various residual soil P levels.

The effect of climate on band vs. broadcast P at various residual soil P levels can be seen in contrasting the Nebraska data cited above to results from dryland spring wheat studies in North Dakota (Alessi & Power, 1980). They conducted a comprehensive study for 6 yr by banding 15 kg P/ha per year on either previously unfertilized plots or on plots that had initially received 20, 40, 80, or 160 kg P/ha broadcast. Results indicated that wheat yields were increased from use of the banded 15 kg P/ha per year of all levels of residual P found in their study. Consequently, they concluded that banded P for spring wheat production in the cool spring climate of the Northern Great Plains was justified.

Similar results were reported from a field study conducted in Ontario (Sheard, 1980). This study was conducted to test the effect of banded P ei-

Table 15-3. Comparison of broadcast and row applied fertilizer P at different soil test residual P levels on winter wheat yields (Peterson et al., 1981).

Bray and Kurtz no. 1 P level	Calculated fertilizer P rates: Broadcast for maximum yield	Calculated fertilizer P rates: Row application for same yield as broadcast	Ratio of row to broadcast
kg/ha	kg P/ha		
18	37	12	0.32
22	45	28	0.62
31	45	30	0.67
38	NS†	NS	1.00
43	NS	NS	1.00
43	27	22	0.82

† No significant difference between broadcast and row rates.

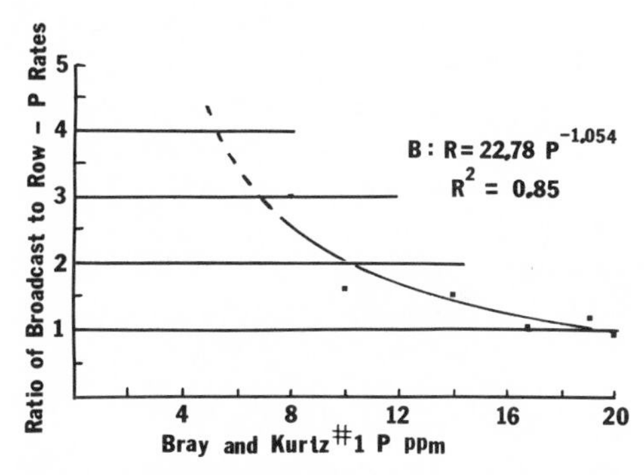

Fig. 15-3. Relationship of soil test P level and the ratio of broadcast and row P fertilizer rates required to obtain an equal grain yield response (Peterson et al., 1981).

ther alone or in the presence of ammonium-N (NH_4^+-N) or nitrate-N (NO_3^--N) on establishment of forage species. Results showed as much as a fivefold increase in seedling growth with up to 30 kg P/ha regardless of the forage species or residual soil P level used in this study.

III. FERTILIZER PLACEMENT WITH REDUCED TILLAGE

Due to concern over erosion control and fuel costs, there is currently much interest in crop production systems requiring less mechanical seedbed preparation than the common methods involving plowing, disking, and spike-tooth harrowing to prepare a smooth, clean seedbed. Several reduced tillage techniques have been developed, but two have become very popular: chisel plowing and no-till planting. Chisel plowing reduces the amount of field surface area inverted during primary tillage; no-tilling results in a soil disturbance of only 2.5- to 7.5-cm width for each planted row and for all practical purposes represents use of no primary tillage. Because of less soil disturbance involved with either no-till or chisel plowing, there has been much interest in how vertical fertilizer distribution affects root zone fertility levels, particularly with respect to fertilizer application methods necessary to obtain best crop production.

Several authors, including Wells (1984), discuss the effect that continuous no-tillage has on soil properties. Phillips et al. (1980) defines the no-tillage system as:

> one in which the crop is planted either entirely without tillage or with just sufficient tillage to allow placement and coverage of the seed with soil to allow it to germinate and emerge. Usually no further cultivation is done before harvesting. Weeds and other competing vegetation are controlled by chemical herbicides. Soil amendments, such as lime and fertilizer are applied to the soil surface.

Review of several papers (Triplett & Van Doren, 1969; Moschler et al., 1973; Bandel et al., 1975; Blevins et al., 1977; Blevins et al., 1978; Doran, 1980; Phillips et al., 1980; Blevins et al., 1980, Randall, 1980; Moncrief & Schulte, 1982; Ketcheson, 1980; Lal, 1979) shows the following effects of no-tillage on soil properties:

1. Soil moisture content. Because of less surface evaporation and better infiltration of rainfall, there is usually 15 to 25% more available soil water during the growing season with no-tillage compared with conventional tillage.
2. Soil temperature. Due to the insulation effect from the surface mulch of killed sod, killed small grain, or previous crop residues, soil temperature changes more slowly under no-till conditions. This results in cooler soil temperatures during the growing season and much less diurnal fluctuation.
3. Organic matter. In continuous no-till systems, organic residues from the previous crop collect on the soil surface with subsequent redistribution of soil organic matter compared with conventional tillage. Greater root growth activity in the top 5 cm of soil resulting from no-tillage also adds to increased organic matter content in the uppermost part of the soil profile. As a result, organic matter content becomes concentrated at the soil surface with continuous no-tillage, whereas it is mixed somewhat uniformly to plow depth in conventional tillage.
4. Microbial activity. As would be expected from more moisture and accumulation of a concentrated layer of organic residues at the soil surface, there is greater surface microbial activity in no-tilled soils compared with conventionally tilled soils. Greater numbers of both aerobic and anaerobic bacteria have been measured under no-tillage compared with conventional tillage. Even though large numbers of aerobes are present, the relatively large presence of anaerobes, together with a high concentration of organic materials and a higher soil moisture content, results in less potential for oxidation in no-tilled soils compared with conventionally tilled soils.
5. Residual soil N. Residual soil N content increases in continuous no-till production despite the increased potential for both leaching and denitrification of NO_3^-. It appears, based on the few studies reported, that this increased residual soil N results from the increased

concentration of organic residues near the soil surface and that it is in organic form. Indications are that mineralization of this labile pool of organic N is sufficiently slower in no-tillage that an increased residual N content develops, even though it is only slowly available to crops in any one year. Because of this, NO_3^- soil tests are not likely to indicate the presence of this labile pool of organic N.

6. Bulk density. Some studies have shown little difference in long-term effect of no-tillage on bulk density, indicating that even though annual seedbed tillage in conventional systems initially lowers the plow layer bulk density relative to no-till, subsequent recompaction of the plow layer through the growing season and overwinter results in bulk densities of near the same value as those measured in no-till. Other studies, notably those from the northern Corn Belt have indicated that unless the plow layer receives some manner of primary tillage, bulk density is high enough to create problems with some nutrients, particularly K. Because of this, it has been suggested that fields should test high in exchangeable K^+ or starter fertilizer containing K should be used for no-tilling in the northern Corn Belt.
7. Soil pH. Studies indicate that after 3 to 4 yr of continuous no-till production of corn, a very acid layer 1 to 5 cm thick develops at the surface of the mineral soil. This accelerated acidification related to no-till has been explained as resulting from decomposition of the concentrated layer of organic residues on the surface with subsequent intensive leaching by the resultant organic acids in a thin layer at the top of the soil profile. As with conventional tillage systems, this acidification process is increased by use of acid-forming fertilizer N.

It should be emphasized that the effects listed above are largely based on continuous, long-term use of no-till techniques. From the practical standpoint of developing crop production systems to utilize no-till techniques, particularly those involving rotations, it is debatable as to whether producers will use no-till continuously in the same field. Most studies reporting on such long-term effects have lasted 5 to 10 yr due to the relatively recent advent of no-till production systems. Although these studies provide the best current knowledge of no-till effects on soil properties, one should keep in mind that 5 to 10 yr is probably not long enough to evaluate long-term effects and at best likely represents only directions of change that take place. Care should be taken in interpolating these effects to individual production systems where the field is not continuously no-tilled for long periods of time.

A. Tillage Effects on Mineralization, Incorporation, and Nutrient Stratification

Randall (1980) reported results of 10 yr of field experiments conducted to study some of the fertility problems involved in shifting from

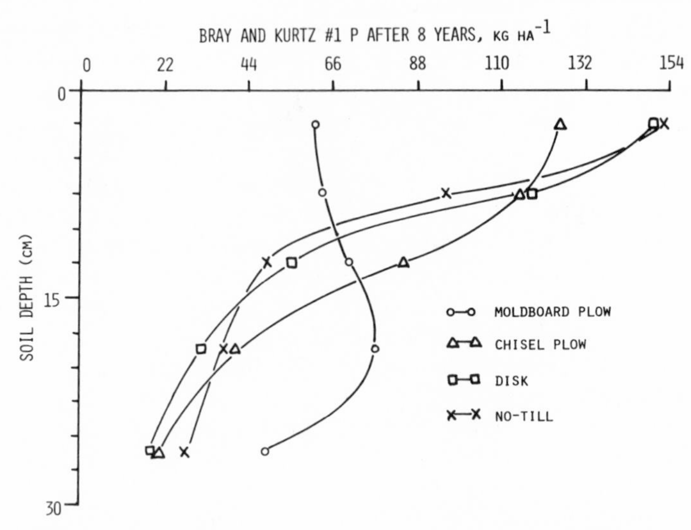

Fig. 15–4. Effect of tillage system on root zone distribution of soil test P after 8 yr (Randall, 1980).

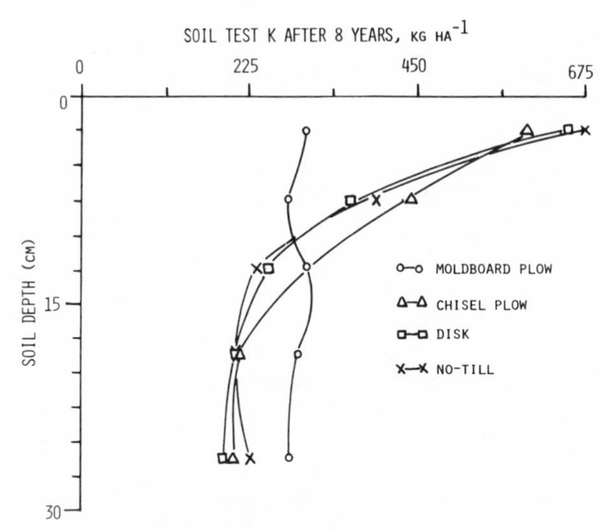

Fig. 15–5. Effect of tillage system on root zone distribution of soil test K after 8 yr (Randall, 1980).

conventional tillage practices to practices involving less tillage on Webster (fine-loamy, mixed, mesic Typic Haplaquolls) soils in Minnesota. Results from some of these studies (Fig. 15–4 and 15–5) show that with reduced primary tillage, much of the added P and K, along with that mineralized from crop residues, accumulates in the top 10 cm of soil. These results contrasted with a fairly uniform distribution within a 30-cm plow depth that resulted from conventional moldboard plow primary tillage each year and mixing to 15 cm with annual chisel plowing. Alternating moldboard plowing every other year (or less often with a reduced tillage system) was suggested as a means to uniformly distribute P and K to greater depths. He reported that even though early growth and nutrient uptake of corn was slower for no-till due to cooler soil temperatures associated with the presence of the surface mulch, the resultant fertility effect on yields was of little importance, with the exception of K when soil test K levels were less than optimum. Moncrief and Schulte (1982) reported similar results from Wisconsin, concluding that unless soil K levels were optimum for fields to be no-tilled or till-planted, a starter K application was necessary to get yields equivalent to those from conventionally plowed fields. They suggested that decreased aeration of the rooting zone resulting from no-till or till-planting resulted in reduced K availability compared with conventional tillage. They attributed the decreased root zone aeration to higher soil bulk density measured in their experiments that had been no-tilled. In their studies, they obtained higher corn yields from no-tillage than from conventional tillage on a well-drained, loessial Fayette (fine-silty, mixed mesic Typic Hapludalfs) silt loam soil when N and K levels were optimized. On a Plano silt loam soil, till-plant and conventional moldboard systems produced better yields than chisel plowing.

Although Randall (1980) reported lower corn yields over a 10-yr period from no-till compared with moldboard plowing, he attributed the difference to poorer weed control, increased bulk density, cooler soil temperature, and delayed emergence rather than to soil fertility or plant nutrition. He also found lower NO_3^- values in the soil profile from no-till compared with moldboard plowing, with chisel plowing being intermediate. This was attributed to greater N losses and poorer fertilizer N efficiency with no-till.

Ketcheson (1980) observed that no-till corn grown on medium- and fine-textured soils in Ontario was producing less than conventionally grown corn. Based on these observations, he tested higher than recommended rates of NPK to determine if no-till yields could be improved. Despite incorporating different P-K rates to a 15-cm depth before conducting the experiment, he was unable to obtain improved no-till corn yields and found tissue concentrations of NPK to be adequate regardless of tillage. He suggested that poor no-till yields were due to higher soil bulk density with no-tillage compared with conventional tillage.

In contrast to some of the results indicated in the foregoing studies conducted in the northern Corn Belt, residual soil P-K levels have not appeared to greatly influence yield of no-till corn grown in the mid- to upper South. Kentucky (Belcher & Ragland, 1972; K. L. Wells, 1977, unpublished data) and Virginia (Singh et al., 1966; Moschler & Martens, 1975) data indicate that no-till corn yields equivalent to or greater than from conventionally tilled corn can be obtained from unincorporated surface broadcast applications of P and K even on low-testing soils. This is attributed to increased rooting activity at or near the soil surface resulting from no-till production.

B. Residue Effects on Nutrient Availability and Efficiency

The foregoing studies indicate rather clearly that the differing degree to which crop residues are incorporated into soil as affected by tillage technique causes variations in available nutrient content within the rooting zone. On one extreme, continuous no-tillage results in greater accumulation of nutrients within the top 5 to 10 cm of soil since crop residues and fertilizer are not incorporated into the rooting zone each year. On the other extreme, continuous moldboard plowing results in a rather uniform distribution of nutrients to plow depth since both crop residues and fertilizers are incorporated into the rooting zone each year. The effect of chisel plowing appears to be intermediate between no-tilling and moldboard plowing.

Studies reported indicate that nonuniform mixing of crop residues and fertilizers within the rooting zone is more likely to unduly influence yields in the northern Corn Belt than in the mid- to upper South. Data tend to suggest that this response is in some way related to the cooler climate of the northern Corn Belt and to higher root zone bulk density of soil measured under no-till in those studies. This has led some states to recommend slightly higher N rates for corn grown under no-tillage compared with conventional tillage.

Despite results of studies that clearly show lower availability of residual soil N under no-till due to immobilization, leaching, denitrification, or some combination thereof, yields of no-till corn have often been shown to be equivalent to or better than conventionally tilled corn, particularly at optimum fertilizer N rates when other nutrients are not limiting. In fact, Lal (1979) reported from long-term no-till corn studies conducted on an Alfisol in the tropics that even though lower yields were measured initially from 0-N no-till treatments, they progressively improved and exceeded yields of the 0-N conventionally tilled treatments after 6 yr. He attributed this to a progressive increase or improvement in soil organic matter content, N and P levels, CEC, soil structure, and soil water–holding capacity in no-till plots and to a decline of these characteristics due to erosion in the conventionally tilled plots.

C. Placement Methods to Maximize Fertilizer Availability in a Reduced Tillage System

Since N is so mobile in soil, there is no great concern about placement method for row crops with respect to positional availability. However, there is concern about N efficiency with reduced tillage systems, just as with conventional tillage systems. The likely routes of inefficiency are the same—leaching, volatilization, denitrification, and immobilization. Soil and climatic characteristics will often be the determining factors as to which route or routes of N inefficiency will be most likely. Such practices as incorporation, split application, surface broadcasting, band application, or use of nitrification inhibitors should be considered depending on the specific cropping systems and the various soil and climatic situations.

Of the reduced tillage systems, it appears that no-till would require more concern about placement with respect to fertilizer N efficiency since fertilizer N is often surface-applied. This is due to greater potential risks for leaching, volatilization, denitrification, and immobilization. Most reported data suggest that potential risk of surface losses from surface application of urea to no-till corn is greater than from NH_4NO_3 or UAN solutions (Bandel et al., 1980; V. A. Bandel, 1981, personal communication; McKibben, 1975; Fox & Hoffman, 1981). Kentucky data reported by Wells et al. (1976) indicated that although results from surface application of either urea, NH_4NO_3, or UAN solution were variable, the field results obtained provided little basis to discriminate among these sources for no-till corn.

Tomar and Soper (1981) reported on field experiments with barley (*Hordeum vulgare* L.) grown in Manitoba, Canada, where urea tagged with ^{15}N had been applied either banded 10 cm deep or broadcast on the surface. They measured more efficient recovery of fertilizer N from the bands (42–53%) than from that broadcasted (23–43%). Measurements also indicated negligible leaching of fertilizer N and nearly the same degree of gaseous losses (15% from bands and 13% from broadcast). On this basis, they concluded that superiority of the banded N was due to less immobilization from the subsurface banded urea.

Results from Indiana studies reported by Griffith (1974) and more recently by Mengel et al. (1982) and Maryland studies (V. A. Bandel 1982, personal communication) indicate that subsurface application of fertilizer N improves its efficiency for use by no-till corn. This practice could lower the risk for volatilization and immobilization. Although these studies show promise for better fertilizer N efficiency from subsurface application, commercially available subsurface application machines have not been available for use with strictly defined no-till systems. However, applicators to inject fertilizer satisfactorily into the soil in very reduced tillage systems became available in the early 1980s and are now showing widespread acceptance.

There is reason for greater concern about availability of the immobile nutrients than for mobile nutrients as influenced by placement method. As previously indicated, efficiency of fertilizer P and K in low-testing soils can be materially increased by positionally placing the fertilizer in such a way that plant roots can intercept the placement zone where soil fixation of the fertilizer P and K is minimized or in soils where root growth is inhibited. As discussed by Randall (1980), possibilities for overcoming the surface stratification of P and K that results from reduced tillage techniques would be to periodically use the moldboard plow to redistribute them more uniformly to greater depths or to inject fertilizer materials or manure below the soil surface. Also, the data reported by Moncrief and Schulte (1982) show a benefit from row or band applications, particularly for K with reduced tillage systems. From studies reported, this appears to be more important in cooler climates with shorter growing seasons than in the upper to mid-South, where no-till studies have shown surface application of P and K to be as effective as row applications, even on low-testing soils.

IV. FERTILIZER APPLICATION WITH IRRIGATION

The application of fertilizers in irrigation water, commonly called *fertigation,* has become a well-established agronomic practice. Over the years the technology of this fertilizer application method has changed markedly. Two major problems encountered with this application method have been the lack of uniform fertilizer distribution and the fixation of P as well as some micronutrients in the soil surface where they are essentially unavailable for root uptake. Remarkable progress on both problems has been made. Irrigation management (i.e., rate of water application, distribution of water, and methods of irrigation) has improved tremendously. New fertilizer sources and methods of introducing the fertilizer into the irrigation water have been developed. All of these developments in technology have led to a vast improvement in fertigation as a method to apply fertilizer and increase efficiency.

Irrigation water is often the most convenient and inexpensive method of applying fertilizers. Some of the advantages for applying fertilizers with irrigation water include (i) lower cost of energy, labor, and equipment; (ii) greater flexibility in choice of time and rate of fertilizer application (particularly with N); (iii) precise application/distribution of nutrients with some low-pressure systems; and (iv) reduced leaching losses of mobile nutrients (e.g., N on coarse-textured soils with proper application techniques). The potential disadvantages are (i) fertilizer distribution is determined by water distribution; (ii) precise application rates are not always possible with some of the water application equipment in current use; (iii) fertilizer can be lost in runoff water; (iv) anhydrous or aqua NH_3 applied in surface or sprinkler irrigation water is subject to loss by volatilization; (v) N may be lost by volatilization from NH_4^+-containing fertilizer applied to the surface of calcareous soils; (vi) there is stratification on the soil surface or localized concentration near the emitter of immobile nutrients (e.g., P,K, and some micronutrients); and (vii) precipitation problems may result when some forms of nutrients are introduced into irrigation waters high in Ca^{2+}, Mg^{2+}, and bicarbonate (HCO_3^-).

In spite of these potential disadvantages, the concept of fertigation is becoming increasingly popular. New developments in irrigation technology along with irrigation and fertilizer management, which has led to improved water and fertilizer efficiency, and the economics of the concept have largely been responsible for the rapidly expanding growth of this fertilizer application method.

A. Irrigation Methods

As the demand for and price of water increase in irrigated agriculture, crop producers must maximize water use efficiency (WUE) while maintaining optimum yields. These improvements in WUE and fertilizer nutrient utilization were accomplished through a number of changes in irrigation methods since the 1960s. Formerly flood irrigation was most popular, especially in the western USA. The development of the center-pivot system has now resulted in a rapid expansion in sprinkler irrigation throughout the USA. Efficiency of water use when applied by center pivot is about 80% compared with 60 to 70% with other sprinkler methods (G. W. Rehm, 1982, personal communication). In areas where agriculture is more intensive and water more scarce or costly, trickle/drip and subsurface irrigation systems have become popular. A thorough discussion of these irrigation methods as they relate to N application has been prepared by Gardner and Roth (1984).

1. Flood Irrigation

Flood irrigation involves applying water from a source (ditch, canal, or pump) at the high end of the field and then allowing the water to flow via gravity as a sheet or in furrows to the lower end of the field. Concern in this system arises from the potential nonuniform distribution of water from the high end to the lower end and, thus, a poor distribution of fertilizer. This can be improved greatly by using proper management and shorter furrow runs.

As with all irrigation systems, the metering of fertilizer into the irrigation water is extremely important. Regardless of the type of fertilizer need, the fertilizer should be applied at a point several meters upstream from the field outlets and preferably at a weir box, check drop, or some point where water turbulence will allow for maximum mixing of the fertilizer (Gardner & Roth, 1984).

2. Sprinkler Irrigation

The sprinkler system involves the transfer of water via a pump to a series of overhead or aboveground sprinklers. These sprinkler systems can be solid set, moved by hand or machine, or moved hydraulically, which is the case with most common center-pivot systems. As with flood irrigation, a major disadvantage of this method is lack of uniform distribution, especially when wind speeds are high. Distribution patterns can be greatly affected by nozzle malfunction (plugged, damaged, improper size, etc.), wind speed that causes drift of the water, location and distribution of the nozzles or sprinkler heads, and speed or consistence of movement of the sprinkler system. All of these would, of course, effect fertilizer distribution as well.

Major advantages of the center-pivot sprinkler system include (i) both water and fertilizer application is easy and timely; (ii) the system is capable of irrigating strongly sloping soils (because erosion can be a problem on severe slopes, the operation must be watched carefully); and (iii) the system does not compact soils greatly, damage much of the crop, or require that noncrop strips be established for movement of the irrigation system.

Fertilizer must be injected continuously with center-pivot, traveling gun, and traveling lateral systems from the start of fertilization until the system has moved across or covered the field. Some fertilizer will remain on the crop after irrigation, but because of extreme dilution, little or no damage to plants by the fertilizer has been reported (Gardner & Roth, 1984).

3. Trickle/Drip Irrigation

With trickle or drip irrigation, water under low pressure is delivered through pipes to drippers or emitters or through tubing with equally spaced emitters at a very slow rate. The soil is near saturation at the emitters, with gradual decrease in soil moisture content in all directions away from the emitter. The amount of water delivered to a plant or row can be controlled by number and spacing of emitters and the rate of flow from each emitter.

Two features are characteristic of this irrigation method: (i) high frequency of irrigation (daily in many cases) and (ii) localized water application to only part of the crop's potential root zone. Thus, this system may be designed for use on row crops as well as trees and orchards. It is especially common among commercial fruit and vegetable growers. California scientists provide a detailed discussion of drip irrigation management (Anon., 1981b).

Clogging of orifices by particles or organic slime is one of the major problems with the trickle or drip system. Bucks and Nakayama (1980) suggest that any fertilizer or chemical added to a drip system meet the following criteria: (i) not clog or corrode any component of the system, (ii) be safe for field use, (iii) be water soluble, (iv) not react adversely with irrigation waters, and (v) not reduce crop yields. Addition of NH_3 or ammonium phosphates into water containing high amounts of Ca or Mg can present extensive precipitation problems. Nitrogen applied continuously through the system can increase microbial growth and thus cause clogging problems. Therefore, the system should be rinsed after each application.

Recognized advantages of applying fertilizers and chemicals through the trickle or drip irrigation system are improved efficiency, labor savings, energy savings, and flexibility of timing nutrient application to crop demand, regardless of growth stages or accessibility of the field to machinery (Anon., 1981a). Another advantage of this system is that the crop producer can maintain part of the root zone at optimal soil water and nutrient content for vigorous plant growth while leaving the unwetted soil volume for storage of rainfall. This advantage would help alleviate leaching losses by decreasing the use of soil as a storage reservoir for nutrients. Sammis (1980) reported highest WUE was obtained with trickle and subsurface systems when potatoes were grown on a sandy loam soil in New Mexico where rainfall contributed a significant portion of the water requirements. Field trials in Israel (Goldberg & Shmueli, 1970) demonstrated that trickle irrigation increased yields by 30% or more than furrow or sprinkler irrigation. Bernstein and Francois (1973) found that trickle irrigation required about one-third less water than furrow or sprinkler irrigation for bell peppers (*Capsicum annum* L.).

4. Subsurface Irrigation

A recent analog to the above trickle/drip system is the placement of the emitter lines and tubing below the soil surface. The obvious advantage of this system is to prevent disturbance of the emitter lines during tillage operations, which would make it possible to extend the system's functional life, thereby reducing cost and making it feasible to irrigate lower value crops. Mitchell (1981) concluded that a properly installed subsurface irrigation system using anhydrous NH_3 as the N source can have a functional life of at least 7 yr. In addition, N use efficiency was increased and P and K concentrations in the corn earleaf were improved compared with broadcast NH_4NO_3. Subsurface irrigation gave the highest potato yield and WUE of four systems tested in New Mexico studies where rainfall was extremely limited (Sammis, 1980).

B. Adding Nutrients to Irrigation Water

1. Introduction into the System

There are a variety of injection equipment and methods to introduce fertilizer into irrigation systems. One of the simplest methods is to introduce the solution at the suction side of a centrifugal pump. However, a pollution hazard due to siphoning is associated with this method. Safety devices such as vacuum relief valves and chemical pumps interlocked with the water pumping system are now recommended (Fischbach, 1973). High-pressure pumps can be used to inject fertilizer solutions directly into the irrigation main lines. Pumps must be constructed of materials that are noncorrosive and should be rinsed to prevent corrosion.

Mixing or agitation devices must be used to predissolve dry fertilizers before injection into the lines. Only those fertilizer sources that are easily dissolved should be used. For this reason, UAN solutions are the most frequently used materials injected into irrigation systems.

It is extremely important that the fertilizer is introduced into an area of the system where turbulence and good mixing between the fertilizer and water occurs. The most ideal injection point will vary with the irrigation system.

2. Precipitation of Insoluble Salts

Insoluble salts can be formed in irrigation water if fertilizers containing free NH_3 or NH_4^+phosphates are used. Anhydrous or aqua NH_3 injected into irrigation water will cause the pH to increase, and in the presence of high Ca^{2+}, Mg^{2+}, and HCO_3^- concentrations, precipitates of Ca and Mg salts (e.g., calcite) will form and may clog the system. If irrigation water is high in Ca and Mg, application of most P fertilizers through sprinkle or trickle irrigation systems is not recommended because of possible precipitation of insoluble Ca and Mg phosphates as dicalcium phosphate or dimagnesium phosphate in the irrigation pipe and emitters (Anon., 1981a).

With certain precautions these problems can be prevented or corrected by acid treatment. The amount of acid necessary to prevent calcium carbonate ($CaCO_3$) precipitation depends on the irrigation water quality, CO_2 partial pressure, and temperature. Miyamoto and Ryan (1976) evaluated the effects of NH_3 and H_2SO_4 applied to ammoniated water on Ca precipitation. Ammonia application increased exchangeable Na^+ and NH_4^+ and reduced water infiltration rates, especially when the irrigation water contained high Na^+ relative to Ca^{2+} and Mg^{2+}. Sulfuric acid applied to ammoniated water neutralizes OH^- produced by the NH_3 injection. Thus, the H_2SO_4 reduced Ca precipitation and exchangeable Na^+ and consequently prevented a decline in infiltration rates. They concluded that H_2SO_4 could be used for correcting excessive lime incrustation or poor water penetration induced by the use of anhydrous or aqua NH_3 without corroding irrigation systems.

Phosphoric acid can be added in high concentrations to trickle irrigation systems to keep the system's pH low enough for the salts to remain soluble. Rauschkolb et al. (1976) encountered no precipitation or clogging problems by injecting orthophosphoric acid (H_3PO_4) into trickle irrigation water relatively high in bicarbonate plus Ca and Mg. However, the H_3PO_4 was injected at a point beyond metal connections or filters to avoid corrosion. If irrigation water is low in Ca and Mg, few problems should be encountered in applying H_3PO_4 through trickle/drip irrigation systems. In Arizona studies, O'Neill et al. (1979) compared spot placement of treble superphosphate (TSP) 5 to 10 cm below the soil surface and directly under the emitters to injection of H_3PO_4 with a trickle system. The H_3PO_4 treatments lowered the pH of the irrigation water enough to eliminate clogging problems of P precipitation. Also, the P was delivered to a greater soil volume when applied as OP rather than as TSP.

3. Ammonia Volatilization

Ammonia can be lost directly from irrigation water or from the soil surface by volatilization. Miyamoto et al. (1975) reported NH_3 losses from water of 45 to 73%. Henderson et al. (1955) reported losses of 10% when NH_4^+ salts were applied through sprinklers, but losses of up to 60% were recorded when anhydrous NH_3 was the source of N. Higher N concentrations, water pH, and temperatures along with greater water turbulence and exposure time increase the potential for NH_3 losses. Losses of NH_3 can be decreased some by the addition of acids to the irrigation water. In any case, choice of N source is important; anhydrous or aqua NH_3 should not be applied with sprinkler or furrow irrigation.

Volatilization losses can also be large once the water has been applied to the soil surface. A number of scientists have reported NH_3 volatilization losses from surface-applied ammoniacal sources of N, especially with high soil pH. This can also be true with ammoniacal sources applied with irrigation water. The loss of NH_3 from furrow irrigation water is also a function of windspeed and crop height (Denmead et al., 1982). They noted a 7% loss of NH_3/h with short crops and only about a 1% loss of NH_3/h with tall crops. The N content of their irrigation water decreased by 84% over a 400-m furrow with a short crop and by 59% with the tall crop. Practical remedies include irrigating at night when volatilization rates were one-half those observed during the day.

Volatilization losses of NH_3 are thought to be relatively small with trickle irrigation but can occur in the vicinity of the emitter when anhydrous or aqua NH_3 is injected into the water. However, saturated or near-saturated soil conditions in the vicinity of each emitter might lead to denitrification of NO_3^--N–containing sources of fertilizer N.

C. Agronomic Factors in Fertigation

1. Nitrogen

The chemistry of various N sources should be considered when evaluating N distribution and application efficiencies. Ammonia or the NH_4^+ salts readily convert to NH_4^+, which attaches to the cation exchange sites in the top 2 to 3 cm of soil. When placed directly on the soil surface, NH_4^+ is subject to volatilization loss. Urea is quite mobile, is not strongly adsorbed by soil, and thus will move deeper than the NH_4^+ salts. Nitrate is extremely mobile, is not adsorbed, and moves freely in the soil water. Both urea and NO_3^- tend to move with the water to the wetting front. These forms, consequently, are more susceptible to leaching losses if excessive amounts of water are applied. Urea is generally hydrolyzed to NH_4^+ within 24 h if the temperature is warm enough.

2. Phosphorus

Phosphorus has not generally been recommended for application in irrigation systems because of the assumed restricted movement into the soil or root zone when placed on the soil surface and precipitation with Ca and Mg, causing plugging in the system. There is not complete agreement on the degree of movement of P in soil with irrigation water, however. Some studies have indicated that a preirrigation mixing of P in the soil supplemented by addition of P to the irrigation solution is necessary because of the limited mobility of phosphate ions (Bar-Yosef & Sheikholslami, 1976). However, Goldberg et al. (1971) found that inorganic P applied at a rate of 20 kg P/ha moved approximately 20 cm both vertically and horizontally from a trickle emitter in a loamy sand. Raushkolb et al. (1976) demonstrated that orthophosphate moved a much greater distance into a clay loam soil when applied through a trickle irrigation system than had been observed for comparable application rates spread uniformly on the soil surface. This is more than likely due to the saturation of soil reaction sites near the point of application (emitter) with P and subsequent mass flow with the soil water. Hergert and Reuss (1976) found sprinkler-applied P to move to a depth of 18 cm on a Haxtun (fine-loamy, mixed, mesic Pachic

Argiustolls) loamy sand but to a depth of only 4 to 5 cm on a Nunn (fine, montmorillonitic, mesic Aridic Argiustolls) clay loam in the year of application. Thus, because of uncertainty of the time required for sprinkler-applied P to move into the rooting zone even in a loamy sand, they suggested that P be preplant applied and incorporated.

Organic P compounds move greater distances into the soil than inorganic phosphates because the organic forms must be enzymatically hydrolyzed to orthophosphate before soil reactions inhibit movement. Rolston et al. (1975) found that the movement of six organic P compounds into a clay loam soil was about 12 cm compared with 2 to 3 cm for inorganic P.

3. *Potassium*

Soluble sources of K such as KCl, potassium nitrate (KNO_3), and K_2SO_4 can be injected into irrigation water and should cause few precipitation problems. Potassium, like P, generally moves to a very limited extent in soils. It is similar to NH_4^+ in that it is a cation and once applied to the soil is adsorbed on the exchange complex until the exchange complex is saturated. Research by Uriu et al. (1980) showed that K could move 60 to 90 cm into the soil in one season when applied with a trickle irrigation system. But in low CEC Oxisols in Puerto Rico, a highly concentrated K zone was noticed around the drip irrigation tubing with the hydrolyzed urea (NH_4^+) being located at some distance away (Keng et al., 1979).

4. *Micronutrients*

Micronutrients such as Fe, Mn, Zn, and Cu can be predissolved and metered into irrigation water. Either chelates or SO_4^{2-} sources may be used, although the latter have been known to cause some precipitation and clogging of trickle lines. The chelated forms are generally more water soluble and thus will cause little precipitation. However, the high cost of chelates has somewhat limited their use in this manner. In the same experiments as with P, Hergert and Reuss (1976) reported that sprinkler-applied Zn as zinc ethylenediaminetetracetic acid (ZnEDTA) moved to a slightly greater depth (10 cm) compared with Zn as zinc sulfate ($ZnSO_4$) (5 cm) on a loamy sand in the year of application. Because movement of Zn into both coarse and fine-textured soils was small, they suggested broadcast application and soil incorporation of Zn. Little other agronomic research has been conducted to test the efficacy of micronutrient application through irrigation water.

5. *Application Timing*

As has been mentioned previously, grower control over the time of nutrient application is one of the main advantages for fertigation. Fertilizers can be applied with the irrigation water when crop damage caused by mechanical methods would limit application. Multiple applications of mobile nutrients can be made to highly permeable soils to overcome losses due to leaching. On sandy soils, preplant applications of N without a nitrification inhibitor are often leached out of the rooting zone before the time of major crop uptake occurs. Smika and Watts (1978) compared single broadcast application of N with five equal increments of N injected into sprinkler irrigation before tasseling of corn grown on sandy Nebraska soils and found that sprinkler application greatly minimized NO_3^- leached below the root zone. When broadcast-applied at seeding, very little residual NO_3^- remained in the 150-cm sampling depth after the cropping season, regardless of N application rate or water application amount used. Increased corn yields and greater recovery of applied N have been obtained when N was added to the irrigation water and applied throughout the season to sandy Nebraska soils without layers of fine-textured materials (Rehm & Wiese, 1975). They concluded that a portion of the N fertilizer requirement should be applied with the irrigation water as a recommended practice on these soils. With limited irrigation on medium- and fine-textured Kansas soils, Anderson et al. (1982) concluded that all N should be applied preplant rather than in-season with irrigation. Roberts et al. (1981) cautioned that applying N too late in July to sugar beets (*Beta vulgaris* L.) grown in Washington significantly reduces sucrose content but added that it should be possible to apply half of the sugar beet N requirement through irrigation during June and early July to supplement preplant N without reducing sugar yields.

6. Nonuniform Infiltration

Infiltration of sprinkler-applied irrigation water is generally assumed to be uniform. However, Saffiigna et al. (1976) reported nonuniform infiltration beneath the hills of irrigated potatoes grown on a sand in Wisconsin. Their data showed that large irrigations (> 2.5 cm) moved water below 25 cm in the furrow and parts of the hill without wetting all parts of the root zone. This can lead to positional unavailability of sprinkler-applied N to crops with limited, shallow root systems that are in widely spaced rows. Nitrogen applied in the irrigation water or broadcast is particularly prone to leaching under the furrows. Frequent, smaller applications of water and N should improve WUE and minimize NO_3^- leaching.

7. Efficiency of Trickle Fertigation Versus Soil Application

California scientists (Anon., 1981a; Feigin et al., 1982) concluded that N applied through trickle/drip irrigation systems is more efficient than when the N is banded and either furrow or trickle irrigated. Uptake of N by celery *(Apium graveolens)* was greater for UAN applied through the irrigation system than for either a slow-release fertilizer or ammonium sulfate [$(NH_4)_2SO_4$] banded in the center of the beds and then trickle irrigated. Greater leaching of the slow-release fertilizer and $(NH_4)_2SO_4$ below the root zone was also noticed. On Oxisols in Puerto Rico, Keng et al. (1979) found that green pepper yields from fertigation with N plus K were 15.8% higher than from broadcast fertilizer and slightly higher than from banded fertilizer. Significantly higher P contents of tomato (*Lycopersicon esculen-*

tum Mill.) seedling leaves were found when P was applied by drip irrigation compared to the same rate banded.

8. Water Management

Good irrigation management remains one of the key factors for improving fertilizer efficiency through fertigation. Excessive rates of water application, whether due to poor distribution within the system, higher than necessary application rate, or an unexpected rainfall immediately after irrigation, can lead to serious N losses. Hergert et al. (1978) showed striking effects due to excessive irrigation and leaching in Nebraska. Significant N loss occurred regardless of the use of the nitrification inhibitor nitrapyrin [2-chloro-6-(trichloromethyl) pyridine], although the effects on corn grain yield were not as great where nitrapyrin was used.

D. Summary

Research and farmer experience have shown quite conclusively that fertilizer can be applied very efficiently through irrigation water. Choice of the fertilizer salt in combination with the type of water and kind of irrigation system used is extremely important in arriving at maximum fertigation efficiency.

These choices then must be coupled with an irrigation system that delivers the proper rate and distribution of water. Water and fertilizer management are the keys to fertigation efficiency and success.

V. COMBINED FERTILIZER-PESTICIDE APPLICATIONS

Crop producers are continually looking for methods to reduce their input costs. A cost-cutting technique that is becoming more commonplace is that of combining pesticides with fertilizers prior to application. Potential advantages of this labor-saving method are obvious. Farmers can improve their timeliness and save fuel because the number of application trips are reduced. This is especially important for larger operators who are using a minimum amount of labor to manage their operation in an efficient manner. Indications of excellent pesticide performance, especially herbicides, when applied with fertilizers is an added bonus.

A. Formulations Available

Pesticides can be combined with both liquid (solution and suspension) and dry fertilizers. Since the early 1970s, the most common technique has been the mixing of herbicides with liquid fertilizers. This practice has assumed the common name of *weed and feed* and has become very popular among both dealers and farmers. More recently, dry fertilizers have become the carrier for certain herbicides that can be impregnated into the fertilizer by the fertilizer dealer. In this case, the dealer usually performs the custom application as well. Some flowable insecticides can be combined with liquid, band-placed starter fertilizers that are applied by farmers. Certain volatile herbicides are being introduced into the NH_3 stream just before the injection knives.

B. Compatibility

Compatibility problems may be due to either chemical or physical causes. There has been some concern that the conditioners (e.g., colloidal materials) in the fertilizer carrier may reduce the activity of the pesticide due to chemical inactivation. Because of the varied chemical nature of pesticides, it is best to check with the manufacturer or the product label for this information. On the other hand, some pesticides are not easily miscible with fluid fertilizers and, therefore, tend to separate. If the ingredients in the tank mix do tend to separate, the end result is nonuniform application of both the pesticide and fertilizer carrier. Separation usually occurs when the combination is stored for even short periods of time without agitation or is applied with equipment that does not have adequate agitation.

Some of the earliest work on the physical compatibility of fluid fertilizer-pesticide mixtures was reported by Achorn et al. (1970). Their results indicated that more pesticides were physically compatible in suspension fertilizers than in clear liquids. Moreover, less agitation was necessary to maintain dispersion with suspensions. Their recommendations suggested constant agitation of clear liquid–herbicide mixtures during application or use of an emulsifying agent to help prevent separation of the herbicide from the liquid fertilizer.

It is extremely important that the pesticide be completely compatible with the fertilizer carrier. This means that it must mix freely with the carrier, stay in suspension with the fertilizer without undue agitation, and be effective and consistent in its control of the specific pest for which it was intended. Before a particular pesticide is mixed with a specific fertilizer, it can be checked for its mixing compatibility by simply adding the appropriate ratios of both materials to a small bottle and mixing them. Many combinations mix well and the pesticide stays dispersed with proper agitation. However, some materials do not mix well without considerable and constant agitation. Pesticide labels should be consulted for information as to their compatibility and recommended use with fertilizers.

C. Potential Use Situations and Limitations

Not all crop systems allow for the efficient use of fertilizer-pesticide combinations. For the combination to be most effective, it must allow the fertilizer nutrients to be effectively applied for that crop while maximizing the effectiveness of the herbicide. For instance, the preemergence application of a mobile nutrient such as N with a herbicide to corn would be much more desirable and efficient than the application of that preemergence herbicide with a solution or suspension containing primarily immobile nutrients that could be left positionally unavailable.

Another limitation is the potential for nonuniform applications. Generally pesticide application requires more precision and uniformity than fertilizer application. For instance, plant roots come in contact with nutrients throughout the rooting zone as they grow. This satisfies the nutrient needs of the plant over the season. On the other hand, germinating weed seedlings must come into contact with the herbicide through root or shoot uptake very early in their growth before their roots are below the herbicide or before the herbicide degrades. Also, insects that damage the root systems must be controlled by insecticides uniformly placed in the root zone where the insects are active. This requirement for specific application positioning and uniform distribution can be a problem with materials that are customarily applied with fertilizer spreaders. One can anticipate greater inconsistency in herbicide performance, especially when applying dry materials using twin spinner types of applicators or if calibration of the spreaders is not precise. The pneumatic spreaders may improve this uniformity and allow greater use of these fertilizer-pesticide combination systems.

Perhaps the present most widespread use of fertilizer-pesticide combination is the weed and feed program using N solutions. This flexible program allows the combination of a number of herbicides with UAN for preplant, preemergence, or even early postemergence application. Herbicide activity and effectiveness with this program are usually good. This is probably due to the additive effect of the N solution that acts as both a surfactant and a desiccant for postemergence application and to more uniform distribution of the herbicide because of greater carrier volume. Application rates of 140 to 190 L/ha of UAN serving as the carrier supplies from 50 to 68 kg N/ha, which because of N mobility in the soil is available to the crop that growing season.

Band applications of UAN-herbicide combinations to row crops at the time of planting or shortly after have proved to be very effective. In addition, the herbicide cost per hectare with this system are reduced because of the narrowed application width. However, mechanical cultivation must then be relied on to remove weeds from the interrow areas.

Dry fertilizers can also serve as the carrier for herbicides. The herbicide (primarily dinitroanilines and thiocarbamates) is sprayed on or impregnated into the fertilizer immediately before application. These combinations are then spread on the soil surface and incorporated into the upper 10 cm of soil with secondary tillage equipment. These preplant, incorporated applications are usually made in the spring prior to planting. In the northern USA, application and incorporation is frequently done in the late fall for sugar beets the next spring. This technique has the advantage of allowing the farmer to plant early the next spring without having to perform additional secondary tillage operations immediately prior to planting on these wet, poorly drained, and fine-textured soils. The major disadvantage is that the fall secondary tillage creates a smooth and level soil surface that is extremely susceptible to wind erosion.

A practice that is not very popular at present is the use of liquid starter fertilizers to carry flowable insecticides. This combination allows the grower to band-apply the fertilizer and insecticide close to the seed at the time of planting. This practice is generally used with continuous corn where corn rootworm (*Diabrotica* spp.) control is necessary. However, even though it is an easy and convenient application technique, compatibility, phytotoxicity, and insecticide performance problems have been reported. Proper application equipment and selection of an appropriate insecticide for the equipment used is critical where insecticides are placed in or near the seed zone.

Beginning in the early 1980s, interest has developed regarding the addition of thiocarbamate herbicides with anhydrous NH_3 by injecting them into the NH_3 stream after it passes through the regulator. The combination is then injected into the soil where the mobile herbicide diffuses throughout the upper layer of soil. Preliminary investigations indicate rather narrow spacing of NH_3 knives (perhaps 25 cm or less) and shallow injection are needed. This method also shows some potential when NH_3 is applied with secondary tillage equipment (disk, field cultivator, etc.).

D. Performance

Relatively few studies have been reported in the literature on the agronomic benefits or limitations of fertilizer-pesticide tank mix combinations. Murphy et al. (1972) concluded that the dual application of atrazine [2-chloro-4-ethylamino-6-isopropylamino-*s*-triazine] and suspension fertilizers seemed to be as feasible as that of applying atrazine with liquid N solutions. Under their Kansas nonirrigated conditions, however, they found a consistent 250 kg/ha corn yield advantage with the preplant incorporated compared with preemergence application of the NPK suspension–atrazine material. This was thought to be due to both improved weed control and improved utilization of the incorporated nutrients. Follow-up studies reported by Murphy et al. (1973) indicate acceptable results with a number of herbicides added to an NPK suspension. However, the herbicides did distribute themselves in varying manners through the solid and liquid phases of the suspension and, thus, required agitation to maintain a homogeneous mixture of fertilizer and herbicide.

Postemergence application of N solutions containing herbicides generally result in leaf burn to the crop unless drop nozzles are used to direct the solution below the leaves. The degree of burn is greater with increasing N and herbicide rate and advanced growth stage of the plant. Leaf burn often is of concern to the grower who immediately suspects a yield loss. Russ et al. (1974), in Kansas studies with 80 kg N/ha applied as 32% N solution combined with various herbicides, found rather significant leaf burn on applications made to corn and sorghum [*Sorghum* bicolor (L.) Moench) at the three- and six-leaf stages, but yields were seemingly unaffected. They concluded that applications of these materials as late as the six-leaf stage are a feasible means of applying N as well as herbicides; however, they should be regarded as a salvage operation. From their studies they suggested that the N-herbicide combination be applied as soon as possible in the cropping system to alleviate the possibility of leaf damage and to enhance the likelihood of weed control. Results from Minnesota studies (Randall, 1984) indicate that UAN (28% N) can be applied over the top of corn through the four-leaf stage without seriously damaging the plants or reducing yields if N rates do not exceed 134 kg/ha. When 2.1 kg atrazine/ha was applied with UAN at the four-leaf stage, yields were not reduced as long as the N application rate did not exceed 68 kg/ha. Weed control with this system was excellent.

E. Legal Aspects

An overriding concern of the fertilizer-herbicide combination technique is the legal aspect associated with the various mixtures. Many tank-mix combinations that may be compatible and effective are not registered for use and consequently are not recommended primarily because of potential incompatibility, phytotoxicity, and performance problem. Therefore, it is of utmost importance that the grower, dealer, or custom applicator check the pesticide label or check with the manufacturer before tank mixing pesticides with fertilizers. If the combination is not registered, the grower has little litigation recourse in recovering damages caused by the combined materials.

VI. FOLIAR FERTILIZATION

A. Basis for Use

Because plants are capable of absorbing nutrients through aerial plant parts, foliar applications of fertilizer solutions can be an effective method of applying fertilizer nutrients (Barel & Black, 1979; Garcia & Hanway, 1976; Vasilas et al., 1980). However, yield increases have not been obtained consistently (Boote et al., 1978; Harder et al., 1982; Poole et al., 1983a; Syverud et al., 1980). The amount of fertilizer applied per application is limited. Severe leaf burn has been shown to result from applications of excessive amounts of fertilizer salts at any one time (Parker & Boswell, 1980). Therefore, foliar fertilization has been primarily for applications of minor elements that are needed only in small amounts (Boynton, 1954; Wittwer & Bukovac, 1969).

Foliar fertilization probably should be used most generally as a supplement to good soil fertilization. As such, foliar applications of fertilizer solutions may be very useful:

1. To quickly correct nutrient deficiencies in plants.
2. To avoid problems such as excessive fixation or leaching, which occurs with some applications in some soils.
3. To provide nutrients to the plants at times when nutrient uptake through plant roots is inadequate.

Foliar applications of such minor elements as Fe, Zn, Mn, B, Cu, and Mo are used extensively to correct deficiencies of these elements in deciduous fruit trees, grape (*Vitis* spp.) vineyards, and citrus and are also used for other crops, such as beans (*Phaseolus vulgaris* L.), tomatoes, peas (*Pisum* spp.), sugar beets, corn, soybeans, grain sorghum, potatoes, and pineapples (*Ananas comosus* (L.) Merr.). These foliar applications are effective and are used especially on crops grown on soils where soil applications are difficult or where even massive doses are not effective (Brady et al., 1948; Wittwer & Bukovac, 1969).

B. Potentials and Limitations

Foliar applications of the major elements N, P, and K, can be effective in colder regions or seasons where low temperatures restrict nutrient uptake by roots. In warmer climates, as in Hawaii, foliar applications of N, P, and K along with minor elements are used regularly for pineapple and sugarcane (*Saccharum* spp.).

Recent research on foliar fertilization with the major elements N, P, K, and S for such field grain crops as soybeans, corn, and wheat has shown that very significant yield increases sometimes result from foliar applications made during the seed-filling period. Although these yield increases have not been obtained consistently, decreases in nutrient concentrations are commonly observed in leaves of grain crops during the seed-filling period. Such decreases occur in crops even on very fertile soils.

Seed quality as indicated by protein content has been improved with foliar application of N–P–K–S fertilizer materials. In the 1950s, Finney et al. (1957) found that foliar application of urea near flowering increased the protein content of wheat grain. Harder et al. (1982) reported a 10% increase in the N concentration and a 4.7% increase in the P concentration of corn grain, and Poole et al. (1983a) found some increase in the percentage of N in soybeans.

More research data are available for foliar fertilization of soybeans during seedfilling than for other grain crops, but the basic principles apply to all grain crops. Soybeans have been studied most extensively because soybean seeds contain much higher concentrations of several nutrient elements, especially N, P, K, and S, than do cereal grains, and the concentration of these nutrients in the leaves are decreased substantially during seed filling.

During seed filling, most of the photosynthate (sugar) produced in the soybean leaves is channeled to the developing seeds. Dinitrogen fixation in the nodules of leguminous plants and root growth stops or becomes very limited. Although nutrient uptake from the soil continues, much of the supply of several nutrients (especially N, P, K, and S) to the developing seeds results from translocation of these elements from the leaves and other plant parts to the seeds. This translocation results in depletion of these nutrients in the leaves. During the seed-filling period, photosynthesis in the leaves gradually slows and then stops. Other nutrient elements (especially Ca, Mg, and Mn) are not effectively translocated from the leaves to the seeds. The concentrations of these elements in the seeds are low, and during seed-filling their concentrations in the leaves increase.

Research to date has supplied some useful guides:

1. Foliar fertilizer applications (especially those containing urea) should not be made when the sun is shining brightly. Applications should be made in the early morning or evening to avoid serious leafburn (Poole et al., 1983b).
2. Where foliar fertilizer applications resulted in the best soybean yield increases, the application included all four nutrient elements—N, P, K, and S. A ratio of 10N–1P–3K–0.5S (10–2.2–3.6–0.5) (similar to the ratio in the soybean seeds) was more effective for soybeans than other ratios (Garcia & Hanway, 1976).
3. Urea is the most effective form of N for foliar application.
4. The amount of urea-N that can be applied in one foliar application without causing serious leaf burn generally should not exceed 22 kg N/ha.
5. Including micronutrients or the less phloem-mobile elements (e.g., Zn, Cu, Mn, B, Fe, Ca, and Mg) or both in the foliar applications have not increased soybean seed yields.
6. Including chelating agents (e.g., citric, malic, and lactic acids) with Ca in the foliar spray solutions has not increased seed yields or the Ca content of the seeds.
7. Excessive amounts of certain contaminants such as cyannate, biuret, and cyanamid in urea used in the foliar spray solution can result in severe leaf burn and decreased yields.
8. Foliar applications of N–P–K–S during seed filling do not delay leaf senescence appreciably.
9. Foliar N–P–K–S application later in the seed-filling period has been more effective in increasing seed yields than applications prior to or early in the seed-filling period.

C. Problems to be Solved

Increased grain crop yields result only from increases in the number or size (or both) of the harvested seeds. These must be associated with increases in the rate of dry weight increase of seeds during seed filling or the length of the seed-filling period.

Abortion of flowers, pods, and seeds within pods normally occurs to the extent that $< 75\%$ of the potential number of seeds is produced. Practices that would limit this abortion would increase the potential number of seeds to be filled. Foliar application of growth regulators that limit continued vegetative growth of the plants and increase seed set offer promise. Other practices can and will be developed to decrease abortion and increase seed set.

Although grain yield increases are most generally associated with increases in numbers of seeds produced, there is great diversity in seed sizes from one variety to another. The large-seeded varieties normally produce fewer seeds for harvest. Increasing seed set of varieties that produce larger seeds would increase the potential for foliar fertilization during seed filling.

Certain nutrient elements, especially Ca, Mg, and Mn, are not translocated effectively in the phloem to the developing seeds. The concentrations of these elements in the seeds are low. The beneficial effect of Ca for seed development separate from Ca available to plants roots was demonstrated for peanuts (*Arachis hypogaea* L.) as early as 1948 (Brady et al., 1948). Treatments to increase the level of available Ca in the phloem may improve translocation of materials in the phloem and may result in improved growth of developing seeds.

Foliar applications of N–P–K–S during seed filling have not maintained the concentrations of these elements in the leaves or significantly increased the time that leaves on the plant stay green and photosynthetically active. Development of genotypes that maintain green leaves longer during seed filling or foliar applications of growth regulators, such as naphthaleneacetic acid (NAA) and benzyl adenine (BA), to delay leaf senescence and extend the seed-filling period offer promise.

Studies are needed to determine the most practical and effective methods for applying the needed fertilizer nutrients with minimum leaf burn damage.

D. Summary

Foliar fertilization is now practiced extensively for application of micronutrients on many tree and vegetable crops. But, foliar fertilization of major field grain crops with the major nutrients generally has not been successful and has received relatively little attention. However, research has shown that during seed filling there is appreciable accumulation of several nutrient elements in the developing seeds, translocation from and depletion of several nutrient elements (especially N, P, K, and S) in the leaves and other vegetative parts, and a decreasing rate of photosynthesis in the leaves with a decrease in soluble sugars in various plant parts. Thus, foliar application of nutrients during this seed-filling period does offer some potential for improving yields.

Although foliar fertilization of grain crops during seed filling shows promise of being an effective and practical method of increasing yields above those obtained with good soil fertilization practices, it currently is not recommended for general use by farmers. Additional research is needed to identify the mechanisms and technology necessary to obtain consistent yield enhancement.

VII. FUTURE IMPROVEMENTS IN APPLICATION TECHNIQUES

Improvements in application technology will need to be made as fertilizer sources and methods of fertilizer application change. Many of these changes will be dictated by the desire to improve plant uptake and yield and thus improve fertilizer efficiency.

The adoption of reduced tillage throughout the USA will necessitate substantial changes in fertilizer application methodology. Subsurface application of all forms of N may become a standard recommended practice. This placement below the surface-accumulated plant residue will reduce potential immobilization and volatilization losses of ammoniacal N. To conserve energy consumed in fertilizer application, we may see widespread adoption of subsurface or deep placement of P, K, secondary nutrients, and micronutrients at the same time that the N is injected. Rather than dual N-P placement as it is now practiced in some areas, we may find multiple nutrient deep placement becoming commonplace.

Projected increased costs of all fertilizers relative to grain prices is also likely to regenerate more interest in banding fertilizer P–K on medium- or lower-testing soils to obtain the economic efficiency resulting from the lower P–K rates necessary for banding. Under these conditions, there will be renewed interest in developing fertilizer handling systems for rapid filling of fertilizer hoppers on planters so that planting will not be unduly slowed.

As crop production system change, developments and improvements must be made in application equipment. The precision of application, whether it be the uniformity among knives of an NH_3 toolbar or the distribution of dry and liquid fertilizers applied with mechanical spin-type spreaders or liquid applicators, has to be improved. The capability to increase or decrease application rates on fertilizer application equipment as a farmer or a custom applicator moves through the field from areas with low production potential to areas with a higher production potential must be developed. In other words, can electronic sensors or similar devices be adapted to application equipment to measure important soil properties (organic matter, moisture, etc.) so that application rates can be adjusted more precisely to meet the crop's need throughout the field? These are some future areas in application methodology that need investigation so that fertilizer efficiency can be improved.

REFERENCES

Achorn, F. R., W. C. Scott, and J. A. Wilbanks. 1970. Physical compatibility tests of fluid fertilizer-pesticide mixtures. Fert. Solutions 14(6):40–46.

Alessi, J., and J. F. Power. 1980. Effects of banded and residual fertilizer phosphorus on dryland spring wheat yield in the Northern Plains. Soil Sci. Soc. Am. J. 44:792–796.

Anderson, C. K., L. R. Stone, and L. S. Murphy. 1982. Corn yield as influenced by in-season application of nitrogen with limited irrigation. Agron. J. 74:396–401.

Anonymous. 1981a. Applying nutrients and other chemicals to trickle-irrigated crops. Univ. of California Div. of Agric. Sci. Bull. 1893.

Anonymous. 1981b. Drip irrigation management. Univ. of California Div. of Agric. Sci. Leaflet 21259.

Bandel, V. A., S. Dzienia, and G. Stanford. 1980. Comparison of N fertilizers for no-till corn. Agron. J. 72:337–341.

Bandel, V. A., S. Dzienia, G. Stanford, and J. O. Legg. 1975. N behavior under no-till vs. conventional corn culture: I. First-year results using unlabeled N fertilizer. Agron. J. 67:782-786.

Barel, D., and C. A. Black. 1979. Foliar application of P: II. Yield responses of corn and soybeans sprayed with various condensed phosphates and P–N compounds in greenhouse and field experiments. Agron. J. 71:21–24.

Bar-Yosef, B., and M. R. Sheikholslami. 1976. Distribution of water and ions in soils irrigated and fertilized from a trickle source. Soil Sci. Soc. Am. J. 40:575–582.

Belcher, C. R., and J. L. Ragland. 1972. Phosphorus absorption by sod-planted corn (*Zea mays* L.) from surface-applied phosphorus. Agron. J. 64:754–756.

Bernstein, L., and L. E. Francois. 1973. Comparisons of drip, furrow, and sprinkler irrigation. Soil Sci. 115:73–86.

Blevins, R. L., G. W. Thomas, and P. L. Cornelius. 1977. Influence of no-tillage and nitrogen fertilization on certain soil properties after 5 years of continuous corn. Agron. J. 69:383–386.

Blevins, R. L., L. W. Murdock, and G. W. Thomas. 1978. Effect of lime application on no-tillage and conventionally tilled corn. Agron. J. 70:322–326.

Blevins, R. L., W. W. Frye, and M. J. Bitzer. 1980. Conservation of energy in no-tillage systems by management of nitrogen. p. 14–20. *In* Proc. of the 3rd Annual No-Tillage Systems Conf. University of Florida, Gainesville.

Boote, K. J., R. N. Gallaher, W. R. Robertson, K. Hinson, and L. C. Hammond. 1978. Effect of foliar fertilization on photosynthesis, leaf nutrition, and yield of soybeans. Agron. J. 70:787–791.

Boynton, D. 1954. Nutrition by foliar application. Annu. Rev. Plant Physiol. 5:31–54.

Brady, N. C., J. F. Reed, and W. E. Colwell. 1948. The effect of certain mineral elements on peanut fruit filling. Agron. J. 40:155–167.

Bucks, D. A., and F. S. Nakayama. 1980. Injection of fertilizer and other chemicals for drip irrigation. Tech. Conf. Proc. Irrig. Assoc. 1980:166–180.

Cope, J. T., Jr. 1981. Effects of 50 years of fertilization with phosphorus and potassium on soil test levels and yields at six locations. Soil Sci. Soc. Am. J. 45:342–347.

Cope, J. T., Jr., and F. E. Khasawneh. 1981. Soil test summaries for phosphorus and potassium. Agron. Abstr. American Society of Agronomy, Madison, WI, p. 235.

Creamer, F. L., and R. H. Fox. 1980. The toxicity of banded urea or diammonium phosphate to corn as influenced by soil temperature, moisture, and pH. Soil Sci. Soc. Am. J. 44:298–300.

Denmead, O. T., J. R. Freney, and J. R. Simpson. 1982. Dynamics of ammonia volatilization during furrow irrigation of maize. Soil Sci. Soc. Am. J. 46:149–155.

Doran, J. W. 1980. Microbial changes associated with residue management with reduced tillage. Soil Sci. Soc. Am. J. 44:518–524.

Engelstad, O. P., and G. L. Terman. 1966. Importance of water solubility of phosphorus fertilizers. Commerc. Fert. Plant Food Ind. 113:32–35.

Feigin, A., J. Letey, and W. M. Jarrell. 1982. Nitrogen utilization efficiency by drip irrigated celery receiving preplant or water applied N fertilizer. Agron. J. 74:978–983.

Fenn, L. B., J. E. Matocha, and E. Wu. 1981a. Ammonia losses from surface-applied urea and ammonium fertilizers as influenced by rate of soluble calcium. Soil Sci. Soc. Am. J. 45:883–886.

Fenn, L. B., and S. Miyamoto. 1981. Ammonia loss and associated reactions of urea in calcareous soils. Soil Sci. Soc. Am. J. 45:537–540.

Fenn, L. B., R. M. Taylor, and J. E. Matocha. 1981b. Ammonia losses from surface-applied nitrogen fertilizer as controlled by soluble calcium and magnesium: General theory. Soil Sci. Soc. Am. J. 45:777–781.

Finney, K. F., J. W. Meyer, F. W. Smith, and H. C. Fryer. 1957. Effect of foliar spraying of Pawnee wheat with urea solutions on yield, protein content, and protein quality. Agron. J. 49:341–347.

Fischbach, P. E. 1973. Anti-pollution devices for applying chemicals through the irrigation system. Neb Guide G73-43. University of Nebraska, Lincoln.

Fixen, P. E., and R. P. Wolkowski. 1981. Dual placement of nitrogen and phosphorus on corn and a Plano silt loam. Agron. Abstr. American Society of Agronomy, Madison, WI, p. 235.

Fox, R. H., and L. D. Hoffman. 1981. The effect of N fertilizer source on grain yield, N uptake, soil pH, and lime requirement in no-till corn. Agron. J. 73:891–895.

Garcia, L., Ramon, and J. J. Hanway. 1976. Foliar fertilization of soybeans during seedfilling. Agron. J. 68:653–657.

Gardner, B. R., and R. L. Roth. 1984. Applying nitrogen in irrigation waters. p. 493–506. *In* R. D. Hauck (ed.) Nitrogen in crop production. American Society of Agronomy, Crop Science Society of America, and Soil Science Society of America, Madison, WI.

Goldberg, D., and M. Shmueli. 1970. Drip irrigation—a method used under arid and desert conditions of high water and soil salinity. Trans. ASAE 13(1):38–41.

Goldberg, D., B. Gornat, and Y. Bar. 1971. The distribution of roots, water and minerals as a result of trickle irrigation. J. Am. Soc. Hortic. Sci. 96:645–48.

Griffith, D. R. 1974. Fertilization and no-plow tillage. *In* Proc. Indiana Plant Food and Agric. Chemical Conf., West Lafayette, IN. 17–18 December. Purdue University Press, West Lafayette, IN.

Harder, H. J., R. E. Carlson, and R. H. Shaw. 1982. Corn grain yield and nutrient response to foliar fertilizer applied during grain fill. Agron. J. 74:106–110.

Henderson, D. W., W. D. Bianchi, and L. D. Doneen. 1955. Ammonia loss from sprinkler jets. Agric. Eng. 36:398–399.

Hergert, G. W., and J. O. Reuss. 1976. Sprinkler application of P and Zn fertilizers. Agron. J. 68:5–8.

Hergert, G. W., K. D. Frank, and G. W. Rehm. 1978. Anhydrous ammonia and N-Serve for irrigated corn. p. 3.1–3.4. *In* University of Nebraska-Lincoln Agronomy Dept. Soil Sci. Res. Rep. 1978.

Keng, J. C. W., T. W. Scott, and M. A. Lugo-Lopez. 1979. Fertilizer management with drip irrigation in an oxisol. Agron. J. 71:971–980.

Ketcheson, J. W. 1980. Effect of tillage on fertilizer requirements for corn on a silt loam soil. Agron. J. 72:540–542.

Lall, R. 1979. Influence of six years of no-tillage and conventional plowing on fertilizer response of maize (*Zea mays* L.) on an Alfisol in the tropics. Soil Sci. Soc. Am. J. 43:399–403.

Lamond, R. E., and J. L. Moyer. 1983. Effects of knifed vs. broadcast fertilizer placement on yield and nutrient uptake by tall fescue. Soil Sci. Soc. Am. J. 47:145–149.

Leikam, D. F., R. E. Lamond, D. E. Kissel, and L. S. Murphy. 1979. More for your fertilizer dollar—dual applications of N and P. Crop Soils Mag. 31(8):16–19.

McKibben, G. E. 1975. Nitrogen for 0-till corn. p. 87–89. *In* Update 75, a research report of the Dixon Springs Agricultural Center. University of Illinois Press, Champaign.

Mengel, D. B., D. W. Nelson, and D. M. Huber. 1982. Placement of nitrogen fertilizers for no-till and conventional till corn. Agron. J. 74:515–518.

Mitchell, W. H. 1981. Subsurface irrigation and fertilization of field corn. Agron. J. 73:913–916.

Miyamoto, S., J. Ryan, and J. L. Stroehlein. 1975. Sulfuric acid for the treatment of ammoniated irrigation water: I. Reducing ammonia volatilization. Soil Sci. Soc. Am. Proc. 39:544–548.

Miyamoto, S., and J. Ryan. 1976. Sulfuric acid for treatment of ammoniated irrigation water: II. Reducing calcium precipitation and sodium hazard. Soil Sci. Soc. Am. J. 40:305–309.

Moncrief, J. F., and E. E. Schulte. 1982. Fertilizer placement in tillage systems—Wisconsin. *In* Proc. of the 34th Annual Fertilizer and Ag. Chemical Dealers Conf., Des Moines, IA. 12–13 January. Iowa State University, Ames.

Moschler, W. W., D. C. Martens, C. I. Rich, and G. M. Shear. 1973. Comparative lime effects on continuous no-tillage and conventionally tilled corn. Agron. J. 65:781–783.

Moschler, W. W., and D. C. Martens. 1975. Nitrogen, phosphorus and potassium requirements in no-tillage and conventionally tilled corn. Soil Sci. Soc. Am. Proc. 39:886–891.

Murphy, L. S. 1981. Placement of nitrogen and phosphorus. *In* Proc. of the Indiana Plant Food and Agric. Chemicals Conf., West Lafayette, IN. 15–16 December. Purdue University Press, West Lafayette, IN.

Murphy, L. S., L. J. Meyer, and O. G. Russ. 1972. Evaluating fluid fertilizer-herbicide combinations. Fert. Solutions 16(2):19–28.

Murphy, L. S., L. J. Meyer, O. G. Russ, and C. W. Swallow. 1973. Update report on evaluating fluid fertilizer-herbicide combinations. Fert. Solutions 17(4):26–36.

Murphy, L. S., and L. M. Walsh. 1972. Correction of micronutrient deficiencies with fertilizers. p. 347–387. *In* J. J. Mortvedt et al. (ed.) Micronutrients in agriculture. Soil Science Society of America, Madison, WI.

Olson, R. A., K. D. Frank, P. H. Grabouski, and G. W. Rehm. 1982. Economic and agronomic impacts of varied philosophies of soil testing. Agron. J. 74:492–499.

O'Neill, M. K., B. R. Gardner, and R. L. Roth. 1979. Orthophosphoric acid as a phosphorus fertilizer in trickle irrigation. Soil Sci. Soc. Am. J. 43:283–286.

Parker, M. B., and F. C. Boswell. 1980. Foliar injury, nutrient intake, and yield of soybeans as influenced by foliar fertilization. Agron. J. 72:110–113.

Parks, W. L., and W. M. Walker. 1969. Effect of soil potassium, potassium fertilizer and method of fertilizer placement upon corn yields. Soil sci. Soc. Am. Proc. 33:427–429.

Parks, W. L., M. E. Walker, and J. T. Cope, Jr. 1981. Residual P-K research in the southeastern United States. Agron. Abstr. American Society of Agronomy, Madison, WI, p. 237.

Peaslee, D. E. 1978. Relationships between relative crop yields, soil test phosphorus levels, and fertilizer requirements for phosphorus. Commun. Soil Sci. Plant Anal. 9:429–442.

Penas, E. J. 1981. Preplant bands of dual placement—a new concept in phosporus application. Soil Science News, 3:14, July 31. Institute of Agriculture, University of Nebraska, Lincoln.

Peterson, G. A., D. H. Sander, P. H. Grabouski, and M. L. Hooker. 1981. A new look at row and broadcast phosphate recommendations for winter wheat. Agron. J. 73:13–17.

Phillips, R. E., R. L. Blevins, G. W. Thomas, W. W. Frye, and S. H. Phillips. 1980. No-tillage agriculture. Science (Washington, D.C.) 208:1108–1113.

Poole, W. D., G. W. Randall, and G. E. Ham. 1983a. Foliar fertilization of soybeans: I. Effect of fertilizer sources, rates, and frequency of application. Agron. J. 75:195–200.

Poole, W. D., G. W. Randall, and G. E. Ham. 1983b. Foliar fertilization of soybeans: II. Effect of biuret and application time of day. Agron. J. 75:201–203.

Randall, G. W. 1980. Fertilization practices for conservation tillage. *In* Proc. of the 32nd Annual Fertilizer and Agric. Chemical Dealers Conf., Des Moines, IA. 8–9 Jan. Iowa State University, Ames.

Randall, G. W. 1981a. Maintaining high P and K levels with reduced tillage systems. p. 22–28. *In* Proc. of the Soils, Fertilizer, and Agric. Pesticide Short Course, Minneapolis, MN. 15-16 Dec. University of Minnesota, St. Paul.

Randall, G. W. 1981b. Residual P-K research in the Midwest. Agron. Abstr. American Society of Agronomy, Madison, WI, p. 237.

Randall, G. W. 1984. Postemergence application of urea–ammonium nitrate (UAN) and atrazine combinations to corn. J. Fert. Issues 1:50–53.

Rauschkolb, R. S., D. E. Rolston, R. J. Miller, A. B. Carlton, and R. G. Burau. 1976. Phosphorus fertilization with drip irrigation. Soil Sci. Soc. Am. J. 40:68–72.

Rehm, G. W., and R. A. Wiese. 1975. Effect of method of nitrogen application on corn (*Zea mays* L.) grown on irrigated sandy soils. Soil Sci. Soc. Am. Proc. 39:1217–1220.

Richards, G. E. (ed.). 1977. Band application of phosphatic fertilizers. Olin Corp., Little Rock. AR.

Roberts, S., W. H. Weaver, and A. W. Richards. 1981. Sugarbeet response to incremental application of nitrogen with high frequency sprinkler irrigation. Soil Sci. soc. Am. J. 45:448–449.

Rolston, D. E., R. S. Rauschkolb, and D. L. Hoffman. 1975. Infiltration of organic phosphate compounds in soil. Soil Sci. Soc. Am. Proc. 39:1089–1094.

Russ, O. G., R. F. Sloan, C. W. Swallow, P. J. Gallagher, and L. S. Murphy. 1974. Effects of application date on crop responses to herbicide-nitrogen solution combinations. Fert. Solutions 18(4):24–34.

Saffigna, P. G., C. B. Tanner, and D. R. Keeney. 1976. Non-uniform infiltration under potato canopies caused by interception, stemflow, and hilling. Agron. J. 68:337–342.

Sammis, T. W., 1980. Comparison of sprinkler, trickle, subsurface, and furrow irrigation methods for row crops. Agron. J. 72:701–704.

Sheard, R. W. 1980. Nitrogen in the P band for forage establishment. Agron. J. 72:89–97.

Singh, T. A., G. W. Thomas, W. W. Moschler, and D. C. Martens. 1966. Phosphorus uptake by corn under no-tillage and conventional practices. Agron. J. 58:147–148.

Smika, D. E., and D. G. Watts. 1978. Residual nitrate-N in fine sand as influenced by N fertilizer and water management practices. Soil Sci. Soc. Am. J. 42:923–925.

Syverud, T. D., L. M. Walsh, E. S. Opplinger, and K. A. Kelling. 1980. Foliar fertilization of soybeans (*Glycine max.* L.). Commun. Soil Sci. Plant Anal. 11:637–651.

Tomar, J. S., and R. J. Soper. 1981. Fate of tagged urea N in the field with different methods of N and organic matter placement. Agron. J. 73:991–995.

Triplett, G. B., Jr., and D. M. Van Doren, Jr. 1969. Nitrogen, phosphorus, and potassium fertilization of non-tilled maize. Agron. J. 61:637–639.

Uriu, K., R. M. Carlson, D. W. Henderson, H. Schulbach, and P. M. Aldrich. 1980. Potassium fertilization of prune trees under drip irrigation. J. Am. Soc. Hortic. Sci. 105:508–510.

Vasilas, B. L., J. O. Legg, and D. C. Wolf. 1980. Foliar fertilization of soybeans: Absorption and translocation of ^{15}N-labeled urea. Agron. J. 72:271–275.

Wells, K. L., L. Murdock, and H. Miller. 1976. Comparisons of nitrogen fertilizer sources under Kentucky soil and climatic conditions. Agron. Notes, Vol. 9, no. 6. p. 1–8. Dep. of Agron., Univ. of Kentucky, Lexington.

Wells, K. L. 1984. Nitrogen management in the no-till system. p. 535–550. *In* R. D. Hauck (ed.) Nitrogen in crop production. American Society of Agronomy, Crop Science Society of Agronomy, and Soil Science Society of Agronomy, Madison, WI.

Westfall, D. G. 1981. Residual P-K research in the western United States. Agron. Abstr. American Society of Agronomy, Madison, WI, p. 239.

Wittwer, S. H., and M. J. Bukovac. 1969. The uptake of nutrients through leaf surfaces. Handb. Pflanzenernalhr. Dueng. 1:235–261.

16

Fertilizer Use in Relation to the Environment

J. W. Gilliam
North Carolina State University
Raleigh, North Carolina

Terry J. Logan
Ohio State University
Columbus, Ohio

F. E. Broadbent
University of California
Davis, California

The effect of fertilizer use on the environment has been a major concern since the 1960s. When this concern first received public attention, there was little scientific evidence to support those who contended that fertilizers were a great threat to the environment or those who believed there was little or no environmental danger of fertilizer application. Fortunately, there is now a significant amount of good data from which some definitive conclusions can be made. The intent of this chapter is to summarize these data and to give the authors' interpretations of current knowledge on this topic.

The concern about fertilizer use with regard to environmental quality has focused primarily on accelerated eutrophication of surface waters and nitrate (NO_3^-) content of drinking water. Eutrophication, the rapid growth and decay of aquatic vegetation, is most often limited by P and sometimes N concentrations in water. There is no question that eutrophication has been a problem in some areas. Because the greatest use of industrially fixed N_2 and mined P is for agricultural fertilizers, it is only natural that concern would be expressed regarding fertilizer use and eutrophication.

Ten micrograms of NO_3^--N/mL has generally been accepted throughout the world as a maximum safe level for drinking water. The danger of higher levels of NO_3^- in drinking water is that NO_3^- in the digestive tract of infant children, particularly those under the age of 6 months, may be reduced to nitrite (NO_2^-), which reacts with the hemoglobin of blood. This reaction reduces the oxygen carrying capacity of the blood and is known as *methemoglobinemia*. The number of documented causes of methemoglobinemia in humans is very small, but the linkage with NO_3^- requires that the potential for NO_3^- pollution of water supplies be minimized. Because the subsurface drainage water from agricultural land, particularly irrigated land, often exceeds 10 μg NO_3^--N/mL, attention has focused on agriculture as a problem with regard to drinking water quality.

Other concerns about the use of fertilizers are the effect on increased emission of nitrous oxide (N_2O) as a result of N fertilization and the increased level of soil Cd as a result of P fertilization. Nitrous oxide participates in a series of reactions in the stratosphere, which may result in the depletion of ozone (O_3), which shields the earth from ultraviolet radiation. Cadmium is a heavy metal that can be absorbed by plants at high soil Cd levels in amounts that may be toxic to animals and humans eating these plants.

I. NITROGEN

The potential for N added to soils as fertilizer to become a pollution problem first received national attention in the late 1960s. Before that time, agronomists had been quite concerned with fertilizer N efficiency in crop production but did not have quantitative information about what happened to the unutilized N, particularly with regard to the amounts lost through denitrification and in drainage waters. When fertilizer N is added to soils, the N may be harvested with the crop, incorporated into the soil organic matter, denitrified, volatilized, or lost to drainage waters. The factors controlling the distribution between each of these possible paths are complex and vary greatly between soils and locations.

Plant uptake is the primary objective of fertilizer application; however, the efficiency of fertilizer N uptake is seldom much above 50%, even with good management. Although perennial forage crops may be much more efficient in utilizing N applications, the large majority of N applications are to annual cultivated crops. The efficiency of N uptake for several crops, based on measurements with isotopically labeled fertilizer, is given in Table 16–1.

Some of the N not used by plants is incorporated into the soil organic fraction through microbiological immobilization, but unless mature crop residues or similar materials have recently been returned to the soil, this seldom accounts for a large fraction of the total. However, as much as half

Table 16–1. Percent uptake of fertilizer N by some important crops.

Crop	N rate	Uptake	Reference
	kg/ha per year	%	
Corn	50–168	23–32	Bigeriego et al. (1979)
Corn	90–360	24–60	Broadbent & Carlton (1979)
Wheat	50–100	21–44	Olson et al. (1979)
Rice	100	38–44	Patrick & Reddy (1976)
Barley	60	14–25	Sapozhnikov et al. (1974)
Sugarbeet†	56–280	12–40	Hills et al. (1978)

† Sugarbeet, *Beta vulgaris* L.

of the applied fertilizer N may eventually find its way into the soil organic matter through decomposition of crop residues. In an experiment with wheat (*Triticum aestivum* L.) followed by corn (*Zea mays* L.), Broadbent and Krauter (1974) reported that 30% of the applied N was present in the soil organic fraction at the end of the experiment. Even though a significant amount of N found in soil organic matter may be N that was applied as fertilizer, few cultivated soils are increasing in the amount of organic N. Thus, the amount of fertilizer N remaining in soils as organic N is of minimal importance in long-term N balances.

Reliable estimates of the quantities of fertilizer N lost through denitrification are very difficult to obtain in a field situation. Using isotopically labeled nitrate fertilizer in a soil maintained at constant soil-water pressure heads ranging from −8 to −70 cm of water, Rolston et al. (1978) estimated denitrification losses to range from 1% in an uncropped plot to 70% in a plot at −15 cm water soil-water pressure head treated with manure. Loss was only 16% in a manure-treated plot at a soil-water pressure head of −70 cm of water. Ryden et al. (1978), using a method based on measurement of N_2O emissions in an environment in which reduction of N_2O to N_2 was blocked by acetylene, found a loss of 51 kg N/ha in a celery (*Apium graveolens* L.) field to which 335 kg N/ha had been applied.

With the exception of instances involving soils under very wet conditions, the limited number of estimates of fertilizer N loss through denitrification under field conditions, which have become available as a result of development of direct measurement techniques in recent years, agree quite well with older estimates based on the difference method.

The N not utilized by crops, incorporated into soil organic matter, volatilized, or denitrified is subject to loss in drainage water. It is this N that is the greatest environmental threat. The N is a potential problem to both ground and surface waters.

A. Movement of Nitrogen into Surface Water

When this concern first received national attention, partially as a result of a presentation in 1968 by Commoner (1970), there was little information available about the effect of N fertilization on N movement into waters. Since that time, there has been extensive work on nonpoint (diffuse) entry of N into surface waters and the factors that control the movement of N from agricultural land to surface waters. Some representative examples of N losses from land to surface waters are given in Table 16–2. These data are intended to illustrate the common range of values that have been reported in the literature from around the world.

1. Pasture Crops

The losses of N from pasture crops are usually quite small as illustrated in Table 16–2. This is partially because most of the transport from these areas is via surface runoff as observed by Olness et al. (1975). Surface runoff losses of N are most frequently in the range of 1 to 10 kg/ha per year,

Table 16–2. Examples of N loss to surface waters from forest, pasture, and cultivated lands.

Location	Crop	N loss: Surface drainage	N loss: Subsurface drainage	N loss: Total	N loss: Range measured	Reference
		kg/ha per year				
New Hampshire	Forest	1.8		1.8		Borman et al. (1968)
West Virginia	Forest	0.8		0.8		Aubertin & Patric (1974)
Oklahoma	Pasture	6		6		Olness et al. (1975)
North Carolina	Pasture			8	3–12	Kilmer et al. (1974)
Great Britain	Pasture		33†		11.55†	Hood (1976)
Iowa	Corn		38		27–48	Baker & Johnson (1981)
Minnesota	Corn		56†		19–120†	Gast et al. (1978)
Oklahoma	Cotton	13				Olness et al. (1975)
California	Citrus		64			Bingham et al. (1971)
North Carolina	Corn	25	21	46	45–48	Gambrell et al. (1975)
Canada	Mixed crops		34†‡		4–64†	Miller (1979)
Canada	Mixed crops		145§		37–245	Miller (1979)
Texas	Mixed crops	8				Kissel et al. (1976)
Netherlands	Mixed crops		30		0–60	Kolenbrander (1969)

† These values include losses from fields fertilized at rates higher than those recommended.
‡ Mineral soils.
§ Organic soils.

although larger values are sometimes observed. Much of the N loss by surface runoff is organic N associated with sediment (Schuman et al., 1973; Kissel et al., 1976; Menzel et al., 1978).

The use of N fertilizers usually does not have a large influence on N movement from pastures. For example, Kilmer et al. (1974) observed an annual average of 12 kg N/ha leaving a fertilized Kentucky bluegrass (*Poa pratensis* L.) watershed in North Carolina, while a similar unfertilized watershed lost 3 kg N/ha. Most of the N lost from these watersheds was in the NO_3^- form with the average N concentration in the drainage water for the fertilized and unfertilized watersheds being approximately 4 and 1 μg/mL, respectively. In the bluegrass region of central Kentucky, Thomas and Crutchfield (1974) found much higher annual losses of NO_3^-. The approximately 25 kg N/ha observed in their study was largely a result of native N loss, because little fertilizer is used for pasture crops in that area. As will be illustrated again later with drainage from organic soils, relatively high NO_3^--N concentrations in drainage water from agricultural land are not necessarily a result of N fertilization.

There are examples of relatively high NO_3^--N in drainage waters from intensively managed forage and pasture crops. Hood (1976) measured an average concentration of 8.4 mg/L (11 kg/ha per year) of NO_3^--N in the lines draining a pasture in Great Britain fertilized at a rate of 250 kg N/ha. Hood indicated that this fertilizer rate was low compared with intensive standards in Great Britain. When the N application rate was increased to 750 kg/ha, a rate higher than the common 400 kg/ha, the NO_3^- loss increased to 54 kg/ha with the average concentration being 33 mg/L. Other examples of high N losses from high N input livestock systems are given by Kolenbrander (1969). However, there are very few pastures in the USA that have both high rates of N fertilizer application and subsurface drainage systems installed so the N losses from U.S. pastures are usually relatively low.

2. *Forests*

Nitrogen losses in drainage water from both fertilized and unfertilized forests are very low as indicated in Table 16–2. Normal losses are in the range of 1 to 2 kg N/ha with little or no increase upon application of N fertilizer. Sanderford (1975) could measure no difference in N concentrations in drainage waters before and after fertilization of a forested watershed in the Piedmont of North Carolina. Neary and Leonard (1978) applied 200 kg/ha of N as urea on a Monterey Pine (*Pinus radiata* D. Don) stand in New Zealand and found 0.33% of the N in the drainage water. Others (Meehan et al., 1975) have also observed small increases in stream N concentration after forest fertilization. Much of the N that enters the water from fertilization of forests is believed to fall directly in the streams during aerial application.

3. *Cultivated Crops*

Nitrogen is much more susceptible to being lost in drainage water from cultivated land than from land in perennial crops. However, there are many factors that influence the quantities of N lost, and the range of values that has been observed is quite large as shown in Table 16–2. These losses may occur by both surface and subsurface drainage, but the large losses are nearly always via subsurface drainage.

a. Surface Runoff—Surface runoff from any land invariably contains some N regardless of land use or cultural practices. Much of the N transported in surface runoff from cultivated and fertilized fields is organic N associated with the sediment (Alberts et al., 1978; Langdale et al., 1979; Burwell, 1977; Romkens et al., 1973). Thus, any factor that influences the amount of erosion will influence the amount of N lost via surface runoff. Burwell et al. (1976) found that a terraced watershed cropped to corn in Iowa lost more N via surface runoff compared with contoured corn watersheds. However, because of increased infiltration on the terraced watershed, more N was lost in subsurface drainage waters. Skaggs and Gilliam (1981) observed similar effects when subsurface drainage was improved to reduce surface runoff. It has long been known that the greater the slope under similar management practices, the greater the loss of sediment and N (Rogers, 1941).

The addition of N fertilizers frequently results in greater loss of N via surface runoff, although the increase is usually small and rather insignificant with regard to fraction of the applied fertilizer lost (Dunigan et al., 1976; Romkens et al., 1973; Timmons et al., 1973; Kissel et al., 1976). The effect of fertilizer application on increased N runoff is usually limited to the first runoff event following fertilization. When a surface application of N fertilizer is followed by an intense rainstorm, significant losses of N in surface runoff can occur (Moe et al., 1967). These losses can be reduced by fertilizer incorporation (Timmons et al., 1973; Romkens et al., 1973). Langdale et al. (1979) also observed that runoff losses could be reduced by shifting fertilizer applications to periods of rapid plant canopy development and periods of less intense rainfall. Viets (1970) has suggested that over the entire year, N application might reduce the total amount of N lost by surface runoff by promoting a good ground cover; application of N fertilizer will also tend to maintain the soil organic matter at a higher level so similar sediment losses will result in greater N losses from the fertilized soils. Gambrell et al. (1975) observed this in North Carolina. In summary, the N losses via surface runoff from cultivated fields are usually relatively small. This is particularly true for inorganic N, even from fertilized fields. The ultimate fate of the organic N lost to drainage waters is poorly understood, but it certainly is less of a eutrophication problem than entry of inorganic N into the same system (Keeney, 1973).

b. Subsurface Drainage—There is a large range in the quantities of N that are lost to surface waters from agricultural fields via subsurface drains (Table 16–2). There is little doubt that this mechanism of entry, particularly NO_3^--N through tile drainage, can make a significant contribution to the total N entering streams. The factors that affect this entry have been studied extensively.

One of the most important factors influencing quantities of N loss through tile lines is rate of fertilizer applied. Any time the rate of fertilizer N applied greatly exceeds the amount of N utilized by the crop to which it is applied, there is a potential for large losses of N through tile outlets. Typical of the type of data that have been obtained in many experiments throughout the world is that of Gast et al. (1978) in Minnesota. These authors observed that losses of NO_3^--N through tile lines after 3 yr of continuous corn grown on Typic Haplaquoll soils were 19, 25, 59, and 120 kg N/ha per year for annual N application rates of 20, 112, 224, and 448 kg N/ha, respectively. For this soil, the recommended N application rate for corn is 112 kg N/ha in Minnesota and this rate increased the N loss through tile lines by only a small amount. However, one can note the large increase in the contribution of N to surface waters when the recommended N rate was exceeded. This type of observation has been noted in many locations (Baker & Johnson, 1981; Miller, 1979; Zwerman et al., 1972; Logan et al., 1980).

It is not necessary for agricultural fields to have subsurface drains installed for N application to increase the entry of N into surface waters. Fellows and Brezonik (1981) observed that excessive fertilization of a citrus grove in Florida resulted in large amounts of NO_3^--N seeping into a nearby lake, whereas normal fertilization practices did not appear to enhance seepage fluxes of N. In Ohio, Chichester (1976) noted that the N contents of spring flows were related to the N regime of the different agricultural practices investigated.

High NO_3^--N in agricultural drainage waters is not necessarily a result of overfertilization or poor management. Some soils, particularly Histosols, may have mineralization rates that exceed crop needs during certain periods with the result that water draining from these soils has a high NO_3^--N content. This has been observed in New York (Duxbury & Peverly, 1978) and Canada (Miller, 1979). Nicholaichuk and Read (1978) have also observed that N concentrations in water from unfertilized fields in Saskatchewan exceed local water quality criteria, so they regard the current guidelines for their area as unobtainable.

Just as it cannot be assumed that high concentrations of NO_3^--N in drainage water is a result of overfertilization, it cannot be assumed that recommended N rates will not have a significant effect on NO_3^--N content of drainage waters. As indicated earlier, the potential loss of fertilizer to drainage water is closely related to the efficiency with which the applied N is used by the crop. Several authors (Frink, 1969; Gilliam & Terry, 1973) have computed pollution potential for their areas utilizing average or assumed yields and N application rates. An example of these computations is given in Table 16–3. It is clear from these calculations that there is a potential for N increases in drainage water as a result of N fertilization unless all of the N is harvested with the crop. Even 100% utilization just transfers the potential problem to the area where the crop is utilized. However, this is an unavoidable problem.

The data in Table 16–3 show the potential for N increases in drainage as a result of fertilization, but this potential is usually not realized because of denitrification. Soils with higher denitrification potentials lose much less

Table 16-3. Possible average N concentration in drainage water under two fertility conditions assuming that all N not harvested in corn crops goes into the 35 cm of drainage water (average for North Carolina) leaving the fertilized fields.

Corn yield	N in grain	N applied (kg/ha per year) 150	200
kg/ha per year		N in drainage water (μg/mL)	
3 136	45	29	44
6 272	90	17	31
9 048	135	4	18
12 544	180	0	6

NO_3^- to drainage waters than well-drained soils with no water restrictive horizons. Drainage water from some soils contains very little NO_3^- because of denitrification. Gentzsch et al. (1974) observed that cultivated soils in Illinois with natric horizons or poorly drained soils contained very little NO_3^- in the profile to be lost in drainage waters. In California, Devitt et al. (1976) found that a smaller fraction of applied N was lost in tile effluent from profiles containing layers of high clay content, compared with coarse-textured profiles. Similar observations about denitrification affecting NO_3^- content of drainage waters have been made in North Carolina (Gilliam et al., 1979) and in Iowa (Hanway & Laflen, 1974).

Nitrate concentrations in irrigation return flows depend significantly on the volume of such flows. Where irrigation water is cheap and abundant, the volume may be quite large. Conversely, surface irrigation return flows are negligible where water is scarce or expensive. Return flows that percolate through the soil profile to reach an aquifer offer the greatest potential for pollution, and factors that influence the transfer of some fertilizer N from near the surface to an underlying aquifer will be discussed in a later section. Since fertilizers are often placed below the soil surface, the contributions of fertilizer N to nitrate concentrations in irrigation tailwater are usually quite low; however, with surface application of fertilizer, considerable degradation of irrigation water can occur during passage down a furrow or over a field. In a study of two irrigation districts in the Central Valley of California, Tanji et al. (1977) found less than 0.2 μg NO_3^--N/mL in drain water from a district where the principal crop is flooded rice (*Oryza sativa* L.), compared with 13 μg N/mL in drain water from another district where furrow-irrigated row crops and some barley (*Hordeum vulgare* L.) and wheat are grown. They attributed the degradation of the applied water in the latter case to the collected subsurface drainage.

Discharge of tailwater into surface waters may not necessarily be harmful. If the water is reused for irrigation the contained NO_3^- may reduce the N requirement for growth of crops.

Among the factors affecting quality of surface irrigation tailwater are leaching fraction; evapotranspiration, which has the effect of concentrating the salts; nature of the crop; rate of water application; and erodibility of the soil.

Of greater significance than concentration of NO_3^- in irrigation return flows is the mass emission of NO_3^-. Low concentrations can be maintained by excessive water application, but in addition to being inefficient in the use of water, this is likely to increase total emissions substantially.

In addition to soil factors that may reduce the fertilizer N input to water via subsurface drainage, NO_3^- losses may be reduced by certain management alternatives. Alfalfa (*Medicago sativa* L.) may be used in crop rotations to absorb unutilized fertilizer N left in the soil profile (Robbins & Carter, 1980; Schertz & Miller, 1972). A more common rotation includes soybeans [*Glycina max* (L.) Merr.], which can utilize much of the inorganic N remaining after a fertilized corn crop (Johnson et al., 1975). Water management can also be utilized in some areas to reduce the NO_3^- leaving cultivated fields. Gilliam et al. (1979) used controlled drainage in North Carolina to induce denitrification of unutilized fertilizer N and reduced N in the drainage water by approximately 50%. Skaggs and Gilliam (1981) and Baker and Johnson (1976) have shown that various combinations of subsurface and surface drainage can have a large influence on the quantities of N in subsurface drainage. It should be noted that the drainage conditions which minimize NO_3^- in drainage waters will maximize the loss of sediment via surface runoff.

In irrigated land, Devitt et al. (1976) found that irrigation management to provide low leaching fractions resulted in relatively higher NO_3^--N concentrations in tile effluents but smaller amounts of total NO_3^--N compared with irrigation management for high leaching fractions.

There is much more quantitative information about N reaching surface water through subsurface flow when the subsurface outlet is a tile line than where the outlet is a seep or spring. However, there certainly is entry of N into surface waters when the outlet is not a tile line as was shown by Chichester (1976). However, there is a much greater potential for denitrification in transit (George & Hastings, 1951) or at the outlet (Gilliam et al., 1974). Several investigations have proposed that riparian vegetation can reduce the NO_3^- in groundwater before it enters surface water (Schlosser & Karr, 1981). Lowrance et al. (1984) in Georgia and Jacobs and Gilliam (1983) in North Carolina have noted decreases in N content as a result of groundwater flowing through a wooded area before entering streams. However, Omernik et al. (1981) observed little relationship between absence or presence of near-stream vegetation and water quality.

4. Fertilizer Effects on Stream Nitrogen Contents

The effect that increased fertilizer N usage is having on N contents of surface waters of the USA varies with the area of the country. In many areas, there apparently has been little change in NO_3^--N in streams since the mid-1930s or 1940s, even though there has been a tremendous increase in fertilizer N applied (Thomas & Crutchfield, 1974 [Kentucky]; Gilliam & Terry, 1973 [North Carolina]; Bower & Wilcox, 1969 [Colorado]). However, many areas have observed significant increases in N content of the stream (Budd, 1972, [Illinois]; Johnson et al., 1976 [New York]; Skogerboe, 1974 [western USA]). Areas that have not shown significant increases in N content of streams are generally those with relatively high rainfall and a low percentage of the land area under cultivation.

B. Movement of Nitrate into Groundwater

Although there is ample evidence that NO_3^- is accumulating in groundwater in some agricultural areas such as in the San Joaquin Valley of California (Nightingale, 1970), the contribution of fertilizers is not well defined. As a consequence of the uncertainties associated with N balance sheets, particularly with respect to denitrification losses, estimates of leaching losses obtained by difference are highly unreliable. Direct measurements of N movement to groundwater usually entail measurement of NO_3^- concentrations in soils or in soil solution extracts and calculation of soil water flux. Both NO_3^- concentrations and soil water properties exhibit a wide range of spatial variability; consequently, calculations of mass flow of N through the soil profile also have a considerable degree of uncertainty associated with them, particularly when based on only a few measurements.

Many papers have been published showing a relationship between fertilizer application levels and NO_3^- concentrations in the soil profile. For example, McNeal et al. (1977) found high levels of dissolved inorganic N in soil solutions taken from potato fields (*Solanum tuberosum* L.) in the Columbia basin area of Washington, and concluded that there was a considerable potential for groundwater and drainage water contamination by soil solution displaced by excessive applications of irrigation water. Increases in dissolved N levels in their studies were consistently observed with increased fertilization rates. They noted that ceramic extraction cups gave samples that varied greatly over short distances, and concluded that extrapolation from a few extraction cup sites to an entire irrigated field was an unreliable technique to determine fertilizer leaching patterns. Rible et al. (1979) analyzed NO_3^- in soil cores taken to a depth of 15 m at 83 locations in central and southern California, and calculated the quantity of N drained past the root zone, N_d, from the following relation:

$$N_d(\text{kg/ha per year}) = \frac{[NO_3^-\text{-N}(\mu\text{g/mL})]\,D(\text{cm/yr}),}{10},$$

where D is the drainage volume. The value of D was calculated from the product of the volume of water applied and the leaching fraction. The leaching fraction was calculated from the ratio of the Cl concentration in the irrigation water to that in the soil water of the unsaturated zone. Considering data from all sites, they found that NO_3^- concentrations were not well correlated with amount of N fertilizers applied, but were negatively correlated with amounts of water moving past the root zone. The total mass flux of NO_3^- moving past the root zone, however, was not well correlated with drainage volume. By selecting data for sites where records of water and N inputs were considered most reliable, they found that mass emissions of NO_3^- were positively correlated with both N applications and drainage volume, whereas NO_3^- concentrations were not correlated with either of these factors. At nonfertilized sites the average NO_3^--N concentration was 19 μg/mL compared with an average of 41 μg/mL for all sites. They regarded soil profile characteristics as being of major importance in influencing the amount of NO_3^- moving past the root zone. In their study they encountered a range of average concentrations of NO_3^--N from 6 to 220 μg/mL.

At an intensively instrumented site in northern California, Nielsen et al. (1979) calculated hydraulic flux from soil water pressure and soil water content measurements. These values, together with NO_3^- concentrations in soil solution samples, were used to calculate annual cumulative NO_3^- seepage, or mass emissions. Scaling factors were used to deal with spatial variability in both water and NO_3^- measurements. Table 16–4 presents some of their data for total mass emissions of N and those due to the applied fertilizer.

It is clear that contributions of fertilizer N to total mass emissions were small except where applications of water or fertilizer or both were excessive in relation to crop requirements. Figure 16–1, based on data from a 5-yr field experiment with isotopically labeled N fertilizer applied to corn (Broadbent & Rauschkolb, 1977) supports the view that in this situation fertilizer and water management practices required to attain maximum production had a low potential for pollution. However, fertilization in excess of that needed for maximum yield sharply increased the pollution potential.

Another management tool that can be used to decrease NO_3^- leaching is the use of slow-release fertilizers and/or nitrification inhibitors. For example, Swoboda (1977) reported that the amount of NO_3^--N leached below

Table 16–4. Annual cumulative NO_3^--N seepage at the 3-m depth over a 4-yr period in a field trial with corn on Yolo fine sandy loam† (from Nielson et al., 1979).

	Fertilizer N applied (kg/ha per year)						
	0	90		180		360	
Year	Total	Total	From Fert.	Total	From Fert.	Total	From Fert.
	NO_3^--N seepage (kg/ha per year)						
	Irrigation = ET						
1974	81	143	1.1	175	4.2	182	21.2
1975	18	21	0.5	25	0.7	49	8.7
1976	−5	10	0.2	4	0.2	70	11.0
1977	36	9	1.0	20	0.7	107	34.6
	Irrigation = 5/3 ET						
1974	149	119	6.9	151	19.3	281	61.6
1975	45	59	3.1	83	6.2	157	57.0
1976	31	36	2.6	57	6.3	83	29.2
1977	48	69	1.8	113	5.5	47	10.2

† Yolo fine sandy loam, Typic Xerorthents.

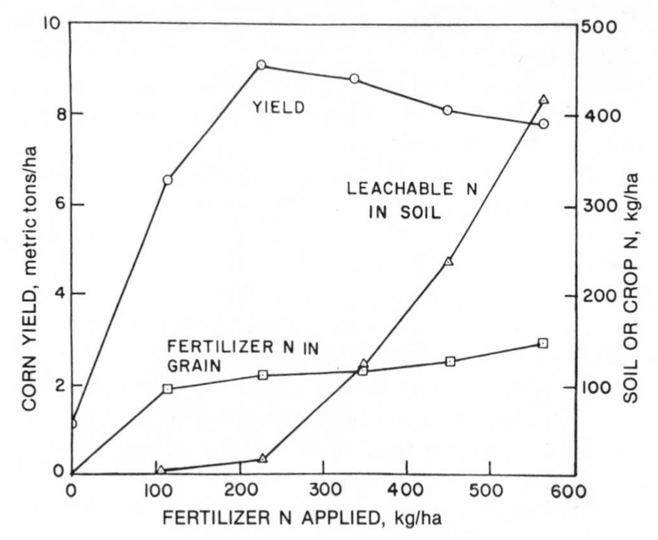

Fig. 16–1. Relationships among corn yield, amount of fertilizer N applied, and N recovered in grain or remaining as leachable N in soil (Broadbent & Rauschkolb, 1977).

60 cm in Houston black clay (fine, montmorillonitic, thermic Udic Pellustert) was decreased 53% where sulfur-coated urea (SCU) was used instead of soluble sources of N. The nitrification inhibitor N-Serve (nitrapyrin, The Dow Chemical Co., Midland, MI) decreased leaching of soluble N by 42% and of SCU by 58%.

II. PHOSPHORUS

The two major sources of P to waters are sewage discharges and agricultural runoff, the relative contributions of each determined by the land use of the drainage basin. In the USA, Canada, and other more technologically developed nations, advanced sewage treatment systems have greatly reduced the P load to waters, and the use of P in detergents has decreased in many areas or has been eliminated by ordinance. Ontario, Canada, New York, and Michigan have partial or total bans on use of P detergents, as do other areas, and the level of P in household detergents in the USA has significantly declined since the 1970s (Shannon, 1980). The U.S. Water Quality Act and its amendments dictate that advanced waste treatment be in use throughout the nation by 1983. In some environmentally sensitive areas such as the Great Lakes drainage basin, P in sewage effluents discharged to rivers and lakes must not exceed 1 μg P/mL.

As the sewage sources of P (referred to as *point* sources because they are discharged at a given point, e.g., a pipe) are reduced, land runoff will constitute an increasing proportion of the remaining P load to water. In instances where the runoff P load is great, significant reduction of this source may be required, in addition to point source reduction, to achieve the desired improvement in water quality. An example is the well-known case of Lake Erie, where it is estimated that, if the P concentrations of all of the point source discharges were reduced to < 1 μg P/mL, a further reduction of the diffuse P load of ~50% would be required to satisfactorily reduce the lake's eutrophication (PLUARG, 1978; U.S. Army Corps of Eng., 1979).

The diffuse source of P to water is comprised of numerous individual and diverse sources that are intermixed in the land drainage process so that their identities at the inlet to the lake or stream are lost, and monitoring their individual contributions is almost impossible. Individual sources include: runoff from storm sewers, livestock facilities, and agricultural and forest land; tile drainage and interflow; and atmospheric inputs. Of these, by far the greatest contribution is runoff from intensely cultivated and fertilized row-crop agricultural land. In the following sections, the magnitudes of P lost from agricultural land, the nature and bioavailability of P in agricultural land drainage, and the mechanisms of P movement in drainage water will be discussed.

A. Movement of Phosphorus into Groundwater

Movement of P to groundwater poses little environmental threat. With few exceptions, the P retention capacity of soil, especially the lower horizons, is great enough to retard the movement of even very heavy applications of P in the form of fertilizer, manure, or sewage sludge. In most instances, most applied P is retained near the soil surface when the fertilizer is not incorporated or placed within the surface layer after plowing. Brown (1935) found that P from biennial applications of superphosphate to permanent pasture penetrated no more than 5.7 to 7.5 cm after 16 yr. Schaller (1940) found that a surface application of 600 kg P/ha of superphosphate only penetrated 3 cm. Oloya and Logan (1980) found that application of 268 kg P/ha to no-till soil over a 3-yr period increased the Bray P-1–extractable P at least 10-fold in the 0- to 5-cm depth, but showed no increase below that zone.

In the classical experiments at Rothamstead, Cooke (1967) reported that P moved to a depth of 45 cm after annual applications of manure of 35 metric tons/ha since 1845. Furrer (1981) studied the movement of P from annual applications of 12 metric tons/ha of sewage sludge. After 5 yr, P had not moved below the 0- to 25-cm sampling depth. Johnson (1981) reported on the movement of P from superphosphate, manure, and sewage sludge in soil at the Market Garden Experiment, Woburn, England. Between 1942 and 1960, 43, 5575, and 11 700 kg P/ha as superphosphate, manure, and sewage sludge, respectively, were applied to a sandy loam soil. There was no movement of P from fertilizer or sewage sludge below the 0- to 23-cm sampling depth, while P from the manure was detected in significant quantity at the 46- to 60-cm depth. Phosphate in sewage sludges appears to be much less labile than that in livestock wastes; this is probably due to the formation of insoluble phosphates with metals commonly found in sludge (Fe, Al, Zn, Pb, etc.).

The greatest potential for movement of P to groundwater is from the application of large quantities of P to soils with low P retention capacity, such as sands or organic soils. Spencer (1957) showed that, when 600 to 2000 kg P/ha as superphosphate was applied to Lakeland fine sand, P was leached to depths up to 200 cm with the greatest accumulation in the 15- to 45-cm zone. Fiskell and Spencer (1964) applied up to 13 000 kg P/ha as monocalcium phosphate to Lakeland fine sand and found that 22% was retained in the surface 15 cm with some P movement up to 4 m. Larsen et al. (1958) showed that 76 cm of water leached 80% of applications of 40 and 200 kg P/ha of phosphate from a virgin muck. Fox and Kamprath (1971) were able to increase the low P retention capacity of an organic soil and a high organic matter sand by addition of $AlCl_3$. Spencer (1957) and Fiskell and Spencer (1964) found that liming increased the P retention capacity of Lakeland fine sand.

Phosphate can also leach from soil if the soil water permeability is high and sufficient water for leaching is applied. This is exemplified by the Pennsylvania State University wastewater study in which 2.5 to 5.0 cm/week of wastewater were applied to clay loam and sandy loam soils under forest vegetation (Kardos & Hook, 1976). There was significantly more leaching of P to the 120-cm sampling depth on the sandy loam soil than on the clay loam. The kinetics of phosphate sorption by soil may be limiting for soils with high water permeability.

The tendency of P to leach from soil may also be due to its chemical form. Several researchers (Spencer & Stewart, 1934; Hannapel et al., 1964; Duxbury & Peverly, 1978; Miller, 1979) have shown that a significant component of dissolved P in tile effluents and leachates is organic P, especially with the application of livestock wastes and on organic soils.

B. Movement of Phosphorus into Surface Water

Most of the P lost from land is by surface runoff. Phosphate is strongly retained by soil, and in the runoff process is transported primarily as eroded sediment with lesser amounts as dissolved P. Natural and anthropogenic processes tend to concentrate P at the soil surface where it is susceptible to runoff and erosion. Fertilizer applied to pasture, hay fields, or no-till land will be adsorbed and retained within a few centimeters of the surface, and the natural cycling of P through senescence and decay of leaf litter and crop residue also tends to concentrate P at the surface.

Figure 16–2 illustrates the processes by which P leaves the land in surface runoff. As precipitation strikes the land surface, some infiltrates and some runs off. Runoff detaches and transports sediment containing P (total particulate, P-TPP) and also removes some dissolved inorganic P (orthophosphate, OP) and organic P. There is also a shallow zone of runoff interaction (< 1 cm) in which some dissolved P is extracted from the soil by the runoff water, which then enters the runoff stream downslope.

The processes that determine the loss of TPP in runoff are illustrated in Fig. 16–2a. During soil detachment and transport, there is a selective

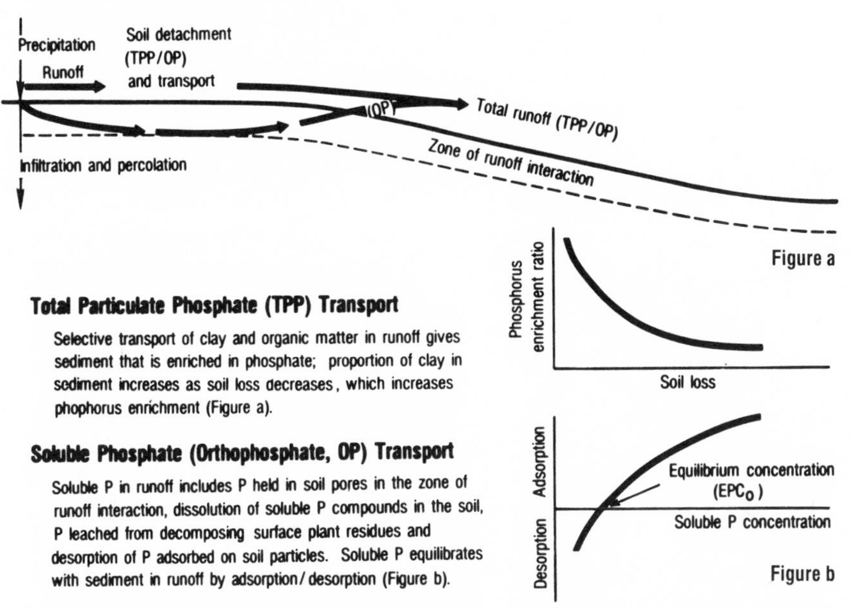

Fig. 16–2. A conceptual view of phosphorus transport in runoff from agricultural land (Logan & Adams, 1981).

removal of clay-size particles including organic matter. These clay particles usually contain higher concentrations of P than the larger silt and sand components, and the result is an enrichment of the sediment with P compared with the original surface soil. As soil loss (or sediment concentration) increases, the P enrichment ratio (PER) decreases and approaches 1.0 when the sediment composition approaches that of the surface soil. Menzel (1980) and Sharpley (1980) have shown that this relationship can be expressed as logarithmic functions:

$$\ln(\text{PER}) = 2.00 - 0.20 \ln(\text{sediment load}), \text{(Menzel, 1980)}$$

$$\ln(\text{PER}) = 2.48 - 9.27 \ln(\text{sediment load}), \text{(Sharpley, 1980)}.$$

The total particulate P fraction of runoff is made up of a number of chemical forms including insoluble minerals of Fe, Al, Mn, and Ca; organic P; sparingly soluble fertilizer reaction products such as dicalcium phosphate; and absorbed P. Only the more labile soil P fractions will be bioavailable and available to algae. A recent review of the literature on bioavailability of sediment-bound P (Lee et al., 1980) indicated that about 20% of the total particulate P of sediments is bioavailable with a range of 5 to 50%.

Phosphorus fertilization in excess of crop removal will increase the total P content of surface soil and will also increase TPP losses in runoff if the level of erosion is unchanged. More significant, however, is the effect of P fertilization on bioavailable P. Lake and Morrison (1975) reported on a

Table 16–5. Runoff losses of TPP and Bray P-1–extractable sediment and P and EPC_o under simulated rainfall (Lake & Morrison, 1975).

Soil	Treatment	Runoff sediment			Bray P-1–Extractable P		EPC_o clay
		TPP	Bray P-1	EPC_o	Whole soil	Clay	
		μg/g		μg/mL	μg/g		μg/mL
Haskins loam	Unfertilized	530	76	0.028	46	155	0.028
(Aeric Ochraqualfs)	Fertilized†	643	213	0.455			
Nappanee clay loam	Unfertilized	267	113	0.074	44	75	0.009
(Aeric Ochraqualfs)	Fertilized	547	199	0.300			
Morley clay loam	Unfertilized	553	25	0.004	12	16	ND‡
(Typic Hapludalfs)	Fertilized	665	196	0.104			
Hoytville silty clay	Unfertilized	1230	231	0.015	117	166	0.044
(Mollic Ochraqualfs)	Fertilized	1375	368	0.284			

† 56 kg P/ha. ‡ Not detected.

simulated rainfall study of fertilized and unfertilized Indiana soils. Their data (Table 16–5) show that, while TPP in runoff increased with fertilization, there was an even greater increase in Bray P-1–extractable P, a measure of sediment P bioavailability. Phosphate fertilizer is strongly adsorbed by the clay fraction of soils, which is selectively eroded and transported, and results in an enrichment of bioavailable P in the sediment.

Figure 16–2b also illustrates the processes that govern the transport of dissolved inorganic P in runoff. As dissolved inorganic P reacts with soil particles, there is strong and rapid adsorption of P until equilibrium or near equilibrium is reached. The result is a partitioning of P between the solid phase (soil or sediment) and the solution phase. The equilibrium dissolved inorganic P concentration (EPC_0) is the concentration at zero net adsorption of desorption and is the concentration of inorganic P in solution that a given soil or sediment will maintain without significant P removal or addition or desorption and is the concentration of inorganic P in solution that a given soil or sediment will maintain without significant P removal or addition. The EPC_0 is not only characteristic of the retention capacity of the soil or sediment, but also reflects the level of previously adsorbed P. This is illustrated in Table 16–5 (Lake & Morrison, 1975), which shows the increase in EPC_0 with fertilization. McDowell et al. (1980) found that EPC_0 was positively correlated with soil test estimates of available P, and was a good indicator of dissolved inorganic P losses in runoff as long as runoff sediment concentrations were high enough to provide adequate buffer capacity. At low sediment concentrations as those found with no-till, EPC_0 was found to underestimate dissolved P losses. Sharpley et al. (1981) have recently shown that the kinetics of P adsorption/desorption may be limiting at low sediment concentrations, and have developed an exponential kinetic model to account for the effects of sediment concentration and previously adsorbed P on concentration of inorganic P in runoff. Logan and Adams (1981) recently reviewed the literature on effects of no-till or conservation tillage vs. plowed systems on losses of total particulate P and dissolved inorganic P in runoff. They found that no-till and conservation tillage reduced TPP losses by as much as 90% when compared with plowing on the same soil, but in all cases dissolved inorganic P losses increased with no-till or conservation tillage. They attributed this increase to the reduction in P buffer capacity as sediment concentrations were reduced and the increase in labile P levels at the soil surface from surface-applied fertilizer and decay of crop residues.

Table 16–6 gives a range of total particulate and dissolved inorganic P losses in surface runoff with different land use conditions. The crop data are for conventional tillage, and the high values for particulate P for corn (Minnesota) and cotton (*Gossypium hirsutum* L.) reflect the high erosion rates on these sites. On alfalfa, dissolved P losses were as high as the particulate load. While erosion was low on this site, there was significant transport of dissolved P from decay of alfalfa residue. Rangeland particulate P transport was high where continuous grazing led to increased erosion. In

Table 16-6. Particulate and dissolved P losses in surface runoff under different land use conditions.

Location	Size	Land use	P in runoff		Reference
			Particulate	Dissolved inorganic	
	ha		—— kg/ha per year ——		
Iowa	3.3	Corn	--	0.09	Alberts et al. (1978)
Minnesota	0.01	Corn	18.19	0.41	Burwell et al. (1975)
Ohio	3.2	Soybeans	1.09	0.13	Logan & Stiefel (1979)
Oklahoma	5.3	Wheat	2.40	0.54	Olness et al. (1975)
Oklahoma	17.9	Cotton (irrigated)	9.22	1.93	Olness et al. (1975)
Oklahoma	10.7	Alfalfa	1.25	1.23	Olness et al. (1975)
Oklahoma	7.8–11.1†	Range—continuous grazing	4.46	0.14	Olness et al. (1975)
Oklahoma	11.0	Range—rotation grazing	1.14	0.13	Olness et al. (1975)
Minnesota	0.01	Native prairie	0.09	0.02	Timmons & Holt (1977)
Mississippi	1.5–2.8‡	Southern pine forest§	0.21	0.09	Duffy et al. (1978)
Minnesota	6.5	Aspen-birch forest¶	0.13	0.13	Timmons et al. (1977)

† Mean of two watersheds.
‡ Mean of five watersheds.
§ Pine, *Pinus* sp.
¶ Aspen, *Populus* sp.; birch, *Betula* sp.

Table 16–7. Particulate and dissolved P losses in subsurface drainage in Iowa, Minnesota, and Ohio (Logan et al., 1980).

Location	Size	Crop†	P in subsurface drainage: Total	P in subsurface drainage: Dissolved inorganic
	ha		—— kg/ha per year ——	
Iowa	0.4	Corn-oats-corn-soybean	0.018	0.003
Minnesota	0.9	Small grain-flax-sorghum-soybeans	0.10	0.04
Minnesota	2.2	Corn	0.20	0.09
Minnesota	97	50% corn, 50% soybeans	0.16	0.09
Minnesota	26	45% corn, 40% soybeans, 15% oats	0.04	0.02
Ohio	0.9	Soybeans	0.39	0.09
Ohio	0.004	Soybeans	0.45	0.10

† Oats, *Avena sativa* L.; flax, *Linum usitatissimum* L.; sorghum, *Sorghum bicolor* (L.) Moench.

comparison, levels of particulate and dissolved P are low for prairie and forest conditions.

C. Movement of Phosphorus by Subsurface Drainage

In a hydrologic sense, subsurface drainage should be considered as part of the surface runoff from a watershed, and as such, P loads from subsurface drains contribute to the eutrophication potential of the watershed drainage. This is in contrast with P movement to groundwater where there is only a potential for algal growth if the groundwater is pumped into surface waters. Unlike surface runoff, subsurface water is generally free of sediments (although some sediment will be discharged from tile drains over clay soils) and the equilibrium dissolved inorganic P concentration (EPC_0) of the subsoil is generally much lower than that of the surface soil.

Table 16–7 gives values for total P (particulate plus dissolved) and dissolved inorganic P for several studies in the North Central USA. These data have been summarized in detail by Logan et al. (1980). The values are about one order of magnitude lower than the surface runoff data given in Table 16–6.

The movement of P to subsurface drains is complicated by the nature of water movement in different soils, especially under unsaturated conditions. Schwab et al. (1973) found significant movement of sediment through tile lines on a deep-cracking lake plain soil in northwestern Ohio. Logan and Stiefel (1979) also observed very high concentrations of dissolved inorganic P in tile flow immediately after broadcast application of P fertilizer in the fall on a deep-cracking clay soil. Hergert et al. (1981) recently found that, while NO_3^- and Cl^- concentrations in tile flow decreased with increased flow, dissolved inorganic P concentrations increased as flow increased. They attributed the P increase to decreased P adsorption by the soil as Ca^{2+} and electrolyte concentrations declined at higher flows. The role of organic matter and organic P on the downward movement of P has already been discussed.

D. Fertilizer Effects on Stream Phosphorus Contents

The phosphorus content of a stream is determined by a number of factors including flow, relative contributions from various sources and biological, chemical and physical sinks in the stream system. When significant amounts of sediment (> 100 μg/mL) are transported in a stream, most of the phosphate (> 75%) occurs as total particulate P and the dissolved inorganic P fraction is buffered by the particulate P. The highest particulate and dissolved P concentrations occur in a stream during storm events, and these high flow periods account for most of the P load. Agricultural sources, including fertilizer, make up a large portion of the phosphate transported during storm events in streams draining watersheds with significant amounts of cropland, while point sources, including effluent discharges from sewage treatment plants, may represent a significant fraction of the P in a stream during low flow.

Fertilizers contribute to the dissolved and particulate P concentrations in streams to the extent that they contribute to P in runoff as discussed previously (section B, Movement of Phosphorus into Surface Water). As the extractable phosphate content of soil and sediment increases with fertilization, the eroded sediment will maintain a higher dissolved inorganic P concentration in the stream and will have less capacity to serve as a sink for point sources of phosphate that have higher P concentrations than normally found in agricultural runoff. Taylor and Kunishi (1971) found that sediments in a stream draining an area of low P fertilization were highly effective in reducing the high phosphate concentration in agricultural waste discharged to the stream.

III. ATMOSPHERIC CONCERNS ABOUT NITROGEN FERTILIZER

The fixation of N_2 from the atmosphere, utilization of fixed N_2 by plants and animals, and return of N to the atmosphere as N_2 and N_2O through denitrification is a well-known cycle. The denitrification step has long been recognized as a problem by soil scientists concerned with the efficient use of soil and fertilizer N by plants because of loss from soils of potentially utilizable N. Only recently has it been recognized that the release of N_2O to the atmosphere through denitrification and nitrification may be causing problems because of the participation of N_2O in reactions that result in loss of O_3 from the stratosphere (Crutzen, 1971; McElroy & McConnell, 1971; Johnson, 1972) and contributes to the "greenhouse" effect (Crutzen, 1981). An excellent review of the stratospheric reactions of N_2O and concern of atmospheric chemists about the loss of protection from ultraviolet radiation because of destruction of O_3 was given by Crutzen (1981).

The concern that increased N_2 fixation for N fertilizer and by legumes results in greater N_2O release to the atmosphere prompted an evaluation and report by a team of Council of Agricultural Science and Technology

(CAST) scientists (CAST, 1976). This group concluded that there was cause for concern and need for more information because there was a very limited amount of data on which to base their calculations. The lack of information was particularly evident for N_2O evolution from fertilized and nonfertilized land areas. Their estimation of N_2O evolution from cropland was based largely on an estimate by Hauck (1969) of an average denitrification loss from cropland of approximately 16 kg/ha per year and a ratio of 16:1 for N_2/N_2O evolved during denitrification. Thus they estimated that approximately 1 kg/ha per year of N_2O-N is evolved from cropland.

One new observation resulting from the interest in N_2O evolution is that much of the evolution from soils is a result of nitrification (Bremner & Blackmer, 1978; Freney et al., 1979; Mosier & Hutchinson, 1981). It was previously thought that essentially all N_2O evolution was a result of denitrification (Focht, 1974).

Nitrous oxide evolution has been measured from a number of cropland soils throughout the world. Most of the estimates of N lost via this pathway fall in the range of 2 to 10 kg/ha per year (Hutchinson & Mosier, 1979 [Colorado]; Burford et al., 1981 [Great Britain]; McKenney et al., 1980 [Canada]; Bremner & Blackmer, 1981 [Iowa]; Spooner & Gilliam, 1978 [North Carolina]; Duxbury et al., 1981 [New York]). However, rates of 19 to 42 kg/ha per year have been measured on some irrigated soils in California (Ryden & Lund, 1980). One factor that the recent studies have very clearly shown is that N_2O evolution from soils is extremely variable, both spatially and temporally. There are many factors known to affect N_2O evolution during both nitrification and denitrification such as moisture content, time, soil pH, oxygen content, temperature, nitrate concentration and reduction-oxidation potential (Firestone et al., 1979; Blackmer & Bremner, 1978; Focht, 1974; Lety et al., 1980, 1981; Lensi & Chalamet, 1979). Thus, any predictions of N_2O evolution from agricultural lands have a large uncertainty factor (Gilliam et al., 1978).

It is clear from recent work that N fertilization will increase N_2O evolution from soils. However, it is also evident that an increase in N_2 fixation through the cultivation of legume crops, or improvement of N_2 fixation by legumes, will also result in an increase in N_2O evolution to the atmosphere. Although the concern for the detrimental effects of increased N_2O evolution seems valid, magnitude of the problem is not clear at this time. The only recommendation regarding N fertilizers and N_2O evolution that can be made at this time is that any fertilizer practice that maximizes fertilizer N efficiency will minimize N_2O evolution problems.

IV. THE ACCUMULATION OF CADMIUM IN SOIL WHEN ADDED AS A CONTAMINANT OF PHOSPHATE FERTILIZERS

Cadmium is a metal element that poses some environmental risk to the human food chain (CAST, 1980). Although the major source of anthropogenic Cd addition to soil is from sewage sludges and other industrial wastes, Cd is also added to soil with P fertilizers. Cadmium occurs naturally in phosphate rock at levels that vary with the source of the rock. Cadmium in U.S. phosphate rock can be as high as 980 μg Cd/g (USDA–USDI, 1977; Mulla et al., 1980) with the highest concentrations in U.S. ores in the west and the lowest in the southeast. Mulla et al. (1980) reported a range of 50 to 200 μg Cd/g in triple superphosphate (TSP) manufactured from western ores, and 10 to 20 μg Cd/g for southeastern sources of TSP. Williams and David (1973, 1976) showed that long-term applications of TSP containing < 50 μg Cd/g increased surface soil total Cd concentrations from 0.046 to 0.212 μg Cd/g soil, and Mulla et al. (1980) found that application of about 175 kg P/ha per year for 36 yr to citrus groves in California increased total soil Cd from a background concentration of 0.07 μg Cd/g to about 1.0 μg Cd/g. They estimated that the Cd content of TSP used on the citrus groves was approximately 174 μg Cd/g.

Mulla et al. (1980) estimated that even long-term applications of high rates of TSP produced from western ores that have higher Cd contents than other U.S. ores would result in less accumulation of Cd in soil than is currently recommended by USEPA (USEPA, 1979).

V. SUMMARY AND CONCLUSIONS

Since the late 1960s there has been a large amount of research designed to assess the environmental hazards of fertilizer use. This research has focused primarily on the nonpoint source contributions of N and P to eutrophication of surface waters and to NO_3^- content of drinking water. Some concern has been expressed about the increase in N_2O emission from N fertilized fields with regard to N_2O contributing to the decrease of the protective O_3 layer in the stratosphere. The contribution of P fertilizers to the Cd levels of soils has also been investigted.

When the annual quantity of N entering surface waters from agricultural land exceeds 10 kg/ha, it is almost always via subsurface flow from cultivated fields. The annual amount of N leaving fertilized or nonfertilized forests in drainage waters is usually 1 to 2 kg N/ha. The amount of N in drainage waters from pastures seldom exceeds 10 kg/ha per year. Nitrogen fertilization usually increases the amount of N lost to surface water from pastures, but this increase is small. The loss of N through surface runoff is usually relatively small and consists primarily of organic N, which does not cause as much of a water quality problem as does excessive quantities of inorganic N.

There are many factors that influence the amount of N lost from cultivated crops to surface water via subsurface flow and to deeper groundwater. Some of the more important factors are fertilizer N rate, the percentage of the fertilizer N absorbed by the crop, soil properties that influence the amount of denitrification that occurs, level of mineralizable soil N, and drainage volume (particularly in irrigated soils). Because of the many factors influencing the quantities and concentrations of N in drainage waters, there has been a large range of values observed. In spite of the complexities involved in determining the amount of N lost and the uncertainties associated with the attempt to measure NO_3^- flux below the root zone, a consideration of the data available from many sources leads to the conclusion that leaching of NO_3^- from cropland represents a significant loss of available N and that this contributes to the NO_3^- load in both surface and groundwater. Whether that contribution is of major or minor magnitude depends on local factors. In many areas, no known problem exists as a result of excessive N content of water and in some localized areas, high N concentrations in groundwater, particularly, is considered a problem.

Because of the general lack of mobility of P in soils, P fertilization poses little or no threat to quality of groundwater. However, P is the element that is most often limiting in the growth of algae in surface water and increases in the amount of P entering surface waters can contribute to the eutrophication of these waters.

Most of the P lost from land is by surface runoff, primarily as eroded sediment with lesser amounts as dissolved P. Most of the P associated with the sediment is not bioavailable so it does not contribute to eutrophication problems. However, an average of about 20% of the total particulate sediment P is bioavailable and this percentage increases when P fertilizers are applied to the soil. Although the total amount of P lost to drainage water from cultivated crops (usually 1 to 10 kg/ha per year) is usually small with regard to fertilizer inputs, this loss can contribute to eutrophication problems, particularly in sensitive areas.

The use of N fertilizer has been shown to increase the N_2O emission from some soils. Whether or not this is a significant threat to the O_3 content of the stratosphere is not known at this time, but the general conclusion seems to be that there is cause for concern but not for alarm. The Cd content of P fertilizers is not considered a problem under normal agricultural practices.

REFERENCES

Alberts, E. E., G. E. Schuman, and R. E. Burwell. 1978. Seasonal runoff losses of nitrogen and phosphorus from Missouri Valley loess watersheds. J. Environ. Qual. 7:203–208.

Aubertin, G. M., and J. H. Patric. 1974. Water quality after clearcutting a small watershed in West Virginia. J. Environ. Qual. 3:243–249.

Baker, J. L., and H. P. Johnson. 1976. Impact of subsurface drainage on water quality. p. 91–98. *In* Proc. 3rd Natl. Drainage Symp. ASAE Pub. 1–77.

Baker, J. L., and H. P. Johnson. 1981. Nitrate-nitrogen in tile drainage as affected by fertilization. J. Environ. Qual. 10:519–522.

Bigeriego, R., R. D. Hauck, and R. A. Olson. 1979. Uptake, translocation and utilization of ^{15}N-depleted fertilizer in irrigated corn. Soil Sci. Soc. Am. J. 43:528–533.

Bingham, F. T., S. Davis, and E. Shade. 1971. Water relations, salt balance, and nitrate leaching losses of a 960-acre citrus watershed. Soil Sci. 112:410–417.

Blackmer, A. M., and J. M. Bremner. 1978. Inhibitory effect of nitrate on reduction of N_2O to N_2 by soil microorganisms. Soil Biol. Biochem. 10:187–191.

Bormann, F. H., G. E. Likens, D. W. Fisher, and R. S. Pierce. 1968. Nutrient losses accelerated by clear-cutting of a forest ecosystem. Science (Washington D.C.) 159:882–884.

Bower, C. A., and L. V. Wilcox. 1969. Nitrate content of the upper Rio Grande as influenced by nitrogen fertilization of adjacent irrigated lands. Soil Sci. Soc. Am. Proc. 33:971–973.

Bremner, J. M., and A. M. Blackmer. 1978. Nitrous oxide: Emission from soils during nitrification of fertilizer nitrogen. Science (Washington D.C.) 199:295–296.

Bremner, J. M., and A. M. Blackmer. 1981. Terrestrial nitrification as a source of atmospheric nitrous oxide. p. 151–170. *In* C. C. Delwicke (ed.) Denitrification, nitrification and atmospheric nitrous oxide. John Wiley & Sons, Inc., New York.

Broadbent, F. E., and A. B. Carlton. 1979. Field trials with isotopes—Plant and soil data for Davis and Kearney sites. p. 433–465. *In* Final Report to the National Science Foundation for Grant nos. GI34733X, GI143664, AEN74-11136 AO1, ENV76-10283 and PRF76-0283, Nitrate in Effluents from Irrigated Lands, University of California.

Broadbent, F. E., and C. Krauter. 1974. Nitrate movement and plant uptake of N in a field soil receiving ^{15}N-enriched fertilizer. p. 236–239. *In* Trace contaminants in the environment, Proc. 2nd Annual NSF–RANN Trace Contaminants Conference, Asilomar, Pacific Grove, CA. Lawrence Berkeley Laboratory, University of California, Berkeley.

Broadbent, F. E. and R. S. Rauschkolb. 1977. Nitrogen fertilization and water pollution. Calif. Agric. 31:24–25.

Brown, L. A. 1935. A study of P penetration and availability in soils. Soil Sci. 39:277–287.

Budd, T. 1972. Are we polluting steams? Prairie Farmer 144(2): 16–17.

Burford, J. R., R. J. Dowdell, and R. Crees. 1981. Emission of nitrous oxide to the atmosphere from direct-drilled and ploughed clay soils. J. Sci. Food Agric. 32:219–223.

Burwell, R. E. 1977. Nutrient loss research. p. 28–34. *In* USDA–ARS Pub. 57. U.S. Department of Agriculture, Columbia, MO.

Burwell, R. E., G. E. Schuman, K. E. Saxton, and H. G. Heinemann. 1976. Nitrogen in subsurface discharge from agricultural watersheds. J. Environ. Qual. 5:325–329.

Burwell, R. E., D. R. Timmons, and R. F. Holt. 1975. Nutrient transport in surface runoff as influenced by soil cover and seasonal periods. Soil Sci. Soc. Am. Proc. 39:523–528.

Chichester, F. W. 1976. The impact of fertilizer use and crop management on nitrogen content of subsurface water draining from upland agricultural watersheds. J. Environ. Qual. 5:413–416.

Commoner, B. 1970. Threats to the integrity of the nitrogen cycle: Nitrogen compounds in soil, water, atmosphere and precipitation. p. 341–366. *In* S. F. Singer (ed.) The changing global environment. D. Reidel Publishers, Hingham, MA.

Cooke, G. W. 1967. The availability of plant nutrients in soils and their uptake by crops. p. 48–69. *In* Annual Report of East Malling Res. Stn. for 1966. Rothamstead Exp. Stn., Harpenden, Herts, UK.

Council for Agricultural Science and Technology. 1976. Effect of increased nitrogen fixation on stratospheric ozone. Rep. no. 53, Iowa State University, Ames, IA.

Council for Agricultural Science and Technology. 1980. Effects of sewage sludge on the cadmium and zinc content of crops. Rep. no. 83, Iowa State University, Ames, IA.

Crutzen, P. J. 1971. Ozone production rates in an oxygen-hydrogen-nitrogen oxide atmosphere. J. Geophys. Res. 76:7311–7327.

Crutzen, P. J. 1981. Atmospheric chemical processes of the oxides of nitrogen, including nitrous oxide. p. 17–44. C. C. Delwicke (ed.) *In* Denitrification, nitrification and atmospheric nitrous oxide. John Wiley & Sons, Inc., NY.

Devitt, D., J. Letey, L. J. Lund, and J. W. Blair. 1976. Nitrate-nitrogen movement through soil as affected by soil profile characteristics. J. Environ. Qual. 5:283–288.

Duffy, P. D., J. D. Schreiber, D. C. McClurkin, and L. L. McDowell. 1978. Aqueous and sediment-phase phosphorus yields from five southern pine watersheds. J. Environ. Qual. 7:45–50.

Dunigan, E. P., R. A. Phelan, and C. L. Mondart, Jr. 1976. Surface runoff losses of fertilizer elements. J. Environ. Qual. 5:339–342.

Duxbury, J. M., A. Bekine, and S. Vogkel. 1981. Emission of N_2O from cropland. Agron. Abstr. American Society of Agronomy, Madison, WI, p. 160.

Duxbury, J. M., and J. H. Peverly. 1978. Nitrogen and phosphorus losses from organic soils. J. Environ. Qual. 7:566–570.

Fellows, C. R., and P. L. Brezonik. 1981. Fertilizer flux into two Florida lakes via seepage. J. Environ. Qual. 10:174–177.

Firestone, M. K., M. S. Smith, R. B. Firestone, and J. M. Tiedje. 1979. The influence of nitrate, nitrite and oxygen on the composition of the gaseous products of denitrification in soil. Soil Sci. Soc. Am. J. 43:1140–1145.

Fiskell, J. G. A., and W. R. Spencer. 1964. Forms of phosphate in Lakeland fine sand after six years of heavy phosphate and lime applications. Soil Sci. 97:320–327.

Focht, D. D. 1974. The effect of temperature, pH and aeration on the production of nitrous oxide and gaseous nitrogen-a zero-order kinetic model. Soil Sci. 118:173–179.

Fox, R. L., and E. J. Kamprath. 1971. Adsorption and leaching of P in acid organic soils and high organic matter sand. Soil Sci. Soc. Am. Proc. 35:154–156.

Freney, J. R., O. T. Denmead, and J. R. Simpson. 1979. Nitrous oxide emission from soils at low moisture contents. Soil Biol. Biochem. 11:167–173.

Frink, C. R. 1969. Water pollution potential estimated from farm nutrient budgets. Agron. J. 61:550–553.

Furrer, O. J. 1981. Accumulation and leaching of phosphorus as influenced by sludge application. p. 235–240. *In* T. W. G. Hucker and G. Catroux (ed.) Phosphorus in sewage sludge and animal waste slurries. Proc EEC Seminar, Groningen, Netherlands. D. Reidel Publishers, Hingham, MA.

Gambrell, R. P., J. W. Gilliam, and S. B. Weed. 1975. Nitrogen losses from soils of the North Carolina coastal plain. J. Environ. Qual. 4:317–323.

Gast, R. G., W. W. Nelson, and G. W. Randall. 1978. Nitrate accumulation in soils and loss in tile drainage following nitrogen applications to continuous corn. J. Environ. Qual. 7:258–261.

Gentzsch, E. P., E. C. A. Runge, and T. R. Peck. 1974. Nitrate occurrence in some soils with and without natric horizons. J. Environ. Qual. 3:89–94.

George, W. O., and W. W. Hastings. 1951. Nitrate in the groundwater of Texas. Trans. Am. Geophys. Union 32:450–456.

Gilliam, J. W., R. B. Daniels, and J. F. Lutz. 1974. Nitrogen content of shallow groundwater in the North Carolina Coastal Plain. J. Environ. Qual. 3:147–151.

Gilliam, J. W., S. Dasberg, L. J. Lund, and D. D. Focht. 1978. Denitrification in four California soils: Effect of soil profile characteristics. Soil Sci. Soc. Am. J. 42:61–65.

Gilliam, J. W., R. W. Skaggs, and S. B. Weed. 1979. Drainage control to diminish nitrate loss from agricultural fields. J. Environ. Qual. 8:137–142.

Gilliam, J. W., and D. L. Terry. 1973. Potential for water pollution from fertilizer use in North Carolina. North Carolina Agric. Ext. Circ. 550.

Hannapel, R. J., W. H. Fuller, S. Bosma, and J. S. Bullock. 1964. Phosphorus movement in a calcareous soil: I. Predominance of organic forms of phosphorus in phosphorus movement. Soil Sci. 97:350–357.

Hanway, J. J., and J. M. Laflen. 1974. Plant nutrient losses from tile-outlet terraces. J. Environ. Qual. 3:351–356.

Hauck, R. D. 1969. Quantitative estimates of N cycle processes: Review and comments. Paper presented at meeting on recent developments in the use of ^{15}N in soil-plant studies. Sponsored by the joint FAC/IAEA Division of Atomic Energy in Food and Agriculture, Sofia, Bulgaria.

Hergert, G. W., D. R. Bouldin, S. D. Klausner, and P. J. Zwerman. 1981. Phosphorus concentration-water flow interactions in tile effluent from manured land. J. Environ. Qual. 10:338–344.

Hills, F. J., F. E. Broadbent, and M. Fried. 1978. Timing and rate of fertilizer nitrogen for sugarbeets related to nitrogen uptake and pollution potential. J. Environ. Qual. 7:368–372.

Hood, A. E. M. 1976. The high nitrogen trial on grassland at Jealott's Hill. Stikstof (Engl. Ed.) 83:395–404.

Hutchinson, G. L., and A. M. Mosier. 1979. Nitrous oxide emission from an irrigated cornfield. Science (Washington D.C.) 205:1125–1126.

Jacobs, T. J., and J. W. Gilliam. 1983. Nitrate loss from agricultural drainage waters: Implications for nonpoint source control. Water Resour. Res. Inst. of the Univ. of North Carolina Rep. no. 209. Water Resources Research Institute of the University of North Carolina, Raleigh, NC.

Johnson, A. E. 1981. Accumulation of phosphorus in a sandy loam soil from farmyard manure (FYM) and sewage sludge. p. 273–290. *In* T. W. G. Hucker and G. Catroux (ed.) Phosphorus in sewage sludge and animal waste slurries. Proc. EEC Seminar, Groningen, Netherlands. D. Reidel Publishers, Hingham, MA.

Johnson, A. H., D. R. Bouldin, E. A. Goyette, and A. M. Hedges. 1976. Nitrate dynamics in Fall Creek, N.Y. J. Environ. Qual. 5:386–390.

Johnson, H. 1972. Newly recognized vital nitrogen cycle. Proc. Natl. Acad. Sci. U.S.A. 69:2369–2372.

Johnson, J. W., L. F. Welch, and L. T. Kurtz. 1975. Environmental implications of N fixation by soybeans. J. Environ. Qual. 4:303–306.

Kardos, L. T., and J. E. Hook. 1976. Phosphorus balance in sewage effluent treated soils. J. Environ. Qual. 5:87–90.

Kenney, D. R. 1973. The nitrogen cycle in sediment-water systems. J. Environ. Qual. 2:15–29.

Kilmer, V. J., J. W. Gilliam, J. F. Lutz, R. T. Joyce, and C. K. Eklunk. 1974. Nutrient losses from fertilized grassed watersheds in Western North Carolina. J. Environ. Qual. 3:214–219.

Kissel, D. E., C. W. Richardson, and E. Burnett. 1976. Losses of nitrogen in surface runoff in the blackland prairie of Texas. J. Environ. Qual. 5:288–292.

Kolenbrander, G. J. 1969. Nitrate content and nitrogen loss in drainwater. Neth. J. Agric. Sci. 17:246–255.

Lake, J., and J. Morrison. 1975. Environmental impact of land use on water quality. p. 98–110. *In* Black Creek Progress Report. USEPA Rep. 905/9-75-006. U.S. Government Printing Office, Washington DC.

Langdale, G. W., R. A. Leonard, W. G. Fleming, and W. A. Jackson. 1979. Nitrogen and chloride movement in small upland piedmont watersheds: II. Nitrogen and chloride transport in runoff. J. Environ. Qual. 8:57–63.

Larsen, J. E., R. Langston, and G. F. Warren. 1958. Studies on the leaching of applied labelled phosphorus in organic soils. Soil Sci. Soc. Am. Proc. 22:558–560.

Lee, G. F., R. A. Jones, and W. Rast. 1980. Availability of phosphorus to phytoplankton and its implications for phosphorus management strategies. p. 259–308. *In* R. C. Loehr et al. (ed.) Phosphorus management strategies for lakes. Ann Arbor Science Publishers, Ann Arbor, MI.

Lensi, R., and A. Chalamet. 1979. Nitrate-nitrous oxide relations during denitrification in a hydromorphous soil. Rev. Ecol. Biol. Sol. 16:315–323.

Letey, J., N. Valoras, D. D. Focht, and J. C. Ryden. 1981. Nitrous oxide production and reduction during denitrification as affected by redox potential. Soil Sci. Soc. Am. J. 45:727–730.

Letey, J., N. Valoras, A. Hadas, and D. D. Focht. 1980. Effect of air-filled porosity, nitrate concentration, and time on the ratio of N_2O/N_2 evolution during denitrification. J. Environ. Qual. 9:227–232.

Logan, T. J., and J. R. Adams. 1981. The effects of reduced tillage on phosphate transport from agricultural land. Tech. Rep. Ser. Lake Erie Wastewater Management Study. U.S. Army Corps of Engineers, Buffalo, NY.

Logan, T. J., G. W. Randall, and D. R. Timmons. 1980. Nutrient content of tile drainage from cropland in the North Central region. North Central Res. Bull. 268. OARDC, Wooster, OH.

Logan, T. J., and R. C. Steifel. 1979. The Maumee River Basin pilot watershed study. Vol. 1. Watershed characteristics and pollutant loadings. USEPA Rep. 905/9-79-005-A. U.S. Government Printing Office, Washington, DC.

Lowrance, R. R., R. L. Todd, and L. E. Asmussen. 1984. Nutrient cycling in an agricultural watershed. I. Phreatic movement. J. Environ. Qual. 13:22–27.

McDowell, L. L., J. D. Schreiber, and H. B. Pionke. 1980. Estimating soluble (PO_4-P) and labile phosphorus in runoff from croplands. p. 509–533. *In* CREAMS: A field-scale model for chemicals, runoff and erosion from agricultural management systems. USDA Conserv. Res. Rep. U.S. Government Printing Office, Washington, DC.

McElroy, M. G., and J. C. McConnell. 1971. Nitrous oxide: A natural source of stratospheric NO. J. Atmos. Sci. 28:1095–1098.

McKenney, D. J., K. F. Shuttleworth, and W. I. Findlay. 1980. Nitrous oxide evolution rates from fertilized soil: Effects of applied nitrogen. Can. J. Soil Sci. 60:429–438.

McNeal, B. L., B. L. Carlile, and R. Kunkel. 1977. Nitrogen and water management to minimize return-flow pollution from potato fields of the Columbia basin. p. 33–43. *In* J. P. Law and G. V. Skogerboe (ed.) Proc. National Conference on Return Flow Quality Management. Colorado State University, Ft. Collins.

Meehan, W. R., F. B. Lotspeich, and E. W. Mueller. 1975. Effects of forest fertilization on two southeast Alaskan streams. J. Environ. Qual. 4:50–55.

Menzel, R. G. 1980. Enrichment ratios for water quality modeling. p. 486–492. *In* CREAMS: A field scale model for chemicals, runoff and erosion from agricultural management systems. USDA Conserv. Res. Rep. U.S. Government Printing Office, Washington, DC.

Menzel, R. G., E. D. Rhoades, A. E. Olness, and S. J. Smith. 1978. Variability of annual nutrient and sediment discharges in runoff from Oklahoma cropland and rangeland. J. Environ. Qual. 7:401–406.

Miller, M. H. 1979. Contribution of nitrogen and phosphorus to subsurface drainage water from intensively cropped mineral and organic soils in Ontario. J. Environ. Qual. 8:42–48.

Moe, P. G., J. V. Mannering, and C. B. Johnson. 1967. Loss of fertilizer nitrogen in surface runoff water. Soil Sci. 104:389–394.

Mosier, A. R., and G. L. Hutchinson. 1981. Nitrous oxide emissions from cropped fields. J. Environ. Qual. 10:169–174.

Mulla, D. J., A. L. Page, and T. J. Ganje. 1980. Cadmium accumulations and bioavailabilities in soils from long-term phosphorus fertilization. J. Environ. Qual. 9:408–412.

Neary, D. G., and J. H. Leonard. 1978. Effects of forest fertilization on nutrient losses in streamflow in New Zealand. N. Z. J. For Sci. 8:189–205.

Nicholaichuk, W., and D. W. L. Read. 1978. Nutrient runoff from fertilized and unfertilized fields in Western Canada. J. Environ. Qual. 7:542–544.

Nielsen, D. R., C. S. Simmons, and J. W. Biggar. 1979. Flux of nitrate from a spatially variable field soil. p. 201–246. *In* Final Report to the National Science Foundation for Grant nos. GI34733X, GI143664, AEN74-11136 AO1, ENV 76-10283 and PER76-10283, Nitrate in Effluents from Irrigated Lands. University of California.

Nightingale, H. L. 1970. Statistical evaluation of salinity and nitrate content and trends beneath urban and agricultural areas—Fresno, California. Ground Water 8:22–28.

Olness, A., S. J. Smith, E. D. Rhoades, and R. G. Menzel. 1975. Nutrient and sediment discharge from agricultural watersheds in Oklahoma. J. Environ. Qual. 4:331–336.

Oloya, T. O., and T. J. Logan. 1980. Phosphate desorption from soils and sediments with varying levels of extractable phosphate. J. Environ. Qual. 9:526–531.

Olson, R. V., L. S. Murphy, H. C. Moser, and C. W. Swallow. 1979. Fate of tagged fertilizer nitrogen applied to winter wheat. Soil Sci. Soc. Am. J. 43:973–975.

Omernik, J. M., A. R. Abernathy, and L. M. Male. 1981. Stream and nutrient levels and proximity of agricultural and forest land to streams. J. Soil Water Conserv. 36:227–231.

Patrick, W. H., Jr., and K. R. Reddy. 1976. Fate of fertilizer nitrogen in a flooded rice soil. Soil Sci. Soc. Am. J. 40:678–681.

Pollution from Land Use Activities Reference Group (PLUARG). 1978. Environmental management strategy for the Great Lakes System. International Joint Commission, Windsor, Ontario.

Rible, J. M., P. F. Pratt, L. J. Lund, and K. M. Holtzclaw. 1979. Nitrates in the unsaturated zone of freely drained fields. p. 297–320. *In* Final Report to the National Science Foundation for Grant nos. GI34733X, GI143664, AEN74-11136 AO1, ENV76-10283 and PRF 76-10283, Nitrate in Effluents from Irrigated Lands. University of California.

Robbins, C. W., and D. L. Carter. 1980. Nitrate-nitrogen leached below the root zone during and following alfalfa. J. Environ. Qual. 9:447–451.

Rogers, H. T. 1941. Plant nutrient losses from a corn, wheat, clover rotation on Dunmore silt loam. Soil Sci. Soc. Am. Proc. 6:263–271.

Rolston, D. E., D. L. Hoffman, and D. W. Toy. 1978. Field measurement of denitrification: I. Flux of N_2 and N_2O. Soil Sci. Soc. Am. J. 42:863–869.

Romkens, M. J. M., D. W. Nelson, and J. V. Mannering. 1973. Nitrogen and phosphorus composition of surface runoff as affected by tillage method. J. Environ. Qual. 2:292–295.

Ryden, J. C., and L. J. Lund. 1980. Nitrous oxide evolution from irrigated land. J. Environ. Qual. 9:387–393.

Ryden, J. C., L. J. Lund, and D. D. Focht. 1978. Direct in-field measurement of nitrous oxide flux from soils. Soil Sci. Soc. Am. J. 42:731–738.

Sanderford, S. G. 1975. Forest fertilization and water quality in the North Carolina Piedmont. M. S. thesis. North Carolina State University, Raleigh, NC.

Sapozhnikov, N. A., N. A. Ivanova, T. K. Livanova, I. P. Rusinova, V. V. Siderova, L. B. Sirota, and T. Tarvis. 1974. Transformation of nitrogen in soil and nitrogen nutrition of plants studied according to the data from research using nitrogen-15. Trans. Int. Congr. Soil Sci. 10th 1974 9:30–38.

Schaller, F. W. 1940. The downward movement of lime and superphosphate in relation to permanent pasture fertilization. Soil Sci. Soc. Am. Proc. 5:162–166.

Schertz, D. L., and D. A. Miller. 1972. Nitrate-N accumulation in the soil profile under alfalfa. Agron. J. 64:660–664.

Schlosser, I. J., and J. R. Karr. 1981. Water quality in agricultural watersheds: Impact of riparian vegetation during base flow. Water Res. Bull. 17:233–240.

Schuman, G. E., R. E. Burwell, R. F. Piest, and R. G. Spomer. 1973. Nitrogen losses in surface runoff from agricultural watersheds on Missouri Valley loess. J. Environ. Qual. 2:299–302.

Schwab, G. O., E. O. McLean, A. C. Waldron, R. K. White, and D. W. Mitchener. 1973. Quality of drainage from a heavy-textured soil. Trans. ASAE 16:1104–1107.

Shannon, E. E. 1980. Water quality and wastewater treatment considerations for detergent substitutes. p. 391–414. *In* Phosphorus management strategies for lakes. R. C. Loehr et al. (ed.) Ann Arbor Science Publishers, Ann Arbor, MI.

Sharpley, A. N. 1980. The enrichment of soil phosphorus in runoff sediments. J. Environ. Qual. 9:521–526.

Sharpley, A. N., L. H. Ahwja, M. Yamamoto, and R. G. Menzel. 1981. The kinetics of phosphorus desorption from soil. Soil Sci. Soc. Am. J. 45:493–496.

Skaggs, R. W., and J. W. Gilliam. 1981. Effect of drainage system design and operation on nitrate transport. Trans. ASAE 24:929–934, 940.

Skogerboe, G. V. 1974. Agricultural impact on water quality in western rivers. p. 12-1–12-25. *In* Hsieh Wen Shen (ed.) Environmental impact on rivers. Hsieh Wen Shen Publisher, Fort Collins, CO.

Spencer, V. E., and R. Stewart. 1934. Phosphate studies: I. Soil penetration of some organic and inorganic phosphates. Soil Sci. 38:65–79.

Spencer, W. F. 1957. Distribution and availability of phosphates added to a Lakeland fine sand. Soil Sci. Soc. Am. Proc. 21:141–144.

Spooner, J., and J. W. Gilliam. 1978. N_2O evolution from high organic matter soils of the Coastal Plain of North Carolina. Agron. Abstr. American Society of Agronomy, Madison, WI, p. 146.

Swoboda, A. R. 1977. Nitrate movement in clay soils and methods of pollution control. p. 19–25. *In* J. P. Law and G. V. Skogerboe (ed.) Proceedings of National Conference on Irrigation Return Flow Quality Management. Colorado State University, Ft. Collins.

Tanji, K. K., J. W. Biggar, R. J. Miller, W. O. Pruitt, and G. L. Horner. 1977. Evaluation of surface irrigation return flows in the central valley of California. p. 167–173. *In* J. P. Law and G. V. Skogerboe (ed.) Proceedings of National Conference on Irrigation Return Flow Quality Management. Colorado State University, Ft. Collins.

Taylor, A. W., and H. M. Kunishi. 1971. Phosphate equilibria on stream sediment and soil in a watershed draining an agricultural region. J. Agric. Food Chem. 19:827–831.

Thomas, G. W., and J. D. Crutchfield. 1974. Nitrate-nitrogen and phosphorus contents of streams draining small agricultural watersheds in Kentucky. J. Environ. Qual. 3:46–49.

Timmons, D. R., R. E. Burwell, and R. F. Holt. 1973. Nitrogen and phosphorus losses in surface runoff from agricultural land as influenced by placement of broadcast fertilizer. Water Resour. Res. 9:658–667.

Timmons, D. R., and R. F. Holt. 1977. Nutrient losses in surface runoff from a native prairie. J. Environ. Qual. 6:369–373.

Timmons, D. R., E. S. Verry, R. E. Burwell, and R. F. Holt. 1977. Nutrient transport in surface runoff and interflow from an aspen birch forest. J. Environ. Qual. 6:188–192.

U.S. Army Corps of Engineers. 1979. Lake Erie wastewater management study: Methodology report. U.S. Army Corps of Engineers, Buffalo, NY.

U.S. Department of Agriculture and U.S. Department of the Interior. 1977. Development of phosphate resources in S. E. Idaho: Final environmental impact statement. 1:52–54. U.S. Government Printing Office, Washington, DC.

U.S. Environmental Protection Agency. 1979. Criteria for classification of solid waste disposal facilities and practices. Fed. Regist. 44:53438–53468.

Viets, F. G. 1970. Soil use and water quality. J. Agric. Food Chem. 18:789–792.

Williams, C. H., and D. J. David. 1973. The effect of superphosphate on the cadmium content of soils and plants. Aust. J. Soil Res. 11:43–56.

Williams, C. H., and D. J. David. 1976. The accumulation in soil of cadmium residues from phosphate fertilizers and their effect on the cadmium content of plants. Soil Sci. 121:86–93.

Zwerman, P. J., T. Greweling, S. D. Klausner, and D. J. Lathwell. 1972. Nitrogen and phosphorus content of water from tile drains at two levels of management and fertilization. Soil Sci. Soc. Am. Proc. 36:134–137.

17

Nutritional Quality of Plants in Relation to Fertilizer Use[1]

David L. Grunes

U.S. Plant, Soil, and Nutrition Laboratory
Agricultural Research Service, USDA
Ithaca, New York

W. H. Allaway

Cornell University
Ithaca, New York

This chapter will discuss the effects of fertilizers, and of mineral nutrients in the soil, on the nutritional quality of plants. The nutritional quality of a plant is controlled by the concentration and bioavailability of essential nutrients and potentially detrimental substances in the plant, with reference to the requirements and tolerances of the person or animal that consumes the plant as part of a specific diet. Special attention will be directed toward instances where variation in plant composition due to soil variation results in an effect on the health and/or growth rate of people and animals.

I. FACTORS INVOLVED IN THE SOIL–PLANT–ANIMAL SYSTEM

In the soil to plant to animal chain the soil supplies the plant with inorganic nutrients and water. Plants capture light energy and use nutrients and water from the soil and carbon dioxide (CO_2) from the air to form reduced organic compounds of C, H, N, S, and a few other elements. Animals ingest these reduced compounds from the plants in their diets, oxidize them, and use the chemical energy contained in them to sustain their life processes. The animals then excrete inorganic and partially oxidized simpler organic compounds.

Epstein (1972) lists the following elements as essential for at least some higher plants: C, H, O, P, K, N, S, Ca, Mg, Fe, Mn, Zn, Cu, Mo, B, Cl, Na, Se, and Si. He adds Co, I, and V as essential for some algae and bacteria. Since Epstein's summarization, evidence for the essentiality of Ni has been presented (Eskew & Welch, 1982). Cobalt is required by the N_2 fixing organisms in legume root nodules. Mertz (1981) lists the elements C, H, O, P, K, N, S, Ca, Mg, Fe, Mn, Zn, Cu, Mo, Cl, Na, Se, Si, Co, I, V, Ni,

[1]Cooperative investigation, USDA–ARS, and Department of Agronomy, Cornell University, Ithaca, NY. Department of Agronomy paper no. 1444.

As, and Cr as essential for animals. He indicated that there are unconfirmed reports that Cd and Sn are also essential.

The dietary concentrations of the different elements required by animals frequently differ from the concentrations found in normal plants. Animals may tolerate dietary concentrations of Zn, Mn, and sometimes Cu that would be indicative of plant toxicity on the basis of analyses of dried plants. Normal plants may contain either too much or too little Se for optimum nutritional status of animals.

The nutritional requirements of people and animals include a number of specific organic compounds, such as amino acids, vitamins, and fats. The requirements of different animal species vary both quantitatively and qualitatively. The most striking differences are between ruminant and nonruminant or monogastric animals. Monogastric animals must obtain all of the required amino acids from their diet, either from plant or animal products. They require Co in the form of vitamin B_{12}, and they are able to utilize only a very small part of the energy in complex carbohydrates, such as cellulose. In ruminant animals, microscopic organisms (bacteria, etc.) in the rumen can utilize energy from a wide range of carbohydrates and can synthesize amino acids from simple N compounds such as urea. The rumen microorganisms can synthesize vitamin B_{12} from inorganic Co, and they can reduce inorganic S compounds and incorporate the reduced S into the essential S amino acids. These and other essential nutrients are then absorbed farther along in the digestive tract of ruminant animals. Therefore, the nutritional requirements of ruminant animals are not so complex as for nonruminants, and ruminants may, under certain specific conditions, obtain nearly all of their requirements from just one or two plant species. On the other hand, nonruminants must have diets containing a wider range of plant species plus some animal products or else have access to dietary supplements.

The nutritional requirements of a given species of animal may vary quantitatively, and in some instances qualitatively, depending on stage of growth, reproduction, lactation, etc. For example, the very young ruminant cannot accomplish some syntheses that occur in adult ruminants; therefore, very young ruminants have nutritional requirements similar to those of monogastric animals. Laying hens *(Gallus gallus domesticus)* require substantially higher levels of Ca than do broilers, and the Ca and Mg requirements of lactating ruminants are higher than for nonlactating animals of the same species.

Different plant species, and even different cultivars of the same species, normally contain very different amounts of the nutrients essential to people and animals. The forage legumes are generally high in Ca in relation to the Ca requirements of ruminant animals, while the cereal grains are nearly always too low in Ca to sustain animal species. However, the cereal grains, if they produce satisfactory crop yields, can provide all or nearly all of the P required by animals.

No single plant species, regardless of the soil on which it grows, can supply all of the nutrient requirements of any animal. Therefore, the nutritional quality of a plant must be evaluated with respect to the entire diet in which the plant is eaten and the specific animal involved. Some of this com-

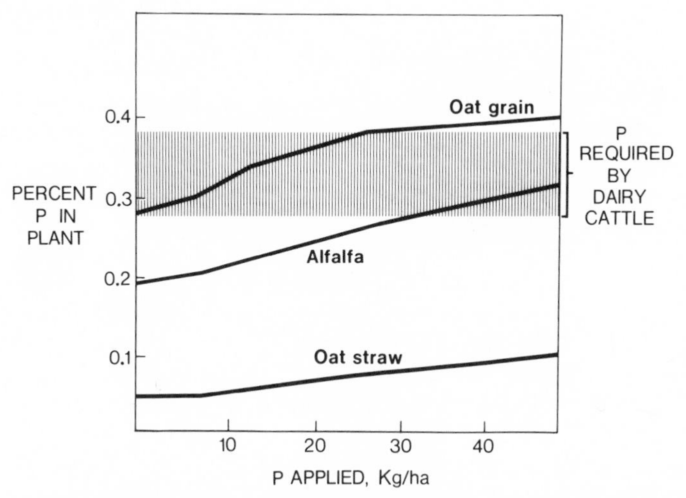

Fig. 17–1. Effect of P fertilization on the concentration of P in different plants grown on Clarion loam (Typic Hapludolls) (data abstracted from Larson et al., 1952).

plexity is shown in Fig. 17–1, which summarizes plant P levels observed in Iowa (Larson et al., 1952) when different rates of P fertilizer were applied to a new seeding of alfalfa (*Medicago sativa* L.) with oats (*Avena sativa* L.) as a nurse crop. The application of 40 kg P/ha increased the P concentration of alfalfa from an inadequate to an adequate level for many milking dairy cattle *(Bos taurus)*. Although P concentrations in oat grain and straw were also increased by P fertilization, the oat grain would have been an almost adequate source of dietary P for cattle, even without fertilization, while the oat straw would have contained too little P to meet the requirements of cattle, even at the highest rate of P fertilizer applied.

II. SPECIFIC EFFECTS OF FERTILIZERS ON THE NUTRITIONAL QUALITY OF PLANTS

A. Nitrogen

1. Protein

The effects of N in the soil on protein in plants, especially in the cereal grains, are of great importance in world food problems. Discussion of these effects can best be based on consideration of the biochemical pathways involved in movement of N from the soil through plants and animals. These pathways are outlined in Fig. 17–2.

Nitrogen enters the plant from the soil or from nodules on the roots of legumes as nitrate (NO_3^-) or ammonium (NH_4^+). Within the plant, NO_3^- is reduced to NH_4^+ and then combined with C skeletons to form about 100 different amino acids. These amino acids contain N in the $-NH_2$ form with the N bonded to the alpha C of an organic acid.

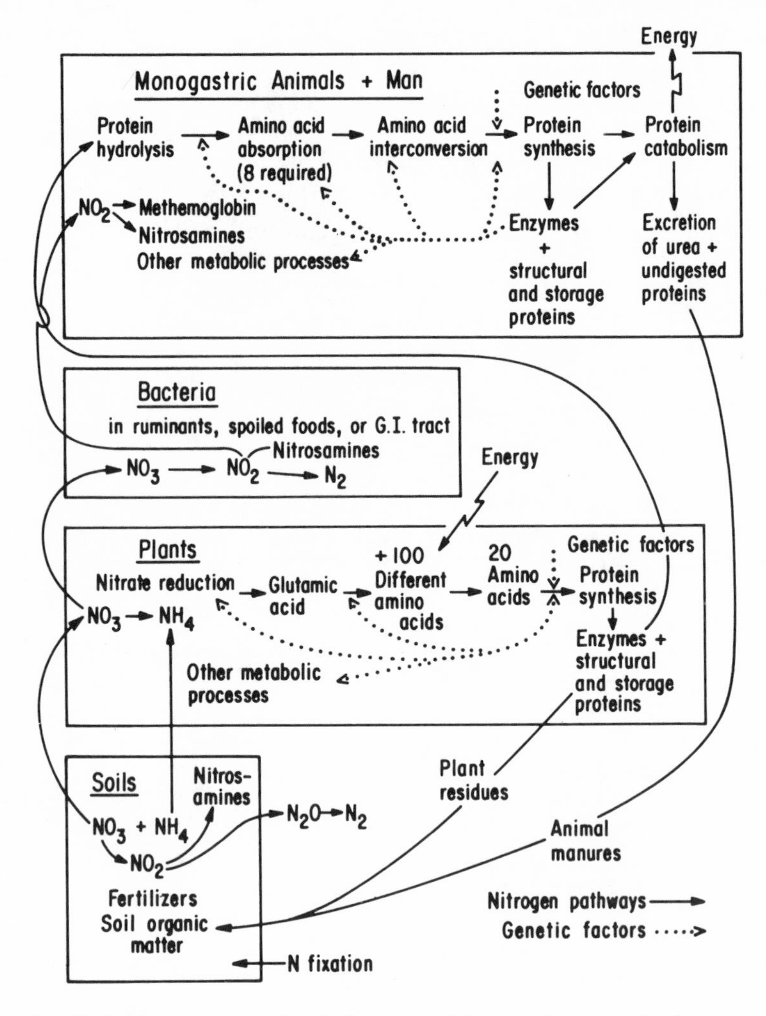

Fig. 17–2. Movement and chemical changes of N in a soil–plant–animal system.

About 20 of the different amino acids are then linked together into long chains called polypeptide chains. These chains may contain several hundred amino acid links. The order in which different amino acids occur along the polypeptide chain, and therefore the ratios of different amino acids in the chains, is controlled by genetic information contained in the nucleic acids in the plant. The polypeptide chains are then folded, coiled, cross-linked and modified in other ways to form proteins. Some of these proteins may be stored in the seed for use by the new seedling after germination. Some of them are enzymes that regulate all the metabolic processes within the plant, and others may become structural components or membranes. In normally growing plants, 75% or more of the total N in the plant is combined into protein. Under conditions where the rate of plant growth is slowed and where NO_3^- uptake from the soil remains rapid, NO_3^-, amides, and nonprotein amino acids may accumulate.

In addition to the genetic control over the order in which amino acids are linked together in the polypeptide chains, there are genetic controls over the total concentration of protein in the plant. When plant growth is slowed due to deficiency of nutrients other than N, lack of water, low light intensity, or low temperatures, protein concentrations in vegetative tissues may be very high. The genetic controls over total protein are most evident in seeds of cereals and legumes producing near-optimum seed yields. Regardless of the level of available N in the soil or amount of N fertilizer applied, seeds of low-protein crops such as rice (*Oryza sativa* L.) will contain lower concentrations of protein than high-protein crops such as durum wheat (*Triticum durum* Desf.) or soybeans [*Glycine max* (L.) Merr.].

When an animal eats a food that contains protein, the protein molecules are partially broken down and the individual amino acids are released from the polypeptide chains. Animals require only 9 or 10 of these amino acids. These are the *essential amino acids*. The amino acids are then absorbed from the gastrointestinal (GI) tract and recombined into animal proteins. The *protein requirements* of people and animals are, in reality, amino acid requirements. The sequence of different amino acids in the polypeptide chains of animal proteins is controlled by the genetic information in the nucleic acids of the animal. The value of a molecule of dietary protein to the animal is thus determined by the degree to which the protein is broken down to its individual amino acids and the frequency with which the essential amino acids are incorporated into the polypeptide chains.

No one protein or even any one plant contains the essential amino acids in exactly the right ratios and amounts to meet the requirements of people and other monogastric animals. To quantify the nutritional value of the different proteins, some protein that is of high nutritional quality—usually whole egg or whole milk protein— is selected as a standard and given a value of 100; the amino acid patterns of other proteins are then compared with the amino acid pattern of the standard protein. The essential acid that is present at the lowest percentage of the concentration in the standard protein is the *limiting amino acid* for the protein being studied. The nutritional quality of the protein is the percent of this limiting amino acid in the protein as compared with this same amino acid in the standard protein. In protein from cereal grains, lysine is frequently the limiting amino acid, while methionine is frequently the limiting amino acid in pulses such as soybeans. Plant breeders have made important progress in producing cereal varieties that contain higher levels of lysine, thus increasing the nutritional quality of cereal proteins.

In addition to the adjustments in nutritional quality due to amino acid balances, adjustments due to the percentage digestibility of proteins are common. In general, proteins from plants are less digestible than proteins from animal foods. In some bioassays of protein quality using growing rats (*Rattus* sp.) (NAS, 1980), wheat (*Triticum aestivum* L.) gluten has been found to be only 20% as effective as whole egg protein. The National Academy of Science (1980) recommended that dietary allowances include a 30% increase in protein consumption to provide for incomplete utilization of the mixed proteins typical of U.S. diets.

The effects of nutrition of the plant on the amino acid composition of protein and nonprotein fractions of turnip (*Brassica rapa* L.) greens have been investigated by Thompson et al. (1960). They concluded that "the amino acid composition of the protein was little, if at all, affected by deficiencies of any of the macronutrient elements." Mineral nutrition had more effects on the amino acids in the nonprotein fraction. Most of the changes in nonprotein amino acids that were due to differences in the mineral nutrition of the plant involved the nonessential amino acids.

It is evident from the above discussion that the level of available N in the soil affects the supply of dietary protein for people and animals by promoting greater growth of plants having somewhat higher total protein content, and thus markedly higher production of protein per hectare. The nutritional value per kilogram of protein is generally not greatly affected by fertilizers, including N fertilizers.

The effects of N fertilizer use on yield and total protein concentrations in cereals has been reviewed by Sander et al. (projected 1987). Their review stresses the importance of the relationship of available soil N to available soil water at different stages in the growth of the plant. This relationship is of great importance in the yield and protein concentration in the hard red wheats, most of which are grown in regions of fairly limited rainfall. High levels of available N at early stages of plant growth, followed by dry periods in late stages of plant growth, may result in low grain yields. Where soil moisture levels are high and available N is low, higher yields and lower protein concentrations in the grain usually result. If both soil moisture and available N levels are high during the entire growth period prior to maturity, high yields of high-protein grain may result. Sander et al. (projected 1982) indicate that "The amount of fertilizer N required to maximize grain yields is not adequate to produce maximum protein concentrations in the grain. Without reimbursement for producing a high protein product, producers will apply N to provide the most economic grain yields."

The effect of N fertilizer on the protein nutrition of humans and animals is therefore primarily to increase total protein production per hectare, per farm, or per nation. The nutritional quality of each unit of this protein is controlled by the genetics of the plant or animal that produces the proteins consumed by humans or other animals. Plant geneticists and breeders are making substantial progress in developing plants with the genetic ability to form proteins of improved nutritional quality. The fullest use of these new plants in meeting the protein needs of people will require that these new plants be supplied with adequate N to permit them to use their genetic capability to form these high-quality proteins.

2. *Toxicity of Nitrates (Nitrites) and Nitrosamines*

As indicated earlier, whenever the process of protein synthesis in plants is interrupted or slowed down, NO_3^- may accumulate in plant tissues. Nitrate is not highly toxic to animals or humans, but if it is reduced to nitrite (NO_2^-), the NO_2^- may combine with hemoglobin in the animal blood and form a compound, methemoglobin, which reduces the capacity of the blood to transport O_2 to all parts of the body. Accumulation of high concentrations of NO_3^- in food or feed crops is, therefore, to be avoided, because of the possibility of formation of NO_2^- from NO_3^-.

Factors affecting the accumulation of NO_3^- in plants and the impact of nitrates and nitrites on ruminant animals have been reviewed by Wright and Davison (1964) and Kemp et al. (1978). Factors affecting the accumulation of nitrates in food for humans has been reviewed by Wright and Davison (1964) and by Maynard et al. (1976). The problem of nitrates, nitrites, and methemoglobin formation, with special emphasis on human health, has been reviewed by Lee (1970). From these reviews, it is apparent that different plant species vary in their tendency to accumulate nitrates. Among the vegetable crops, kale [*Brassica oleracea* L. (*Acephala* Group), spinach (*Spinacia oleracea* L.), celery [*Apium graveolens* L. Var. dulce (Mill.) Pers.], beets (*Beta vulgaris* L.), and radishes (*Raphanus sativus* L.) are often high in nitrates. As far as the forages are concerned, annual grass species, such as sudangrass [*Sorghum sudanense* (Piper) Stapf], cereal grains used at an immature stage for pasture or hay, and corn (*Zea mays* L.) at the silage stage, are among crops that have been implicated in cases of NO_3^- (NO_2^-) poisoning in animals. Perennial grasses are less likely to contain high concentrations of NO_3^-, but cases of NO_3^- (NO_2^-) toxicity have been observed in cattle grazing on Kentucky bluegrass (*Poa pratensis* L.), tall fescue (*Festuca arundinacea* Schreb.), and bromegrass (*Bromus inermis* Leyss.). Certain weed species, especially pigweed (*Amaranthus* sp.), are noted accumulators of NO_3^- and have been implicated in deaths of ruminant animals. High concentrations of NO_3^- are rarely found in perennial forage legumes. In most plant species the NO_3^- concentration declines with approach to maturity, and high concentrations of NO_3^- are rarely found in the seeds.

Environmental factors favoring the accumulation of nitrates in plants include high levels of NO_3^- in the soil, cloudy periods, shading, drought, excessive temperatures, damage to plants from insects or weed-control chemicals, and nutrient imbalance in the soil (Wright & Davison, 1964; Maynard et al., 1976).

Nitrate (nitrite) toxicity is primarily a problem of ruminant animals. It has been the cause of numerous deaths of cattle in the USA. Occurrences in sheep *(Ovis aries)* have been less common. Affected animals show discoloration of the light areas of exposed skin, rapid breathing, and a brown discoloration of the blood. Abortions may occur, and deaths are frequent. If affected animals are treated promptly with methylene blue (tetramethylthionine chloride) recovery is likely.

Nitrate (nitrite) toxicity that can be traced to high levels of NO_3^- in food crops has been rare among humans in the USA. In his review, Lee (1970) cites evidence from Europe of NO_3^- (NO_2^-) poisoning in infants consuming improperly stored high-nitrate vegetables.

There has been poisoning of infants from nitrates in well water (NRC, 1972). Apparently, the reaction of NO_2^- with hemoglobin is inconsequential in adults, but it can be troublesome in infants, particularly in those under 3 months of age.

Lee (1970) also reviewed the concern that nitrites formed in contaminated high-nitrate vegetables may react with secondary amines to form nitrosamines. Some of the nitrosamines are potent carcinogens. There are recorded fatalities resulting from the evolution of toxic gaseous oxides of N when high-nitrate crops are ensiled. These gases may be inhaled by those working with this silage.

It is very difficult to set a critical level of NO_3^- in either food or feed crops. Since the toxicity of the NO_3^- itself is low, the effect of NO_3^- in plants on animals and people depends on the rate that NO_3^- is converted to NO_2^-. This conversion depends on the types of organisms that attack the high-nitrate plants, other components in the diet, rate of food intake, the gastrointestinal flora of the animal or human, storage conditions, etc.

Nitrate, nitrite, and secondary and tertiary amines are precursors of nitrosamines (NRC, 1972). The N-nitroso compounds, and in particular N-nitrosamines, are an important class of chemical carcinogens that have induced tumors in all species of laboratory animals tested, and in virtually every vital tissue (NRC, 1978a). However, few studies provide adequate data to define detailed dose-response relationships. The inadequacy of dose-response data, from either epidemiologic or toxicologic studies with people, precludes a determination of whether the exposure to N-nitroso compounds that might result from dietary NO_3^- or NO_2^- is "safe." Also unknown is how large a risk of human cancer such exposures may represent. In the absence of more definitive information on the magnitude of the health hazard, the most prudent course of action is to take reasonable measures to minimize human exposures to N-nitroso compounds (NRC, 1978a).

Information on the occurrence of nitrosamines in food is accumulating (Maynard et al., 1976). Too few food samples have been analyzed to comment on the frequency of nitrosomine contamination of the human food supply; also, the analyses have been confined largely to the volatile nitroso compounds and will have to be extended to include those of lower volatility (NRC, 1978a).

In acid solution, NO_2^- is converted to nitrous acid (HNO_2), which subsequently forms nitrous acid anhydride (N_2O_3). The N_2O_3 subsequently reacts with nonionized secondary amines to form N-nitrosamines (NRS, 1978a). Nitrosation of secondary and tertiary amines can occur in air, soil, and water, as well as in vivo in humans or animals. Too little information is available to estimate the extent to which nitrosation may take place in soil, water, and air, or in humans, but research on these topics is active and such information should be forthcoming.

3. The Prevention of Nitrate (Nitrite) Toxicity

It should be recognized that the high levels of NO_3^- in the soil that are associated with accumulation of NO_3^- in plants are not always due to use of

N fertilizers. High-nitrate plants and outbreaks of NO_3^- (NO_2^-) toxicity in ruminant animals have been recorded in situations where no N fertilizers have been used. The source of the NO_3^- in these instances has been from decomposition of soil organic matter, from animal manures, from previous legume crops, or from municipal wastewater and sludge. As far as food crops are concerned, there is evidence that NO_3^- levels in commercially produced vegetables in the USA are no higher today than they were in the years prior to extensive use of N fertilizers. This evidence was reviewed by Lee (1970) and by Maynard et al. (1976).

Any accumulation of NO_3^- in plants at harvest (or grazing) time can be regarded as undesirable, even though critical limits for NO_3^- (NO_2^-) toxicity have not been definitely established. Plants high in nitrates are a more probable source of nitrates than low-nitrate plants. Furthermore, since the NO_3^- must be reduced to NH_4^+ before it is used by the plant, any accumulation of NO_3^- at harvest time represents wasted N as far as crop growth, yield, and protein content are concerned.

The following steps may be useful in various situations to prevent accumulation of NO_3^- in plants and to minimize the plant nitrates impact on human and animal health:

1. Nitrogen fertilizers should be used in the amounts required to provide optimum crop yields and protein concentrations, but not in excess of this amount. The overuse of N fertilizers, merely to provide insurance of high yields, should be discouraged.
2. Systems of soil and crop management that provide the required, but not excessive, levels of N to the plant should be developed. Information to be used includes release rate of soil organic N and the NO_3^- content of plants at preharvest stages in relation to content at harvest time. Also, crops should be irrigated to eliminate drought and to leach excess NO_3^- from the root zone when needed.
3. A system should be developed for warning cattle and sheep producers of the danger of NO_3^- (NO_2^-) poisoning when weather conditions associated with NO_3^- accumulation have occurred. Stockmen should have access to rapid testing of forages and pastures for NO_3^- concentrations. Colorimetric methods of analyses for NO_3^- can be used. Also, the "NO_3^- electrode" currently offers promise for the use in these rapid testing services.
4. Where leafy vegetables are processed for human consumption, the NO_3^- concentration should be monitored, and materials very high in NO_3^- should be rejected before moving food into distribution channels.
5. The water used to cook food plant species that are known to accumulate nitrates should be discarded. When these species are preserved by canning, they should be consumed soon after the cans are opened, without storage of opened material prior to consumption. This precaution is especially important in vegetables canned for infant diets.

4. Nitrogen Interactions With Other Elements

Earlier research on the effect of N fertilization on the absorption of soil and fertilizer P by plants was summarized by Grunes (1959). Fertilization with N frequently increased root growth, with a resultant increase in root foraging capacity for soil and fertilizer P. In addition, residually acidic banded N fertilizers may decrease the soil pH, increase soluble soil P, and thus increase the uptake of P by plants.

Cole et al. (1963) found that corn plants absorbed P from nutrient solution much more rapidly when they had been previously grown in nutrient solutions high in N. Translocation of P from roots to tops was also greatly increased by N pretreatment. The presence of NO_3^- or NH_4^+ ions in the culture solution during a 2-h uptake period did not affect the rate of P uptake. The stimulation of P uptake rates with higher plant levels suggests a connection between N metabolism and P uptake processes.

Miller et al. (1970) found that ammonium sulfate [$(NH_4)_2SO_4$] lowered the pH in the vicinity of a band of monocalcium phosphate in the soil, and that the $(NH_4)_2SO_4$ prevented precipitation of P at the root surface. Absorption of P by corn plants was greatly increased by the addition of the $(NH_4)_2SO_4$. Miller (1974) later reviewed the effects of N on P absorption by plants.

In a greenhouse experiment on a soil with an initial pH of 7.2, Viets et al. (1957) found that $(NH_4)_2SO_4$ markedly reduced the soil pH, ammonium nitrate (NH_4NO_3) reduced the pH somewhat, and sodium nitrate ($NaNO_3$) increased the pH. Milo (*Sorghum vulgare* Pers.), a grain sorghum, was cut at the preheading stage, and ladino clover (*Trifolium repens* L.) was cut at the start of bloom. Zinc concentrations in both crops were increased most by $(NH_4)_2SO_4$, increased somewhat less by NH_4NO_3, and decreased by $NaNO_3$. In a later field study, Boawn et al. (1960) found that $(NH_4)_2SO_4$ increased Zn uptake as compared with calcium nitrate [$Ca(NO_3)_2$] when the crop was grain sorghum [*Sorghum bicolor* (L.) Moench] or potatoes (*Solanum tuberosum* L.), but this effect was not shown with sugarbeets (*Beta vulgaris* L.).

Schnug and Finck (1982) worked with cereals and oil seed rape (*Brassica napus* L.) in field trials in northern Federal Republic of Germany. On carbonate-free soils the Mn and Zn concentrations increased as much as 70 and 30%, respectively, following the application of residually acidic N fertilizers. The B concentrations in oilseed rape increased as much as 30%.

Grunes (1973), Mayland and Grunes (1979), and Grunes (1983) reviewed the effects of N fertilizer sources on concentrations of Ca and Mg in plants. Added NH_4^+-N resulted in lower concentrations of Mg and Ca in plants than occurred with NO_3^--N sources. This effect would be especially marked in culture solutions, or when nitrification is low, as would occur with low soil temperatures in the spring. In a greenhouse study, Mathers et al. (1982) used nitrapyrin (N-Serve, The Dow Chemical Co., Midland, MI)[2] to depress nitrification of $(NH_4)_2SO_4$ added to the soil. The nitrapyrin

[2]Mention of a trade name does not constitute a recommendation for use by the USDA or Cornell University.

depressed the concentrations of K, Ca, and Mg in wheat forage, but this depression effect was greater for the divalent cations Ca^{2+} and Mg^{2+} than for the monovalent cation K^{+}.

B. Phosphorus

1. Direct Effects of Phosphorus

The problem of P deficiency in animals has been reviewed by Underwood (1981). Animal problems now known to be associated with the grazing of low P forages have been recognized since the 1800s and there is no doubt that P deficiency is the most widespread and economically important of all the mineral deficiencies affecting grazing livestock.

The successful use of P fertilization to correct P deficiency in grazing animals is dependent on the soil and the plant species involved. On some soils, addition of P fertilizers will not bring about increases in the P concentrations in the plants grown, even though substantial increases in yield may be noted. However, in most forages, the application of P fertilizers slightly increased P concentrations, particularly if the concentration of P is at the minimum level for the normal growth of a particular forage type (Beeson & Matrone, 1976). The effect of P fertilization on concentrations of P in alfalfa, oat grain, and oat straw in Iowa has been referred to in the discussion of Fig. 17–1 in section I.

When mixed pastures containing both grasses and legumes are fertilized, the increase in percent of legume in the mixture resulting from P fertilization may have important effects on the animals grazing the pasture, independent of the effect of fertilization on the P concentration in the pasture plants. In general, decisions concerning use of P fertilizers are based on the expected response in crop production. Where little or no response in plant growth can be expected, animal diets are supplemented to meet P requirements.

Phosphorus is present in all foods, and dietary deficiency is extremely unlikely to occur in humans (NRC, 1980). Animal products, such as milk, meat, and eggs are major sources of P in human diets.

2. Interactions of Phosphorus and Calcium

The recommended nutritional requirements for beef cattle are: 0.46% Ca and 0.36% P for 150-kg steer calves and yearlings gaining 0.7 kg/day; 0.28% Ca and P for 450-kg cows of average milking ability with nursing calves; and 0.40% Ca and 0.37% P for cows of superior milking ability with nursing calves (NRC, 1976).

The recommended nutritional requirements for dairy cattle are: 0.37% Ca and 0.26% P for dry pregnant cows; 0.43% Ca and 0.31% P for 500-kg cows producing $<$ 11 kg milk/day; 0.48% Ca and 0.34% P for 500-kg cows producing 11 to 17 kg milk/day; 0.54% Ca and 0.38% P for 500-kg cows producing 17 to 23 kg milk/day; and 0.60% Ca and 0.40% P for 500-kg cows producing $>$ 23 kg milk/day (NRC, 1978b).

A dietary Ca/P ratio between 1:1 and 2:1 is assumed to be ideal for growth and bone formation, since this is approximately the ratio of the two minerals in bone (Underwood, 1981). The Agricultural Research Council (1980) agreed that this is a generally safe range. Ruminants can tolerate a wide range of Ca:P ratios, particularly when their vitamin D status is high (Underwood, 1981). Ratios between Ca and P as high as 2:1 have been shown to be beneficial in reducing urinary calculi. When calculi problems are encountered, higher levels of Ca may be advisable. Ratios between Ca and P of 7:1 in the diet have been reported to be satisfactory for cattle, provided P is adequate (NRC, 1976).

Available moisture and stage of growth may markedly affect concentrations of P in forages, while having little effect on Ca concentrations. Spring rain promotes new growth and relatively high P concentrations. As plants mature and P concentrations decrease, the P available to animals would decrease (Beeson & Matrone, 1976).

Calcium concentrations in pasture plants can be increased by increasing the proportion of legumes and herbs in the forage. High rates of K fertilizer, or high K levels in soil, can reduce Ca concentrations, particularly on sandy soils (Grunes et al., 1970; Comm. Miner. Nutr., 1973).

Poultry and pigs *(Sus scofra domesticus)* appear to be less tolerant of high dietary Ca/P ratios. The optimum ratio for growing chicks and pigs lies between 1:1 and 2:1. Ratios below this are well tolerated, and ratios greater than 3:1 are poorly tolerated. For laying hens, the permissible ratio is considerably higher due to their greater requirements for Ca than for P. The P requirements of poultry and pigs, but not of ruminants with a functioning rumen, increase as the proportion of dietary P occurring as phytate increases (Underwood, 1981).

The National Research Council (1980) indicated that humans appear to be able to tolerate ratios of Ca/P between 1:2 and 2:1. They stated that some have concluded that the Ca/P ratio ideally should remain at or above 1. In setting their recommended allowances for both Ca and P, a Ca/P ratio of unity has been maintained. They also indicated that for adult humans, wide variation of the Ca/P ratio in the diet is tolerated, provided the amount of vitamin D is adequate (NRC, 1980). A recent survey (McCarron et al., 1982) indicates that low Ca intake may be related to increased hypertension in people in the USA.

3. Interactions of Phosphorus and Other Elements

In a field experiment in which available Zn in the soil was not high, P fertilization depressed concentrations of Zn in table beet roots, petioles, and blades; in cabbage heads *(Brassica oleracea* var. *capitata)*; in snap bean pods (*Phaseolus vulgaris* L.) and immature seeds; and in immature pea seeds (*Pisum sativum* L.). However, when Zn fertilizer was added, P fertilization increased the Zn concentrations (Peck et al., 1980). In solution culture, low Zn levels greatly increased P concentrations in okra [*Abelmoschus esculentus* (L.) Moench] (Loneragan et al., 1982). It is known whether this also occurs in the field.

High levels of P sometimes decrease Fe concentrations in plants (Brown, 1961). In their review, Murphy et al. (1981) indicated that high levels of P frequently depress uptake of Zn, Fe, and Cu by plants.

In pigs, phytate has been shown to increase levels of Zn required in feed. For monogastric animals, high levels of dietary phosphate reduce Fe absorption, presumably by the formation of insoluble ferric phosphate (Underwood, 1977, 1981). There have been some reports that high levels of dietary phytate reduce Fe absorption in monogastric animals (Underwood, 1981). However, Welch and Van Campen (1975) found that Fe absorption by rats was not related to phytic acid levels in soybean seeds.

High levels of phosphate have also been shown to reduce Fe absorption in man. Severe Zn deficiency in Middle East countries has occurred in young males eating diets consisting predominantly of unrefined cereals high in phytate (Underwood, 1977).

C. Calcium, Magnesium, and Potassium

Uncomplicated deficiencies of Ca, Mg, and K are apparently rare in both animals and humans. These nutritional disorders are usually due to an imbalance among these elements, to imbalance of one or more of these with other elements, or to interferences with the utilization of dietary Ca or Mg.

A conditioned Ca deficiency called *milk fever* is fairly common in heavily producing, heavily fed milk cows soon after the birth of the calves (Underwood, 1981). Milk fever apparently results from deranged Ca metabolism within the animal body. This disease has not responded to increase in dietary levels of Ca. A deficiency of vitamin D may lead to poor absorption of dietary Ca and Mg.

The range of Ca concentrations in plants is predominately a characteristic of the plant species, and is less dependent on the level of available Ca in the soil. Most legumes contain more Ca than do grasses. Cereal grains are generally low in Ca, and Ca supplements are frequently provided to grain-fed animals to provide desired Ca/P ratios.

Milk is a major source of Ca in many human diets. The Ca concentration in milk is nearly constant and is not altered by changes in the level of Ca in the cow's ration.

As far as direct nutritional effect of K fertilizers is concerned, Wilde (1968) has stated: "If you can induce the plant to grow, it will always contain enough K for animal nutrition. Man and animals suffer K deficiencies only when ill for reasons other than improper food composition.

A conditioned deficiency of Mg, known as *grass tetany* or *grass staggers,* is a major problem to cattle producers under some circumstances. This problem has been reviewed by Grunes et al. (1970), Grunes (1973), Mayland and Grunes (1979), and other articles in the publications by Rendig and Grunes (1979) and Littledike et al. (1981). The level of Mg in forages has been related to soil parent material in several regions of the USA

(Kubota et al., 1980). Grass tetany affects, and is frequently fatal to, lactating cows that graze on certain grass or small grain forages during the cool seasons of the year. Similar animals that graze on legumes at these same times rarely show evidence of this disease, and there are very few instances of the disease during the warm, summer months. The level of Mg, and often the level of Ca, in the blood of the animals is very low at the time the disease strikes, and affected animals will recover if they are given timely intravenous injections of the proper forms of Mg and Ca.

A wide K/(Mg + Ca) or K/Mg ratio within the forage grass has been associated with this problem. High levels of available K in the soil tend to depress concentrations of Mg in grasses, and increased incidence of grass tetany has been observed following heavy applications of K in fertilizers or manure. However, as pointed out by Grunes et al. (1970), the etiology of the problem is complex and serious outbreaks of grass tetany have occurred on pastures that have not been fertilized with K. High levels of N in the soil also increase the incidence of grass tetany.

On acid, coarse-textured soils, the use of Mg fertilizers or dolomitic limestone has been effective in increasing the Mg concentration in the grass and in decreasing the incidence of grass tetany. This practice has been less successful on finer-textured soils or soils of higher pH. Incorporation of the dolomitic lime into the soil increases its effectiveness.

Whenever liberal application of K and/or N fertilizers are applied for the grass species that grow well in the cool seasons and whenever these grasses are grazed by lactating cows or ewes, special attention should be given to maintenance of a high level of daily Mg intake by the animals. On some soils, Mg fertilization of the soil will be effective for this purpose, but on other soils, dusting of pastures with magnesium oxide (MgO) or the use of a mineral supplement high in Mg may be necessary.

Adding magnesium sulfate ($MgSO_4$) to drinking water is helpful but some diarrhea may occur. Magnesium acetate [$Mg(C_2H_3O_2)_2$] may be used instead of $MgSO_4$. To be effective, the drinking trough must be the only source of water.

On wheat pasture, fertilization with N has increased frothy bloat of younger cattle (Stewart et al., 1981). Bloat-producing pastures were shown to be low in dry matter and soluble carbohydrates and high in crude protein, soluble N, soluble protein N, and soluble nonprotein N. Interestingly, these indices occur in the same relative standings in forages that tend to produce grass tetany. Bloat-producing pastures were lower in total fiber, but this constituent has not been related to grass tetany.

Magnesium deficiencies of humans and monogastric animals have been observed, but there are no recorded cases where these deficiencies have been traced to a deficiency of available Mg in the soil. However, as indicated by Grunes et al. (1970), severe Mg deficiency has occurred in children in developing countries eating Mg-poor foods such as cornstarch, yam (*Dioscorea* sp.), and cassava (*Manihot esculenta* Crantz). Recently, there has been a good deal of interest in Mg in relation to human nutrition (Cantin & Seelig, 1980; NRC, 1980; Seelig, 1980; Wacker, 1980). The use of stable isotopes of Mg (Schwartz et al., 1980) should markedly increase Mg research and understanding in humans.

D. Sulfur

Sulfur moves from the soil into the plant as the sulfate ion (SO_4^{2-}). Within the plant the S is reduced to sulfide (S^{2-}) and incorporated into the essential amino acid cysteine. Some of the cysteine is converted to methionine, another essential S amino acid, and from methionine into other organic S compounds in the plant. Most of the cysteine and methionine formed is combined into the polypeptide chains of protein according to the genetics of the plant and so cysteine and methionine do not accumulate in the free or uncombined form in normal plants. Cysteine, methionine, and some of the other organic S compounds are essential to all living cells, both plant and animal. The metabolism of S in plants has been reviewed by Thompson et al. (1970).

When SO_4^{2-} is taken up by the plant in amounts greater than needed for protein synthesis, SO_4^{2-} will accumulate in the plant. The presence of SO_4^{2-} in the tops of plants may be a useful indicator of an adequate supply of available S in the soil, unless some factor other than S supply is limiting the rate of protein synthesis (Ensminger & Freney, 1966).

When a monogastric animal eats a plant food, the protein is hydrolyzed to release the essential S amino acids and these are then absorbed from the GI tract and incorporated into animal proteins. Monogastric animals cannot reduce SO_4^{2-} and so they must obtain the essential S amino acids from their diet. Monogastric animals can convert methionine to cysteine and their requirement for both of these can be met by an adequate level of methionine in the diet. Monogastric animals also require two S-containing vitamins. Very small amounts of dietary SO_4^{2-} can be utilized by animals in the formation of SO_4^{2-}esters such as the chrondroitin sulfate of connective tissue. However, most of the SO_4^{2-} ingested by monogastric animals is excreted in the urine within a few days of ingestion (Dziewiatkowski, 1970). Excess dietary SO_4^{2-} is not detrimental to the animal, except as it may accentuate a Mo–Cu interaction that will be discussed in the sections on trace element fertilizers.

The effect of S fertilizers added to a S-deficient soil on the nutrition of monogastric animals, including people, is to provide higher yields and higher production per hectare of the essential S amino acids and protein.

The nutritional value of each unit of protein is determined by the genetic control over the ratios of S amino acids to other amino acids in the plant protein. Plant proteins, and particularly the proteins in the seeds of some of the legumes such as soybeans, are usually low in S amino acids in relation to the requirements of monogastric animals. Supplementation of diets of pigs and chickens, fed largely corn plus soybean protein, with methionine or high methionine proteins such as fish meal is a common practice. The S (or more accurately S amino acid) requirements of humans have been reviewed by Swendseid and Wang (1970). Deficiency of S amino acids is a very important problem in developing countries, where much of the dietary protein comes from legume seeds.

The effects of S fertilization on the nutrition of ruminant animals may be very different from those for monogastric animals because the microorganisms in the rumen can synthesize cysteine and methionine from inorganic S in the animals' diet. Levels of inorganic S in the plant normally increase much more than levels of S amino acids when a S fertilizer is added to a soil low in available S. For the ruminant, the nutritional quality of the plant, with respect to S, is controlled by total S content or by the N/S ratio. The S nutrition of ruminant animals and the effects of S supplementation on the growth, rate of wool production, and feed efficiency of ruminants has been reviewed by Garrigus (1970). According to this review, growth rates and wool production of sheep are optimized when the N/S ratio in the diet is about 10:1. The optimum ratio for cattle is slightly higher, about 15:1. When the N/S ratio in the diet is higher than these, the dietary N is used less efficiently and the ruminant responds to marginal S deficiency by wasting dietary N.

Allaway and Thompson (1966) have pointed out that the N/S ratios required for optimum growth of most plants are slightly higher, ranging from about 15:1 to 18:1, than those required for ruminant animals. This disparity between optimum ratios for plants and for ruminants indicates that the level of available S in soils should be maintained somewhat above that needed for optimum plant growth to ensure optimum S nutrition of ruminant animals.

E. Trace Element Fertilizers

1. Boron

Boron is required by plants, but no requirement on the part of animals has been established. Any nutritional quality effects from the use of B fertilizers, or of variations in level of available B in soils, are indirect. Boron added to low-B soils has permitted the plant to grow normally and synthesize organic compounds of importance in human or animal nutrition. A potential effect of this type has been described by Kelly et al. (1952). In their experiments, B fertilization was found to increase the concentration of carotene (pro-vitamin A) in carrot (*Daucus carota* L. 'Red Cored Chantenay') roots. The increased concentration of carotene in the roots was attributed to enhanced translocation of carotene from the tops in the B-adequate plants. Other effects of B on carotene concentrations have been noted in forage where use of B fertilizers corrected chlorosis in the leaves. It seems possible that any fertilizer treatment that corrects a chlorotic condition in the plant might increase carotene concentration in the leaves.

2. Cobalt

Potential effects of Co fertilizers on human and animal nutrition are associated with the requirement of monogastric animals for Co as part of the molecule of vitamin B_{12}.

Legumes require Co for the N_2 fixation process conducted in root nodules. No requirement for Co in nonleguminous plants has been established. Legumes may grow normally and produce optimum or near-optimum yields even though Co levels in the tops are less than 0.05 μg/g (dry weight). Field cases of growth response of legumes to Co fertilizers have been reported from Australia but are apparently very rare in the USA.

When a ruminant animal eats a plant containing Co, some of the Co is converted to vitamin B_{12} by the rumen organisms. The process of converting inorganic Co in the diet to vitamin B_{12}, which is then transmitted in milk and meat to the next animal in the food chain, is one of the important roles of ruminants. The dietary requirement of ruminants for Co is between 0.05 and 0.10 μg/g of the dry diet. Therefore, on some soils legumes may make nearly optimum growth and yield, even though they contain too little Co to meet the requirements of ruminant animals. There are large areas in Australia where production of sheep and cattle was severely restricted until it was discovered that the problem was due to Co deficiency (Underwood, 1977). Areas of the USA where forage legumes contain $<$0.05 to 0.10 μg Co/g have been described by Kubota and Allaway (1972). In areas where legumes do not contain sufficient Co to meet dietary requirements of ruminants, the use of Co-fortified fertilizers may result in higher levels of plant Co and improved Co nutrition of ruminants. Dietary supplements or salt mixtures containing inorganic Co may also be used. Pellets containing inorganic Co in a slowly degradable matrix may be placed in the rumen of cattle and sheep, and these will dissolve slowly to release the Co needed by the animal over long periods of time. Grasses rarely contain adequate concentrations of Co to meet ruminant requirements and use of some method of Co supplementation of diets of grass or cereal silage is very common.

The use of Co in fertilizer can improve human and animal nutrition only if the Co taken up from the soil by the plant is processed through the rumen to form vitamin B_{12} for use by the ruminant animals, and for transmission in meat and milk to humans and other monogastric animals.

3. Copper and Molybdenum

Copper and Mo are discussed together because most of the nutritional effects of variations in the concentration of these two elements in plants are associated with their interactions with each other. Molybdenum in the diet may interfere with the utilization of dietary Cu, and bring about Cu deficiency when dietary Mo is high in relation to dietary Cu. Where dietary Cu is high and Mo low, an increase in the level of dietary Mo may alleviate Cu toxicity. These effects are especially evident in ruminant animals. Mongastric animals are much less likely to show effects of Cu deficiency or toxicity than are ruminants. There are excellent reviews of the Cu–Mo problem in animal nutrition (Underwood, 1977, 1981).

Both Cu and Mo are required for plants. Molybdenum is required for the enzyme NO_3^- reductase. It is also required for the symbiotic bacteria that fix N_2 in the nodules of legumes. Of all the required elements, Mo is required in the lowest concentrations in plant tissue and nearly all of the critical concentrations suggested to be indicative of Mo deficiency in plants are < 0.5 μg/g dry weight of plant tissue (Allaway, 1976). In spite of this low requirement, there are vast areas in the world where the addition of extremely small amounts of Mo in fertilizers have permitted economic crop and pasture production on otherwise nearly wasteland areas. These dramatic responses to Mo bearing fertilizers have been reviewed by Stout (1972). In 1956 an entire issue of *Soil Science* was devoted to reports of Mo deficiency and crop responses to Mo fertilizers. These reports and others more recent have been summarized by Allaway (1976). Responses to Mo fertilizers have been most common on well-drained, acid, weathered soils. In some cases, liming to increase soil pH will increase the availability of native soil Mo and minimize the need for Mo fertilizers. In other cases plants have responded to Mo fertilizers on well-limed and naturally neutral or alkaline soils. The amounts of Mo required for increased yield are very small, usually measured in grams per hectare. In some cases, treatment of the seed with Mo has been effective. Use of Mo in fertilizers and as seed treatments has been reviewed by Murphy and Walsh (1972). A report from Australia (Freney & Lipsett, 1965) documents increases in yield of wheat from Mo fertilization of soils very high in available N. Molybdenum-deficient wheat accumulated NO_3^--N to the point of toxicity to the wheat plant. These results would indicate a possibility that Mo fertilization might be useful in preventing accumulation of NO_3^- in plants and consequent NO_3^-–NO_2^- toxicity in ruminants, as well as undesirable NO_3^- concentrations in human foods. However, there are also many reports of high levels of NO_3^- in plants that cannot be attributed to Mo deficiency and where NO_3^- levels were not reduced following Mo fertilization.

Molybdenum is not very toxic to plants; legumes, growing normally and containing well over 100 mg Mo/kg, have been reported by several investigators of the problem of Mo-induced Cu deficiency in livestock. Areas in the USA where plants contain high concentrations of Mo are described by Kubota (1976). The naturally occurring areas of high-Mo plants occupy poorly drained soils derived from high Mo parent materials. High Mo plants have also been found in several places where the soils have received Mo bearing industrial discharges and fumes.

According to Underwood (1976), "under naturally occurring conditions a true Mo deficiency has never been reported in man or farm animals and nutritional interest in the element is overwhelmingly concerned with its toxic effects on animals and its interactions with Cu and S."

Copper, like Mo, is required by both plants and animals. The "normal" levels of Cu in plants have been considered to be between 5 and 20 μg/g (Chapman, 1966). Copper deficiency in plants has been very common in many countries. Some of the acid Histosols of the eastern USA could scarcely be used for economic crop production until Cu fertilization was a common practice. The use of Cu fertilizers has been summarized by Murphy and Walsh (1972). On some acid soils, one application of Cu may last for several years, while repeated applications may be needed on alkaline soils. Copper has also been a component of many fungicidal sprays and use of these sprays has alleviated Cu deficiency in plants as well as providing protection from disease. Copper toxicity in plants has resulted from excessive use of Cu fertilizers and Cu fungicides. Plant concentrations of $> 20\ \mu g$ Cu/g have been considered indicative of Cu toxicity.

Both Cu deficiency and toxicity have been important practical problems in ruminant animal nutrition in many parts of the world. The dietary concentrations of Cu needed to prevent deficiency, and those likely to cause toxicity in ruminants, cannot be specified without consideration of the levels of dietary Mo and specific forms of S. There is also evidence that the Cu in different forages varies in its digestibility by ruminants. The occurrence of Cu deficiency and toxicity in Northern Ireland and its relationship to Cu–Mo interactions and the use of Cu fungicides has been described by Todd (1976).

The use of a Cu fertilizer on a soil where the Cu concentration in the plant is increased by such fertilization can have the effect, if Mo in the diet is moderate or high, of correcting Cu deficiency in ruminants. However, where dietary Mo is very low, this same Cu fertilizer application may accentuate Cu toxicity in ruminants. Similarly, the use of a Mo fertilizer in amounts that will increase plant Mo may aggravate Cu deficiency if dietary Cu is low. The probability that Mo fertilizers would have a long-lasting effect of accentuating Cu deficiency is not great if the Mo is added to well-drained soils where added Mo can be expected to revert to unavailable forms at a rate that will prevent high concentrations of Mo in the plant. These are the soils where Mo deficiency is most apt to occur. The addition of Mo fertilizers to poorly drained soils could have much more lasting effects on plant concentrations of Mo, but Mo fertilization is rarely needed on poorly drained soils.

4. Iron

Iron is required by both plants and animals, and deficiencies of Fe have seriously limited crop yields in many areas. Iron deficiency in people is a very important nutritional problem, especially in countries where the diet contains large amounts of processed cereals. However, use of Fe fertilizers to increase crop yields may not be useful in ameliorating Fe deficiencies in people.

The concentration of Fe in plants is controlled not only by the supply of available Fe in the soil but by a number of interactions with other elements, as well as by homeostatic controls. Many of these interactions have been described by Olsen (1972) and Murphy et al. (1981). The use of Fe fertilizers or sprays may or may not increase the concentration of Fe in the leaves, and are even less likely to increase Fe concentrations in seeds, even though seed yield may be substantially increased. Different plant species, and different cultivars of the same plant species, may vary markedly in their Fe concentrations, even when growing on the same soil.

Experiments on the effect of variations in Fe concentration in culture solutions on Fe in 'Shogoin' turnip leaves has been reported by Wein et al.

(1975). Increasing the concentration of chelated Fe in the culture solution in which the turnips were grown generally, but not always, increased Fe concentration in the leaf blades, although the increases were not large. Plants grown in culture solutions deficient in the major elements N, P, and K contained higher concentrations of leaf Fe than plants grown where levels of N, P, and K were more adequate. When the leaves were fed to Fe-depleted rats, the 13-day retention of Fe by the rats was not significantly affected by the Fe levels in the plant culture solutions. The Fe in leaves grown in solutions deficient in N, P, and K was of lower bioavailability to the rats than Fe from leaves grown where N, P, and K were adequate. Leaf Fe was less available to the rat than ferric chloride ($FeCl_3$) added to the homogenates of leaves immediately prior to feeding. In a somewhat similar experiment using 'Amsoy' soybean seeds, Welch and Van Campen (1975) found that the Fe concentration in the seeds was not affected by the concentration of Fe in the culture solution in which the plants were grown. The Fe contained in immature soybean seeds was less available to rats than Fe contained in mature seeds, even though the level of phytic acid ($C_6H_{18}O_{24}P_6$), a suspected inhibitor to digestibility of Fe, was higher in the mature seeds. Iron in the mature seeds was similar to $FeCl_3$ added to the seed homogenates as regards bioavailability to rats.

From these experiments it appears that the role of Fe fertilizers and sprays in human nutrition is to increase crop yields and food production from soils on which plant growth is limited by Fe deficiency. Correction of human Fe deficiency will require use of foods that are high in available Fe, use of Fe supplements to the diet, and correction of losses of body Fe due to parasites and diseases.

Farm animals, except for young pigs maintained in confinement on concrete floors, rarely show Fe deficiency. Among grazing animals, Fe from ingested soil may be an important dietary source. The supplementation of diets of young pigs with Fe, or the use of Fe injections, is very common in modern pig production.

5. Manganese

Manganese deficiency and toxicity are important problems in crop production. Manganese deficiencies are most often observed on well-drained neutral or alkaline soils, while Mn toxicity is most common in strongly acid or waterlogged soils (Murphy & Walsh, 1972). Due to interactions with other elements, especially Fe, precise limits of Mn concentration indicative of deficiency or toxicity are difficult to establish.

Manganese deficiency in poultry, resulting in a leg defect called "slipped tendon" has been a very serious practical problem (Underwood, 1977, 1981). Manganese deficiencies in grazing animals have not been confirmed. Manganese toxicity has been a problem in miners who come into prolonged contact with Mn ores and dusts.

There are no reports of beneficial or detrimental effects on humans or animals from the use of Mn fertilizers.

6. Silicon

Silicon is the second most abundant element in the earth's crust (Lindsay, 1979). Growth responses to applications of Ca silicate have been reported for rice and sugarcane (*Saccharium officinarum* L.) in the field, and for other graminaceous and for several nongraminaceous crops in pots (Silva, 1971, 1973). These responses have occurred on highly weathered volcanic soils of the tropics and on old alluvial terraces in Japan. There is not uniform agreement concerning the essentiality of Si for plant growth. Jones and Handreck (1967) indicated that Si does not yet deserve a place among the essential elements for plants. Lewin and Reimann (1969) concluded that although Si seems not to be required for vegetative growth of most plants, it appears to be necessary for the healthy development of many plants and may be considered an essential element for those plants with high Si contents, such as rice and certain other grasses and horsetails (*Equisetum arvense* L.). Silva (1973) indicated that Si is considered essential only for diatoms and horsetail.

The Gramineae contain 10 to 20 times the Si concentrations found in legumes and other dicotyledons (Jones & Handreck, 1967). Lowland rice has been reported as containing 7 to 9% Si in the straw.

Van Soest (1982) stated that the main form of Si in plants is opaline silica. He indicated that Si is used as a structural element complementing lignin to strengthen and rigidify plant cell walls. Silica is also deposited in hairs on the plant surface and cuticular edges and can contribute to the harsh character of certain plants. In grasses the Si level is highly dependent on soil type and the availability of Si, as well as on transpiration and the nature of the plant species. Consequently, a wide variation in Si content occurs among forages of the same plant species grown on different soils.

Van Soest (1982) indicated that the relationship of Si to forage digestibility is often affected by a compensatory association with lignin. The decrease in forage digestibility is more closely related to the sum of the percent Si and percent lignin than to either one. Ruminants grazing in arid regions, and where grasses are high in Si, often suffer from siliceous kidney stones.

Silicon is required by the animal for collagen synthesis and bone formation, with a requirement of approximately 50 μg/g in the diet (Carlisle, 1974). Although some tropical soils are highly depleted of Si, clinical animal deficiencies are as yet unknown in the field (Van Soest, 1982).

7. Zinc

Zinc has been known to be required for both plants and animals since about 1930, and interest in Zn fertilization of soils and Zn nutrition of people has substantially increased since the mid-1970s.

The recognition of new soil areas, where plants show a growth response to Zn fertilization, has led to an expanded use of Zn fertilizers since 1945. The use of Zn fertilizers has been described by Murphy and Walsh

(1972) and by Viets (1966). Expanded use of Zn fertilizers on beans, rice, corn, and grain sorghum has been especially notable. Almost all of the commercially produced citrus fruit trees have been fertilized with Zn.

Zinc concentrations in plants normally range from 20 to 60 or 100 μg/g, with values much below 20 in the leaves indicative of Zn deficiency. Plants are very tolerant of excess Zn and may contain 100 to 200 μg Zn/g before showing signs of Zn toxicity (Jones, 1972). Zinc concentrations in plants are very much dependent on interactions with other elements, especially P and Fe (Olsen, 1972). When a Zn fertilizer is applied to a Zn-deficient soil, Zn concentration in the crop leaves is usually increased more than Zn concentration in the seeds.

The role of Zn in human and animal nutrition was reviewed by Underwood (1977). Zinc deficiency in pigs fed diets of corn plus soy protein was recognized in about 1940. Beef cows have responded to Zn supplementation when grazed on seimarid rangelands in the western USA (Mayland et al., 1980). Recent documentation of severe Zn deficiency in people has generated additional interest in the need for increasing human dietary intake of this element. The human Zn deficiency symptoms include dwarfism and failure of sexual development in young males (Prasad, 1966), delayed wound healing (Pories & Strain, 1966), and difficulties in childbirth (Jameson, 1981).

Due to variability in the bioavailability of Zn from various sources, it is difficult to establish very precise limits for the required dietary concentration of Zn for animals. However, Underwood (1977, 1981) indicated that dietary levels of up to 32 μg Zn/g may be required for optimum sexual development of young rams. Thus, the dietary Zn requirement of some animals is probably in excess of the Zn concentrations in plant leaves required for good plant growth. Animals are very tolerant of high levels of Zn, and diets containing 500 to 1000 μg Zn/g have been fed to experimental animals without evidence of toxicity.

Inasmuch as Zn deficiency is an important problem in both human and animal nutrition, it is appropriate to consider whether increased use of Zn fertilizers would be useful in increasing dietary Zn, and especially in cases where the diet is primarily based on seeds of cereals and pulses. The work of Welch et al. (1974) indicates that the concentration of digestible Zn in 'Early Market' pea seeds can be substantially increased by adding Zn to the culture solution in which the peas were grown, even though the Zn in the peas was not quite as available (to rats) as Zn from zinc sulfate ($ZnSO_4$). Similarly, Lantzsch et al. (1981) increased the level of available Zn in 'Jubilar' wheat and 'Carina' barley (*Hordeum vulgare* L.) by fertilizing the soil with 50 kg $ZnSO_4$/ha. In the experiments of Welch et al. (1974), immature pea seeds grown in the low-Zn culture contained 21.5 μg Zn/g. In the experiments of Lantzsch et al. (1981), the unfertilized wheat seeds contained 21.3 μg Zn/g and unfertilized barley seeds contained 24.8 μg Zn/g. In both of these studies, the rates of Zn fertilization used to obtain the increases in bioavailable Zn were probably greater than that needed to obtain optimum plant yields. If growers were to fertilize crops with Zn beyond levels necessary for optimum yield to improve Zn intakes of people, some procedure for monitoring the increased Zn concentrations and reimbursing the growers for high concentrations would be needed. In the absence of such a monitoring system, direct supplementation of diets with inorganic Zn will be more effective.

8. Nutritional Effects of Accessory Compounds and Impurities in Fertilizers

Cadmium is found in certain phosphates and occurs as a contaminant in P fertilizers prepared from them. Concern for the fate of this Cd stems from a possible relationship between Cd and human hypertension (Underwood, 1977) and from evidence of severe Cd toxicity in Japanese women resulting from Cd pollution of river water (Kobayashi, 1971). The studies of Williams and David (1976) in Australia indicate that most of the Cd applied to medium-textured soils as an impurity in phosphates remains in the surface soil. In one of their experiments, the addition of 1680 kg/ha of superphosphate over a period of 5 yr led to an increase in the Cd content of wheat grain from 0.008 to 0.028 μg Cd/g. Where NH_4NO_3, an acid-forming source of N, was used, Cd concentrations were increased by 50%. In other experiments, they showed that plant uptake of Cd, both from phosphate fertilizer residues and that added as cadmium chloride ($CdCl_2$), was greatly decreased when the soil was treated with calcium carbonate ($CaCO_3$) and magnesium carbonate ($MgCO_3$) to increase the soil pH from 5.0 to 6.8.

In the USA and Europe, concern over Cd is centered on the possibility that Cd contained in municipal sewage sludges could lead to undesirably high levels of Cd in food plants grown on sludge-amended soils. Problems concerned with Cd in sewage sludge and other organic fertilizers have been reviewed by Allaway (1977).

It appears that whenever Cd has been added to soils, either in organic fertilizers and sludges or in phosphates, there should be a regular monitoring of the Cd concentrations in the crops produced, and the pH level of these soils should be maintained at 6.5 or above to minimize Cd uptake by plants.

Fluorine is a common constituent of phosphate rocks and some F, in the form of calcium fluoride (CaF_2) is present in fertilizers, especially single superphosphates, made from these rocks. Fluorine is not required by plants and, except in a few known F accumulators, is rarely found in high concentrations in plants. Fluorine is possibly required by animals and is frequently added in very low concentrations to water supplies to minimize dental caries. High intakes of F are toxic to animals, and are also detrimental to normal teeth development in children. The occurrence of F toxicity in animals has been reviewed by Underwood (1977). Ingestion of toxic levels of F has occurred where drinking water is high in F and in areas where F-containing dusts have deposited on plant leaves from industrial exhausts. Other instances of F toxicity have occurred when F-bearing phosphates have been inadvertently used as feed supplements. There are no records of F toxicity that can be traced to the application of F-bearing phosphate fertilizers to soils. Calcium fluoride and most other F compounds are of very low solubility, except in very acid environments and the near absence of Ca. It is probable that plant growth would be severely restricted on any soil that was sufficiently acid to contain appreciable levels of plant-available F.

Iodine is present in the N fertilizers prepared from "Guano." *Guano* is obtained from naturally occurring deposits of bird droppings that accumulate in very arid regions, primarily close to the Pacific Ocean in South America. Iodine is essential to people and animals, and goiter was at one time a very serious human health problem in regions where soils and rainfall are low in I (Underwood, 1977). Iodine has not been found to be essential for higher plants. Human and animal requirements for I are currently provided through use of iodized salt, and movement of I from fertilizers into food chains is of minor importance in contributing to human and animal requirements for this element.

Nickel is ubiquitous in nature, occurring in air, water, rocks, and soils, and is incorporated into all biological materials. Concentrations of Ni in plants may be enhanced by the use of Ni-containing fungicides, fertilizers, soil amendments, and sewage sludges in the production of agronomically important crops. High concentrations of Ni in soil are phytotoxic (Nielsen et al., 1977).

Welch (1981) indicated that while the essentiality of Ni has not been established for higher plants, many beneficial effects of Ni on plant growth have been reported. He indicated that urease is a Ni-containing metalloenzyme, and Ni may be an essential element for higher plants if urease performs an essential function in plant growth and development. He suggested that Ni may be required by nodulated legumes that translocate large amounts of N from roots to tops via ureide compounds.

Eskew and Welch (1982) have shown that soybeans grown in purified nutrient solution developed necrosis of soybean leaflet tips. Leaflet tip necrosis was completely prevented by adding either 1 or 10 μg Ni/L. Nickel supplementation (1 μg/L) prevents leaflet tip necrosis in soybeans grown in nutrient solutions purified using 8-hydroxyquinoline–controlled pore glass chromatography. Eskew and Welch (1982) suggest that the absence of necrotic leaflet tips when Ni was added indicates strong evidence that Ni is an essential nutrient for higher plants.

Evidence from several laboratories suggests that Ni is essential for animals, but its function has not been defined (Welch, 1981). Nickel is relatively nontoxic orally, and Ni contamination when ingested probably does not present a serious health hazard (Nielsen et al., 1977). Some forms of Ni (e.g., metallic Ni dust and Ni carbonyl) are carcinogenic when inhaled.

Selenium has been the cause of economically important toxicities and deficiencies in farm animals, and Se deficiency has very recently been confirmed as the cause of severe human disease in parts of China (Zhu, 1981). Selenium is a component of some phosphate rocks and occurs at low levels in fertilizers prepared from them. It has been suggested that Se in certain phosphate fertilizers might be useful in prevention of Se deficiency in farm animals (Robbins & Carter, 1970).

The soil chemistry of Se and pathways of its movement into plants and on into animals has been compiled (NAS, 1976). Selenium is not required by most plants. However, the use of soil applications of Se to increase Se concentrations in plants up to levels required to prevent Se deficiency in livestock has been investigated (Cary & Allaway, 1973; Grant, 1965). Selenium in the form of selenate is quite soluble in soils and is readily taken up by plants. Rates of annual application of selenate Se must be approximately 5 g/ha and inadvertent overuse might pose a hazard of environmental contamination or Se toxicity to animals fed the forage. Selenite Se is tightly bound in most soils and is of very low availability to plants. One application of 2 to 4 kg Se/ha has been found to produce forages having protective but nontoxic concentrations of Se for 3 to 5 yr after application. However, the recovery of the added Se is very low, and thus soil application of selenite Se represents an uneconomic use of an expensive element.

The application of foliar sprays of selenite Se to silage and grain crops has been shown to be an effective method of ensuring protective concentrations of Se in these crops (Cary & Rutzke, 1981; Gissel-Nielsen, 1981). This procedure is more efficient in terms of recovery of added Se in the animal feed than is soil application.

Selenium deficiencies in livestock are treated or prevented by use of Se injections, Se supplements to mixed feeds, Se containing rumen "bullets," or by Se additions to salt in the animal diet. The human Se deficiency in China has been substantially reduced by weekly use of Se tablets by the population at risk and by addition of Se to table salt. Foliar application of Se to food crops is under investigation in China.

Sodium is essential for humans, animals, and for some plants. It is present in sodium nitrate ($NaNO_3$) fertilizers and in small amounts in some K fertilizers. The concentrations of Na found in plants other than certain halophytes are low, and movement of Na from soils and fertilizers into food and feed crops is of secondary importance in meeting Na requirements of people and animals. The direct consumption of common salt (NaCl) has been used since ancient times to meet human and animal requirements for Na. In the Netherlands, use of Na fertilizer on pastures is suggested where salt licks have been inadequate (Comm. Miner. Nutr., 1973).

In addition to the elements discussed above, there are a number of other elements present in low concentrations in some fertilizers. Some of these may be essential to plants or animals. Among the elements possibly essential to people and animals, Mertz (1981) lists Cr, V, and As in addition to those discussed earlier in this chapter. The effect of the traces of Cr, V, and As that may occur as impurities in fertilizers on the nutritional quality of food and feed crops is not known, but is probably very minor.

III. GENERAL ASPECTS OF FERTILIZER USE AND HUMAN AND ANIMAL NUTRITION

Fertilizers are normally used by farmers to provide high yields of crops or to permit them to grow some new crop that could not be produced on their unfertilized soils. Use of fertilizers primarily for the purpose of improving the nutritional quality of the crop has been rare, even though the preceding sections of this chapter indicate that some improvements in the nutritional quality of crops have been an added benefit from use of fertilizers for increasing crop yields. The preoccupation of farmers and agronomists with the yield increases stemming from fertilizer use has led some people to question whether or not the high-yielding crops produced on fertilized soils are of inferior nutritional quality.

In the second edition of *Fertilizer Technology and Use,* certain public health statistics such as longevity and infant mortality in different countries were compared with per-hectare and per-capita use of fertilizers (Allaway, 1971). These comparisons indicate that countries that use high amounts of fertilizer usually have better public health statistics than do countries that use very little fertilizer, but these comparisons cannot be taken as evidence that fertilizer use promotes public health. Countries that use large amounts of fertilizers generally have better medical services, improved food safety standards, and many other features that contribute to improved public health statistics than do countries where fertilizer use is low.

Some people have suggested that food crops produced with organic fertilizers were in some way of better nutritional quality than those produced with chemical fertilizers. This question can only be considered with reference to specific organic or chemical fertilizers and the concentration and plant availability of the different elements they contain, plus knowledge of the soil to which they are applied and the crop to be grown. Then the kind of animal consuming the crop and its entire diet must be considered. Misuse of organic fertilizers containing high levels of available N has been involved, along with misuse of chemical N fertilizers, in causing NO_3^-–NO_2^- toxicity in ruminant animals. Similarly, overuse of organic fertilizers high in K and N and low in Mg has been involved, along with overuse of inorganic K and N fertilizers, in "grass tetany" in cattle.

The earlier sections of this chapter provide evidence that the impact of fertilizers, organic or chemical, on the nutritional quality of crops can be clarified by knowledge of the nutritional requirements of plants in relation to the chemistry of the nutrients in the fertilized soils. When this type of information is related to the known nutritional requirements of people and animals and to the composition of the entire diet consumed, one or more techniques for avoiding undesirable effects on human and animal nutrition usually become apparent.

Fertilizers are used to supply the specific nutrients that are not present in adequate supply in the soil but are needed for optimum crop growth. Other nutrients required by plants, and also by people and animals, must be furnished to the food chain from naturally occurring reserves in the soil. With the continued application of fertilizers containing only those nutrients to which the plants respond, and removal of abundant harvests, these naturally occurring reserves can be depleted. If the reserves of an element required by plants become depleted, evidence of plant deficiencies of this element should lead to its inclusion in the fertilizer to maintain crop yields at the desired level. Where the element is required by animals but not by plants, or the dietary concentration required by animals exceeds the plant tissue concentration indicative of adequate supply to the plant, the native supply of the element in the soil may become depleted. In this case, the animal deficiency remains uncorrected until animal health problems occur. Some instances of this type have been described by Rooney et al. (1977). In the USA it is possible that the increasing recognition of Se-deficiency diseases in livestock may be due to depletion of supplies of native available Se in the soils of some areas. Lack of information on the Se concentrations in forages that were present some years ago in the areas where Se deficiency in livestock now occurs precludes a definite answer to this possibility. The increasing recognition of several micronutrient deficiencies in livestock may be due to improved diagnostic services. It is evident, however, that continuous surveillance of the composition of food and feed crops for the nutrients required by both plants and animals is a necessary feature of modern agriculture.

A major point to consider in evaluating the effects of fertilizers on problems of human nutrition is that, on a worldwide basis, deficiencies of calories and protein are the most crucial human nutritional problems. It is unquestioned that modern use of fertilizers has been one factor that has led to the high yields of food crops that are necessary to combat the problem of protein-calorie malnutrition among people.

IV. SUMMARY AND FUTURE NEEDS

The most important consideration in any discussion of the effects of fertilizers on nutritional quality of plants is that these discussions can only be meaningful if they are confined to specific elements, specific soils, specific plants, and to specific human or animal diets. Generalizations and statements attributing nutritional effects, either good or bad, from the use of "commercial fertilizers, organic or inorganic fertilizers, depleted soils, etc." are invalid, but the study of well-defined specific cases has been very useful. The preceding sections of this chapter indicate a number of instances where use of specific fertilizers has affected the nutritional quality of certain plants growing on certain soils, when these plants are consumed in certain diets. It is very probable that additional specific cases will come to light with the progress of future research.

It is essential that the nutritional quality of food and feed crops be subjected to continuous monitoring. This monitoring must be designed with respect to the growing information on human and animal nutritional requirements, nutrient bioavailability, and interactions. When changes in food production practices, including changes in fertilizer use, take place, these monitoring programs should be especially vigilant.

REFERENCES

Agricultural Research Council. 1980. The nutrient requirements of ruminant livestock. Commonwealth Agricultural Bureau, Farnham Royal, Slough, UK.

Allaway, W. H. 1971. Feed and food quality in relation to fertilizer use. p. 533–556. *In* R. A. Olsen (ed.) Fertilizer technology and use, 2nd ed. Soil Science Society of America, Madison, WI.

Allaway, W. H. 1976. Perspectives on Mo in soil and plants, Vol. II. p. 317–339. *In* W. R. Chapell and K. K. Peterson (ed.) Molybdenum in the environment. Marcel Dekker, Inc., New York.

Allaway, W. H. 1977. Food chain aspects of the use of organic residues. p. 282–298. *In* L. F. Elliot and F. J. Stevenson (ed.) Soils for the management of organic wastes and wastewaters. American Society of Agronomy, Soil Science Society of America, Crop Science Society of America, Madison, WI.

Allaway, W. H., and J. F. Thompson. 1966. Sulfur in the nutrition of plants and animals. Soil Sci. 101:240–247.

Beeson, K. C., and G. Matrone. 1976. The soil factor in nutrition, animal and human. Marcel Dekker, Inc., New York.

Boawn, L. C., F. G. Viets, Jr., C. L. Crawford, and J. L. Nelson. 1960. Effect of nitrogen carrier, nitrogen rate, zinc rate, and soil pH on zinc uptake by sorghum, potatoes, and sugar beets. Soil Sci. 90:329–337.

Brown, J. C. 1961. Iron chlorosis in plants. Adv. Agron. 13:329–369.

Cantin, M., and M. S. Seelig (ed.). 1980. Magnesium in health and disease. SP Medical and Scientific Books, New York.

Carlisle, E. 1974. Silicon as an essential element. Fed. Proc. Fed. Am. Soc. Exp. Biol. 33:1758–1766.

Cary, E. E., and W. H. Allaway. 1973. Selenium content of field crops grown on selenite-treated soils. Agron. J. 65:922–925.

Cary, E. E., and M. Rutzke. 1981. Foliar application of selenium to field corn. Agron. J. 73:1083–1085.

Chapman, H. D. (ed.). 1966. Diagnostic criteria for soils and plants. University of California Division of Agricultural Science, Riverside, CA.

Cole, C. V., D. L. Grunes, L. K. Porter, and S. R. Olsen. 1963. The effects of nitrogen on short term phosphorus absorption and translocation in corn *(Zea mays)*. Soil Sci. Soc. Am. Proc. 27:671–674.

Committee on Mineral Nutrition. 1973. Tracing and treating mineral disorders in dairy cattle. Center for Agricultural Publication and Documentation, Wageningen, The Netherlands.

Dziewiatkowski, D. D. 1970. Metabolism of sulfate esters. p. 97–125. *In* O. H. Muth (ed.) Sulfur in nutrition. AVI Publishing Co., Westport, CT.

Ensminger, L. E., and J. F. Freney. 1966. Diagnostic techniques for determining sulfur deficiencies in crops and soils. Soil Sci. 101:283–290.

Epstein, E. 1972. Mineral nutrition of plants: Principles and prospectives. John Wiley & Sons, Inc., New York.

Eskew, D. L., and R. M. Welch. 1982. Nickel supplementation (1 μg/1) prevents leaflet tip necrosis in soybeans grown in nutrient solutions purified using 8-hydroxyquinoline-controlled pore glass chromatography. Plant Physiol. 69(4):43.

Freney, J. R., and J. Lipsett. 1965. Yield depression in wheat due to high nitrate applications and its alleviation by molybdenum. Nature (London) 205:616–617.

Garrigus, U. S. 1970. The need for sulfur in the diet of ruminants. p. 126–152. *In* O. H. Muth (ed.) Sulfur in nutrition. AVI Publishing Co., Westport, CT.

Gissel-Nielsen, G. 1981. Foliar application of selenite to barley plants low in selenium. Commun. Soil. Sci. Plant Anal. 12:631–642.

Grant, A. B. 1965. Pasture topdressing with selenium. N. Z. J. Agric. Res. 8:681–690.

Grunes, D. L. 1959. Effect of nitrogen on the availability of soil and fertilizer phosphorus to plants. Adv. Agron. 11:369–396.

Grunes, D. L. 1973. Grass tetany of cattle and sheep. p. 113–140. *In* A. G. Matches (ed.) Anti-quality components of forages. Spec. Pub. 4. Crop Science Society of America, Madison, WI.

Grunes, D. L. 1983. Uptake of magnesium by different plant species. *In* J. P. Fontenot et al. (ed.) John Lee Pratt Int. Symp. on the Role of Magnesium in Animal Nutrition. Blacksburg, VA. 13–16 May. Virginia Polytechnic Institute Press, Blacksburg, VA.

Grunes, D. L., P. R. Stout, and J. R. Brownell. 1970. Grass tetany of ruminants. Adv. Agron. 22:331–373.

Jameson, S. 1981. Zinc nutrition and pregnancy in humans. 243–248. *In* J. McHowell et al. (ed.) Trace element metabolism in man and animals (TEMA-4). Australian Academy of Science, Canberra, Australia.

Jones, J. B. 1972. Plant tissue analysis for micronutrients. p. 319–346. *In* J. J. Mortvedt (ed.) Micronutrients in agriculture. Soil Science Society of America, Madison, WI.

Jones, L. H. P., and K. A. Handreck. 1967. Silica in soils, plants, and animals. Adv. Agron. 19:107–149.

Kelly, W. C., G. J. Somer, and G. H. Ellis. 1952. The effect of boron on the growth and carotene content of carrots. Am. Soc. Hortic. Sci. 59:352–360.

Kemp, A., J. H. Geurink, A. Malestein, and A. Th. van'T Klooster. 1978. Grassland production and nitrate poisoning in cattle. 7th General Meeting European Grassl. Fed., Ghent, Belgium.

Kobayashi, J. 1971. Relation between "itai-itai" disease and the pollution of river water by cadmium from a mine. p. 1–7. *In* Proc. 5th Int. Water Pollution Res. Conf. July–August, 1970. San Francisco, CA.

Kubota, J. 1976. Molybdenum status of United States soils and plants, Vol. 2. p. 545–582. *In* W. Chappell and K. Peterson (ed.) Molybdenum in the environment. Marcel Dekker, Inc., New York.

Kubota, J., and W. H. Allaway. 1972. Geographic distribution of trace element problems. p. 525–554. *In* J. J. Mortvedt (ed.) Micronutrients in agriculture. Soil Science Society of America, Madison, WI.

Kubota, J., G. H. Oberly, and E. A. Naphan. 1980. Magnesium in grasses of three selected regions in the United States and its relation to grass tetany. Agron. J. 72:907–914.

Lantzsch, H. J., S. E. Scheuermann, and H. Marschner. 1981. Improvement of zinc bioavailability in cereal grains. p. 107–110. *In* J. McHowell et al. (ed.) Trace element metabolism in man and animals (TEMA-4). Australian Academy of Science, Canberra, Australia.

Larson, W. E., L. B. Nelson, and A. S. Hunter. 1952. The effect of phosphate fertilization upon the yield and composition of oats and alfalfa grown on phosphate deficient Iowa soils. Agron. J. 44:357–461.

Lee, D. H. K. 1970. Nitrates, nitrites and methemoglobinemia. Environ. Rev. no. 2. Natl. Inst. of Environ. Health Sci., Department of Health, Education, and Welfare, Washington, DC.

Lewin, J., and B. E. F. Reimann. 1969. Silicon and plant growth. Annu. Rev. Plant Physiol. 20:289–304.

Lindsay, W. L. 1979. Chemical equilibria in soils. J. Wiley & Sons, Inc., New York.

Littledike, E. T., J. W. Young, and D. C. Beitz. 1981. Common metabolic diseases of cattle: Ketosis, milk fever, grass tetany, and downer cow complex. J. Dairy Sci. 64:1465–1482.

Loneragan, J. F., D. L. Grunes, R. M. Welch, E. A. Aduayi, A. Tengah, V. A. Lazar, and E. E. Cary. 1982. Phosphorus accumulation and toxicity in leaves in relation to zinc supply. Soil Sci. Soc. Am. J. 46:345–352.

Mathers, A. C., B. A. Stewart, and D. L. Grunes. 1982. Effect of a nitrification inhibitor on the K, Ca, and Mg composition of winter wheat forage. Agron. J. 74:569–573.

Mayland, H. F., and D. L. Grunes. 1979. Soil-climate-plant relationships in the etiology of grass tetany. p. 123–175. *In* V. V. Rendig and D. L. Grunes (ed.) Grass tetany. Spec. Pub. 35. American Society of Agronomy, Madison, WI.

Mayland, H. F., R. C. Rosenau, and A. R. Florence. 1980. Grazing cow and calf responses to zinc supplementation. J. Anim. Sci. 51:966–974.

Maynard, D. N., A. V. Barker, P. L. Minotti, and N. H. Peck. 1976. Nitrate accumulation in vegetables. Adv. Agron. 28:71–115.

McCarron, D. A., C. D. Morris, and C. Cole. 1982. Dietary calcium in human hypertension. Science (Washington, D.C.) 217:267–269.

Mertz, W. 1981. The essential trace elements. Science (Washington, D.C.) 213:1332–1338.

Miller, M. H. 1974. Effects of nitrogen on phosphorus absorption by plants. p. 643–668. *In* E. W. Carson (ed.) The plant root and its environment. University Press of Virginia, Charlottesville.

Miller, M. H., C. P. Mamaril, and G. J. Blair. 1970. Ammonium effects on phosphorus absorption through pH changes and phosphorus precipitation at the soil-root interface. Agron. J. 62:524–527.

Murphy, L. S., R. Ellis, Jr., and D. C. Adriano. 1981. Phosphorus-micronutrient interaction effects on crop production. J. Plant Nutr. 3:593–613.

Murphy, L. S., and L. M. Walsh. 1972. Correction of micronutrient deficiencies with fertilizers. p. 347–387. *In* J. J. Mortvedt (ed.) Micronutrients in agriculture. Soil Science Society of America, Madison, WI.

National Academy of Science. 1976. Selenium. Committee on medical and biologic effects of environmental pollutants. Division of Medical Sciences, National Research Council, Washington, DC.

National Academy of Science. 1980. Protein and amino acids. p. 39–54. *In* Recommended dietary allowances. Committee on Dietary Allowances, Food and Nutrition Board. National Academy of Science, Washington, DC.

National Research Council. 1972. Accumulation of nitrate. National Academy of Science, Washington, DC.

National Research Council. 1976. Nutrient requirement of domestic animals. No. 4, Nutrient requirements of beef cattle, 5th revised ed. National Academy of Science, Washington, DC.

National Research Council. 1978a. Nitrates, an environmental assessment. National Academy of Science, Washington, DC.

National Research Council. 1978b. Nutrient requirements of domestic animals. No. 3, Nutrient requirements of dairy cattle, 5th revised ed. National Academy of Science, Washington, DC.

National Research Council. 1980. Recommended dietary allowances, 9th revised ed. National Academy of Science, Washington, DC.

Nielsen, F. H., H. T. Reno, L. O. Tiffin, and R. M. Welch. 1977. Nickel. *In* Geochemistry and the environment, Vol. 2. National Academy of Science, Washington, DC.

Olsen, S. R. 1972. Micronutrient interactions. p. 243–264. *In* J. J. Mortvedt (ed.) Micronutrients in agriculture. Soil Science Society of America, Madison, WI.

Peck, N. H., D. L. Grunes, R. M. Welch, and G. E. MacDonald. 1980. Nutritional quality of vegetable crops as affected by phosphorus and zinc fertilizer. Agron. J. 72:528–534.

Pories, W. J., and W. H. Strain. 1966. Zinc and wound healing. p. 378–391. *In* A. S. Prasad (ed.) Zinc metabolism. Charles L. Thomas, Publisher, Springfield, IL.

Prasad, A. S. (ed.). 1966. Zinc metabolism. Charles C. Thomas, Publisher, Springfield, IL.

Rendig, V. V., and D. L. Grunes (ed.). 1979. Grass tetany. Spec. Pub. 35. American Society of Agronomy, Madison, WI.

Robbins, C. W., and D. L. Carter. 1970. Selenium concentrations in phosphorus fertilizer materials and associated uptake by plants. Soil Sci. Soc. Am. Proc. 34:506–509.

Rooney, D. R., N. C. Uren, and D. D. Leaver. 1977. Some aspects of fertilizer use in relation to animal health and the environment. Aust. Vet. J. 53:9–16.

Sander, D. H., W. H. Allaway, and R. A. Olson. Projected 1987. Modification of nutritional quality by environment and production practices. *In* R. A. Olson (ed.) Nutritional quality of cereal grains. Agronomy Monograph. American Society of Agronomy, Crop Science Society of America, Soil Science Society of America, Madison, WI.

Schnug, E., and A. Finck. 1982. Trace element mobilization by acidifying fertilizers. p. 582–587. *In* A. Scaife (ed.) Plant nutrition 1982. Int. Plant Nutr. Colloq., Proc. 9th, Warwick Univ., England. 22 August. Commonwealth Agricultural Bureau, Farnham Royal, Slough, UK.

Schwartz, R., D. L. Grunes, R. A. Wentworth, and E. M. Wien. 1980. Magnesium absorption from leafy vegetables intrinsically labeled with the stable isotope ^{26}Mg. J. Nutr. 110:1365–1371.

Seelig, M. S. 1980. Magnesium deficiency in the pathogensis of disease. Plenum Publishing Corp., New York.

Silva, J. A. 1971. Possible mechanisms for crop response to silicate applications. *In* J. S. Kanwar and S. P. Ray Chaudhuri (ed.) Proc. Int. Symp. Soil Fert. Eval. (New Delhi, India) 1:805–814.

Silva, J. A. 1973. Mineral nutrition of plants. p. 338–340. *In* D. N. Lapedes (ed.) McGraw-Hill yearbook of science and technology. McGraw-Hill International Book Co., New York.

Soil Science. 1956. Soil Sci. 81(1). (Entire issue.) Soil Science, Baltimore, MD.

Stewart, B. A., D. L. Grunes, A. C. Mathers, and F. P. Horn. 1981. Chemical composition of winter wheat (*Triticum aestivum* L.) forage grown where grass tetany and bloat occur. Agron. J. 73:337–347.

Stout, P. R. 1972. Introduction. p. 1–5. *In* J. J. Mortvedt (ed.) Micronutrients in agriculture. Soil Science Society of America, Madison, WI.

Swendseid, M. E., and M. Wang. 1970. Sulfur requirements of the human. p. 209–221. *In* O. H. Muth (ed.) Sulfur in nutrition. AVI Publishing Co., Westport, CT.

Thompson, J. F., C. J. Morris, and R. K. Gering. 1960. The effect of mineral supply upon the amino acid composition of plants. Qual. Plant Mater. Veg. 6:261–275.

Thompson, J. F., I. K. Smith, and D. P. Moore. 1970. Sulfur requirement and metabolism in plants. p. 80–96. *In* O. H. Muth (ed.) Sulfur in nutrition. AVA Publishing Co., Westport, CT.

Todd, J. R. 1976. Problems of copper-molybdenum imbalance in the nutrition of ruminants in Northern Ireland. p. 33–49. *In* W. R. Chapell, and K. K. Peterson (ed.) Molybdenum in the environment, Vol. 1. Marcel Dekker, Inc., New York.

Underwood, E. J. 1976. Molybdenum in animal nutrition. p. 9–31. *In* W. R. Chapell and K. K. Peterson (ed.) Molybdenum in the environment, Vol. 1. Marcel Dekker, Inc., New York.

Underwood, E. J. 1977. Trace elements in human and animal nutrition, 4th ed. Academic Press, Inc., New York.

Underwood, E. J. 1981. The mineral nutrition of livestock. Commonwealth Agricultural Bureau, Farnham Royal, Slough, UK.

Van Soest, P. J. 1982. Nutritional ecology of the ruminant. O & B Books, Inc., Corvallis, OR.

Viets, F. G. 1966. Zinc deficiency in the soil-plant system. p. 90–127. *In* A. S. Prasad (ed.) Zinc metabolism. Charles C. Thomas, Publisher, Springfield, IL.

Viets, F. G., Jr., L. C. Boawn, and C. L. Crawford. 1957. The effect of nitrogen and types of nitrogen carrier on plant uptake of indigenous and applied zinc. Soil Sci. Soc. Am. Proc. 21:197–201.

Wacker, W. E. C. 1980. Magnesium and man. Harvard University Press, Cambridge, MA.

Wein, E. M., D. R. Van Campen, and J. M. Rivers. 1975. Factors affecting the concentration and bioavailability of iron in turnip greens to rats. J. Nutr. 105:459–466.

Welch, R. M. 1981. The biological significance of nickel. J. Plant Nutr. 3:345–356.

Welch, R. M., W. A. House, and W. H. Allaway. 1974. Availability of zinc from pea seed to rats. J. Nutr. 104:733–740.

Welch, R. M., and D. R. Van Campen. 1975. Iron availability to rats from soybeans. J. Nutr. 105:253–256.

Wilde, W. A. 1968. Role of K in human and animal nutrition. p. 203–220. *In* V. J. Kilmer et al. (ed.) Role of potassium in agriculture. American Society of Agronomy, Madison, WI.

Williams, C. H., and D. J. David. 1976. The accumulation in soil of cadmium residues from phosphate fertilizers and their effect on the cadmium content of plants. Soil Sci. 121:86–93.

Wright, M. J., and K. L. Davison. 1964. Nitrate accumulation in crops and nitrate poisoning in animals. Adv. Agron. 16:197–247.

Zhu, L. 1981. Keeshan disease. p. 514–517. *In* J. McHowell et al. (ed.) Trace element metabolism in man and animals (TEMA-4). Australian Academy of Science, Canberra, Australia.

SUBJECT INDEX